Helmut Vonhoegen

Excel

The Comprehensive Guide

Editor Megan Fuerst
Acquisitions Editor Hareem Shafi
German Edition Editor Erik Lipperts
Translation Torsten T. Will
Copyeditor Doug McNair
Cover Design Graham Geary
Photo Credits iStockphoto: 2197568960/© akinbostanci, 175489698/© 400tmax
Layout Design Vera Brauner
Production Kelly O'Callaghan
Typesetting SatzPro, Germany
Printed and bound in the USA, on paper from sustainable sources

ISBN 978-1-4932-2780-8
1st edition 2026

Rheinwerk Publishing, Inc.
2 Heritage Drive, Suite 305
Quincy, MA 02171
USA
info@rheinwerk-publishing.com
+1.781.228.5070

Represented in the E.U. by:
Rheinwerk Verlag GmbH
Rheinwerkallee 4
53227 Bonn
Germany
service@rheinwerk-verlag.de
+49 (0) 228 42150-0

Library of Congress Cataloging-in-Publication Control Number: 2025041791

Contents at a Glance

Contents

4 Designing Worksheets 265

10 Refining Worksheets with Graphics 465

22 Automating Routine Tasks with Macros 807

23 Visual Basic for Applications 823

Preface

You're expecting a useful book about Excel, but what application is it about, exactly? For several years, Microsoft has been pushing the shift from one-time purchases to subscription-based services. These services are called Microsoft 365, and include Microsoft Excel. This approach has clear advantages, especially when working with Microsoft's cloud services like OneDrive. Subscription versions can be continuously expanded, and you can control the update pace.

Excel Standalone and Excel Subscription

Microsoft would certainly like to attract all potential customers to this subscription model, but it has recognized that not everyone is satisfied with it. The cloud still raises concerns for some, as not everyone feels comfortable entrusting their data to sprawling server farms around the globe.

The compromise is that Microsoft has also released a new one-time purchase version: Excel 2024. Its features and user interface mostly match the subscription version of Excel as of September 2024, with a few exceptions.

If you choose the purchase version, this book offers a reliable foundation for the coming years. While it will receive security updates as needed, its features will remain stable. If you're a subscription customer, this book can help too, but you'll likely see new features and some changes at least every six months. Microsoft usually documents these updates well within the program through help sections and embedded tips, so based on the knowledge in this book, they shouldn't be a major issue.

Updates

The updates Microsoft introduced to Excel since 2021 are generally included in Excel 2024 as well. Without going into detail, the main points are:

- A new default font and subtle changes to the interface design
- Additional text functions: `DETECTLANGUAGE()`, `TEXTAFTER()`, `TEXTSPLIT()`, `TEXTBEFORE()`, `TRANSLATE()`, `REGEXEXTRACT()`, `REGEXREPLACE()`, `REGEXTEST()`
- A series of new functions in the lookup and reference category: `IMAGE()`, `EXPAND()`, `GROUPBY()`, `HSTACK()`, `PIVOTBY()`, `WRAPCOLS()`, `CHOOSECOLS()`, `TAKE()`, `VSTACK()`, `DROP()`, `WRAPROWS()`, `CHOOSEROWS()`, `TOCOL()`, `TOROW()`. Except for `IMAGE()`, these functions focus on handling data in matrix form.
- This also applies to the new functions in the logic category: `MAP()`, `SCAN()`, `REDUCE()`, `MAKEARRAY()`, `BYCOL()`, `BYROW()`.

- Analyzing data from images
- Enhanced autocomplete
- Usage of checkboxes
- Inserting images into cells
- Usage of data types in PivotTables

What This Book Offers

This book describes Excel's capabilities as of 2024. Most examples are explained step by step, allowing you to follow along to the final result. They are selected to showcase the program's power across all areas.

In **Chapter 1** we cover the basic principles of using Excel, and a first example gives you a clear look at Excel's workflow.

Chapter 2 through **Chapter 4** focus on the core spreadsheet functions, formula usage, and table design. **Chapter 5** and **Chapter 6** introduce tools for analyzing existing data. **Chapter 7** through **Chapter 10** demonstrate how to visualize data with charts, sparklines, and free graphics in worksheets.

Chapter 11 to **Chapter 14** cover all ways to publish calculation models and charts, from printing locally and emailing to presenting data on the internet or a company intranet. They also explain the collaboration features Excel supports on internal networks and the web.

Chapter 15 to **Chapter 18** provide complete lists of spreadsheet functions with many examples, plus advanced features like building and querying data tables. They also cover analyzing these tables with PivotTables and charts and managing extensive data models using Power Pivot.

Chapter 19 to **Chapter 21** focus on importing and exporting data between Excel and various applications and data sources.

Chapter 22 and **Chapter 23** offer a concise introduction to automating repetitive tasks and show how to customize Excel with Visual Basic for Applications to fit your needs.

Chapter 24 provides tips on using the Copilot AI extension in Excel.

Books like this are usually not read straight through like novels. If you want to access a specific topic directly, you'll find a detailed index at the end of the book.

Helmut Vonhoegen
hv@helmut-vonhoegen.de

Chapter 1
Basic Knowledge for Working with Excel

Often, the user interface of a program can be most rapidly grasped through an initial attempt to produce something useful within it. For this reason, this chapter starts with an example. It follows an explanation of basic Excel terms and concepts to give you a solid foundation for learning from the rest of the book.

1.1 Starting with a Cost Comparison

Programs like Excel are designed to create documents, and a very common way to use Excel is to create a spreadsheet document that will help you calculate the total prices of purchases. Imagine you want to set up a small wireless network—including a desktop PC, five notebooks, and two tablets—in your company or workgroup. You'll also need an access point with a built-in router for networking. You want to know how much you can afford, what the essentials will cost, and what extras are available. For the notebooks, you'll compare costs for features like 13.5- or 17-inch monitors and 500-GB or 1-TB hard drives, and for the tablets, you'll compare options with and without keyboards.

1.1.1 Starting with the Labels

Begin by setting up the calculation framework and labeling the table's rows and columns.

1. Open Excel by clicking the program's icon. After the Excel logo appears, the Excel window loads. First, a start screen appears, and it offers a button for a blank workbook, quick access to recently opened workbooks, and links to templates. You can disable this start screen under **File • Options • General • Start up options.** If disabled, Excel opens a new workbook immediately.

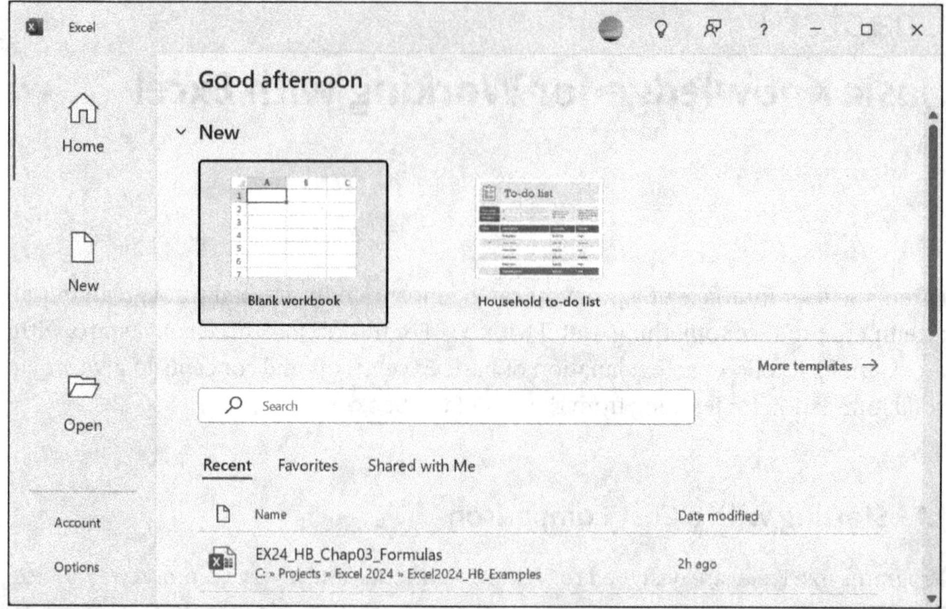

2. Unlike older versions, the current version of Excel opens a separate application window for each workbook. The title bar initially displays the default name **Book1**.

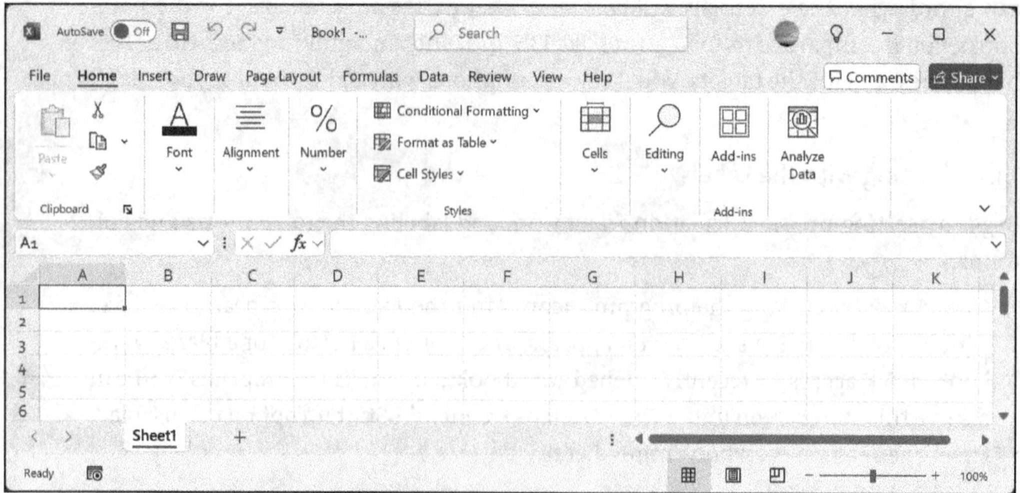

3. You should start by entering your shopping list, preferably in the first column. To do this, you move the mouse pointer to the second cell in the first column and click the left mouse button, or you can use the arrow keys instead of the mouse. On a touchscreen, just tap the cell you want. However you do this, the cell will be outlined with a thicker frame to show it's selected.

4. You proceed by entering into this cell a heading for the entire table, such as "Hardware costs for a workgroup." First, click the **Home** tab on the ribbon, which is always visible. Then, in the **Font** group, click the **Bold** icon, type the heading, and press `Enter` to finish. At that point, the cell pointer will automatically move down one cell.

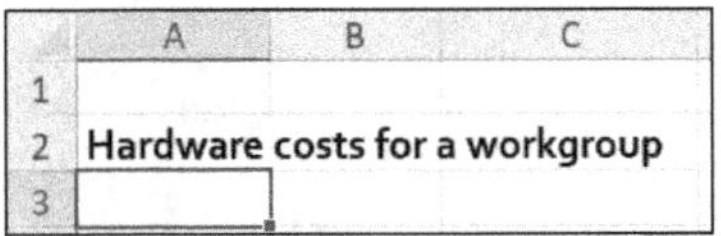

5. Select cell A5, then enter the names of the components needed, one cell at a time, in column A. Press `Enter` after each entry, and to record surcharges for additional features, add the necessary rows.
6. To clearly distinguish the notebook configurations, use an indent. Select cells A7 to A8 by holding down the mouse button, then click the **Increase Indent** icon in the **Alignment** group ❶.

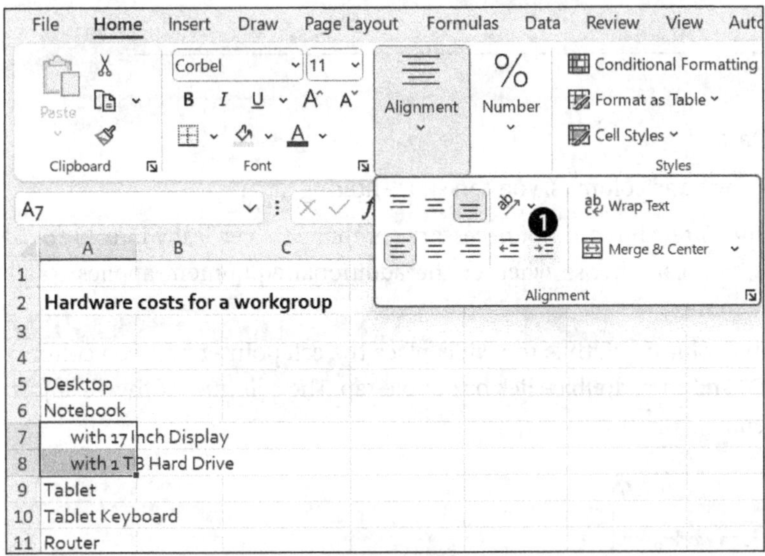

7. To finish, move the mouse pointer to the line between column headers **A** and **B** ❷. Then, click and drag the pointer to the right until all component texts fit into the column.

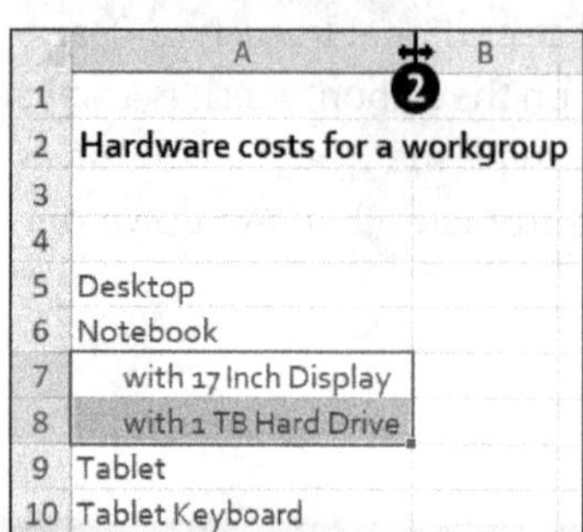

8. Select cell A4, begin labeling the columns from there, and finish each entry by pressing [Tab] or [→].
9. Select the range from A4 to F4 and click the **Bold** icon again.

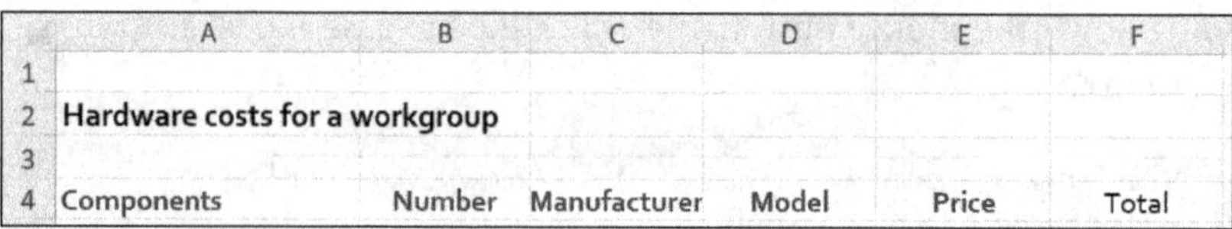

A
B
C
D
E
F
Hardware costs for a workgroup
Components
Number
Manufacturer
Model
Price
Total

1.1.2 Entering Data

After labeling the rows and columns, you can start entering data:

1. Select cell B5 and enter the count of necessary components vertically in column B. For notebooks, you can choose whether the additional equipment applies to all devices or only some.
2. Since the default column width is too wide, place the cell pointer between column headers B and C and then double-click or double-tap. The column width will adjust to fit the content.

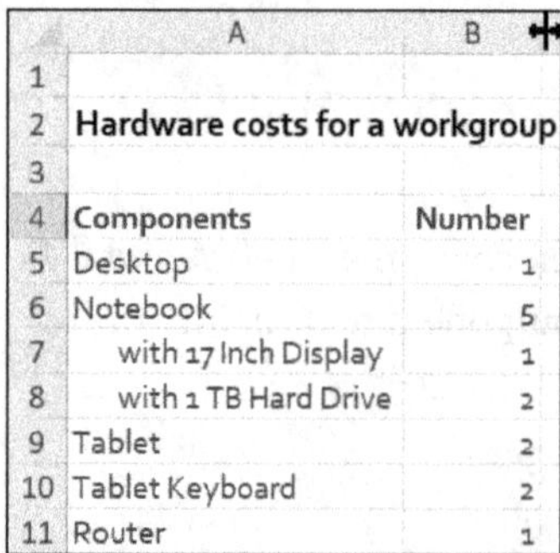

3. Select cell C5 to start entering the manufacturers' names and then press Enter to complete each entry in column C. If components come from a manufacturer that's already listed, simply accept Excel's input suggestion once the first letter matches.

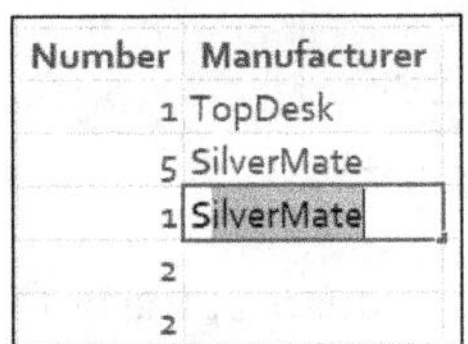

Number	Manufacturer
1	TopDesk
5	SilverMate
1	SilverMate
2	
2	

4. Under **Model**, enter the type designations, and in the **Price** column, enter the amounts provided by each supplier. Next, select cells E5 through E12 and click the **Accounting Number Format** icon in the **Number** group.

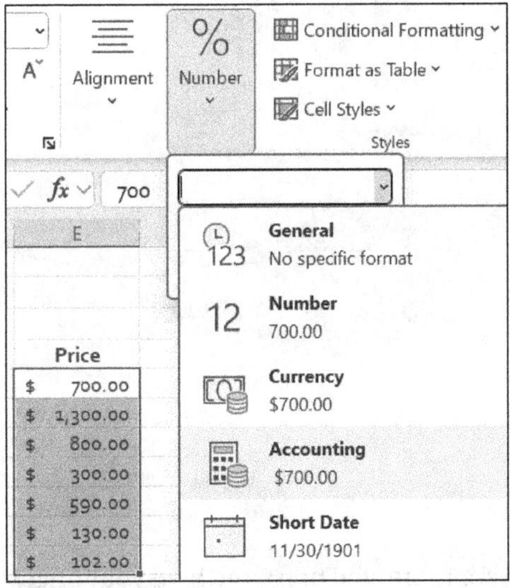

1.1.3 Calculating What It Costs

After entering all prices, calculate the cost of the minimum amount of equipment you need and the price of additional equipment by multiplying each price by the given quantity.

1. Select cell F5, type an equal sign, and select cell E5. Then, enter an asterisk for multiplication and select cell B5. Finish the formula by clicking the checkmark in the formula bar ❶ while keeping the cell selected. The cell will then display the formula's result, which is the value that would normally appear on the other side of the equal sign in math.

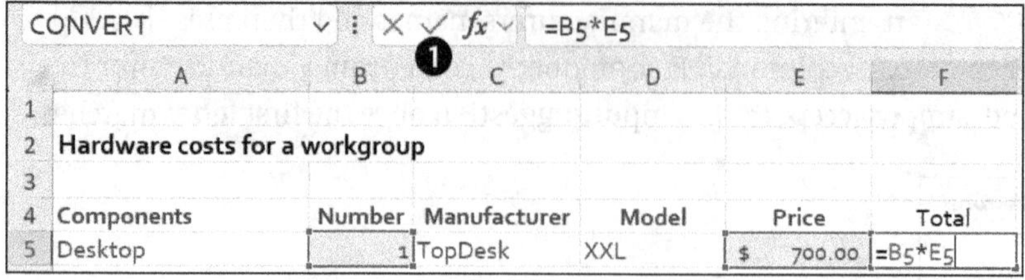

2. Hover the mouse pointer over the fill handle—which is the small square at the bottom right corner of the cell border—and double-click. Excel will then fill the column with copies of the formula from cell F5 down to cell F11, based on the length of column D where values exist. Cell references will adjust row by row, and Excel will show the **Auto Fill Options** button at the bottom. Clicking the small arrow reveals different fill options, and in this case, the default **Copy Cells** option is the right choice. You don't need to make any changes.

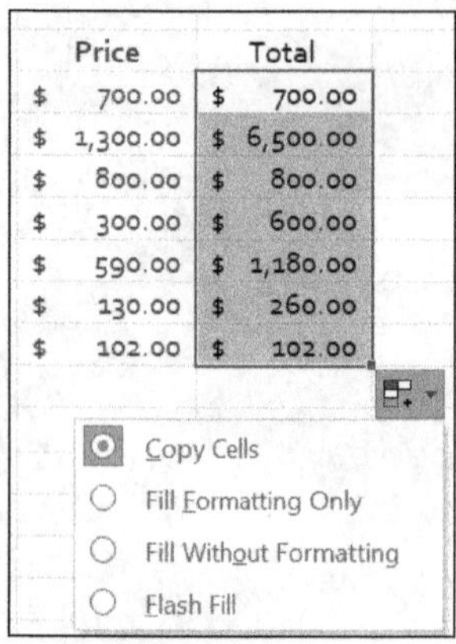

The process is a bit different on a touchscreen. There, you press and hold your finger or stylus on cell F5 briefly and a horizontal context menu appears. When you tap the **AutoFill** icon, Excel will display a fill handle at the bottom right corner of the selected cell, and you can drag it down to cell F11.

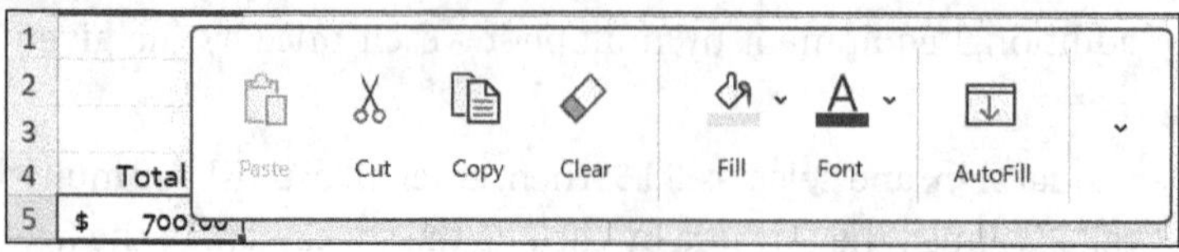

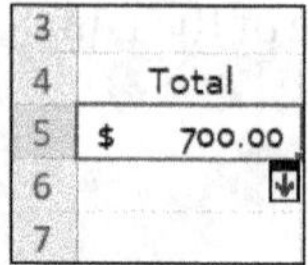

3. Column F automatically applies the dollar format here because a currency amount was multiplied. The column width may now be too narrow to display all values correctly, and values that don't fit might appear as # signs. This only happens if the column width was previously adjusted manually, but if you widen the column as described previously, the full values will reappear.
4. Now, the total sum is still missing. To retrieve it, place the cell pointer in cell F12 and click the **Sum** icon on the **Home** tab in the **Editing** group. Excel is smart enough to detect that you want to sum the range above, and it highlights the range with a dashed border. Confirm by pressing `Enter` or clicking the **Enter** icon in the formula bar.

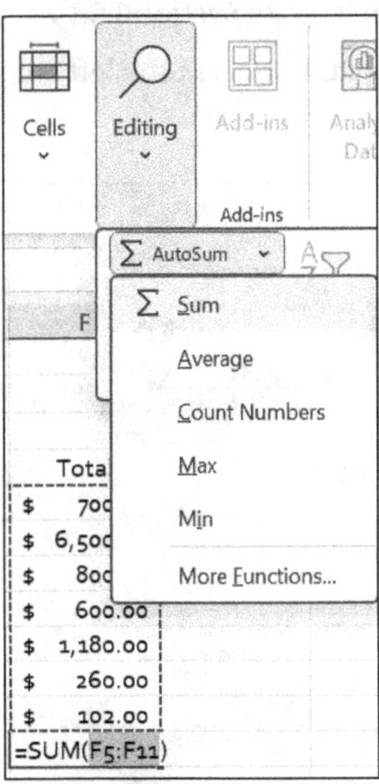

5. To finish the table, enter **Total** in cell E12 and slightly widen column E.

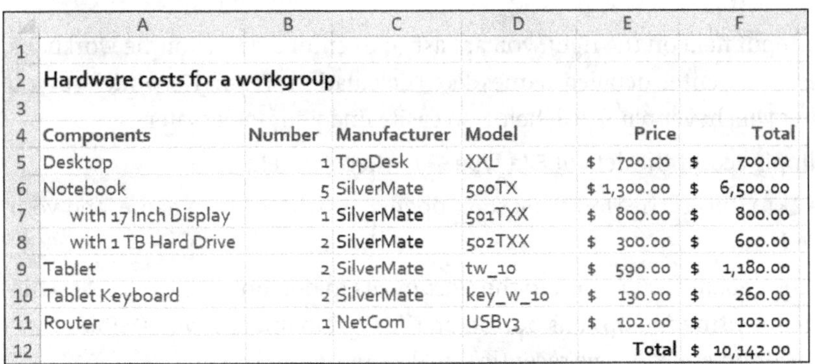

	A	B	C	D	E	F
1						
2	**Hardware costs for a workgroup**					
3						
4	**Components**	**Number**	**Manufacturer**	**Model**	**Price**	**Total**
5	Desktop	1	TopDesk	XXL	$ 700.00	$ 700.00
6	Notebook	5	SilverMate	500TX	$ 1,300.00	$ 6,500.00
7	with 17 Inch Display	1	SilverMate	501TXX	$ 800.00	$ 800.00
8	with 1 TB Hard Drive	2	SilverMate	502TXX	$ 300.00	$ 600.00
9	Tablet	2	SilverMate	tw_10	$ 590.00	$ 1,180.00
10	Tablet Keyboard	2	SilverMate	key_w_10	$ 130.00	$ 260.00
11	Router	1	NetCom	USBv3	$ 102.00	$ 102.00
12					**Total**	**$ 10,142.00**

The first step is now complete. The table works as intended, even if it still looks a bit plain.

1.1.4 Saving the Results

You should save the calculation model now—ideally, right after entering the labels—since, until now, the final version has existed only in your system's main memory.

1. To do this, select the **Save As** command from the **File** tab.
2. On the page that appears, Excel lets you save locally or to internet servers like OneDrive or other web storage. Also, the button with the left-pointing arrow ❶ always returns you to the other tabs. We'll cover saving to the web later, in Section 1.9.
3. If you want to stick with saving locally at first, click the **This PC** button in the left column of the **Save As** page.

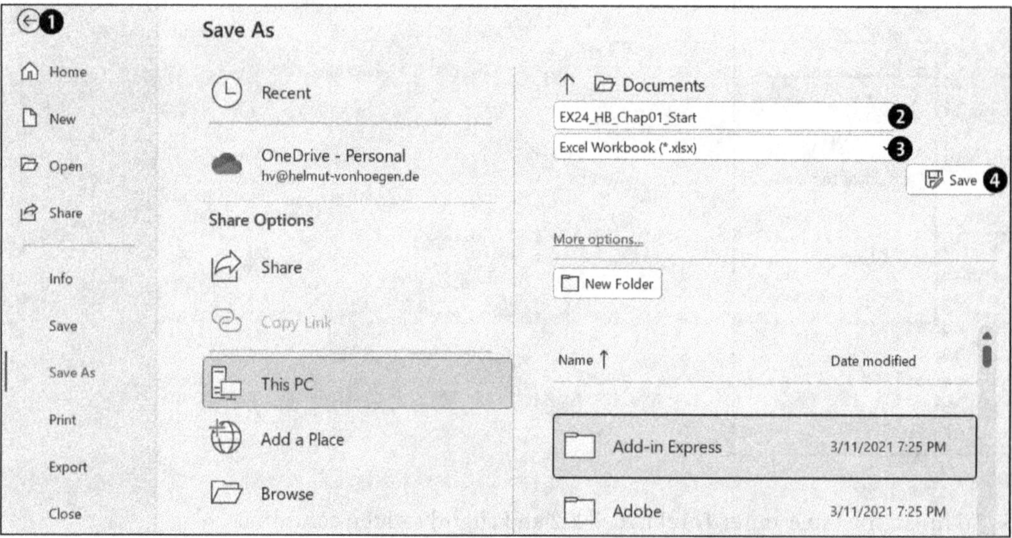

4. In the first input field on the right, you are asked to enter a name for the workbook ❷. Excel lets you enter detailed names like "Calculation_Team_Network," so you should enter whichever name will help you easily find your tables later.
5. You can simply accept the default **File type** ❸ in the next field.
6. Click **Save** ❹ to finish. The workbook stays open, now with its own name, and your data is saved.
7. Excel sets the default file location to the *Documents* folder, unless you choose a different folder by using the **Options** button on the **File** tab under **Save • Default local file location,** or unless you have recently saved to another folder.
8. If you select **This PC**, the right column will show folders or workbooks you've used before. Selecting an item fills in the folder or file name in the fields above.

9. Using the **More options** or **Search** link opens the **Save As** dialog box, which offers additional saving options. This is covered in detail in Section 1.8.2.
10. If you want to save the table again later, click **File • Save,** click the **Save** icon on the Quick Access Toolbar, or press Ctrl+S. By now, your work has a name, so Excel knows where to save the data.
11. When you finish working on the workbook for this session, you can close both the workbook and its application window. To do this, use the **Close** icon in the title bar or press Alt+F4. To close only the workbook, click **File • Close** or press Ctrl+F4.
12. To access the calculation sheet later, use the **Open** command on the **File** tab. As happens when you use the **Save As** command, Excel first shows a page with various locations for workbooks, plus lists of recently edited workbooks and recently used folders.

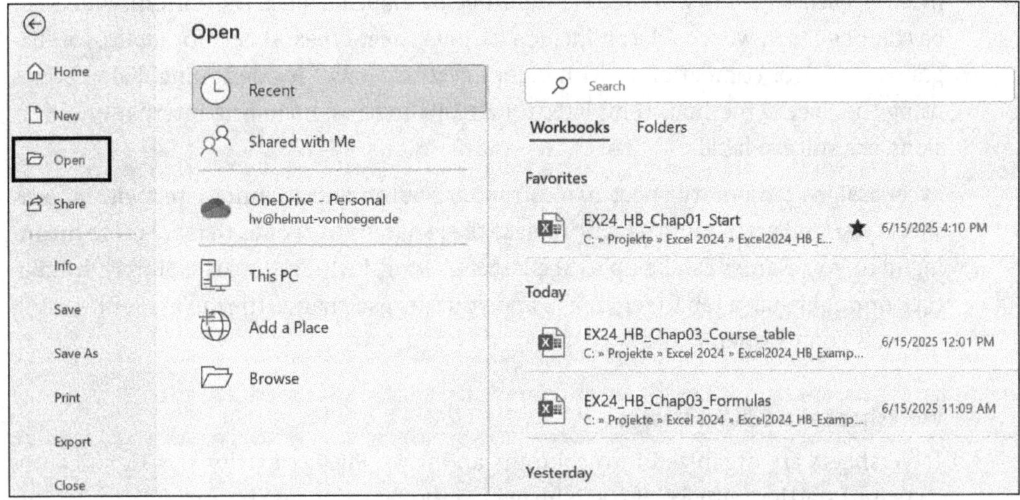

13. If you choose the **This PC** option, Excel displays the current folder and recently used folders, along with the **Search** button, which opens the **Open** dialog box at the last used folder.
14. The **Open** dialog box lists the available workbooks, so you can use it to select the desired name and then click **Open**.

1.2 Basic Concepts

Excel's workspace is modeled on the familiar setup of a physical desk. On your desktop, Excel provides a work area in which multiple tasks can be handled simultaneously in separate windows.

1.2.1 Workbook, Worksheet, and Cell

Excel offers workbooks for individual tasks to work with. A workbook, much like a binder or file folder, can, for example, gather all tables and charts you need to manage an inventory. One key advantage of using workbooks is that you can use them to perform certain operations—such as formatting specific table areas—across multiple sheets at once. You can also use workbooks to run calculations that draw data from several tables.

A workbook opens with a user-selected number of sheets. By default, Excel opens a new workbook with a single sheet, but you can increase the default number of sheets up to 255 (the maximum number of sheets allowed in a workbook) via menu path **File • Options • General** under **Include this many sheets**.

Sheet Types

In Excel, each sheet in a workbook is identified by a labeled tab at the bottom, which can be color coded. A workbook can include various sheet types: sheets for tables, for diagrams, and—for compatibility with older Excel versions—for designing dialog boxes using the Excel 5 method. Templates for Excel 4 macros, including international versions, are still available.

Excel assigns temporary sheet names that are automatically numbered within each sheet type, and you should usually replace these names with ones that are more meaningful to you. Names can be up to 31 characters long, including spaces. Simply double-click or double-tap a tab to rename it, and you can also change the tab's color by using the tab's context menu.

Worksheet and Cell Addresses

Table sheets are organized into columns and rows. Sheet capacity was significantly increased starting with Excel 2007: Instead of the previous maximum of 256 columns, Excel now supports up to 16,384 columns, which are labeled with letters from A to XFD. The row capacity has also increased, from 65,536 to 1,048,576 rows, which are numbered starting at 1. However, when working with data models, you can handle up to two million table rows.

The extent to which you can actually leverage these vast capacities depends heavily on the size of your system's main memory. To manage such large data sets, the 32-bit version of Excel now allows up to 4 GB of memory when saving a workbook. Excel also supports multicore processors, so if you need larger workbooks, you can use the 64-bit version of Excel—which removes this 4 GB limit and is now Excel's default version. However, you'll need to keep in mind that users with 32-bit Excel cannot open such large workbooks.

At the intersection of a column and a row is a *cell*, which is where you enter data. Each cell can hold up to 32,767 characters, and formulas can contain up to 8,192 characters. Since Excel 2007, formula nesting levels have increased from 7 to 64.

Like each square on a chessboard, each cell in a sheet has a position that is defined by its column and row labels. Each cell also has a unique address, like this one:

```
[Report3]Costs!F6
```

The meaning of each component of this address is summarized in the following table:

Address Component	Meaning
[Report3]	The cell is in the workbook file named *Report3*.
Costs!	The cell is on the sheet named *Costs*.
F	The cell is in column F.
6	The cell is in row 6.

The workbook name is always enclosed in square brackets, and an exclamation mark follows the name to show it refers to a sheet, not a cell range. When entering a formula, you can omit the file name if the cell is in the current workbook—that is, in the file shown in the active application window. If the cell is in the sheet where the cell pointer currently is, then you can also omit the sheet name.

Since its first version, Excel has allowed an alternative notation for cell references that uses row and column numbers. For example, instead of B2, you can use R2C2. To enable this alternative reference style, go to the **File** tab, click the **Options** button, and select the **Formulas** tab. There, under **Working with formulas**, check the **R1C1 reference style** option. However, note that this book does not use the alternative reference style.

The Active Cell

A single cell in the table is highlighted with a thicker border, called the *cell pointer*, that marks it as the active cell. In the lower-right corner of the cell border, you'll also see a small box that's called the *fill handle*. A selected cell is ready for manual data entry, and if the fill handle isn't visible, then the user's ability to fill the cell is disabled. This feature is useful when creating applications that restrict user input to specific areas. To display the fill handle again, go to **File • Options • Advanced**, and under **Editing options**, check **Enable fill handle and cell drag-and-drop**.

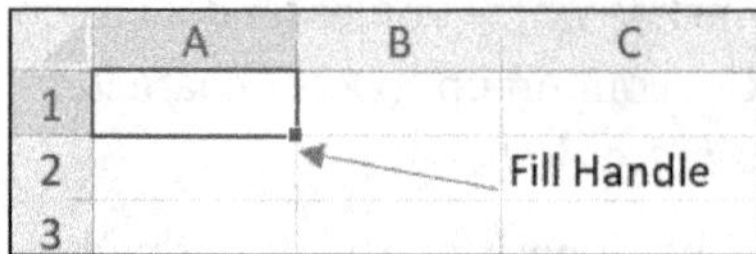

Figure 1.1 Cell and Fill Handle

Commands affecting a cell apply only to the selected cell. You can move the cell pointer by using the mouse, arrow keys, or commands like **Go To**.

On a touchscreen, you tap a cell to select it. To fine-tune the selection, you double-tap the cell border on the side where you want to move it. The fill handle also appears on the Surface Tablet touchscreen when you use the Microsoft Pen instead of your fingers.

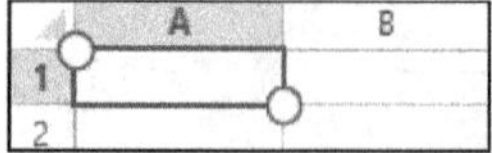

Figure 1.2 Selected Cell on Touchscreen with Two Handles

1.2.2 Cell Content and Cell Format

Each cell has at least two aspects. One aspect is the cell's content: A cell can contain data, or it can be empty. The other aspect is the cell's format, which determines how the content is displayed: the format, font, color, and other visual characteristics of the data in the cell.

Excel treats the two aspects separately. You can clear a cell's contents while keeping its format, or you can copy a cell's format without copying its contents. This aspect also covers properties that protect the cell from changes or set data validation rules.

A third aspect can include a comment or note with hints about a value or formula. Cells containing *sparklines* have a special status, and they are stored in a separate aspect of the cell. Learn more in Chapter 9.

1.2.3 Cell Ranges

Excel commands apply not only to individual cells but also to groups of cells called ranges. A *range* is a rectangular group of cells that you either select manually or reference by a range address. When you select a cell range, Excel outlines it with a border and fills the cells with a background color. It also highlights the headers of the rows and columns where the range appears, and the fill handle appears in the lower-right corner. The active cell in the cell range is marked by a white background.

A cell range includes at least one cell and can extend to all cells in a sheet or workbook. Ranges within a sheet are two-dimensional, but they can also span cells across adjacent sheets. Ranges are defined by the addresses of two opposite or diagonally paired corner cells, and range addresses are specified using a "from: to" notation. For example, the range shown in Figure 1.3 has the address A1:D5.

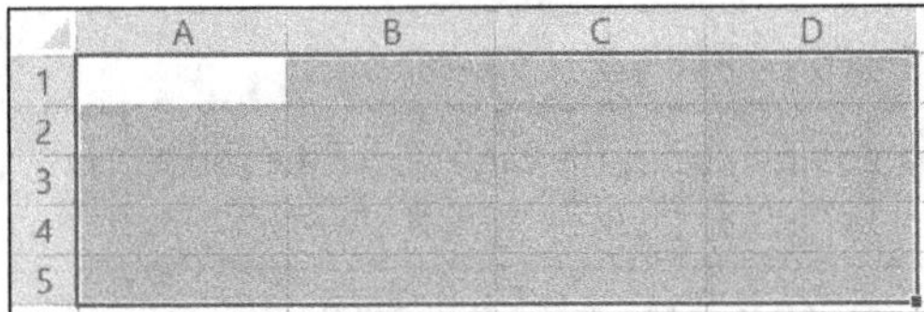

Figure 1.3 Cell Range A1:D5 with A1 as Active Cell

On a touchscreen, the selected range appears slightly different when you work in **Touch input** mode.

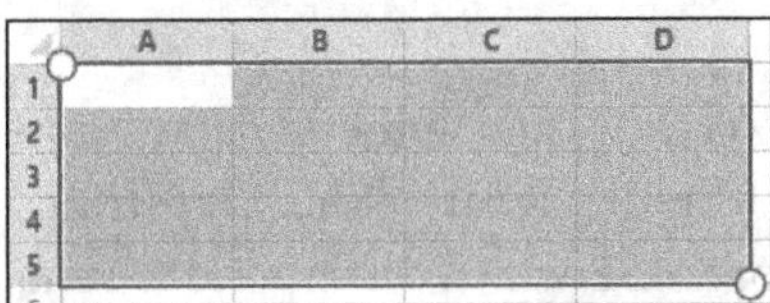

Figure 1.4 Cell Range A1:D5 on Touchscreen

When you tap a cell with the pen, the pen tip acts like a mouse pointer, allowing you to drag the range's frame and fill handle just as you would with a mouse. Excel automatically detects whether you select a cell with your finger or the pen, so it's easy to select a cell sometimes with your finger and other times with the pen. To use this functionality, you need to have the **Use pen to select and interact with content by default** option enabled under **Advanced** in the **Excel Options** dialog.

You can also name cells and ranges, which allows you to reference them by those names.

Multiple Ranges

Excel lets you select multiple nonadjacent cell ranges at the same time. Most operations that are available for single ranges also work on multiple ranges, but cutting, deleting, or inserting cells is only allowed for single ranges.

Data Type

Cells in sheets can contain two basic types of data:

1. **Constants** are fixed values that are entered directly into the cell: numbers; text like labels, names, and product descriptions; dates such as 10/12/2025; or times like 12:33.
2. **Formulas** are calculation rules or data manipulation instructions that tell the application to display the result in the cell. Formulas always begin with an equal sign and can include constant values, references to cells and ranges, named cells or ranges, operators, and functions.

Formulas	Functionalities
=b2	It returns the value of cell B2.
=b3+b4	It returns the sum of the two cell values.
=NetAmount*VAT	It calculates the gross amount by using the values from the two named ranges.
=LEFT(b3, 15)	It returns the first 15 characters of the content in cell B3.

When a cell contains a formula, the formula appears in the formula bar when the cell is selected, while the cell itself shows the result.

Text Strings and Numeric Values

The second important distinction here is between text strings and numeric values. Constants are either text strings or numeric values, and formulas return either text strings or numeric values. Logical expressions return TRUE or FALSE, which correspond to 1 and 0. (Logical values can also be used in simple calculations.)

Text strings can be words or sequences of numbers, so for example, item numbers can be stored as text strings. When you enter data into a cell, the program determines whether to treat it as a text string or a value and displays it accordingly. Text strings are left-aligned, while values are right-aligned.

1.3 Starting and Closing Excel

Excel should start quickly, and you have several options to start and close the program.

1.3.1 Startup Options

Excel requires Windows 11 or 10 as the operating system. You can customize the **Start** menu in **Windows Settings** under **Personalization • Start**, and you can also choose to create a **Start** menu item for Excel to launch the app. A more convenient option is **Pin to Taskbar**, which is available in the element's context menu. In some cases, it's useful to open the program with administrator rights by using the **Run as administrator** option. If the program icon is pinned to the taskbar, you can use just one click or tap to open Excel.

When you start Excel by clicking the entry in the **Start** area or the icon on the taskbar, the **Start Screen** appears first—unless you've disabled it. This screen offers a template for a new workbook along with a variety of other templates. You can click the **More templates** link to open the **New** page.

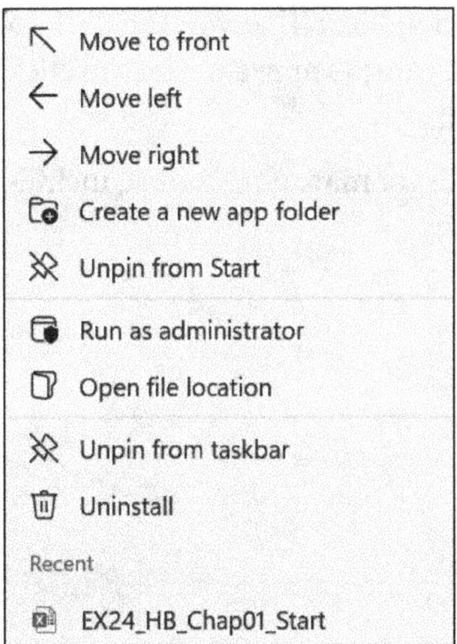

Figure 1.5 Context Menu for Excel Icon

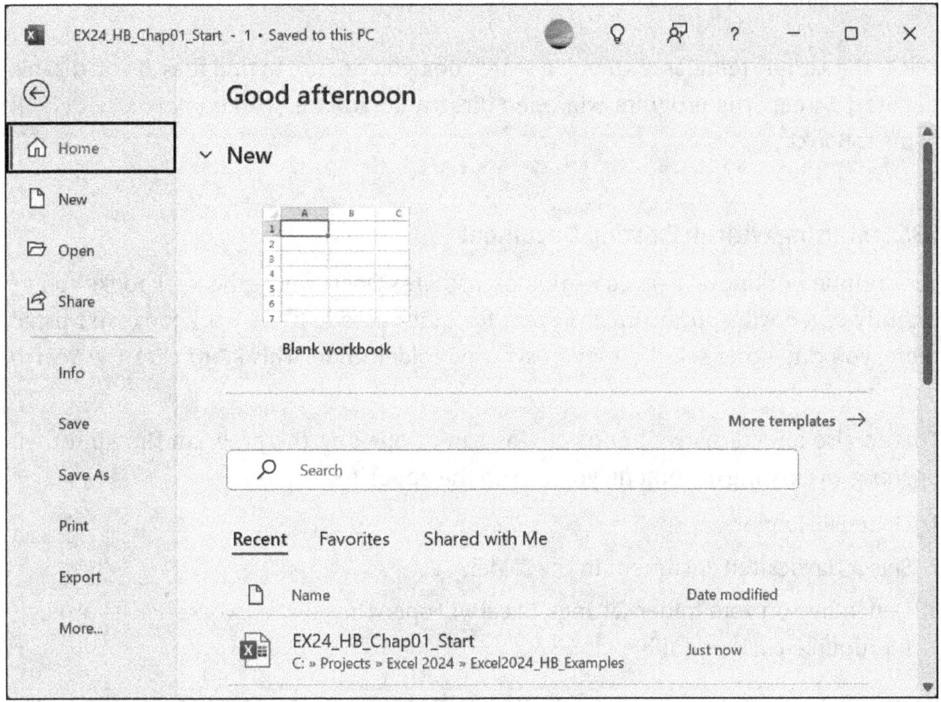

Figure 1.6 Start Screen, Which Facilitates Access to Your Documents

Below that, you'll see a list of recently edited workbooks, pinned workbooks that always show up, and—if available—the names of shared workbooks. This speeds up access to the workbooks you use every day.

To the right of each selected file name, you'll see a small pin icon that you can click to add the file to the **Pinned** list and thus ensure that the file always appears. You can click the other icon to share the workbook without opening it.

The context menu for a list entry offers useful commands for managing the list, including **Open a copy** and **Copy path to clipboard**.

Figure 1.7 Context Menu for File on Start Menu List

Below the sample templates, there's a search box you can use to find files. If you disable the **Start Screen**, the program will open directly to a blank workbook whose default name is **Book1**.

1.3.2 Starting with an Existing Document

To continue working on a saved workbook, use **File • Open**, where the workbooks you've recently saved will appear under **Recent** for quick access. If the workbook isn't listed there, you can either select a previously used folder under **This PC** or click the **Search** button to find it.

There's also an alternative approach. Instead of opening the program first and then opening an existing document, you can do the opposite:

1. Open File Explorer.
2. Select the desired document in the folder.
3. Depending on your folder settings, Excel will open the selected workbook if you click, tap, double-click, or double-tap it.

If you don't remember the exact file name or the folder where it's saved, use the search function to find it.

The context menu for the Excel icon in the taskbar also gives you quick access to your recently used workbooks. These menus are called *jump lists*, and each entry has a context menu that lets you, for example, print the workbook immediately.

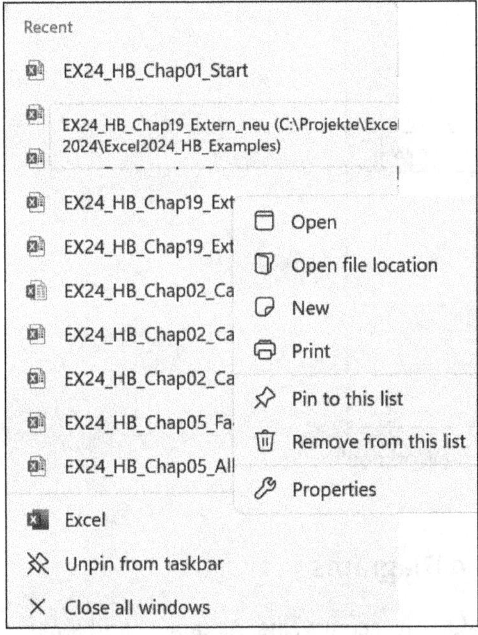

Figure 1.8 Context Menu for Excel Taskbar Icon

[+]

Pin Workbooks to Jump Lists

A jump list usually shows only the workbooks you've used most recently because Excel removes workbooks that you haven't used for a long time. To keep a workbook on a jump list, click the pin icon at the end of its row or select **Pin to this list** from the context menu. That will add the workbook to the pinned workbooks area.

1.3.3 Exiting Excel

To end your Excel session properly, you need to make sure to close it correctly so that your session settings will be available the next time you start Excel. You can close an Excel application window in one of two ways:

1. Click or tap the **Close** button in the title bar.
2. Press [Alt]+[F4].

Excel automatically checks for unsaved changes, and if there are any, it will prompt you to save them (see Figure 1.9).

You can choose **Save** to keep your changes—or you can choose **Don't Save** to discard them, which can be useful if you've tried things you want to discard. The **Close** command on the **File** tab closes only the current workbook, leaving the application window open.

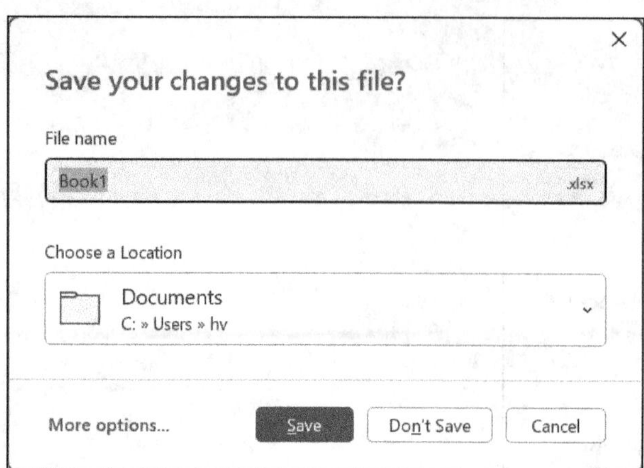

Figure 1.9 Prompt to Save Changes When Closing Workbook

1.4 Construction Site for Tables and Diagrams

The device landscape in IT has changed rapidly in recent years. As tablets and smartphones featuring touchscreens have become available, fingers have become a newly recognized input tool, and the mouse and keyboard now face competition from fast fingers and their gestures. And while it's true that people who enter large amounts of data will likely still prefer a physical keyboard over a virtual one, in many situations—especially on the go—accessing workbooks via a tablet is convenient and can quickly become a habit.

1.4.1 Two Operating Modes

Excel offers two operating modes: ❶ the classic mouse-and-keyboard mode and ❷ a mode for fingers or a pen, as used on tablets or touchscreens. To better support these devices, Excel also includes a special touch mode. If your device supports touch, the **Draw** tab appears automatically in Excel, and you can also switch manually to this mode by using the **Touch/Mouse** mode button in the Quick Access Toolbar.

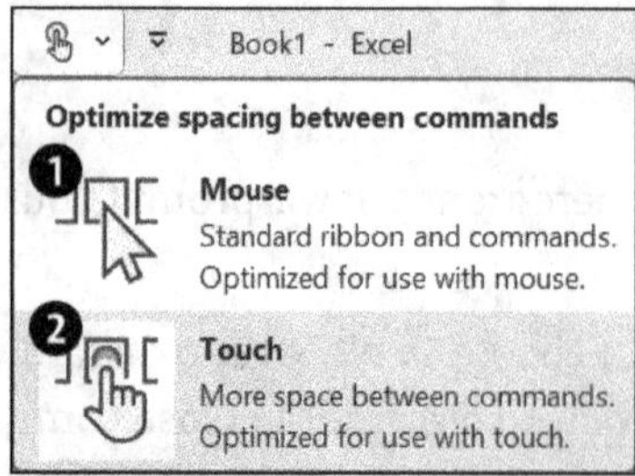

Figure 1.10 Menu for Changing Input Modes

In **Touch input** mode, the Quick Access Toolbar and ribbon display buttons are spaced farther apart, making it easier for you to tap the correct icon with your finger. Even in context menus, like the one for the status bar, there's wider spacing between options. Some context menus have been redesigned so the most frequently used options appear side by side and less common ones are arranged vertically, and palettes, such as the color picker, are enlarged to help you select individual patterns more easily.

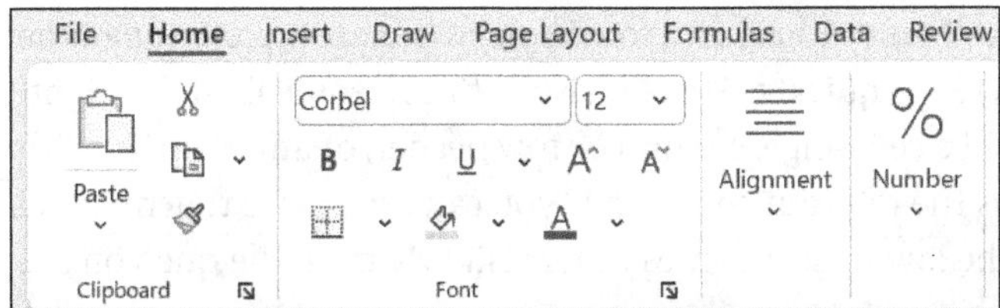

Figure 1.11 Expanded Ribbon for Touch Input

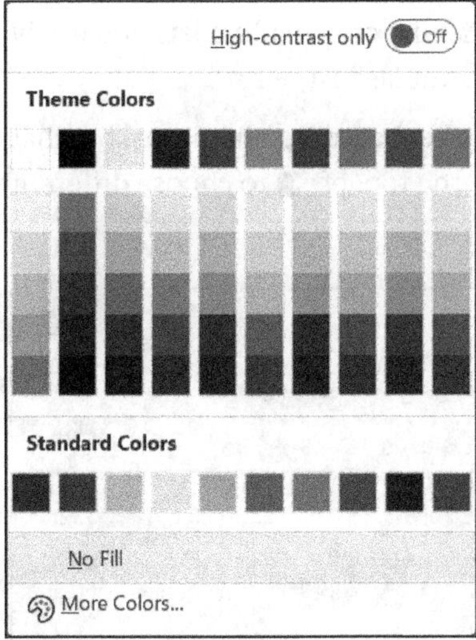

Figure 1.12 Enlarged Color Palette for Mouse and Touch Input

To select a cell, simply tap it. You can also select the insertion point in text the same way. Dragging with your finger mimics dragging with your mouse to select cell ranges in Excel, and small round handles appear at the corners of ranges on touchscreens to make this easier.

Figure 1.13 Selecting Text in Cell by Using Touch Input

To select part of the text inside a cell, tap the text and then drag the circle that appears to expand the selection. You can adjust the selection in both directions, and you can use the circles that appear at each end of the selection to fine-tune it.

To clear a cell's contents, press and hold the cell briefly and then tap the **Clear Selection** button in the horizontal context menu. To move graphic objects out of a cell, tap and drag them. To resize an object, tap it and drag the handles.

To mimic double-clicking a cell on a mouse on a touchscreen (for example, to open the cell for editing), you tap it twice quickly. There's also a finger equivalent for right-clicking. If you press and hold a cell range, column or row header, or any element with a context menu, Excel shows the relevant commands you can use on the touchscreen. However, as noted earlier, the layout sometimes differs slightly from the one you see when using a mouse.

Swiping is a special gesture where direction matters. Swiping up on the cell grid moves the screen area down, while swiping left reveals columns further to the right. Using two fingers enables additional gestures. For example, pinching or spreading lets you easily zoom in or out, which is often useful when editing formulas in a cell.

On tablets that use only a virtual keyboard, you'll see a keyboard icon in the taskbar that you can tap to open the keyboard. Use the button at the top left ❶ to choose different keyboard layouts.

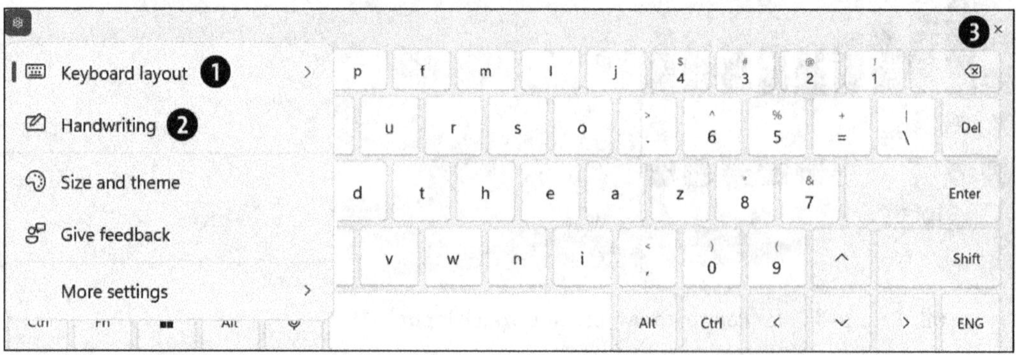

Figure 1.14 On-Screen Keyboard with Menu to Select Layout

The **Traditional** layout makes it easier to enter symbols and numbers. Other layouts separate letters and numbers, and you can switch between them by using the **&123/abc** key. You can also tap the **Close** button with the cross ❸ to hide the on-screen keyboard. When the keyboard is open, Excel automatically adjusts the screen area so you can always see the input field.

When using Excel on a touchscreen, you can also enter characters with your finger or the pen by using the **Handwriting** ❷ feature. It recognizes your input as digits and letters and inserts them into the active cell when you press `Enter`.

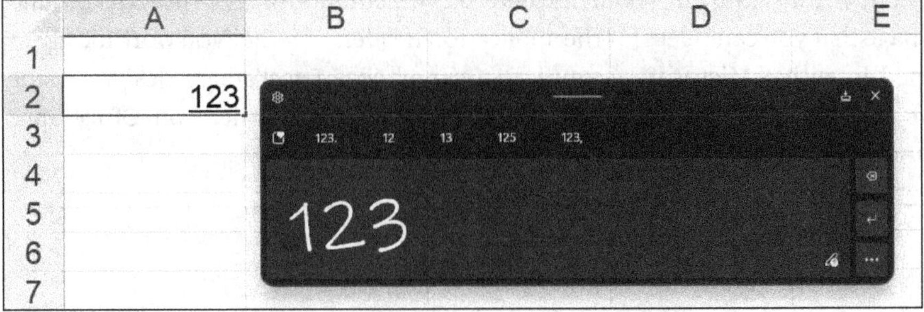

Figure 1.15 Data Entry with Finger or Pen

You can also operate the Surface tablet with clickable cover keyboards. (Or, you can easily connect a USB mouse or keyboard to the USB port; this is fully supported.)

1.4.2 Local and Network Users

Since Excel 2013, Microsoft has distinguished between (1) local users who work only on their own device and (2) those who sign in with an account to Microsoft to use cloud services like OneDrive and then sync their activity across multiple devices. You can sign in by using the link in the top right corner of the application window, as long as the program isn't linked to an account yet. Just click the provided link.

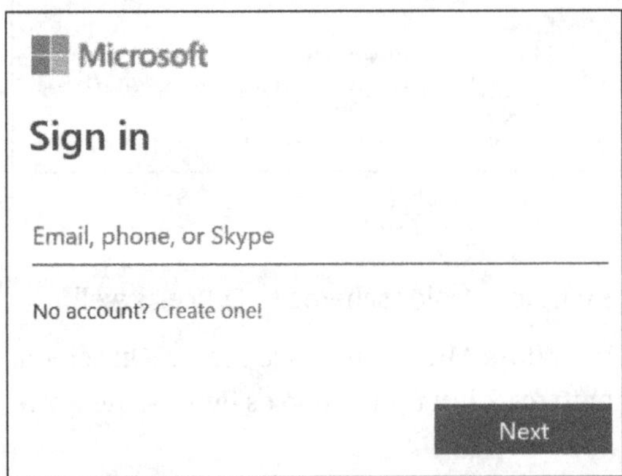

Figure 1.16 First Step to Creating Account

You can sign in as an individual with an account for other Microsoft services, or for an organization, school, or business that's subscribed to Windows 365.

When you sign in to Microsoft, your account details appear in the top right corner of the application window and on the **Account** page, which you can access via the **File** tab. You can then connect to various web services like OneDrive or OneDrive for Business by entering the appropriate credentials. Use the **Add a service** button to do this.

As an incentive for signing in, you'll find decorative elements under **Office Background** on this page that you can select for the application header. Whether you work locally or are signed in with a Microsoft account, you can choose Excel's base design on the **Account** page under **Office Theme**. This setting applies to all Microsoft Office apps simultaneously.

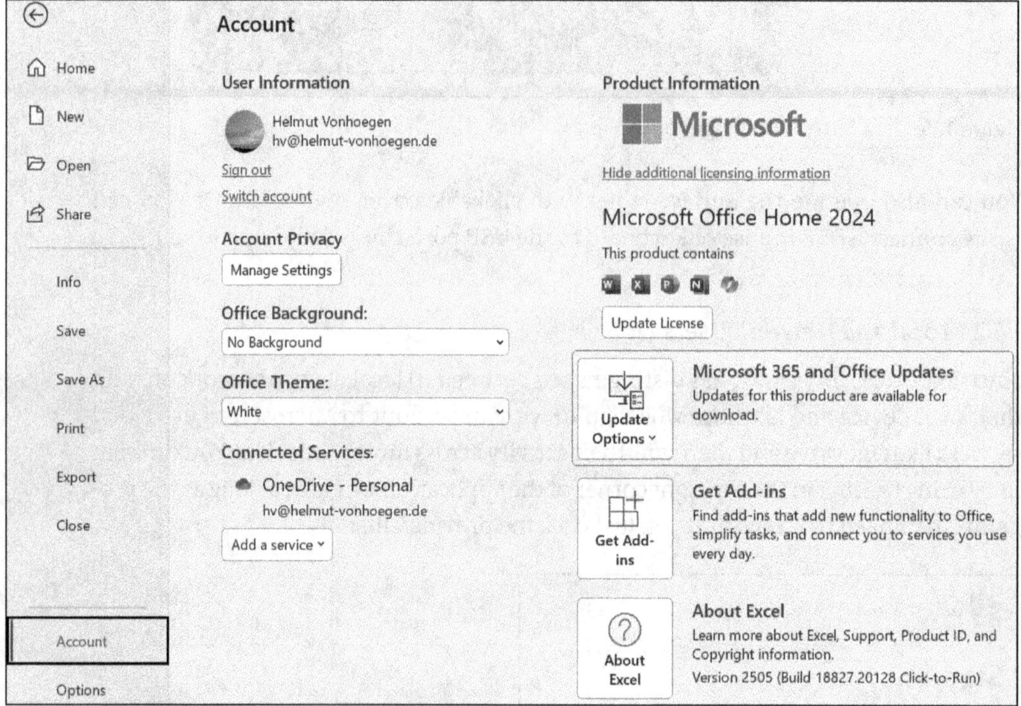

Figure 1.17 Account Page Showing Details

The **Use system setting** option applies the Windows color scheme to Office as well.

The figures in this book consistently use the **White** Microsoft Office theme. On screen, dark or black themes often make text easier to read, but for the book's figures, the white design is more suitable.

Office Theme:
White
Dark Gray
Black
White
Use system setting
Colorful

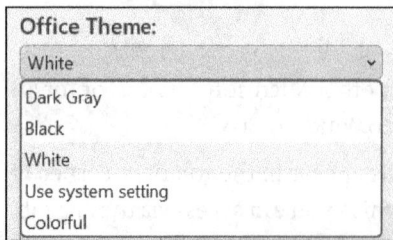

Figure 1.18 Menu You Can Use to Select Base Design for Your Office Apps

1.4.3 The Application Window

When you start Excel, a program window opens for a single workbook, as mentioned earlier. You cannot manage multiple workbooks in one application window, so to work on multiple workbooks simultaneously, simply open as many application windows as you need.

Each window has a title bar with tree buttons on the right and a minimal frame to save space.

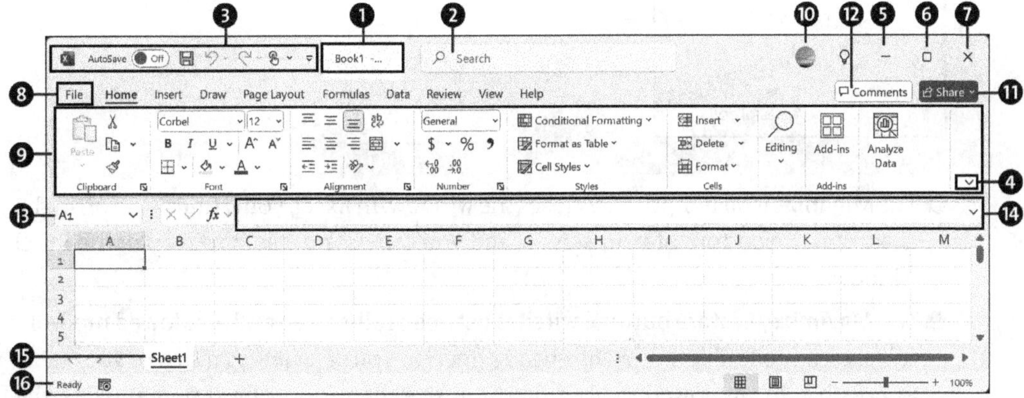

Figure 1.19 Screen Elements in Excel

Here's a brief overview of the different elements that are called out in the figure:

❶ The title bar shows the workbook name in the center.

❷ Next to it is a search box where you can look up a column label or a cell value, for example. You can also use the search box to directly access certain actions, like increasing the font size of the selected cell. Clicking or tapping the search box lets you type a question and offers access to help pages.

Using the **More search results for ...** menu opens Excel's **Search** task pane, which provides internet resources.

❸ The Quick Access Toolbar holds frequently used icons and can be placed above or below the ribbon.

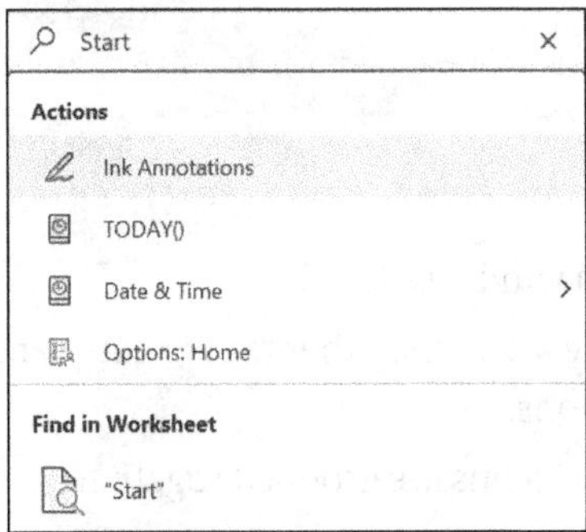

❹ Click the **Ribbon Display Options** icon to choose what appears on the ribbon. **Full-screen mode** hides the ribbon completely until you need it, when you can click or tap the top edge to show the ribbon. **Show tabs only** shows only the tab names at first, and you can click or tap a tab to open it. **Always show Ribbon** keeps all tabs visible permanently, and **Hide Quick Access Toolbar** does what it says.

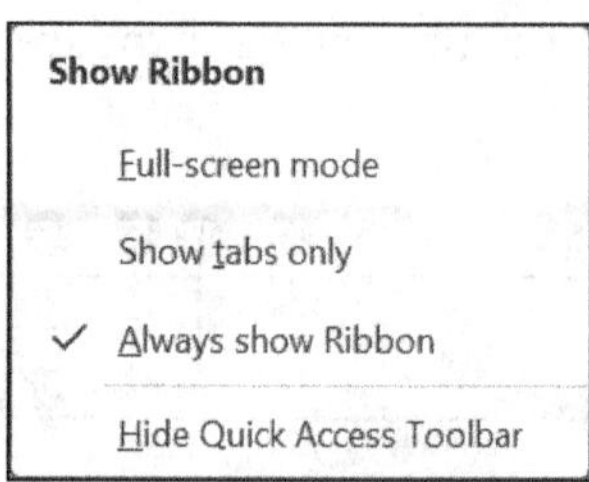

❺ The **Minimize** icon temporarily closes the window. In Excel, you'll find a button in the taskbar at the bottom of your screen, and you can click or tap it to reopen the Excel window.

❻ The **Maximize/Restore** button switches between full-screen and windowed mode. In full-screen mode, Excel uses the entire screen so you can view as much of your table as possible. In windowed mode, you can resize the window by simply moving the mouse pointer to the window's edge and then dragging it in the desired direction while holding down the left mouse button. This is useful when you want to view multiple program windows on the screen simultaneously. Also note that the mouse pointer changes to a double-headed arrow when you're resizing the window.

To view alternative window sizes, hover the mouse pointer or a pen over the icon. You can select alternative sizes by clicking or tapping them, and this is especially handy on large screens when you want to keep multiple windows open simultaneously.

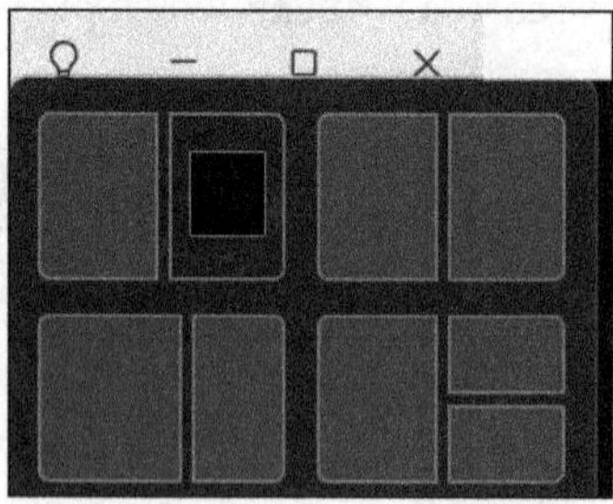

❼ Use the **Close** button to exit the current application window.

❽ Use the **File** tab to open the program's Backstage view and the other tabs in this bar to access commands and options for specific functions.

❾ In each section of the ribbon, related commands and icons are grouped together.

⑩ Use the **Sign In** button to set up and manage your user account. After signing in, your username or photo will appear, and from that link, you can edit account settings or switch accounts if multiple users share the program. You can also access these commands via **File • Account**.

⑪ Use the **Share** button to share the workbook.

⑫ Use the **Comments** button to open the **Comments** task pane.

⑬ The name box shows the current address or range and recently used functions.

⑭ Use the formula bar to edit formulas and cell values.

⑮ Use sheet tabs to switch between the sheets in a workbook.

⑯ The status bar displays the application's current status and provides quick summaries of selected cell ranges.

1.4.4 The Ribbon

The tools for controlling the program are organized in a ribbon made up of overlapping tabs, with the tabs arranged in a bar below the title bar. You can customize the ribbon elements to fit your work habits and needs, and you can remove functions you don't use from the ribbon. For example, if you don't need the suggestions under **Data Analysis**, you can remove the **Analysis** group. This removal isn't permanent; you can undo it anytime.

Each tab also offers task-related groups of icons or commands. These groups can include various controls: buttons that open submenus or task panes, buttons with selection palettes, checkboxes, spin boxes for choosing values, and more. Sometimes, you'll find a *dialog box launcher* ❶ in the lower right corner of a group, which—as the name suggests—opens a dialog box with more detailed options (see Figure 1.20).

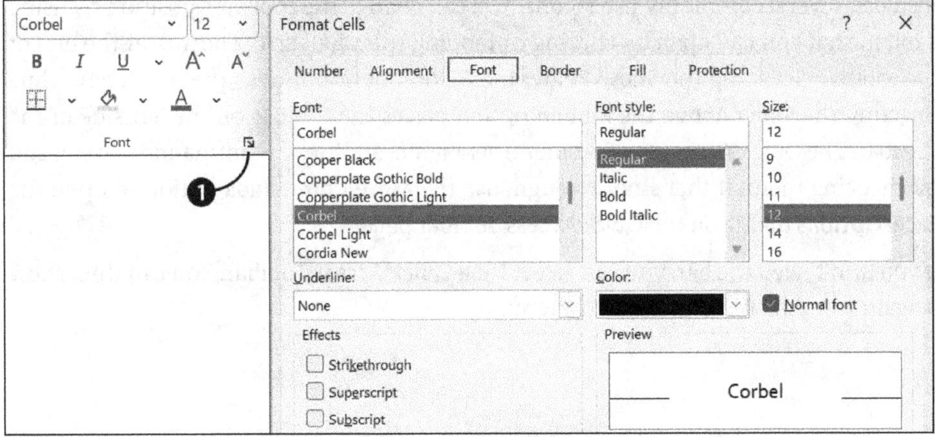

Figure 1.20 Group with Dialog Box Launcher and Its Associated Dialog

For example, in the **Home • Number** group, you can use the dialog box launcher to access the familiar **Format Cells** dialog box from earlier versions.

You bring tabs to the front by clicking or tapping them, unless Excel automatically shows the needed tab based on your current task.

The number of visible tabs depends on your current task, so functions that are irrelevant to your current task remain hidden. This reduces clutter in the user interface. The ribbon adjusts to the program window's width, and if there isn't enough space to display all groups on a tab, then some groups are gradually replaced by narrower buttons. The arrow lets you show hidden groups, and in rare cases, the ribbon temporarily hides completely.

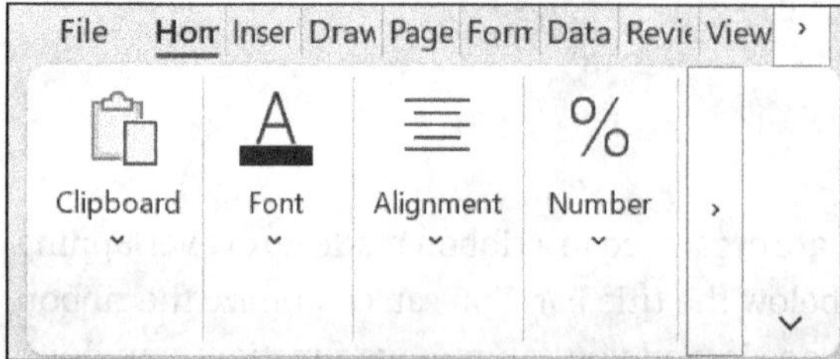

Figure 1.21 Highly Compressed Ribbon

You can also minimize the ribbon via the context menu by using **Minimize Ribbon**, which leaves only the tabs visible. You can then use the same command to restore the ribbon, and you can also double-click or tap a tab to minimize or restore it.

1.4.5 The Quick Access Toolbar

Excel can display the *Quick Access Toolbar* above or below the ribbon, which by default includes several commonly used icons. You can change the toolbar's position by using a menu that you can open by clicking or tapping the small arrow button at the end of the toolbar. Selecting the **Show Below the Ribbon** option moves the bar down, while selecting the **Show Above the Ribbon** option places the toolbar on the left side of the title bar. The Quick Access Toolbar menu also offers additional commands you can add by checking them. If that's not enough, use the **More Commands** option to open the **Excel Options** dialog on the **Quick Access Toolbar** page.

If you don't need the bar, you can select **Hide Quick Access Toolbar**. You can then show it again from the ribbon's context menu.

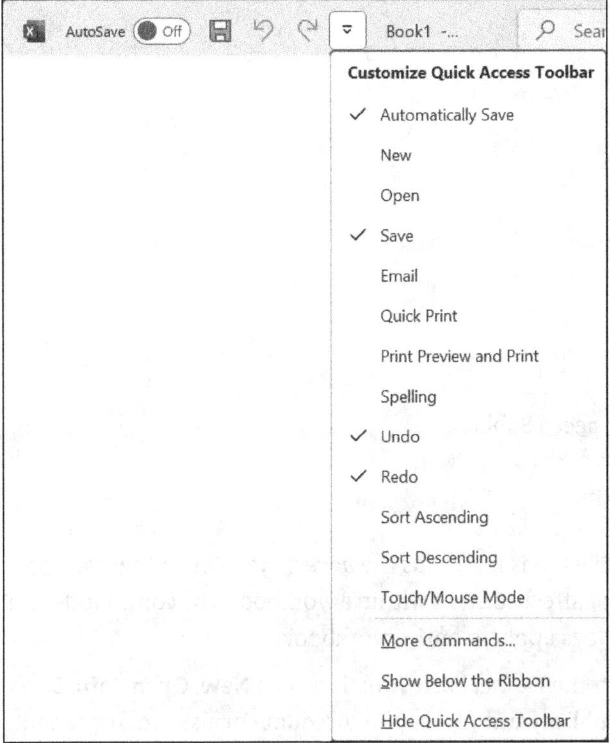

Figure 1.22 Quick Access Toolbar and Its Customization Menu

1.4.6 The Quick Analysis Tool

The Quick Analysis tool is a useful feature that consists of a button that appears when you select a range containing values in more than one cell. Clicking or tapping the button opens a kind of speech bubble with a menu bar and a row of icons for the selected menu item. You can also open the speech bubble by pressing `Ctrl`+`Q`, and if you do, you won't need to select the range first. Instead, you can just place the cell pointer anywhere within the range. If the Quick Analysis tool doesn't appear, check the **Show Quick Analysis Options** setting under **File • Options • General**.

This tool offers a variety of common analysis options for data areas, such as applying conditional formatting or creating charts. Each option is explained in the corresponding chapters of this book.

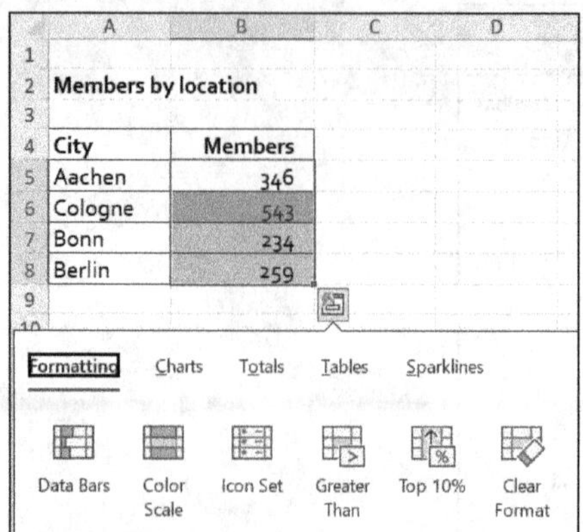

Figure 1.23 Quick Analysis Tool's Speech Bubble

1.4.7 The Backstage View

The area accessed through the **File** tab is known as the *backstage*. While other tabs provide commands and options that affect content within a workbook, the commands and options "behind the scenes" always apply to entire workbooks.

The **File** tab includes, along with the **Home**, basic functions like **New**, **Open**, **Info**, **Save**, **Save As**, **Print**, **Share**, **Export**, **Publish**, and **Close**, plus **Account** for user management, **Feedback**, and **Options**. Each button opens a page with a submenu you can use to choose different options, and some buttons reveal additional options on the right side of the window (see Figure 1.24).

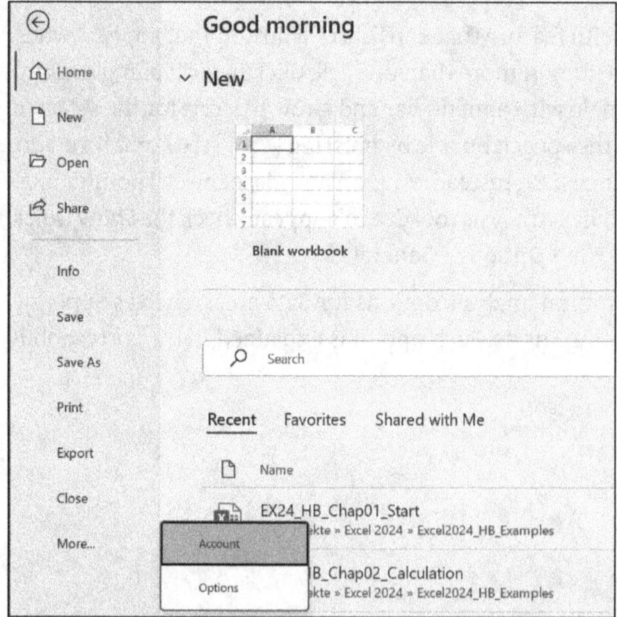

Figure 1.24 File Tab with Homepage

Use the **Feedback** function to contact Microsoft if you notice any issues with how the program works. The **Options** command opens a detailed dialog box with numerous settings you can use to customize the program to your needs, and the left-arrow button at the top of the menu always returns you to the program's normal view. Above the file lists, you'll also find a search box you can use to quickly locate workbooks—just enter the first letters of the name of the workbook you're looking for.

Under **Recent**, you'll see links to your recent workbooks if you've set a value for **Show this number of Recent workbooks**. You can find this value under **File • Options • Advanced • Display**.

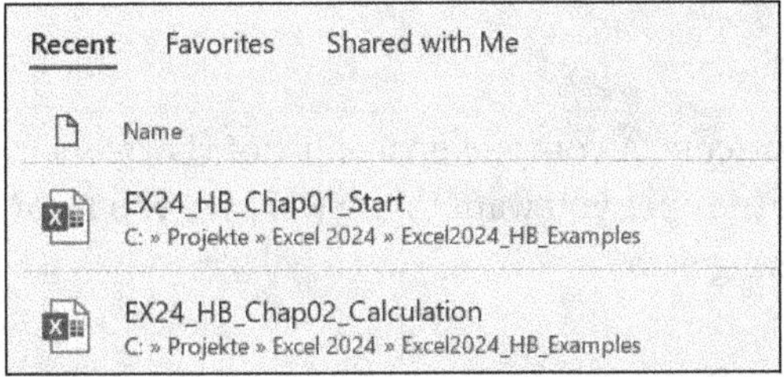

Figure 1.25 Quick Access to Workbooks via Recent List

When you select a workbook from the list, icons appear that you can use to share it or add it to the **Pinned** list. Shared workbooks appear under **Shared with Me**.

1.4.8 Keyboard Shortcuts and Key Sequences

Most users will likely use a mouse or tap with a finger or stylus to operate the ribbon and Quick Access Toolbar, but Excel doesn't leave mouse-phobics or those with tendon injuries stranded. You can definitely master the program by using the keyboard, and you can use special key sequences, all starting with Alt. Unlike as with most keyboard shortcuts, you don't have to press the keys simultaneously. When you press Alt, letters for the different tabs appear on the ribbon.

These letters are based on the command names—for example, N for **Insert**. Once you enter one of these letters, letters for the various options in the groups on the selected tab appear. You can also press Esc to hide the letters if you need to. At first, this feature may seem unappealing, but to see how useful it is, try creating a column chart from a selected table with the key sequence Alt, N, C, and Enter. Then, select palette patterns or list options by pressing Tab or the arrow keys and press Enter to confirm.

In dialog boxes you've opened via the dialog box launcher, hold down Alt and press the underlined letter to choose an option. To justify text in the **Alignment** tab of the **Format Cells** dialog, press Alt+V, use the arrow keys to select **Justify**, and then press Enter. Press Ctrl+Tab to bring tabs in dialog boxes to the front. Press Tab to move to the next item within a tab, the spacebar to check a box, the arrow keys to select from a list, and Enter to confirm.

You can also access Quick Access Toolbar commands by pressing the numbers that correspond to their position on the toolbar.

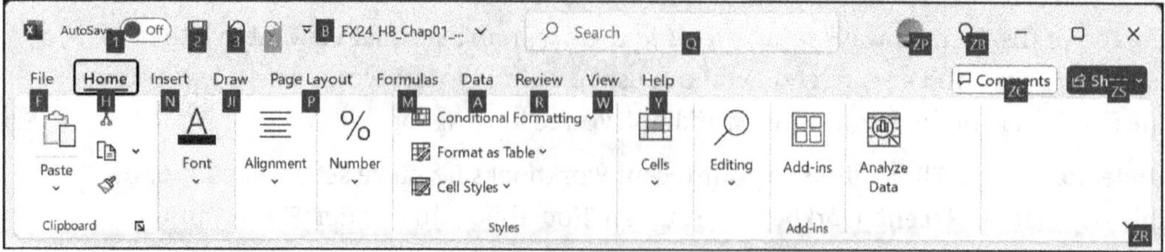

Figure 1.26 Display of Characters for Possible Key Sequences

Using Keyboard Shortcuts

Many commands also have keyboard shortcuts. A comprehensive list of shortcuts is available in Appendix A, and memorizing the ones you want to use will save you from having to switch input devices while entering data.

1.4.9 The Formula Bar

Using the formula bar for data entry is optional since you can enter and edit data directly into the cell, and you can hide the formula bar via the **View** tab in the **Show** group to create more space for workbook windows. On the other hand, while you're developing the worksheet and especially when entering or correcting formulas, using the formula bar is usually better because it offers greater clarity.

The Name Box

The formula bar has several components. The first one, called the *name box*, shows the reference of the active cell, the selected object's reference, or the size of the selection. If ranges have been named, their names appear in the name box when the corresponding cells are fully selected. To select a range of fifty rows and twelve columns when formatting a planned table, you can monitor the range size here while dragging. If **50R × 12C** appears in the name box, release the mouse button or withdraw your finger or pen.

Clicking the arrow button opens the list of previously named ranges, and clicking or tapping a name moves the cell pointer directly to that range. You can also enter names here for selected cells or cell ranges.

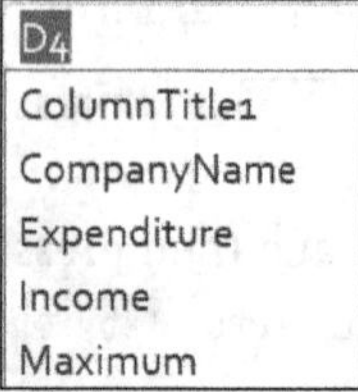

Figure 1.27 List of Names

The Input Icons

The right component in the formula bar stays empty until you enter or edit data in a cell. Once you enter a character in the active cell, the first two buttons to the left of the field activate: the cross button you can use to cancel the entry and the checkmark button you can use to confirm the entry. The **Insert function** button remains visible at all times, and you can click it to open the dialog box that you can use to insert functions into a formula.

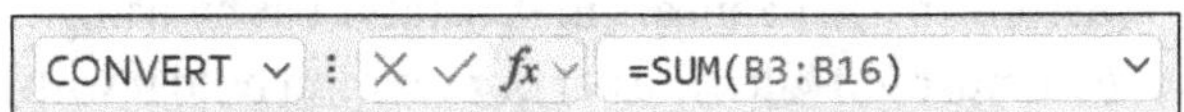

Figure 1.28 Components of Formula Bar

The Input Area

The right component is the input area where you edit the active cell's content. If a cell contains data, then the data appears here, stripped of any formatting, when you select the cell. If the cell contains a formula, then the formula appears here while the cell shows the formula's result. When you create a formula by typing an equal sign, the name box is replaced by a button displaying your most recently used functions. Clicking the arrow next to it opens the recently used functions palette.

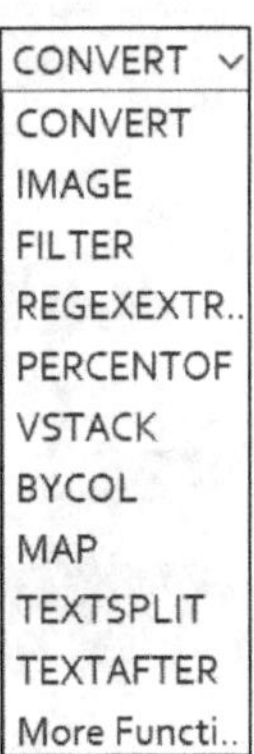

Figure 1.29 Open Functions Palette

When entering complex formulas, it's often helpful to expand the formula bar to multiple rows by dragging its bottom edge. Clicking the small button at the end of the bar toggles the multiline display on and off, and you can adjust the space between the name box and the rest of the formula bar by dragging the divider that's marked with a vertical dotted line.

```
fx  ="There is a new regular expression" & "You can extract part of a string"
    & "The result can be the first match" & "Alternatively you can exstract
    all matches" & "or you can capture groups from the first match."
```

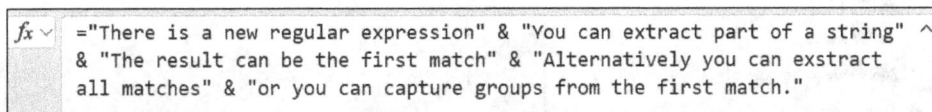

Figure 1.30 Example of Especially Long Formula (for Demonstration Purposes Only)

1.4.10 The Status Bar

The status bar appears at the bottom of the Excel window, except in full-screen mode. It displays the application's current messages, and it also shows context-sensitive tips, such as **Select the target range...** when you've copied a range. So, it's always worth checking the status bar from time to time. Otherwise, the program's current status is displayed on the bar. You can also use the status bar to see if the program is ready to receive data, and the bar can also display different states where certain operations are either allowed or not. You can also customize what appears in the status bar by using a detailed context menu. To access the menu, right-click the bar or press and hold it with your finger.

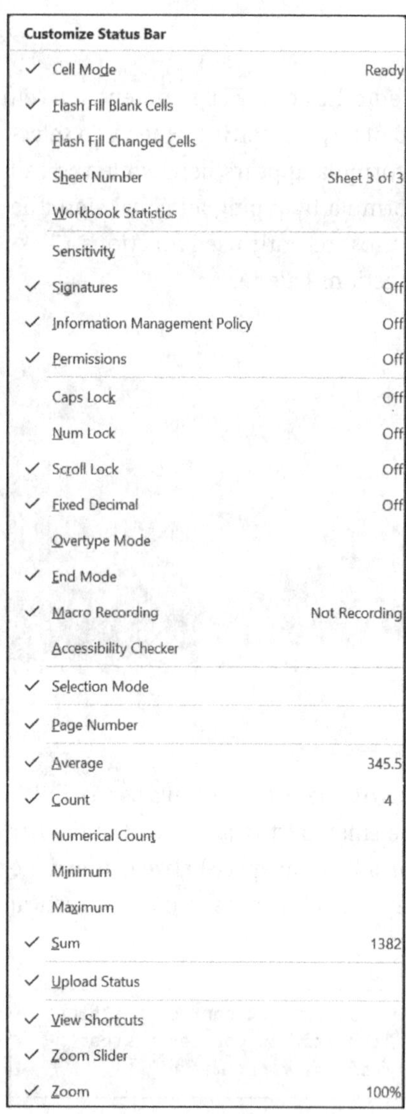

Figure 1.31 Status Bar Context Menu

If you enable **Workbook Statistics**, a button with the same name appears in the status bar and opens a related dialog.

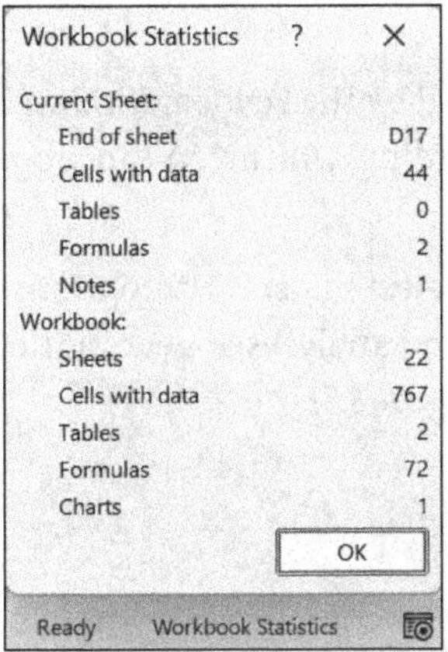

Figure 1.32 Workbook Statistics

The following table explains the meanings of the mode indicators on the status bar.

Indicators	Meanings
Ready	The program is waiting for data or commands.
Enter	You are re-entering the content of a cell.
Edit	You are editing the content of a cell.
Show	You can move the cell pointer to select a cell or range.
Calculate	The workbook needs recalculation because automatic recalculation is off and data has changed.

The following keyboard settings can also be displayed in the fields to the right.

Keyboard Settings	Descriptions
Add to selection	Press `Shift`+`F8` to select additional ranges.
Expand selection	Press `F8` to expand the selection.
Num	The number lock key is on.
Fixed decimal place	In **Options • Advanced • Automatically Insert Decimal Point,** a fixed number of decimal places is set. You don't need to enter the decimal point.
Scroll	Scroll lock is on. Arrow keys move the screen area, not the cell selection.
Caps Lock	Caps lock is on.

Figure 1.33 Status Indicators in Status Bar

On the far left of the status bar is the mode indicator, followed by the keyboard indicators. Clicking the button with the list icon starts macro recording, which you can stop by clicking the same button.

To the right of these displays, you can quickly analyze the selected data with options like **SUM**, **COUNT**, or **AVERAGE**. You control which displays and analyses appear in the status bar through the context menu shown previously in Figure 1.31.

Figure 1.34 Quick Analyses in Status Bar

The three view buttons on the right let you switch quickly between the **Normal** view, the **Page Layout** view, and the **Page Break Preview**. The slider is handy for smoothly zooming into or out of the worksheet, or you can use the plus and minus buttons. To the right, the current zoom level is always shown as a percentage. Clicking or tapping it opens a small dialog with various zoom options.

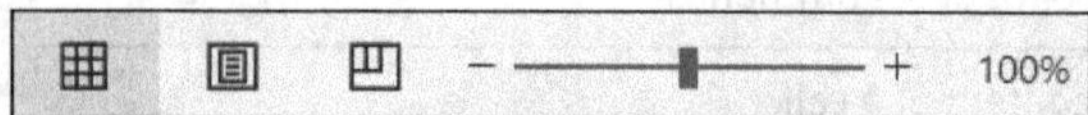

Figure 1.35 View and Zoom Buttons

Below that, you'll find the **Fit Selection** option, which resizes a cell range you've selected to fit the current window.

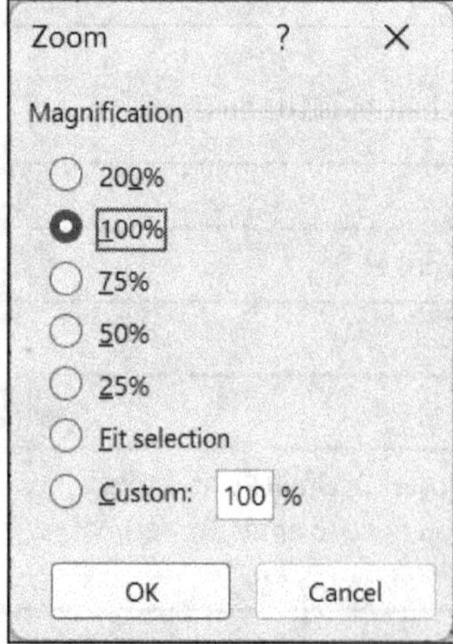

Figure 1.36 Zoom Menu

1.4.11 The Workbook Area

Below the formula bar is the space for the workbook's sheets. When you start Excel, it usually opens a blank workbook named **Book1**—or up to **Book<*N*>** if you've already saved other workbooks under default names. Even when you open a new window via the **File** tab with **New • Blank Workbook**, Excel initially assigns default names.

Also by default, Excel starts with a single sheet in a workbook. Clicking the **+** button adds new sheets and activates them immediately, and pressing `Shift`+`F11` works similarly but inserts the new sheet before the active one and activates it. You can create different types of sheets in a workbook: worksheets, chart sheets, macro sheets, and more. Each type is numbered separately.

Each worksheet also has column headers at the top that are labeled with letters ❶ and row headers on the left that are labeled with numbers ❷. In the upper-left corner ❸ is the **Select All** box, which you can click to select all cells in the sheet at once. The worksheet also has scroll bars ❹ that you can use to access table areas outside the current viewport. These scroll bars also have a context menu.

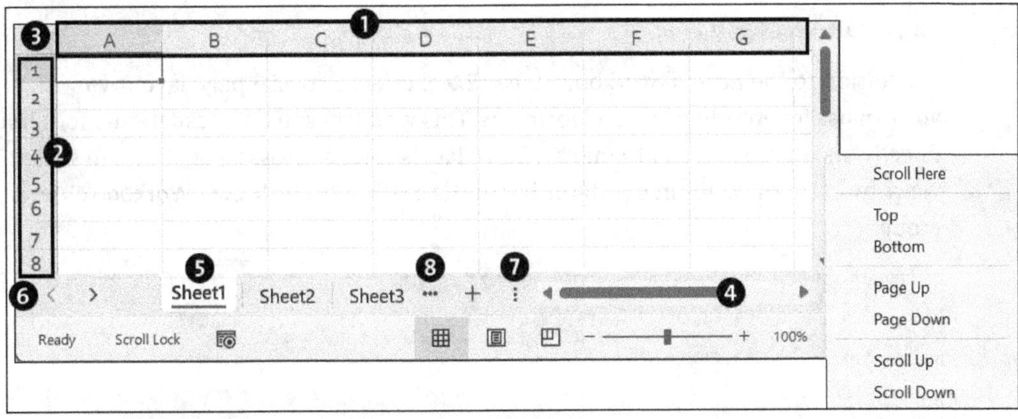

Figure 1.37 Workbook Window Controls

Each sheet in the workbook also has a tab along the bottom edge ❺. Clicking or tapping a tab selects that sheet, and the active sheet is indicated by bold, underlined text in its tab. To access tabs that aren't visible, use the small arrow buttons to the left of the tabs ❻, which activate only when not all tabs visible. Holding down `Ctrl` while clicking these buttons moves you to the first or last sheet in the workbook. You can also drag the tab splitter with three vertical dots ❼ to show more or fewer tabs, and clicking buttons with three horizontal dots ❽ always activates the adjacent sheet.

In **Touch input** mode, you have another way to change the visible tabs. You can drag tabs in either direction or quickly swipe to bring distant tabs into view, and you can right-click or press and hold the small tab buttons to opens a list of tabs you can select from.

Descriptive Sheet Names

To change a default sheet name, you can double-click or tap the sheet tab and enter a new name, or you can use the **Rename** command in the tab's context menu. Names can be up to 31 characters long, including spaces. Sheet names cannot include square brackets, exclamation points, question marks, colons, asterisks, slashes, or backslashes, as these characters can cause errors in cell references. (Renaming sheets automatically updates existing cell references.)

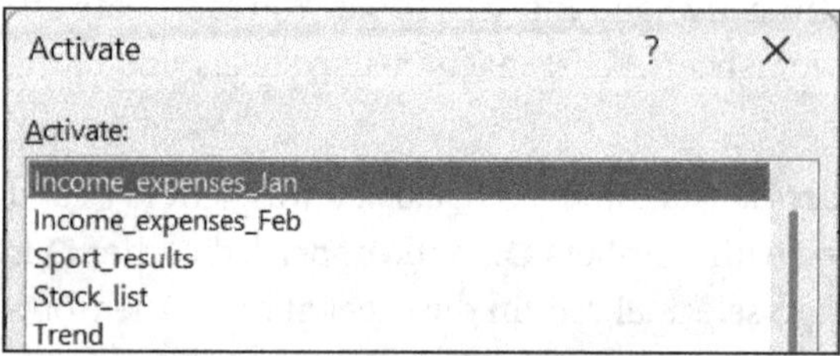

Figure 1.38 Menu Showing Sheet Names in Workbook

1.4.12 Alternative Views

In addition to the normal workbook view, Excel offers a special page layout view that you can use for entering data and formulas. This view lets you edit headers and footers directly and makes it easier to manage how data is spread across pages. You can switch views by using the icons in the status bar or the buttons in the **View • Workbook Views** group.

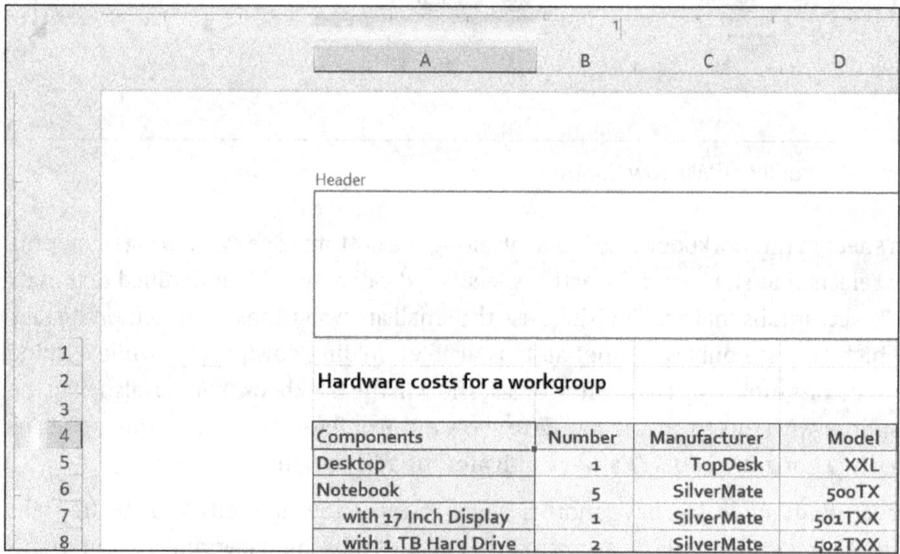

Figure 1.39 Page Layout View

The Side-by-Side View

Excel lets you arrange windows for two workbooks with the same structure side by side, which makes it easier to compare their contents. If both workbooks are open, you can use the **View • Window • View Side by Side** ❶ command in one of the application windows. Clicking the **Synchronous Scrolling** ❷ button in the same group lets you sync the cell pointer movement between the two tables. This setting is on by default, so turn it off if you want the cell pointers to move independently. You can also use the **Arrange All** ❸ button's dialog to choose whether the windows are arranged vertically or horizontally.

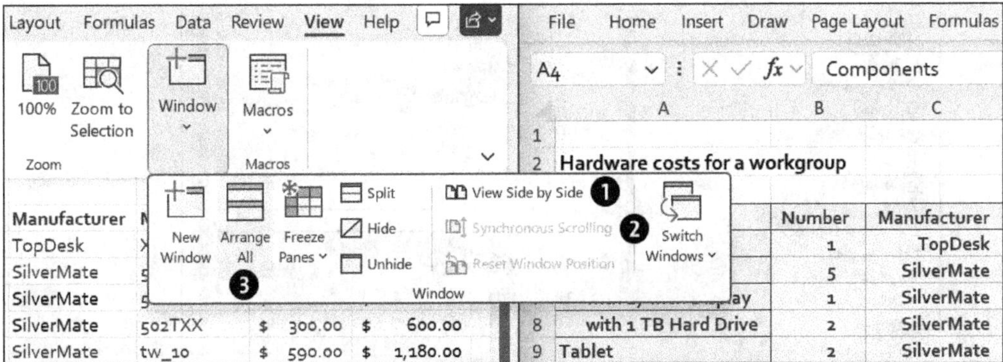

Figure 1.40 Side-by-Side Workbook Windows

1.4.13 Customizing the Ribbon

To modify the ribbon, select the **Customize the Ribbon** command from the ribbon's context menu. This opens the **Excel Options** dialog and displays the **Customize the Ribbon** page.

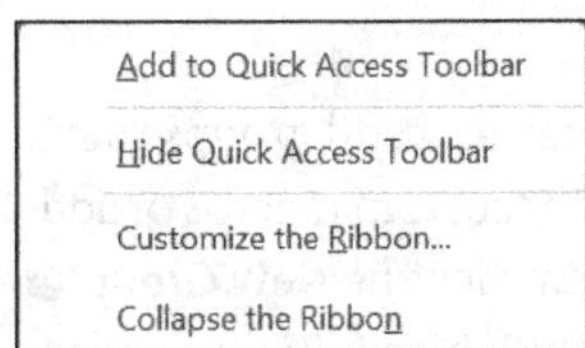

Figure 1.41 Context Menu for Ribbon

You can limit customization to the **Main Tabs** or the **Tool Tabs** by using the list box at the top right ❹.

You can also hide tabs by clearing the checkmark next to them, and if you don't plan to develop Excel-based applications, you can leave the **Developer Tools** tab unchecked. You can't remove the default tabs, but you can hide them to restrict editing options on certain workstations.

You can also remove groups within tabs, such as the **Sparklines** group from the **Insert** tab. Here's how: In the dialog box, open the **Insert** tab ❺, select the **Sparklines** group ❻, and click the **Remove** button ❼. Repeat these steps for each group you want to remove.

To add a group back in later, select **All Tabs** under **Choose commands** and check the group you want. Then, click the **Add** button ❽.

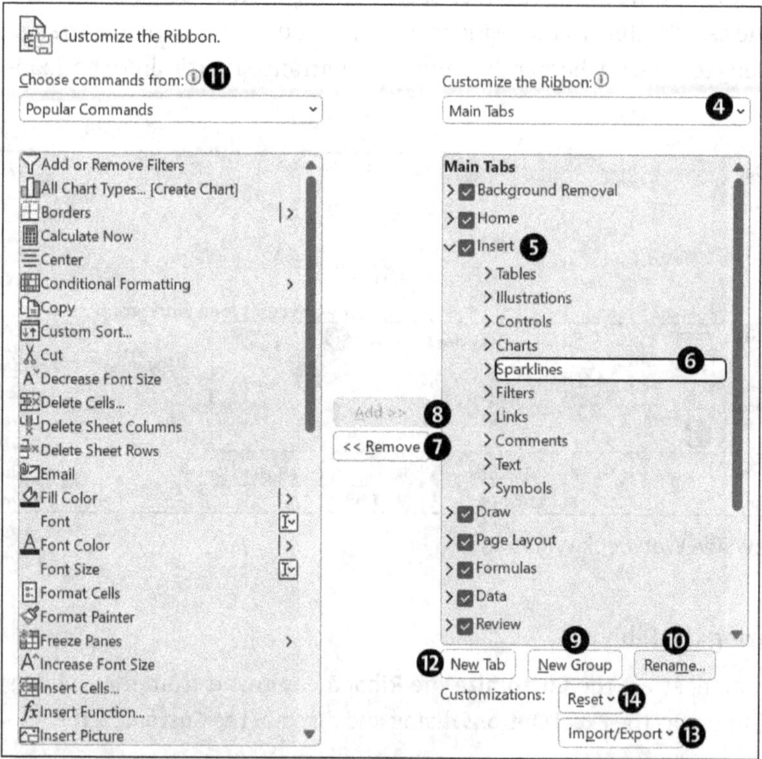

Figure 1.42 Removing Group from Tab

You can also add a new group to a tab or create a brand-new tab and add groups there. In the new groups, you can insert existing commands or calls for custom macros or add-ins. To add a group to a tab, first select the tab name and then click the **New Group** ❾ button. You can also use the small arrow keys to move the group, which initially appears at the end, to any position you want. Click the **Rename** ❿ button to give the group a suitable name, and while the new group is selected, you can also add commands to it, including your own macros.

Under **Choose commands from** ⓫, select the command category first. If you're unsure where a command is, simply select **All Commands**. You can also display commands as a separate group if they don't appear on the ribbon. To do this, select the commands or macros you want from the list box on the left, then click **Add** to assign them to the selected group. To remove icons from the ribbon group, click **Remove**, and to reorder commands within the group, use the arrow keys.

To create a new tab—such as one you'll use to access certain macros faster—click the **New Tab** button ⓬, which will create a new group automatically.

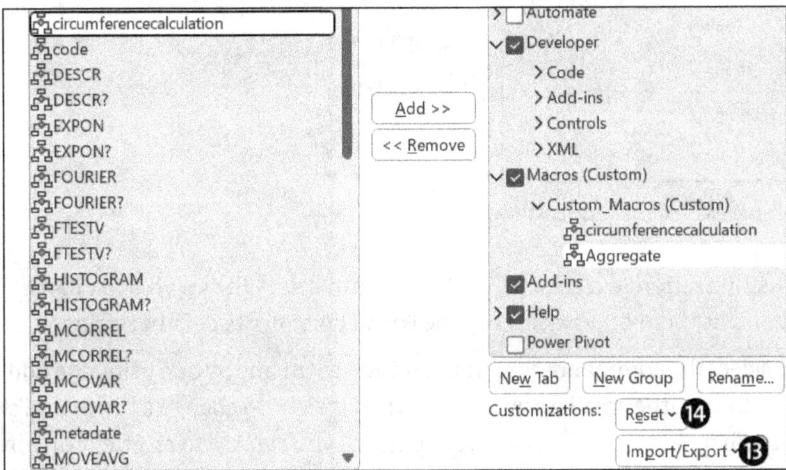

Figure 1.43 Screen for Creating New Tab for Macros

You can click **Rename** to change the names of tabs and groups and to create new groups, and you can fill in the groups as described previously.

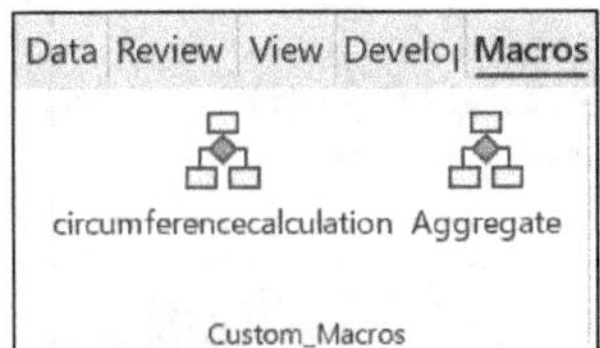

Figure 1.44 New Group on New Tab

You can also export ribbon changes to a small file, which makes it easy to import them to another computer. To do this, use the **Import/Export** button ⓭ and the **Export All Customizations** command and then enter a suitable file name. The data will be saved in an Office UI file with the *.exportedUI* extension, and you can import this data to a different file by using the same button and the **Import Customization File** command.

Use the **Reset** button ⓮ to restore the ribbon to its factory default at any time. The button's menu lets you reset all customizations or just the changes on the currently selected tab.

1.4.14 Customizing the Quick Access Toolbar

You can also customize the Quick Access Toolbar in the **Excel Options** dialog. To do this, either click the small arrow at the end of the toolbar and select **More commands** or choose **Customize...** from the toolbar's context menu.

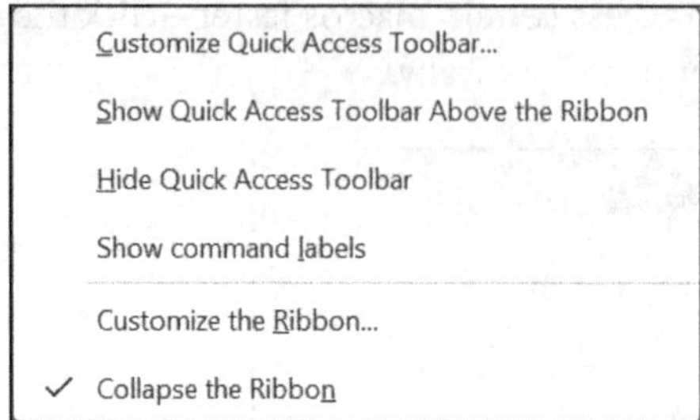

Figure 1.45 Quick Access Toolbar Context Menu

To apply this customization to a specific workbook only, select **For <workbook name>** from the righthand list box; otherwise, keep the **For all documents** default setting.

In the left pane, select the commands you want and add them one by one using the **Add** button. You can also use the small arrow keys on the far right to change a command's position in the toolbar. This dialog also includes buttons you can use to reset, export, or import by using a **.exportedUI* file.

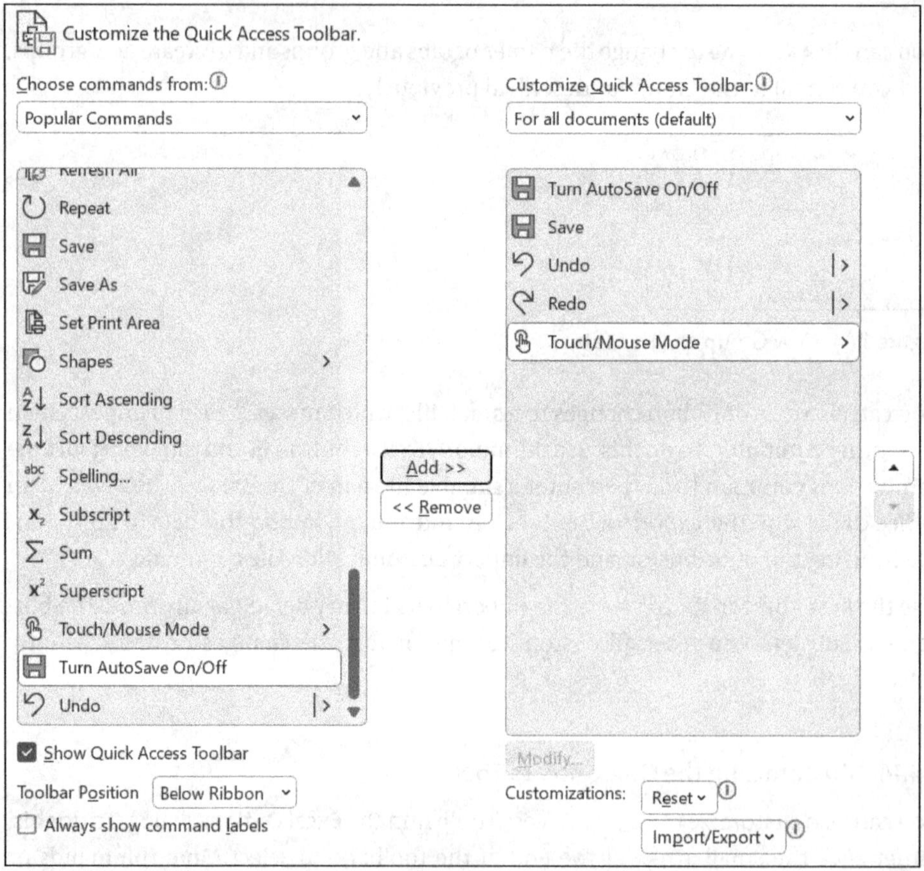

Figure 1.46 Customizing Quick Access Toolbar

There's also a faster way than using the dialog. When you find yourself frequently using a button in the ribbon, right-click it and choose **Add to Quick Access Toolbar** from the context menu.

1.4.15 Task Panes

Some Excel commands open task panes, such as **Review • Spelling • Thesaurus**, **Review • Language • Translate** or **Review • Insights • Smart Lookup**. *Task panes* are commonly used to format charts and graphic objects, and when you're working with external data sources, the **Queries and Connections** task pane will open.

A task pane docks to either the left or right side of the workspace. To move it, drag the title bar in the desired direction; the mouse pointer will change to crosshairs while you're doing that.

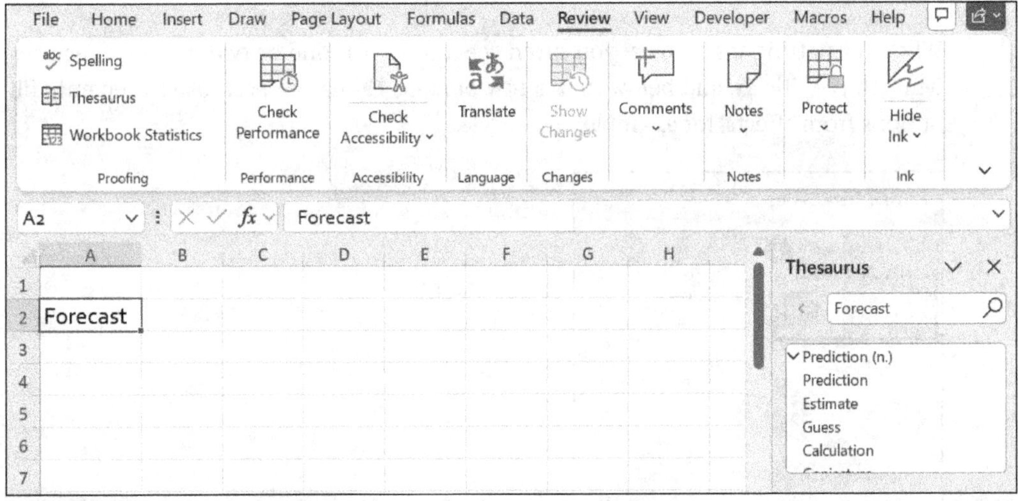

Figure 1.47 Task Pane for Thesaurus

Opening the menu by clicking the small arrow in the upper righthand corner gives you options to resize or close the pane. Resizing starts from the left, with the mouse pointer changing to a double arrow. You can also move the task pane outside the application window, and double-clicking or double-tapping the title bar returns the pane to its last docked position. You can also open multiple task panes at the same time if needed, and when you open two panes in succession, the last one initially covers the first. Use the icons to switch between panes.

Figure 1.48 Icons for Two Task Panes

If you choose the **Move out of Tab** option in the small menu, the panes appear side by side. The menu also offers the reverse option, **Move to Tab**.

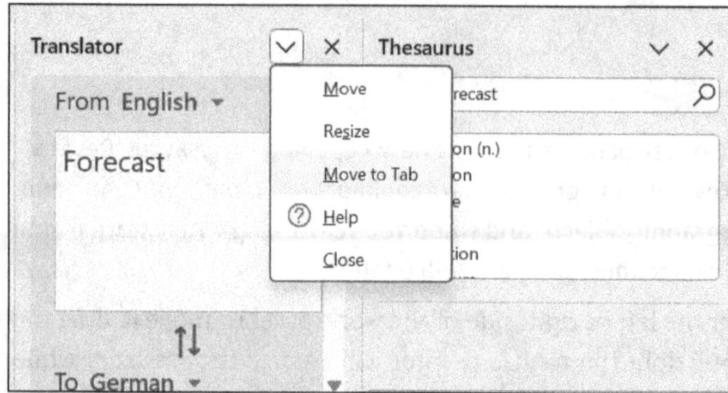

Figure 1.49 Two Task Panes Appearing Simultaneously

When formatting task panes, you often see a row with one or two option categories beneath the title ❶, and below that, a row of icons ❷ that you can use to separate fill options from effects, for example.

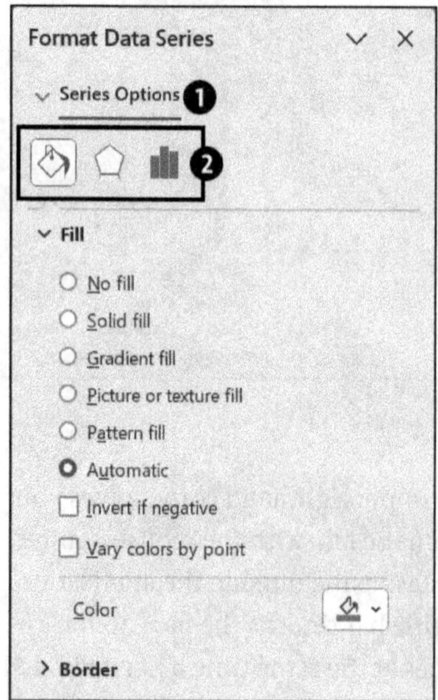

Figure 1.50 Task Pane for Chart Element

1.4.16 The Undo and Redo Commands

Excel remembers the last commands or inputs it executed. Specifically, it saves the state before those commands or inputs so that you can restore it if needed. You use the **Undo** ❸ button in the **Quick Access Toolbar** to do this. To undo multiple steps, click the small arrow next to the icon to open the list of recorded actions, select the steps you want to undo by dragging over them, and then click or tap to undo. The most recent steps appear first. You can also press Ctrl+Z instead of clicking **Undo**.

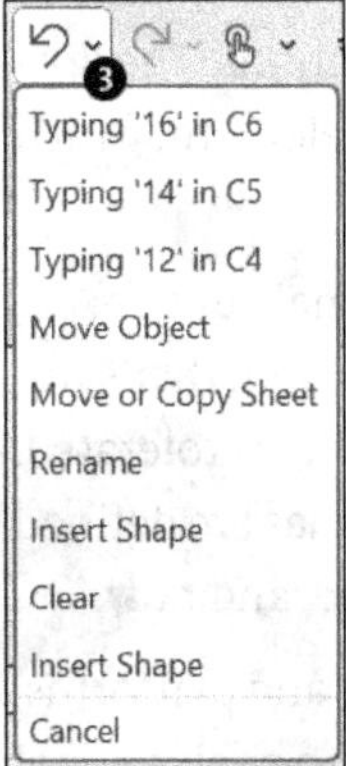

Figure 1.51 List of Recent Steps

Note that not all commands can be undone. For example, you can't undelete an entire sheet, so if a sheet contains data and you try to delete it, Excel will display a warning before it will let you proceed. Otherwise, after undoing an action, you can reapply it by using the **Redo** ❹ button. If you accidentally undo inputs, you can restore them step by step by using the **Redo** icon.

Figure 1.52 Redo Icon

1.5 File Formats

Microsoft first introduced an XML-based file format as a secondary option with Excel XP, and it expanded XML support significantly in Excel 2003. Then, with Excel 2007, the focus shifted. XML-based file formats became the standard, while the binary workbook format was offered as the secondary option. This change reflects Microsoft's acknowledgment that XML documents have become the global, cross-platform standard for data storage.

1.5.1 The XML Family

Extensible markup language (XML) was the first standard for describing structured data. Controlled by the World Wide Web Consortium (W3C), this language uses simple element tags and attributes that are similar to HTML. However, unlike HTML, XML defines the semantic meaning of data elements rather than their visual presentation, and the content is separated from its form. On the other hand, like HTML documents, XML files are plain text. Because the format is platform- and application-independent, it works well for exchanging data between different systems and applications. The documents consist of a hierarchy of elements that always start with a root element, which allows them to be represented as a tree.

An XML document is considered well-formed if it follows specific formal rules, especially the rule that every start tag must have a matching end tag—something that's not always true in HTML, where every start tag not having a matching end tag is tolerated. Well-formed XML documents can be further validated by parsers if schemas are defined that specify which tags and attributes are allowed, where they're allowed, and how.

Because XML is a data format that's independent of specific applications and platforms, it greatly expands Excel's capabilities. Supporting XML turns Excel into an analysis tool for information from any source and lets it share results with applications on any platform that also support XML.

1.5.2 The Standard Open XML Format

Since Office 2007, Microsoft has used XML-based file formats consistently in its core Office programs: Excel, Word, and PowerPoint. However, unlike in Office 2003, Microsoft Office now combines its XML files with ZIP technology, which has become a quasi-standard. When you create a workbook, Office generates a compressed ZIP archive, even though the *.xlsx* file extension doesn't make this obvious. Then, when you open the file with Excel, the archive is expanded automatically, and when you save it, it is recompressed.

If you extract the file with a ZIP program into a folder, you'll find a multilevel hierarchy of components, most of which are XML documents you can view and edit with any text editor. Because these are mainly plain text files, they compress well, resulting in small file sizes, sometimes even smaller than a binary format.

Excel separates workbooks without macros from those with macros. Typically, a workbook contains no executable code, so it can be safely shared via email or over networks. However, macros cannot be added to such files afterward. Workbooks with macros or Object Linking and Embedding (OLE) objects use their own unique file extension, making them easy to identify and enabling necessary security measures. The file format is essentially a container that defines the relationships among its components.

1.5.3 Advantages of Container Formats

Another benefit of this seemingly complex process is that the document remains usable even if some components are damaged, such as by network transmission errors.

In binary files, such errors often make the entire file inaccessible—but individual file components, like an inserted image, can be replaced without opening Excel. Document properties—such as the author's name, topics, or keywords—can also be modified externally, such as by a small batch application that searches for and replaces names across multiple documents.

Microsoft has published the schemas used in the various XML files, but the standard's documentation spans six thousand pages. The specification for the formats and schemas is published under a royalty-free license, and since August 2008, Open XML has been recognized as an ISO standard.

1.5.4 Strict Open XML Workbooks

However, Microsoft has taken its time to fully implement this self-initiated standard in its own file formats. This was only achieved with the Strict Open XML workbook variant Excel offers as an alternative format that also uses the *.xlsx* file extension. This format finally supports date values in the ISO 8601 format and fixes the leap year calculation error for 1900 that previous Excel versions always carried. When you open such a file in Excel 2010, you'll see a notice that it's a newer file format, but you can open it using the provided **Open** button. (When you save a file in the new format, any cell containing the incorrect date 2/29/1900 will be converted to text.)

1.5.5 The Structure of the Open XML Formats

An Open XML file consists of multiple components, which are organized in a list. Most components are XML files, but the container can also include non-XML parts, such as binary files for embedded images. Images are therefore not encoded in XML, and special relationship components define how the individual parts are connected. By packaging the components in a ZIP container, the document remains a single file that users can save or open as usual, and the complex background remains hidden. Figure 1.53 shows the main components of a workbook container and how they connect.

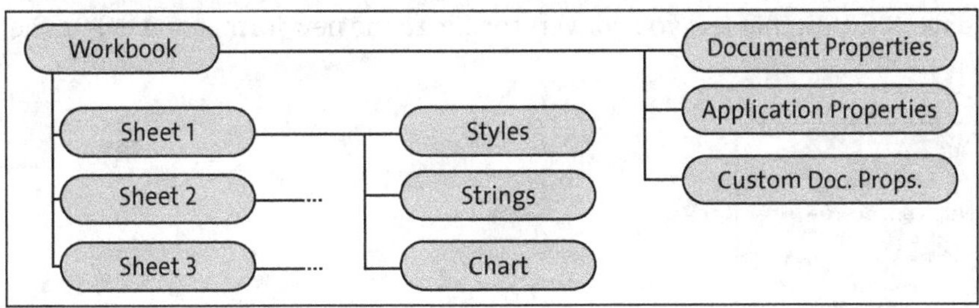

Figure 1.53 Components in Excel Container

Excel also creates a predefined folder structure, as shown in following the illustration for a sample file. However, this folder hierarchy is not mandatory. You can change the arrangement and names of components within the ZIP container, but you must also update the defined relationships accordingly. Components for different content types are created based on their XML schemas.

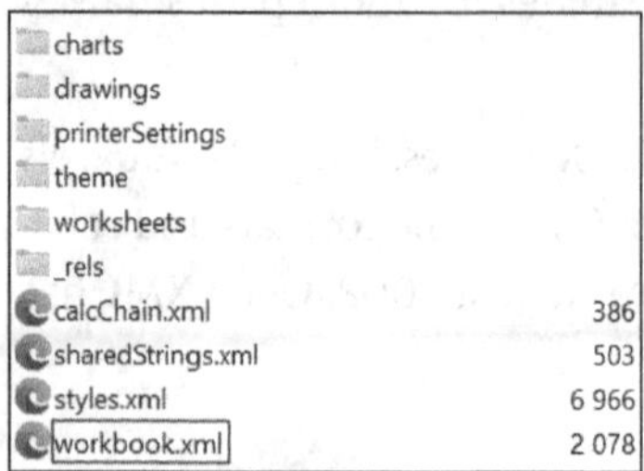

Figure 1.54 Example Workbook Folder Hierarchy

1.5.6 File Extensions

Since version 2007, Excel's standard file format has used the *.xlsx* extension. The familiar*.xls* file extension has been extended with an *x* to indicate it's an XML document. If the workbook contains macros, it uses the *.xlsm* extension, and for template files, the file type is *.xltx*. *Templates* are files used as design patterns for tables, diagrams, and macro templates, and examples include invoice or order forms. If these templates can contain macros, use the*.xltm* file type.

Visual Basic for Applications (VBA)-based add-ins that are integrated into Excel use the*.xlam* file extension. Alternatively, you can still save workbooks in the binary format as before, and in that case, you'll use the *.xlsb* file extension. These files may also contain macros.

1.5.7 File Conversion

When you open a workbook that was saved in an older file format, Excel switches to a special compatibility mode, which is indicated in the title bar. In this mode, newer Excel features are disabled and cannot be enabled manually. On the **File** tab, under **Info**, you'll find the **Convert** button that lets you convert the file to the new format and delete the old one.

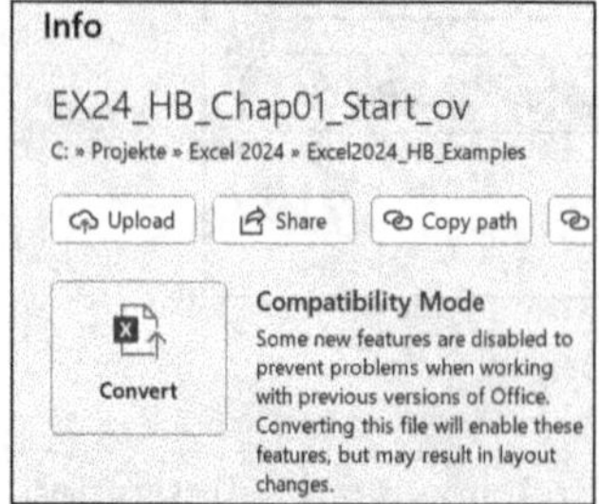

Figure 1.55 Convert Button

When you convert a file, the file extension automatically changes to *.xlsm* or *.xlsx*, depending on whether the file contains macros. You might also see warnings about possible accuracy loss during conversion. You need to save the workbook once and reopen it to use the new format, so to be safe, consider saving the file under a different name in the new format before converting it.

1.5.8 An Alternative: OpenDocument Spreadsheets

For several years, the OpenDocument Format (ODF), which is another XML-based file format, has been used by free office suites. ODF was standardized by ISO in November 2006 as ISO/IEC DIS 26300 Open Document Format for Office Applications (OpenDocument) v1.0. Excel supports the advanced ODF 1.4 format, which is the default when saving, but older formats can still be opened.

General Features

OpenDocument combines XML technologies with ZIP compression. Documents are saved either as a single XML file or as a ZIP package containing subdocuments and binary components, and the structure is consistent across all applications.

Each OpenDocument package includes a *META-INF* folder containing the *manifest.xml* file, and this file lists the archive's contents along with their media types. Most components are XML files describing the document's structure, content, styles, and settings, and as in Open XML, any media files are embedded in binary format. The *meta.xml* file contains metadata such as the title, description, author, creation date, and keywords, the *content.xml* file holds the workbook data, and the *styles.xml* file stores all formats used in the document. These documents use application-specific file extensions; the OpenDocument spreadsheet (ODS) extension is *.ods*.

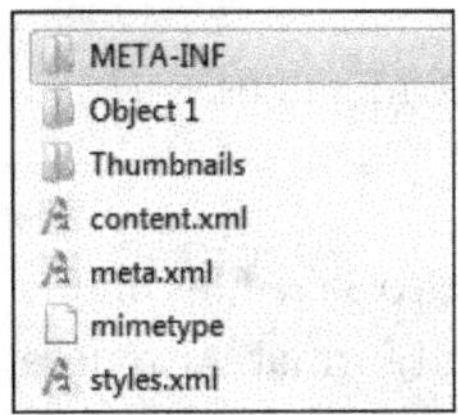

Figure 1.56 Folder Structure of Workbook in ODS Format

ODF or Open XML?

ODF was created as a competing product to Open XML. The structures of the two are similar in that they both combine a compressed archive with XML documents. ODF's schemas are more compact but do not support all the options available in Microsoft Office applications. ODF was designed largely from scratch, while Microsoft aimed to fully replicate all features found in the extensive Microsoft Office document collections within the XML schemas.

Rumors of debates over which of the two formats deserved support have been greatly exaggerated, especially during the still-ongoing process of ISO standardization of Open XML, in which resentment toward Microsoft has often stood in the way of objective appraisal. Ultimately, end users would do well to observe this format rivalry with a degree of equanimity. The development of markup languages for Office documents is likely far from complete, even with these two standards available. Each one has strengths and weaknesses, and expectations for what a document should do remain fluid. The key point is that the XML format breaks the fixed link between documents and their source applications. Moreover, XML documents are much easier to convert from one XML dialect to another than are documents in other formats.

Saving as an ODS File

To save a single workbook in ODS format initially, you can select the file type in the **Save As** dialog. To save it in ODS format after having saved it in a different format, you can go to the **File** tab, choose **Export**, click **Change File Type**, select **OpenDocument Spreadsheet**, and then click **Save As**. This opens the **Save As** dialog with the selected file type.

If you save an Open XML workbook in the ODS format, you can generally expect that the data and formulas will survive the conversion intact. Formatting, however, is another story: You may encounter minor or even significant discrepancies since some formatting features supported by Open XML are only partially—or not at all—supported in ODS.

Some Excel operations are also unsupported in ODS, so you'll want to avoid using them if you plan to save as ODS:

- The **Repeat rows at top** and **Repeat columns on left** print functions are not supported.
- Encrypted file saving and workbook sharing for collaboration are also unsupported.
- Tables lack support for total rows and table style templates.
- PivotTables don't support style templates, calculated fields, or OLAP tables.
- OLAP formulas and the **Consolidation** data tool are not supported.

If you want to work with ODS files, you can set **File • Options • Save** to **Save files in this format** and select **OpenDocument Spreadsheet (*.ods)** as the default format. Also note that when you first start Excel, you'll see a dialog that allows you to choose the default file type.

1.6 Options for Working with Excel

How you use Excel may be very different from how others use it. Some people use Excel mainly as a calculation tool, while others focus on its table functions. Some users are most interested in turning numbers into charts, while developers use Excel's tools to build programmed solutions. Excel is designed to adapt to different needs, and you can customize your Excel workspace to fit your requirements.

1.6.1 Customizing Excel to Suit Your Needs

When you start using Excel, you'll find it helpful to adjust the options in the **Excel Options** dialog under **General**.

The first setting under **User Interface options** deals with multiple monitors. If display issues arise, you can switch the default **Optimize for best appearance** setting to **Optimize for compatibility.**

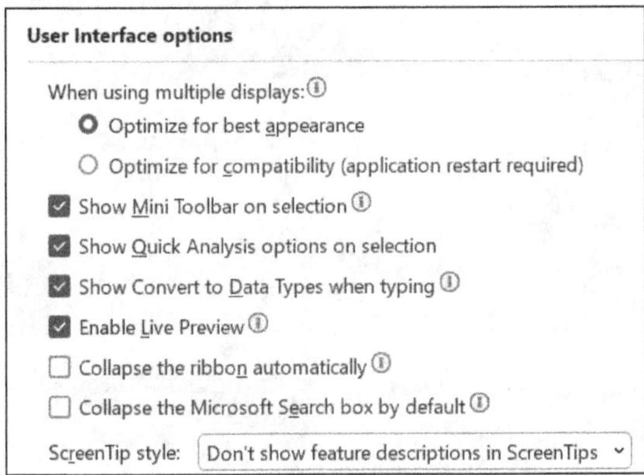

Figure 1.57 General Excel Options

The next option is **Show Mini Toolbar on selection.** When this setting is enabled, the most common formatting icons appear above selected text in a cell or the formula bar. This option doesn't affect the context menu for selected cell ranges; you can't disable the mini toolbar there.

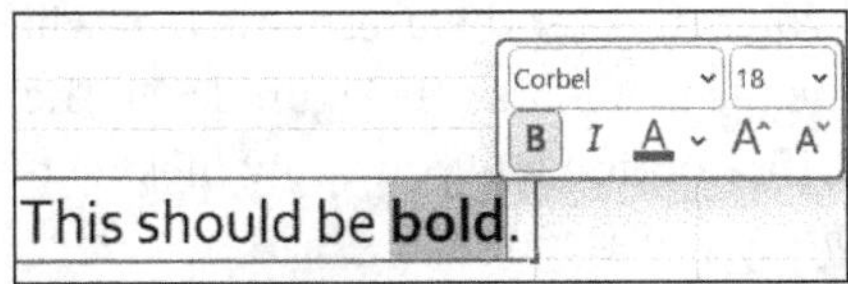

Figure 1.58 Mini Toolbar for Text Selected Inside Cell

If you need to, you can disable the button for the Quick Analysis tool mentioned earlier under **Show Quick Analysis options on selection**, but you can still access this feature by using the Ctrl+Q keyboard shortcut.

If you want to use advanced data types like stocks, enable **Show Convert to Data Types when typing.**

A useful feature is **Enable Live Preview**, which instantly shows how many settings—like color or font size—affect the selected area before you confirm your choice. However, some individuals might find the occasional screen flicker distracting, so they can disable the **Enable Live Preview** option here. You can also use the **ScreenTip style** option to control how screen tips appear for commands and options in the menu bar. You can choose detailed tips, brief tips, or no tips.

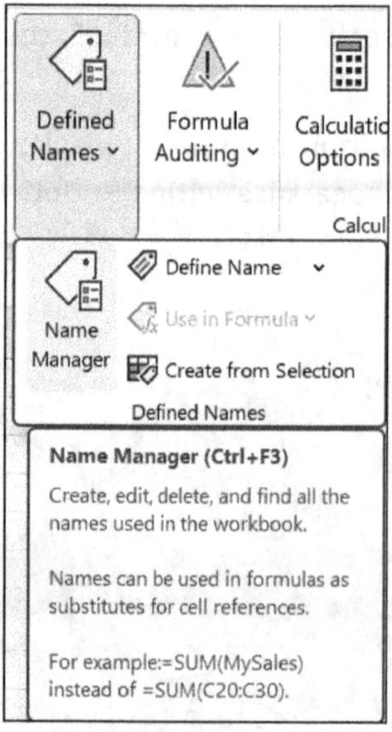

Figure 1.59 Example of Detailed ScreenTip

You can use the second group of options to set general defaults for new documents. Here, you can choose a default font and font size for data entry. Excel uses **Aptos Light** as the default font for headings and 11-point **Aptos** for body text. You can also choose whether to work by default in **Normal** view, **Page Layout** view, or **Page Break Preview**, and you can also set how many sheets a new workbook should include. These changes take effect only after you restart Excel.

Under **Personalize your copy of Microsoft Office**, you can enter the username that will automatically be added to the document properties as metadata (see Figure 1.60). The **Always use these values regardless of sign in to Office** option lets you apply these settings regardless of whether you sign in with an *Office.com* account.

There are two more list boxes you can use to customize Excel's appearance to your liking. The first, called **Office Background,** offers various designs that affect the header area of the application window (see Figure 1.61). As mentioned earlier, this option is unavailable to local users.

Under **Office Theme**, you'll find options that allow you to color the application background. The default option is **Colorful**, and other options are **Dark Gray**, **Black**, **White**, and **Use system setting** (which refers to Windows settings).

In the **Privacy Settings**, you can control how much diagnostic data Microsoft may collect while you work on your device. If you use the Microsoft-owned professional network

LinkedIn, you can activate services that let you connect directly to LinkedIn from Excel. Under **Home Options**, you'll find a **Default Programs** button that opens the **Control Panel** dialog, where you can assign or change file type associations for Excel. The **Start Screen** setting has already been mentioned, and you can disable the **Show at startup** option for this application there.

Figure 1.60 Personalizing Options

Figure 1.61 Background Called "Clouds," Displayed with Slightly More Contrast

Under **Language**, Excel provides a dedicated page where you can select editing languages as alternatives to your primary language. To add a language, open the list box below the selected editing languages, pick the language you want, and click **Add a Language**. In the lower section, you can also choose which languages to prioritize for program displays and get help when multiple languages are available.

Figure 1.62 Choosing Editing Language

1.6.2 Showing and Hiding Screen Elements

If you need to, you can hide some elements that usually appear on the Excel screen. If you only need to view data, you can skip the screen elements required for editing. The **View** tab offers several settings for this in the **Show** group. After tables are fully developed, the formula bar is often unnecessary, and you can deselect it here to instantly gain more space for displaying tables.

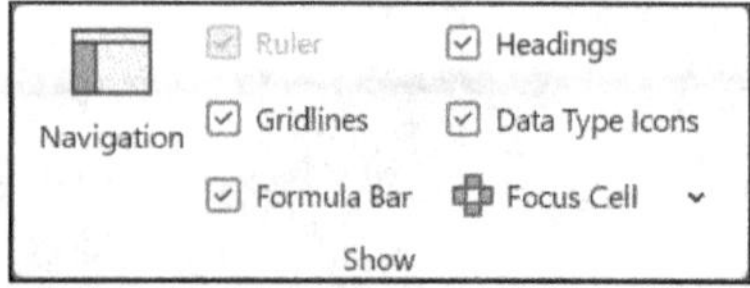

Figure 1.63 Show Group

Alternatively, instead of the options on the **View** tab, you can use options offered in the **Excel Options** dialog box specifically on the **Advanced** page. These settings have different scopes, and settings under **Display** apply to Excel in general and remain active until you change them.

To change these settings, under **Display**, you can start by setting how many files appear in the **Recent** list on the **File** tab under **Open**. Additionally, under **Quickly access this number of Recent Workbooks**, you can set how many files appear directly under the File menu, regardless of the selected page on this tab. You can also set the number of recently used folders here, and under **Ruler units**, you can set the scale used in the ruler shown in the **Page Layout** view.

You can also disable the formula bar and function tooltips here. If cells contain notes or comments, you can toggle their display and the small indicators on or off.

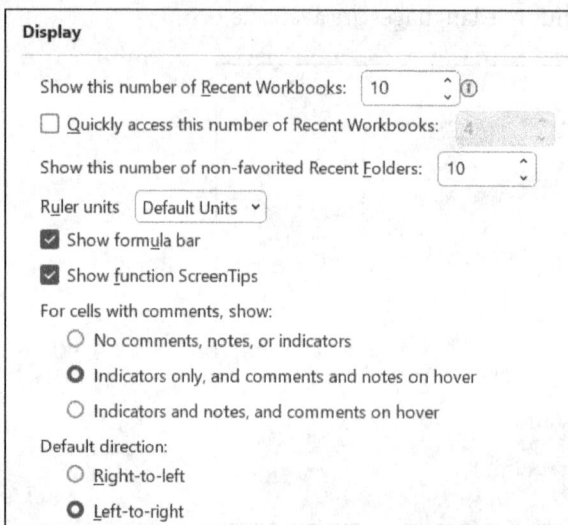

Display
Show this number of Recent Workbooks: 10
Quickly access this number of Recent Workbooks: 4
Show this number of non-favorited Recent Folders: 10
Ruler units Default Units
Show formula bar
Show function ScreenTips
For cells with comments, show:
No comments, notes, or indicators
Indicators only, and comments and notes on hover
Indicators and notes, and comments on hover
Default direction:
Right-to-left
Left-to-right

Figure 1.64 Display Options

Display Options for Individual Workbooks

Under **Display options for this workbook**, you'll find settings that apply only to specific workbooks. If multiple workbooks are open, use the small list box to choose one. These settings are saved with each workbook, and they begin with options to show or hide scroll bars and sheet tabs. The **Group dates in the AutoFilter menu** option adds groups for months or years to the filter's context menu. For more details, see Chapter 17.

As the **For objects, show:** setting, you can choose either **All** or **Nothing (hide objects)**. This applies to charts, graphic objects, buttons, text boxes, and objects embedded via **Insert • Text • Object**. The **Nothing (hide objects)** setting can be temporarily useful on large sheets with many charts when you only need to focus on the table data. It speeds up navigation on the sheet by hiding all graphic objects and preventing them from printing. There's also a keyboard shortcut to quickly toggle between these two settings: Ctrl+6.

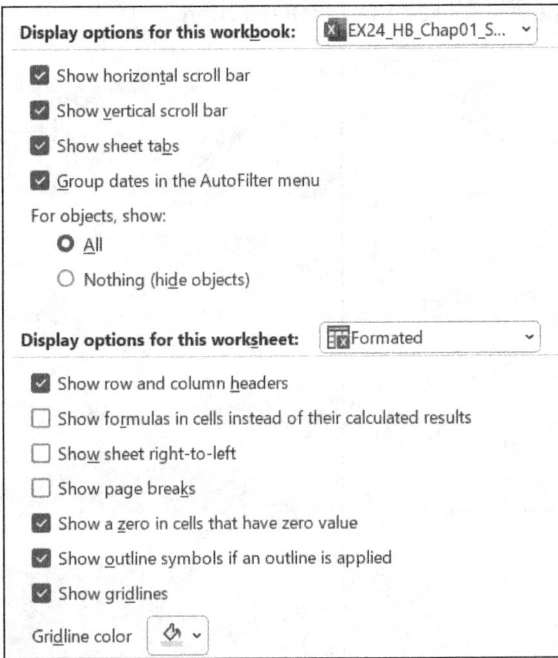

Figure 1.65 Display Options for Sheets or Workbooks

Sheet Options

Under **Display options for this worksheet,** you can configure settings for individual sheets that you select from the list box. You can show or hide the row and column headers, and to view all formulas at once in a worksheet for checking a calculation model, you can check **Show formulas in cells instead of their calculated results.** Displaying formulas also helps document calculation models. When **Show page breaks** is checked, lines appear to mark page breaks in large tables.

You can also uncheck the **Show a zero in cells that have zero value** option, which is helpful when configuring tables where many formulas initially return zero because values are missing. However, keep in mind that this setting applies to the entire worksheet, where a zero may be distracting in one place but important in another. So, it's usually better to control the display of zeroes through the number format. You can learn more about this in Chapter 4, Section 4.3.6.

With the **Show outline symbols if an outline is applied** option, you can enable or disable the display of outline symbols. You can choose a different color for the gridlines, since once you've fully designed a worksheet, it's often best to turn off the gridlines and use custom borders to organize the table areas. This usually makes the table easier to read.

1.6.3 Editing Options

A large set of options on the **Advanced** page relates to worksheet editing. Under **Editing options**, you can set how pressing the `Enter` key affects cell selection.

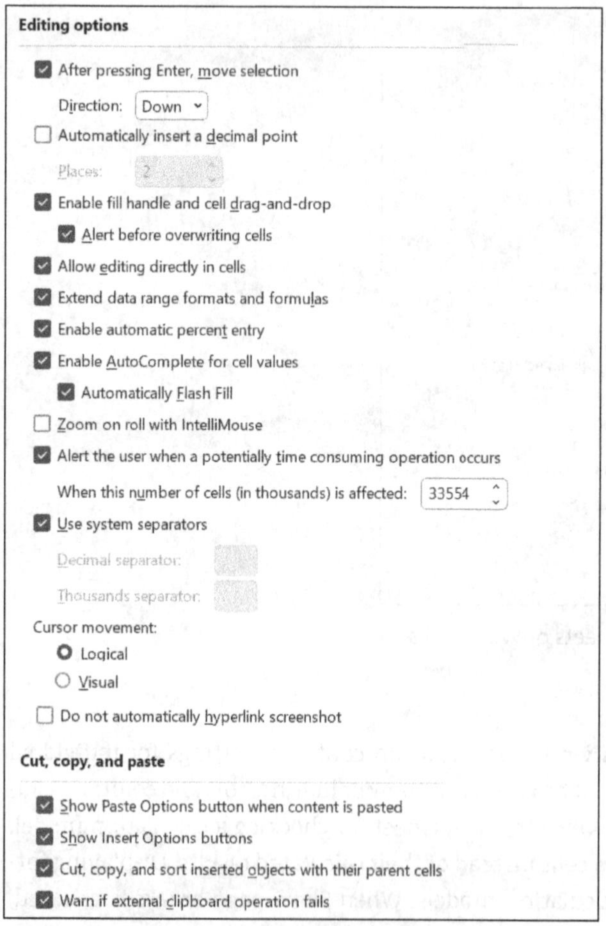

Editing options

- [x] After pressing Enter, move selection
  Direction: Down
- [ ] Automatically insert a decimal point
  Places: 2
- [x] Enable fill handle and cell drag-and-drop
  - [x] Alert before overwriting cells
- [x] Allow editing directly in cells
- [x] Extend data range formats and formulas
- [x] Enable automatic percent entry
- [x] Enable AutoComplete for cell values
  - [x] Automatically Flash Fill
- [ ] Zoom on roll with IntelliMouse
- [x] Alert the user when a potentially time consuming operation occurs
  When this number of cells (in thousands) is affected: 33554
- [x] Use system separators
  Decimal separator:
  Thousands separator:

Cursor movement:
- (•) Logical
- ( ) Visual
- [ ] Do not automatically hyperlink screenshot

Cut, copy, and paste

- [x] Show Paste Options button when content is pasted
- [x] Show Insert Options buttons
- [x] Cut, copy, and sort inserted objects with their parent cells
- [x] Warn if external clipboard operation fails

Figure 1.66 Editing Options on Advanced Page

The cell pointer moves with the **Direction** options, either to the next cell in the row (to the right) or to the next cell in the column (down). Depending on the order of data entry, one setting or the other will prove to be more practical.

Usually, it's best to keep the **Enable fill handle and cell drag-and-drop** default option enabled. The same goes for **Allow editing directly in cells** and **Extend data range formats and formulas**, the latter of which ensures that formulas and formats carry over when ranges expand. You should also generally keep **Enable AutoComplete for cell values** turned on.

Checking the **Automatically Flash Fill** option activates an advanced fill feature for table columns. Checking the **Do not automatically hyperlink screenshot** option prevents a screenshot of a webpage that's been inserted into a workbook with the **Screenshot** function from automatically acting as a button that navigates to the webpage.

You should also keep the default options under **Cut, copy, and paste**, which ensure that after these actions, buttons with small menus will appear so you can choose exactly what happens.

In the **Chart** group, you'll find settings for managing data points. You can choose a setting for all future workbooks, or you can select a workbook under **Current workbook** and apply the **Properties follow chart data point for current workbook** option.

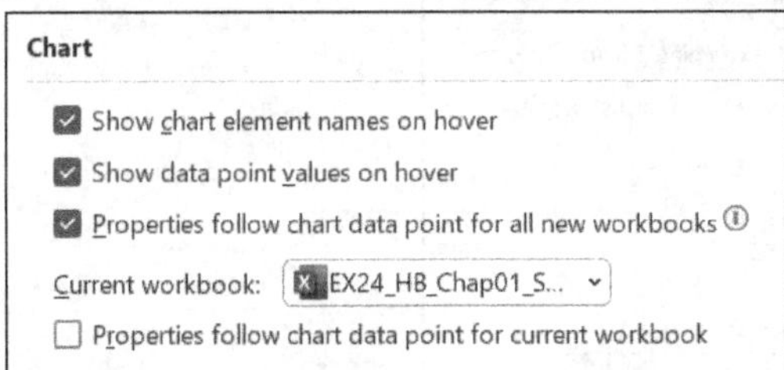

Figure 1.67 Chart Options

In the **Formulas** group, you'll find an option to speed up calculations in large workbooks. Excel can distribute calculation tasks across multiple threads, and if you have a multi-core processor, you should enable the **Enable multi-threaded calculation** option.

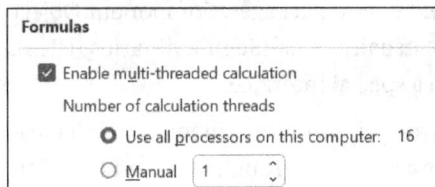

Formulas
Enable multi-threaded calculation
Number of calculation threads
Use all processors on this computer: 16
Manual 1

Figure 1.68 Calculation Speed Options

On the **Data** tab, you'll find options for managing PivotTables and data models. For large data sets, you can disable the **Undo** feature.

Other settings on the **Advanced** tab will be covered later, in sections that are relevant to each topic. This also applies to options on the **Formulas** tab, which we'll discuss in Chapter 3, Section 3.9.

1.6.4 Save Options

On the **Save** page, you can set local saving to always be offered first. If you enable **Save to Computer by default,** the **This PC** location will always be the default on the **Save As** page.

The default save location is also set in the dialog; for example, on Windows 10 and 11, it's *C:\Users\<username>\Documents*. You can also change the default file format under **Save files in this format,** if needed. If you save table data as a CSV file and **Show data loss warning ...** is enabled, you'll get a warning that some data might be lost. If you use this format often, you can disable this warning.

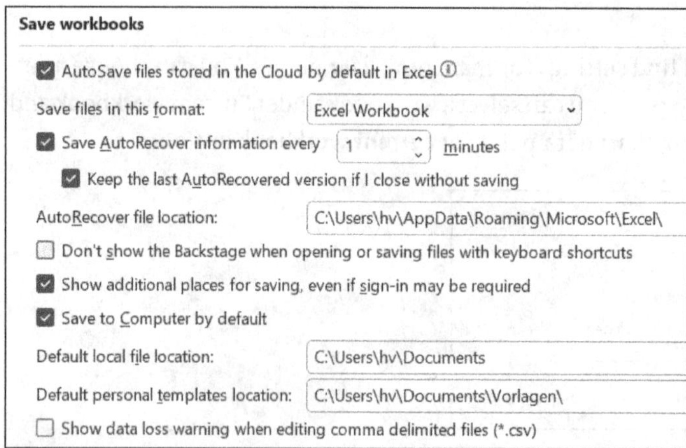

Figure 1.69 Options for Saving Workbooks

1.6.5 Integrating Add-Ins

Some of Excel's more complex functions are delivered through add-ins. These include Excel add-ins, COM add-ins, macros, and XML extension packages. Component Object Model (COM) add-ins are ActiveX components, usually dynamic-link libraries (DLLs), that communicate with the host application via a special interface.

When an add-in is loaded, you'll usually see extra options on the ribbon to use its features. Other add-ins offer table functions that are accessible through the **Insert Function** dialog. You can view the list of enabled or disabled add-ins by going to **File • Options** and selecting the **Add-ins** page. Because not every add-in is needed in every work scenario, Excel uses a startup list to load the necessary add-ins for each session. To enable or disable add-ins, choose the add-in type from the **Manage** dropdown list and then click the **Go** button.

Add-ins

Name ^	Location	Type
Active Application Add-ins		
Analysis ToolPak	C:\Program Files\Microsoft Office\root\Offi	Excel Add-in
Analysis ToolPak - VBA	C:\Program Files\Microsoft Office\root\Offi	Excel Add-in
Microsoft Power Pivot for Excel	C:\Program Files\Microsoft Office\root\Offi	COM Add-in
Solver Add-in	C:\Program Files\Microsoft Office\root\Offi	Excel Add-in
Inactive Application Add-ins		
Date (XML)	C:\Program Files\Common Files\Microsoft	Action
Eigene_Funktionen	C:\Users\hv\AppData\Roaming\Microsoft\	Excel Add-in
Euro Currency Tools	C:\Program Files\Microsoft Office\root\Offi	Excel Add-in
Microsoft Actions Pane 3		XML Expansion Pack
Microsoft Data Streamer for Excel	C:\Program Files\Microsoft Office\root\Offi	COM Add-in
Microsoft Power Map for Excel	C:\Program Files\Microsoft Office\root\Offi	COM Add-in
PDFMaker.OfficeAddin	C:\Program Files\Adobe\Acrobat DC\PDFM	COM Add-in

Add-in: Analysis ToolPak
Publisher: Microsoft Office
Compatibility: No compatibility information available
Location: C:\Program Files\Microsoft Office\root\Office16\Library\Analysis\ANALYS32.XLL

Description: Provides data analysis tools for statistical and engineering analysis

Manage: Excel Add-ins | Go...

Figure 1.70 Add-Ins Page

To enable an add-in, check the box next to its name in the dialog, and to remove it from the startup list, uncheck the box. If the add-in doesn't appear under **Add-ins available**, click the **Browse** button since it might be located in a different folder. When you find the file you want, select it and confirm by clicking **OK**.

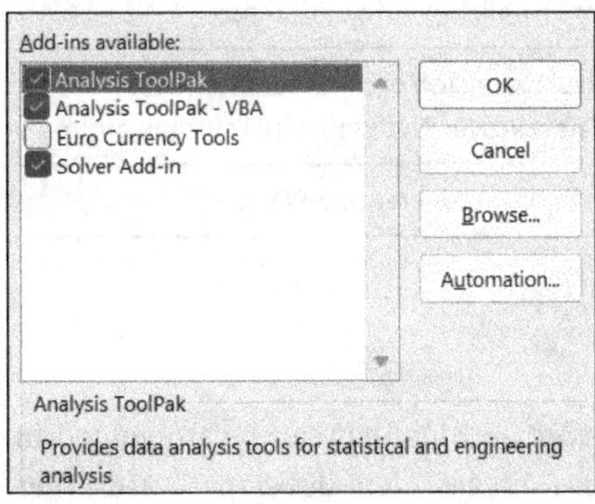

Figure 1.71 Add-Ins Dialog

If you only need an add-in for one session, you can load it through the **Open** dialog. Excel add-ins use the *.xlam* file extension, and older add-ins use *.xla*. If you have the **Developer Tools** tab enabled, you'll find an **Add-Ins** group on that tab with buttons to load add-in components.

You can also create your own VBA add-ins to integrate custom macros into Excel. If a workbook contains modules, you must choose **Excel Add-In (*.xlam)** as the file type in the **Save As** dialog box when saving it.

1.6.6 List of Built-In Excel Add-Ins (VBA)

The following Excel add-ins come with all editions:

Name	Description
Analysis Functions	This offers many additional financial and technical functions, along with tools for statistical analysis.
Analysis Functions VBA	This provides many additional financial and technical VBA functions.
Euro Currency Tools	This includes a button for the Euro format and the `EUROCONVERT()` function to convert between currencies.
Solver	This solves systems of equations and inequalities with multiple variables.

1.6.7 List of Built-In COM Add-Ins

Depending on the edition, Excel may or may not include the following built-in COM add-ins:

Name	Description
Microsoft Data Streamer for Excel	This performs data analysis for external sensors.
Microsoft Power Map for Excel	This is a tool for visualizing data with maps.
Microsoft Power Pivot for Excel	This is a powerful extension of pivot functions that adds to the ribbon a new **Power Pivot** tab with multiple groups.
PDFMaker.OfficeAddIn	This is an Adobe add-in for creating PDFs.

1.7 Office Add-Ins

Another way to extend Excel is with special Office add-ins that can be embedded in workbooks. Microsoft provides a store for these but also lets developers use other stores. Technically, these apps consist of web components controlled by script code.

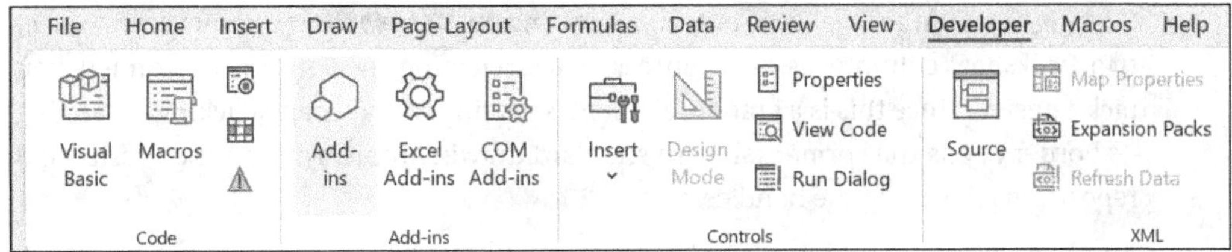

Figure 1.72 Add-Ins Group

Under **Developer Tools • Add-Ins**, you can access the **Store** and use the **My Add-Ins** button to open your downloaded add-ins. Add-ins like **People Graph** add their own layers to the worksheet.

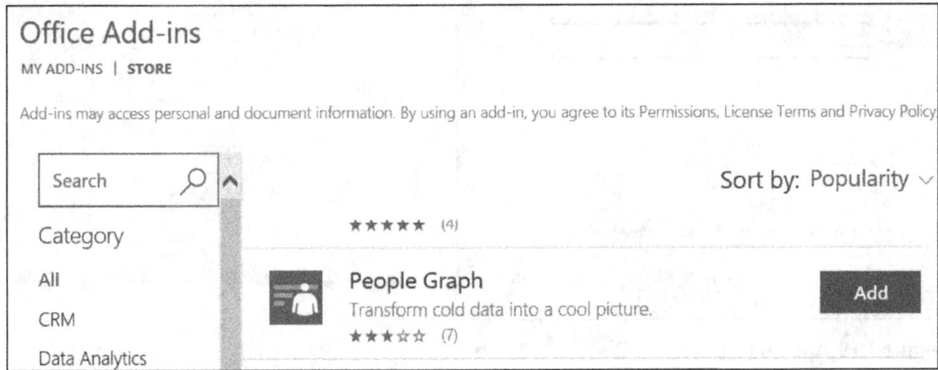

Figure 1.73 Access to App Store

Figure 1.74 shows an example that visualizes member distribution across several cities.

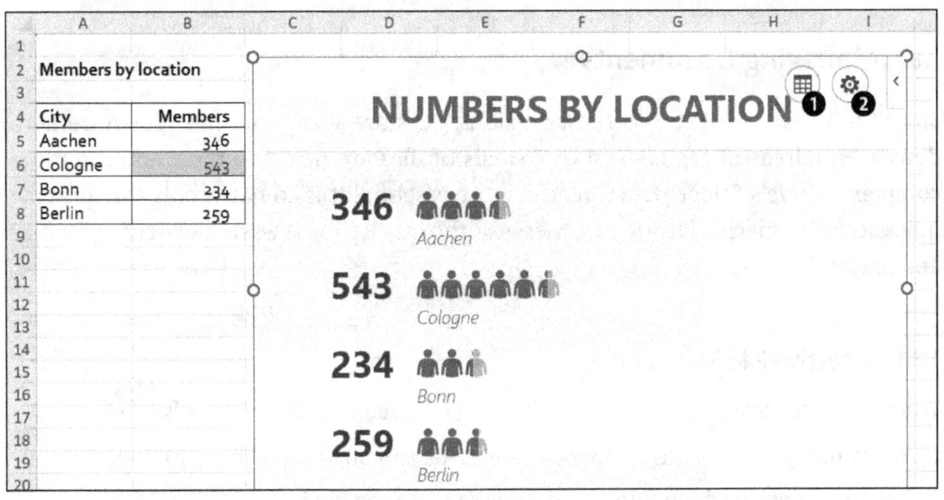

Figure 1.74 People Graph Example

Clicking or tapping reveals two buttons. Selecting the **Data** ❶ icon prompts you to select two worksheet columns as data sources, while selecting the **Settings** ❷ icon lets you pick a design. Since this is a graphic element, selecting the layer and clicking or tapping its border opens the **Format** tab with standard drawing tools. You can then resize and reposition it by using the handles.

Another type of add-in doesn't add its own layer to the sheet but opens a new task pane instead. An example of this is **Wikipedia**.

Figure 1.75 App That Inserts Its Own Task Pane

These examples show that such add-ins can create attractive extensions for your workbooks and thus offer you valuable opportunities for Office development.

1.8 Managing Documents

Since the release of Excel's earliest versions, PC storage capacity has grown tremendously. Hundreds of folders and thousands of files are now common, and the more complex a drive's folder structure, the more essential it is to have tools that provide quick access to specific locations. Otherwise, too much time is wasted searching for files and navigating to specific folders.

1.8.1 Effective File Management

Whenever you save data to a storage device, you need to make two decisions:

1. You must give every file an appropriate name so you can find it again.
2. You must decide which folder and drive to store the file in.

Using the same file name multiple times is allowed if the files are stored in different folders, but giving files unique names usually make them easier to find. This, of course, does not apply to backup copies. To retrieve stored data, you need to know the file name and its location—the folder and drive. That sounds simpler than it often is. After all, it's not easy to remember exactly where any given piece of urgently needed data is stored, which is why there are methods available to help you find data if you don't know its file name and location.

1.8.2 Saving Documents

Only saving to a storage device ensures that your data remains permanently accessible. Main memory storage is always temporary, which becomes painfully clear when the power goes out and unsaved data is lost. When you save for the first time using **Save As**, you must assign a file name and location—and later, **Save** will assume that the name and location will stay the same.

When you select the **File • Save As** command, Excel offers the default location, standard folder, and file type you chose under **Options • Save**. Afterward, it always suggests the folder that's currently in use.

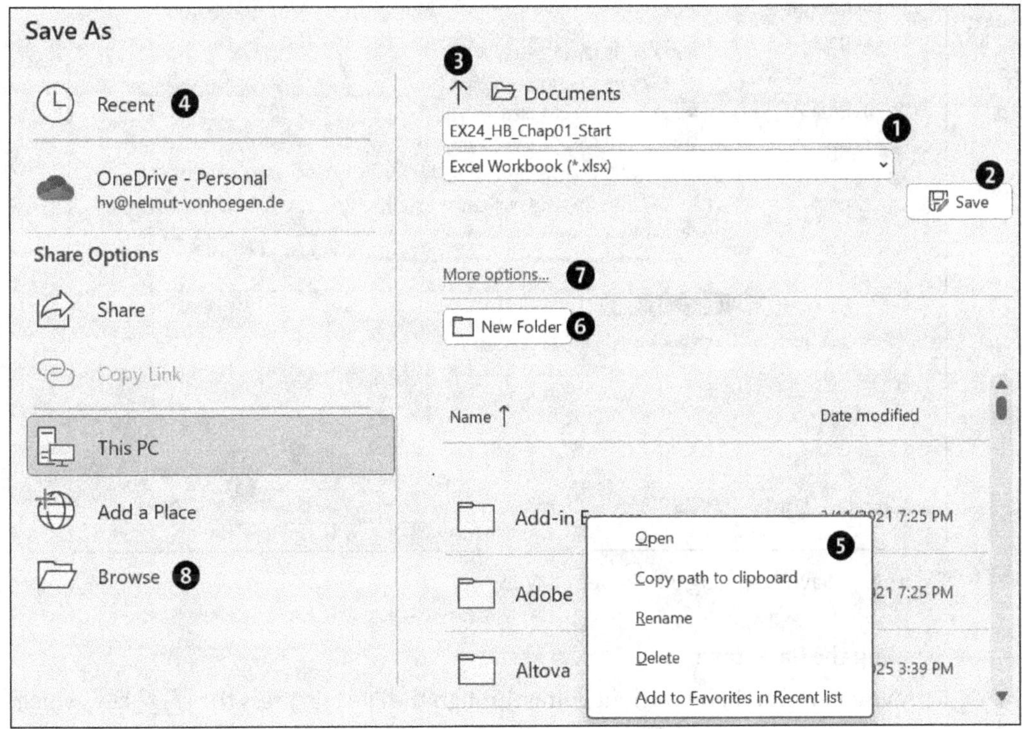

Figure 1.76 Save As Page

If you want to accept these default options, simply enter the file name ❶ and click the **Save** button ❷. To save the file in a different folder, click the arrow ❸ before the current folder to go up one level, then choose from the available folders. To reuse a folder, select

Recent ❹ to access the list of recently used folders. You can manage this list through the context menu ❺ on any folder in the list. You can create a new folder at the current level by clicking the **New Folder** ❻ button. Both the **More options** ❼ link and the **Browse** ❽ icon open the **Save As** dialog box, where you'll have other options for saving your data.

The dialog box layout suggests you first choose the folder where you want to save the file. If the displayed folder contains subfolders or subfiles, they will appear in the large list box. You can toggle the folder contents display on and off using the **Hide Folders** or **Search Folders** button ❾.

This is useful if you always save a series of workbooks to the same folder. To have a preview available when opening the workbook later, select the **Save Thumbnail** option ❿ and then click **Save** ⓫ to save the workbook.

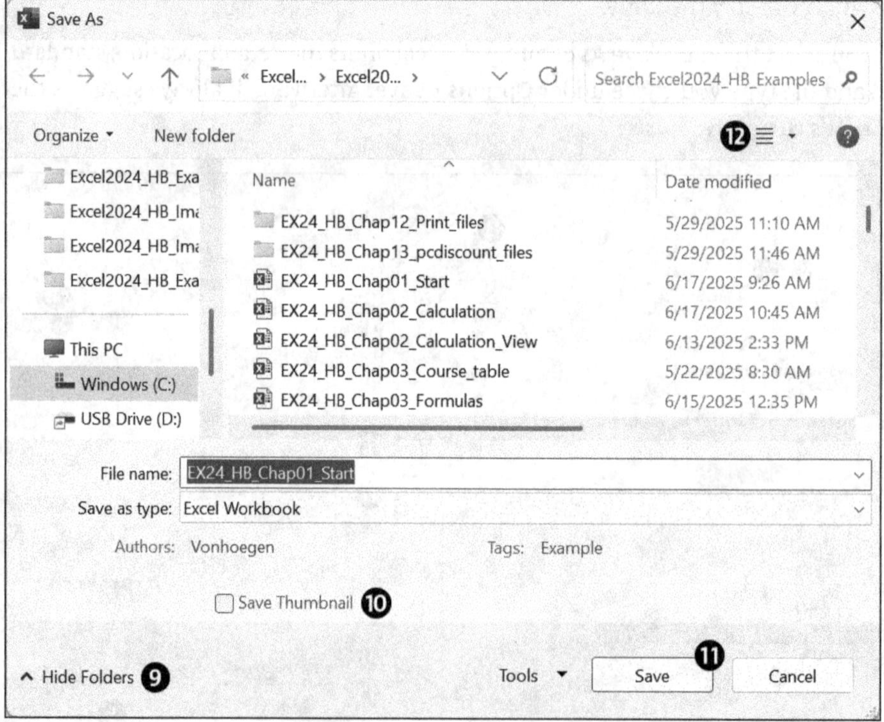

Figure 1.77 Save As Dialog Box in Details View

Skipping the Backstage

To save a new workbook without going through the **File** tab, press the [F12] key, which will immediately open the **File Save As** dialog box. Using the [Ctrl]+[S] keyboard shortcut will still take you to the tab the first time you save, but if you uncheck the **Don't show the Backstage when opening or saving files with keyboard shortcuts** option under **File • Options • Save** for workbooks, then the **File** tab won't appear in future when you save a workbook with [Ctrl]+[S] or open it with [Ctrl]+[O].

Another way to save a file is by clicking or tapping the file name in the title bar.

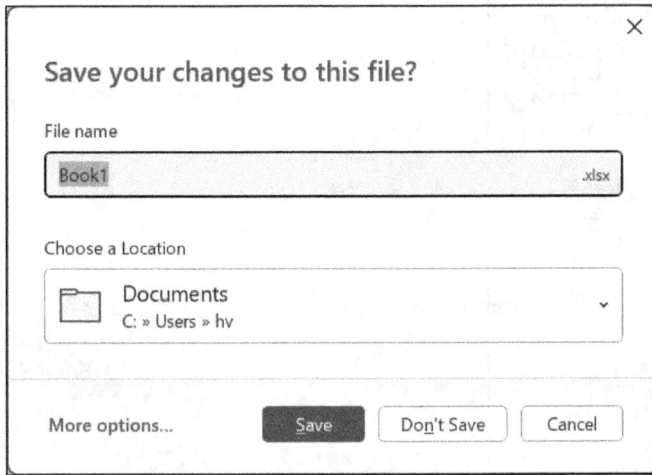

Save your changes to this file?

File name

Book1 .xlsx

Choose a Location

Documents
C: » Users » hv

More options... Save Don't Save Cancel

Figure 1.78 Quick Access to Saving

1.8.3 Choosing the View

The amount of information shown about each file in the **Save As** or **Open** dialogs depends on the selected view. You can click the **Change View** icon (refer to Figure 1.77 ⓬) to cycle through the available views, and you can also access these options via the arrow next to the **Change View** button. You can either click directly on a view or use the slider to preview different views before choosing the best option. The **Details** view offers the most information.

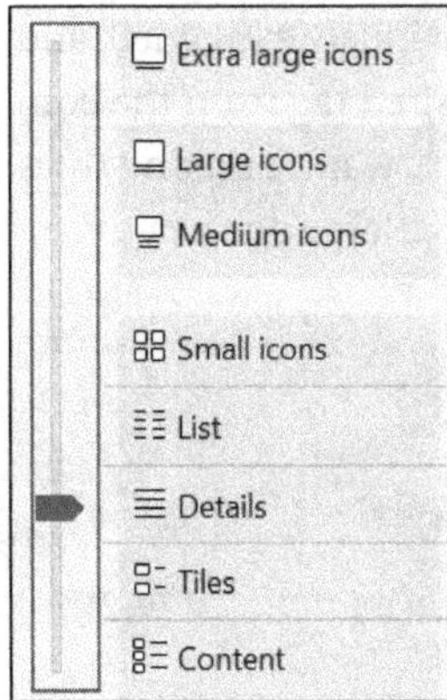

Figure 1.79 Slider for Choosing View

It shows extra details like **Size**, **Type**, and the last **Date modified**. You can select which columns to display via the context menu on the column header bar or via the **Select Details** dialog box, which you can open in the **More** option.

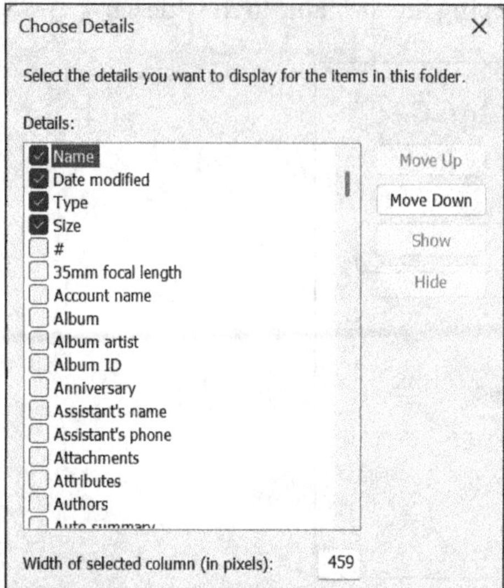

Figure 1.80 Select Details Dialog Box

The column labels appear on small, convenient buttons. Clicking or tapping the **Name** button sorts the column alphabetically—in ascending order on the first click and descending order on the second. Sorting by modification date in descending order always shows the newest document first.

Tags

You can also assign documents categorizing tags, which appear in a separate column and can be used as filters. For example, in Figure 1.81 the workbooks are tagged with the **Example** keyword. You can add tags in the **Save As** dialog, as previously explained, and you can then easily group these workbooks by selecting the desired tag from the context menu of the **Tags** column header. Simply check the tag you want, and only files with that tag will then be listed. Tagging is an easy way to organize large document collections, even after they're created.

Name	Type	Tags
EX24_HB_Chap01_Start	Microsoft Excel Work...	Example
EX24_HB_Chap02_Calculation	Microsoft Excel Work...	Example

Figure 1.81 Organizing by Using Tags

You can resize the dialog box by dragging its edges to show more items or get a better preview before opening. The size you choose will be saved and will be the same the next time you open the dialog box.

Viewing the files that are already in the folder will help you name your file. Since each file needs a unique name, this helps you prevent duplicates, and it can also help you avoid names that are too similar to existing ones, which can make files harder to find.

Of course, you can intentionally choose an existing name if you want to overwrite that file's current content, and in that case, you enter the name you've selected into the File name field. Excel will then ask for confirmation to ensure you really want to proceed. To number names sequentially, select the last name in the series from the list box in **File name** and then simply change the digit instead of retyping the entire name each time.

1.8.4 Choosing the Desired Folder

What if the target folder you want doesn't appear initially in the dialog box? You can use the small arrow buttons in the first list box, and the first button on the left ❶ will give you quick access to the desktop, the computer, the network, and other frequently used resources.

If the selected level above has sublevels, they'll appear under the next arrow button ❷. (Click it to view the list.) Using this method lets you quickly switch between folders. It also complements the navigation options provided by the arrow buttons in the top lefthand corner, which allow you to go back one step ❸ or forward ❹, display recently used folders ❺, or navigate to the parent folder ❻. The first three options activate only after you've moved at least one step within the folder hierarchy.

Clicking the arrow button at the end of the list box ❼ opens a list of recently used paths. Next to it is a button you can click to refresh the view of the selected path ❽, which is useful if, for example, a file has been added or changed by another application or through network access.

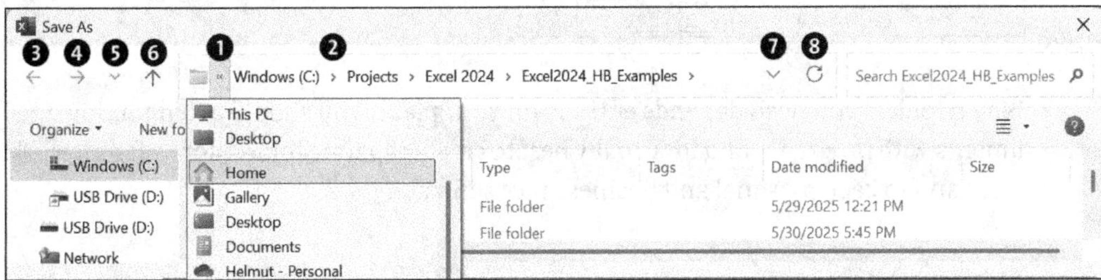

Figure 1.82 Quick Selection of Storage Location

1.8.5 Quick Access

By clicking the links in the navigation pane, you can find the right storage location faster. You can show or hide this functionality by using the **Organize** button and the **Layout • Navigation Pane** option. Click the small triangle next to a name to open the list of favorite folders, then click or tap **Home** to view frequently used folders.

This button also has a context menu, where you can use **Pin current folder to Quick access** to add the selected folder to the links.

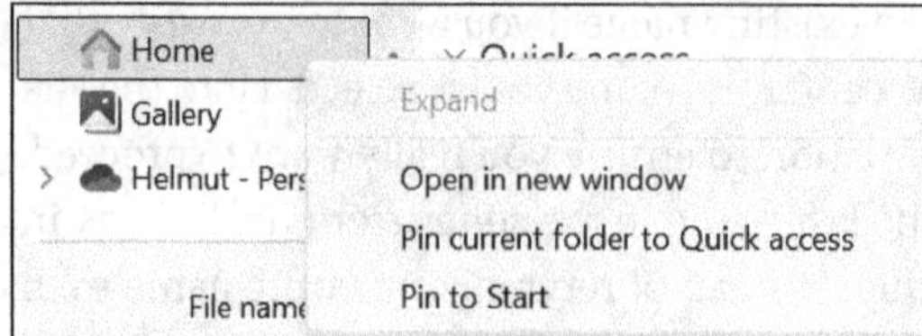

Figure 1.83 Navigation Pane with Context Menu for Quick Access

1.8.6 Creating New Folders

Creating a new folder is quick and easy. First, you should decide where the new item fits best within the folder structure, and then, do the following:

1. Make sure the dialog box shows the drive and folder where you want to create the new folder.
2. Click the **New folder** button.
3. Replace the default name with the name you want.
4. Click to set the new folder as the current folder so you can save files there immediately.

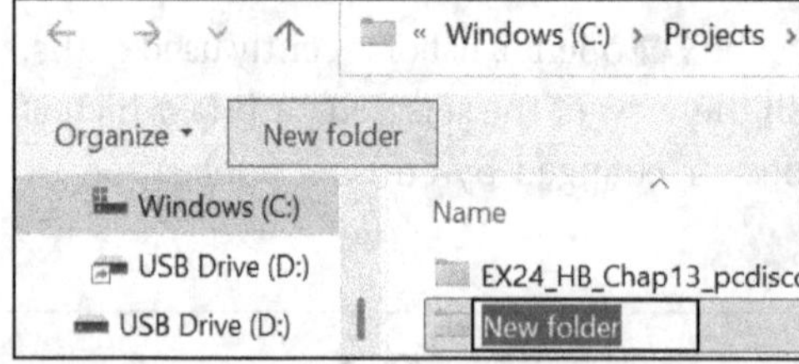

Which folders you need depends entirely on your plans. You should avoid mixing documents with programs, and it's usually best to create separate folders for different work areas and to keep personal and business files separate.

1.8.7 Choosing a File Name and File Type

After choosing the storage location, you need to name the document. Every document needs a name that's easy to recognize, and creating one for each document isn't easy, given the massive amounts of data now stored on typical drives. However, you can use longer filenames, including ones with spaces, to help create unique and memorable names.

Each name can be up to 255 characters, but overdoing it on name length is rarely helpful. In the **File name** text box, a new or empty workbooks initially shows a temporary default name, and you should replace it with a clear, unique name.

Under **File type**, you usually don't need to change anything. However, if you want to save the data in a different format than before so another program can process it, you can open the single-line list box and select the format you want.

Excel Workbook (*.xlsx)
Excel Macro-Enabled Workbook (*.xlsm)
Excel Binary Workbook (*.xlsb)
Excel 97-2003 Workbook (*.xls)
CSV UTF-8 (Comma delimited) (*.csv)
XML Data (*.xml)
Single File Web Page (*.mht, *.mhtml)
Web Page (*.htm, *.html)
Excel Template (*.xltx)
Excel Macro-Enabled Template (*.xltm)
Excel 97-2003 Template (*.xlt)
Text (Tab delimited) (*.txt)
Unicode Text (*.txt)
XML Spreadsheet 2003 (*.xml)
Microsoft Excel 5.0/95 Workbook (*.xls)
CSV (Comma delimited) (*.csv)
Formatted Text (Space delimited) (*.prn)
Text (Macintosh) (*.txt)
Text (MS-DOS) (*.txt)
CSV (Macintosh) (*.csv)
CSV (MS-DOS) (*.csv)
DIF (Data Interchange Format) (*.dif)
SYLK (Symbolic Link) (*.slk)
Excel Add-in (*.xlam)
Excel 97-2003 Add-in (*.xla)
PDF (*.pdf)
XPS Document (*.xps)
Strict Open XML Spreadsheet (*.xlsx)
OpenDocument Spreadsheet (*.ods)

Figure 1.84 List of Available File Types

When you confirm your entries in the dialog box by clicking **Save**, the file will be saved with the name you chose in the selected folder. The file will then remain open on the screen so you can continue working on it.

1.8.8 Adding Metadata to a File

Besides the file name and type, you can enter various metadata directly into the **Save As** dialog. If you hide the folders and drag the lower-right corner down slightly, the dialog will provide several text fields for this: **Authors**, **Tags**, **Title**, **Subject**, and more.

File name: EX24_HB_Chap01_Start
Save as type: Excel Workbook
Authors: Vonhoegen
Tags: Example
Title: Introductory_Example
Subject: Local_Net
Manager: hv
Company: Helmut Vonhoegen
Categories: IT_Equipment
Comments: Plan for a small network
Save Thumbnail

Figure 1.85 Entering Metadata Directly into Dialog

The input fields provide prompts that you can simply highlight and overwrite. To enter metadata, you can also use an alternative method that's independent of the **Save As** dialog. To do this, select **Info** on the **File** tab. File properties will appear on the right side of the page, where you can expand or collapse the data using the **Show All Properties** or **Show Less Properties** links.

For metadata that's not set automatically—like file size or modification dates—you can edit all entries by clicking or tapping the corresponding text field. You can also add more authors to the text field by using the **Add Author** prompt.

Above the metadata, the **Properties** button offers the **Advanced Properties** option, which you can select to open the **Properties** dialog box. The entries on the **Summary** tab mainly record the document's author and topics, plus keywords that describe its content.

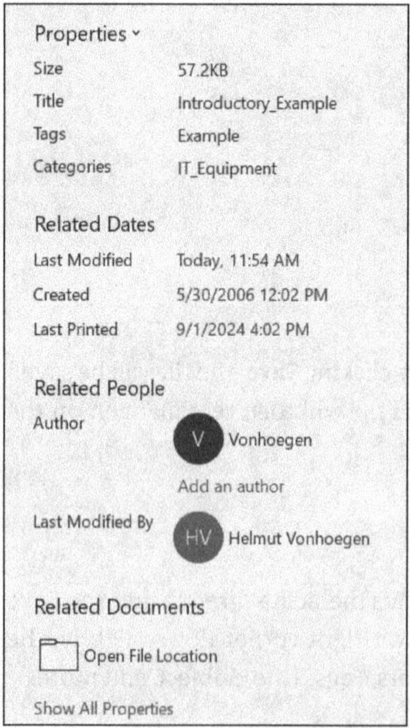

Figure 1.86 Metadata in Information Section

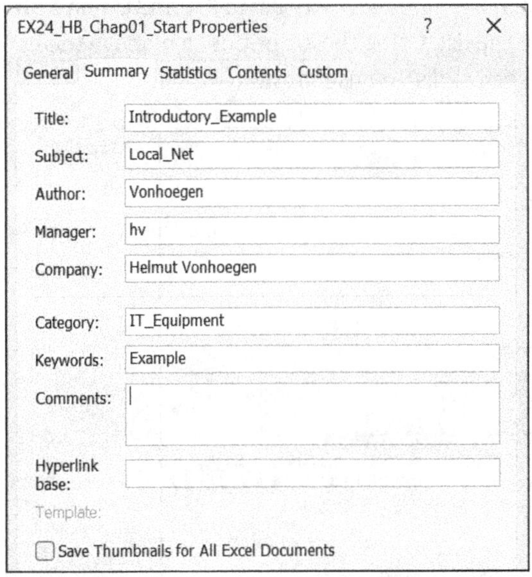

Figure 1.87 Summary Tab

1.8.9 Opening Recently Used Files

To resume work on a document or view its information, you must first reload the file. There are many ways to open a recently edited file, and as mentioned earlier, the **Recent** section on the start screen shows your most recently edited files. Click or tap a file name to open it, and if the Excel icon is pinned to the taskbar, then its context menu will provide a jump list of recently used workbooks (see Figure 1.88).

1.8.10 Creating New Workbooks

To create a new workbook, open the **File** tab and click the **New** button. The right pane will show a **Blank workbook** button along with various templates, and the Excel start screen will offer a similar selection when you launch the program. Clicking or tapping **Blank workbook** lets you start working immediately, and Excel will assign a default name, which you should overwrite right away using by the **Save As** dialog. The quickest way to create a new workbook is with the `Ctrl`+`N` shortcut.

Sometimes, it's useful to create a workbook based on an existing one. The recently used workbooks lists on the homepage and the **File • Open** page always offer an **Open a copy** option in the context menu. If you use that option, Excel will create a copy of the original and append a number to the file name. You should then assign a final name to the new workbook as soon as possible.

The context menu for a recently used file offers several useful options. **Copy path to clipboard** can be helpful when you need a complex path elsewhere. You can also use **Remove from list** to delete selected workbooks from the list, **Remove resolved items** to clear all files that are not pinned to the persistent section with **Pin to list**, and **Unpin from list** to remove a file from this section of the list. Using these options to manage the list can often speed up your access to the workbooks you use most.

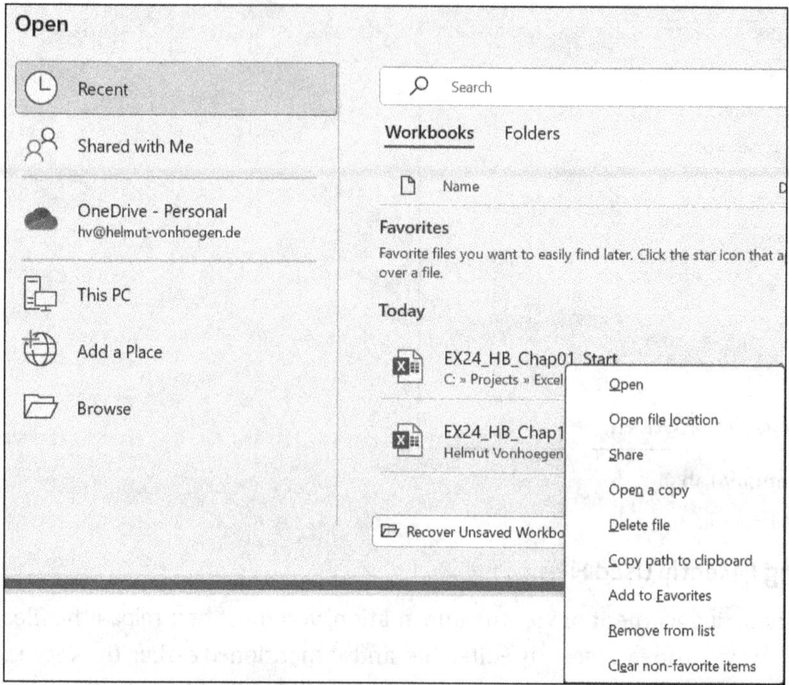

Figure 1.88 Context Menu for Recently Used Workbook

1.8.11 Working with Online Templates

As mentioned earlier, the **New** page offers many online templates. For example, to create a calendar in a workbook, you can select the **Daily Planner** template to view some information about the template. Then, you can click the **Create** button to download the template.

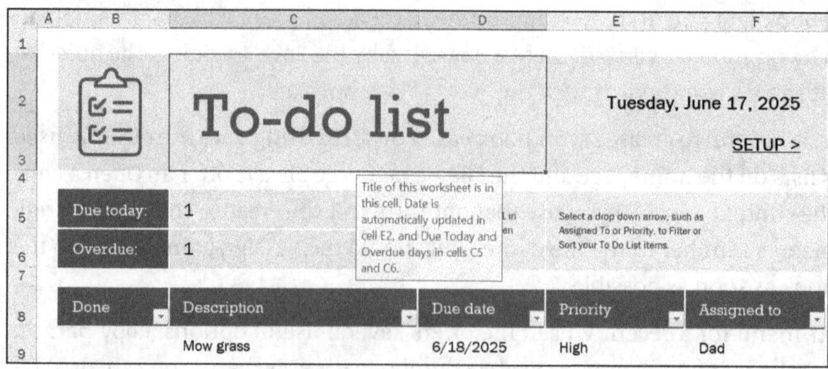

Figure 1.89 Daily Planner Template

To find a template on a specific topic, enter a keyword in the search box with the magnifying glass. For example, if you enter "To-do list," you'll see more than a dozen results.

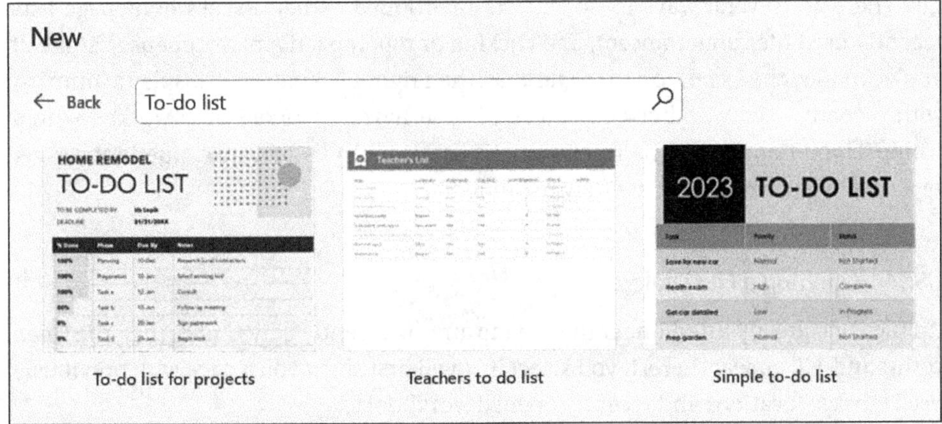

Figure 1.90 Results for "To-Do List" Search Term

1.8.12 Storing Custom Templates

In earlier versions of Excel, files with the *.xltx* or *.xltm* extensions were automatically saved in the user-specific default templates folder at ... *\AppData\Roaming\Microsoft\Templates*. But in the current version, you can choose the folder where your templates are saved. To do this, go to **File • Options • Save** and set the **Default personal templates location**. This folder will then be selected automatically in the **Save As** dialog when you save a file with the *.xltx* extension, and templates stored in this folder will appear under **Personal** on the **File • New** page.

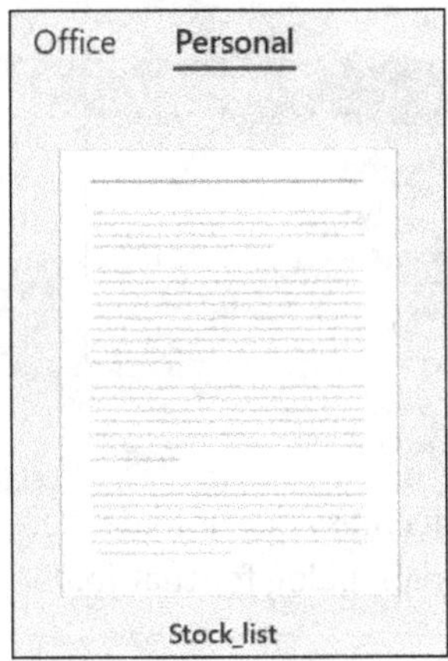

Figure 1.91 Custom Template That Appears Under Personal

1.8.13 Opening Existing Files

Before you can resume work on a document or view its contents, you need to reload the file. There are several ways to do this. As mentioned earlier, Excel's homepage lists recently used files under **Recent**, and clicking or tapping a file name opens that file. If you've pinned the Excel icon to the taskbar, then right-clicking it will display a jump list with recently used workbooks. However, if you haven't worked on a workbook in a while, it won't appear in the recently used list or jump list because older entries get replaced by newer ones unless they're pinned.

1.8.14 Opening a File Dialog

If a file isn't already listed under the **File** tab among recently edited files, then the **Open** command will appear there. If you select it, it will first show you a page with previously used storage locations and recently opened workbooks.

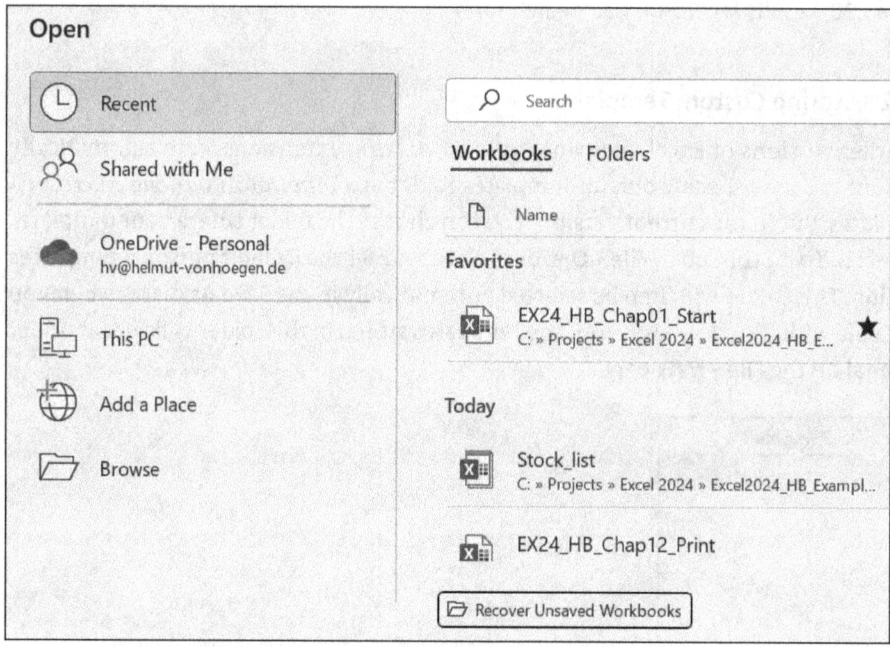

Figure 1.92 File • Open Page

Selecting **This PC** under **Open** displays the current folder with its files and subfolders on the right side of the page, and choosing a folder opens the **Open** dialog for that folder (see Figure 1.93).

To use the **Search** box above the list to find folders or files, you enter at least the first few letters of the name. If the desired file isn't in one of the listed folders, then click the **Search** button, which will open the **Open** dialog to the most recently used folder.

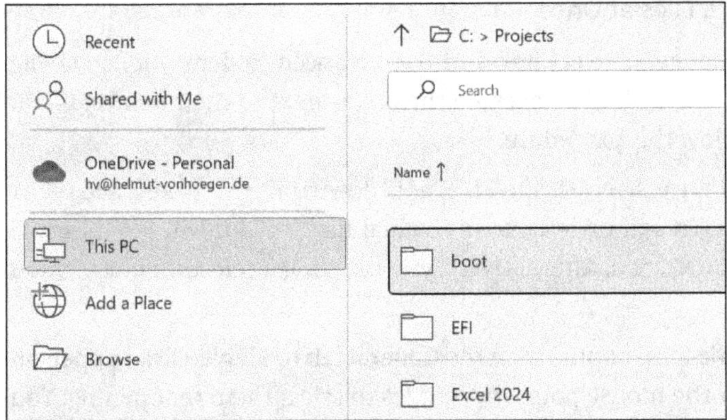

Figure 1.93 List of Subfolders in Current Folder

You can control what appears in the large list box by using the **Change View** button, which we also described for the **Save As** dialog. Before you open a workbook, it's often helpful to preview its contents—and you can do this by using the **Extra large icons** view, which will show you a snippet of the table if preview images were created when it was saved. This means you don't always have to open a file to see if it contains the data you need.

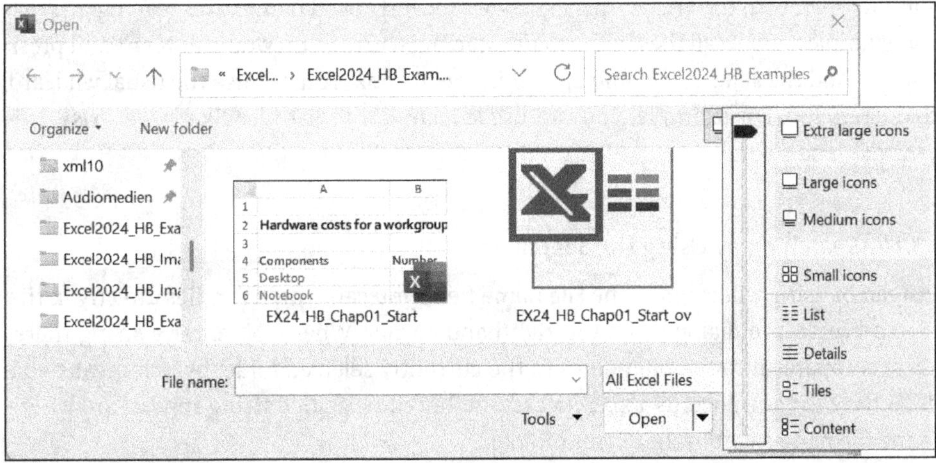

Figure 1.94 Open Dialog Box with Extra-Large Icons

The Prompt You May See When Opening a File

When you open a file that's already open, you'll receive a warning prompt, and if you continue, you'll lose any changes you've made to the open file. Usually, you should choose **No** to stop opening it, but sometimes, you'll want to roll back changes—like when you've tried reformatting a workbook but aren't happy with the results.

1.8.15 Opening Multiple Files at Once

If you enable **Use checkboxes to select items** in the advanced folder options via File Explorer, you can easily select files by checking the boxes next to their names. If you don't have it enabled, follow this procedure:

To open a group of files at once, select the first file, hold down the `Shift` key, and select the last file in the group. To add or remove individual files, hold down `Ctrl` while selecting files and then click **Open**. Alternatively, you can drag a selection box around the files you want to open.

If you've set File Explorer's folder options on the **General** tab to **Single-click to open an item**, then you can hover the mouse pointer over files to select them for opening. You can also press the `Shift` or `Ctrl` key to manage multiple selections here, or you can type the file name directly into the **File name** text box and confirm by clicking **Open**. The list box activates when you open the dialog box.

1.8.16 Finding Files by Using Search Patterns

You can use part of a file's name—called a search pattern—instead of the full name, and you can enter multiple search patterns separated by semicolons. Also, instead of typing a name or search pattern in the **File name** box, you can often select one from a list that you can open with the arrow. This list shows recently used names and search patterns. For example, to view all files in the folder starting with *M*, you would enter `m*` and press `Enter`, and the selection would appear in the list box. You can use the usual wildcard characters too. For example, you can use *ledger?.xlsx* to list files like *ledger1.xlsx*, *ledger2.xlsx*, and so on.

1.8.17 Searching by Using the Search Box

Instead of using wildcards in the **File name** field, you can search for files directly in the search box in the dialog box's top righthand corner. When you enter a string there, Excel compares it to the file names in the currently selected folder, which means you don't have to use wildcards. Excel checks whether the specified string appears anywhere in the file names within the current folder.

Excel uses the search term in the search box to find files whose names include that string. You can use file properties that Excel records automatically—like modification date, type, or size—along with the metadata you've entered into the search box to find files that match specific criteria. For example, if you've assigned the IT equipment category to several workbooks, you just type "Categories: IT-Equipment" in the search box. Remember to include the colon.

If matching workbooks exist in the selected folder, then they will appear. Excel uses a special query language based on key-value pairs. In Figure 1.95, *Categories* is the key and *IT-Equipment* is the value. The colon separates the key from the value.

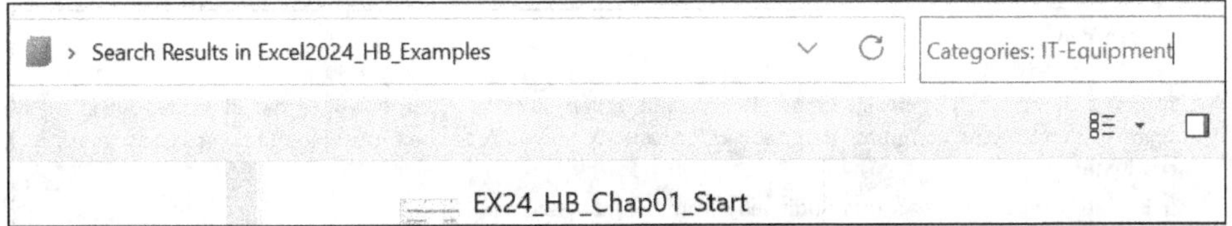

Figure 1.95 Searching with Metadata

In these search expressions, you can use all of these common comparison operators: =, <, >, <=, and >=. You can also specify multiple criteria at once, separating them with spaces. This lets you create very precise search patterns.

1.8.18 Opening Finished Documents as Read-Only

It's best to open some files as read-only from the start. Files that include final data should generally not be modified, so in the **Open** dialog box, you should open them with the **Open Read-Only** option, which appears when you click the arrow on the **Open** button. This also skips the write-access password prompt if the file is locked.

Write protection prevents accidental changes to the original file, so if you modify the opened file, you can only save it under a new name. You can also open the file as a copy via the menu for the **Open** button, after which, *Copy(1)...* will appear in the title bar. This ensures that the original file always remains unchanged. Another option is to use **Open in Browser**, which is only available for HTML files.

Another option is **Open in Protected View**. It's a special mode for handling files that might pose security risks, and it's particularly useful for opening files you've downloaded from the internet. To use it, go to **File • Options • Trust Center** and click the **Trust Center Settings** button.

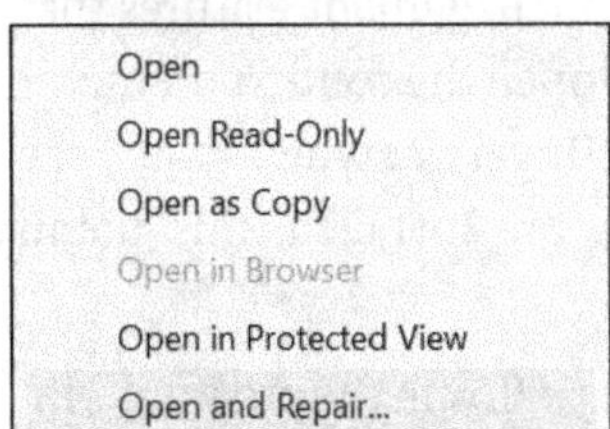

Figure 1.96 Menu for Open Button

Then, in the dialog under **Protected View**, you can choose which document types should open in protected view by default (see Figure 1.97).

These include workbooks downloaded from the internet or from local network locations that are not already marked as safe. After reviewing the workbook, if you want to enable normal editing, click the **Enable Editing** button (see Figure 1.98).

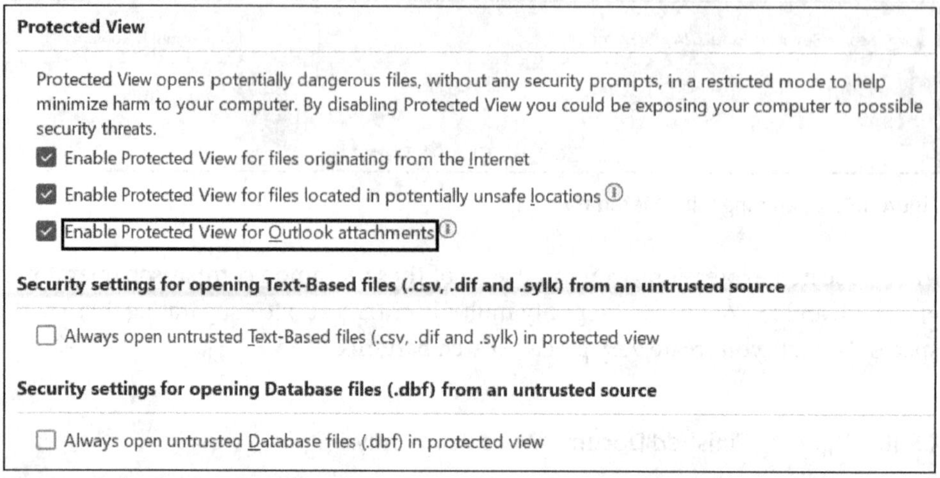

Protected View

Protected View opens potentially dangerous files, without any security prompts, in a restricted mode to help minimize harm to your computer. By disabling Protected View you could be exposing your computer to possible security threats.

☑ Enable Protected View for files originating from the Internet

☑ Enable Protected View for files located in potentially unsafe locations

☑ Enable Protected View for Outlook attachments

Security settings for opening Text-Based files (.csv, .dif and .sylk) from an untrusted source

☐ Always open untrusted Text-Based files (.csv, .dif and .sylk) in protected view

Security settings for opening Database files (.dbf) from an untrusted source

☐ Always open untrusted Database files (.dbf) in protected view

Figure 1.97 Protected View Options

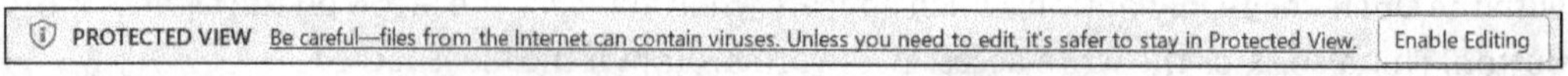

PROTECTED VIEW Be careful—files from the Internet can contain viruses. Unless you need to edit, it's safer to stay in Protected View. Enable Editing

Figure 1.98 Warning Received When Opening Downloaded File

1.8.19 Selecting a Folder

If your file is in a different folder from the one currently shown or is on another drive, you can start by selecting the drive or folder where your file is stored. The buttons available there are the same as those in the **Save As** dialog box.

1.8.20 Local File Management

The **Save As** and **Open** dialog boxes include many basic file management features that are otherwise accessed through the File Explorer. When you open the context menu for a selected file or files—by right-clicking or holding it with a finger or stylus—you'll see many options. If tools like WinZip or antivirus software are installed, this menu expands accordingly.

For example, you can print a file immediately without opening it first in the program window. With **Send to**, you can quickly copy the file to a compressed folder or a Bluetooth device, or you can send it directly as an email. You can also move a file to the **Documents** folder or create a shortcut on the desktop.

You can delete files here as well. For safety, you'll be asked if you want to move the file to the Recycle Bin, which serves as a safety net when deleting files, unless it's disabled via the **Properties** tab. A file is only permanently deleted once it's removed from the Recycle Bin, and once deleted, it no longer exists (though some tools might still help you retrieve it). If you accidentally move a file to the Recycle Bin and want to get it back, you can select the **Open** command in the Recycle Bin's context menu, select the file, and choose **Restore** from the context menu.

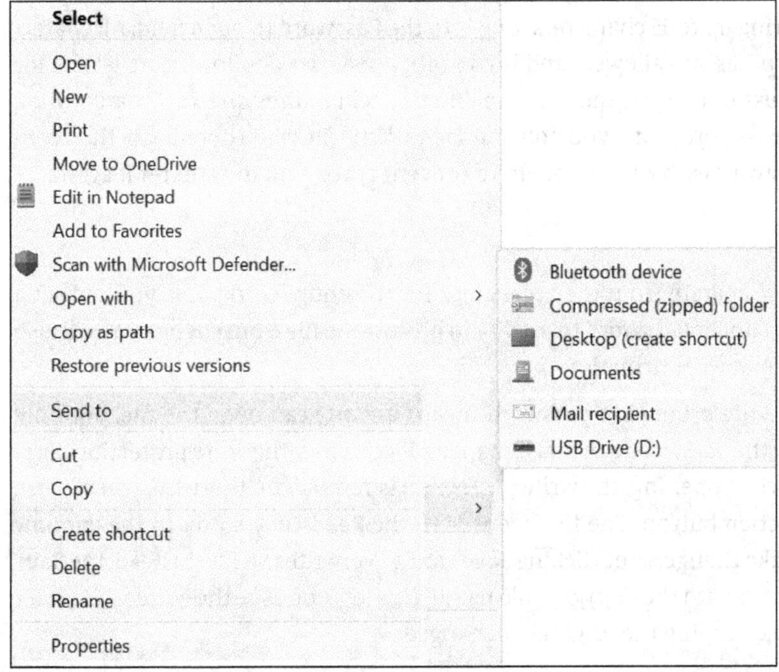

Figure 1.99 Context Menu for File Selection

Renaming a File

You can rename a file in place if its current name no longer fits. To do this, select **Rename** from the context menu, change the name, and confirm by pressing `Enter`.

1.8.21 Security Options: Password Protection and Encryption

If your documents contain personal or sensitive data, use the available options to protect files from unauthorized access or changes.

Excel offers simple data protection for workbooks in the **Save As** dialog box through the **Tools** button. The **General Options** dialog box lets you set two passwords.

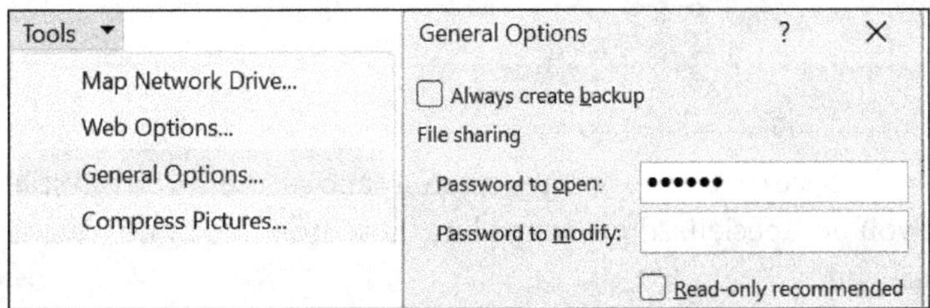

Figure 1.100 Setting Passwords Before Saving

Access Protection

The **Password to open** completely restricts access to a workbook, and only users who enter the correct password when opening the file can access the data. To create a password, enter a string up to 15 characters long into the **Password to open** field. All characters, including spaces, are allowed, and if you mix uppercase and lowercase letters for security, you must remember the exact spelling. Excel requires an exact match when re-entering the password, but if you only want to lock the file when opening it, then confirming the password with **OK** is enough. To prevent typos, you must enter it again.

Write Protection

In addition to or separate from the access lock for the entire workbook, you can set a second password under **Password to modify** to protect the file from changes. You'll also be required to input a confirmation.

Using only the write protection password means anyone can open the file, but only those who know the password can make changes. If you enter the write protection password correctly when opening, the write protection is removed. Otherwise, you can use the **Write Protection** button. The file opens with the **Read-Only** suffix in the window title. You can make changes, but clicking **Save** won't overwrite the file; instead, the **Save As** dialog box opens with the *Copy of*... file name. This lets you save the edited file under a new name while keeping the original unchanged.

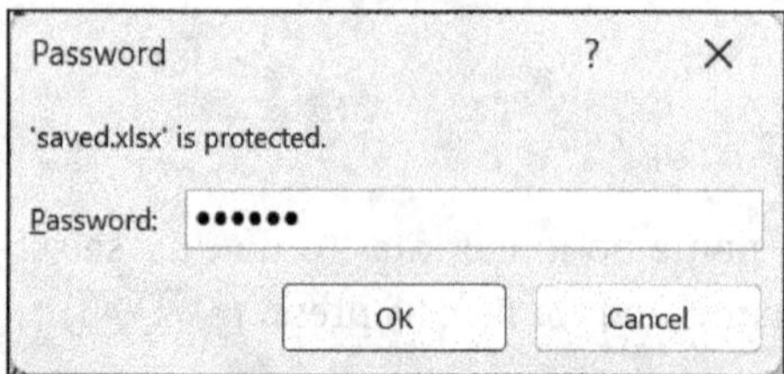

Figure 1.101 Write Protection Prompt

You can use both passwords at the same time to set a lock before opening and before editing. If you check the **Read-only recommended** box, Excel will display a prompt when opening the file if you haven't used the **Write Protection** button. Alternatively, in the **Open** dialog, you can click the small **Open** button and choose **Open as read-only.** You can also enable the **Always create backup** option in this dialog to create a backup file every time you save.

Forgot Your Password?

Even if you follow all good advice to store passwords carefully and securely, you can still lose or forget them. If you do, specialized programs like those from Passware (*www.passware.com*) can assist you.

Removing or Changing Passwords

If you no longer need to have password protection on a workbook, then if you know the password, you can remove it at any time. Delete the entry in the dialog box by pressing [Delete] and then confirm by clicking **OK**. You'll be asked if you want to overwrite the file's previous version, so confirm to proceed. Alternatively, instead of removing the password, you can follow the same steps to enter a new one.

1.8.22 Automatic Backup and Recovery

Despite improvements in system security, the operating system or a program may crash or a power outage may cause an unexpected shutdown. Normally, if Excel crashes, all changes made to a workbook since the last save are lost.

To minimize potential loss, Excel enables the **AutoRecover** feature by default. While it doesn't replace regular saving, this feature helps you recover work after a crash and reduces data loss by automatically and intermittently saving information about changes since the last save. You can enable or disable **AutoRecover** in Excel by going to **File • Options**, going to the **Save** tab, and then adjusting the **Save AutoRecover information every ... minutes** setting.

The default backup interval is ten minutes. That may be too long if you're entering large amounts of data, so you should select the interval that best suits your workflow. You can also specify a dedicated folder for backups. In addition, make sure the **Keep the last autosaved version if I close without saving** option remains enabled. If you work with a workbook where AutoRecover isn't needed, you can add it to **AutoRecover exceptions** to disable recovery for that file.

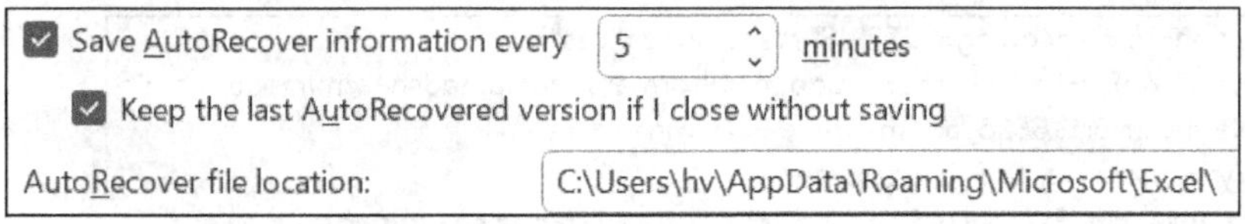

Figure 1.102 AutoRecover Options in Excel

If there's an unexpected shutdown and you reopen the program, the start screen will display a **Recover Unsaved Workbooks** link, which you can click to open the **Document Recovery** pane that lists the affected files. Each file has a button with a small menu that lets you open the file for viewing or save it under a new name. If the file has been repaired, you can view the changes.

If a file labeled **Recovered** is offered alongside the one marked **Original**, it's usually best to open or save the recovered version because it contains a more recent state. If you're in a hurry, you can save all versions under different names and compare them later.

You can also start repairing a file through the **Open** dialog. If you suspect a file wasn't saved correctly, open the **Open** dialog from the menu, click the **Open** button, and select

Open and Repair. You can choose to attempt a full repair with **Repair** or extract the data and formulas by using **Extract Data**. If you choose the latter, you'll need to specify how Excel should handle references in formulas (such as named cell ranges) that can't be restored. You can select **Convert to Values** to save the current cell value; if you don't, Excel will try to restore the formulas.

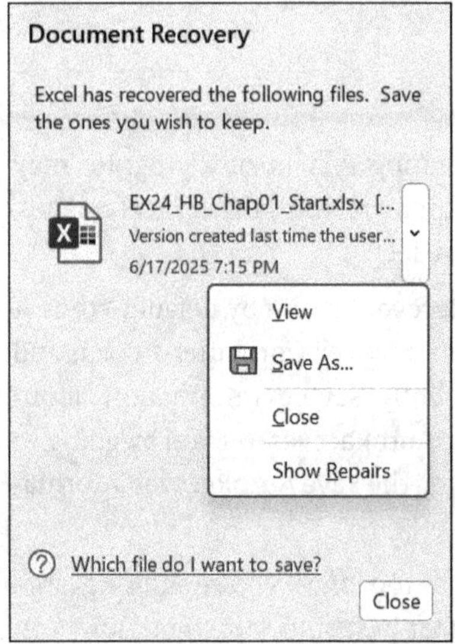

Figure 1.103 Menu for Recovered File

Excel also creates an XML log file for repairs and provides a link to it in the **Repairs in...** dialog.

```
<?xml version="l.o" encoding="UTF—8" standalone="yes"?>
<recoveryLog xmlns="http://schemas.openxmlformats.org/spreadsheetml/2006/main">
<logFileName>error116640_01. xml</logFileName>
<summary>Error in file 'C:\Projects\Excel
2024\Excel2024_HB_Examples\EX24_HB_chap19_extern.xlsx'</summary>
```

Figure 1.104 Example of Error Log

1.8.23 Version Control

Excel lets you manage multiple versions of a workbook when the file is shared and saved on OneDrive. These versions are created automatically as long as automatic caching is enabled in the **Excel Options** dialog. If multiple versions exist for an open file, you can view them on the **File** tab under **Info** by clicking the **Manage Workbook** button. The exact version is shown in the bar above the formula bar; you can use the **Restore** button to make the selected version the current one.

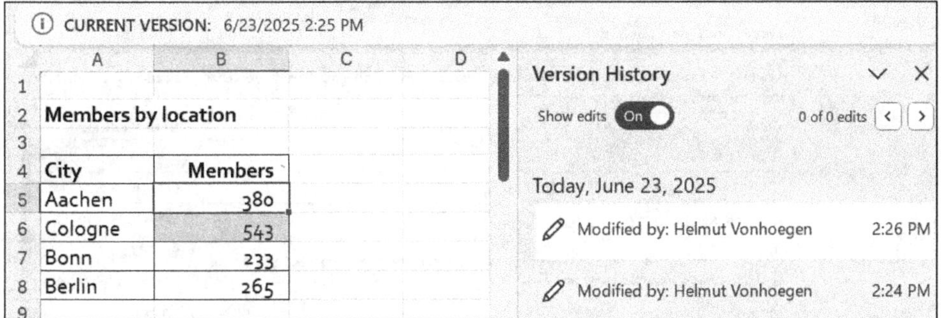

Figure 1.105 Different Versions of Same Workbook

The **Manage Workbook** button also includes the option to **Recover Unsaved Workbooks**, and you can select it to open the **Open** dialog to a folder that lists any files that weren't saved during previous Excel sessions. This only works if the option **Keep the last autosaved version if I close without saving** is enabled under **File • Options • Save**.

The folder contains files as they were at the time of the last automatic backup, and it uses the binary data format. When you open one of these files, a notice appears, and you can then use the **Save As** button to save the file.

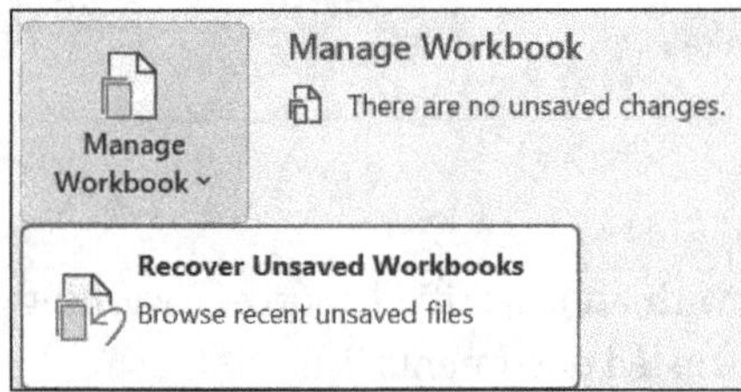

Figure 1.106 Manage Workbook Button

1.8.24 The Security Center

In the Security Center, Excel groups security measures that are designed to enhance protection. You can access it via **File • Options • Trust Center**, and you can click the **Trust Center Settings** button to open a detailed dialog box.

The **Trusted Locations** page offers various options for allowing exceptions to the security policies that are set in this dialog. One of the most important options is designating certain paths as trusted locations. Use the **Add new location** button to specify these paths. Some trusted locations are predefined, like the *Templates* folder and the *XLStart* startup folder. The last two options on this page also appear separately on the **Trusted Documents** page, and they let you enable or completely disable trust settings for documents on a network. The **Delete** button clears all trust assignments for documents at once.

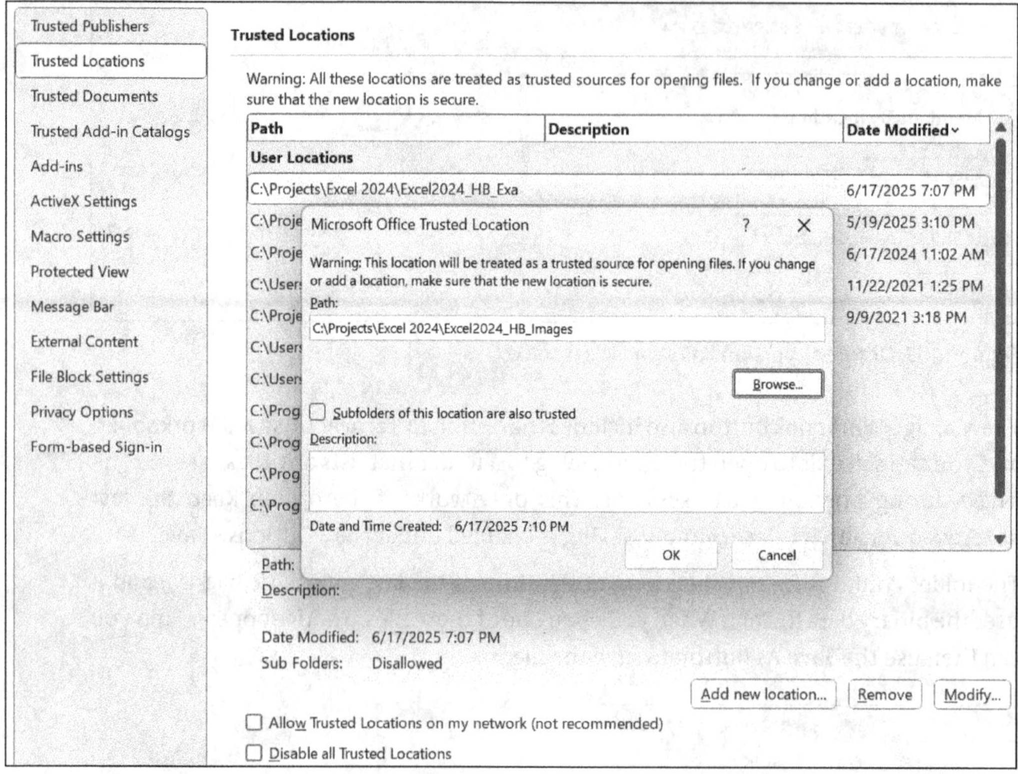

Figure 1.107 Choosing Trusted Locations

When you open a document from the internet with an unverified source, a warning appears that lets you add the document to your trusted documents list.

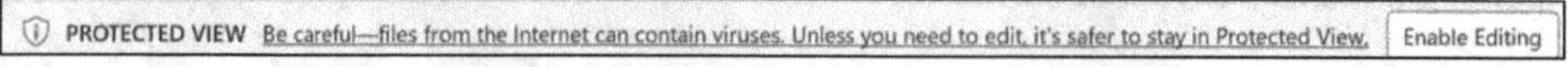

Figure 1.108 Security Warning for Files from Internet

When you open a document containing macros, a security warning usually appears on a bar above the formula bar, indicating that macros have been disabled. If you click the **Enable Content** button because you trust the file, the document will be added to your trusted documents list, and you won't be prompted again the next time you open it.

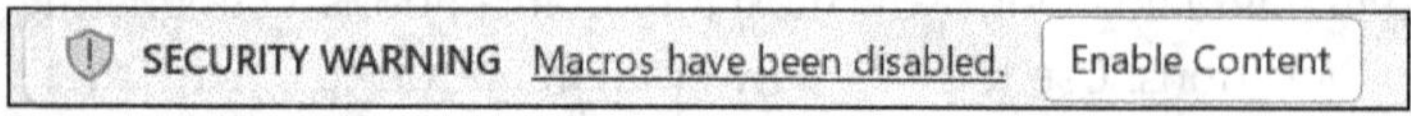

Figure 1.109 Macro Warning

A similar warning appears under **Information** on the **File** tab. Clicking the **Enable Content** button lets you activate a file's macros only for the current session through **Advanced Options**.

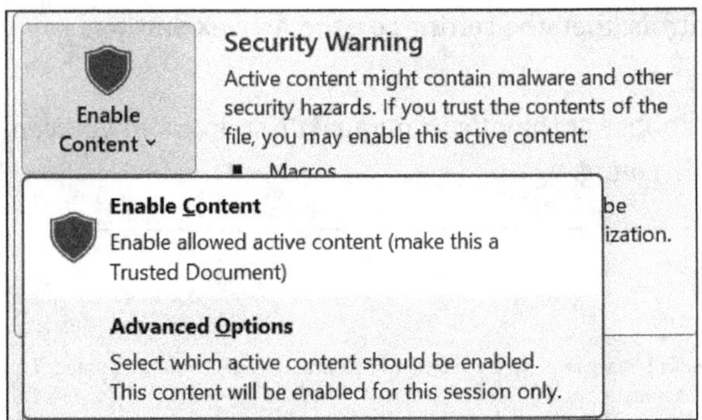

Figure 1.110 Sharing for a session

On the **Trusted Web Add-in Catalogs** page, you can decide how to manage the add-ins that are offered for Office programs.

The first option on the list is a strict one: **Don't allow any web add-ins to start**. Alternatively, you can exempt add-ins from Microsoft's Office Store if you trust them, and instead of a blanket refusal, you can specify the web addresses of trusted add-in catalogs here. Click the **Add catalog** button to specify the appropriate URL.

Trusted Web Add-in Catalogs

Use these settings to manage your web add-in catalogs.

☐ Don't allow any web add-ins to start.

☐ Don't allow web add-ins from the Office Store to start.

☐ Next time Office starts, clear all previously-started web add-ins cache.

Trusted Catalogs Table

You should only add a catalog if you trust its owner. You may also select one of each catalog type to show in the insert add-in menu. We will automatically start web add-ins from your insert add-in menu catalogs when opening documents.

Catalog Url: [ ] Add catalog

Trusted Catalog Address	Catalog Type	Show in Menu

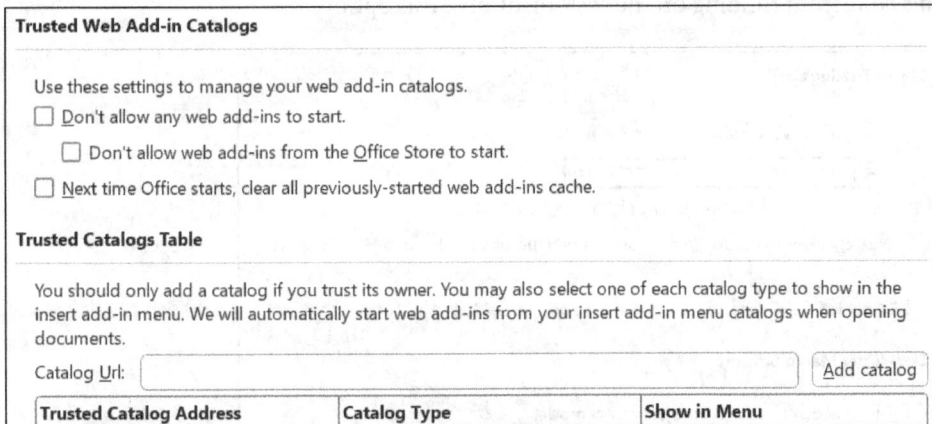

Figure 1.111 Security Policies for App Catalogs

The **Add-ins** page gives you options to block unsigned add-ins or disable all add-ins entirely.

Add-ins (COM, VSTO and Other)

☑ Require Application Add-ins to be signed by Trusted Publisher

☐ Disable notification for unsigned add-ins (code will remain disabled)

☐ Disable all Application Add-ins (may impair functionality)

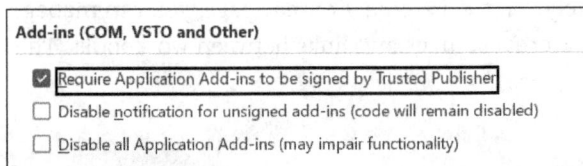

Figure 1.112 Add-In Settings

If documents contain ActiveX controls, then the settings on the **ActiveX Settings** page apply.

The **Macro Settings** page lets you choose automatic or manual disabling, and you can also exclude signed macros from disabling.

ActiveX Settings for all Office Applications

- Disable all controls without notification
- Prompt me before enabling Unsafe for Initialization (UFI) controls with additional restrictions and Safe for Initialization (SFI) controls with minimal restrictions
- Prompt me before enabling all controls with minimal restrictions
- Enable all controls without restrictions and without prompting (not recommended; potentially dangerous controls can run)

Safe mode (helps limit the control's access to your computer)

Figure 1.113 ActiveX Control Settings

If you enable the **Trust access to the VBA project object model** option, you'll be able to program automated Excel workflows with VBA code using the object model. To prevent this code from running on the system, disable this option.

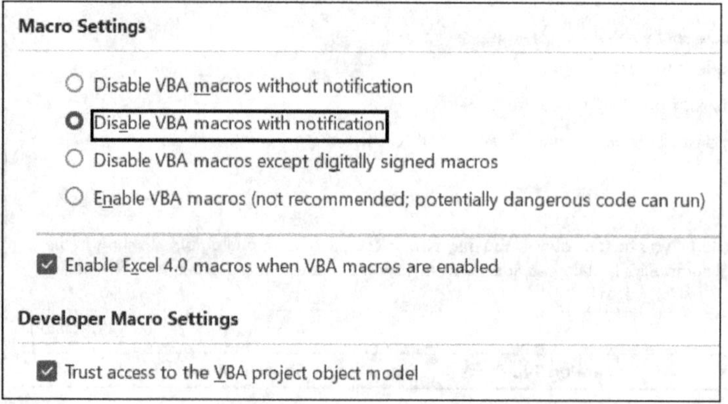

Figure 1.114 Macro Settings

The **Protected View** option has already been discussed. These three options specify which file types will open in this restricted mode, and you can enable opening text or database files in this view separately. On the **External Content** page, you can manage settings for data connections to external sources and links between workbooks. This also covers settings for linked data types.

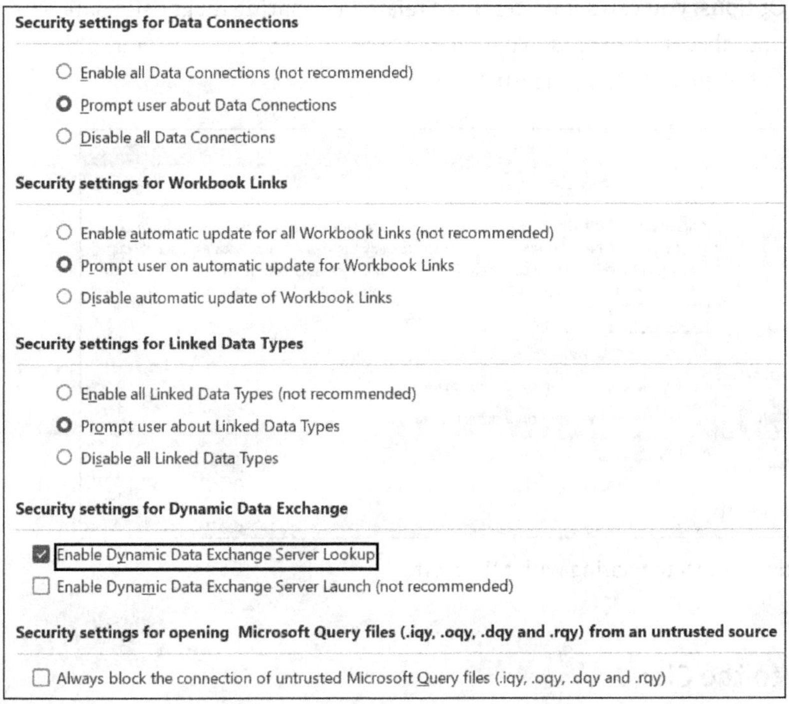

Security settings for Data Connections

○ Enable all Data Connections (not recommended)
◉ Prompt user about Data Connections
○ Disable all Data Connections

Security settings for Workbook Links

○ Enable automatic update for all Workbook Links (not recommended)
◉ Prompt user on automatic update for Workbook Links
○ Disable automatic update of Workbook Links

Security settings for Linked Data Types

○ Enable all Linked Data Types (not recommended)
◉ Prompt user about Linked Data Types
○ Disable all Linked Data Types

Security settings for Dynamic Data Exchange

☑ Enable Dynamic Data Exchange Server Lookup
☐ Enable Dynamic Data Exchange Server Launch (not recommended)

Security settings for opening Microsoft Query files (.iqy, .oqy, .dqy and .rqy) from an untrusted source

☐ Always block the connection of untrusted Microsoft Query files (.iqy, .oqy, .dqy and .rqy)

Figure 1.115 Settings for External Content

To set general restrictions for specific file types, use access protection through the **File Block Settings** page (see Figure 1.116). For example, older file formats like Excel 4 always open first in protected view.

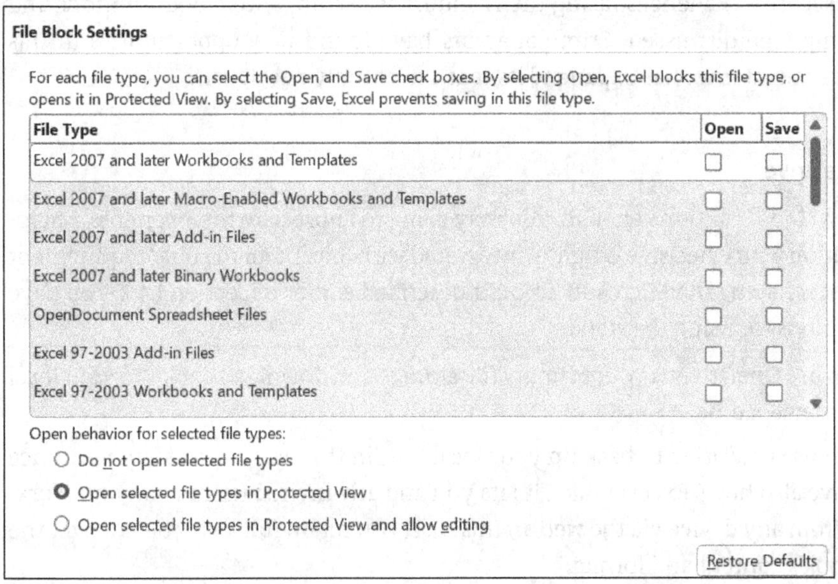

File Block Settings

For each file type, you can select the Open and Save check boxes. By selecting Open, Excel blocks this file type, or opens it in Protected View. By selecting Save, Excel prevents saving in this file type.

File Type	Open	Save
Excel 2007 and later Workbooks and Templates	☐	☐
Excel 2007 and later Macro-Enabled Workbooks and Templates	☐	☐
Excel 2007 and later Add-in Files	☐	☐
Excel 2007 and later Binary Workbooks	☐	☐
OpenDocument Spreadsheet Files	☐	☐
Excel 97-2003 Add-in Files	☐	☐
Excel 97-2003 Workbooks and Templates	☐	☐

Open behavior for selected file types:
○ Do not open selected file types
◉ Open selected file types in Protected View
○ Open selected file types in Protected View and allow editing

Restore Defaults

Figure 1.116 Access Protection for Specific File Formats

Under **Privacy Options**, you can set preferences related to sharing diagnostic data with Microsoft. This mainly concerns trust. The last two buttons let you enable or disable language directions for the translation feature and reference tools.

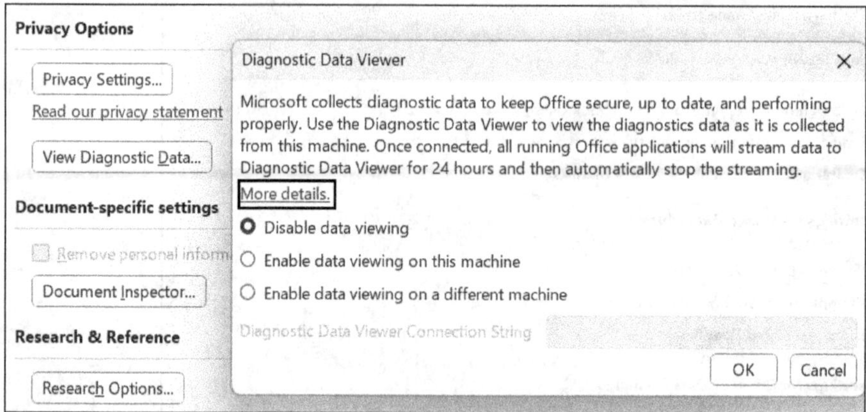

Figure 1.117 Options for Data Sharing with Microsoft

1.9 Saving to the Cloud

Storing data on servers around the world has become common, despite the inherent security concerns. Generally, people are more willing to trust an external provider with less-sensitive data than with data that must remain inaccessible to unauthorized users. On the other hand, data stored in the cloud is often better protected than data on a user's own system, especially if they neglect backups and system updates. However, the common security challenges facing the IT industry also apply to cloud solutions and cannot simply be dismissed. Criminal actors have found new opportunities in this industry as the internet has expanded its reach to entirely new levels.

1.9.1 OneDrive

If you trust cloud solutions enough, you may come to appreciate their benefits. Microsoft's solution with OneDrive is tightly integrated with Excel, and the only requirement for using it is having the Microsoft account described earlier in Section 1.4.2. You have several options for using the cloud:

- You can use OneDrive as your primary file storage and download copies to your local device as needed.
- You can use OneDrive to back up your local files in the cloud as a safety net. Since OneDrive also hosts Excel Online, it lets you and any invited users view your workbooks from any device via the web. In this case, you retain full local control over the workbook's content and format.

- You can not only copy your workbooks to the cloud for viewing but also allow changes to be made there, and you can use the cloud to facilitate teamwork or support mobile work.

First, we'll show you how to save files to OneDrive and open them locally from OneDrive when needed. We'll also briefly cover the teamwork feature in Chapter 14.

If you use your local Excel version with a Microsoft account, **OneDrive** ❶ is automatically offered as the save location when you choose **Save As** from the **File** tab. You can find previously used folders ❷ and documents in the right pane of the window. You can click or tap a folder to select it for saving. Click the **New Folder** ❸ button to create a folder at the current level, and then, just enter the file name ❹.

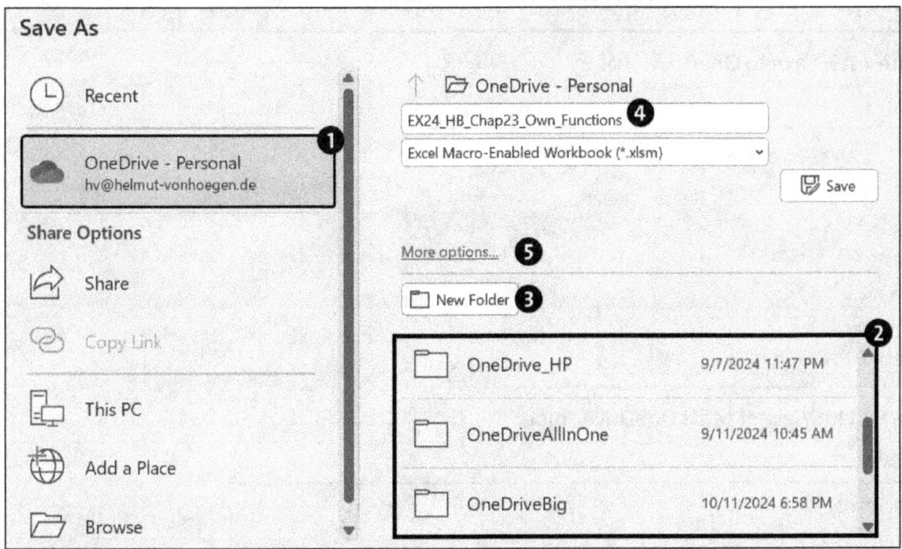

Figure 1.118 OneDrive as Storage Location on File Tab

Clicking More options ❺ opens the **Save As** dialog (see Figure 1.119) with the selected storage location, and the address bar shows your chosen OneDrive location. Excel uses local OneDrive storage that automatically syncs with the OneDrive cloud, as long as you're connected to the internet (see Figure 1.120). If you're not, then the data is available offline, and you can save it as you would to a local folder.

Start saving by clicking the **Save** button. The **AutoSave** button will then activate in the title bar of the saved workbook, and from that moment on, all changes to the workbook will be saved immediately. This is especially useful when multiple people need to work on the workbook over a network.

In the local OneDrive folder, the **Status** column shows icons indicating whether the data is synced with the cloud or still syncing. To open a workbook stored in the cloud, go to **Open**, select **OneDrive**, and find the workbooks in the recently used folder. If you've recently accessed a workbook, you'll find it in the **Recent** list (see Figure 1.121).

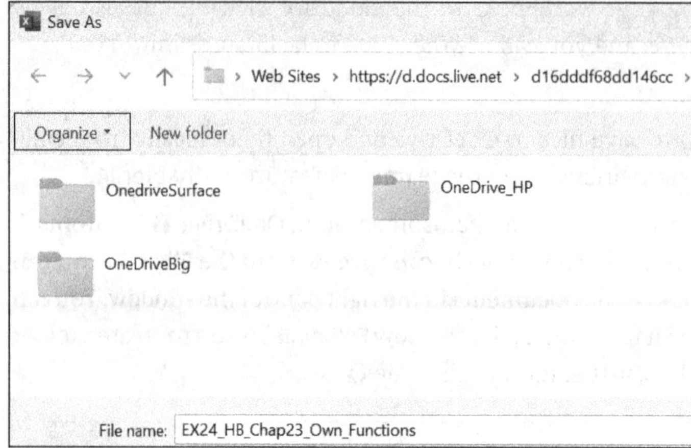

Figure 1.119 Save to OneDrive Dialog

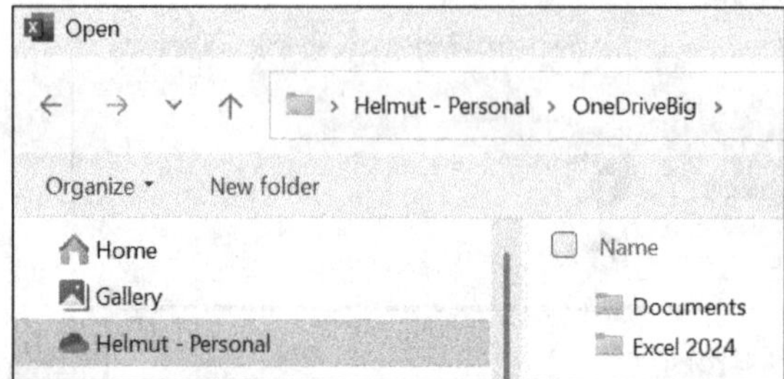

Figure 1.120 View of Local OneDrive Folder

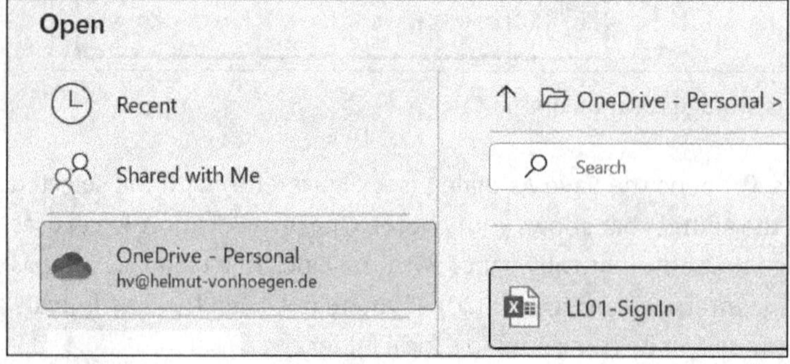

Figure 1.121 Opening Workbook from OneDrive

1.10 Excel Help

Because of its complexity and many features, Excel needs easily accessible support tools. One way to get help is via the search box in the title bar, which you can also open by pressing [Alt]+[M]. Type what you want to do or where you need help. Previous requests will appear under **Recently Used Actions**, along with some suggested actions.

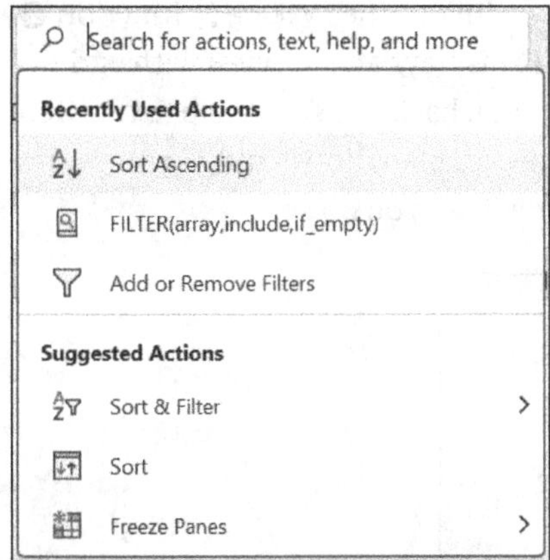

Figure 1.122 Search Bar Display of Recent and Suggested Actions

1.10.1 The Help Assistant

The help system does more than just provide information about your question. Often, the commands to perform a specific step appear directly as options, so the help system functions like an assistant handling routine tasks for you. This is easiest to understand via an example. If you have a long data list on a sheet and you want to sort it, do the following:

1. Click any cell in the list and type "define filter" into the search box.
2. A menu will appear with several options, including **Sort Ascending**.
3. Select this option to immediately sort the column containing the active cell in ascending order.

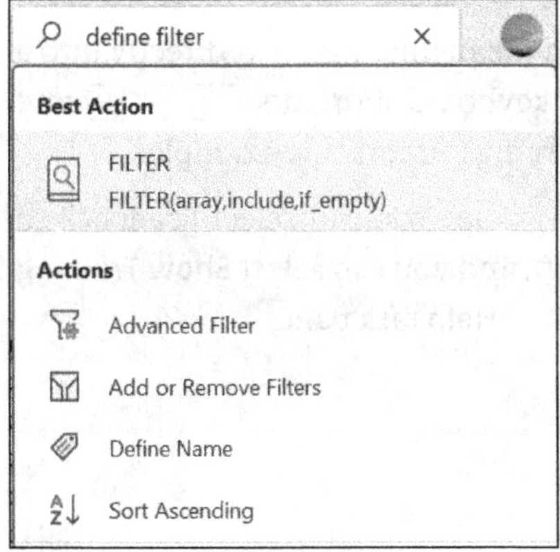

1.10.2 The Help Tab

Another way to get help is through the **Help** tab and its button of the same name. Clicking the button opens the **Help** task pane with a search box ❶ at the top, where, for example, you can type "calculate average" to show a link to the AVERAGE() function ❷ and links to related topics. Clicking the home icon ❸ displays an overview with various access points, and clicking the arrow ❹ always takes you back one step. Help texts often include additional links to related topics or detailed explanations, and if you prefer reading help in your browser, you'll find a link that will let you do so.

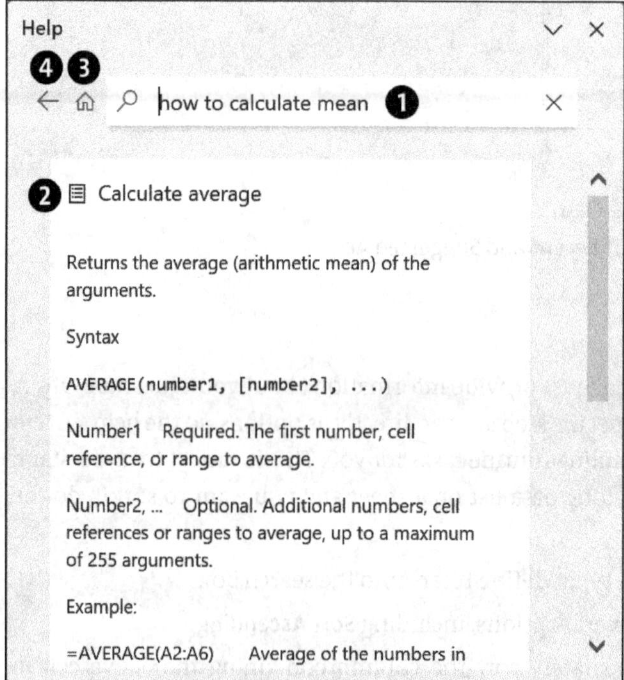

Figure 1.123 Example of a Query in the Help Task Pane

To copy text, simply highlight the section and drag the highlighted text directly into a Word file. Although there's no context menu, the keyboard shortcuts Ctrl+C and Ctrl+V work and are very useful when you're trying out formula examples.

One final note about the other icons in the **Help** group: You can select **Feedback** to report errors or suggest improvements to Microsoft, and you can select **Show Training** to access video tutorials and solution examples in the **Help** task pane.

Chapter 2
The Structure of Spreadsheets

This chapter begins by explaining how to best define a table's structure to solve a specific problem. As an example, it uses an income and expense statement that tracks your finances. Next, this chapter teaches you how to enter your data most efficiently, and finally, it covers all the basic operations that are essential for working with calculation models.

2.1 Planning and Designing Calculation Models

Programs like Excel tend to tempt you to dive straight in and add labels and data the moment you open them. All too often, the result is a worksheet that, on closer inspection, just doesn't hold up. Although correcting these problems is fairly easy in Excel, it's usually more effective to clarify a few things beforehand. You can do this on a separate sheet in the workbook that's used to document the calculation model.

2.1.1 What to Consider When Building Tables

Before you enter data into a table, it's important to consider several questions, such as these:

- What is the table's purpose?
- What information should it provide to support which decisions?
- What data is needed to fulfill that purpose?
- Which information should be emphasized?
- Who will enter the data, and who will analyze it?
- Which data will be entered manually, which data will come from existing sources, and are there links to other tables?
- What time periods, regions, items, or people does the table cover?
- How long is the data valid, which data is static, and which data changes—and how often?
- What level of accuracy is required, and what is realistically achievable?
- What calculations are planned?
- How is the accuracy of the data and calculations ensured?

- Which data will be analyzed graphically?
- How are the data published, in what format, and for whom?

Tables are generally used to present specific facts, relationships, and trends in an organized way. A table should include all the information necessary for it to achieve its purpose. For example, our income-expense statement should list all income and expense items to accurately represent the current situation. Unnecessary data should be excluded from a table as it reduces clarity. Often, it's better to spread data across multiple tables than to cram all the material on a topic into one.

On the income and expenses sheet, for example, you could list the amounts for each individual insurance policy separately—but that would actually make things less clear. Therefore, it's smarter to use a dedicated worksheet for insurance costs, and to help you access that sheet quickly when you need to, you can add a hyperlink from one table to the other.

2.1.2 Labels, Values, and Calculation Rules

Usually, a table's content consists of three main parts:

- Labels for the columns and rows used
- Values
- Calculation rules (formulas)

These often reflect three stages in developing a table. First, you define the table's structure with labels, then, you enter the values (the numerical data for calculations), and finally, you create the formulas. If you have forms you want to fill out later, you can enter the formulas in advance. It's also often helpful to test formulas with sample values that you can delete later.

Labeling Rules

You should follow these guidelines for table labels to ensure clarity:

- The temporal, spatial, factual, or personal scope should be clearly identifiable.
- Keep column labels as brief as possible to display more data columns in the window. Place related items next to each other, and always position row sums either before or after the value columns for consistency.
- Use uniform row heights for row labels to avoid unintentionally highlighting items.
- When possible, let large tables expand vertically rather than horizontally. Use column labels with fewer items.
- Arrange values vertically when calculating sums, and calculate row sums by using a manageable number of values whenever possible.

2.1.3 Defining the Structure of an Income and Expense Table

The best structure to use when creating a table usually depends on the key characteristics of the data you'll be entering into the table.

Labels

A two-part table works well for an income and expense statement: income is listed on one side and expenses on the other. Using indents helps distinguish main labels from those for specific income types.

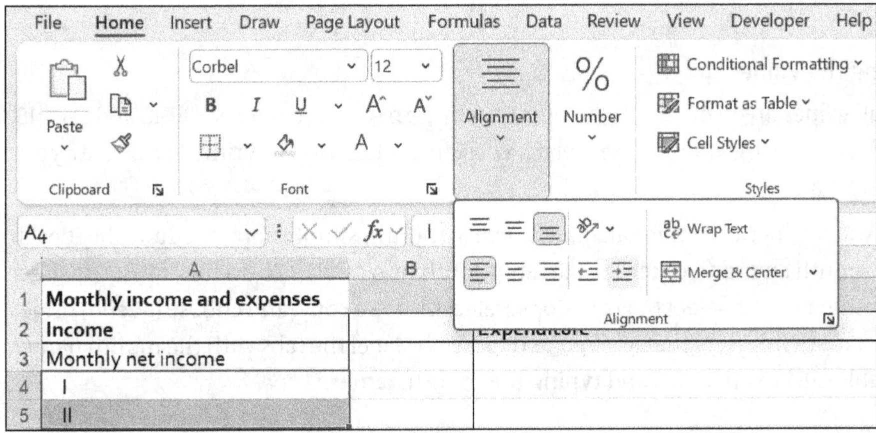

Figure 2.1 Labels for Income, Organized with Indents

Classifying expenses is generally more challenging than classifying income because you need to find categories that clearly assign each expense amount. The example solution in Figure 2.2 is just a suggestion.

	A	B	C	D
1	**Monthly income and expenses**			**Jan-25**
2	**Income**		**Expenditure**	
3	Monthly net income		Rent and other housing expenses	
4	I		Expenses for car	
5	II		Public transport expenses	
6			Insurance premiums	
7			Contractually fixed savings rates	
8			Interest and repayment rates	
9			Food	
10	Other regular income		Expenses for the children	
11	Capital gains		Expenses for private purchases	
12			Expenses for information, education, etc.	
13			Expenditure on health and sports	
14			Expenses for entertainment	
15			Donate	
16			Other expenditure	
17	**Total**		**Total**	

Figure 2.2 Fully Labeled Income and Expenses Table

Setting the Calculation Rule

This table requires minimal calculation. You only need to sum the two columns for income and expenses, so after entering the labels, you can input the formulas as follows:

1. Select cell B18.
2. Click the **AutoSum** icon on the **Home** tab in the **Editing** group. Because Excel can't automatically detect the range due to empty cells, select the range from B3 to B17 and press `Enter` to confirm the formula.
3. Repeat the same steps for cell D18.

Entering the Values

Since all values are in euros, apply the currency format to the ranges B3:B18 and D3:D18. Then, if you save the table as a template, you can use the form anytime to calculate your monthly expenses.

If you want to perform this calculation every month, simply copy the first sheet each time a month ends and enter the current month into cell D1. To do this, open the sheet tab's context menu, select **Move or Copy**, select **Create a copy**, and choose **(at end)** under **Before sheet**. On sheets that are updated monthly, label the tabs with the month names by double-clicking the tabs and typing the month names.

	A	B	C	D
1	**Monthly income and expenses**			**Jan-25**
2	**Income**		**Expenditure**	
3	Monthly net income		Rent and other housing expenses	$3,000.00
4	I	$5,000.00	Expenses for car	$500.00
5	II	$4,000.00	Public transport expenses	$250.00
6			Insurance premiums	$600.00
7			Contractually fixed savings rates	$400.00
8			Interest and repayment rates	$340.00
9			Food	$1,200.00
10	Other regular income		Expenses for the children	$500.00
11	Capital gains	$400.00	Expenses for private purchases	$500.00
12			Expenses for information, education, etc.	$400.00
13			Expenditure on health and sports	$1,000.00
14			Expenses for entertainment	$300.00
15			Donate	$100.00
16			Other expenditure	$100.00
17	**Total**	$9,400.00	**Total**	$9,190.00

Figure 2.3 Completed Income and Expense Statement

2.2 Navigation and Selection

In Excel, whether you're entering data and formulas or formatting and editing cells, cell ranges, and objects, you should always select first and then act. Selecting cells, cell ranges, sheets, or other objects is one of the most common tasks you'll perform, and you'll save time by mastering the selection methods and choosing the most effective one for each situation. This is especially important when you're working on large tables, where only a small portion is visible on screen.

If you're using a computer, you can select by using either a mouse or keyboard commands—and which option is better depends on the situation. It's often easiest to select the first cell by clicking it and to then use the arrow keys to move through neighboring cells. If you're using a touchscreen, then finger gestures replace the mouse; or if your device supports it, you can use a stylus instead of your fingers. Finally, if you're using a tablet, you might use an on-screen keyboard instead of a physical one. In the following sections, we mainly describe the classic mouse-and-keyboard operation, but we also include touch mode instructions where needed. If you're using an on-screen keyboard, keep in mind that you can't combine finger presses with key presses directly. In many cases, you can tap a key first in **Touch input** mode and then perform an action with your finger, but be aware that not all mouse and keyboard functions have exact equivalents in this mode.

2.2.1 Sheet Selection and Group Editing

Imagine you've completed the income and expense report from the last section for several months. To view May's data, you can just click or tap the tab of the corresponding sheet, and you can also switch sheets by using `Ctrl`+`Page ↓` or `Ctrl`+`Page ↑` on the keyboard.

If you need to have multiple sheets share the same formatting or print them together, you can select them all at once. If the sheets are adjacent, click the first one, hold down the `Shift` key, and then click the last tab in the group. To add or remove individual sheets from the group, use `Ctrl` and click or tap the tab. However, you cannot remove the currently active sheet—which is marked by an underline and white background—from the group. The tabs of the other sheets in the selected group also have a white background, and the workbook title displays the word **Group**.

You can also use this group mode to, for example, add a common title row to all sheets at once. Any entry in a cell on a sheet within a group appears in the same position on all other sheets in that group, and clicking or tapping a tab outside the group cancels the multiple selection. To select all sheets at once, choose **Select All Sheets** from the tab's context menu. To cancel this selection, choose the **Ungroup** option.

2.2.2 Selecting Cells and Cell Ranges

If a cell is visible in the window, you can select it with a mouse click or a tap with your finger or stylus. Selecting cells far from the current cell pointer takes a bit more effort, so you'll need to use the scroll bars to move the screen area and then select the cell that appears.

Dragging the vertical scroll bar moves the screen area row by row, and dragging the horizontal scroll bar works the same for columns. Holding down `Shift` speeds up the scrolling, and clicking or tapping the arrows on the scroll bar moves the window by one row or column. Holding down the arrows causes continuous scrolling, and clicking or tapping above or below the scroll bar shifts the screen area by about the size of the current window.

On a touchscreen, you can also adjust the screen area by sliding the grid horizontally or vertically with your finger. It's even faster if you swipe in the opposite direction within the grid.

Selecting Cell Ranges in a Sheet

To select cell ranges, drag over the desired area while holding down the left mouse button. The direction you drag in doesn't matter, and the cell where you start dragging is always the active cell of the range. To extend the range beyond the visible area, drag past the window edge until the desired table area appears and then release the mouse button once you reach the correct position. A counter at the moving edge helps you accurately define the range, and if you go too far, just drag a little in the opposite direction.

When you're selecting a large area with the mouse, it's usually more effective to use a different method. Click the first corner cell of the area and then hold down the `Shift` key while clicking the opposite corner cell. If the selection isn't quite right, click the correct spot again while holding down the `Shift` key. Using this method is especially helpful when you need to scroll the view to make the last cell visible. If you select an incorrect area, click a single cell to cancel the selection. However, running a command that affects the area doesn't cancel it, so you can use the same selection for multiple commands.

Another way is to select the first corner cell, press `F8` (which will make the status bar display **Extend Selection**), and then select the opposite corner cell. Then, press `F8` again to exit extension mode.

To select a range using **Touch input** mode, tap the top-left corner cell and then drag the round handles that appear in the direction you want. Even in **Touch input** mode, you can use `F8` as described.

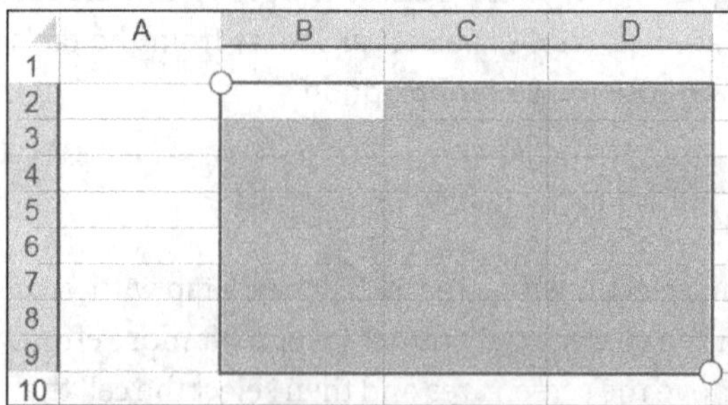

Figure 2.4 Range Selected by Dragging Round Handles

Selecting Rows and Columns

You can select an entire column or row by simply clicking or tapping its header. To select multiple rows or columns at once, drag the mouse over the row or column headers. The number of selected items is always shown for easy reference.

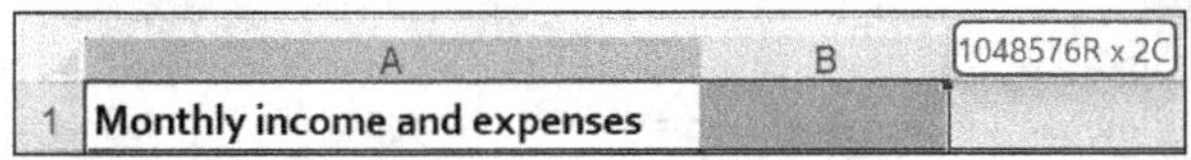

Figure 2.5 Selecting Two Income Columns

In **Touch input** mode, drag the round handles that appear on both sides of the selected column. To select the entire table at once, click or tap the **Select All** box in the upper-left corner where the column and row headers meet.

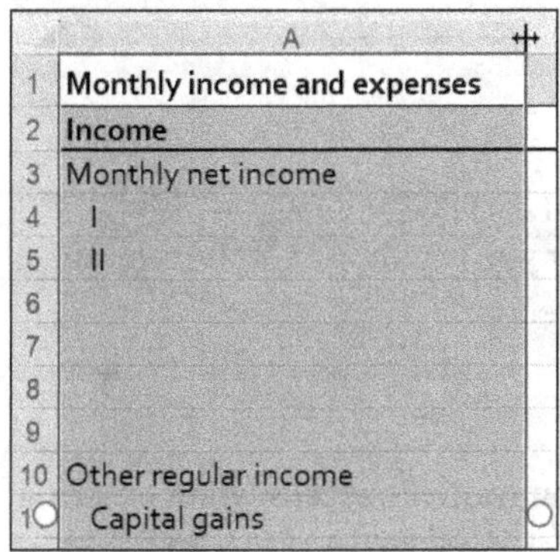

A
1 Monthly income and expenses
2 Income
3 Monthly net income
4 I
5 II
6
7
8
9
10 Other regular income
Capital gains

Figure 2.6 Multicolumn Range with Round Handles on Touchscreen

Multiselection

Although a range must always be rectangular, Excel lets you combine multiple rectangles into a multiselection. This allows you to select nonadjacent cells, for example, to format them consistently. The areas can even overlap.

For example, to highlight the two number columns in the income-expense table with different colors, follow these steps:

1. Define the first area in column B by dragging as usual.
2. To include the expense values, hold down the Ctrl key while dragging in column D.

The active cell is always in the most recently selected area (see Figure 2.7).

To include entire columns or rows in a multiple selection, hold down Ctrl while clicking the column or row headers.

You can also use the function key method. After selecting the first area, press Shift+F8, select the start of the second area, press F8, and then click or tap the end of the second area. Press F8 again to finish the multiple selection.

	A	B	C	D
1	**Monthly income and expenses**			**Jan-25**
2	**Income**		**Expenditure**	
3	Monthly net income		Rent and other housing expenses	$3,000.00
4	I	$5,000.00	Expenses for car	$500.00
5	II	$4,000.00	Public transport expenses	$250.00
6			Insurance premiums	$600.00
7			Contractually fixed savings rates	$400.00
8			Interest and repayment rates	$340.00
9			Food	$1,200.00
10	Other regular income		Expenses for the children	$500.00
11	Capital gains	$400.00	Expenses for private purchases	$500.00
12			Expenses for information, education, etc.	$400.00
13			Expenditure on health and sports	$1,000.00
14			Expenses for entertainment	$300.00
15			Donate	$100.00
16			Other expenditure	$100.00
17	**Total**	$9,400.00	**Total**	$9,190.00

Figure 2.7 Two Selected Number Columns

To remove individual cells or cell ranges from a multiple selection, you can hold down `Ctrl` and click or tap the cells you want to exclude, or you can drag over the cell range. This method doesn't work with a finger on a touchscreen.

Navigating within a Range

Sometimes, it's better to select a cell range rather than individual cells when entering data. Keep in mind that you can't use arrow keys to finish data entry because they cancel the range selection, and you also can't use the mouse inside a selected range since clicking will cancel the selection. The following key functions apply:

Key Combinations	Effects
`Enter`	Move cell down
`Shift`+`Enter`	Move cell up
`Tab`	Move cell right
`Shift`+`Tab`	Move cell left

All key functions in this area work cyclically, meaning after the last cell at the bottom right, the selection returns to the first cell at the top left.

Using Data Blocks

Data blocks are closed cell ranges filled with data, and when multiple tables are set up as data blocks in a worksheet, additional navigation options become available.

If you double-click the border of an active cell on one side, the selection jumps in that direction—either to the end of the data block the cell belongs to or to just before the next data block in that direction. This is helpful when tables are arranged side by side or above each other, separated by empty columns or rows. In very wide tables, it lets you quickly jump from the row label on the far left to a column with the row sums on the far right. When combined with the Shift key, double-clicking the selection frame extends the range to the end of the block or to just before the next one. You can quickly select a long column filled entirely with data and no empty cells this way.

Click the top cell, then double-click the bottom edge of the selection while holding the Shift key. To include adjacent columns, double-click the right edge of the selection while holding the Shift key.

	A	B	C	D	E	F	G	H
1								
2			Data block 1				Data block 2	
3				1. Jump →	2. Jump →			3. Jump →
4			100 m Results				200 m Results	
5		Day	Rank	Time		Day	Rank	Time
6		6/10/2015	3	12.34		6/10/2015	2	22.83
7		6/11/2015	2	12.45		6/11/2015	4	23.03
8		6/12/2015	4	12.23		6/12/2015	1	22.63
9		6/13/2015	1	12.10		6/13/2015	2	22.39

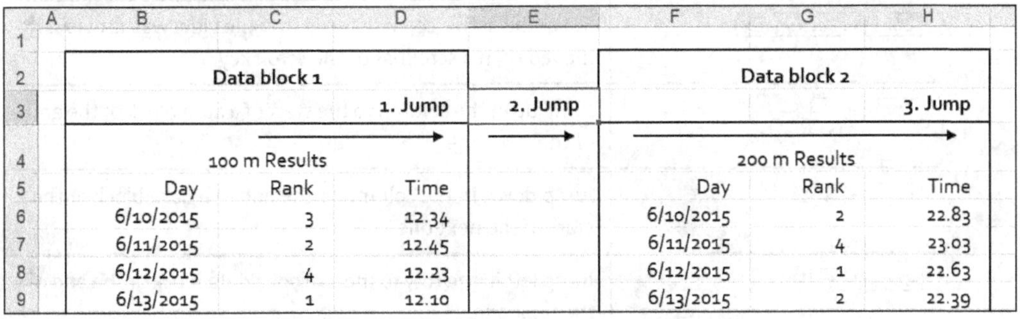

Figure 2.8 Navigating Within Data Blocks

2.2.3 Moving and Selecting with the Keyboard

Use the ←, ↓, ↑, and → keys to select cells with the keyboard. Each keystroke moves the cell pointer one column or row, and holding down an arrow key scrolls the view continuously in that direction. For larger moves, keyboard shortcuts are more efficient.

Navigating within a Table

There are many keys and keyboard shortcuts you can use to make small and large movements within the cell grid. There are also some handy methods to make working with large workbooks easier.

Key Combinations	Effects
→ or Tab	Move one column to the right.
← or Shift+Tab	Move one column to the left.
↓	Move down one row.
↑	Move up one row.

Key Combinations	Effects
Home	Jump to the beginning of the row.
Ctrl + Home	Jump to the beginning of the table.
Ctrl + End	Jump to the end of the used table area.
Page ↓	Move down one window.
Page ↑	Move up one window.
Ctrl + Page ↓	Jump to the next sheet.
Ctrl + Page ↑	Jump to the previous sheet.
Ctrl + Backspace	Make the active cell visible again if the view was previously moved with a scrollbar or the scroll key.
Ctrl + ↑	Jump up in the column to the start of a data block or the end of the next one.
Ctrl + ↓	Jump down in the column to the end of a data block or the start of the next one.
Ctrl + ←	Jump left in the row to the start of a data block or the end of the next one.
Ctrl + →	Jump right in the row to the end of a data block or the start of the next one.

[+]

If Scrolling in the Sheet Feels Slow

If your table contains many charts and graphic objects, navigating it can become sluggish. You can speed up scrolling by going to **Options • Advanced • Display options for this workbook** and selecting **For objects, show: Nothing (hide objects)**. You can speed it up even more by pressing Ctrl + 6.

Range Selection

Many keyboard shortcuts can simplify selecting cell ranges.

Key Combinations	Effects
Shift + ←↓↑→	Extend the selection by one row or column in the direction of the arrow key.
Shift + Home	Extend the selection to the start of the row.
Shift + Spacebar	Select the entire row.

Key Combinations	Effects
Ctrl+Spacebar	Select the entire column.
Ctrl+Shift+Home	Extend selection to the start of the table.
Ctrl+Shift+End	Extend selection to the end of the table.
Ctrl+Shift+Spacebar	Select the entire table.
Shift+Page ↓	Extend selection down by one screen.
Shift+Page ↑	Extend selection up by one screen.
Ctrl+Shift+← ↓ ↑ →	Extend selection to the end or start of the block in the direction of the arrow key or to the start or end of the next data block.
Ctrl+*	Select the entire data block containing the active cell. Instead of the key combination, you can use the button with four arrows (**Select Current Region**), which you can add to the Quick Access Toolbar.
Shift+Backspace	Reduce the selection to the active cell.
Ctrl+A	Select the entire worksheet. If the cell pointer is inside a cell range, then that range is selected; pressing the shortcut again selects the whole sheet. In the formula bar, this selects all content.

When selecting with your keyboard, you can also pick the first corner cell, press F8, select the last corner cell, and press F8 again.

Keys in End Mode

Like keyboard shortcuts for moving and selecting data blocks, some keys work when the End key activates **End mode**, which appears in the status bar. You don't need to hold down the End key, and the mode deactivates immediately afterward.

Key Combinations	Effects
End	Toggle **End mode** on or off.
End+← ↓ ↑ →	Jump to the start or end of a data block in a column or row.
End+Home	Jump to the end of the table.
End+Enter	Jump to the last cell in the current row.
End+Shift+← ↓ ↑ →	Extend the selection to the start or end of a data block in the direction of the arrow key.

Key Combinations	Effects
End + Shift + Home	Extend selection to the end of the table.
End + Shift + Enter	Extend the selection to the last cell in the current row.

The table end is the cell where the lowest row and the rightmost column that contain values or formatting intersect.

Keys in Scroll Mode

Pressing the scroll key toggles scroll lock on and off. In this mode—which is indicated by **Scroll** in the status bar—the keys shown in the following table perform functions that are different from their usual ones.

Key Combinations	Effects
← ↓ ↑ →	Move the window view by one column or row in the direction of the arrow key without changing the active cell's position.
Home	Move to the top-left cell in the window.
End	Move to the bottom-right cell in the window.
Shift + Home	Extend the selection to the top-left cell in the window.
Shift + End	Extend the selection to the bottom-right cell in the window.

Selecting Multiple Ranges by Using the Keyboard

To select multiple table ranges with the keyboard, select the first range as usual by using Shift and the arrow keys. Before adding another range, you must freeze the current selection so the next range doesn't cancel it. You do this by pressing Shift + F8, after which, the status bar will show the message **Add or remove section**.

Selecting with the Name Box

If you know the address of a distant cell you want to select, you can use the name box that's located in the formula bar. For example, to jump to cell F400 by using the name box, just enter F400 and press Enter. You can also enter a range like F200:G400 to select that range, and by entering addresses like F200:G400; H200:I400, you can select nonadjacent ranges.

If a range already has a name, simply click or tap it in the name list to select it. The list shows names from the active sheet and the entire workbook. You can also use the name box to name a range. To do this, select the range, click or tap the arrow next to the name box, enter the name, and press Enter. Chapter 3 covers named ranges in detail.

The Go To Command and the F5 Key

You can use **Home • Editing • Find and Select • Go To** as an alternative to the name box. Excel provides a small dialog box where you can enter the address of the cell to activate under **Lookup**, and you can also specify cell addresses in other open workbooks by prefixing the full file name in square brackets.

Instead of a cell address, you can select the name of a cell or range if the target cell or its range has been named beforehand. When you use the command, Excel places the address of the last active cell—meaning the jump address—at the top of the name list. This makes it easy for you to return quickly to that spot, which you do by simply pressing F5 and Enter. This lets you easily jump back and forth between two distant cell ranges.

You can also use the **Go To** command to select a range. For example, you can enter A1:Z300 under **Reference** to select that range. You can also place the cell pointer on A1, press F5, and then confirm the Z300 entry by pressing Shift+**OK** or Shift+Enter. You can also enter multiple ranges here by separating each range address with a semicolon.

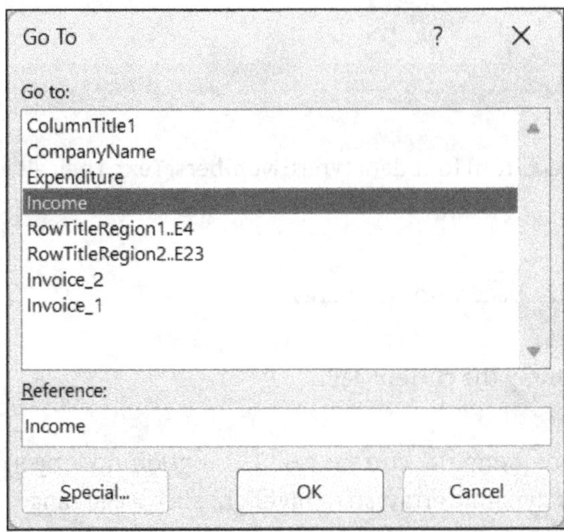

Figure 2.9 Go To Dialog Box

2.2.4 Selecting Specific Content

Sometimes, you need to select cells that share certain properties. For example, if you need to find empty cells in a large table because their values are missing, you can select the entire range and then open the **Go To** dialog box by pressing F5. Clicking the **Special** button opens a dialog box where you can select the **Blanks** option to highlight all empty cells in the range, and after you confirm this, the table will highlight all empty cells as a multiselection. You can easily enter missing data by pressing Tab after each entry, and the cell pointer moves to the next empty cell after each entry you make.

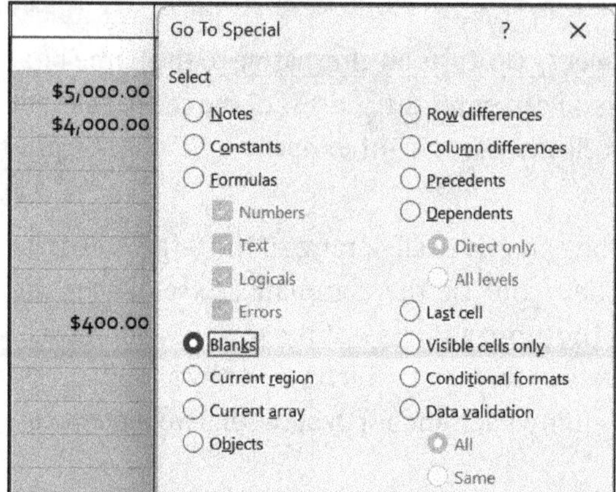

Figure 2.10 Selecting Blank Cells Using Select Special Dialog Box

This feature lets you search the entire table, selected sheets, or a marked range by using various criteria, but you can only apply one criterion at a time. You can select from the following criteria:

- **Comments**
- **Constants**
- **Formulas**

For constants or formulas, you can select from four data types: **Numbers**, **Text**, **Logicals**, and **Errors**.

- **Blanks**
  This highlights empty cells within the used worksheet area.
- **Current region**
  This highlights the cell block containing the current cell.
- **Current array**
  This highlights the entire array containing the current cell. This option no longer applies to cell ranges that contain a dynamic array, so to highlight such a cell range, use the **Current region** option instead.
- **Objects**
  This highlights all objects in the file, if they haven't been protected using **Review • Protect • Protect Sheet** and the **Edit Objects** option is unchecked.
- **Row differences**
  This identifies cells in a selected range of data rows whose content differs from the cell in the comparison column for the same row. The comparison column is where you place the active cell, and this function lets you compare columns.
- **Column differences**
  This identifies cells in a selected range of data columns whose content differs from the cell in the comparison row for the same column.

- **Precedents**
  This highlights cells referenced by the selected formulas. You can extend the search to all levels or limit it to directly preceding cells.
- **Dependents**
  This highlights cells whose formulas are affected by the selected cells. You can extend the search to all levels or restrict it to directly dependent cells.
- **Last cell**
  This highlights the cell at the intersection of the lowest row and the rightmost column that contain content or formatting.
- **Visible cells only**
  This excludes hidden cells from the selection. When you select a range, hidden cells are usually included. (Hidden cells in the print area are automatically excluded only during printing.)
- **Conditional formats**
  This highlights cells that have conditional formatting applied.
- **Data validation**
  This highlights cells with specific validation rules. To select all cells, choose the **All** option, and to select only the cells that follow the same rule as the currently selected cell, choose the **Same** option.

The locations are highlighted as a range or multiple ranges. When multiple sheets are selected, this function only works if the sheets were created in group mode. The program highlights the same cells on each sheet, based on the active sheet's context.

2.3 Efficient Data Entry and Editing

Moving the cell pointer lets you control where to enter data or formulas in the grid, and the cell grid helps you organize your data effectively. This section explains how to enter different types of data into a worksheet and how to edit cell contents when you need to. When you enter data into a cell, Excel evaluates the characters you typed, and it determines whether the input should be interpreted as a number, text string, or formula.

2.3.1 Text and Character Strings

If Excel can't interpret any given input as a number or formula, it treats the input as a text string. Text strings can include any characters and can be up to 32,767 characters long. Each character can be formatted individually, while numbers are always formatted consistently. To treat a number as text, you must explicitly tell Excel to do so, and you can do it in two ways:

1. Format the cell range by using the **Text** format. You can do this either before or after entering data.
2. Place a single quotation mark before the digits. This character won't appear in the cell.

Excel also displays error notifications related to text and character strings if error checking is enabled. If you open the menu for the error icon by clicking or tapping, you can select **Ignore Error** or, under **Error Checking Options**, deselect the **Numbers formatted as text ...** rule entirely. Thereafter, this notification won't appear again.

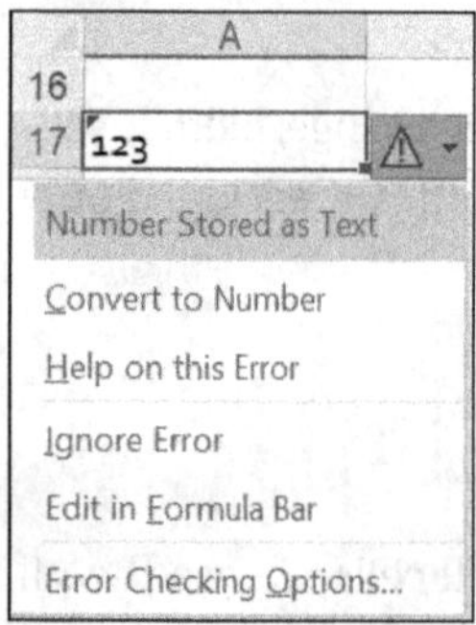

Figure 2.11 Menu for Error Notification

Wrapping Text within a Cell

Labels often require longer entries to clearly identify their content. Wrapping text within the cell helps here, and you can do this by using the **Home • Alignment • Wrap Text** button. However, Excel's text wrapping is a bit rough. While the row height adjusts automatically, words that don't fit in the column are simply cut off without hyphens or proper breaks. Widening the column can improve the result, but after you do that, the row height will no longer adjust automatically. To fix this, double-click or tap the bottom border of the row header. Then, if you shorten the text later so it fits in fewer rows, the row height will adjust automatically.

You can also control line breaks manually. To do this, press Alt + Enter when entering data where you want a new line to start, and Excel will automatically enable text wrapping for that cell. Another option is to suggest breaks while typing. For example, if you enter "Pre-vious year," Excel will break at the hyphen when wrapping is enabled. However, you should avoid this method if you want to name the column with the header or use the header as a field name in a PivotTable. In those cases, Excel won't ignore the break character, which means that in range names, the hyphen will be replaced by an underscore. In PivotTables, the separator is part of the field name.

Accepting Entries (AutoComplete)

Sometimes, certain entries—like product group names or suppliers in an item list—appear multiple times in a column. To help you keep track of these, Excel remembers existing entries for each column. For example, if the product group name *Jalousie* is already present, Excel will suggest completing *J* to *Jalousie* the next time you type it. It doesn't matter if the cell you're typing in is above, below, or between previous labels in the column. Because the suggestion is highlighted, every new keystroke replaces all highlighted characters—so if you don't want to use the suggestion, you won't be slowed down. If you want to use part of the suggested characters, add the remaining characters in the formula bar. Click or tap where the entry looks different.

	B	C
3	**Product name**	**Category**
4	Louvre Ccxs	Louvre
5	Rollo BT 33	Rollo
6	Awning Blue Sk	Awning
7	Louvre VVx	Louvre
8	Rollo Dark	
9	Louvre Louise	

Figure 2.12 Example of Input Suggestion

You can also disable this functionality by going to the **Excel Options** dialog, selecting **Advanced**, and unchecking **Enable AutoComplete for cell values** under **Editing options**.

If you often use the same labels in a column, there's another way to save time typing. If you're using a mouse and keyboard, right-click a cell directly below a filled cell in a column to access existing entries via the dropdown list. If you're using a touchscreen, hold your finger on the cell, tap the arrow at the end, and choose the command.

	B	C
3	**Product name**	**Category**
4	Louvre Ccxs	Louvre
5	Rollo BT 33	Rollo
6	Awning Blue Sk	Awning
7	Louvre VVx	
8	Rollo Dark	Awning Louvre
9	Louvre Louise	Rollo

Figure 2.13 Displaying Dropdown List for Column

2.3.2 Entering Numbers

While text in a spreadsheet is mainly used for labels, numbers provide the data for calculations. Follow these rules when entering numbers:

- Format values of the same type consistently.
- Omit decimal places when they don't matter to keep the table clear.
- If values are missing initially, decide whether to leave the cell empty, enter a zero, or use the `NA()` function (the last of which shows `#NA` to indicate that the value is not yet available).

You can use digits and certain special characters: plus or minus signs, decimal points, comma separators, slashes, parentheses, percent signs, *E* or *e* for scientific notation, and currency symbols like € and $. If you enter only digits, Excel right-aligns the input by default. If the input contains special characters along with numbers, then Excel checks whether the entire entry matches a valid number format—and if it does match, then Excel treats the input as a number.

2.3.3 Input and Output Formats

Excel initially applies the standard format to numbers. It assigns this minimal format automatically, before you apply more specific number formats. Numbers reset to this standard format when you clear the formatting in a cell range, but you can go beyond this minimal format when entering data. If the entry matches a valid number format, Excel applies that format to the cell and displays the number accordingly, and this remains until you assign a different format to the cell. For example, say you entered the following number value in cell B1:

```
1,000,000
```

The number would display the same way in the cell because a number with comma separators is a valid format. The cell adopts a fixed number format when you enter data into it, and you'll notice this if you overwrite the value in cell B1 with this number, which doesn't include comma separators:

```
2000000
```

After you enter the number, the cell will display this:

```
2,000,000
```

So, you can set the number format this way with the first entry you make in a cell. This is useful if you need to enter hundreds of numbers into a column, because it would be tiresome to type comma separators and currency symbols each time you make an entry. It's more practical to start by setting a format that automatically adds these characters so you only have to type the raw numbers. Chapter 4 explains these formatting options in detail.

Excel also offers another option for this: If you set the format in the first cell of a column, you can drag the fill handle down to the end of the column and click the **Auto Fill Options** button. Then, from the menu that you can open by clicking the small triangle, you can select **Fill Formatting Only** (see Figure 2.14). The copied numbers initially shown will disappear, but the cells will then match the format of the first cell.

To do this when using a touchscreen, tap the cell and then tap **AutoFill** in the context menu. Next, drag the arrow icon down and choose **Fill Formatting Only** from the **AutoFill** button menu (see Figure 2.15).

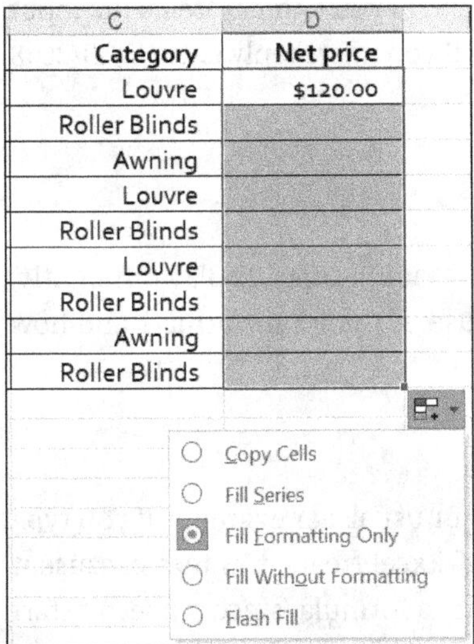

Figure 2.14 Extending Range's Formatting with Fill Handle

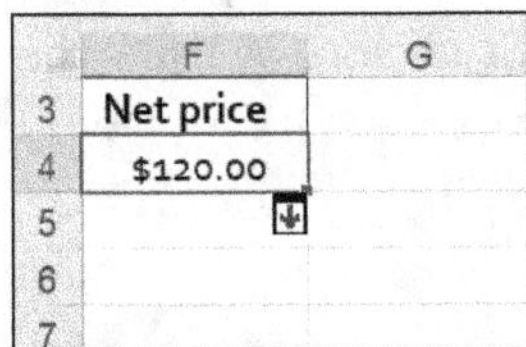

Figure 2.15 AutoFill in Touch Input Mode

Number Size and Column Width

If a number entered is wider than the column, Excel first drops decimal places from the number—unless the standard format has been replaced. If that doesn't work, it switches to scientific notation with an exponent, like 3.45E+12.

For all other number formats, Excel automatically widens the column if there isn't enough space. This only happens if the column width hasn't been manually changed. If it has, Excel shows a series of # characters instead of the number that won't fit. However, Excel still displays the value when you hover the mouse pointer over the cell. In that case, either adjust the column width or choose a more suitable format.

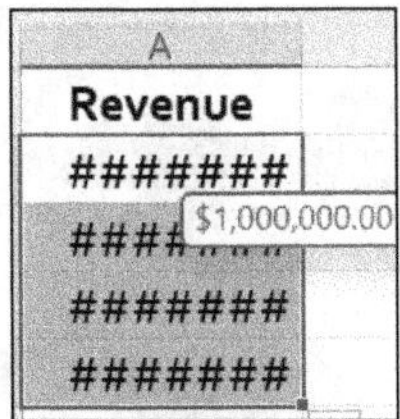

Figure 2.16 Values That Are Wider Than Column

If Excel can't find a valid number format for the characters you enter, it treats the input as text and aligns it to the left in the cell. For example, if you mistakenly enter 13/13/2018 as a date, Excel will simply treat it as text.

2.3.4 Fractions, Leading Zeros, Dates, and Times

If you're confused by how fractions or numbers with leading zeros display differently, the following sections will help. This happens because of the input formats and how Excel interprets your entries.

Entering Fractions

Excel's way of handling fraction entries is somewhat unusual. For example, if you type "1/3," the cell shows `01-Mar`—but if you type "4/100," Excel treats it as text because it can't be converted into a valid date and because it isn't a formula (since it doesn't start with an equal sign). In contrast, Excel accepts your typing "2 3/4" without issue and interprets it as the value `2.75`, as the formula bar shows. And if you type "0 1/3," Excel will show the desired result of `1/3`. So, you can memorize all of these Excel quirks and be careful how you type fractions, or you can format the range in advance by using the single-digit fraction format.

Entering Leading Zeros

Excel also poses a small challenge when you enter numbers that begin with zeros. Normally, Excel removes leading zeros. For example, 007 becomes 7, both in the cell and the formula bar, which reduces the input to its numeric value.

However, leading zeros are often needed in numbering systems—common examples include customer numbers, article numbers, or document numbers. So, one way to keep Excel from deleting these zeroes is by entering the number as text, as explained in the previous section. However, this causes the number to lose its numeric properties, and that's a drawback for document numbers, which often need to be incremented.

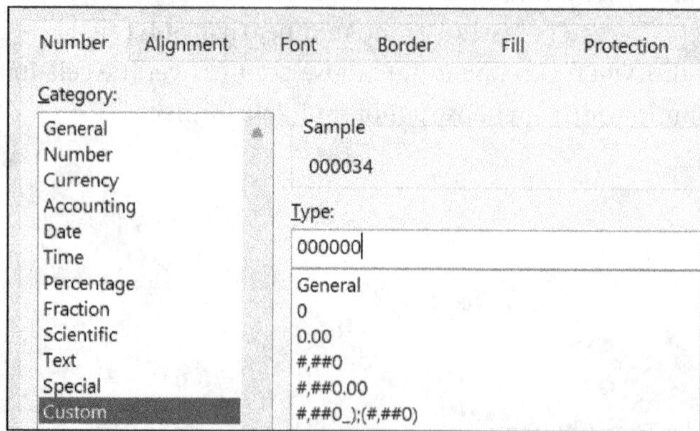

Figure 2.17 Format Pattern for Numbers with Leading Zeros

If you want to enter a number with leading zeros as a number, your best option may be to define a custom format. You can do this by opening the list under **Home • Number** to **Number Format** and selecting **More Number Formats**. Then, on the **Number** tab in the **Custom** category, enter a number pattern under **Type**, using as many zeros as the number's digits (see Figure 2.17).

Input with Fixed Decimal Places

If a workbook contains only numbers with the same number of decimal places, such as dollar amounts, you can use a function that removes the need to enter the decimal point. For best results, use the numeric keypad to enter numbers. Press the [Num] key to activate the function, and in **Excel Options**, go to the **Advanced** tab and check **Automatically insert decimal point** under **Editing options**. Then, set the number of decimal places to 2, for example, and after you do that, Excel will automatically treat the last two digits of every number as decimals and insert the decimal point. You'll no longer need to enter a zero before the decimal point. However, because this setting applies to the entire workbook, you must add two extra zeros at the end of numbers in areas where you don't want decimals, such as counts.

Entering Dates and Times

Excel treats dates and times as numeric values. This makes it easy to calculate values like the date 20 days after today by simply adding 20. You can enter date and time values into a cell either by using date and time functions (formulas) or directly as constants. The allowed input formats for dates as constants vary depending on your regional settings, and some examples of valid input formats are as follows:

```
12-10-25
12/10/25
12.12.24
12.12.2025
December 12, 2025
12/10
12-10
Jan 12, 25
Feb 4
Jan 08
February 4
```

The default display format depends on your system's regional settings, which you can change in the Windows Control Panel. Excel automatically completes incomplete date entries. If the year is missing, Excel adds the current year, and if the day is missing, Excel adds the first day of the month. The full date always appears in the formula bar and forms the basis for any calculations.

You can enter time in any of the following formats:

```
10:30       10:30 PM
10:30 AM    10:30:30
```

If you omit PM/AM, Excel assumes and enters AM. Lowercase letters are automatically converted to uppercase. The formula bar always shows time in 24-hour format, so for example, 7:12 PM appears as 19:12:00. You can also insert the current date and time from your device by using `Ctrl`+`:` and `Ctrl`+`;`, respectively. You can also enter the time and date consecutively in one cell, but you must separate them with a space.

[»]

Serial Numbers for Dates and Times

Excel stores entries it recognizes as dates or times internally as serial numbers, counting the days since 1/1/1900—or alternatively, since 1/1/1904.

Time is represented by the decimal portion of the serial number, so for example, 12 PM equals 0.5. The serial number appears when you clear the cell's format.

Input Control

By default, pressing `Enter` moves the cell pointer down to the next cell. However, this doesn't apply if you finish data entry by clicking or tapping the **Enter** icon in the formula bar. You can disable or redirect the cell pointer movement through **Excel Options**, as explained in Chapter 1.

2.3.5 Changing, Searching, and Deleting Content

You can overwrite any entry unless the worksheet is protected against changes. To replace an incorrect or outdated number, select the cell, enter the new number, and confirm. For longer entries like labels or formulas, it's inefficient to retype the entire cell content after making a typo or minor change. Editing is more efficient here.

Editing Within a Cell

Excel lets you edit cell content directly in a cell or in the formula bar. To edit cell content directly, you can double-click or tap the cell, or you can press `F2`. The program will then enter edit mode, and for formulas, it will show the formula instead of the result. A blinking cursor marks the insertion point.

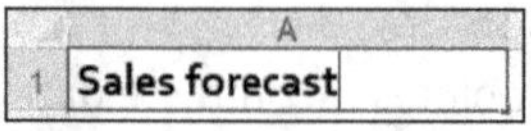

Figure 2.18 Cell Open for Editing

On a touchscreen, the insertion point appears as a line with a round handle at the bottom.

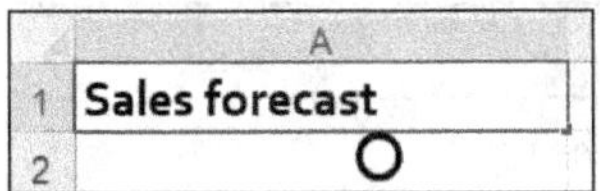

Figure 2.19 Insertion Point with Handle in Touch Input Mode

You can move the insertion point by clicking, tapping, or using the arrow keys, and you can confirm your edits by clicking another cell, tapping it, or pressing `Enter`.

Editing in the Formula Bar

To edit cell content in the formula bar, start by selecting the cell so its content appears there. To change a specific spot, click or tap directly on it in the formula bar. You can also highlight characters to replace by dragging with the mouse, or on a touchscreen, you can drag the round handle to select the characters.

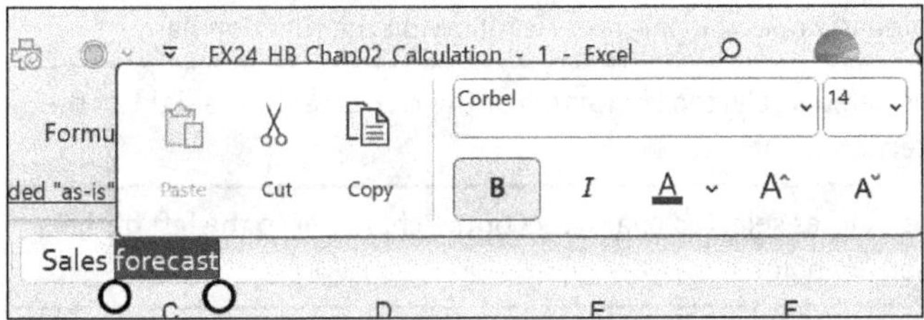

Figure 2.20 Highlighting Cell Content on Touchscreen by Dragging Handles

You can highlight entire words or arguments in a formula by double-clicking or double-tapping them, and holding down the `Shift` key lets you extend a selection to where you place the mouse pointer. The next character you type replaces the selected characters, and any further characters are then inserted at that position. In other words, the program normally operates in insert mode.

You can also place the insertion point on the first character to overwrite and then enable overwrite mode by pressing `Insert`. The insertion point is shown by a gray background on the character under the cursor. To switch back to insert mode after overwriting, press `Insert` again.

Key Functions for Editing Cells

In the formula bar or an open cell, certain keys and keyboard shortcuts perform specific actions.

Key Combinations	Effects
`Home`	This moves to the beginning of the entry.
`End`	This moves to the end of the entry.

Key Combinations	Effects
Ctrl+→	This moves one word or argument to the right.
Ctrl+←	This moves one word or argument to the left.
Shift+→	This extends the selection by one character to the right.
Shift+←	This extends the selection by one character to the left.
Ctrl+Shift+→	This extends the selection by one word or argument to the right.
Ctrl+Shift+←	This extends the selection by one word or argument to the left.
Ctrl+Shift+Home	This extends the selection to the beginning of the entry.
Ctrl+Shift+End	This extends the selection to the end of the entry.
Ctrl+Shift+A	This inserts the arguments for a function, and you should use it after typing the opening parenthesis following the function name.
Delete	This deletes selected characters or the character to the right of the insertion point.
Backspace	This deletes selected characters or the character to the left of the insertion point.
Ctrl+Delete	This deletes from the insertion point or selection to the end of the entry.
Ctrl+A	This selects all content in the formula bar or the cell being edited.
Ctrl+C	This copies the selected characters to the clipboard.
Ctrl+X	This cuts the selected characters and moves them to the clipboard.
Ctrl+V	This pastes data from the clipboard at the insertion point or replaces the selected characters.
Ctrl+Shift+V	This pastes only the value from the clipboard, ignoring its format.

Cut or copied text can replace a selected string or be pasted into another cell or the formula bar of a different cell. When working with longer texts or complex formulas, use the editing commands Excel provides. The **Cut**, **Copy**, **Paste**, and **Delete** commands are also available in the formula bar and during direct cell editing.

Finding and Replacing

Sometimes, you need to change names, labels, or numeric values after you enter them, and you can use the **Find and Replace** dialog box to do this. You can find it on the **Home** tab, in the **Editing** group, under **Find & Select** and the **Find** options as well as the **Replace** options. The dialog box has separate tabs for find and replace, with mostly identical options.

Keep three points in mind when using this functionality:

1. Excel compares the search criterion with the content of the cell ranges included in the search, and it finds a match if the search string appears anywhere in the cell content. For example, a search for the *Hat* criterion will also find cells containing *Summer hat* or *Hat box*. This also applies to numbers: the *5* criterion matches *5, 15, 57*, and so on. To match the entire cell content exactly, enable the **Match entire cell contents** option ❶.
2. When you're searching for numbers, dates, or times, Excel compares the search term by default to the cell's stored content as shown in the formula bar, not the displayed content in the cell. To find a date displayed as *12-Dec*, you can enter a search term like "12/12/2024." Excel will also find the date if you enter just "2024" because it matches part of the cell's content.
3. If you need to, you can expand the search for numbers and text to the entire workbook by clicking **Options** ❷ and choosing the options under **Within:** ❸, so the search isn't limited to the sheet. The list under **Search:** ❹ lets you choose whether to search by rows or columns, and you can also limit the search to formulas, values, notes, or comments by using **Look in:** ❺.

You can click **Find All** ❻ on the **Search** tab to list all matches in the table or workbook, and you can jump to each location you find by clicking or tapping the table rows or using the arrow keys.

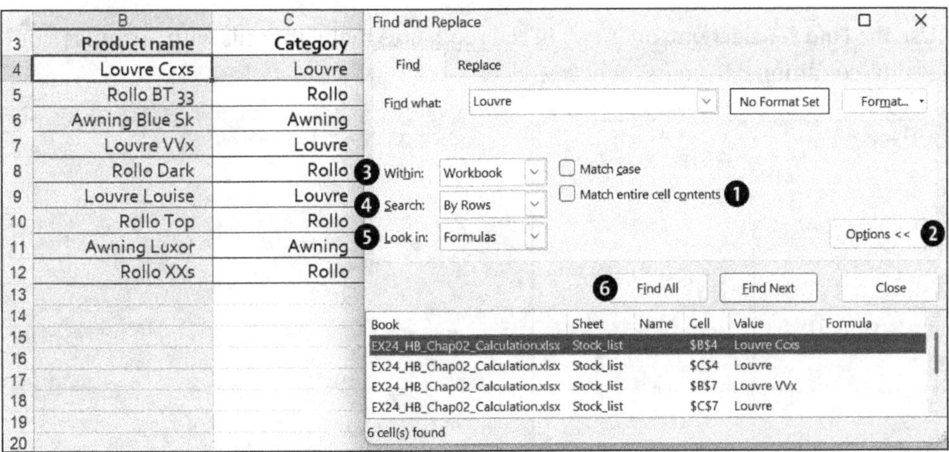

Figure 2.21 Search Results on Find and Replace Screen

You can press Ctrl+F to open the **Search** dialog box quickly, and you can use the Alt+F4 and Ctrl+Alt+F4 shortcuts to search forward or backward with the same term without opening the dialog box again. Press Ctrl+H to open the dialog box directly on the **Replace** tab.

In Excel, you can search for and replace not only numbers and text but also formats—for example, to update cells showing two-digit years to display full years. After you press the **Options** button, additional buttons appear, one being **Format...**, whose arrow you can click to open a small menu. Clicking the first option opens the **Find Format** dialog box, which corresponds to cell formats.

Clicking the second option, **Choose Format From Cell**, is especially useful when you're searching for cells that match a format that's already used in the workbook. You can select the sample cell with the specified format by clicking or tapping it, and to stop searching for formats and search only by content again, you can use the **Clear Find Format** option.

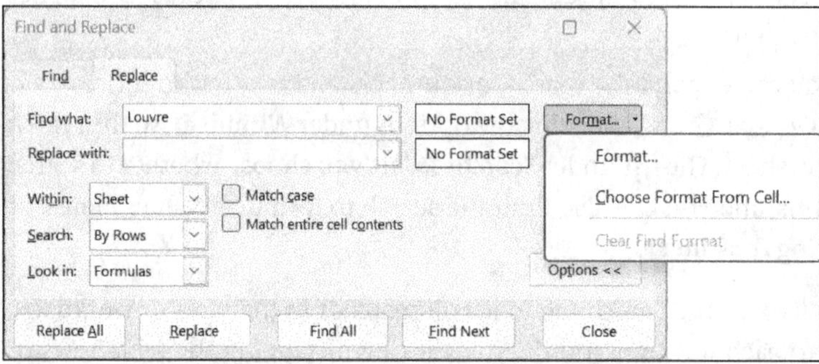

Figure 2.22 Find and Replace Options

Use the **Find & Select** button menu in the toolbar to highlight cells with formulas, constants, conditional formats, or notes.

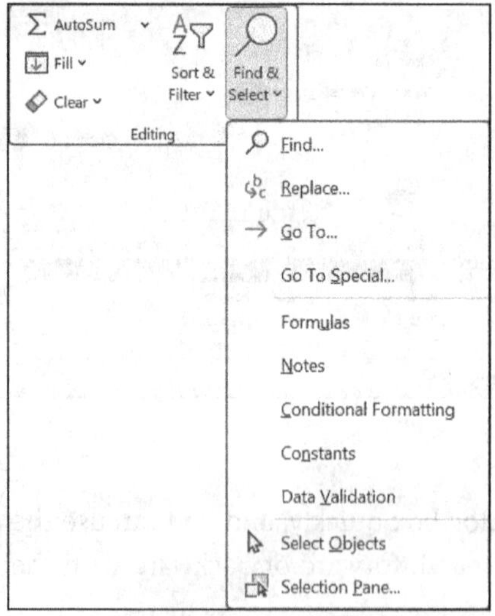

Figure 2.23 Find & Select Button Menu

2.3.6 Clearing Methods

The term *clearing* can refer to different actions related to a cell or cell range's properties:

- Clearing cell content
- Clearing cell format
- Clearing notes or comments
- Clearing links
- Clearing sparklines

You need to distinguish this from removing cells, which is called *deleting*. This will be explained further later in this section.

When you clear a cell's content, the cell becomes empty. This may seem obvious, but it's important to stress because a cell can look empty without actually being empty. This happens when a cell contains spaces, and if you double-click such a cell, the cursor won't be at the far left but just after the last space. These hidden spaces can cause confusion, especially when querying tables.

Clearing the cell format doesn't mean the cell loses all formatting. It simply resets to the default format. For example, if you've chosen a specific font, the cell will revert to the default font. The same applies to number formatting.

Pressing the `Delete` key only removes the cell's content; the formatting stays intact, and notes, comments, and sparklines remain unchanged. To clear cell contents with a mouse, select the cell or range, drag the fill handle inward until the entire area is covered by the grid, and release the mouse button to clear the selected range.

In **Touch input** mode, you can do the same thing by opening the context menu for the range, showing the arrow icon for the **AutoFill** command, and dragging it into the range. You can also quickly clear by tapping the cell or range and selecting **Clear Selection** from the context menu.

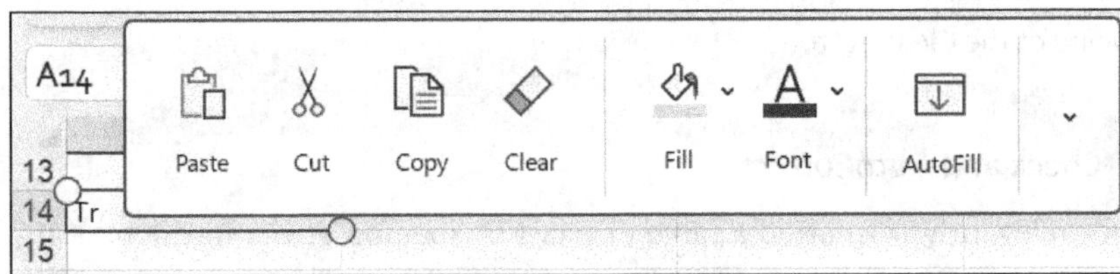

Figure 2.24 Clearing Cell Contents on Touchscreen

To clear entire columns or rows, select them by clicking or tapping the column or row headers, and to clear multiple columns or rows, drag across the relevant headers then drag the fill handle inward. Then, when you release, the contents of the selected columns or rows will be cleared. If you realize you selected the wrong range while dragging, simply drag the mouse pointer back to the fill handle and release the button.

To clear formatting and any comments along with the contents, hold Ctrl while dragging the fill handle—after which, a plus sign will appear next to the crosshairs. To do this on a touchscreen, tap the header of the first column or row, then expand the selection by dragging the round handles. After that, hold your finger briefly on the selected area and choose **Clear Selection** from the context menu.

2.3.7 Clearing Large Areas

Using the fill handle is a handy way to clear single cells or small ranges. For large areas, it's easier to select the range first and then use Delete or **Home • Editing • Clear**.

The button opens a menu where you can choose what to clear: all, only formats, only cell contents, only hyperlinks, or only comments and notes. If you select a link, it will also enable the **Remove Hyperlinks** option, which removes the cell's formatting as well.

You can also use the **Clear Contents** command from a range's context menu, and you can undo any clear action with the **Undo** button if you accidentally select the wrong range. Cells containing sparklines (see Chapter 9) have a special status. They aren't part of the cell content or format, so to clear them, choose **Clear All** from the **Clear** button menu.

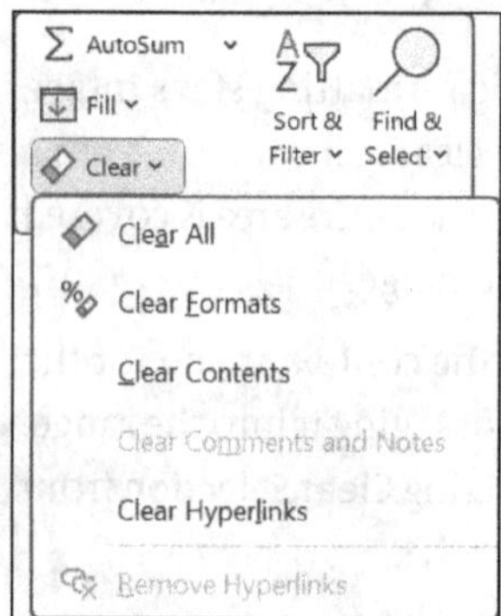

Figure 2.25 Menu of the Clear Button

2.3.8 Spell-Check and AutoCorrect

No matter how elegantly formatted a table or chart is, spelling errors in the text are always frustrating. Spelling mistakes can cause information to be incorrect, such as when a product name is wrong in an order.

Checking with Dictionaries

Excel provides dictionaries for multiple languages, allowing you to easily check workbooks with English, French, or Spanish labels. To use a dictionary in another language when checking a workbook or selected range, select the desired language from the **Dictionary language** list in the **Spelling** dialog box.

Because dictionaries can't cover every word combination or specialized term in your workspace—like technical terms, product names, abbreviations, company names, and customer names—it's a good idea to create your own custom dictionaries (known as *user dictionaries*) that you can use to check the spelling of such terms. Protected documents, formulas, and text generated by formulas will be excluded from such spell-checks.

To check text in the formula bar or the cell you're editing, you have two options:

1. To check a specific word or part of a cell's content, just select it.
2. To check all of a cell's contents, simply start the spell check. The quickest way is to press F7 or click the **Spelling** button in the **Review • Spelling** group. Excel will check the worksheet starting from the active cell, and if you previously selected a range, it will only check that range. If Excel finds no errors, it will simply confirm that the check is complete.

If the pesky typo pixie strikes, call on Excel's **Spelling** dialog. Whenever possible, Excel will offer correction suggestions under **Suggestions** that you can accept by clicking **Change** ❶. If no useful suggestions appear, you can type the correct value in the first text field and accept it by clicking **Change**. Selecting **Change All** ❷ ensures that the error will be automatically corrected whenever it appears again.

While the dialog remains open, you can undo any correction by clicking the **Undo Last** ❸ button. If you click the **Undo** button on the Quick Access Toolbar after closing the dialog, all corrections made within the dialog will be undone.

If the word is spelled correctly but isn't in the dictionary or user dictionary, you can click **Ignore Once** ❹ and Excel will leave the word unchanged and move on to the next error. Clicking **Ignore All** ❺ will leave all other instances of this word unchanged as well, and the program will maintain a list for this. Alternatively, you can add the flagged word to a user dictionary by clicking **Add to Dictionary** ❻.

If you don't want to use the default standard *RoamingCustom.dic* user dictionary, click the **Options** button ❼.

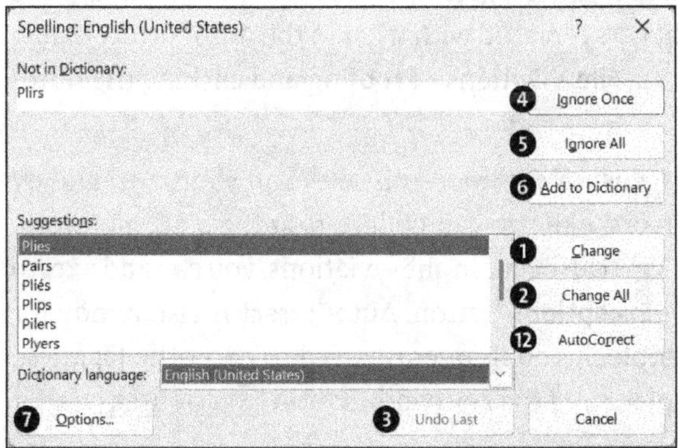

Figure 2.26 Correction Suggestions in Spelling Dialog

This opens the **Excel Options** dialog, and on the **Proofing** page, you'll find a **Custom Dictionaries** button ❽ next to several spell-check settings. Select the dictionary you want from the list, or if it's not listed, click **Add** ❾ and enter the dictionary's path. To create a new dictionary, click the **New** button ❿ and provide a name.

To edit a custom dictionary's word list here, select the dictionary and then click **Edit Word List** ⓫.

The **AutoCorrect** button ⓬ in the **Spelling** dialog (see Figure 2.26) also lets you add both incorrectly and correctly spelled words to the list used by the related **AutoCorrect** feature. Whenever you mistype a word, it will automatically be replaced with the correctly spelled one, as described in the next section.

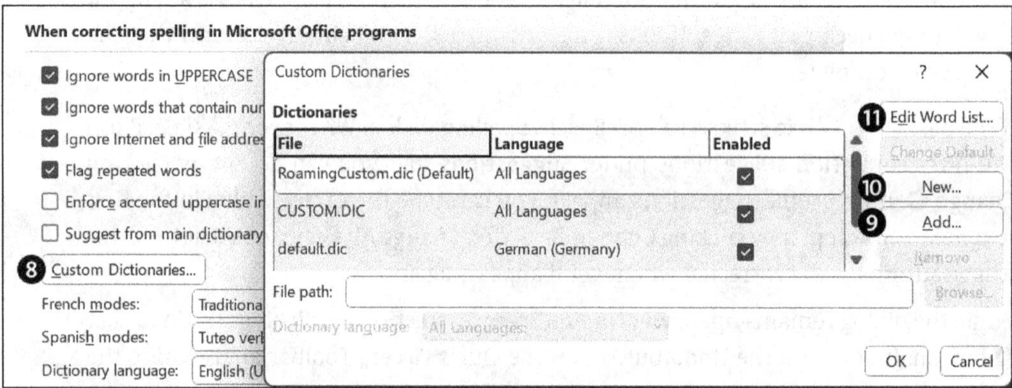

Figure 2.27 Document Proofing Settings

2.3.9 Automatically Replacing Entries

The spellchecker is enhanced by the **AutoCorrect** feature, which catches errors as you type and corrects them automatically. For example, if you accidentally type *comapny* instead of *company*, AutoCorrect will correct it to *company* when you leave the cell. AutoCorrect can also create shortcuts for frequently used long entries, so if, for example, if you often need to enter the term *Finance Office* into your tables, you can use the shortcut *FO*. You can also define what gets replaced with what in the **AutoCorrect** dialog box, and you do this by accessing it via **File • Options • Proofing** and clicking the **AutoCorrect Options** button.

The **AutoCorrect** dialog box includes a list of common mistakes and shortcuts, and by default, it corrects some standard errors, like two capital letters at the start of a word. However, since such letter combinations can occur in abbreviations, you can add exceptions to autocorrection by using the **Exceptions** button. **AutoCorrect** is also handy for entering special characters. For example, if you frequently need to enter the Japanese yen symbol, you can set *yy* under **Replace** and the key combination `Alt`+`0165` under **With** in the **AutoCorrect** dialog box.

Figure 2.28 The AutoCorrect Dialog Box

Mathematical AutoCorrect

Since Excel 2010, this dialog has included the **Math AutoCorrect** tab. It also contains several pairs of replacement strings for inserting mathematical symbols, like `\phi` for the symbol φ. However, these symbols are only available when you're entering formulas with the formula editor, which you can find under **Insert** in the **Symbols** group.

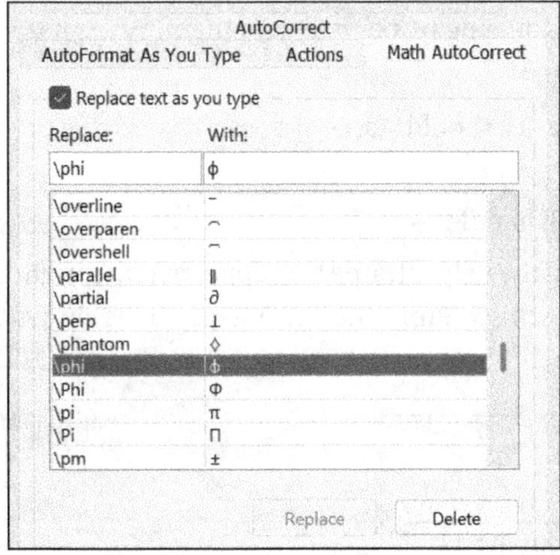

Figure 2.29 Placeholder Strings for Mathematical Symbols

2.4 Automatically Generating Data

Whenever you need to fill a cell range with data that forms a clear sequence, you can automate the entry. Creating a series makes it easier to label rows and columns and to number items or storage locations continuously. If you need test data for simulations or to test formulas or macros, the fill feature can save you a lot of time.

You can use two methods to work with the fill handle: the **AutoFill** icon on the touchscreen or the **Home • Editing • Fill** button, the latter of which opens a menu with several options for creating a series. The first method lets you create small series quickly, while the second method gives you more control over the elements you create large series.

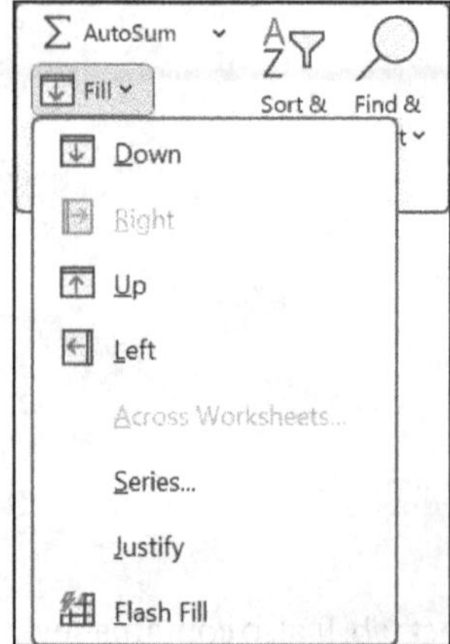

Figure 2.30 Fill Button Menu

2.4.1 Dragging Data Series with a Mouse

Quarter labels are common data series, and dragging works well with them. You can do this as follows:

1. Enter the label for the quarter where the series should start—it doesn't have to be the first quarter.
2. Place the mouse pointer exactly on the fill handle.
3. Hold down the left mouse button and drag the helper frame that appears, either right or left within the row or up or down within the column. The current quarter is always displayed for confirmation.

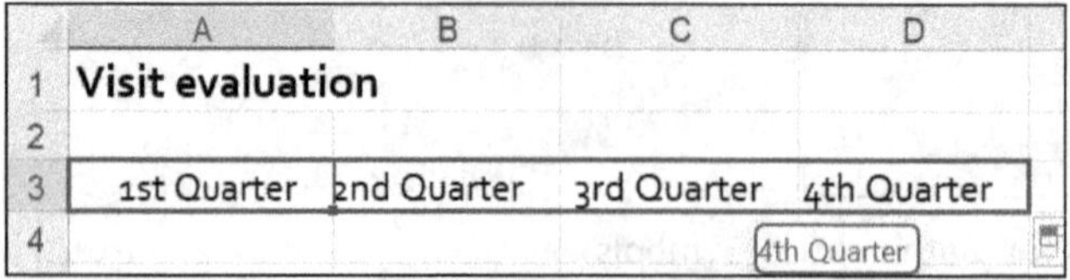

4. If a series is too long, drag the fill handle in the opposite direction until the extra cells are removed.
5. Release the mouse button and Excel will fill the range selected by the helper frame.

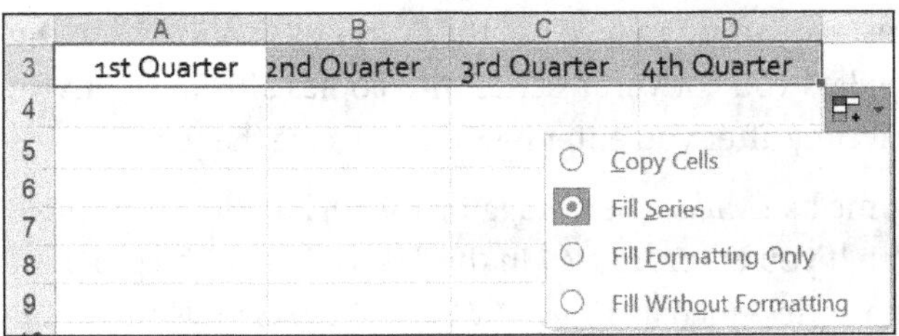

By default, Excel copies the source cell's formatting to the new cells when filling. However, Excel usually shows the **Auto Fill Options** button, which you can click to open a menu where you can specify that Excel shouldn't copy the source cell's formatting in specific cases.

2.4.2 Working with Series on a Touchscreen

Selected cells on a touchscreen don't show a fill handle unless you use a Surface Pen and have enabled **Use pen by default to select and interact with content** under **File • Options • Advanced • Pen**. As mentioned earlier, you can display a larger box that lets you create series by using your finger. Here's how you do that:

1. If you briefly tap the cell for the first quarter, the **AutoFill** button appears in the context menu.
2. This button displays a larger icon with an arrow instead of the fill handle.
3. Dragging this icon to the right with your finger inserts a series with the other quarters.

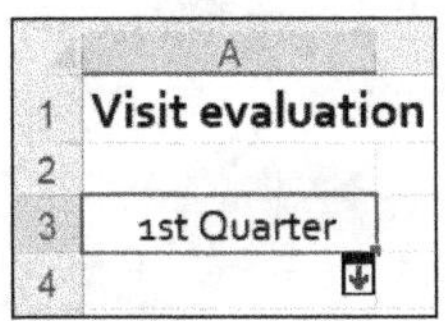

You can also drag the **AutoFill** icon down to create series that run down columns. Dragging the icon up or to the left shortens an existing series or creates a new descending series.

2.4.3 Series or Copies?

What happens when you drag to fill depends on the content of the starting cell(s) you initially select. Excel checks whether it can create a continuous series from those contents or not, and if it can't, it simply copies the cell contents. To prevent Excel from creating a series when you only want copies, hold down `Ctrl` while dragging, which forces Excel to copy the cells instead. Alternatively, use the **Copy Cells** option from the **Auto Fill Options** button, which lets you convert a series into copies afterward or vice versa. The button will disappear only after you enter new data or save the file.

If the original cell contains a name like *Miller*, then dragging the fill handle or the **AutoFill** icon fills the cells with *Miller*. If you enter *House 1* in the first cell, then the program assumes you want to number a series, such as *House 1, House 2, House 3*. Note that it doesn't matter whether the number is at the start or end of the cell entry, but if it's at the start, then it must be separated from the text by a space. However, in some places (like Germany), a period can follow a number, so an entry like *1.Day* (with no space after the period) will always remain *1.Day* while an entry like *1. Day* (with a space after the period) will create a series. The initial text *1 MB* creates a numbered sequence, no matter the locale. You can also use *1st* to generate an ordinal number sequence like *1st, 2nd, 3rd, 4th*, and so on.

To create a continuous numbering, simply enter "1" or another starting number in the first cell and then select the **Fill Series** option under **Auto Fill Options**. Otherwise, Excel will just copy the number. Alternatively, you can hold down the `Ctrl` key while dragging with the mouse. If the sequence shouldn't increase by one, then you should first fill and select two cells. Put the starting value in the first cell and the next value in the second cell so Excel can detect the step size (**Increment**).

2.4.4 Ascending and Descending Sequences

To create ascending sequences, drag the fill handle or the **AutoFill** icon into the right or lower neighboring cells, and to create descending sequences, drag it into the left or upper neighboring cells. Be sure to drag beyond the initially selected range; otherwise, it will be overwritten!

2.4.5 Time Series

Some temporal series, like dates, can be linear. Others, like times, weekdays, or months, can be circular. Intervals between items in a series are either set automatically in steps of one based on the first value or defined by the content of a second cell. Here are some examples:

Starting Values	Series Continuation
8:15	9:15 10:15 11:15
Tue	Wed Thu Fri Sat Sun Mon Tue
Monday	Tuesday Wednesday Thursday

Starting Values	Series Continuation
May-01 May-15	May-29 Jun-12 Jun-26
January	February March April
May-13 Sep-13	Jan-14 May-14 Sep-14
1st quarter	2nd quarter 3rd quarter

Quarter can be written out or abbreviated as *Q1, Q2, ...* and month and day names can be written out or abbreviated.

2.4.6 Arithmetic Series

Arithmetic series have a constant interval, so to create such a series with a mouse, you just select two cells containing the first and second values.

Starting Values	Series Continuation
4 7	10 13 16 19
6 3	0 –3 –6
1.2 1.6	2 2.4 2.8
1st	2nd 3rd 4th
1st 5th	9th 13th 17th

2.4.7 Geometric Series

In a geometric series, each value is multiplied by a constant factor from the previous one. These series can only be created by using the **Series** dialog box, which is accessed via **Home • Editing • Fill • Series**, using the **Type: Growth** option. The options here correspond to those that are available with the fill handle or the **AutoFill** icon.

Starting Values	Series Continuation
2 (Step value 2)	4 8 16
3 (Step value 5)	15 75 375

2.4.8 Creating a Trend Analysis

By using the fill handle or the **AutoFill** icon, you can easily perform a simple trend analysis. For example, you can enter visitor numbers for the last six months in a column, select those six cells, and then drag the fill handle or **AutoFill** icon down to fill six more cells. You'll then receive an arithmetic trend calculation for the second half of the year using the least squares method, and the original values will remain unchanged. Figure 2.31 shows an example with an area chart that analyzes the number series.

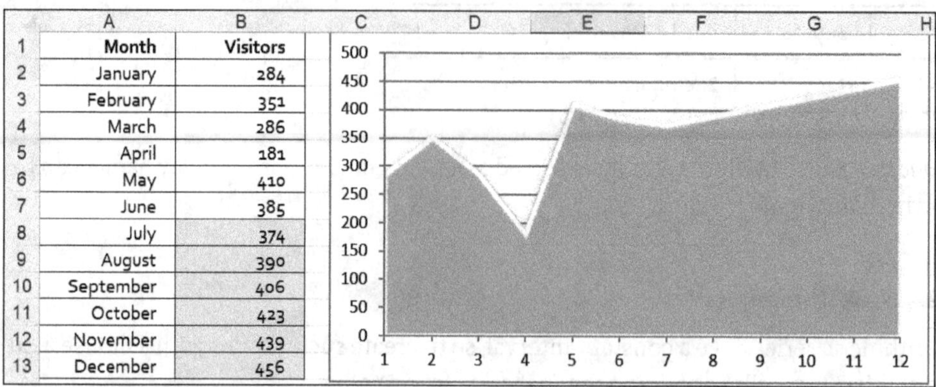

Month	Visitors
January	284
February	351
March	286
April	181
May	410
June	385
July	374
August	390
September	406
October	423
November	439
December	456

Figure 2.31 Linear Trend Analysis of Visitor Numbers

2.4.9 Special Options for Date Values

If you need a list of weekdays, select the **Fill Weekdays** option from the **Auto Fill Options** menu, which automatically fills only the working days.

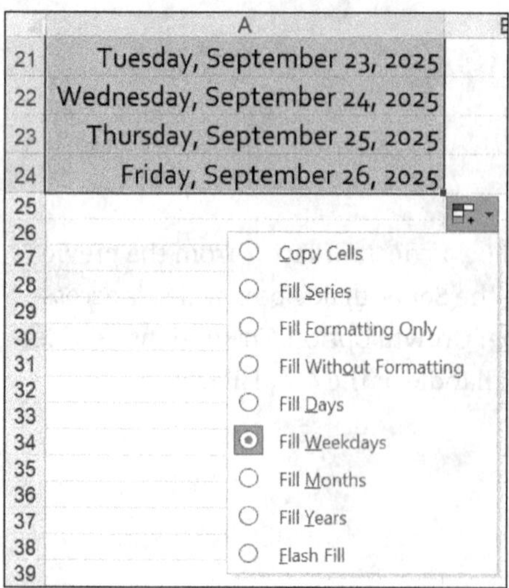

Figure 2.32 Special Series Options for Date Values

Dragging the fill handle with the right mouse button opens additional options in the **Auto Fill Options** menu. For number values, you can calculate both arithmetic and exponential growth trends here while keeping the original values unchanged.

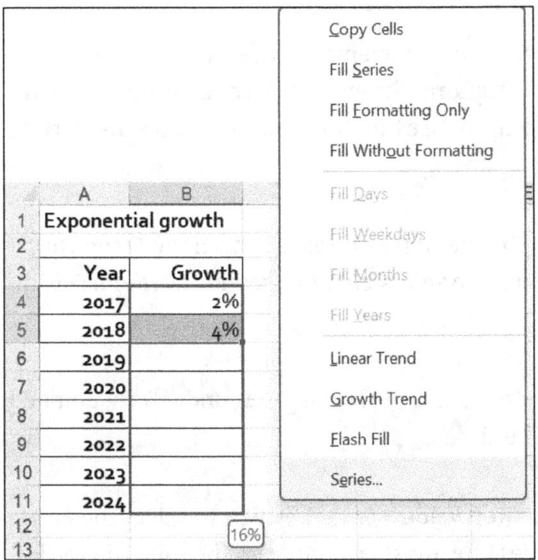

Figure 2.33 Calculating Exponential Trend Using Fill Handle

2.4.10 Creating a Series in the Dialog Box

The dialog at **Home • Editing • Fill • Series** offers additional options for creating a series, but the process works a bit differently. You have two options:

1. Before running the command, select the entire range to fill, including the starting cell. You can also use multiple cells as starting points, as in Figure 2.34.
2. Alternatively, select only the starting cell(s) and control the series length by setting the end value in the dialog box.

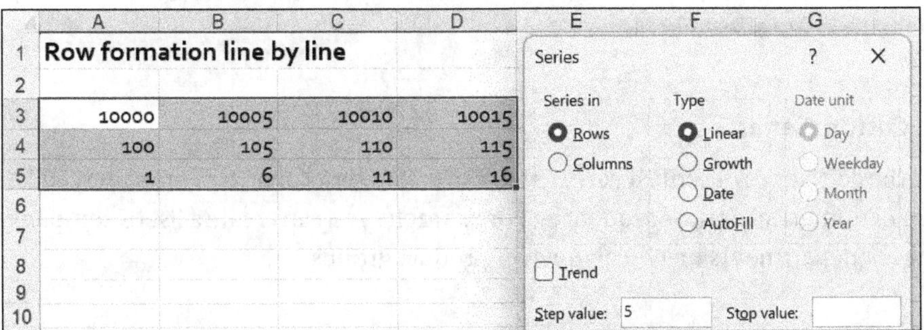

Figure 2.34 Selecting Entire Series Range in Advance

If you have only selected the starting cell(s), choose under **Series in** whether the series should fill across the row (to the right) or down the column. In the second options group, select the **Type** of the data series from the list. Here are the effects of selecting each type:

- **Linear**
  This setting adds the step value entered under **Increment** to the last value each time. If the **Trend** checkbox is checked, Excel ignores the entry in **Step value** and calculates a linear trend. Unlike the corresponding mouse function, this overwrites the starting values.
- **Growth**
  This setting multiplies the last value by the step value each time. If the **Trend** checkbox is checked, Excel ignores the step value and calculates a geometric trend. The initial values will be overwritten.
- **Date**
  If you select this setting, you can choose the unit for the series under **Date unit** and the step value will then apply to this unit.
- **AutoFill**
  This setting ignores **Step value** and **Stop value**. It requires you to select the entire range for the series first, and the series type must be clear from the values in the initial cells. It works the same way as dragging the fill handle.

Excel sets the type based on the **Home** cell you select.

Except for the **AutoFill** option and the **Trend** setting, you can enter the increment for a data series under **Step value**. You only need to enter the **Stop value** if you don't select the entire range for the series. The stop value lets you specify exactly where the series should stop, and you should enter the value in a format that matches the series type—so for time values, use a valid date or time format.

In the **Fill** button menu, you'll also find an **Align** option that applies only to cell ranges containing text strings. This function consolidates text strings spread across multiple cells in a column or row into as few cells as possible, with the column width determining how many rows are affected.

2.4.11 Custom Series

Besides the data series described earlier, you can create series that are completely independent of numerical or time sequences. For example, you can create a list of company branches or departments or a breakdown of product groups.

Defining a List of Branches

To define branch locations as a data series, follow these steps:

1. In the **Excel Options** dialog, go to the **Advanced** page, and under **General**, click the **Edit Custom Lists** button to edit.
2. You can enter the list directly here or import it from a column or row in a worksheet. To do this, select **NEW LIST** in the left pane.

3. If you want to enter the list manually, type each value under **List entries** on a separate line and press `Enter` after each entry. When you're finished, click **Add**.
4. For long lists or when you can use values from existing tables, importing the list entries is more convenient. To do this, select the appropriate range under **Import list from cells** and click **Import**.

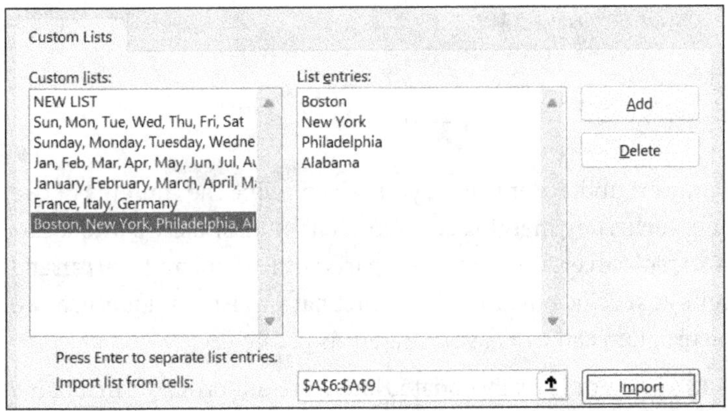

To use a custom list in the worksheet, enter a valid list entry into the first cell of the range you want to fill and then drag the fill handle or the **AutoFill** icon right or down to fill in the remaining entries. The lists cycle continuously. Alternatively, instead of dragging the fill handle, you can select the area to fill in first, go to **Home • Editing • Fill • Series**, and choose the **AutoFill** option.

2.4.12 Input Using Pattern Recognition

A newer way to fill columns is called **Flash Fill**, though the name can be a bit misleading because it uses pattern recognition to work. This method can help when you're working with addresses, so here's an example to illustrate:

1. Start by importing a column of last and first names separated by commas from an external data source.
2. Next, you want to store the last name separately in the adjacent column as a username. To do this, enter the first username in the first cell of the second column.
3. Then, in the second cell below, use the **Home • Editing • Fill • Flash Fill** command—or to do it faster, press `Ctrl`+`E`—and the program will automatically fill in the last name for the entire column.
4. Finally, a **Flash Fill Options** button appears in the first cell, offering commands you can use to undo the fill or select the filled cells if you need to.

This method works with many other patterns as well. For example, if you have a column of ISBN numbers and want to extract the second number, you can do it if the number groups are separated by dashes.

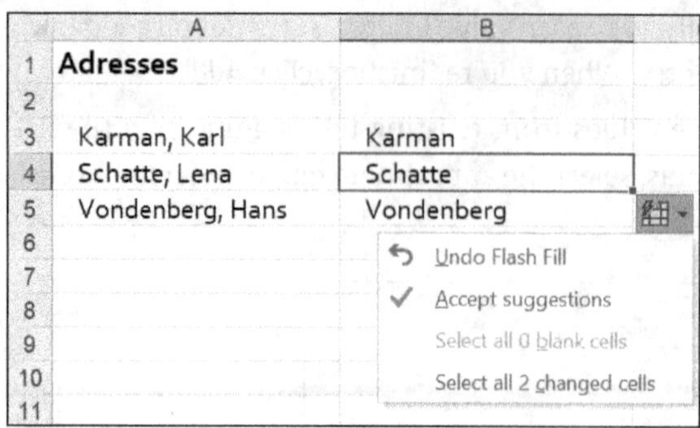

It also works the other way around. For instance, you can combine the contents of two columns into a new one, such as by merging separate first and last names into a customer name. You can also replace certain identifiable parts with others, and you can add character strings or digits at specific positions. If the first pattern isn't unique, use two or three patterns so the program can work as accurately as possible.

A series you create can also be automatically updated later. If you shorten the first name in the first cell of the row in the example with first and last names, then the other first names in the column will also be shortened. Figure 2.35 shows several examples of this.

	A	B	C
1	Helmut Vonhoegen (Hrsg.)	H. Vonhoegen (ed.)	
2	Helmut Kraus (Hrsg.)	H. Kraus (ed.)	
3	Andreas Maslo (Hrsg.)	A. Maslo (ed.)	
4			
5	049-221-888888	+49(221) 888888	
6	049-211-77777	+49(211) 77777	
7	001-77777-999999	+01(77777) 999999	
8			
9	(6/7) + (7/5)	(6 / 7) & (7 // 5)	
10	(3/6) + (5/3)	(3 / 6) & (6 // 3)	
11	(2/5) + (6/3)	(2 / 5) & (5 // 3)	
12			
13	3-540-653566-2	35406535662	
14	34-77-666-555	3477666555	
15	345-7777-66-7777	345777667777	
16			
17	House 1	Boston	House in Boston
18	House 2	New York	House in New York
19	House 3	Buffalo	House in Buffalo

Figure 2.35 Examples of Flash Fill Function

Flash Fill often eliminates the need for complex text transformations using macros and text functions, and the command is also available on the **Data** tab in the **Data Tools** group.

2

2.5 Data Entry Validation

When you enter data into a cell, there won't be any checks on data type, value, or length unless macros control the input. Excel lets you set data rules for specific cell ranges without any programming.

2.5.1 Validation Rules for a Price Column

To prevent accidental entry of a negative price or a price that's higher than that of the most expensive item on a price list, you can do the following:

1. Select the cell range where you want to apply validation rules.
2. On the **Data** tab in the **Data Tools** group, click the **Data Validation** button.

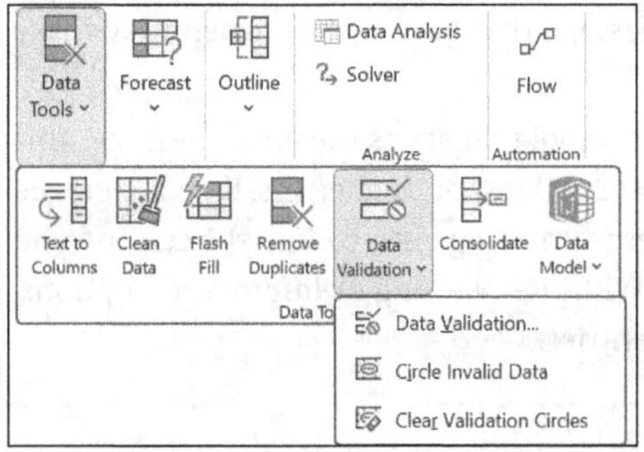

3. On the **Settings** tab, under **Allow,** choose **Decimal,** and under **Data,** select **between.**
4. Under **Minimum**, enter "0," and under **Maximum**, enter, for example, "300."
5. Check the **Ignore blank** box to prevent the system from issuing error messages when the price hasn't been entered yet.

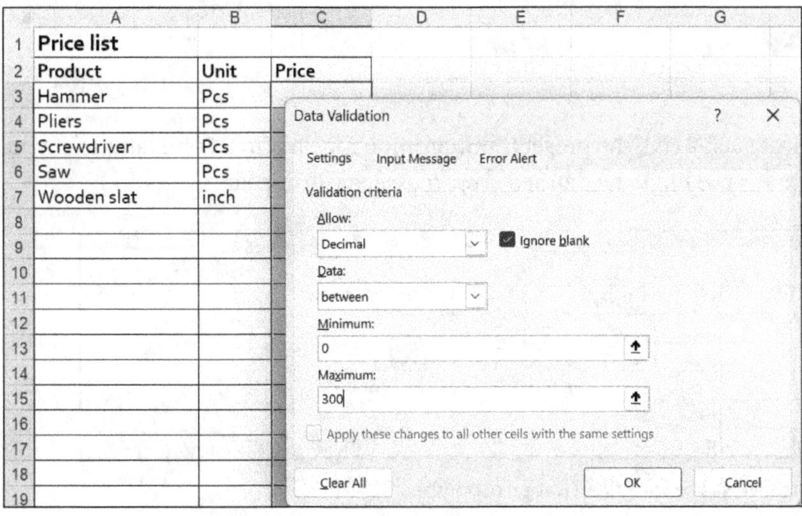

6. On the **Input Message** tab, enter a note that appears when a cell is selected. Make sure the **Show input message when cell is selected** box is checked.
7. Enter a title, like "Price," along with the input message text.

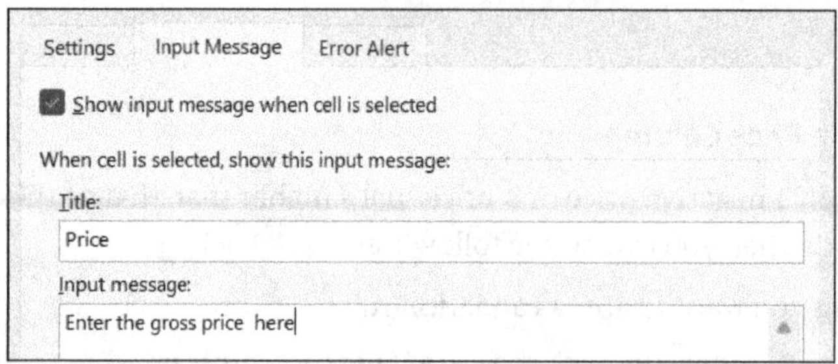

8. Finally, on the **Error Alert** tab, enter a message that guides users when they make invalid entries.
9. From the **Style** dropdown list, select the symbol that starts the error message. This choice isn't just visual; it also controls how Excel responds to errors. If you select the **Stop** option, Excel blocks incorrect entries from being entered into the cell and you must either re-enter the data or cancel. With the **Warning** or **Information** options, Excel can either allow or reject incorrect entries.

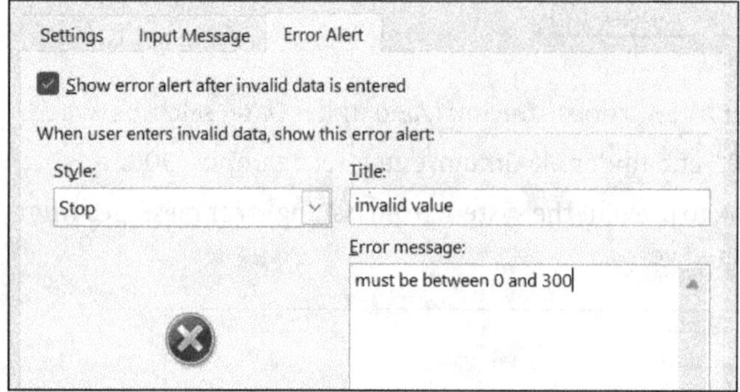

When you select such a cell, the preset input prompt appears first. For example, if you enter a price that's too high, the defined error message will appear.

	A	B	C
1	**Price list**		
2	**Product**	**Unit**	**Price**
3	Hammer	Pcs	400
4	Pliers	Pcs	
5	Screwdriver	Pcs	
6	Saw	Pcs	
7	Wooden slat	inch	

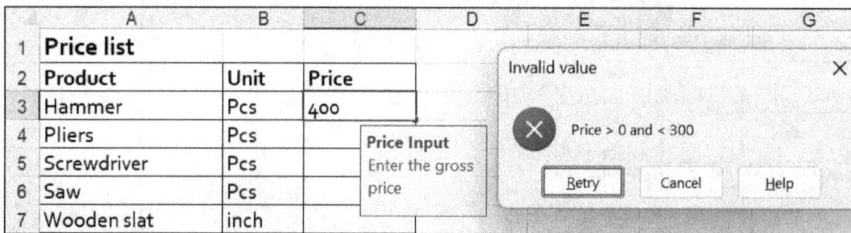

Figure 2.36 Error Message for Price That's Too High

2.5.2 Highlighting Incorrect Data

If you allow invalid data by using the **Warning** or **Information** options, then the menu under the **Data • Data Tools • Data Validation** button will offer the **Circle Invalid Data** option, which highlights the corresponding cells in red. The next option, **Clear Validation Circles**, removes these error markings.

You can also apply validation rules to cells that are used for calculations rather than input. If a calculated value exceeds a set maximum, the cell will be circled in red during validation.

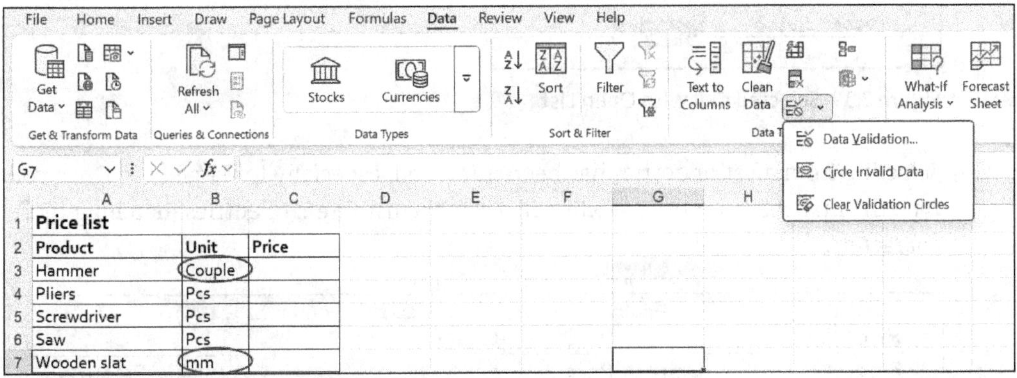

Figure 2.37 Highlighted Invalid Data

2.5.3 Input Lists

Data validation can also use lists that let users select entries without typing. In the displayed price list, only *inch*, *m*, *cm*, and *Pcs* are usually allowed as packaging units, and you can set this by choosing **List** under **Allow** on the **Settings** tab. You can enter the list itself under **Source**, using commas to separate items: *inch, m, cm, pcs*. This is sufficient here. Instead of listing possible values, you can also reference a cell range containing the list. In this case, it's best to use a named range so you don't have to update the reference when the list expands.

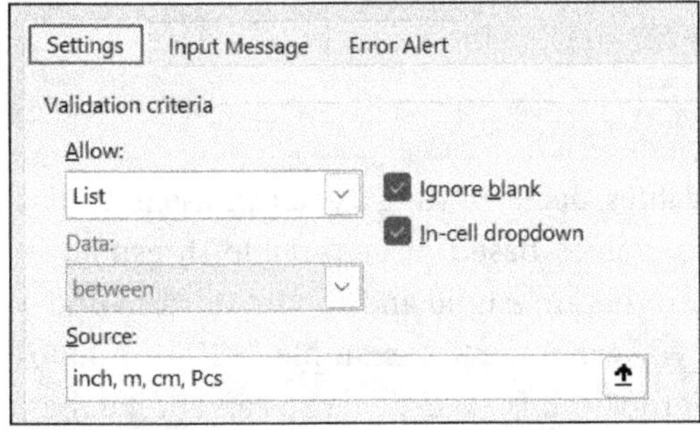

Figure 2.38 Entry of List of Allowed Values

When you select a cell with this validation rule, a small button appears that opens the list of allowed entries, and you can just click or tap the desired entry to select it. This requires that the **In-cell dropdown** option stays checked on the **Settings** tab.

	A	B	C
1	**Price list**		
2	**Product**	**Unit**	**Price**
3	Hammer	Pcs	
4	Pliers	Pcs	
5	Screwdriver	inch	
6	Saw	m	
7	Wooden slat	cm Pcs	
8			

Figure 2.39 Selected Cell with Open List

Selecting from a list of entries has been improved, especially for large lists. Once you type the first character or characters in a cell, the list filters to entries matching those characters.

	A	B	C	D	E	F
1	**Price list**					
2	**Product**	**Unit**	**Price**			**Available Products**
3	Hammer	Pcs				Hammer
4	Pliers	Pcs				Pliers
5	Screwdriver	Pcs				Screwdriver
6	Saw	Pcs				Saw
7	Wooden slat	inch				Wooden slat
8						Drill
9	Hammer					Nails
10	Pliers Screwdriver					Metal file
11	Saw					
12	Wooden slat					
13	Drill					
14	Nails Metal file					
15						

Figure 2.40 Selection from Larger List

2.5.4 Validation with Formulas

In addition to limiting entries to specific values, dates, or times, you can use formulas to validate input. For example, to control values based on a variable threshold in another column, you can name the cell with that threshold and choose **Custom** under **Allow** on the **Settings** tab. Under **Formula**, you can enter, for example, "=SUM(B$3:B$7) <= Maximum" if the column starts at B3. If the limit is exceeded, you can enter a message like "Value too high" under **Error Alert** to prompt a new entry.

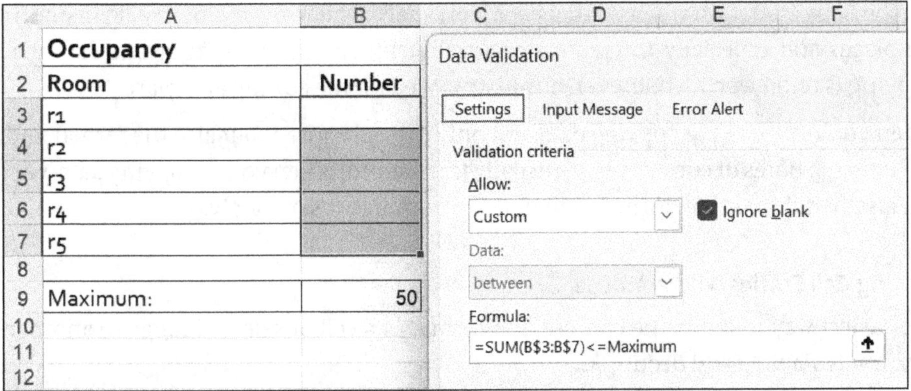

Figure 2.41 Data Validation with Formula

2.5.5 Editing or Deleting Rules

To change a validation rule, you don't need to select all cells that use it. Do this instead:

1. Select one affected cell.
2. Go to **Data • Data Tools • Data Validation** and update the entries in the dialog box.
3. Check the box labeled **Apply these changes to all cells with the same settings.**

Excel will apply the new rule to all cells that previously had the old rule. To delete all rules, click the **Clear All** button in the dialog box.

2.5.6 Applying Rules

To extend a validation rule to additional ranges, select the cells where the rule currently applies along with the new cells you want to include. Then, when you use **Data Validation** again, you'll be prompted to extend the validation to the new cells. Finally, select **Yes** and confirm the data in the dialog box.

Existing rules are also preserved when you move or copy cells, and you can extend a validation rule to adjacent cells by using the fill handle or the **AutoFill** icon.

2.6 Reorganizing and Restructuring Sheets

Unlike characters written on paper, characters in a worksheet don't stick to a cell. You don't need scissors or glue to rearrange the structure of a table.

2.6.1 Rearranging and Copying Cell Ranges

To help you rearrange labels, values, or formulas, Excel offers the **Move** operation, which clears data from the old location and places it in the new one. You'll also need to use the **Copy** operation when working in a worksheet, whenever you need data in multiple places and especially to create structurally similar formulas from one formula. Copying is often worthwhile, even if the copy needs some editing afterward.

When you're moving or copying data, not only the cell content but also the format and any existing notes or comments are usually transferred. When you're copying data, you can also handle content, format, and notes or comments separately.

Moving Cell Entries with a Mouse

The easiest way to move the content and format of a cell or selected range to another location is via drag and drop:

1. Place the cell pointer on the selection frame until it changes into an arrow with crosshairs.
2. Hold down the left mouse button and drag the frame to the target position. The current cell address always appears for confirmation.
3. Release the mouse button at the target position.

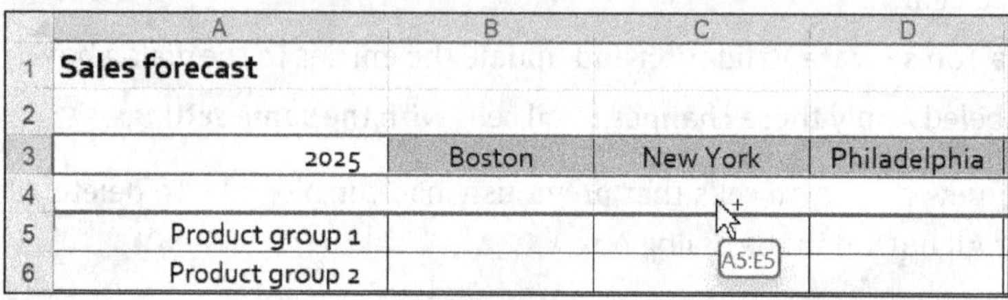

A
B
C
D
1 Sales forecast
2
3 2025 Boston New York Philadelphia
4
5 Product group 1
6 Product group 2
A5:E5

If you drag beyond the window edge, the window scrolls in that direction. You can also move entire rows or columns with this method by simply selecting the row or column headers and dragging the frame. This lets you reorder columns in a table within seconds. If data already exists at the target position, Excel will ask if you want to overwrite those cells.

Moving and Inserting

If you want to move cell contents to a spot where you must first create space by inserting new cells, then you can do it with a mouse in a single step.

Suppose you have a table where the first quarter values for two years are initially arranged side by side. To compare them more easily, you can rearrange the columns so the values for the comparison month are next to each other. Here's how you do it:

1. Select the month column you want to move.
2. Place the cell pointer on the selection border, hold down the Shift key, and drag the mouse pointer to the target position.
3. An I-beam will appear on the mouse pointer, and you should position it exactly on the column line where you want to insert the cells.

	A	B	C	D	E	F
1	Sales					
2						
3		Jan-23	Feb-23	Mar-23	1st Quarter 2023	Jan-24
4	Product 1	10000	12000	12000	34000	11000
5	Product 2	12000	14400	17280	43680	13200
6	Product 3	13000	15600	18720	47320	14300
7	Product 4	11000	13200	15840	40040	12100
8	Product 5	10000	12000	14400	36400	11000
9	Product 6	12000	14400	17280	43680	13200

When you release the mouse button, the data moves from its original spot to the insertion area and the existing data shifts right.

Moving to Other Sheets or Workbooks

To move a range to another sheet, follow these steps:

1. Select the range.
2. Hold down the Alt key and drag the range by its border to the tab of the sheet where you want to place it.
3. The sheet becomes active, and you can drop the selected range where you want it.

In Excel, you can use a mouse to move cells beyond the active workbook. To drag a range from one workbook to another, arrange both workbook windows so their sheet tabs are visible and then drag the selected cell range by its border directly into the other workbook.

Moving with the Clipboard

It's usually easier to move large ranges with **Cut** and **Paste**. These commands use the Windows clipboard, and the quickest way to use them is with the **Cut** and **Paste** buttons in the **Home • Clipboard** group or by pressing Ctrl+X and Ctrl+V:

1. Select the cell range you want to move and use **Cut** ❶. Excel will copy the selected data to the clipboard and outline the selection with a moving border.
2. Select the upper-left cell of the paste range or the entire paste range. Overlaps with the source range are allowed if the target is the corner cell or a range of the same size. If the target range is on another sheet, select that sheet tab, and if it's in another file, choose the appropriate window. You can adjust your selection until it's correct. To cancel the operation, press Esc.
3. Complete the move by pressing **Paste** ❷ or Enter.

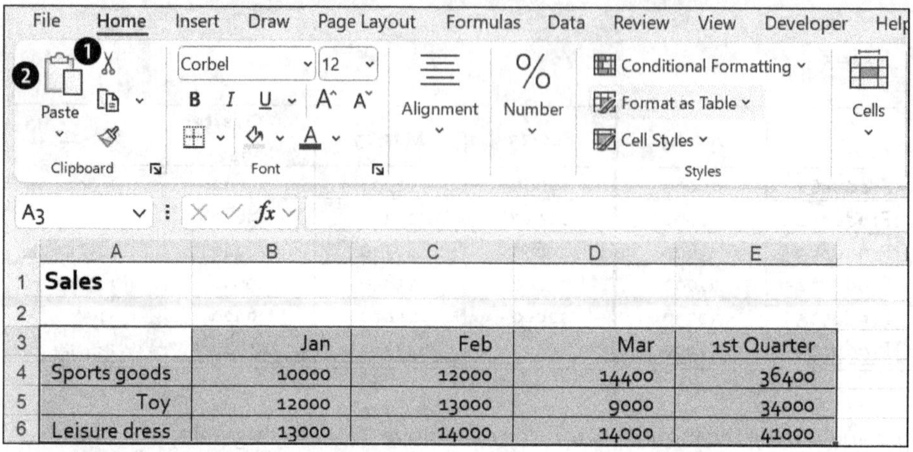

	A	B	C	D	E
1	**Sales**				
2					
3		Jan	Feb	Mar	1st Quarter
4	Sports goods	10000	12000	14400	36400
5	Toy	12000	13000	9000	34000
6	Leisure dress	13000	14000	14000	41000

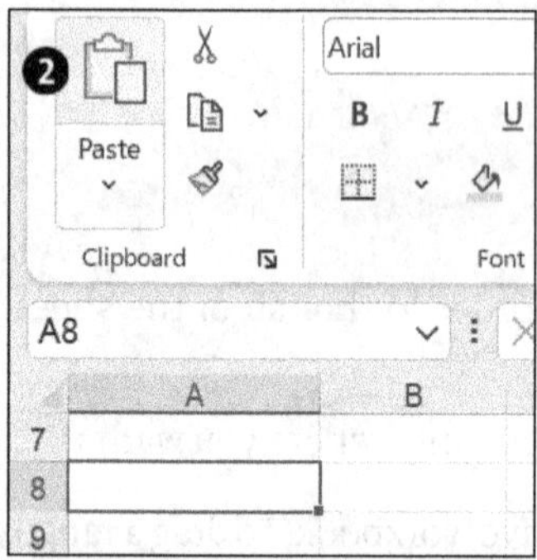

If you need to make space in the paste range, right-click the target range and select **Insert Cut Cells**. This command is only available when cut data is in the clipboard. If there's more than one way to paste, a small dialog box appears, and you can use it to choose whether to shift the existing cell contents right or down at the paste location.

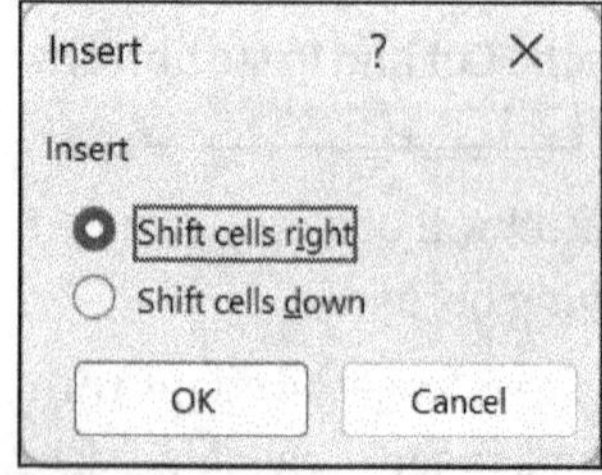

Figure 2.42 Insert Cells with Content Dialog Box

Moving with the Keyboard

If you don't use a mouse, you can move cells by using the keyboard. To do this, select the area and press `Shift`+`Delete`, then set the active cell to the top-left corner of the paste area and confirm with `Enter`.

Moving on the Touchscreen

In **Touch input** mode, you can easily stretch or resize ranges. However, you can't move them with your finger, but you can use the clipboard method as follows:

1. Touch the selected range until the context menu appears.
2. Tap **Cut**.
3. Hold your finger on the corner cell of the target area and then select **Paste** from the context menu.

The target range stays selected thereafter, so you can quickly apply more commands.

If you're using a Surface Pen, you can use the tip like a mouse pointer to drag a selected cell range to a new location. At the target spot, a context menu appears offering **Move here** along with other options, and this menu is the same as the one that appears when you drag a selection while holding the right mouse button (look ahead to Figure 2.53).

2.6.2 Copying to Adjacent Cells

There are several ways to copy cells, which you'll often need to do when copying a value or formula into neighboring cells.

We've already covered copying with the fill handle or the **AutoFill** icon on the touchscreen, but you can also use the icons under **Home • Editing • Fill**. To do this, first select the range with the content to copy and the target range as a block, then pick the icon with the desired copy direction from the **Fill** menu. If you want to copy a single cell's value only into the cell to the right or below, just select the empty target cell and then click the corresponding arrow icon.

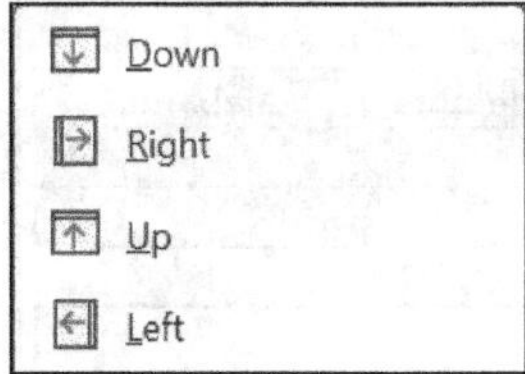

Figure 2.43 Direction Arrows in Fill Menu

To copy with a mouse, select the cell you want to copy and drag the fill handle across the cells in the row or column you want to fill. You can also copy multiple cells at once, whether they're in a row, column, or cell range.

The section on data series explained that when you drag the fill handle or the **AutoFill** icon, Excel looks for a chance to create a series—and if it can't, it simply copies the content. This happens with formulas, single numbers, and text—unless the text includes a counter at the start or end. Even if you can create a series, you can still use the **Copy Cells** auto-fill option to copy only; alternatively, you can hold down the `Ctrl` key while dragging.

2.6.3 Copying with Reference to the Adjacent Column

When copying formulas, it's especially useful to copy a cell's content down as far as the adjacent column extends. A typical example is shown in Figure 2.44. Here, you add the values from three months each time. You create the first formula, keep the cell selected, and double-click the fill handle—and the formula will copy down as far as there are values in column D.

	A	B	C	D	E
1	**Sales**				
2					
3		Jan	Feb	Mar	1. Quarter
4	Sports goods	10000	12000	14400	36400
5	Toy	12000	13000	9000	34000
6	Leisure dress	13000	14000	14000	41000
7					

Figure 2.44 Copying First Row Sum by Double-Clicking

2.6.4 Copying to Nonadjacent Cells

To copy data to nonadjacent cells, you can use the mouse within the worksheet. Select the original cells and drag the border while holding down the `Ctrl` key, and then, when you release, the copy appears in the new location. If the target area extends beyond the visible screen, drag slightly past the scroll bar and the screen will scroll until you release the mouse button.

	A	B	C	D	E
1	**Sales forecast**				
2					
3	2025	Boston	New York	Philadelphia	Alabama
4	Product group 1				
5	Product group 2				
6					
7					
8					
9					
10				A7:E9	
11					

Figure 2.45 Copying Labels

2.6.5 Copying and Pasting

By default, copying overwrites the target range, but you can also create space for the copy beforehand. While dragging the selection frame, hold down both the `Shift` and `Ctrl` keys and the mouse pointer will change to an I-beam. Then, position it precisely between the columns or rows where you want to insert new cells filled with data from the source range. If the I-beam is on a column boundary, existing data shifts right; if it's on a row boundary, the data shifts down.

2

2.6.6 Copying to Other Sheets or Workbooks

In Excel, you can copy cells beyond the active worksheet by using a mouse. To copy a range to another sheet, follow these steps:

1. Select the range.
2. Hold down the `Alt` and `Ctrl` keys, then drag the range's frame to the tab of the sheet where you want to place the copy.
3. The sheet will become active, and you can move the selected range to the desired location.

To copy a range from one workbook to another, drag the selected cell range by its border into the other workbook. Just hold down `Ctrl` while doing this.

[+]

Disabling Drag and Drop

You can disable the **Enable fill handle and cell drag-and-drop** option under **File • Options • Advanced • Editing options**. This is helpful if you want to prevent inexperienced users from experimenting in completed workbooks.

2.6.7 Copying via the Clipboard

As with moving, you can choose to use the clipboard when copying. Typically, you use the **Copy** ❶ and **Paste** ❷ buttons in the **Home • Clipboard** group or press the `Ctrl`+`C` and `Ctrl`+`V` keyboard shortcuts. The `Ctrl`+`Shift`+`V` shortcut pastes only the value without formatting.

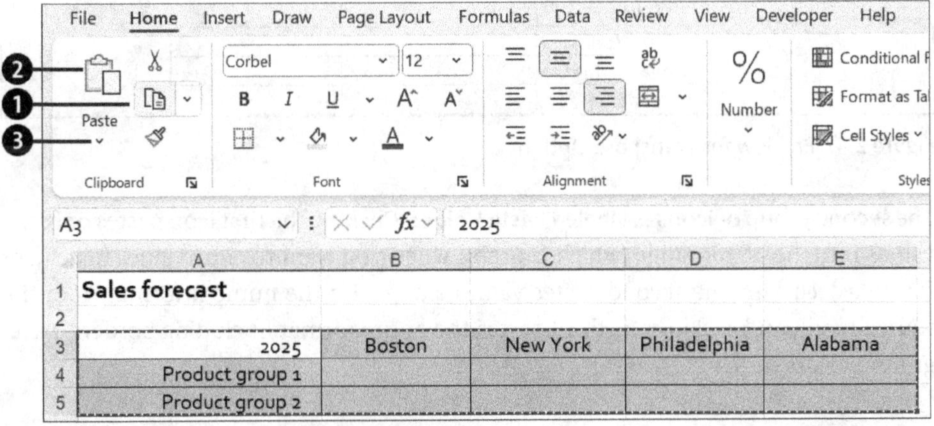

Figure 2.46 Copying Selection to Clipboard

Clicking the arrow ❸ at the bottom of the **Paste** button opens a detailed menu of paste options in Excel. The unique feature here is that the icons not only visually indicate the paste options but also show a preview of the result at the paste location.

To review which paste option works best, hover the mouse pointer over the icons or hold the pen above them. The paste action only happens when you click or tap an icon. If you do that, then at the end of the target area, an icon appears offering the same paste options, thus allowing you to correct the action if you accidentally choose the wrong one. Press Enter to finish and clear the selection of the source area with the marching-ants border.

Clicking the first icon under **Paste** inserts the copied cells entirely, clicking the second inserts only formulas, clicking the third inserts formulas and number formats, and clicking the fourth preserves the original formatting. In the second row, there's an icon that doesn't carry over existing borders from the source range, an icon that applies the source range's column width to the target range, and an icon that "flips" the table by swapping columns and rows, as shown in the preview in Figure 2.47.

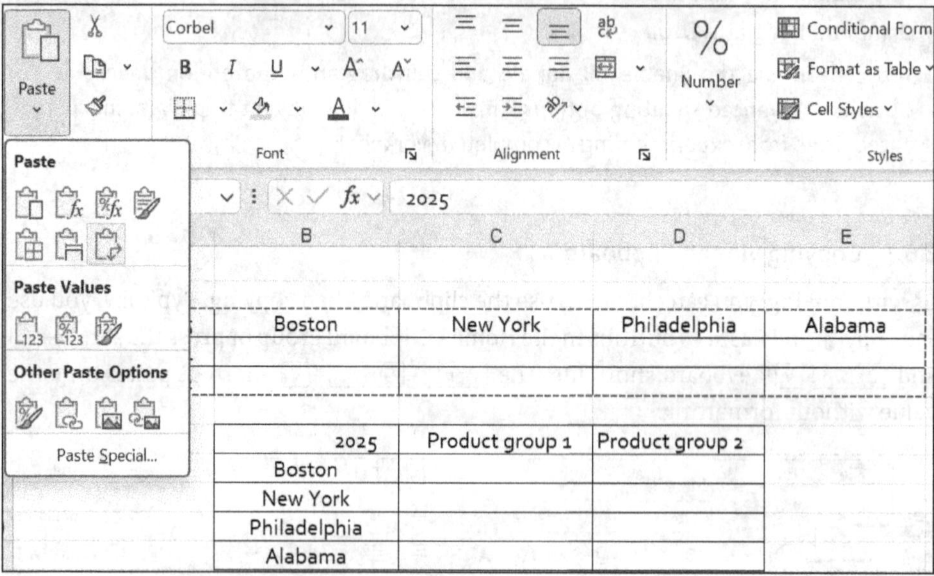

Figure 2.47 Preview for Transpose Option

The second group of icons is labeled **Paste Values**. Clicking the first icon pastes only the values into the target range (which is useful when you want to copy values instead of formulas), clicking the second pastes values along with the number format from the source range, and clicking the third pastes the entire format (including borders, background colors, and more).

The third group of icons, which is labeled **Other Paste Options**, initially offers an icon that applies only the formatting. The second icon links the source and target areas, so changes in the source are reflected in the target. (This topic is covered in detail in Chapter 3.) The third icon takes a snapshot of the copied area and inserts it as a graphic object into the worksheet, while the fourth icon creates a linked image of the source area. We will revisit this topic in Chapter 10.

Also included is the **Paste Special** option, which offers some of these choices in the dialog box. This dialog also offers a few useful operations, like applying calculations to the cells in a target range. You'll find more details in Chapter 5.

The **Skip blanks** option is also noteworthy. A common scenario is copying values from column B to column A, so for example, column B might contain some correction values for the current values in column A, but only for certain cells. When you copy values from B to A, the values in A remain unchanged where column B has no data.

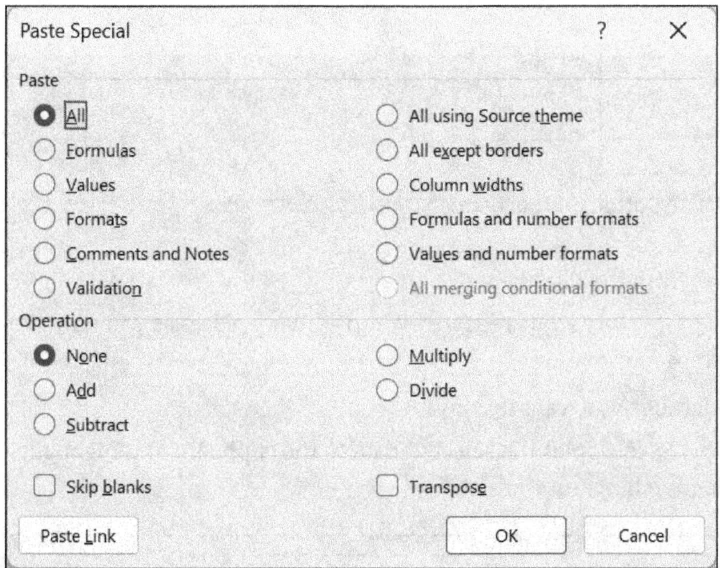

Figure 2.48 Paste Special Dialog

When you open the **Insert** button menu on a touchscreen in **Touch input** mode, Excel shows a vertical menu with larger spacing between options to make selecting commands easier (see Figure 2.49). However, live preview isn't available here, and since the **Paste Options** button usually appears after you paste with one of the options, you can quickly fix any paste mistakes and confirm the paste by pressing `Enter`. This menu is also arranged vertically.

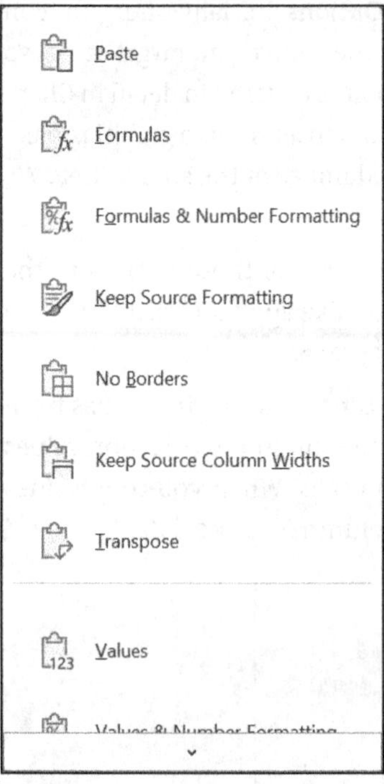

Figure 2.49 Menu for Paste Button on Touchscreen

Pasting Multiple Times from the Clipboard

The clipboard lets you paste stored data repeatedly into as many places as you want. To do this, follow these steps in order:

1. Select the cell or cell range you want to copy.
2. When you click **Copy**, Excel copies the selected data to the clipboard and highlights the selection with a marching-ants border.

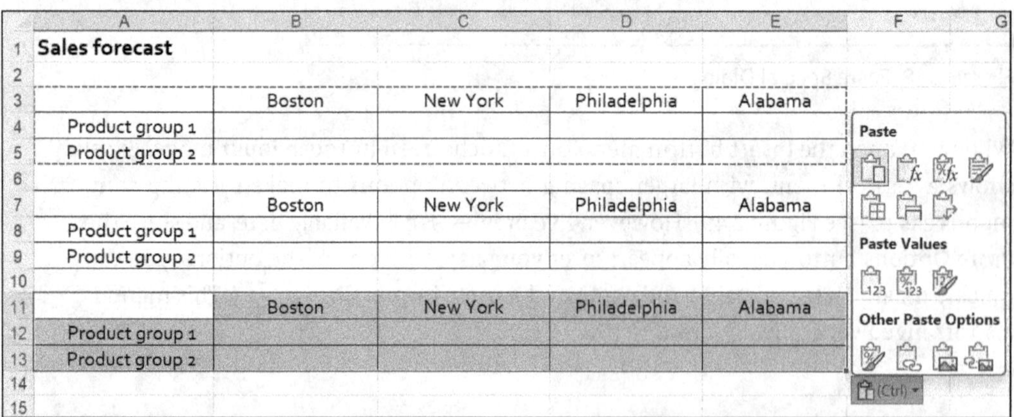

3. Select the top-left cell or the entire range where you want to paste. If it's on another sheet, first select the appropriate sheet tab, or if it's in another file, switch to that window. You can adjust your selection until it's correct. To cancel, press [Esc].
4. Paste the data from the clipboard by clicking the **Paste** button—without opening its menu. Excel will paste the entire cell range from the clipboard but still show the **Paste Options** button at the end of the pasted range, and clicking or tapping it will open a menu to adjust the paste method.
5. To paste another copy, select the next target range and paste again with **Paste**; otherwise, press [Enter] to finish.

If you find the **Paste Options** button distracting, you can disable it under **Cut, Copy, and Paste** in **File • Options • Advanced**.

When the target is a range rather than a single cell, Excel's behavior depends on the shape of the original range. Figure 2.50 shows some examples, with the original highlighted in gray.

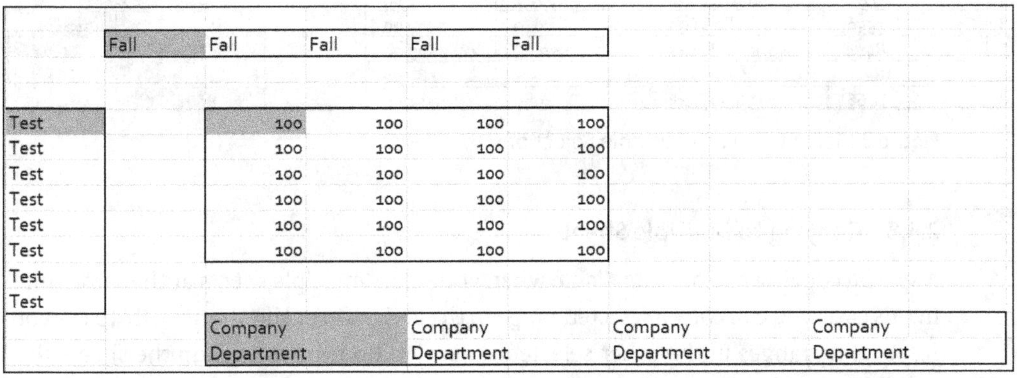

Fall Fall Fall Fall Fall

Test
Test
Test
Test
Test
Test
Test
Test

100 100 100 100
100 100 100 100
100 100 100 100
100 100 100 100
100 100 100 100
100 100 100 100

Company Company Company Company
Department Department Department Department

Figure 2.50 Original and Copy/Copies

If the source area includes more than one cell, the target area must meet one of the following conditions:

- Only the top-left cell is specified as the target area. For single-column source areas, multiple start cells in a row are allowed, and for single-row source areas, multiple start cells in a column are allowed.
- The number of rows is the same.
- The number of columns is the same.
- The target range is the same size as the source area.
- The target range is large enough to be filled multiple times by the source range.

If none of these conditions are met, you will get an error message and should adjust the target range.

Creating Space When Inserting

If the target range already contains data, then you can shift the data at the insertion point to the right or downward. Instead of the **Insert** command, use the **Copied Cells** option from the target range's context menu. In the dialog box, select whether to shift the existing data at the insertion point to the right or down.

Copying Multiple Ranges

Although the **Cut** command isn't available for multiranges, you can copy them using the clipboard. However, there are limitations. The parts of the multirange must share a common dimension—either width or height. They must also be aligned parallel, which means that they must start in the same row if they share the same height or in the same column if they share the same width. When you paste them, the subareas are inserted side by side or stacked vertically. Figure 2.51 shows an example of this.

	A	B	C	D	E	F	G
3	Product number	Product name	Category	Net price		Product name	Net price
4	7777	Louvre Ccxs	Louvre	$120.00		Louvre Ccxs	$120.00
5	5556	Rollo BT 33	Rollo	$100.00		Rollo BT 33	$100.00
6	8444	Awning Blue Sk	Awning	$99.00		Awning Blue Sk	$99.00
7	8443	Louvre VVx	Louvre	$88.00		Louvre VVx	$88.00
8	8666	Rollo Dark	Rollo	$110.00		Rollo Dark	$110.00

Figure 2.51 Copying with Multiple Selections

2.6.8 Copying to Multiple Sheets

A special copying option is available when you select multiple sheets at the same time. In this case, you can copy a selected range in the active sheet simultaneously to the corresponding ranges in the other selected sheets. To do this, you group the sheets that belong to the sheet group, select the cell range to copy, copy it to the clipboard, and then simply use the **Home • Editing • Fill • Across Sheets** command. In the small dialog box, you can choose to copy the cell content, the format, or both.

Moving and Copying Using the Context Menu

The options for moving, copying, and pasting are also available through the context menus of the selected cell ranges. The menu for the selected source range lets you disable the paste commands, and the menu for the target range shows visual paste options and includes the previously mentioned **Paste copied cells** option.

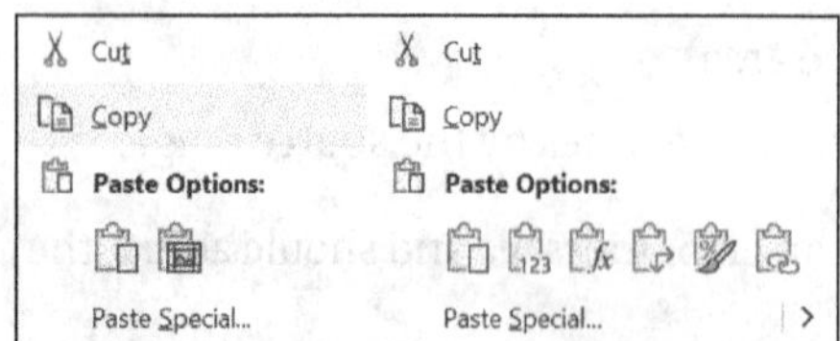

Figure 2.52 Context Menus for Source and Target Ranges

An advanced context menu is also available for mouse users. To access it, select the cells you want to move or copy, place the cell pointer on the frame, and then drag the auxiliary frame to the target position while holding the right mouse button. The menu shown in Figure 2.53 will then appear there.

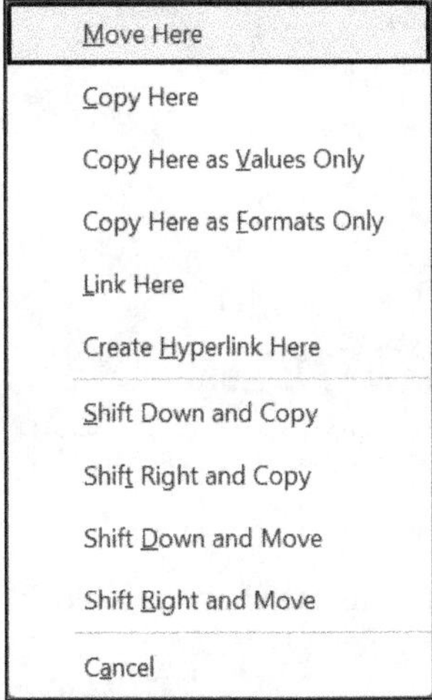

Figure 2.53 Context Menu for Moving and Copying

On the touchscreen, the **Paste**, **Cut**, and **Copy** commands are found in the horizontal section of the context menu.

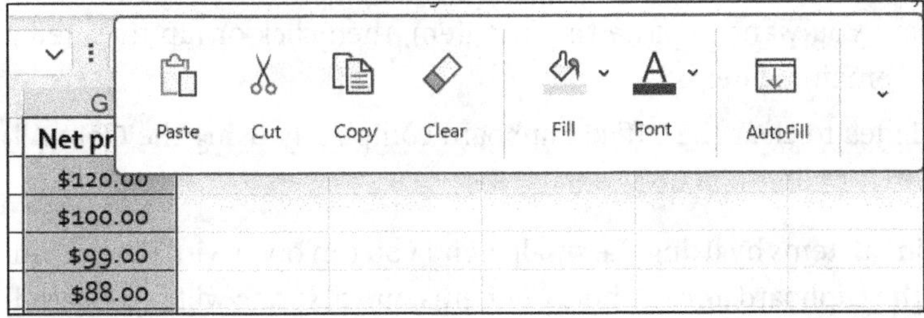

Figure 2.54 Commands for Copying, Cutting, and Pasting on Touchscreen

2.6.9 Copying Multiple Ranges at Once

Another helpful feature when building calculation models is the advanced Office clipboard, which can store up to 24 items. It complements the Windows clipboard, which holds only one item at a time.

If you keep the **Clipboard** task pane open—a click or tap on the dialog box launcher in the lower right corner of the **Clipboard** group will open it—all the items you copy or cut will be listed there. You can also hide the window by using the **Options** button in the task pane and selecting **Collect Without Showing Office Clipboard**, and you can track copied items via an icon in the taskbar.

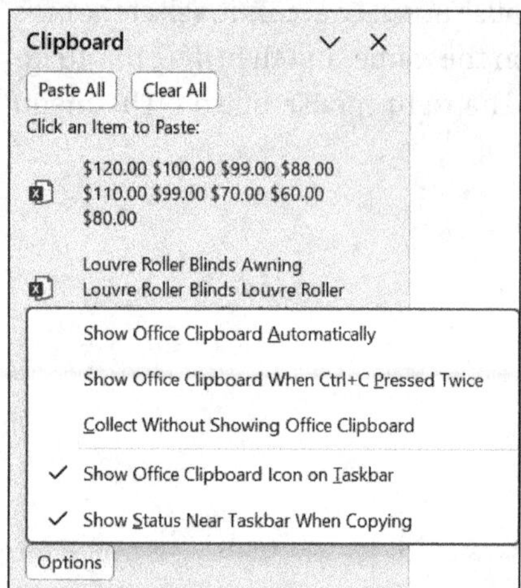

Figure 2.55 Clipboard Task Pane and Its Options

Suppose you want to copy columns from a table and paste them elsewhere as a block. Follow these steps:

1. Select the columns one by one, using the **Copy** command each time.
2. If you use the **Copy** command multiple times, Excel automatically opens the **Clipboard** pane based on the setting you choose via the **Options** button. This pane displays the current items on the Office Clipboard, showing the first part of each snippet along with a small menu for each item.
3. Select the cell where you want to paste the first item, then click or tap that item. Repeat this for the remaining items.
4. Finally, it's a good idea to clear the Office Clipboard completely using the **Clear All** button to start fresh for your next tasks.

You can delete individual items by using the small menu you can open with the button on the right edge. If the clipboard already holds 24 items, the first one will be replaced by the new item. The **Paste All** button lets you combine multiple stored items in a new location. Also note that Excel stacks the cell ranges vertically, so you might need to do some editing afterward.

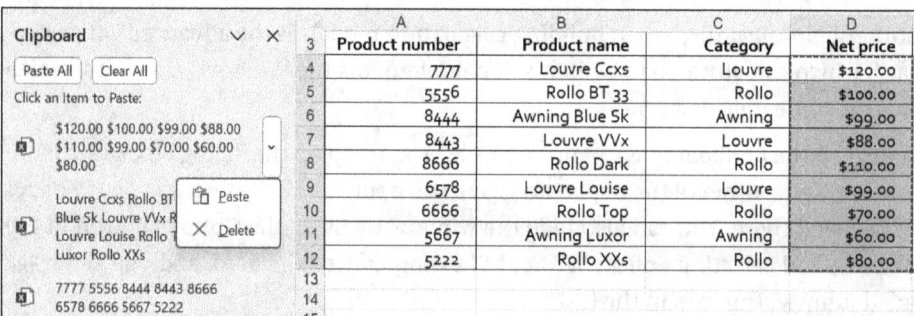

Clipboard

Paste All | Clear All

Click an Item to Paste:

$120.00 $100.00 $99.00 $88.00 $110.00 $99.00 $70.00 $60.00 $80.00

Louvre Ccxs Rollo BT Blue Sk Louvre VVx R Louvre Louise Rollo T Luxor Rollo XXs

Paste

Delete

7777 5556 8444 8443 8666 6578 6666 5667 5222

	A	B	C	D
3	Product number	Product name	Category	Net price
4	7777	Louvre Ccxs	Louvre	$120.00
5	5556	Rollo BT 33	Rollo	$100.00
6	8444	Awning Blue Sk	Awning	$99.00
7	8443	Louvre VVx	Louvre	$88.00
8	8666	Rollo Dark	Rollo	$110.00
9	6578	Louvre Louise	Louvre	$99.00
10	6666	Rollo Top	Rollo	$70.00
11	5667	Awning Luxor	Awning	$60.00
12	5222	Rollo XXs	Rollo	$80.00
13				
14				

2.6.10 Copying Formats

As mentioned earlier, when copying cell ranges, you can handle content and format separately. By default, Excel copies content and format together, which works well in most cases. But sometimes, it's helpful to copy a format separately—either to empty cells before entering data or to cells that already contain data.

If you want to expand a table by three columns and apply the format of the last column, then select that column and drag the fill handle or the **AutoFill** icon to the right across the columns you want to format. After that, click the **Auto Fill Options** button and choose **Fill Formatting Only**. This removes any copied content and keeps only the formatting. Similar options will appear when you're inserting rows.

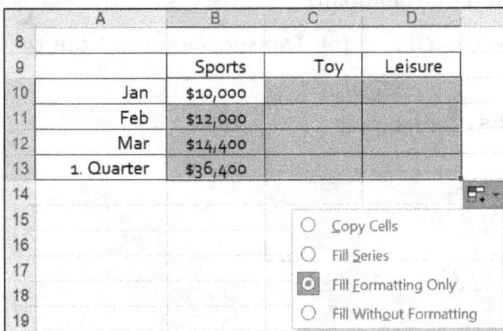

Figure 2.56 Copying Formats

Using the Format Painter

A convenient way to copy formats anywhere is by using the brush button, which is located in the **Home • Clipboard** group under **Format Painter**. You can click this button to apply the active cell's format to the next cell or cell range you select, and the mouse pointer will change to a cross with a paintbrush as long as you hold it. You can also double-click the button to use the format painter repeatedly until you click it again, and if you use this command often, you should consider adding it to the Quick Access Toolbar. You can also use the paintbrush icon in **Touch input** mode by tapping the button and then tapping the cell that should adopt the format of the selected cell or cell range.

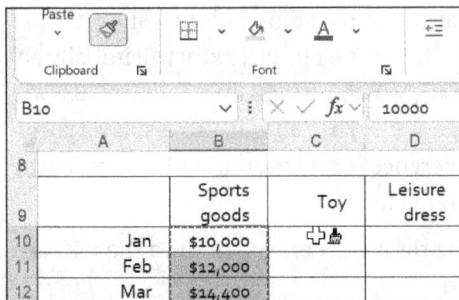

Figure 2.57 Formatting with Paintbrush Icon

2.6.11 Transposing When Copying

Transposing is a special way to copy data that we briefly mentioned earlier. It converts data arranged in columns into rows and data in rows into columns, which effectively flips or mirrors the table. The mirroring point is always the upper-left corner of the selected range. While Excel's PivotTables offer powerful ways to rearrange data, the **Transpose** option remains useful. For instance, if a table has become too wide, you can easily flip it this way.

To demonstrate this feature, we'll use a simple quarterly report as an example. The data for each product group is arranged in a single row, and you can flip this layout so the values for each product group appear in columns. Here's how you do it:

1. Select the entire table range and copy it to the clipboard.
2. Select the corner cell of the new table and choose the **Transpose** option from the **Paste** menu.
3. Press Esc to cancel the selection of the original range.

	A	B	C	D	E
1	**Sales**				
2					
3		Jan	Feb	Mar	1. Quarter
4	Sports goods	$10,000	$12,000	$14,400	$36,400
5	Toy	$12,000	$13,000	$9,000	$34,000
6	Leisure dress	$13,000	$14,000	$14,000	$41,000
7					
8		Sports goods	Toy	Leisure dress	
9	Jan	$10,000	$12,000	$13,000	
10	Feb	$12,000	$13,000	$14,000	
11	Mar	$14,400	$9,000	$14,000	
12	1. Quarter	$36,400	$34,000	$41,000	

Figure 2.58 Transposing Tables

2.6.12 Deleting and Inserting Cells

You can not only clear a cell's content and formatting but also delete the cell itself. However, this description isn't precise: The contents of surrounding cells just shift to close the gap left by the deletion. So if you delete cell B3 in the example, Excel will initially act as if B3 no longer exists—and the formula in C3, which previously referenced B3, will lose that reference and show the error message #REF!. The cell will also display a corresponding error indicator, which means the reference is broken, but if you re-enter the reference to B3, the formula will work correctly again.

When you delete a column, all columns to the right shift one column to the left. The same applies to rows: rows below shift upward.

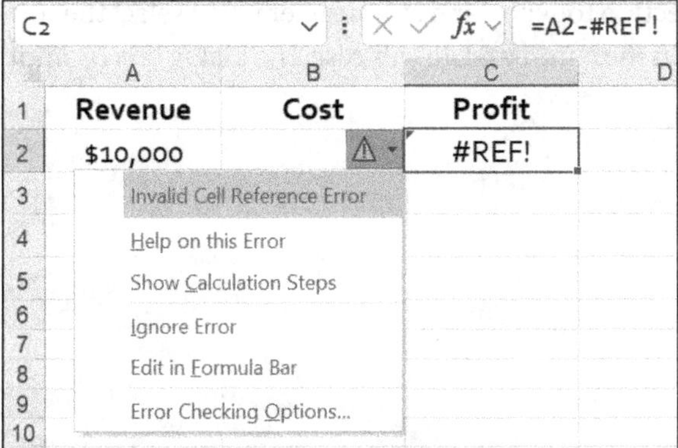

Figure 2.59 Note After Losing Reference

You can also use a mouse to delete cells. Select the range of cells you want to delete, position the mouse pointer over the fill handle, and hold down the `Shift` key. The mouse pointer will change to a double-headed arrow with two crossbars, and you can drag it into the selection until the entire area is filled with the grid. Be mindful of the direction you drag. If you drag upward, the cells below the selection move up, and if you drag to the left, the cells to the right of the selected area shift left. You can also delete entire rows or columns this way. To do this, select the row or column headers, hold down the `Shift` key, and drag the fill handle into the selection.

To delete selected cell ranges with your finger, tap the selection, tap the arrow at the end of the horizontal section in the context menu, and choose **Delete Cells** in the vertical section.

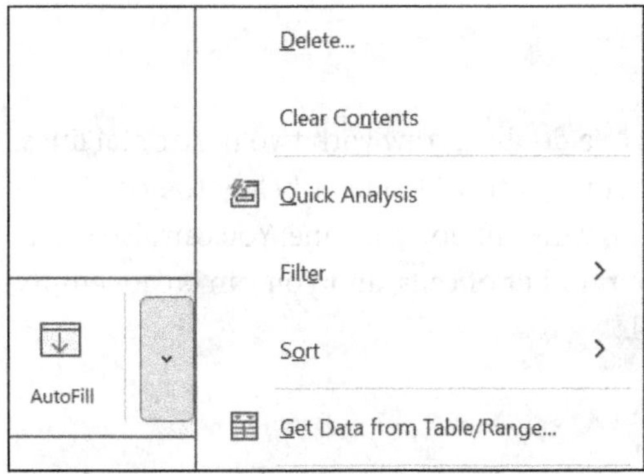

Figure 2.60 Two-Part Context Menu on Touchscreen

Delete Command for Cells

You can also find the delete command for cells under **Home • Cells • Delete • Delete Cells** in the ribbon. Select the range and then use the **Delete Cells** command to opens a dialog box where you can choose to shift surrounding cells left or upward. You can also delete entire rows or columns that the selection touches, and this is faster if you select the row or column headers first and then run the command. In this case, the dialog box prompt is skipped.

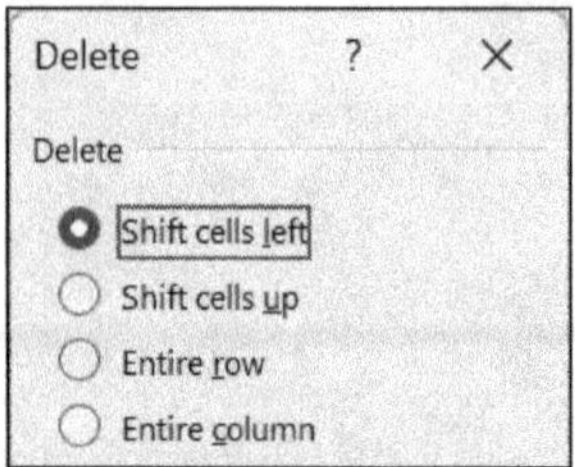

Figure 2.61 Deleting Cell Ranges

Inserting Cells

Inserting cells, rows, and columns works much like deleting them with a mouse, and the main difference is that you drag the mouse pointer out of the selected area. Suppose you enter values in a column and realize after a few entries that you accidentally missed one. All the following values will be one cell too high, so you fix the problem by selecting the last correct value and then dragging the fill handle down one cell while holding down the `Shift` key.

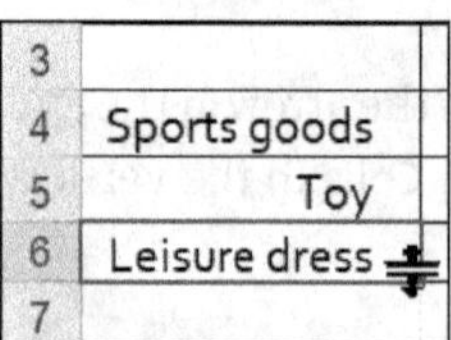

Figure 2.62 Inserting Cells with Mouse

The mouse pointer will then change back to a double arrow with two horizontal lines. When you release the mouse button, an empty cell will appear below the originally selected cell and all other cells in the column will shift down by one. You can also insert multiple cells by dragging down the same number of cells, and you can create empty cells to the right of the selected cell as well.

Adding Rows and Columns

The straightforward way to insert entire rows or columns is to use the row or column headers, just like when deleting them. First, select the row below where you want to insert new rows or the column next to where you want to insert new columns to the right. Then, hold down the `Shift` key and drag the fill handle down or to the right. Finally, extend the frame across multiple rows or columns to insert several rows or columns. By default, Excel applies the formatting of the selected cells to the new cells when you paste.

After you finish pasting, the **Paste Options** button appears and lets you choose whether the new cells should adopt the formatting from the cells above or below or if the formatting should be cleared entirely (see Figure 2.63). 2

3		Jan
4	Sports goods	$10,000
5	Toy	$12,000
6		$13,000
7		
8		
9		
10	Leisure dress	

- Format Same As Above
- Format Same As Below
- Clear Formatting

Figure 2.63 Context Menu for Inserting Rows

Commands, icons, and keyboard shortcuts for pasting are also available. In this case, unlike when you're pasting with a mouse, when you first select the cell below or to the right of the paste location, you select the exact cell range to clear for the new cells. If you select **Home • Cells • Insert • Insert Cells**, a small dialog box appears (see Figure 2.64). You can also open this dialog box by using the Ctrl+[+] keyboard shortcut.

In this dialog box, you can choose whether to shift existing cells right or downward at the insertion point, and you can use the **Entire row** and **Entire column** options to insert full rows or columns at the insertion point. Alternatively, you can select row or column headers and then choose **Insert Cells** from the context menu.

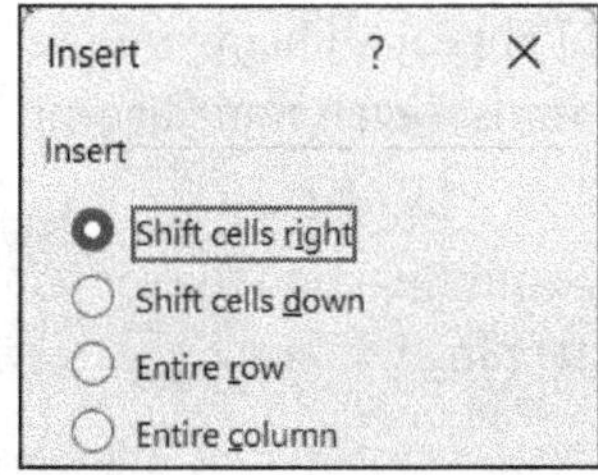

Figure 2.64 The Insert Cells Dialog Box

Also, instead of choosing the **Insert Cells** option, you can choose the **Insert Sheet Rows** or **Insert Sheet Columns** options directly from the **Insert** button menu. As you did when deleting rows and columns, you need to consider how inserting affects hidden parts of the table. Usually, formulas are unaffected and range addresses adjust automatically—and when multiple sheets are selected, cells are inserted into all of them.

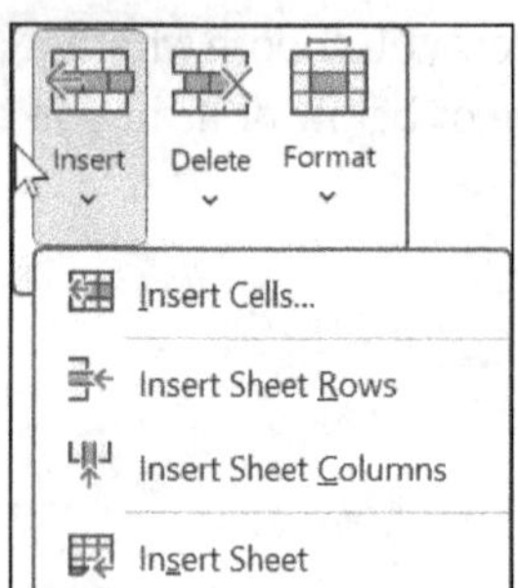

Figure 2.65 Insert Commands Menu

2.6.13 Adjusting Column Width

Excel creates new sheets with a default column width, and if that width isn't enough, you can adjust the width of individual columns, groups of columns, or all columns at once. You can also change the default width so that all columns with unchanged widths get a new default width. Alternatively, you can let Excel automatically adjust the width of the selected columns, and Excel will base this adjustment on the longest entry in each column.

You can also use column width changes to temporarily hide columns, and you do this by setting the column width to zero. Hiding columns is safer than deleting them because Excel keeps the data and allows you to unhide it. To change the width of one or multiple columns, simply select one cell per column in any row. You can also select the column headers.

Resizing by Dragging

A quick but less precise way to adjust column width is to drag the right edge of the column header left or right with your mouse. A guide line will appear to help you see whether the cell contents fit, and if you select multiple columns—even nonadjacent ones—all selected columns will resize evenly.

To do this on a touchscreen, drag the column boundary using your finger on the round handles or use a pen on the edge of the column header after tapping it.

	A	B
1	Sales	
2		
3		Jan
4	Sports goods	$10,000
5	Toy	$12,000
6	Leisure dress	$13,000
7		
8		

Figure 2.66 Widening Columns on Touchscreen

To adjust all columns at once, click or tap the **Select All** box and drag the edge of any column header with your mouse. All columns will adopt the same width, regardless of their previous size. On a touchscreen, open the context menu on a column header and select **Column Width**, which will appear when you tap the arrow at the end.

The Format Command

To set exact widths, use the **Home • Cells • Format** command and select **Column Width**. Enter a value between 0 and 255 here to specify how many characters should fit in the column. This value always refers to the selected default font and size. You can also work with decimal places here.

You can use the **Format • Default Width** command for two different purposes. First, you can reset the width of columns that have been changed to the current default width or a new default width. To do this, include those columns in your selection. Second, if you select a column that hasn't been resized instead of one that has, you can adjust all untouched columns to a new default width while leaving the others unchanged.

To set the optimal column width, double-click or tap the right edge of the column header. This method corresponds to the **Format • AutoFit Column Width** command, and please note that it can cause unwanted results if the column contains headers that are longer than the other entries. If it does, then it's usually better to let the header extend into the adjacent cell and adjust the column width manually.

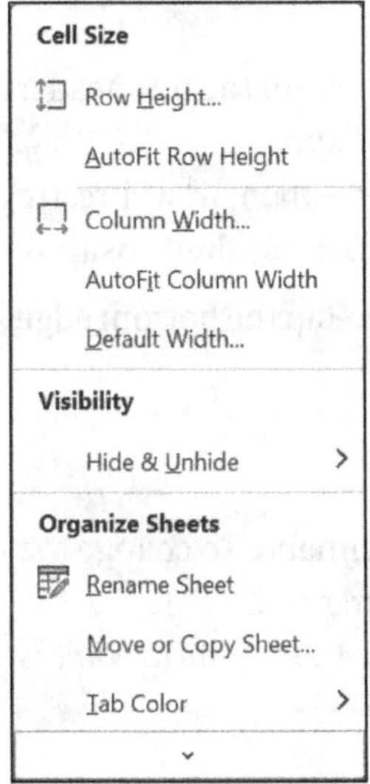

Figure 2.67 Changing Column Width Using Format Command

2.6.14 Hiding and Unhiding Columns

If certain columns aren't relevant to a specific task or if some users shouldn't see data in a given column, then you can hide those columns from view and printing. To do this, drag the column header border left until the column disappears. You can also do this with a finger or stylus on a touchscreen, and a double line between column headers indicates the spot.

Alternatively, you can use the context menu. Right-click or press and hold the selected column headers to open a menu with the **Hide** command. The touchscreen menu arranges most commands horizontally, with only those behind the arrow shown vertically. If you select two column headers with hidden columns between them, you can unhide those columns via the context menu. You can also find these commands under **Format • Visibility • Hide & Unhide**.

2.6.15 Changing the Row Height

Excel sets the default row height to fit the chosen default font at the selected size. When a larger font is used in a row, the row height adjusts automatically, and if the font size is reduced, then the row returns to its default height.

You can also change the height of individual rows, groups of rows, or all rows at the same time. You can also let Excel automatically optimize the height of selected rows, and the program will adjust them based on the largest font in each row. You can also use row height changes to temporarily hide rows. To change the height of one or more rows, simply select one cell per row in any column. You can also select the row headers.

Dragging Row Borders

The quickest way to change the row height is to drag the bottom edge of the row header up or down with your mouse. The current point and pixel count appears as you drag, and if you select multiple rows—including noncontiguous ones—then all will resize evenly. On a touchscreen, tap the row header and drag its border to the desired position.

To set the optimal height for a selected row, double-click or double-tap the bottom edge of its header.

Commands for Adjusting Row Height

The last step corresponds to the **Format • AutoFit Row Height** command. To change the height of all rows, click or tap the **Select All** box. To ensure uniform row heights, use the **Format • Row Height** command and enter a value between 0 and 409 points (a *point* is a unit of measure used for font size).

2.6.16 Hiding and Unhiding Rows

Rows can be hidden just as easily as columns. To hide a row, drag its bottom border upward until it disappears.

Alternatively, you can use the context menu to hide rows. Right-click or press and hold selected row headers with your finger or pen to open a menu with the **Hide** command,

and if you select two row headers with hidden rows between them, you can unhide those rows from the context menu.

2.7 Efficient Workbook Management

To make a workbook a useful tool and organizational aid, you need to arrange the sheets in a clear order. This order might be chronological, with cost sheets for each month; spatial, by department or sales region; or type based, by personnel category or topic. Multiple well-organized workbooks are better than one large, cluttered one, and tables linked by formulas should be grouped into the same workbook whenever possible.

2.7.1 Workbooks as Organizational Tools

One advantage of using workbooks is that you don't need to load the related documents in them separately. Workbooks also make it easier to share documents because it's much simpler to share one workbook than a collection of separate files. Workbooks also support collaborative editing of tables, which makes it easy to create and maintain a consistent table layout across multiple tables in a workbook.

You can create workbooks in various ways. You can either start with a set number of sheets or let the workbook grow by adding sheets over time, and you can also insert sheets from other workbooks at any time.

2.7.2 Adding Worksheets

If the sheets in your workbook aren't enough, you can add new ones anytime. To add standard sheets at the end of your workbook, simply click the [+] button to the right of the sheet tabs.

Figure 2.68 New Sheet Button

Alternatively, you can use the **Home • Cells • Insert • Insert Sheet** command. To insert a sheet between existing ones or choose a different sheet type, right-click the sheet before which you want to insert the new sheet or press and hold the tab with the finger until the context menu appears. Then, choose **Paste** from the context menu, which will open a dialog box with two tabs.

To add a new worksheet, select **Worksheet** on the **General** tab. You can also insert charts as separate sheets, and if you select data for the chart first, it will appear immediately on the new sheet.

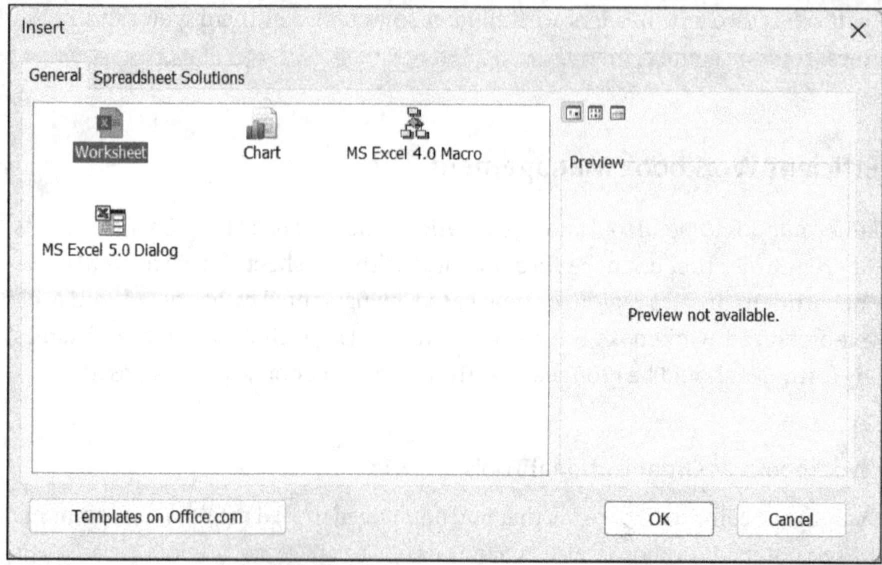

Figure 2.69 Insert Sheet Dialog

The dialog box also offers templates on the **Spreadsheet Solutions** tab. For example, you can insert an invoice form that's available as a template. Click the **Templates on Office.com** button to access Microsoft's template collection.

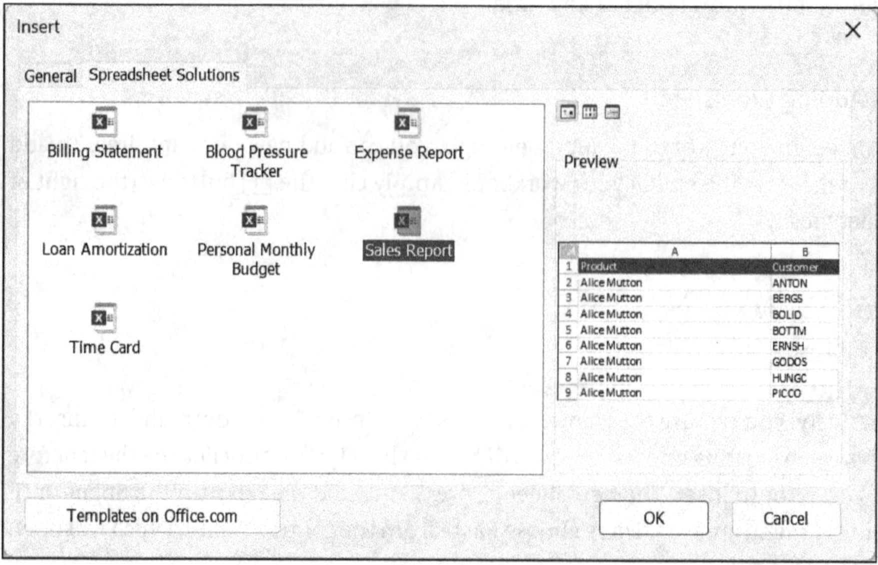

Figure 2.70 Spreadsheet Solutions Tab for Inserting Worksheets

Here's an example of a new sheet based on a template.

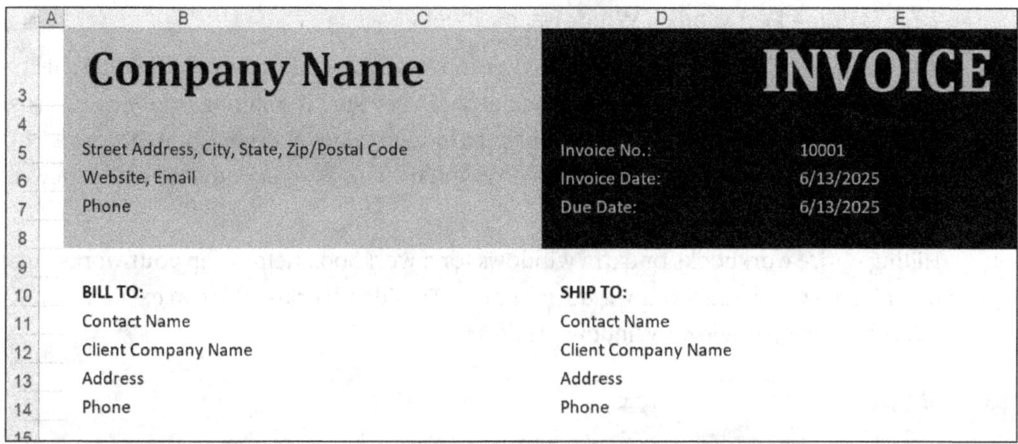

Figure 2.71 New Sheet Based on Template

2.7.3 Inserting Windows

When you're working with large workbooks, it can sometimes be helpful to use multiple windows to view sheets, when you're doing things like comparing data across different sheets. To let you do this, Excel provides the **View • Window • New Window** command. When you select it, Excel opens a second application window and displays the current workbook there as well.

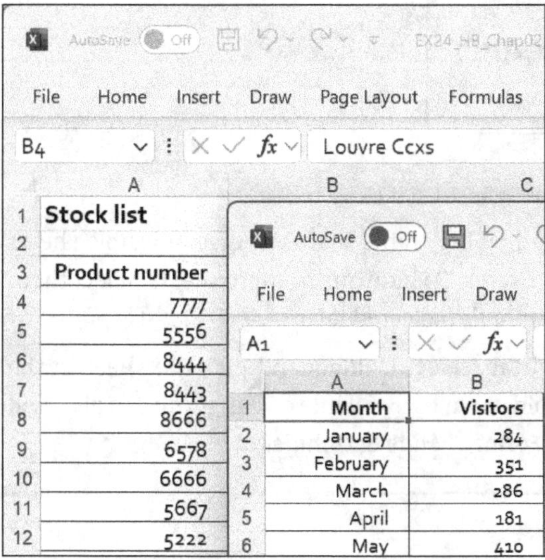

Figure 2.72 Example of Workbook Showing Different Sheets in Separate Windows

You can navigate independently in this new window, separately from the first. To differentiate them, Excel adds a number to the file name in the title bar. When you no longer need the extra window, close it using the **Close** icon.

2.7.4 Hiding Workbooks, Windows, or Sheets

You can hide individual workbooks, extra windows for a workbook, or individual sheets within a workbook during a session as needed. They stay in main memory so macros can still use them, and formula references to hidden tables also continue to work without any issues. However, you can't make manual changes to documents while they're hidden.

Hiding entire workbooks or extra windows for a workbook helps keep your workspace organized by showing fewer windows at once. To hide a workbook or an extra window, select it, then go to **View • Window • Hide ❶**.

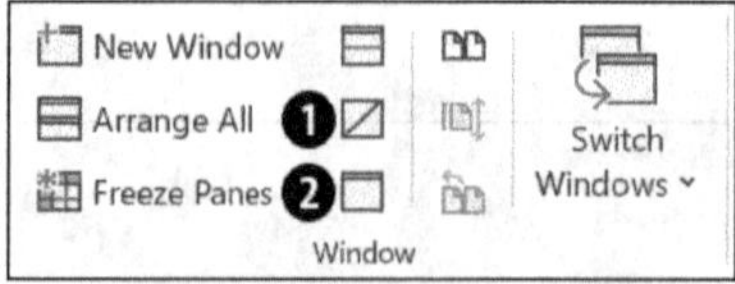

Figure 2.73 Window Group on View Tab

If hidden workbooks or additional windows are open during a session, you can open a dialog box via **View • Window • Unhide ❷** to select from a list of hidden workbooks or windows. You can unhide hidden workbooks or windows individually from there.

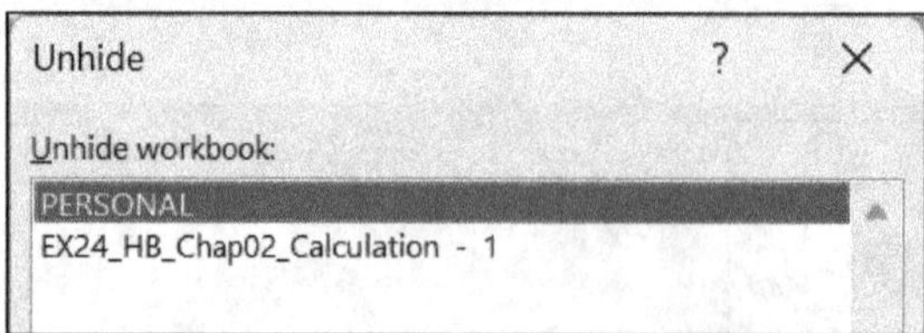

Figure 2.74 Unhide Hidden Workbooks

Hidden sheets are useful for storing named constants that formulas in visible sheets reference, such as VAT rates or exchange rates. To hide one or more sheets, select them and choose **Hide** from the context menu in the selected tabs.

To unhide sheets, right-click any sheet tab, select **Unhide**, and pick the sheet in the dialog box. To select multiple sheets, hold down `Ctrl` while clicking. You can also find these commands under **Home • Cells • Format • Hide & Unhide**.

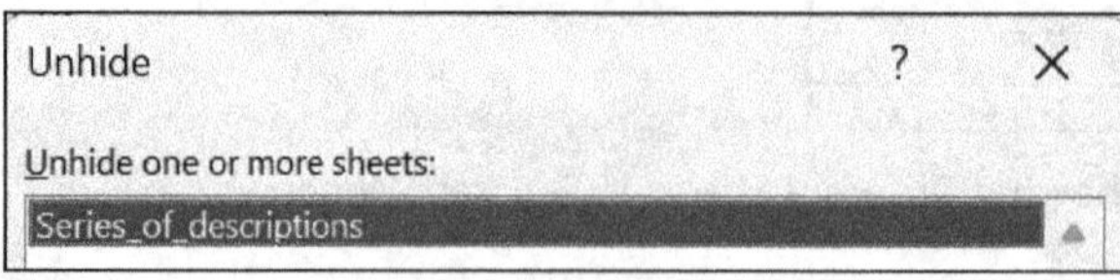

Figure 2.75 Unhiding Hidden Sheets

2.7.5 Deleting Unnecessary Sheets

If you no longer need certain sheets, you can delete them anytime. However, deleting sheets may break cell references in formulas on other sheets, so you should only delete sheets that aren't referenced by formulas elsewhere in the workbook. Here's how you delete sheets:

1. Activate the sheet or select the sheets you want to delete.
2. From the context menu, choose **Delete**, or on the **Home** tab, select **Cells • Delete • Delete Sheet**.
3. If the sheets contain data, you'll be warned before deletion that this action is final. Click **Cancel** if you selected the wrong sheet by mistake.

2.7.6 Rearranging the Order of Sheets

To rearrange sheets, drag the sheet tabs with your mouse to the desired position. This works for a single sheet or a group of selected sheets. The mouse pointer changes to a small sheet or stack icon, and a small triangle shows where the sheets will be placed. When you select multiple nonadjacent sheets, they will be placed consecutively at the new location.

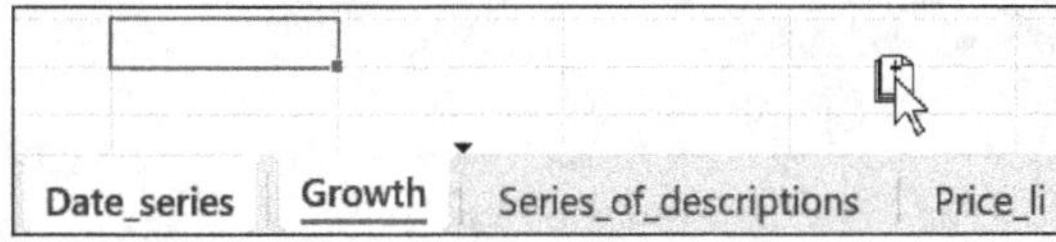

Figure 2.76 Moving Stack of Sheets

On a touchscreen, you can move sheets in **Touch input** mode by using the context menu and the **Move or Copy** command. The pen works like the mouse pointer.

2.7.7 Copying Sheets

Copying a table within a workbook is often useful, especially if it contains a form you want to reuse. First, select the sheet or sheets again, and then, while holding down the `Ctrl` key, drag the selection to where you want the copy to appear.

Excel automatically adds index numbers to new sheet names to avoid duplicates in the workbook, so it's best to replace these with appropriate names right away.

To move or copy sheets, you can use the tab's context menu and select **Move or Copy**, or you can go to **Home • Cells • Format** and choose **Move or Copy Sheet**. In the dialog box, select where to insert the copy under **Insert before:**. To create a copy, check **Create a copy.** You can also select other workbooks or a new workbook as the destination from the **To book:** list box.

Move or Copy
Move selected sheets
To book:
EX24_HB_Chap02_Calculation.xlsx
(new book)
EX24_HB_Chap02_Calculation.xlsx
Income_expenses_Feb
Sport_results
Stock_list
Trend
Rows
Date_series
Growth
Create a copy

Figure 2.77 Dialog Box for Copying or Moving Sheets

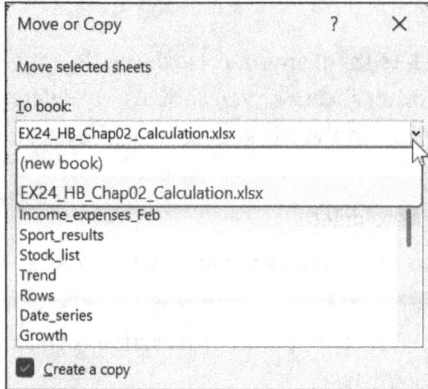

Quickly Rearranging Workbooks

Suppose you've created several workbooks but you don't like their organization anymore because many sheets are in the wrong workbook. To quickly solve this problem, open two or more workbooks and then select one of the options under **View • Window • Arrange All** so all workbook tabs are visible at once. Then, simply drag the tabs with your mouse into the workbook where the sheets belong.

2.7.8 Navigating Large Worksheets

When you compare a table on paper to one on a screen, the paper table—if large enough—has an advantage that's been hard to replicate on screen. Specifically, when you're viewing a large paper table, your gaze can shift to any spot on it in a fraction of a second—but on a screen, you often need to scroll the visible area across the entire table. Despite the availability of scroll bars, mouse movements, and useful keyboard shortcuts, scrolling can be a tedious task, especially on small monitors.

Zooming

The **Zoom** feature offers a solution, with a dedicated slider at the bottom right corner of the window. Dragging the slider or clicking the **+** or **–** icon uniformly enlarges or reduces the grid and its contents within the window. In addition, you can click the percentage display to opens the **Zoom** dialog box, and you can also open the dialog by selecting **View • Zoom • Zoom**.

Excel offers various zoom levels to choose from (see Figure 2.78). You can use **Fit selection** to adjust a previously selected area to fill the entire active window, and you can also zoom continuously. To do the latter, under **Custom**, set zoom percentages between 10% and 400% of the original size.

You can also zoom by using the mouse wheel if the **Zoom on roll with IntelliMouse** option is enabled under **File • Options • Advanced** in **Editing options**. Once you've zoomed the table, you can continue editing it as usual. This function doesn't affect the table's printout because print scaling is controlled through the settings under **File • Print** or the options in the **Page Setup** dialog box.

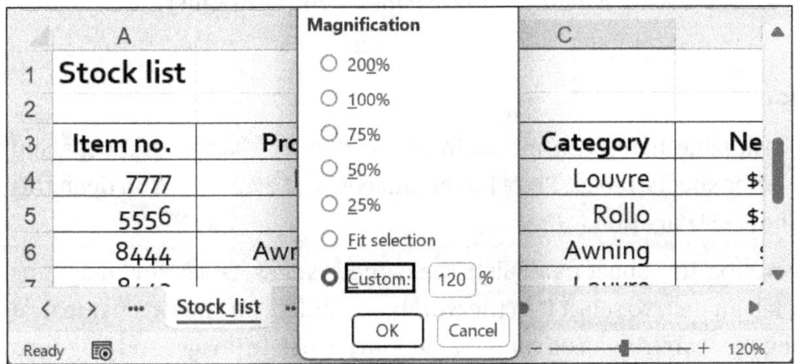

Figure 2.78 Zoom Slider and Zoom Dialog

On a touchscreen, you can zoom in or out by touching the screen with two fingers and spreading them apart or pinching them together.

Freezing Labels

When you're working with large worksheets, you might not see the column and row labels when you scroll the view. Freezing labels helps here. In this example, you'll freeze the first column and the first three rows. Place the cell pointer in cell B4, just to the right of the column(s) and below the row(s) you want to keep visible when scrolling, and then choose **View • Window • Freeze Panes • Freeze Panes**. Clicking this button opens a menu that also lets you freeze only the top row or the first column (see Figure 2.79).

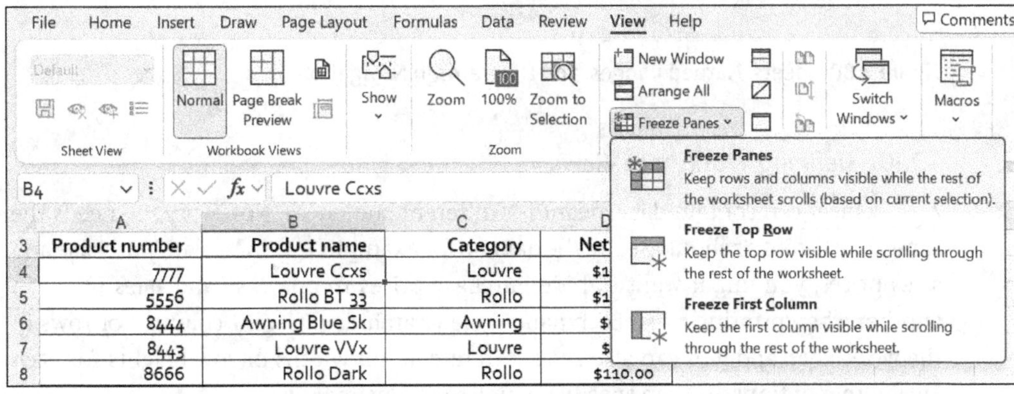

Figure 2.79 Freezing Table Headers

You can apply freezing only once per worksheet, and if you work with two or more large tables, it's best to spread them across multiple sheets. You can still move the cell pointer to the frozen areas to make changes. Note that freezing columns or rows doesn't affect printing. To achieve a similar effect when printing, set the options under **Page Layout • Page Setup • Print Titles** and specify the rows or columns to repeat on each printed page. To unfreeze panes, go to **View • Window • Freeze Panes • Unfreeze Panes.**

2.7.9 Navigation

The **Navigation** task pane helps you maintain an overview in large workbooks and quickly access specific sheets, tables, PivotTables, images, and diagrams. You open this area via **View • Show • Navigation**.

Clicking the arrow next to a sheet name lists the named ranges, tables, and diagrams inside it. Click or tap an item to select it in the workbook, and if you know its name, you can find it by using the search box. You can rename, hide, or delete images and diagrams through the element's context menu.

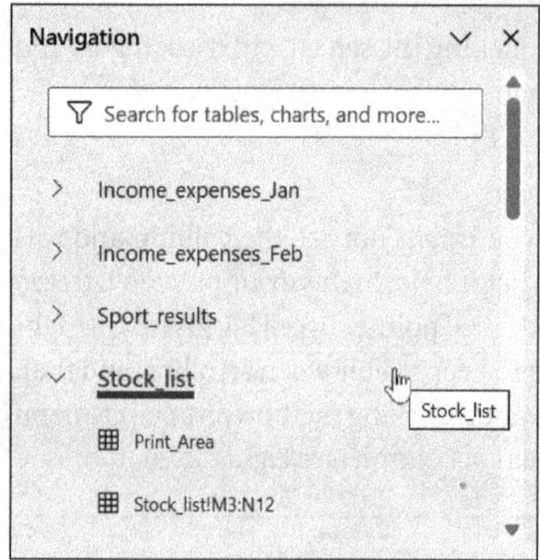

Figure 2.80 Sheets, Named Ranges, and Diagrams in Navigation Area

2.7.10 Defining Worksheet Views

A worksheet can contain data meant for different audiences. Not everyone needs the same data layout or should see all the data, so for example, in a sheet with purchase and sales prices, you might want to hide purchase prices in one case and sales prices in another when printing a list. Each named view can hide different columns or rows in the worksheet, and you can also set different column widths or row heights for each view. Other options you can specify include the following:

- Screen settings, like turning off gridlines or page break lines
- Highlighted cells

- Window size and position
- Window panes
- Fixed rows and columns
- Filter settings for data lists
- Defining a print area or other print settings

However, there's a major limitation: The workbook cannot contain any areas that are formatted as tables. This significantly reduces the feature's usefulness. The **Custom Views** button will then be disabled.

2.7.11 Defining a View

To define a view, follow these steps:

1. Activate the sheet for which you want to create a view. (Views for groups of sheets are not supported.)
2. Choose the display configuration and print settings you want to apply to this view.
3. Click the **View • Workbook Views • Custom Views** button, which will bring up a dialog box that displays a window on the left showing any previously defined views. Clicking the **Add** button opens another dialog box.
4. Enter a name for the current view. Multiword names are allowed.
5. You can also check or leave unchecked two boxes that specify whether to save the print settings and the hidden rows/columns and filter settings that are included in the current view.

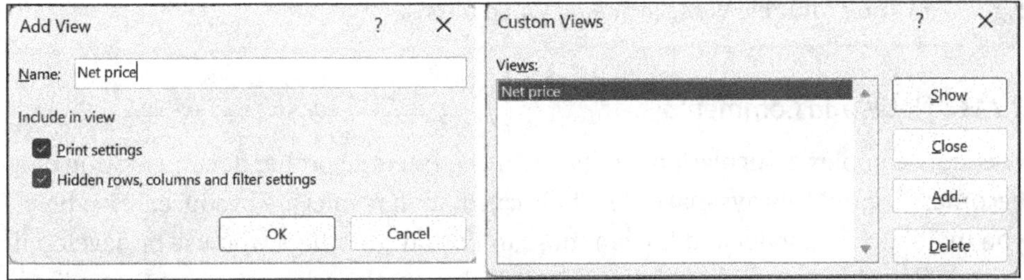

Once you define a view, you can display it anytime. To do this, select the name in the dialog box under **Views** and then click **Show**. When working with views, it's usually best to define the table's "Normal View" as one of the available views first.

2.7.12 Views in Shared Workbooks

When users are collaborating on a workbook, Excel lets each user customize table views individually—applying filters, sorting, conditional formatting, or hiding columns.

To create your first view, prepare the table accordingly, go to **View • Table View,** and select **New Table View** ❶. The text box will show **Temporary View** instead of **Standard View**, and if you're sure the view is useful, you can use **Keep Table View** ❷ to save it in the workbook, where it will be named **View1**. Otherwise, choose **End Table View** ❸. After saving the first view, use the **Table View Options** button to manage and switch between views ❹. Row and column headers will display in a different color ❺ to highlight the views, and the sheet tab will show an eye icon ❻.

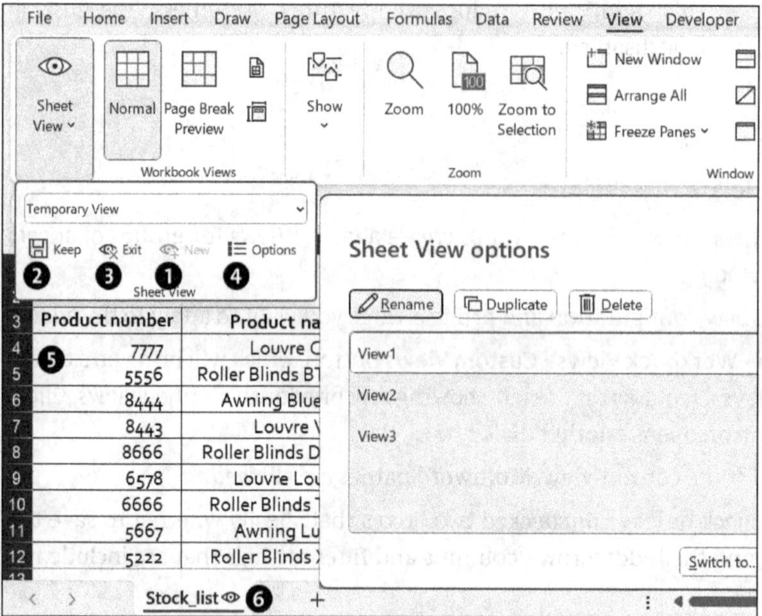

Figure 2.81 Tools for Table Views in Shared Workbooks

2.7.13 Notes and Comments

The more complex a calculation model, the more essential proper documentation of it becomes. It should always make clear how calculation results are produced. This helps you maintain the model and is even more important to others who use or develop it further. You also need to consider regulations that ensure calculations can be audited, such as in accounting.

By consistently naming key table areas clearly and using range names in formulas, a workbook largely documents itself. If that's not enough, you can add necessary information about the entire table, specific calculation methods, and notes on data maintenance and updates in an easily accessible part of the table.

Text Passages and Text Boxes

Excel offers many options for this, and a common method is to add explanatory text in free cell ranges or on a separate sheet in the workbook. Since a cell can hold over 30,000 characters and cells can be merged, it's easy to insert explanatory text directly into the worksheet. Figure 2.82 shows a text passage next to a table where a block of cells has been merged into a single cell.

	A	B	C	D	E	F	G
1	**Customer Classification**						
2							
3	Customer	Revenue	Classification				
4	Berger	600	C		Customers with a revenue of more than $ 5000 are classified in B, customers over $ 50000 in A..		
5	Wehner	1200	B				
6	Brech	220000	A				
7	Schub	60000	A				
8	Schleimin	20000	B				
9	Vosken	900	C				
10	Erben	1300	B				

Figure 2.82 Text in Expanded Cell

For shorter texts, you can also use a text box you create with the **Text Box** button in the **Insert • Text** group. Drag the box to the desired size with your mouse, and when you release the button, you can type text directly into the box. The text automatically wraps within the fixed frame, and you can click a cell outside the text box to end text input.

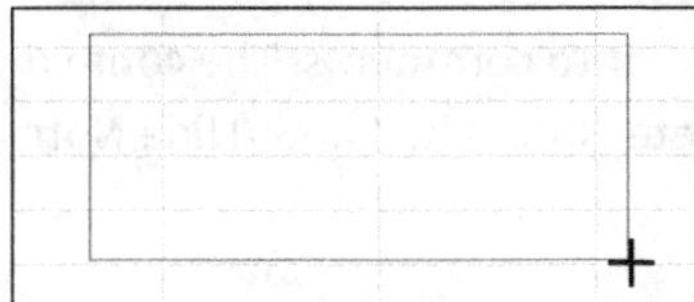

Figure 2.83 Drawing Text Box

You can move these text boxes independently of the table data. If a text box contains instructions for editing the table, like rules for correct data entry, you can move it anywhere on the worksheet where you're working.

Working with Notes and Comments

For individual cells, notes and comments are ideal for explaining the table. Notes attached to a cell usually appear when you hover your mouse pointer over that cell. Here's a brief example of the income-expense calculation from Section 2.1.

	Jan-25	
Expenditure		
Rent and other housing expenses	$3,000.00	
Expenses for car	$500.00	**Helmut Vonhoegen:** Insurance Tax Consumption
Public transport expenses	$250.00	
Insurance premiums	$600.00	
Contractually fixed savings rates	$400.00	
Interest and repayment rates	$340.00	

Figure 2.84 Note on Entered Value

Entering a Note

To add extra information to your table, enter a new comment as follows:

1. Select the cell where you want to add a note. (If you select a range, the note indicator will attach to the first cell.)
2. Go to **Review • Notes • New Note** or press [Shift]+[F2]. Excel will open a text box and insert the user's name at the top. This is especially helpful when collaborating on tables.
3. Enter your note directly into the cell, the formula, or elsewhere.
4. Close the note by clicking or tapping outside the field.

To correct something, select **Edit Note** from the cell's context menu. Then, place the cursor where you want to insert text or drag to select text you want to replace. To finish editing, choose **Exit Text Editing** from the note's context menu. The note box will remain selected.

To move the note box away from the cell, drag its frame in the desired direction. The arrow pointing to the small red triangle will adjust accordingly. Resize the note box by dragging the handles on its frame.

To help you work efficiently with notes, the **Review • Notes** group offers options to navigate existing notes, show or hide them, or convert them into comments. The context menu of a cell with a note provides the **Edit Note**, **Delete Note**, and **Show/Hide Note** commands.

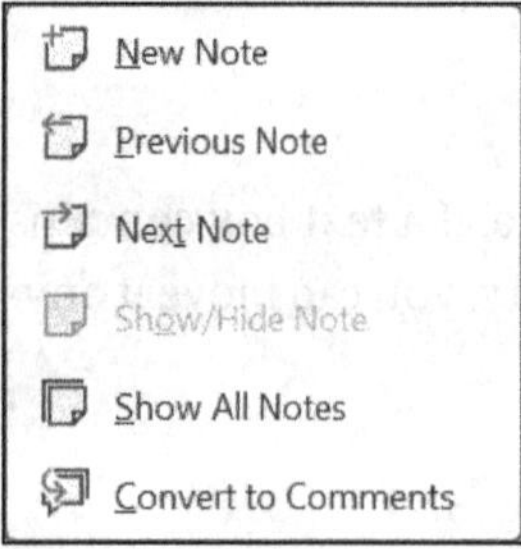

Figure 2.85 Notes Group

Working with Comments

Comments are especially useful when users are collaborating on spreadsheets—when multiple people contribute data to a sheet or review it, or when they must consider specific details during data entry. Unlike simple notes, comments are designed to start a *thread*—which is a conversation with others. They're especially useful when you share a workbook and collaborate with others. Read more about this in Chapter 14.

Adding comments to a cell or cell range works much like adding notes. Follow these steps:

1. Select the cell where you want to add a comment. (If you select a range, the comment indicator will attach to the first cell.)
2. Go to **Review • Comment • New Comment**, where Excel opens a text box and inserts the user's name at the top. This is especially helpful when you're collaborating on tables.
3. Enter your comment directly in the cell, formula, or elsewhere, and close the comment by clicking or tapping the arrow.
4. A reply box will immediately appear where you can respond to the original comment. The ellipses menu lets you delete the thread entirely or resolve it by closing the discussion without deleting it. The pencil icon lets you edit the comment.

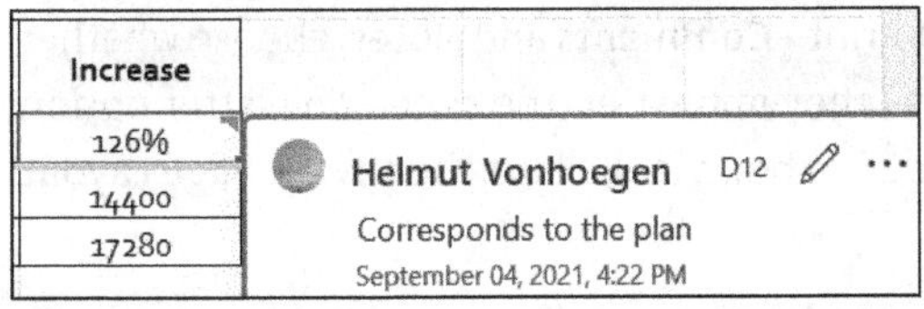

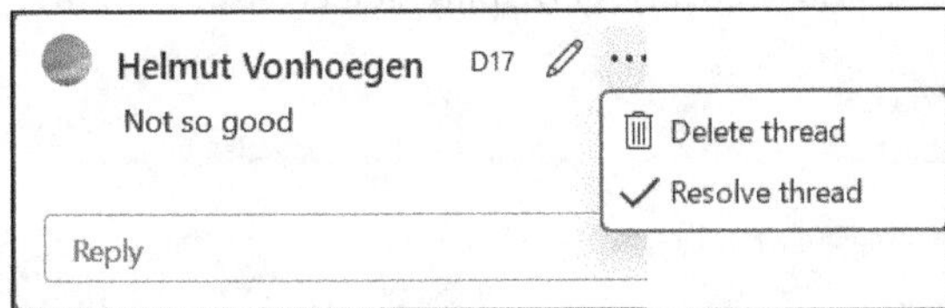

5. Click the **Comments** button above the ribbon to open the **Comments** pane, where all comment threads appear separately from the sheet (see Figure 2.86). Comments stay hidden on the sheet until you close the pane.

The context menu of the cell with a comment then offers the **Reply** and **Delete Comment** commands.

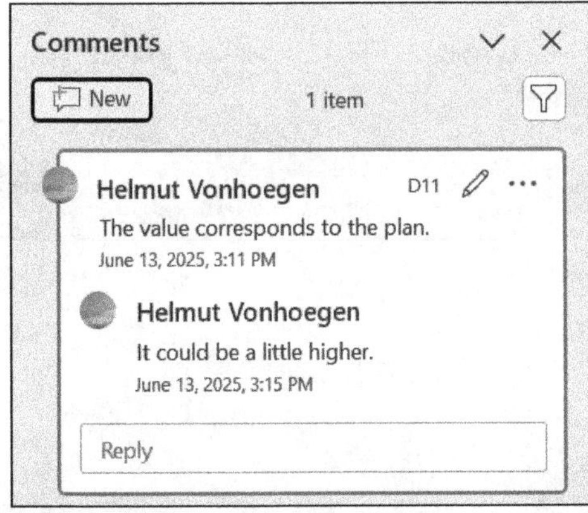

Figure 2.86 Comments Pane

The **Review** group's **Comments** section offers several icons to help you easily navigate, show, hide, or delete comments.

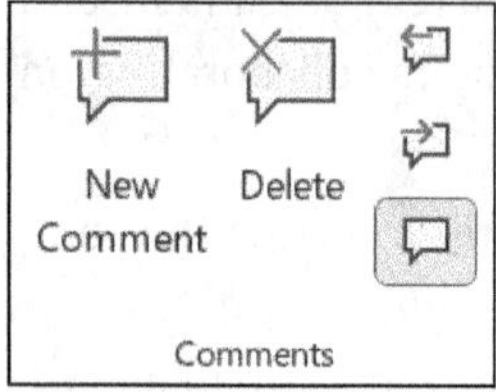

Figure 2.87 Comments Group

You can also print comments and notes along with your table data. In the **Page Setup** dialog box, go to the **Sheet** tab, and under **Print • Comments and Notes,** choose whether to print them at the end of the sheet or as they appear on the sheet. The latter option applies only to notes. You can open this dialog box via the launcher in the **Page Layout** group • **Page Setup.**

You can easily disable the display of notes and comments if you no longer need them. To do this, go to **File • Options • Advanced**, and then, under **Display,** select **No comments, notes, or indicators.**

Chapter 3
Working with Formulas

Excel is useful even when you just want to organize information neatly, as its tabular layout clearly presents data. Whether you need to create a checklist for a business trip, a project calendar, or a staffing plan for a department, you can handle all these tasks efficiently in a single worksheet. Excel's true power shows when you let it perform the calculations for you, and this mainly happens when you enter formulas.

3.1 Formula Structure

Before diving into how to build formulas, we'll highlight an Excel feature that shows calculation results without formulas.

3.1.1 Automatic Calculations in the Status Bar

If you want to quickly find the total of a group of values, you can do it without any formulas. When you select more than one cell, Excel automatically displays values like the average, count, and sum of those cells in the status bar. This works with both contiguous and nonadjacent selections.

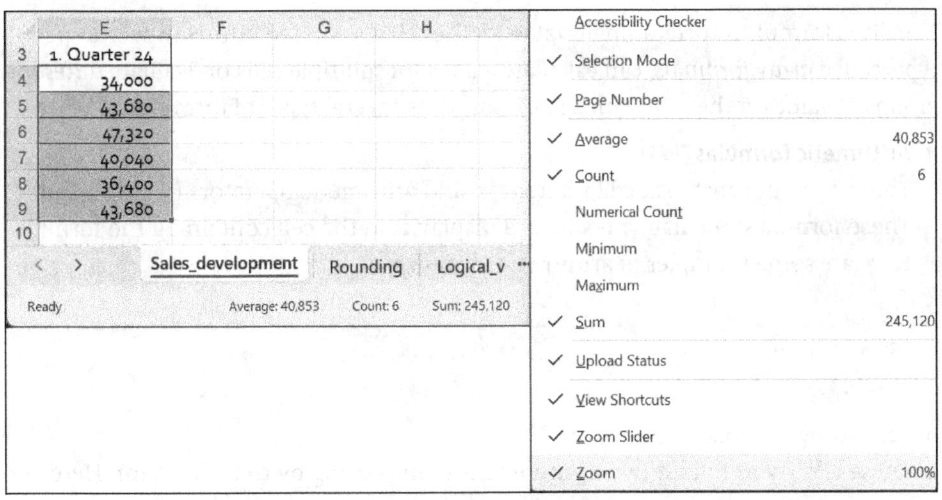

Figure 3.1 Calculations of Selected Cells in Status Bar and Choosing Calculation Type

Instead of the default calculations, you can use the **Customize Status Bar** menu to display a different result. To do this, right-click the status bar and select the calculation type you want, and the menu will remain open until you select a cell or press `Esc`. On a touchscreen, press and hold the status bar with your finger or stylus until the context menu appears, then tap the options to toggle them on or off. Tap any cell to close the menu.

3.1.2 The Role of Formulas

A formula in Excel acts like a standing order to the program. It instructs Excel to repeatedly perform specific tasks. This might be a calculation like *sum a column of numbers* or *multiply the amount by 1.19*, or it could be an instruction like *create a label from the current values of two cells*. Once a formula is created—perhaps with some initial effort—it's then available on demand. If the numbers in the column, the amount, or any of the text strings change, the formulas instantly update with the current result.

Formula Chains

Formulas can use not only numbers and text strings but also other formulas as input. For example, the result of a formula in cell B12 can be used for further calculations in cell F18, and cell F18 will then pass its result to a formula in cell H15. This creates actual formula chains—and the longer the chain, the more likely it is to lose track of how calculations flow through the worksheet. You'll learn how to handle this issue in Section 3.6.

3.1.3 Types of Formulas

Usually, a formula returns a single value for the cell it's entered into. However, a special type, called *array formulas*, can calculate values for multiple cells or evaluate different groups of values at the same time. Excel supports several types of formulas:

- **Arithmetic formulas**
  These contain constants, cell references, and arithmetic operators (+, –, *, /, and ^). These formulas calculate a result and display it in the cell containing the formula. Here are some examples of arithmetic formulas:

  ```
  =B7+C7
  =8000 * 1,19
  =C5/5
  ```

- **Text string formulas**
  These allow you to join two or more text strings using by the & operator. Here are some examples of text concatenation:

```
="Audio"&2011
="Color "&H4
="Margo "&"Lewin"
```

- **Logical formulas**
  These compare constants or cell references by using comparison operators. Formulas like =Z8 > H5 return TRUE if the relationship between the two cell values is true; otherwise, the formula returns FALSE. The displayed TRUE value also has a numeric value of 1, while FALSE equals 0. (And =TRUE + 1 equals 2!)
- **Formulas that include functions**
  These include the sum function:

```
=SUM(F1:F30)
```

3.1.4 Data Types

Different data types can be mixed within a formula, but the results from different parts of a formula must produce compatible data types. For example, a formula that returns a text string cannot be multiplied. That makes little sense, but the & operator accepts numbers as operands and automatically converts them to text:

```
=1111&2222
```

This results in the text string 11112222.

3.1.5 Operators and Their Precedence

The simplest formulas that are stored in a cell link numbers by using basic arithmetic operators. The notation, except for the equal sign, follows the familiar format:

```
=13+25-5
=25*4/3
=3^2
```

Instead of numbers or constants, you can use variables in formulas. In algebra, letters are commonly used to represent variables, and when you're working with Excel, cell or range addresses let you include variables in a formula. Formula operators are processed according to their priority, and the following list shows the priority order, with 1 being the highest.

3.1.6 Operator Table

Operators	Examples	Meanings	Priority
Reference Operators			
:	B3:B7	Range	1
Empty	B3:E8 C4:F12	Intersection	2
@	=[@Column1]	Implicit intersection	2
,	B3:B12,C3:C12	Union	3
#	=D9#	Reference to a dynamic array	3
Arithmetic Operators			
–	-(5*2)	Sign	4
%	10 %	Percent	5
^	5^2	Exponentiation	6
*	5*6	Multiplication	7
/	6/3	Division	7
+	7+4	Addition	8
–	4-2	Subtraction	8
Concatenation Operator			
&	B2&B3	Text concatenation	9
Comparison Operators			
=	B5=B7	Equal	10
<>	B5<>B7	Not equal	10
>	B5>B7	Greater than	10
<	B5<B7	Less than	10
<=	B5<=B7	Less than or equal to	10
>=	B5>=B7	Greater than or equal to	10

Priority determines the order in which Excel evaluates operators. Consider this example: A numerical expression reads this way:

```
=5+6*3-7
```

If the program simply evaluated from left to right, the result would be this:

```
5+6    = 11
11*3   = 33
33-7   = 26
```

But Excel correctly performs multiplication first, since it takes precedence over addition and subtraction:

```
6*3    = 18
5+18-7 = 16
```

The operation with higher priority is executed first, and for operators with the same priority, evaluation proceeds from left to right. To override the priority order, you can use parentheses as usual, like this:

```
=(5*6)*(3-7)
```

Then, the contents inside both parentheses are calculated, and these are the results:

```
=30*-4
=-120
```

3.1.7 Addition and Subtraction

The following list shows examples of correctly entered formulas to demonstrate the syntax. As you can see, date values can also be used as operands in calculations:

```
=12+27
=-15+33
=C17+F18-J21
=B14
=$B16-$C$16
=-(B13+B14+C17)
=SUM(B12:F12)+1000
=SUM(B12:B17)+SUM(C12:C19)
=DATE(2025,12,01)+90
=SUM00+SUM01
```

The last example uses range names, but this is only allowed if each name refers to a single cell. If a range refers to multiple cells, the formula returns the `#VALUE!` error.

3.1.8 Multiplication and Division

When multiplying and dividing, you need to consider how the result handles decimal places. In multiplication, the decimal places add up:

```
=33.33*33.33 results in 1110.8889
=33.33*33.33*33.33 results in 37025.927037
```

If cell formatting limits the result to two decimal places, then the cell displays a rounded value, such as 1,110.89 or 37,025.93. However, the exact result remains stored internally, and when another formula references this cell, it uses the full four or six decimal places for calculations. So, if these values are multiplied again, differences may arise depending on whether the calculation uses the stored precise values or the rounded ones.

Figure 3.2 shows a simple example of this. Column C contains results from column B, which were processed using the ROUND() function. In E4, the unrounded total is multiplied by 17, and in E5, the rounded total is multiplied by the same amount. The difference is noticeable.

	A	B	C	D	E
1	**Example of rounding differences:**				
2					
3	**Formula:**	**Result**	**rounded:**		
4	=12.13*12.6	152.838	152.84	=17*B8	10598.71
5	=13.87*32.77	454.5199	454.52	=17*C8	10598.82
6	=3.66*2.66	9.7356	9.74		
7	=2.12*3	6.36	6.36		
8	=SUM(B4:B7)	623.4535	623.46		
9					

Figure 3.2 Example of Rounding Differences

Preventing Division by Zero

The number of decimal places in a quotient varies widely during division. Also, in addition to the rounding issue that occurs in multiplication, another important problem arises: division by zero is not allowed. If the divisor is entered directly, this error is relatively easy to avoid. But what if the divisor is a cell address or a function with an unpredictable result—meaning the divisor is a variable or unknown?

You need to take precautions in this case. You should prevent the system from dividing by zero, and you can do it with the IF() function, which tells the program what to do instead of dividing when the divisor is zero. It might look like this:

```
=IF(B3<>0,ROUND(A3/B3,2),"")
```

If cell B3 is not zero, it divides A3 by B3; if B3 is zero, the division is skipped, and the cell stays empty. You could also show a relevant message instead.

3.1.9 Concatenating Text

Sometimes it's useful to create a string in a cell by linking different characters or strings with a formula. For example, if you want to expand item numbers by two characters that include the product group, you can enter this into cell C9:

```
=C5&C8
```

Then, the contents of cell C5 will be joined with those of cell C8. If C5 contains product group code *PX* and C8 contains item number *3370086*, then the result in cell C9 will be *PX3370086*. It doesn't matter if the item number in C8 is entered as a number rather than text. The operator automatically converts the number to text. You can then convert this formula to its value. Copy the cells to the clipboard, then paste them back in place by using the context menu's **Paste Special** option and selecting **Values**. To insert a space between two text elements, enter this:

```
=C5&" "&C8
```

Remember to start the formula with an equal sign so Excel recognizes it as a formula. For example, say you just enter this into a cell:

```
C5&C8
```

If you do, Excel will treat it as plain text.

3.1.10 Testing Logical Formulas

Logical formulas check whether specific facts or conditions are true. For example, if you want to decide on an investment based on whether the costs have paid off within five years, you can express this condition as a logical formula:

```
=Costs < Savings
```

This assumes you have named the cell summing the **Costs** and the cell containing the total savings over the last five years' worth of **Savings**. When you enter this formula, Excel compares the values of the two cells, and if the costs are less than the total savings, then the cell displays TRUE. Although it shows TRUE, the cell also holds the numeric value 1. If the costs aren't covered, then the cell shows FALSE, which corresponds to the numeric value 0.

This is handy for checksums. For example, if you enter logical formulas as conditions in three cells that are stacked vertically, you can easily verify whether all conditions are met. Just add the truth values, and if the total is =3, then all conditions are met. Logical formulas have only two possible results: TRUE or FALSE. (In Figure 3.3, logical formulas are also shown as text in column D for documentation.)

	A	B	C	D	E
1	**Example of logical values**				
2					
3	Weight in lbs	264	TRUE	=B3>176	
4	daily Cigarette consumption	40	TRUE	=B4>5	
5	daily Beer consumption in fl oz	100	TRUE	=B5>(17)	
6			3	Highly endangered	

Figure 3.3 Adding Logical Values

A logical formula can include not just one condition but several at once. These conditions can be expressed either alternatively or additively.

3.1.11 Functions

Functions are special expressions that are used either directly as formulas or as parts of formulas. Even if a formula contains only one function, it must still start with an equal sign. Alternatively, functions can be combined with other parts of a formula by using operators.

The following formulas are valid expressions:

```
=A3-SUM(A10:A20)
=B4/ROUND(C6,2)
```

However, this formula is missing an operator:

```
=AVERAGE(costs)G6
```

Functions can replace complex calculations, so you don't need to worry about how they work. You simply provide the necessary arguments, and the program takes care of the rest. A function can have up to 255 arguments, each of which is often cell range containing many values. For example, a sum function can easily add one million values at once by using a range like A3:A1000003.

Functions can also be *nested*, meaning a function's argument can be another function. You can nest up to 64 levels. Chapter 15 provides a brief reference to all Excel functions.

3.2 Entering Formulas and Functions

Follow these rules to create error-free formulas:

- Always start a formula with an equal sign. While you can begin a formula with a plus or minus sign, Excel automatically inserts an equal sign at the start and will ignore the plus or minus sign.
- Always enclose text strings in double quotes.
- Never insert a space between the function name and the opening parenthesis (even though inserting spaces can improve formula readability).

Excel simplifies entering complex formulas by automatically expanding the formula bar to fit the required number of rows. You can use the arrow button at the right end to show or hide these extra rows, and you can also drag the bottom edge of the formula bar downward to create more space. To make very long formulas easier to read, you can insert line breaks by using `Alt` + `Enter`.

=IF(C6>1,"Highly endangered",IF(C6=1,
"endangered",""))
Expand Formula Bar (Ctrl+Shift+U)
B C D E F G

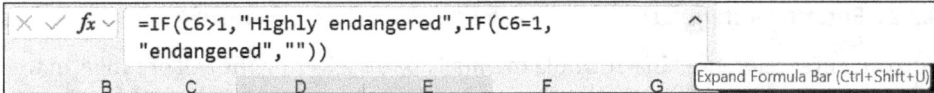

Figure 3.4 Advanced Formula Bar with Complex Formula

3.2.1 Constants in Formulas

When numbers appear as constants in a formula, remember that unlike when you're entering numbers directly into a cell, you cannot use parentheses for negative numbers, commas for separators, or currency symbols. However, you can use the percent sign as a percent operator.

For example, to calculate 10% of the value in cell A1, enter this:

```
=A1*10%
```

You can generate a date or time in a formula by using a function, or you can enter it as a constant. In the latter case, you must enclose the values in double quotation marks—so to calculate the date 100 days after 10/6/24, you enter this:

```
="10/6/24"+100
```

If background error checking is enabled, then because the entry uses a two-digit year, Excel immediately displays a button with a menu that alerts you and lets you convert the year to a four-digit format that clearly assigns it to a century.

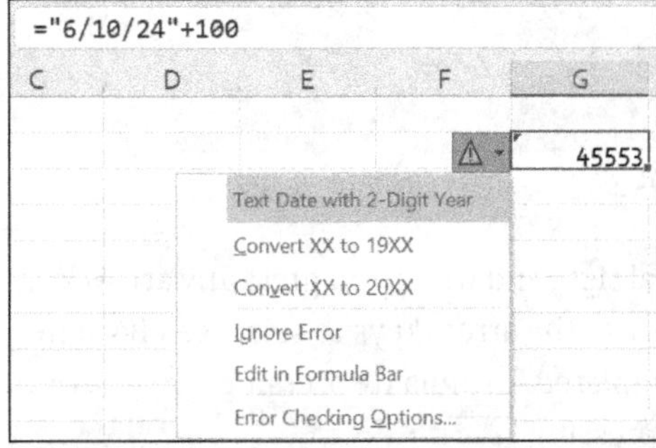

Figure 3.5 Context Menu for Two-Digit Year

Text constants must always be enclosed in double quotation marks within a formula. However, if you use the **Insert Function** dialog, Excel automatically adds the quotation marks when you enter a text string in an argument field. Using the & operator, you can concatenate numbers with text like this:

```
="Founding year " & 2004
```

This is valid syntax and displays `Founding year 2004` in the cell.

3.2.2 Entering References

You can enter references in a formula manually or by selecting the cell or range. In the latter case, Excel automatically switches to **Point** mode, as indicated in the status bar. In this mode, you can keep selecting cells until you enter the next operator.

For example, consider this formula to calculate a simple ratio:

```
=B4/B5
```

To use the **Point** mode while entering a formula, start by typing the equal sign and then click or tap cell B4. The program will switch to **Point** mode ❶, and the cell address will appear in both the cell and the formula bar. If you accidentally select the wrong cell, simply select again, type the division sign, select cell B5, and confirm the formula by clicking the checkmark in the formula bar ❷. If the formula is incorrect, cancel the entry by using the **Cancel** button ❸.

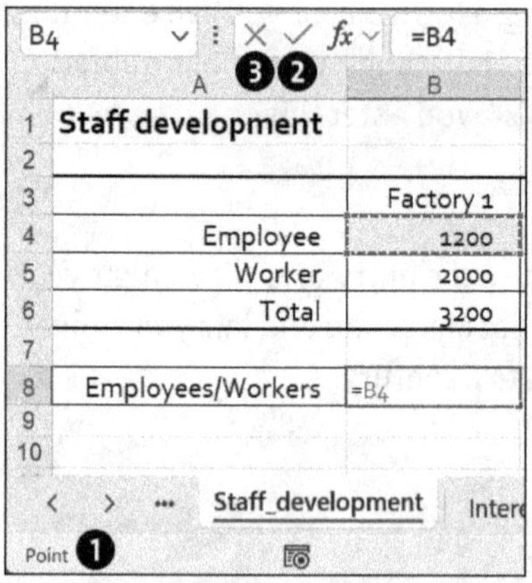

Figure 3.6 Pointing to Range During Formula Entry

The keyboard method is similar. Type an equal sign, and when you press any arrow key, the program will switch to **Point** mode. Then, use the arrow keys to select cell B4, enter the division sign, and so on. Confirm the completed formula by pressing Enter, or if the formula is incorrect, cancel the entry by pressing Esc.

Note that while in **Point** mode, you cannot end formula entry by using the arrow keys or clicking or tapping another cell, as you can when entering values. You can only complete a formula by clicking, tapping, or using the arrow keys if it ends with a closing parenthesis, since entering the parenthesis automatically switches the program back to **Enter** mode.

3.2.3 Range References

If you need a range reference as a formula argument, you have two options.

For option one, if the range isn't named, proceed as you would when entering a cell reference. Instead of clicking a single cell, drag the mouse to select the desired range. If you're using your finger, tap the first cell of the range and drag the round handle to the last cell. If you're using the keyboard, press an arrow key first to enter **Point** mode. Start by selecting the cell where the range should begin and then hold down the `Shift` key while selecting the range with the arrow keys. End the range selection by pressing the `Enter` key, and if you want to continue the formula, end the range selection with a closing parenthesis or an operator. Instead of the `Shift` key, you can also use `F8` to extend the range. Finally, select the starting cell, press `F8`, and select the last cell of the range.

For option two, if a range is named using the methods we'll explain in detail in this section, start by typing something like this:

```
=SUM(
```

Then, you can select the desired range name, like *Factory_1*, via **Formulas • Defined Names • Use in Formula.** Alternatively, you can insert it by pressing `F3` to open the **Insert Name** dialog box, where you can double-click or tap the desired name. Then, you can confirm the formula. The final parenthesis will be added automatically, but this only applies if there's only one closing parenthesis—that is, if the formula contains just one function.

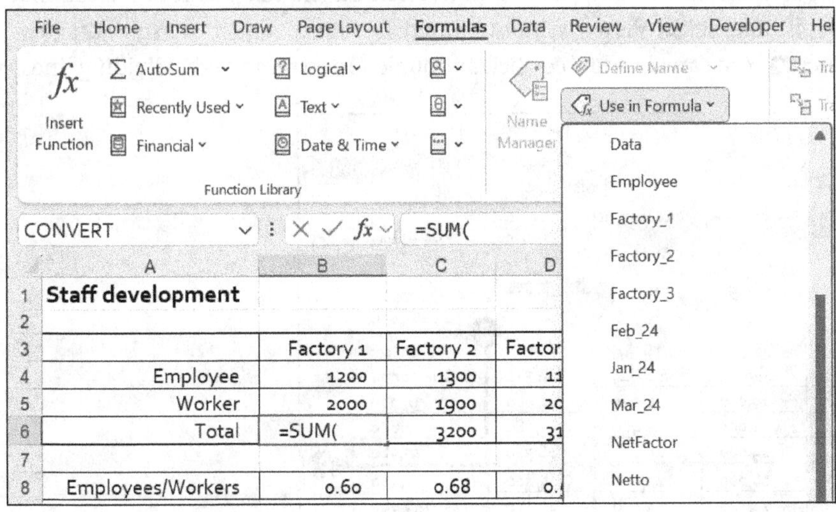

Figure 3.7 Inserting Range Name into Formula

Excel also provides you with extra support for range names when you're entering a formula. As soon as you type the first letter, possible names starting with that letter appear in the function list, as Figure 3.8 shows. Double-click or double-tap a name to insert it into the formula, and if the program selects the correct name, you can insert it by pressing `Tab`.

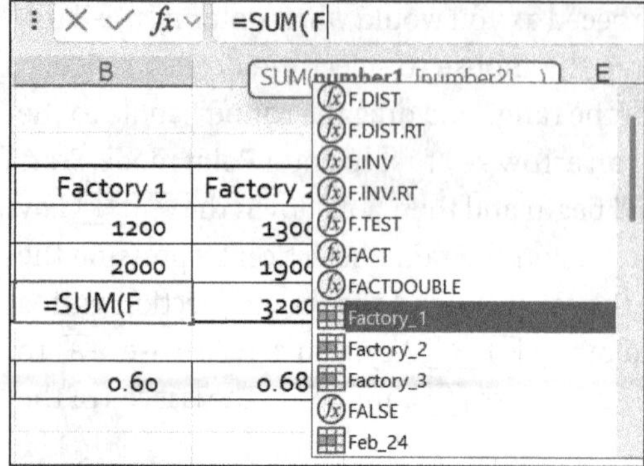

Figure 3.8 Support for Manually Entering Range Names

3.2.4 Tips for Entering References

If an argument refers to a cell or cell range that's far from the formula cell, you can use the scroll bars to adjust the window view and then click the target cell or drag over the desired range. After you complete the formula, the formula cell will return to the view. You can also drag the grid or swipe the screen area with your finger to reveal distant cells.

When cell references are in a distant table area, it's helpful to split the window before entering the formula so that area stays visible. Position the cell pointer on a cell that marks the split line, select **View • Window • Split** ❶, and to adjust the split, drag the dividing lines ❷. You can also double-click or double-tap to remove the dividing lines.

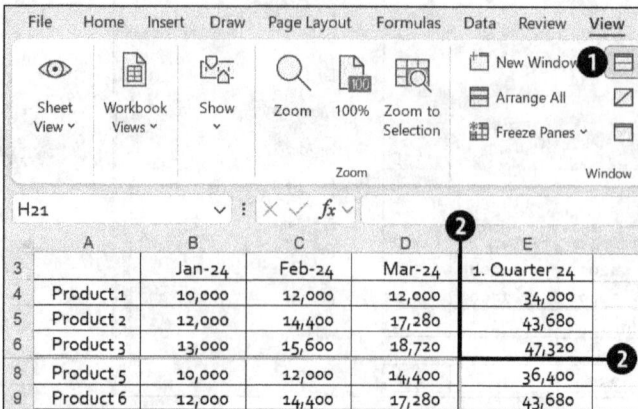

Figure 3.9 Split Window

To view multiple areas of the table simultaneously, you can create additional windows of the active file via **View • Window • New Window**, and you can then arrange them by using **Window • Arrange All • Tiled** so all areas are visible. It's also a good idea to check

Windows of active workbook to hide any other open workbooks and thus maximize the workspace. If it's not checked, the screen will split to show all open workbooks.

References to Other Sheets

To refer to other sheets within the same workbook, place the sheet name before the range address or name and separate them with an exclamation mark. In **Show** mode, if you first click or tap the tab to select the sheet and then select the range, Excel automatically adds the sheet name to the formula.

```
=SUM(Absatzentw!E4:E9)
```

3.2.5 3D References

3D formulas are a special feature of workbooks. Imagine a table that summarizes the results from four similarly structured tables, like the production outputs of four plants. These tables are arranged consecutively in the workbook and share the same structure, and the result for Item A in January appears at the same sheet address in all four tables. Given this, you can create a fifth table that summarizes the results from the four plants, and this table will have the same structure as the others.

In this summary table, you can use a sum formula with a three-dimensional range, which means "Sum the values in cell B7 across the four tables." This is the formula:

```
=SUM(Plant1:Plant4!B7)
```

You can copy this formula into any cell where you want to see the total from the four plants. To enter a 3D reference by pointing with a mouse, first select the sheets and then select the cell or range. Then, click the tab of the first sheet in the range, hold down the `Shift` key, and click the last sheet. Finally, select the cell or range.

Excel automatically adjusts 3D references when sheets are deleted from or added to the sheet area. If sheets are moved out of the area or the entire sheet group is relocated, the references update accordingly. When the last sheet in the group moves further right, the sheets in between become part of the area, and the same applies when the first sheet moves further left.

3.2.6 Entering External References

If a formula refers to values in another workbook, it's best to open that workbook before you enter the formula. To reference cells in the second workbook within a formula, you must switch between windows. The easiest way to do this is to arrange both windows so they're visible at the same time by using **View • Window • Arrange All**. Then, while typing in the formula bar, you can switch to the other window and select the cell references.

When creating a formula in a table that refers to data in another workbook, use this notation:

```
=[WORKBOOK1]Sheet1!$E$2
```

The workbook name must be enclosed in square brackets, and the sheet name must be separated from the cell address or range name by an exclamation mark.

3.2.7 Help with Entering Functions

Excel requires you to use exact syntax when using functions, though it doesn't matter if you use uppercase or lowercase letters. With nearly 500 functions available for table formulas, Excel must support entering these functions. No one can remember all these functions or the many arguments they require, so to save you from constantly checking which arguments a function needs, Excel offers help through on-screen tooltips or the **Insert Function** dialog.

The help section provides function examples and usually provides additional details about how to use the function (see Figure 3.10). You can add a function to a formula in several ways, and Excel assists you with each method.

Examples

Copy the example data in the following table, and paste it in cell A1 of a new Excel worksheet. For formulas to show results, select them, press F2, and then press Enter. If you need to, you can adjust the column widths to see all the data.

Data	Description	
0.06	Annual interest rate	
10	Number of payments	
-200	Amount of the payment	
-500	Present value	
1	Payment is due at the beginning of the period (0 indicates payment is due at end of period)	
Formula	**Description**	**Result**
=FV(A2/12, A3, A4, A5, A6)	Future value of an investment using the terms in A2:A5.	$2,581.40

Figure 3.10 Function Example in Help Section

3.2.8 Manually Entering Functions

If you know the functions well, you can enter them directly into the cell. Excel helps you do this with the **AutoComplete** feature for formulas, as long as it's not disabled under **File • Options • Formulas** in **Working with formulas.** You can also toggle this feature on or off in edit mode by using [Alt]+[↓].

You must start each manual entry with an equal sign, and when you type it and the first letter in a cell, Excel shows functions starting with that letter for you to choose from. Each additional letter you type narrows the list. You can bring up a brief description for every function you select by clicking, tapping, or using the arrow keys. Double-clicking, double-tapping, or pressing the [Tab] key inserts the selected function.

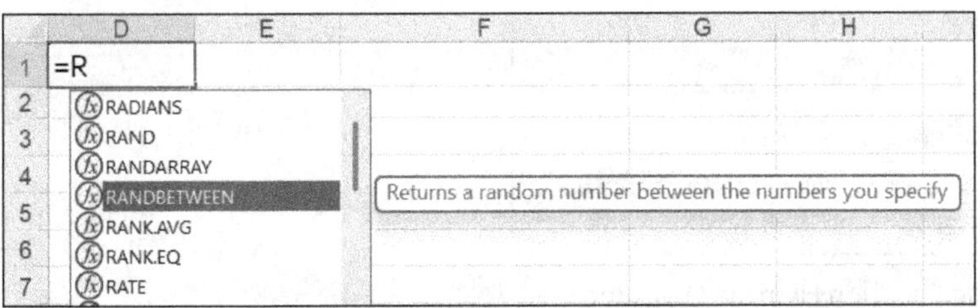

Figure 3.11 List of Possible Functions to Choose From While Typing

If you don't know the exact spelling of a function, you can enter part of its name and the program's autocomplete feature will show you a list of functions that have those letters in their name. For example, typing "=week" will make the program show all functions related to week calculations.

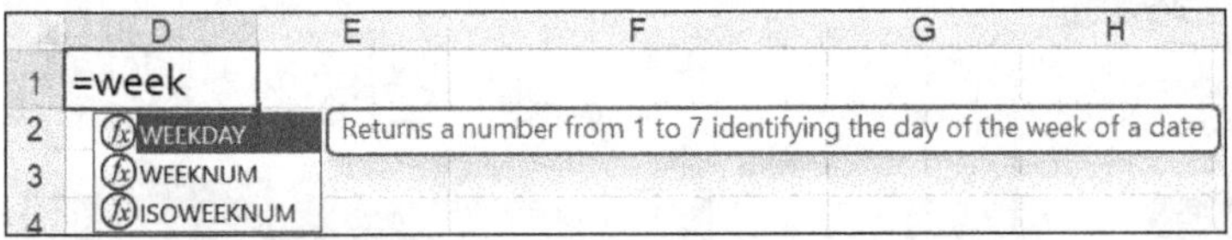

Figure 3.12 AutoComplete Feature Showing Functions with =week in Their Name

When you enter the full function name and the opening parenthesis for its arguments in the cell or formula bar, tooltips appear showing the required arguments, as long as the **Show function ScreenTips** option is enabled under **File • Options • Advanced • Display.**

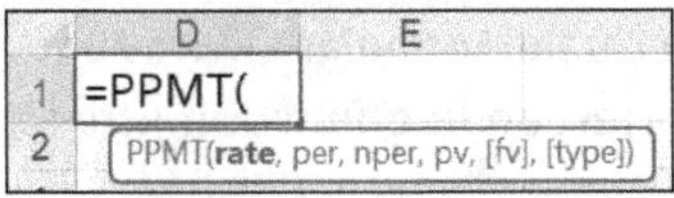

Figure 3.13 Quick Info on PPMT() Function

The elements in the quick info act like hyperlinks, and you can click or tap a placeholder for an argument to select it in the formula and correct it if you need to. The value or values of the argument appear above the formula bar for verification.

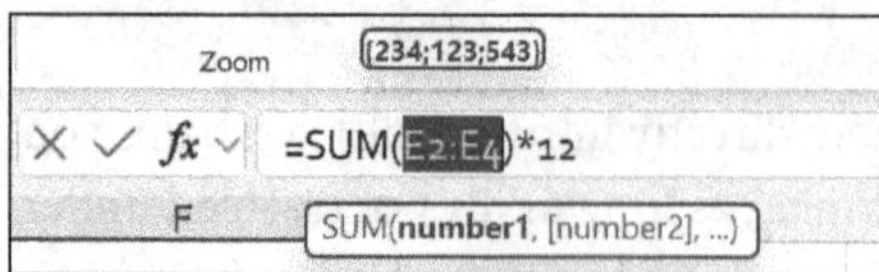

Figure 3.14 Displaying Argument's Values

Clicking or tapping the function name opens the help for that function.

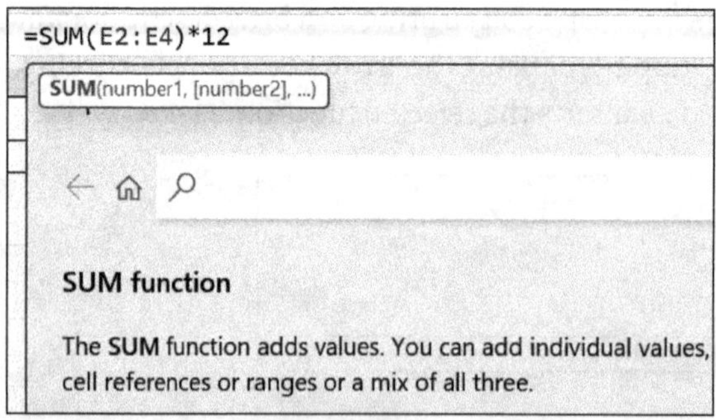

Figure 3.15 Accessing Function Help via Quick Info

If a function expects specific argument values, then these will also be suggested. For example, the AGGREGATE() function requires certain number codes for its first argument, and a list will show the possible codes and their meanings.

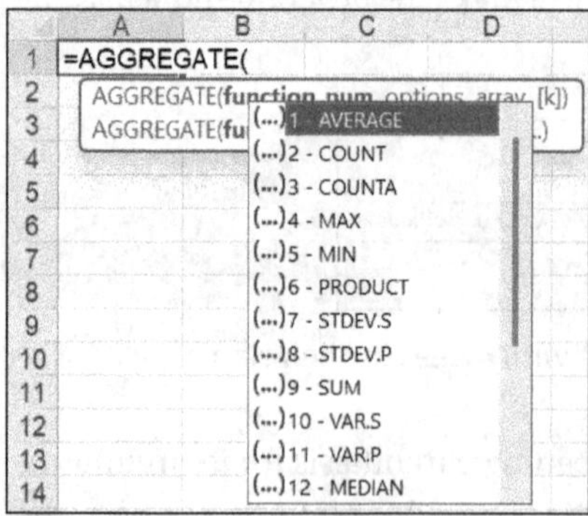

Figure 3.16 Available Codes for First Argument of AGGREGATE() Function

For quick access to specific functions, Excel provides icons in the **Formulas • More Functions** group for key function categories. These show corresponding function lists to choose from, such as the one in Figure 3.17.

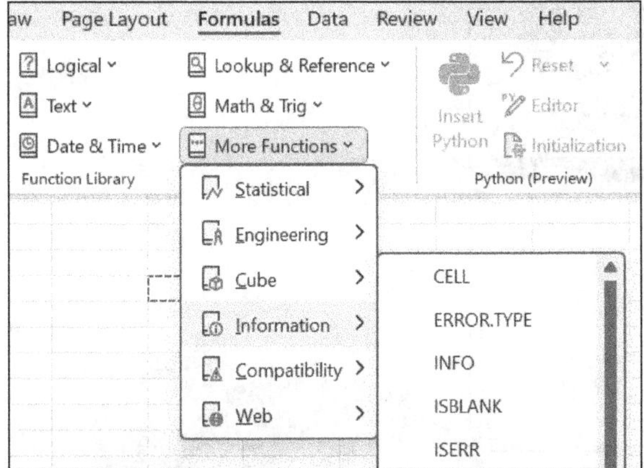

Figure 3.17 Function Library Options

Clicking or tapping a function inserts it into the current cell and immediately opens the dialog box described in the next section for entering its arguments. Recently used functions are accessible via a dedicated button.

3.2.9 Entering Formulas with the Insert Function Dialog

Instead of typing functions directly into a cell, you can use the **Insert Function** dialog. You can select the cell and then click or tap the **fx** icon in the formula bar. Alternatively, you can select **Formulas • Insert Function** or press `Shift`+`F3`. Excel automatically inserts the equals sign at the start of the formula.

Functions are useless if no one knows which one solves a particular problem. For example, which function performs present value calculations? To find out, when you open the **Insert Function** dialog box, you can use the **Search for a function** option. Type "Present Value" and click **Go**, and Excel will list functions for present value calculations under the **Recommended** category.

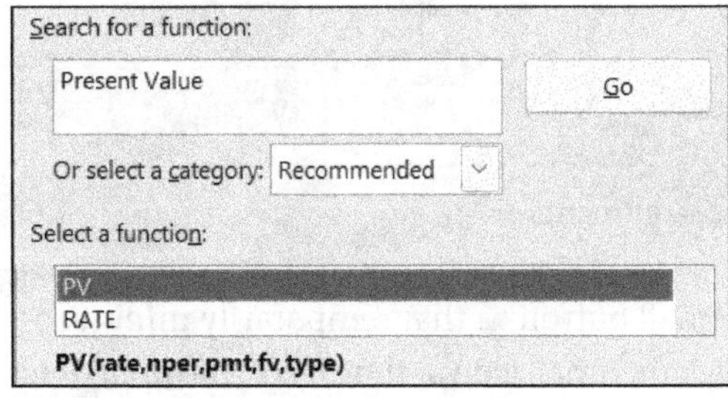

Figure 3.18 Searching for Functions for Specific Purpose

If you already know which function you need, then select the **Financial** category under **Or select a category,** choose the function itself, and use the scroll bars to browse the list of functions. Typing the first letter of the function usually helps you find it faster.

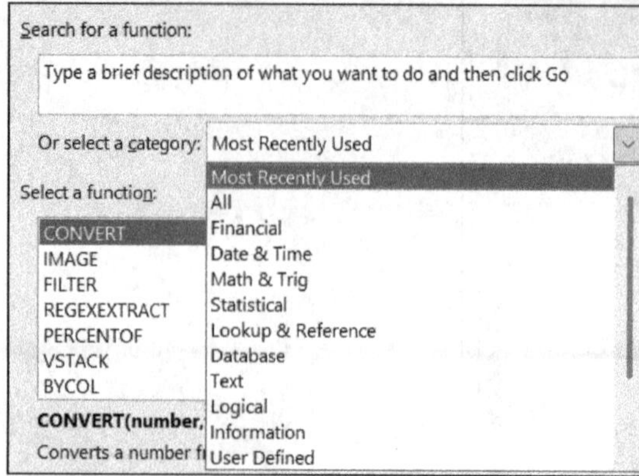

Figure 3.19 Category List

If you're unsure which category the function belongs to, just select **All**. When you highlight a function, a brief description and its syntax appear below, and if it's the right function, click the **OK** button or double-click or tap it to open the **Function Arguments** dialog box. This dialog box provides an input field for each argument, and when the cursor is in a field, you get hints about the type of data needed. Required arguments appear in bold, and you might need to use the scroll bar to see all the arguments.

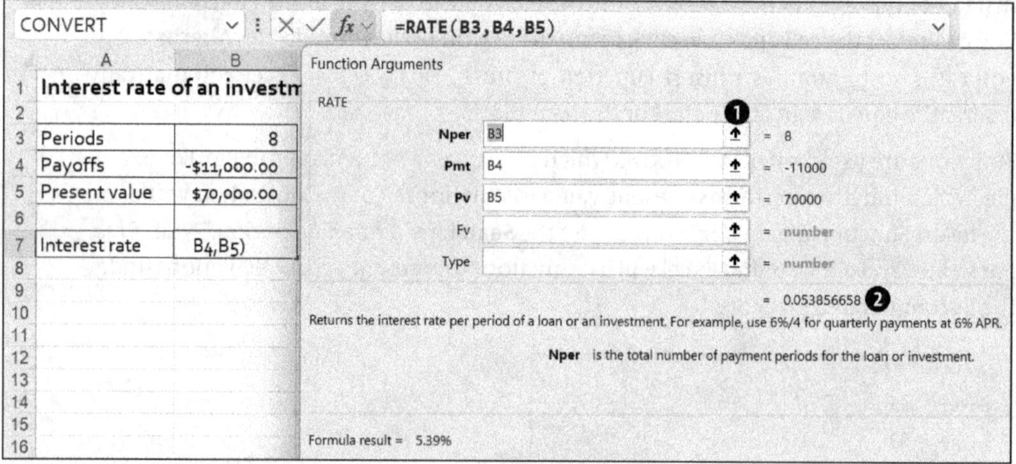

Figure 3.20 Dialog Box for Function's Required Arguments

To enter range references, click or tap the small button ❶ that temporarily minimizes the dialog box, select the range, and then click the small icon at the end of the input field again. It's even easier if you move the dialog box aside and simply select the range on the worksheet containing the needed values. While you're typing, Excel displays the expected formula result ❷ whenever possible and thus helps you catch input errors right in the dialog.

	A	B	C	D	E	F	G
1	**Interest rate of an investment**						
2							
3	Periods	Function Arguments					
4	Payoffs	B5					
5	Present value	$70,000.00					
6							
7	Interest rate	B4,B5)					

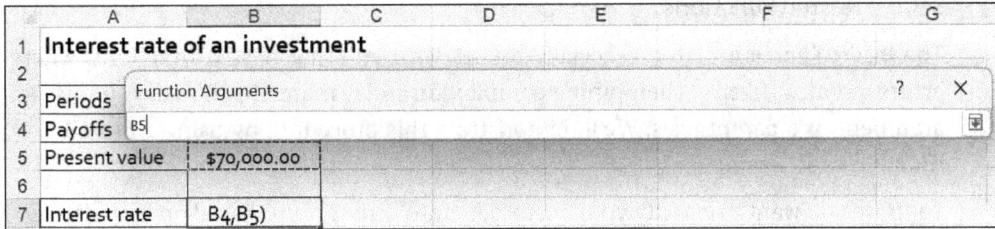

Figure 3.21 Selecting Value for Function Argument

Employing Frequently Used Functions

To employ a function you use often, enter an equal sign in the cell and then click the arrow next to the name box in the formula bar to open the list of recently used functions and choose the one you want.

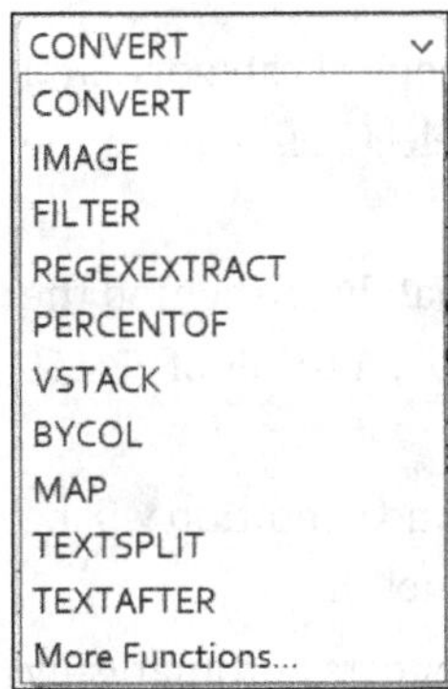

Figure 3.22 List of Recently Used Functions

3.2.10 Editing Functions

Click or tap the function icon and then a function in a formula to edit the function's arguments in the dialog. If a formula contains multiple functions, simply select them one by one. If you need a function at a specific point within a formula, click the **Insert Function** button again.

When a formula mixes one or more functions with other operators, a shortened **Function Arguments** dialog box appears, and it shows only the overall result unless you select a function.

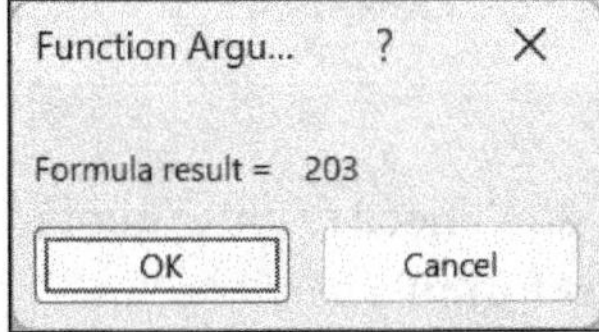

Figure 3.23 Shortened Dialog Box

3.2.11 Nested Functions

The **Insert Function** dialog especially assists with entering nested functions, where errors are most likely. When typing complex formulas manually, it's easy to miss an argument or a parenthesis. We'll demonstrate this procedure by using a nested `IF()` function as an example.

Suppose you want to classify your customers into three groups based on their sales:

- Small customers with sales under $1,000
- Medium customers with sales up to $50,000
- Large customers with sales over $50,000

Follow these steps to accomplish this:

1. In the column next to the sales figures, add a label for the classification. You can do this with an `IF()` function.
2. Click or tap the **Insert Function** button, choose `IF()` from the **Logical** category, and click **OK** to confirm. In the first argument field, **Logical_test**, select cell B4 with the first sales value and then enter ">=1000."

   This condition filters out small customers, and you can immediately see behind the input field whether the condition is met for the selected revenue. A result of `=FALSE` means the condition is not met in the first case.
3. Select the **Value_if_true** field. If the first condition is met, a second condition will be checked, so the second argument must include another `IF()` function.
4. Use the **IF** button in the name box of the formula bar, which appears automatically after you first use the `IF()` function. The dialog box for entering the three arguments of the `IF()` function will also appear again.
5. Enter the second condition: "B4>=50000."

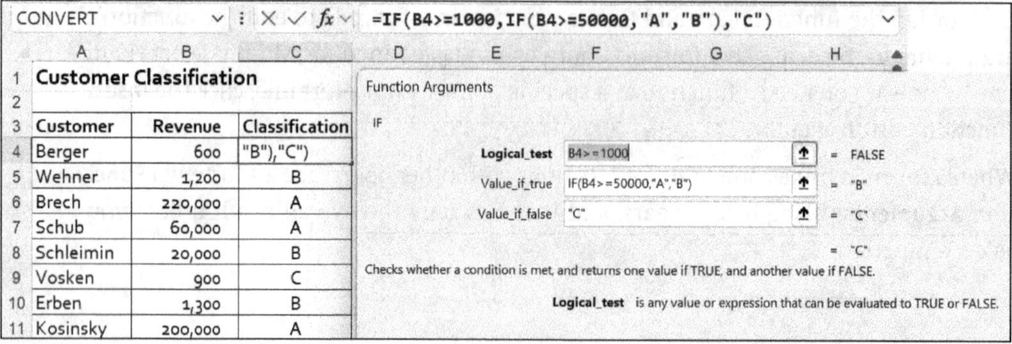

6. Enter the letter A in the **Value_if_true** field. Note that you don't need to enter quotation marks, unlike when typing a function directly into the formula bar.
7. Enter the letter B in the **Value_if_false** field.

8. When you click or tap the first IF in the formula bar, the parent function's dialog box appears again. Enter the letter C as the **Value_if_false** and confirm with **OK** or [Enter].
9. The formula is now complete, so copy it to the other cells.

If you want to replace an argument with a function later in a finished formula, you can use the function icon.

Notes on the SUM() Function

When you're working with the most commonly used function, sum, the **Sum** icon in the **Home • Editing** group or the [Alt]+[=] keyboard shortcut does almost all the work. The icon also offers some basic statistical functions, so if you want to perform a calculation other than sum, you can use a similar approach. Click the arrow next to the icon to access functions like AVERAGE() or to display the minimum or maximum values. You can also find this icon on the **Formulas** tab.

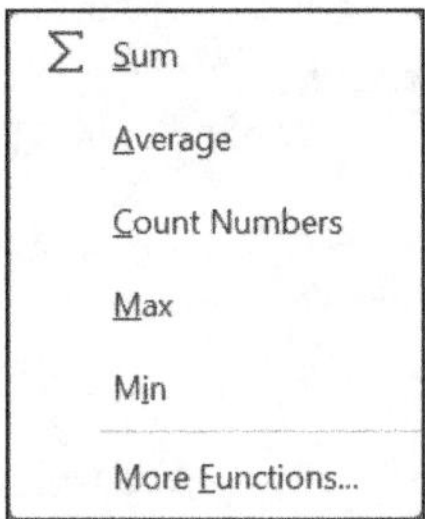

Figure 3.24 Sum Button Menu

If you've entered a continuous range of numbers in a column, just place the cell pointer in any cell below the range and double-click the icon. It doesn't matter how many empty cells are in between, and the program sums only the range containing numbers. On a touchscreen, tap the icon, select the **Sum** option, and press [Enter] or [Tab] to complete the formula.

	A	B	C	
1	**Sales**			
2				
3		Sports goods	Toys	Le
4	Jan	10000	12000	
5	Feb	12000	14400	
6	Mar	14400	17280	
7	=SUM(B4:B6		43680	
8		SUM(**number1**, [number2], ...)		

Figure 3.25 Selected Range for SUM() Function

The same applies to a range of values in a row. Just place the cell pointer in a cell to the right of the range to get the corresponding sum function.

If a selected cell has ranges of values both above and to the left, Excel sums the range closer to the formula cell (with fewer empty cells in between). If the distances are equal, it sums the column range, assuming that's what you intend. If you don't want to use the cell range Excel suggests when you click the **Sum** icon, just drag your mouse over the cell range you want to use instead.

Calculating Row and Column Sums Simultaneously

As mentioned earlier, you can enter multiple formulas at the same time by using the **Sum** icon. This works especially well when you have a closed data block because it allows you to calculate row and column sums at the same time. Here's an example: Select the entire number range, including the cells where the sum formulas will go, and then click the **Sum** icon. The result appears in Figure 3.26.

	A	B	C	D	E
1	**Sales**				
2					
3		Sports goods	Toys	Leisure dress	Total
4	Jan	10000	12000	13000	35000
5	Feb	20000	14400	15600	50000
6	Mar	24000	17280	18720	60000
7	Total	54000	43680	47320	145000

Figure 3.26 Calculating Row and Column Sums in One Step

3.2.12 Calculating Total Sums

Another feature of the **Sum** icon is that it can calculate subtotals in a table with multiple groups of values within a single column. To combine these subtotals into a grand total, just select the cell where you want the total to appear and the **Sum** icon will return the overall total from the subtotals in the column.

	A	B	C
1	**Weekly results**		
2			
3	Mo	120	
4	Tu	130	
5	We	110	
6	Th	130	
7	Fr	140	
8	1. Week	630	
9	Mo	120	
10	Tu	130	
11	We	110	
12	Th	130	
13	Fr	140	
14	2. Week	630	
15	=SUM(B14,B8)		
16	SUM(number1, **[number2]**, [number3], ...)		
17			

Figure 3.27 Calculating Subtotals with Sum Icon

This method also works for subtotals in a row. For example, if you want to combine four quarterly results from three monthly columns each, the **Sum** icon will give you the annual total.

3.3 Relative and Absolute References

To help you better understand how Excel handles formulas, let's start with a simple example. The following table summarizes a company's sales and costs for various product groups.

	A	B	C	D
3	**Product group**	**Revenue**	**Cost**	**Profit**
4	Furniture	$2,000,000.00	$1,240,000.00	$760,000.00
5	Carpets	$3,000,000.00	$1,900,000.00	$1,100,000.00
6	Housing need	$1,500,000.00	$1,050,000.00	$450,000.00
7	Garden furniture	$250,000.00	$150,000.00	$100,000.00
8	**Total**	$6,750,000.00	$4,340,000.00	$2,410,000.00

Figure 3.28 Sales and Costs for Different Product Groups

3.3.1 Working with Relative References

This table requires two calculations: totaling the columns and calculating gross profit by subtracting costs from sales. Instead of entering numbers directly into a formula, you reference the cells by their addresses, and each time, the formula retrieves the value stored at that address. So, you use variables instead of constants.

Calculating the Totals

To help you calculate totals, Excel offers the `SUM()` function, which adds all numeric values within a specified range of cells. Excel only needs to know the size of the range, and the range is passed to the function as a range address. A range address specifies a range like "From–To," which is defined by the address of the starting cell and the address of the ending cell, separated by a colon. For example, if you select cell B8 in the sample table, you could enter the following:

```
=SUM(B4:B7)
```

As mentioned, it's more practical here to double-click the **Sum** icon. Instead of summing each column separately, you can do it all at once by selecting cells B8 through D8 and using the **Sum** icon.

Calculating Profits

To calculate gross profits, simply subtract costs from revenue. The formula for the first row is this:

```
=B4-C4
```

Cell D4 shows the desired result, and if D4 is still selected, just double-click the fill handle to apply the formula to the other product groups. Alternatively, you can drag the fill handle down with a mouse. Double-clicking is especially handy for long columns. In **Touch input** mode, use the context menu to show the **AutoFill** icon and drag it to the bottom of the column.

When you hover your cursor over one of the new formulas, you'll see that Excel updates the row number in the cell references each time. When you're copying, Excel adjusts the addresses of revenue and cost cells instead of treating them as fixed. This adjustment is usually helpful but not always. For example, calculating the percentage of individual product groups in total profit would produce errors if addresses were adjusted. Excel fundamentally distinguishes between two types of references: relative and absolute. Mixed references combine both relative and absolute parts.

What Does Relative Mean?

For a *relative reference*, Excel records the referenced cell's position as the distance from the cell containing the reference. Take another look at the first profit formula in cell D4:

```
=B4-C4
```

Excel interprets this formula as "Subtract the value in the cell two columns to the left from the value in the cell one column to the left." When you copy this formula down, the calculation description stays the same but now applies to cells B5 and C5, and so on. Using a relative reference tells Excel to adjust the reference automatically when the formula is moved or copied.

3.3.2 Absolute and Mixed References

An *absolute reference* always points to the exact same cell. Using an absolute reference ensures that the reference doesn't change when you copy or move the formula, so it always refers to that specific cell. Relative and absolute references can also be mixed within an address. When you're copying mixed or partially absolute references, the absolute part stays the same, while the relative part adjusts.

3.3.3 Types of References

Reference Types	Meanings
B2	This is a relative reference to cell B2.
$B$2	This is an absolute reference to cell B2.
$B2	This is a mixed reference with an absolute column and a relative row.
B$2	This is a mixed reference with an absolute row and a relative column.

When Should You Use Each Type of Reference?

Relative references are Excel's default option, and any simple cell reference that uses cell addresses is relative. References to named ranges, however, are usually absolute. When you delete or insert rows or columns, the named range automatically adjusts its addresses to shrink or expand accordingly, and if you move a named range, its references update automatically as well. When should you use absolute or mixed references? We have already discussed percentage calculations. In our example, say you enter the following formula into cell E4 to calculate the group sales' share of total sales:

```
=B4/B8
```

You'll get a useful result, and then, you can use the **Percent Format** button in the **Home • Number** group to format it as a percentage. However, a problem arises when you try to copy the formula downward. Instead of results, Excel shows error values. When you hover the cell pointer over cell E5, a button appears that provides a hint about the error. Clicking or tapping the button offers several options to address the error or get more details.

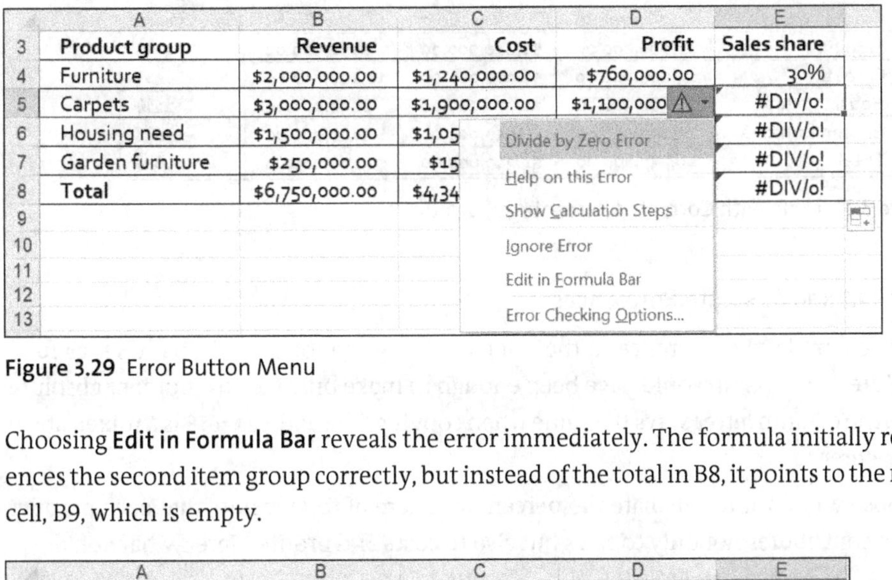

	A	B	C	D	E
3	Product group	Revenue	Cost	Profit	Sales share
4	Furniture	$2,000,000.00	$1,240,000.00	$760,000.00	30%
5	Carpets	$3,000,000.00	$1,900,000.00	$1,100,000	#DIV/0!
6	Housing need	$1,500,000.00	$1,05		#DIV/0!
7	Garden furniture	$250,000.00	$15		#DIV/0!
8	Total	$6,750,000.00	$4,34		#DIV/0!
9					
10					
11					
12					
13					

Divide by Zero Error
Help on this Error
Show Calculation Steps
Ignore Error
Edit in Formula Bar
Error Checking Options...

Figure 3.29 Error Button Menu

Choosing **Edit in Formula Bar** reveals the error immediately. The formula initially references the second item group correctly, but instead of the total in B8, it points to the next cell, B9, which is empty.

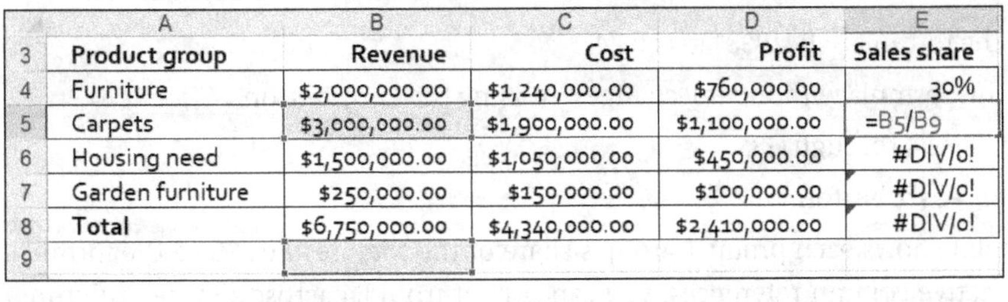

	A	B	C	D	E
3	Product group	Revenue	Cost	Profit	Sales share
4	Furniture	$2,000,000.00	$1,240,000.00	$760,000.00	30%
5	Carpets	$3,000,000.00	$1,900,000.00	$1,100,000.00	=B5/B9
6	Housing need	$1,500,000.00	$1,050,000.00	$450,000.00	#DIV/0!
7	Garden furniture	$250,000.00	$150,000.00	$100,000.00	#DIV/0!
8	Total	$6,750,000.00	$4,340,000.00	$2,410,000.00	#DIV/0!
9					

Figure 3.30 Formula Reference Error

Excel treats empty cells as having a value of 0, and because the reference to the total result was entered as a relative reference, it causes the error `#DIV/0!`, since division by zero is not allowed. To fix this, set the reference to the total result as absolute. After that, you can copy the formula.

Changing the Reference Type

When you're changing the reference type, the F4 key is very useful because it removes the need to manually add dollar signs. Here's how you can use it in our example:

1. Select cell E4 and enter the address "B8" in the formula bar.
2. Press F4 to convert the reference to an absolute reference.
3. Now, you can copy the modified formula `=B4/$B$8` up to cell E8 without any issues.

It doesn't matter if the insertion point is directly before or after the reference or if the entire reference is selected. Pressing F4 repeatedly cycles through the different reference types.

	A	B	C	D	E
3	**Product group**	**Revenue**	**Cost**	**Profit**	**Sales share**
4	Furniture	$2,000,000.00	$1,240,000.00	$760,000.00	30%
5	Carpets	$3,000,000.00	$1,900,000.00	$1,100,000.00	44%
6	Housing need	$1,500,000.00	$1,050,000.00	$450,000.00	22%
7	Garden furniture	$250,000.00	$150,000.00	$100,000.00	4%
8	**Total**	$6,750,000.00	$4,340,000.00	$2,410,000.00	100%

Figure 3.31 Table with Corrected Percentage Formula

3.3.4 Mixed Absolute References

In the example, the reference to the cell containing the total was initially set as fully absolute. However, it would have been enough to make only the row number absolute since the column letter stays the same when copying. The address `=B$8` is a mixed absolute reference.

Suppose you want to calculate the percentage share of total results that each product group contributes not only to sales but also to costs and profits. Here's what you do:

1. Enter the formula "=B4/B$8" in cell E4.
2. Copy this formula without any changes by using the **Copy** button.
3. Select cells E4 through G8.
4. Click the **Paste** button.
5. The result shows each product group's share of the total results. Since the formula uses relative column references, you can copy it from E4 across all three columns. This example also demonstrates that you can copy a single cell to an entire range, with the original cell copying onto itself.

	A	B	C	D	E	F	G
3	**Product group**	**Revenue**	**Cost**	**Profit**	**Sales share**	**Cost share**	**Profit share**
4	Furniture	$2,000,000.00	$1,240,000.00	$760,000.00	30%	29%	32%
5	Carpets	$3,000,000.00	$1,900,000.00	$1,100,000.00	44%	44%	46%
6	Housing need	$1,500,000.00	$1,050,000.00	$450,000.00	22%	24%	19%
7	Garden furniture	$250,000.00	$150,000.00	$100,000.00	4%	3%	4%
8	**Total**	$6,750,000.00	$4,340,000.00	$2,410,000.00	100%	100%	100%

Figure 3.32 Shares of Product Groups in Total Results

3.3.5 Summation with Mixed References

By cleverly combining range references, you can perform summations. Here's a simple example of a weekly summary with a column of running totals that include the previous day's results. The first running total in cell G4 uses this formula:

```
=SUM($B$4:F4)
```

The first argument sets a fixed starting point, and you can copy these formulas down the column. In cell G5, the formula is this:

```
=SUM($B$4:F5)
```

It gives the cumulative total for both rows.

G4 =SUM($B$4:F4)

	A	B	C	D	E	F	G
1	**Weekly results**						
2							
3		Type 1	Type 2	Type 3	Type 4	Type 5	Accumulated total
4	Monday	200	300	250	340	230	1320
5	Tuesday	200	300	250	340	230	2640
6	Wednesday	200	300	250	340	230	3960
7	Thursday	200	300	250	340	230	5280
8	Friday	200	300	250	340	230	6600

Figure 3.33 Example of Summation

3.3.6 Range Unions and Intersections

Excel can combine range references or create references to their union or intersection. To create range unions, you use a comma. For example, to sum two nonadjacent rows, use a formula like this:

```
=SUM(B4:B12,D4:D12)
```

To get the intersection of two ranges in a formula, use the intersection operator, which is a space. Figure 3.34 shows an example of the use of an intersection. The intersection is highlighted in a different color, and the ranges are outlined. Intersection references are also useful for accessing subsets within a table, and working with intersections is especially effective when the ranges have been named in advance. Section 3.4.4 explains this in more detail. Alternatively, instead of the formula used in Figure 3.34, you could use a formula that references only the row numbers and column labels.

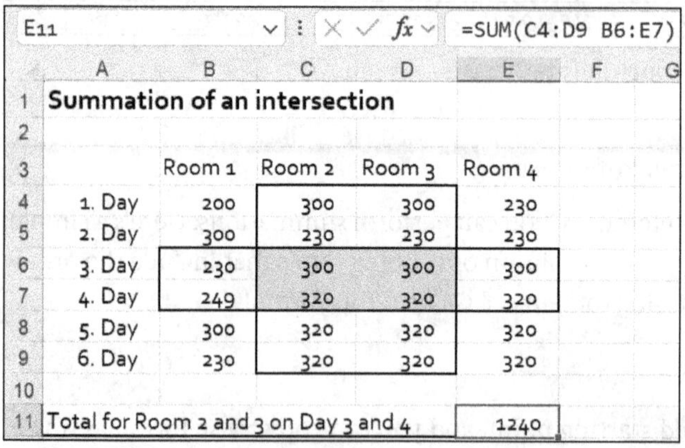

E11	=SUM(C4:D9 B6:E7)				
	A	B	C	D	E
1	Summation of an intersection				
2					
3		Room 1	Room 2	Room 3	Room 4
4	1. Day	200	300	300	230
5	2. Day	300	230	230	230
6	3. Day	230	300	300	300
7	4. Day	249	320	320	320
8	5. Day	300	320	320	320
9	6. Day	230	320	320	320
10					
11	Total for Room 2 and 3 on Day 3 and 4				1240

Figure 3.34 Example of Summing Intersection

Here, the range is the intersection of two entire columns and rows:

```
=SUM(C:D 6:7)
```

3.3.7 Calculating with the Quick Analysis Tool

The Quick Analysis Tool offers you a convenient way to make many calculations involving cell ranges. When you select a range, click the **Quick Analysis** button and then choose **Totals**.

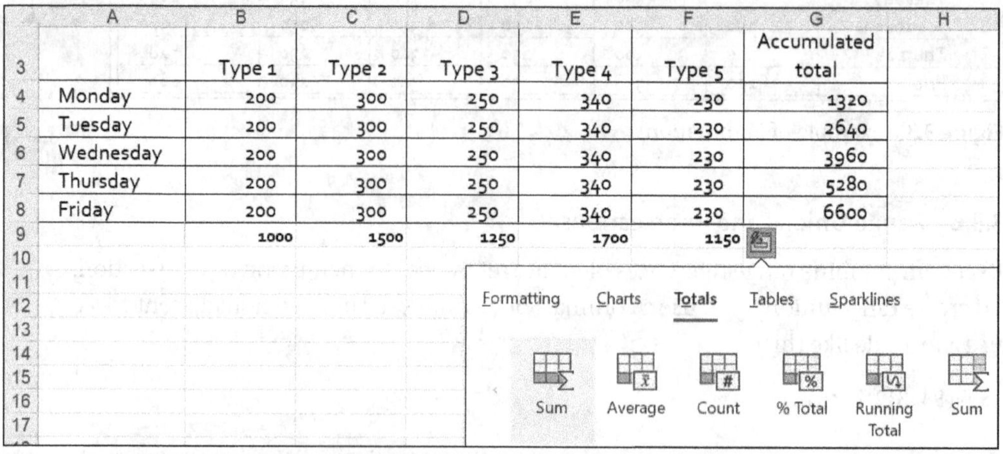

	A	B	C	D	E	F	G
3		Type 1	Type 2	Type 3	Type 4	Type 5	Accumulated total
4	Monday	200	300	250	340	230	1320
5	Tuesday	200	300	250	340	230	2640
6	Wednesday	200	300	250	340	230	3960
7	Thursday	200	300	250	340	230	5280
8	Friday	200	300	250	340	230	6600
9		1000	1500	1250	1700	1150	

Formatting | Charts | Totals | Tables | Sparklines

Sum | Average | Count | % Total | Running Total | Sum

Figure 3.35 Creating Column Sums for Cell Range B4:F8

The menu will show five icons for row-wise analysis and five for column-wise analyses. Use the small arrows to reveal hidden icons.

The tool provides a live preview, displaying results as soon as the mouse pointer hovers over an icon. The calculated values appear in the next row or column, and if you apply the tool again to the same range, a message will inform you that calculated values already exist. Clicking **OK** will overwrite them.

3.4 Descriptive Range Names

References to cells and ranges aren't limited to addresses. You can also name cells and ranges and allow references by their assigned names. Excel also allows you to link formulas and values that are not entered in a cell to a name. We describe this in this section.

3.4.1 Benefits of Range Names

Using range names isn't essential, but it makes working with Excel much easier, especially since names are easier to understand and remember than abstract addresses. Many tasks—such as designing worksheets, copying or moving data, printing, or creating charts—involve referencing or defining ranges. Most formulas use range references as arguments.

Reusability

The main advantage of using names is that you can reuse them repeatedly. Once you assign a name to a range, you can refer to the range by that name without having to specify or select the addresses each time.

While selecting ranges with a mouse, finger, stylus, or arrow keys takes only seconds in small tables, it's a different story when the range spans several screen pages. By the time you reach the bottom-right corner, you've likely lost sight of the other corners of the range and you might even be unsure whether you started the selection in the correct cell. In short, you'll be relieved when the range is finally correctly defined on all sides. In such cases, it's definitely easier to call up a predefined range from a name list. Whenever a command requires a range reference, you can specify the range name instead of the cell addresses. The same applies to ranges used in formulas.

Clarity

When you use simple sum formulas that you place directly below the column being totaled, it's easy to understand what the formula calculates. But this changes once you want to use the result of that formula elsewhere on the worksheet. Then, you might see this:

```
=G20*1.05
```

But what does the value in cell G20 represent? If you've named cell G20 *Profit2024*, the formula becomes this:

```
=Profit2024*1.05
```

This makes it easy for you and anyone else working on the worksheet to understand what the formula means.

Alternatively, compare these two pairs of formulas:

First pair:

```
=J50-M50-P50
=Sales-MaterialCosts-PersonnelCosts
```

Second pair:

```
=IF(SUM(F5:F50) > (SUM(E5:E50) * 1.10), "Goal reached", "")
=IF(Profit2024 > (Profit2023 * 1.10), "Goal reached", "")
```

Clearly, the second formula in each pair is easier to understand.

Named ranges become even more important when you're working with multilevel references. This benefit is especially helpful when you're referencing ranges in other workbooks, where you can't always immediately see what's stored in, say, range B10:B30.

Flexibility

If you use the same range name in multiple formulas, then you don't need to update each formula individually when the range's addresses change. In many cases, range changes happen automatically, such as when you insert or delete a row within a defined range.

The same applies when a value, like a specific conversion factor or tax rate, is defined as a named value. Formulas using the name don't need to be changed if the name's value is updated, such as when sales tax rates change. The same applies when formulas are linked to a name. Any cells using that name, directly or within a formula, don't need to be changed if the named formula is updated.

Easier Navigation

Another handy feature allows you to select named cells and ranges directly by their names. The quickest way to do this is through the name box in the formula bar. You can also access the list of defined names via **Home • Editing • Find and Select • Go To** or by pressing Ctrl+G or F5. This method often saves you from spending time on tedious searching, especially in large sheets.

3.4.2 Naming

The program offers several ways to assign names:

- You can name the current selection directly by using the name box in the formula bar.
- You can use the **Formulas • Defined Names • Define Name • Define Name** ❶ command to assign names to cells, areas, values, and formulas.
- You can use the **Formulas • Defined Names • Create from Selection** command ❷. You can also link cells or cell ranges to names that are already entered as labels in the worksheet, but the cell with the name must border the cell or cell range to be named.
- You can assign names using the command **Formulas • Defined Names • Name Manager** ❸.

Usually, each workbook can use its own list of names, but each name must be unique. The name box shows all names in the workbook, no matter which sheet the named range is on. You can also assign names that apply only to the specific sheet where they're defined. This is useful when the same name, such as *Total_staff*, is used for similar values across multiple sheets. To do this, select **Formulas • Defined Names • Define Names • Define Names** each time. Then, in the dialog under **Scope** ❹, replace the default **Workbook** with the sheet name, which you can choose from the list box.

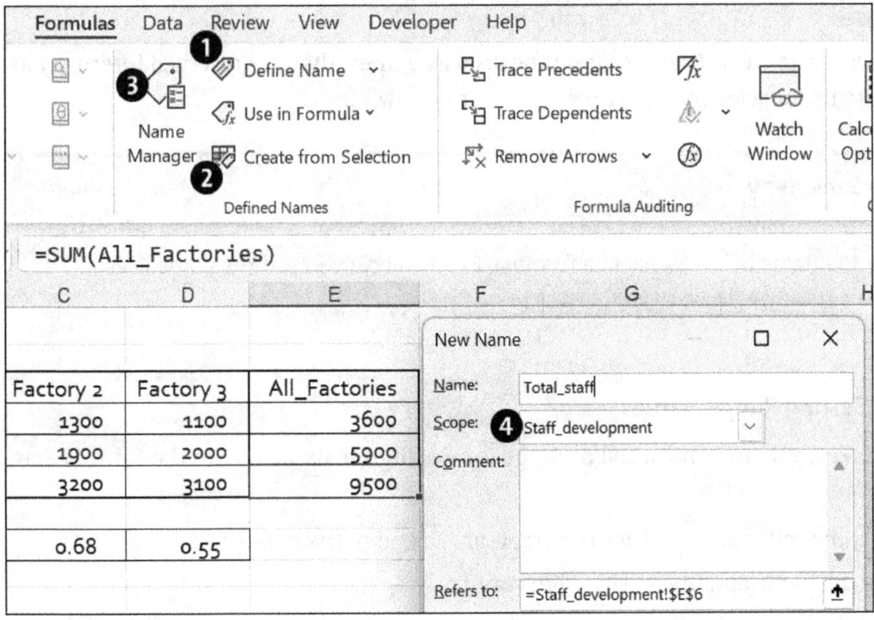

Figure 3.36 Defining Sheet-Specific Name

To use these names in formulas, enter the sheet name before the actual name and separate the two names with an exclamation mark. For example, you can enter "Staff_development!Total_staff." The sheet-specific name appears in the name box only when that sheet is active, and when you use a sheet-specific name in a formula on that sheet, you can omit the sheet name. If the name *Total_staff* exists for the entire workbook, it will be ignored on the sheet where the sheet-specific name is defined. However, if a formula on another sheet refers to the sheet-specific name, you must use the qualified name, including the sheet name prefix.

Naming Syntax Rules

When assigning range names, you must follow these rules:

- Range names can be up to 255 characters long.
- Names are not case sensitive.
- The first character cannot be a number. Allowed characters include letters, underscores (_), and backslashes (\).
- Special characters like !, {, +, *, –, /, @, <, >, &, and # are not allowed.
- Spaces, colons, semicolons and commas are not allowed because they serve as range operators. Use an underscore (_) or a period to separate names.
- Range names must not look like valid cell addresses. For example, B10 or AA3 are not allowed.
- The characters *C*, *c*, *R*, and *r* cannot be used as names; they are reserved internally as shortcuts for selecting the current column or row.

[»]

Case Sensitivity

Although formula references ignore case, using consistent case is still helpful. If you enter the name in lowercase in a formula, Excel will convert it if the name is valid. If it doesn't, there might be a typo.

3.4.3 Setting Range Names

The quickest way to name a cell or range is by using the name box in the formula bar, as follows:

1. Select the cell range and click the arrow at the end of the name box.
2. Type the name and press Enter to confirm.
3. The new name will appear the next time you open the name list, so click or tap the name to select the range.

Quarter_1_24 | fx 10000

	A	B	C	D	E
1	**Sales development**				
2					
3		Jan-24	Feb-24	Mar-24	1. Quarter 24
4	Product 1	10,000	12,000	12,000	34,000
5	Product 2	12,000	14,400	17,280	43,680
6	Product 3	13,000	15,600	18,720	47,320
7	Product 4	11,000	13,200	15,840	40,040
8	Product 5	10,000	12,000	14,400	36,400
9	Product 6	12,000	14,400	17,280	43,680

3.4.4 Defining a Name

The **Define Name** command in the **Formulas • Defined Names** group gives you more control:

1. To name individual cell ranges, it's best to select them first.
2. Click the **Define Name** button and select **Define Name** from the menu.
3. If the active cell in the selected area borders a cell with a label, then that label appears in the name box when the area contains numbers; otherwise, the content of the active cell is suggested. To assign a new name, enter it directly in the **Name** field.
4. Under **Area**, you can keep the default **Workbook** if the name is unique within the workbook.
5. If the correct reference is already selected under **Refers to**, confirm by clicking **OK** or pressing `Enter`. Excel automatically writes the reference using absolute addresses. (Relative addresses are possible for named ranges but are generally impractical.)

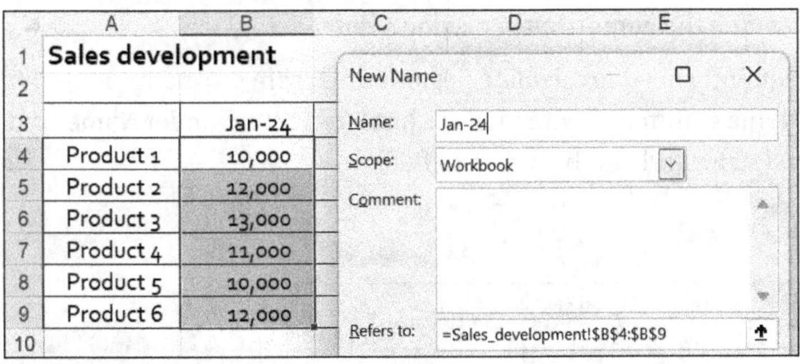

Otherwise, you can correct or reenter the reference. References must start with an equal sign. You can also name multiple ranges, and you separate each range address from the next with a comma. To name the intersection of two ranges, insert a space between their addresses.

3.4.5 The Name Manager

Excel includes a **Name Manager** to simplify working with extensive name lists. You can open it via the **Name Manager** button in the **Formulas • Defined Names** group or by pressing `Ctrl`+`F3`. This dialog clearly lists all names in a workbook and links to the name definition dialog box, which you can access by using the **New** or **Edit** button.

To remove redundant names, click the **Delete** button. The filter function is especially useful here because it offers options like restricting names to a specific sheet.

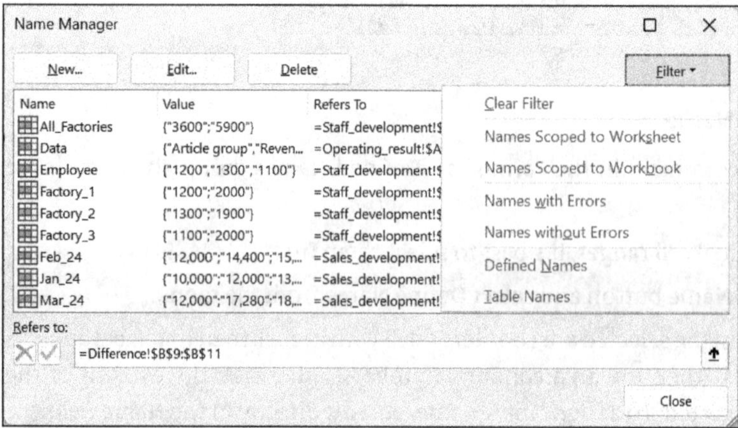

Figure 3.37 Managing Extensive Name Assignments with Name Manager

3.4.6 Defining Named Formulas

Linking a formula to a name works much like naming cells or ranges. Excel stores the definition with the workbook, but not in a cell. You enter the formula directly in the dialog box where you define the name, thus replacing a range reference.

For example, if you frequently need to calculate net amounts from gross amounts, you can select the **Define Name** command, enter a name like "NetFactor" under **Name**, and then enter this conversion formula in the **Refers to** field:

```
=1/107*100
```

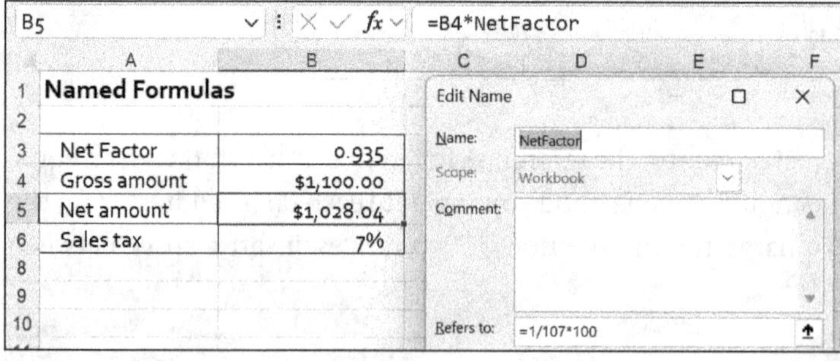

Figure 3.38 Example of Named Formula

If B4 contains the gross amount of an invoice, you can calculate the net amount in B5 with this formula:

```
=B4*NetFactor
```

3.4.7 Named Values or Text Elements

You can also assign names to specific values or text constants. For example, to store a sales tax rate, enter "=6%" in the reference field. If you name this value *SalesTax1*, use the following formula to calculate sales tax for an amount in B10:

```
=B10 * SalesTax1
```

If you enter a text constant under **Refers to**, then it must be enclosed in quotation marks, and if you store the full company name under the name *Company*, then entering "=Company" into a cell will return that name. Both named formulas and named constants appear in the function list during formula entry, as Figure 3.39 shows.

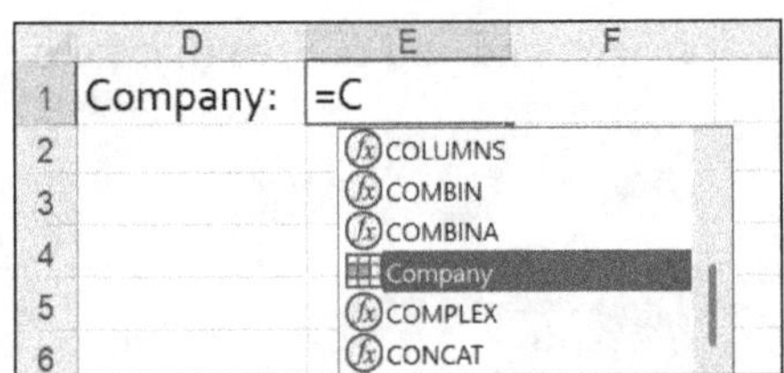

Figure 3.39 Named Constant in Function List

3.4.8 Importing Names from Labels

In uniformly structured tables with complete column and row labels, names for value columns and rows can be generated automatically. Excel simply uses the existing labels as names and adjusts them according to the rules described previously. If the first character is a number, it adds an underscore before it, and it also replaces spaces and most symbols with underscores. It also converts date values into strings using the specified format.

The cells to be named can be above, below, to the right, or to the left of the labeled cells. Cell ranges always refer to groups of cells in a single row or column, and the following example shows how Excel handles this.

To correctly apply the names, start by selecting the range from A3 to D5. You need to initially exclude the totals row so that you can use the names for the column values in the sum formulas in the next step. You can then go to **Formulas • Defined Names • Create from Selection** ❶ or press Ctrl+Shift+F3 to open it.

Then, you'll see four checkboxes ❷, with the first two checked based on the previously selected area. The top row and left column provide the names, so you don't need to change anything here. When you confirm, Excel creates named ranges: two row ranges and three column ranges.

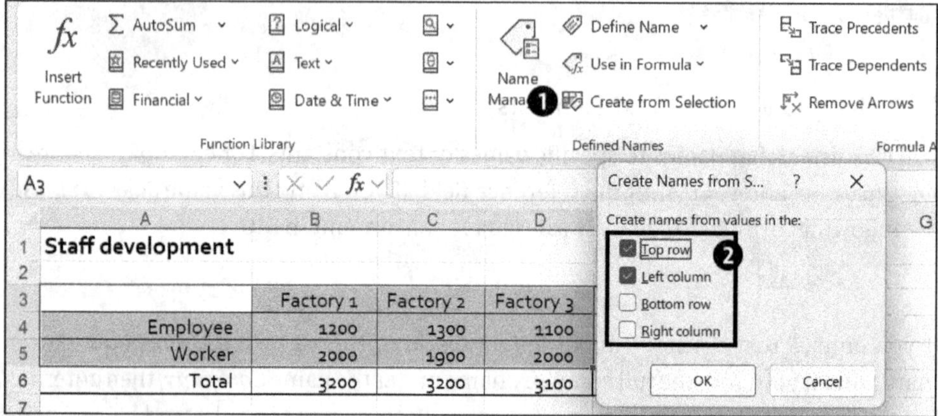

Figure 3.40 Using Existing Labels as Range Names

Figure 3.41 shows the range names in the Name Manager, and you can see that Excel automatically changed *Factory 1* to *Factory_1*.

Name	Value
Employee	{"1200","1300","1100"}
Factory_1	{"1200";"2000"}
Factory_2	{"1300";"1900"}
Factory_3	{"1100";"2000"}
Worker	{"2000","1900","2000"}

Figure 3.41 List of Adopted Range Names

You can now quickly select areas by using the name box. Press F5 to select the intersection of two areas, select the first area's name, and then type the second area's name after separating it from the first area's name with a space. *Employees Factory_2* will then select exactly the number of employees in Factory 2.

3.4.9 Using Names in Formulas

Naming ranges does not automatically update references in existing formulas. You must explicitly do this by using the **Apply Names** command, which is the second option under **Formulas • Defined Names • Define Name**. It doesn't matter where the cell pointer is at the time you do this.

You can choose which names to use in formulas by clicking or tapping, and this choice also affects formulas you enter later. Excel automatically replaces matching range references with their range names.

In the dialog, the names *Employees* and *Workers* are selected. If you enter a row sum for rows 4 and 5 now, *Employees* and *Workers* will be used as range names in the formula. However, keep in mind that range names refer to absolute addresses, so you can't simply copy the row sum for Employees to the next row. It's better to create and copy the row sums using cell addresses first and then replace them with names.

You can control how Excel applies names in detail through **Options**. You should usually check the **Ignore Relative/Absolute** box to ensure that a reference to an area will be replaced, regardless of the reference type used in the formula. If you leave this box unchecked, only absolute cell references will be replaced by area names.

If you uncheck the **Use row and column names** box when applying names, you can decide whether formulas that evaluate intersections of ranges can omit the corresponding names for references within a column or row to simplify the formula.

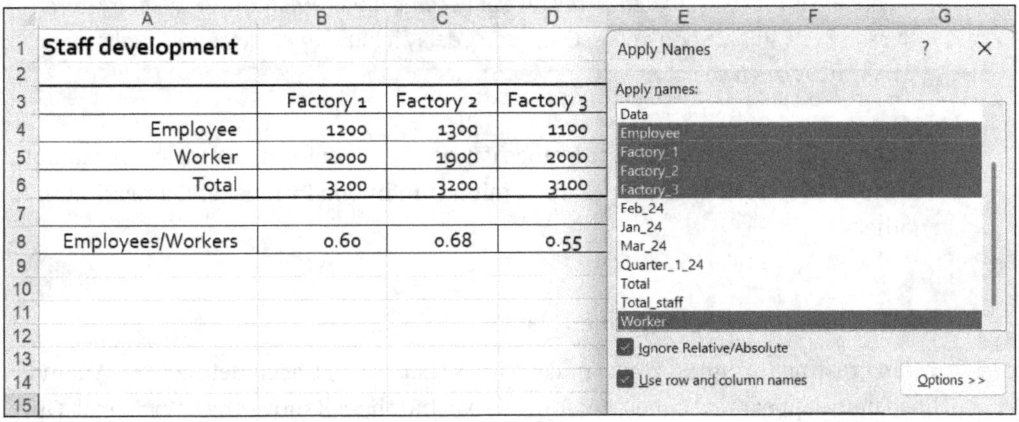

	A	B	C	D
1	**Staff development**			
2				
3		Factory 1	Factory 2	Factory 3
4	Employee	1200	1300	1100
5	Worker	2000	1900	2000
6	Total	3200	3200	3100
7				
8	Employees/Workers	0.60	0.68	0.55

Apply Names

Apply names:

Data
Employee
Factory_1
Factory_2
Factory_3
Feb_24
Jan_24
Mar_24
Quarter_1_24
Total
Total_staff
Worker

Ignore Relative/Absolute

Use row and column names

Options >>

Figure 3.42 Apply Names Dialog Box

You can also choose whether to enter the row or the column first when referencing cells that require both. For example, to calculate the ratio of employees to workers per plant, simply enter this formula in cell B8:

```
=Employees/Workers
```

You enter it instead of this formula:

```
=Employees Factory_1/Workers Factory_1
```

And you do that because both references share the same column label. Using the shorter formula also lets you copy it to cells C8 and D8 without changes, which you can't do with the longer formula.

3.4.10 Correcting Name Definitions

If you need to update a name definition because your sheet's layout has changed, open the Name Manager, select the name from the list, and click the **Edit** button to modify the range addresses. After making changes, check whether any formulas referencing this range are affected. The program automatically updates formulas to the new area addresses, and if this works for you, no further action is needed.

If you find during the development of a complex worksheet that some names you've used are confusing or not distinctive enough, you can change them anytime in the Name Manager without redefining the references. To do this, select the old name in the dialog box then correct it by using the **Edit** button. The reference addresses will remain unchanged.

[+]

Delete Unnecessary Names!

Delete areas that you created by mistake or became unnecessary after you restructured the sheet. Do this as soon as possible—don't delay it indefinitely since you might forget which areas can be deleted.

You can assign multiple names to a range if it clarifies your formulas' logic. For example, a named cell's value might represent a minimum in one formula and a maximum in another.

3.4.11 Formulas with Undefined Names

When creating formulas, you can use names that haven't been defined yet. An undefined name appears in the cell and formula bar, but the cell shows the #NAME? error. Once the name is defined, the error disappears and the formula works correctly. This also means you no longer need to assign the formula's name explicitly through the **Define Name** dialog.

3.4.12 Inserting Names into a Formula

To insert a name while entering a formula or in the **Function Arguments** dialog, you can either insert the name directly at the cursor or replace a selected argument with a name. Use F3 or the **Formulas • Use in Formula** command to select the name you want.

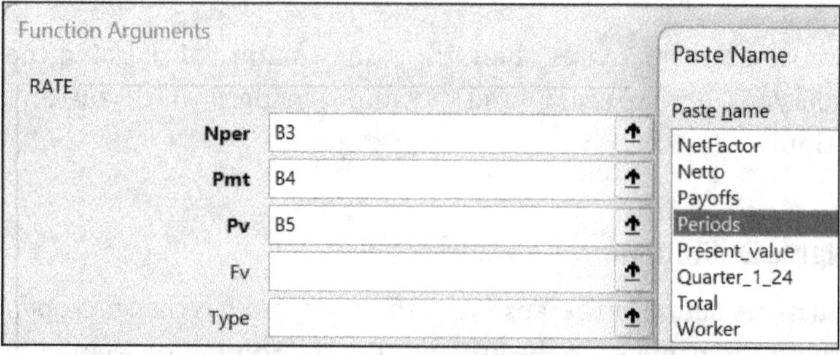

Figure 3.43 Using Name as Function Argument

[«]

References to Labels Are No Longer Supported

Instead of explicitly assigning names, Excel 2003 allowed the use of names from row or column labels directly in formulas. However, this option has not been supported since Excel 2010. When opening older workbooks that use this method, you'll be notified that those references in formulas will be replaced with standard cell references.

3.5 Array Formulas

A simple formula like =B2*B3 multiplies a single value in one cell by a single value in another, displaying the result only in the cell with the formula. Excel also lets you use multiple values as arguments in a formula, and formulas can produce multiple values as results, as seen with certain functions. In both cases, these values are arranged in a structure called a *matrix* or an *array*.

3.5.1 Array Ranges

The simplest way to understand an array is as a contiguous range of cells containing values. The smallest array consists of two adjacent cells, either side by side or stacked vertically, and its size is only limited by available memory. An array is a rectangular cell range where each cell holds a value, so it corresponds to what algebra calls a matrix, provided the values are numerical.

A simple example of an array is the following number pattern with two rows and three columns.

23	26	65
16	10	45

The key point is that Excel treats an array as a single unit, so you can't handle its individual elements separately in many ways. When using an array as an argument in a formula, you have two options.

The first option is to enter a reference to a rectangular cell range containing the array values directly into the formula. Take the function INDEX() as an example, which requires an array as its first argument. In this formula, the array is defined by a simple cell range:

```
=INDEX(B4:D6,2,2)
```

The second option is an array constant, and the preceding small array is an example of such a constant. In a formula, the entire array is enclosed in curly braces. You must enter array constants manually, and you must separate values within a row with commas and separate rows with semicolons. The number pattern in the previous array example appears in the formula bar like this:

```
{23,26,65;16,10,45}
```

This example is a 2 × 3 array, meaning it has two rows and three columns. You enter the data row by row, and a matrix constant must always have rows or columns of equal length. So, this expression is not allowed:

```
{1,2,3;1,2}
```

That's because the second row contains only two values, while the first has three.

Array constants can include numbers, text, logical values, and error values, but you must follow the correct input format. Text must always be enclosed in quotation marks, and using a comma as separator in numbers is not allowed because the period separates the matrix elements. Currency symbols, parentheses, and percent signs are also prohibited, and formulas cannot be elements of an array constant, even if they contain only constants. Therefore, the following expression is invalid:

```
{4+12,4,67}
```

Different data types can be mixed in an array as long as they match the operators or functions used, and individual cell references or range names cannot be part of an array constant. Therefore, this expression is not allowed:

```
{4,B3,16}
```

3.5.2 Working with Array Formulas

Besides array ranges and constants used as formula arguments or results, special array formulas can be applied. These formulas include array ranges or constants as their arguments.

Excel distinguishes between single-value formulas and array formulas. A single-value formula returns one specific value based on the referenced cells, and an array formula can return multiple results across a range of cells simultaneously. How is this to be understood?

Figure 3.44 shows a simple example. The number of nights in column B should be multiplied by the corresponding room price in column C, and instead of each value being multiplied individually, the two cell ranges in columns B and C are multiplied together. You only need to enter the formula in the highlighted array once; it then automatically applies to all cells in the array. The formula performs the specified operations using the corresponding data each time.

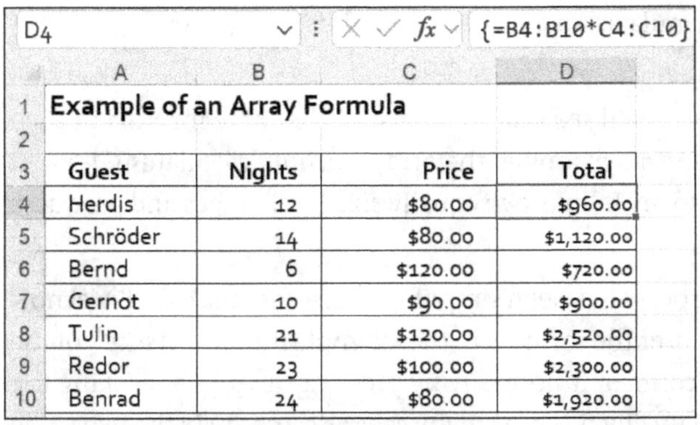

D4 {=B4:B10*C4:C10}

	A	B	C	D
1	**Example of an Array Formula**			
2				
3	**Guest**	**Nights**	**Price**	**Total**
4	Herdis	12	$80.00	$960.00
5	Schröder	14	$80.00	$1,120.00
6	Bernd	6	$120.00	$720.00
7	Gernot	10	$90.00	$900.00
8	Tulin	21	$120.00	$2,520.00
9	Redor	23	$100.00	$2,300.00
10	Benrad	24	$80.00	$1,920.00

Figure 3.44 Example of Array Formula

Array formulas save you time when you need to use certain formulas repeatedly. Alternatively, you could create multiple single-value formulas by copying, but besides incurring extra effort, a group of single-value formulas uses more memory than the equivalent array formula. Making changes later is also easier with an array formula since you only need to update one formula.

Entering an Array Formula

Excel 2021 changed how array formulas are entered, but older formulas remain valid. The following example lays out the steps you perform to use the older method of entering an array formula:

1. Select the cell range that will contain the result matrix. In Figure 3.45, this is cell range D4:D6. Performing this step is crucial because selecting only one cell will cause the array formula to return a single result. The result array should exactly match the number of results the array formula can generate based on the value group(s) it references. If the result array is smaller, some results will be omitted, and if it's larger, the extra cells will show `NA` to indicate that no values are available for the formula to process.
2. Enter the formula into any cell in the result array. It doesn't matter which cell you choose. Like any single-value formula, the formula you enter can include constants, operators, cell references, and functions. What makes it unique is that the formula references groups of values, which the array formula evaluates by row or column.
3. In this example, the values in the 2023 column are subtracted from those in the 2024 column. The formula is this:

   ```
   {=C4:C6-B4:B6}
   ```

Don't manually type the curly braces surrounding the entire formula, which designate it as an array formula. They are added automatically when you confirm the formula, but unlike with single-value formulas, this doesn't happen when you click the checkmark or press [Enter]. To enter a formula as an array formula, finish it by pressing [Ctrl]+[Shift]+[Enter].

If you check the results column, you'll see that Excel placed the same formula in each cell but that the results vary. In this example, the array formula calculates C4–B4 in result cell D4, C5–B5 in D5, and so on. It links two equally sized cell ranges and produces a result matrix of the same size.

This somewhat cumbersome process has been simplified since version 2021. Excel now uses dynamic arrays, so you'd enter the formula in the example into the first cell of the expected result array, D9, and confirm it normally by pressing [Enter] or clicking the checkmark. The formula automatically fills in as many cells as there are in the two value ranges used, and in the formula bar, the formula appears without curly braces. If you need to edit the formula later, you must do so in the first cell of the result range. For the cells below the first result cell, gray-formatted formulas appear in the formula bar, but they are not locked from editing.

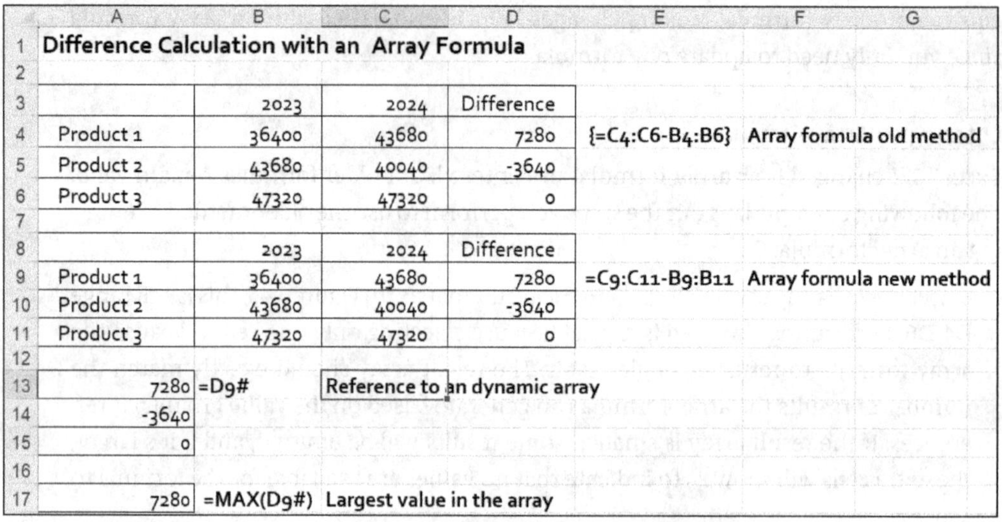

	A	B	C	D	E	F	G
1	**Difference Calculation with an Array Formula**						
2							
3		2023	2024	Difference			
4	Product 1	36400	43680	7280	{=C4:C6-B4:B6}	**Array formula old method**	
5	Product 2	43680	40040	-3640			
6	Product 3	47320	47320	0			
7							
8		2023	2024	Difference			
9	Product 1	36400	43680	7280	=C9:C11-B9:B11	**Array formula new method**	
10	Product 2	43680	40040	-3640			
11	Product 3	47320	47320	0			
12							
13	7280	=D9#	**Reference to an dynamic array**				
14	-3640						
15	0						
16							
17	7280	**=MAX(D9#)**	**Largest value in the array**				

Figure 3.45 Old and New Array Input Methods

You can easily reference a dynamic array by using the # operator. Using =D9# returns the entire result array generated by the formula in D9, so to find the largest value in this array, use =MAX(D9#).

Make sure the range where the result array will be placed is empty, because if it's not, the formula cell will show the #SPILL! error. This error also appears if you try to edit a cell within the result array. Once you delete the obstructing cell content, the array will refill with the results.

When working with dynamic arrays, it's helpful to use named areas. If the areas change in size, the result array adjusts automatically. For tables, as explained in Chapter 16, it's practical to use structured references within array formulas, and these also update automatically when the table size changes.

3.5.3 Simplifying Calculations

The following example is a bit more complex. It demonstrates how array formulas can simplify a table model. This works by combining calculation steps in the formula that would normally require intermediate results. For example, to calculate the total weight of a shipment, the usual method multiplies each item's weight by its quantity and then sums the results, as shown in column E in Figure 3.46.

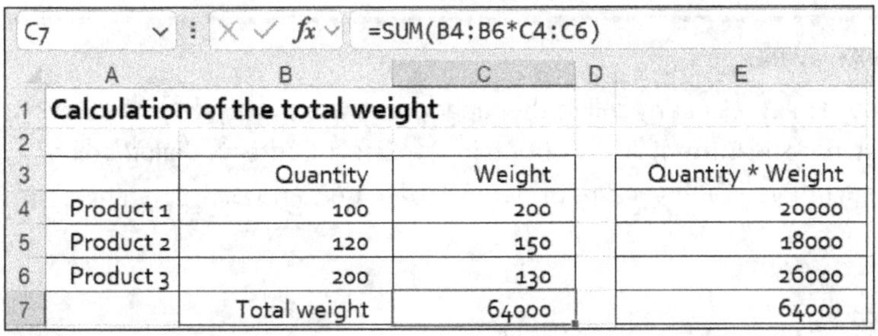

C7 =SUM(B4:B6*C4:C6)

	A	B	C	D	E
1	**Calculation of the total weight**				
2					
3		Quantity	Weight		Quantity * Weight
4	Product 1	100	200		20000
5	Product 2	120	150		18000
6	Product 3	200	130		26000
7		Total weight	64000		64000

Figure 3.46 Weight Calculation

Using an array formula, however, you don't need to display the multiplication results in the table. This formula stores the intermediate results internally and shows only the total in cell C7:

```
=SUM(B4:B6*C4:C6)
```

Note that you cannot access the intermediate results, so this method only makes sense if you don't need the intermediate results for other calculations. It's also important to remember that a calculation model must remain verifiable.

3.5.4 Modifying an Array Formula

When you're modifying a matrix formula, the two methods differ. For an "old" matrix formula enclosed in curly braces, you can select any formula cell within the result matrix, and once you activate the formula bar, the curly braces disappear. After editing the formula, you must confirm it again by pressing Ctrl+Shift+Enter, which updates all formulas within the array.

This step is essential; if you don't perform it, you'll get an error saying part of the array can't be changed. This is because Excel treats a formula array as a single unit, which means you cannot edit individual elements of an array separately. Only operations that affect formatting are exempt from this rule.

All content changes always apply to the entire array, and that's why you can't delete or move individual cells. You can only delete or move the entire array. Inserting cells into an array with array formulas triggers an error, but you can copy the content of individual cells to another part of the table. References adjust automatically during this process. You can only edit array formulas that use the new method—without curly braces—in the first cell of the result array, and you confirm the change by pressing Enter or clicking the checkmark.

To convert array formulas into individual values—that is, their results—select the entire array, use **Home • Clipboard • Copy** to copy the data to the clipboard, and then paste it back in place with **Home • Clipboard • Paste** by using the **Values** option.

How to Select an Array Range

To select a full array range, select any cell in the range and press Ctrl+/. This keyboard shortcut matches the **Current Array** option in the **Select Contents** dialog box of the **Go To** command, but you can't use this option with dynamic arrays.

3.6 Ensuring Quality and Preventing Errors

This section highlights key quality aspects to consider when building calculation models, and it then presents various Excel tools you can use to detect and prevent errors.

3.6.1 Verifiability

With so many commands and functions available, no one can predict the solutions that users will create for their tasks. Generally, the process used to produce specific results in a worksheet should remain clear and verifiable, and a formula system that even its developer can't understand after six months is clearly undesirable. Other users of a table—such as supervisors or colleagues covering during vacations—also deserve clarity, and for tables with tax-related data, relevant regulations must be followed.

It's often helpful to leave comments on cell contents when complex calculations are involved, and you'll find it useful to add direct notes to the table as well. Instead of silently placing a conversion factor in a cell, you should at least label it with the factor's name above or before it. Otherwise, someone might delete the number that appears orphaned in the table as unnecessary, which in extreme cases could cause entire formula systems to fail.

3.6.2 Flexibility

Besides clarity, flexibility is crucial. Rather than using a fixed sales tax rate in an invoice's formulas, it's better to reference the address of a cell containing the local sales tax rate. If the rate changes, you only need to update that one cell—not every formula that uses the value.

3.6.3 Error-Free Operation

The most important quality criterion for a spreadsheet is error-free operation. This applies both to the accuracy of the input data and, more importantly, to whether the formulas and functions work correctly and deliver flawless results.

3.6.4 Avoiding Errors in Formulas

The developers have put special effort into preventing errors. They've continuously improved the error-checking tools and thus given users greater confidence that they'll obtain accurate results. However, various types of errors can still occur when you're working with formulas, and logical errors are the most troublesome because Excel can't detect them automatically. For example, if you enter "AND(ZIP<50000, ZIP>60000)" instead of "AND(ZIP>49999, ZIP<60000)" in a condition, Excel won't flag it, but you might wonder why the program returns no data.

Formula Entry Error

The situation improves if you don't follow the prescribed formula syntax. Excel often automatically corrects formula entry errors, so if you enter an expression with an opening parenthesis but forget the closing one, then Excel will suggest that you add it. You can accept or reject the suggestion if the parenthesis is misplaced. Figure 3.47 shows a simple example. If you reject the correction, Excel provides a more detailed error description.

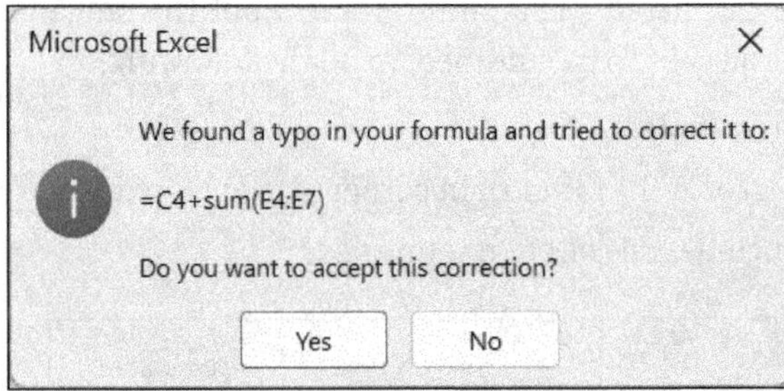

Figure 3.47 Correction Suggestion After Formula Error

Excel offers correction suggestions for many similar errors. For instance, if you were to enter "6B" instead of "B6" in a cell address, meaning you'd entered the row number before the column letter. Excel also detects misplaced spaces, such as those between function names and the opening parenthesis, as well as missing or misplaced quotation marks and colons.

Excel can suggest useful fixes for missing parentheses in simple cases, but not when multiple placements are possible. To help you locate missing or extra parentheses, Excel temporarily highlights matching pairs in bold when you navigate over them with the arrow keys. For example, if an opening parenthesis isn't bolded, then the matching closing parenthesis is missing. Matching pairs also appear in different colors when the cell is in edit mode.

Correcting References

If a formula refers to the wrong cell or range, you can adjust the references later by using a mouse. Double-click the formula cell and Excel will highlight each cell or range reference in the formula with a different color. The corresponding cells and ranges in the table will be outlined with matching colored borders, and you can drag these highlights with the mouse to fix an incorrect cell or range reference in a sum formula.

The following figure shows an example in which the first formula multiplies the **Price** by the **Discount Level1** of **70%**. Use the **Discount Level2** of **80%** and perform the following steps:

1. Double-click the cell containing the formula.
2. Drag the colored frame from the cell with **70%** to the cell with **80%**.
3. Press [Enter] to complete the correction.

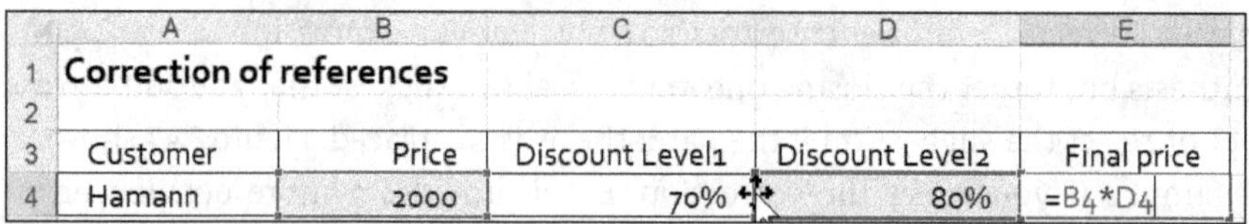

	A	B	C	D	E
1	**Correction of references**				
2					
3	Customer	Price	Discount Level1	Discount Level2	Final price
4	Hamann	2000	70%	80%	=B4*D4

Here's another example. Say you've calculated a row sum for a row but the range is incorrect. The range does not include all cells to be summed, so do the following:

1. Double-click the cell with the sum formula.
2. Place the mouse pointer on the marker at the lower right corner of the colored range.
3. Drag the box to the right until it includes all cells to be summed.

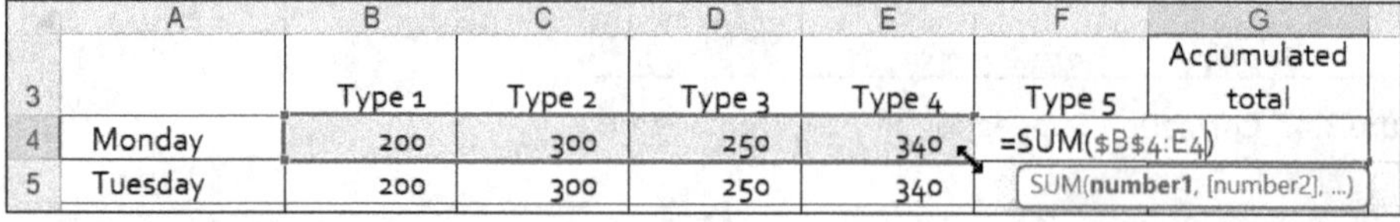

	A	B	C	D	E	F	G
3		Type 1	Type 2	Type 3	Type 4	Type 5	Accumulated total
4	Monday	200	300	250	340	=SUM($B$4:E4)	
5	Tuesday	200	300	250	340	SUM(**number1**, [number2], ...)	

You can do the same on a touchscreen. Double-tap the cell to edit it, and cells referenced in the formula will be outlined in different colors that match the argument text colors. Then, use the handles to fix any incorrect cell references.

	A	B	C	D	E	F	G
3		Type 1	Type 2	Type 3	Type 4	Type 5	Accumulated total
4	Monday	200	300	250	340	230	1090

Figure 3.48 Correcting Range by Dragging Handles

3.6.5 Syntax Checks

When you enter a formula, Excel automatically checks its syntax before accepting the result in the cell. For example, if an argument is missing in the PV() function, an error message will indicate the missing arguments. After you acknowledge the message, Excel stays in the formula bar and highlights the error location. And thanks to the tooltips that appear when editing functions, managing function arguments is easy. When you place the cursor or selection on an argument within a function, the tooltip highlights the placeholder that matches that argument.

	A	B
1	**Present value calculation**	
2		
3	Rate	2.5%
4	Nper	20
5	Pmt	$10,000.00
6	=PV(B3,B4,B5,,0)	
7	PV(rate, nper, **pmt**, [fv], [type])	
8		

Figure 3.49 Controlling Arguments with Tooltips

If you enter one argument too many, you'll get an error message and won't be able to leave the formula bar until you fix it.

Bypassing the Syntax Check

Despite all this help, complex formulas—especially nested ones—can sometimes hide errors you just can't find right away. Since Excel won't let you leave the formula bar unless you press the **Cancel** button—and risk losing the entire entry—you need to outsmart Excel. So, move to the start of the entry by using Home or by tapping and then type a single quotation mark. This will turn the formula into text that is displayed in the cell, and it will give you a chance to pause or consult someone helpful. Once you know the solution, simply remove the quotation mark and the error.

Common Syntax Errors

Some errors happen frequently, so we list them here to help you avoid them from the start:

- The function name is misspelled.
- The formula contains spaces where they are not allowed. There must be no space between the function name and the opening parenthesis. Spaces are allowed elsewhere.
- The logical expression is incomplete. For example, a comparison like this is not allowed:

  `=AND(B1>C1, >D1)`

  You must write this instead:

  `=AND(B1>C1, B1>D1)`

  Otherwise, the second part of the comparison is not a valid logical expression.
- The function has an incorrect number of arguments—either too few or too many.
- An argument separator is missing between two arguments.
- A parenthesis is missing. This error often occurs with nested functions.

You can largely avoid these errors by using the **Insert Function** dialog for input.

3.6.6 Errors Caused by Values

In addition to syntax errors, Excel can detect errors related to the values or arguments used in a formula. If these values are invalid for the formula, Excel displays an error value in the cell, and it also provides an explanatory note when you hover over the hint button. For example, the formula in Figure 3.50 uses the undefined name *Threshold*.

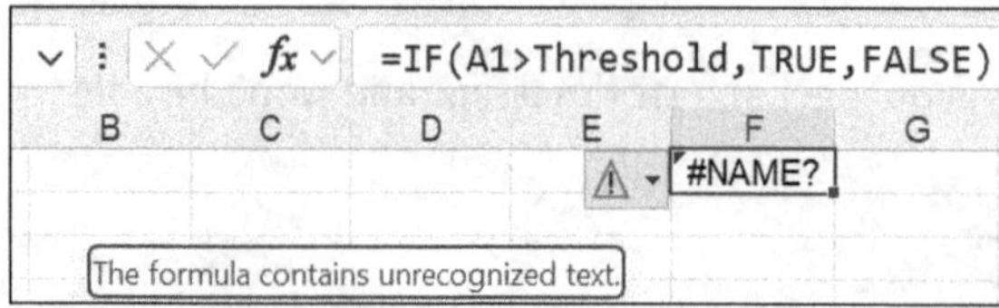

Figure 3.50 Note on #NAME? Error Value

The following table lists the error values and their meanings. The error type number appears in the right column, and you can retrieve it by using the `=ERROR.TYPE` function.

Error Values	Meanings	Type Numbers
#NULL!	There's been an attempt to find the intersection of two non-overlapping areas by using the intersection operator (space).	1
#DIV/0!	The formula tries to divide by zero.	2
#VALUE!	An argument or operand has the wrong data type.	3
#REF!	The formula refers to a deleted or moved cell range.	4

Error Values	Meanings	Type Numbers
#NAME?	The name used in the formula hasn't been defined yet.	5
#NUM!	An estimated value in a function that requires one is invalid.	6
#N/A	The expected value is missing.	7
#GET_DATA	Data retrieval failed.	8
#OVERFLOW!	Not all array results can be returned.	9
#CONNECT!	The program cannot establish the connection.	10
#BLOCKED	Access to the resource is blocked.	11
#UNKNOWN	There's an unknown error.	12
#FIELD!	The data type field does not exist.	13
#CALC!	The calculation cannot be completed.	14
Other		#N/A

3.6.7 Background Error Checking

Excel offers automatic error checking that runs in the background and is controlled on the **Formulas** page in the **Excel Options** dialog, where it can also be disabled if needed. Under **Error Checking Rules**, you can specify when the error indicator should appear.

The check not only detects error values in a cell but also warns about formulas with characteristics that might indicate a problem, such as sum formulas that don't include all cells in the adjacent range. So, you should decide how strict you want the check to be and adjust the criteria to match your workflow. For instance, if you always create formulas before entering values, you can deselect the **Formulas referring to empty cells** option.

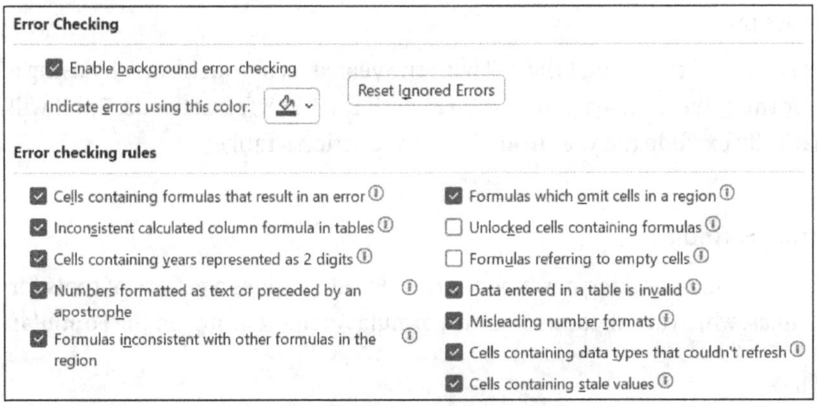

Error Checking

- [x] Enable background error checking

Indicate errors using this color:

Reset Ignored Errors

Error checking rules

- [x] Cells containing formulas that result in an error
- [x] Inconsistent calculated column formula in tables
- [x] Cells containing years represented as 2 digits
- [x] Numbers formatted as text or preceded by an apostrophe
- [x] Formulas inconsistent with other formulas in the region
- [x] Formulas which omit cells in a region
- [ ] Unlocked cells containing formulas
- [ ] Formulas referring to empty cells
- [x] Data entered in a table is invalid
- [x] Misleading number formats
- [x] Cells containing data types that couldn't refresh
- [x] Cells containing stale values

Figure 3.51 Error Checking Settings

While background error checking is enabled, small colored error indicators appear in any cell that returns an error value or meets any of the selected conditions. You can choose the color by using the button under **Error Checking**, and when you select the cell, an error icon appears. Hovering the mouse pointer over it displays a hint about the error, clicking or tapping opens a menu with possible actions you can take to fix the error (see Figure 3.52).

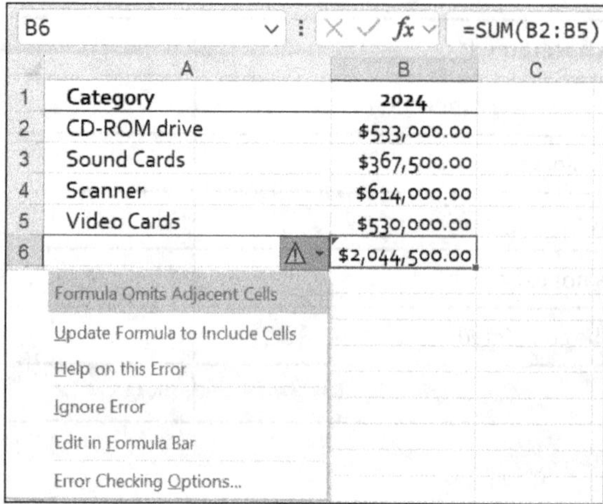

Figure 3.52 Options when Violation Occurs

The first line always provides a brief description of the problem, and you can use **Help on this Error** to find details about the issue. If available, an option lets you correct the error directly. For example, it lets you expand the reference range of the sum function.

However, the example in Figure 3.52 is a false alarm. The warning indicates that the cell containing the year is not included in the sum range, and it appears because Excel can't always recognize a number as a year. You can disable this warning by selecting the **Ignore Error** option, and you can undo this action anytime by using the **Reset Ignored Errors** button under **Error Checking** in the **Excel Options** dialog. You can also reverse undoing this action.

To avoid this issue, you can format the cell with the year as a column header by using a custom date format that displays only the year, with the YYYY format code. Excel will then automatically exclude the year from the sum function's range.

3.6.8 Auditing Formulas

Besides the background checking described earlier, Excel provides a variety of tools for auditing formulas, which are all located in the **Formula Auditing** group on the **Formulas** tab.

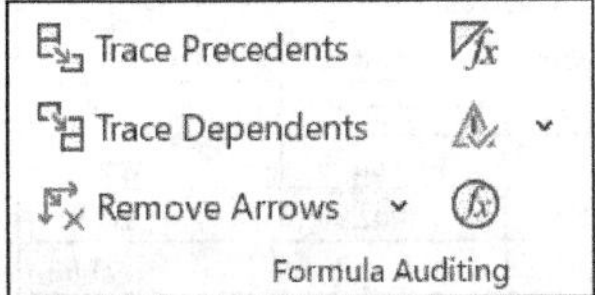

Figure 3.53 Tools for Auditing Formulas

Manual Error Checking

You can manually start checking for errors in any worksheet by using the **Error Checking** button or the same option in its menu, whether or not background error checking is enabled. Use **Next** and **Previous** to move through cells with errors or warning indicators one at a time, and the dialog will display a description of each error and offer buttons for next steps.

E5 =SUM(C5:D5)

	A	B	C	D	E
1	**Sales**				
2					
3		Sports goods	Toys	Leisure dress	Total
4	Jan	10000	12000	13000	35000
5	Feb	12000	14400	15600	30000
6	Mar	14400			
7	Total	36400			

Error Checking

Error in cell E5
=SUM(C5:D5)

Inconsistent Formula

The formula in this cell differs from the formulas in this area of the spreadsheet.

Copy Formula from Above
Help on this Error
Ignore Error
Edit in Formula Bar
Options...
Previous
Next

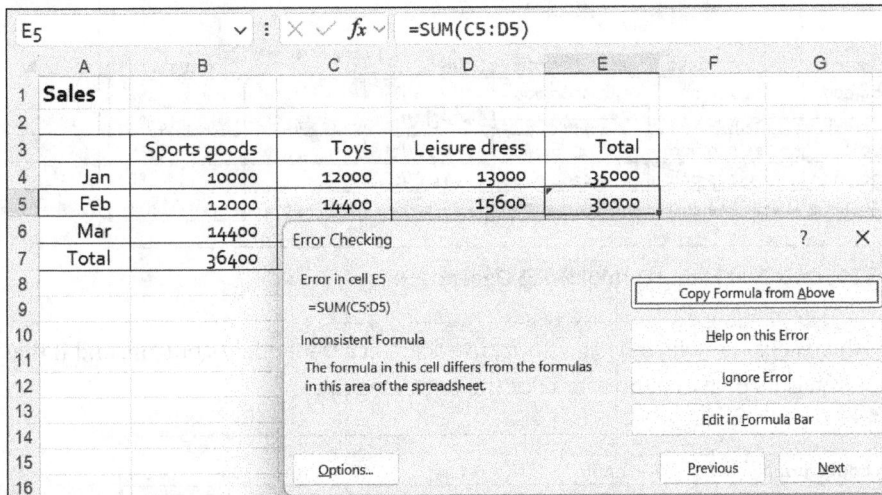

Figure 3.54 Checking for Errors Step-by-Step on Sheet

Tracing Precedents

The most troublesome errors are those that go unnoticed, like logical mistakes or references to the wrong cell. As a calculation model grows and formula results feed into other formulas, it can become difficult to keep track of them, but buttons can help by temporarily creating visual links between cell ranges.

To see which cells a selected formula depends on, use the **Trace Precedents** button in the **Formula Auditing** group. Excel shows connections between cells with arrows, and each arrow begins with a dot. If a predecessor has its own predecessor, repeat the command for that cell. You can do this for multiple cells. Marking a range of cells doesn't help; the watch only works for the active cell. To mark predecessor cells, you double-click the trace arrow, and you can double-click it again to highlight the cell at the arrow's tip. This is especially handy when the arrow's starting point and tip are far apart.

	A	B	C	D	E	F	G
3	Product group	Revenue	Cost	Profit	Sales share	Cost share	Profit share
4	Furniture	$2,000,000.00	$1,240,000.00	$760,000.00	30%	29%	32%
5	Carpets	$3,000,000.00	$1,900,000.00	$1,100,000.00	44%	44%	46%
6	Housing need	$1,500,000.00	$1,050,000.00	$450,000.00	22%	24%	19%
7	Garden furniture	$250,000.00	$150,000.00	$100,000.00	4%	3%	4%
8	Total	$6,750,000.00	$4,340,000.00	$2,410,000.00	100%	100%	100%

Figure 3.55 Cells That Affect Value in G4

If you select a cell with an error, use the **Error Checking** button's **Trace Error** option to find cells that might be causing the problem.

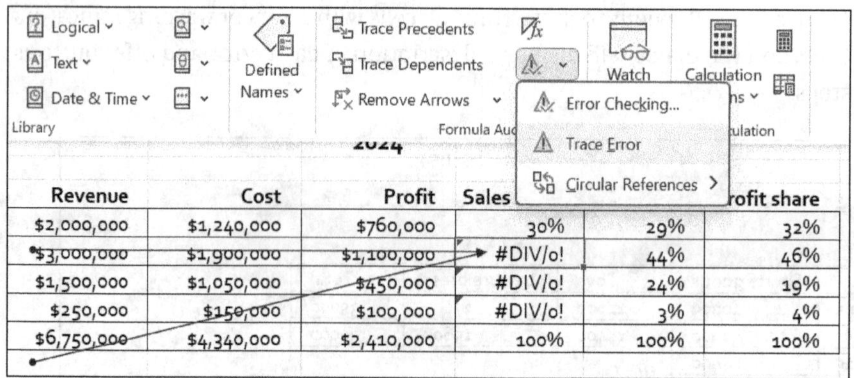

Revenue	Cost	Profit	Sales		rofit share
$2,000,000	$1,240,000	$760,000	30%	29%	32%
$3,000,000	$1,900,000	$1,100,000	#DIV/0!	44%	46%
$1,500,000	$1,050,000	$450,000	#DIV/0!	24%	19%
$250,000	$150,000	$100,000	#DIV/0!	3%	4%
$6,750,000	$4,340,000	$2,410,000	100%	100%	100%

Figure 3.56 Cells That Faulty Formula in E5 Depends On

To see which cells the selected cell affects, use the **Trace Dependents** button, and if the dependent has its own dependents, click the button again.

	A	B	C	D	E
3	Product group	Revenue	Cost	Profit	Sales share
4	Furniture	$2,000,000.00	$1,240,000.00	$760,000.00	30%
5	Carpets	$3,000,000.00	$1,900,000.00	$1,100,000.00	44%
6	Housing need	$1,500,000.00	$1,050,000.00	$450,000.00	22%
7	Garden furniture	$250,000.00	$150,000.00	$100,000.00	4%
8	Total	$6,750,000.00	$4,340,000.00	$2,410,000.00	100%

Figure 3.57 Cells Affected by B8

It looks different when a trace leads to another sheet or workbook. (That workbook must be open as well.) In both cases, an arrow points to or from a small table icon. Double-clicking or double-tapping the arrow opens the **Go To** dialog box with the address of the dependent or precedent cell, and double-clicking or double-tapping that address takes you to the cell.

Removing Tracers

If you no longer need the tracer arrows, you can remove them by using the **Remove Arrows** button in the **Formula Auditing** group. Alternatively, instead of removing all arrows at once, you can delete the two types of tracers separately from the button's menu.

3.6.9 Value Monitoring in the Watch Window

Excel includes a watch window where you can track the values of selected cells. You can open it via the **Formulas • Formula Auditing • Watch Window** button. Then, you can use the **Add Watch** button to select any cells or cell ranges whose current values you want to monitor in the watch window.

Any change in values is immediately shown in this window, and this is especially helpful when you're checking cells that are far apart in the worksheet or on different sheets. Double-clicking or tapping a list entry instantly selects the corresponding cell.

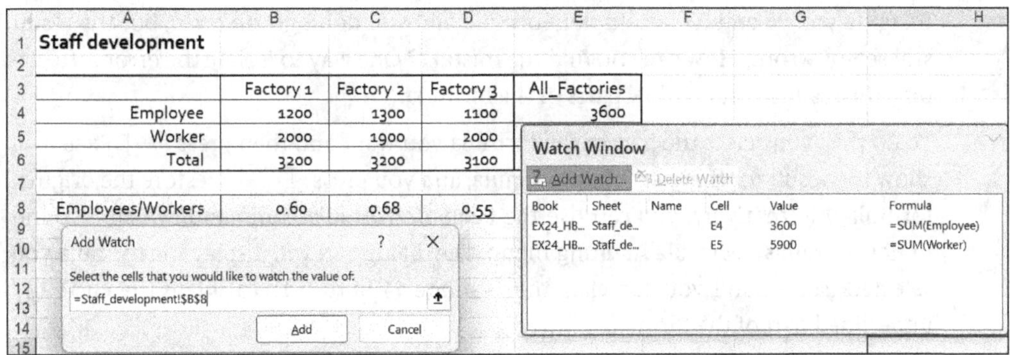

Figure 3.58 Value Checking in Watch Window

3.6.10 Circular Formulas

Excel also notifies you when a formula is circular. For example, the following function is entered in cell B9:

```
=SUM(B4:B9)
```

Cell B9, which should display the sum, is part of the range being summed, so if you try to confirm the sum formula, you'll get an error message indicating a circular reference in the workbook. You'll also see this warning when opening a workbook that contains such a reference.

The status bar shows the address of the cell causing the circular reference, and the cell with the circular formula is set to zero. For more on handling circular formulas, refer to Section 3.9.

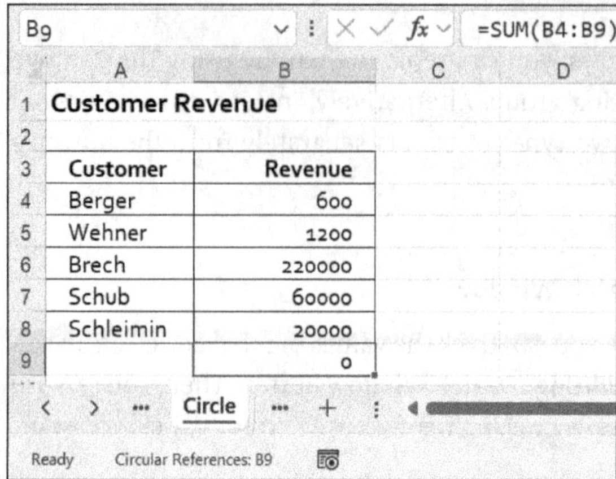

	A	B
1	Customer Revenue	
2		
3	Customer	Revenue
4	Berger	600
5	Wehner	1200
6	Brech	220000
7	Schub	60000
8	Schleimin	20000
9		0

Figure 3.59 Circular Reference Message in Status Bar

3.6.11 Stepping Through Formulas

Imagine you've entered a complex formula and you don't get an error, but the result still seems wrong. How can you find the mistake? One way to isolate the error is to display the results of individual parts of the formula.

To do this, you select the part of the formula you want and then press F9. Excel will show the result for that part of the formula, and you press Esc to restore the original formula. Alternatively, you can use the **Evaluate Formula** command, which is found under **Formulas • Formula Auditing** menu. The dialog box will display the formula you selected earlier, and you can click the **Evaluate** ❶ button to calculate the currently underlined part of the formula.

Clicking **Step In** ❷ shows the value of the highlighted section, and clicking **Step Out** ❸ inserts this value into the formula. Unlike pressing F9, this process doesn't change the formula in the cell.

C4 =IF(B4>=1000,IF(B4>=50000,"A","B"),"C")

	A	B	C
1	Customer Classification		
2			
3	Customer	Revenue	Classification
4	Berger	600	C
5	Wehner	1,200	B
6	Brech	220,000	A
7	Schub	60,000	A
8	Schleimin	20,000	B
9	Vosken	900	C
10	Erben	1,300	B
11	Kosinsky	200,000	A
12	Toker	70,000	A
13	Marchat	23,000	B
14	Berend	70,000	A

Evaluate Formula
Reference: Classification!$C$4
Evaluation: = IF(B4>=1000,IF(B4>=50000,"A","B"),"C")
To show the result of the underlined expression, click Evaluate. The most recent result appears italicized.
❶ Evaluate ❷ Step In ❸ Step Out Close

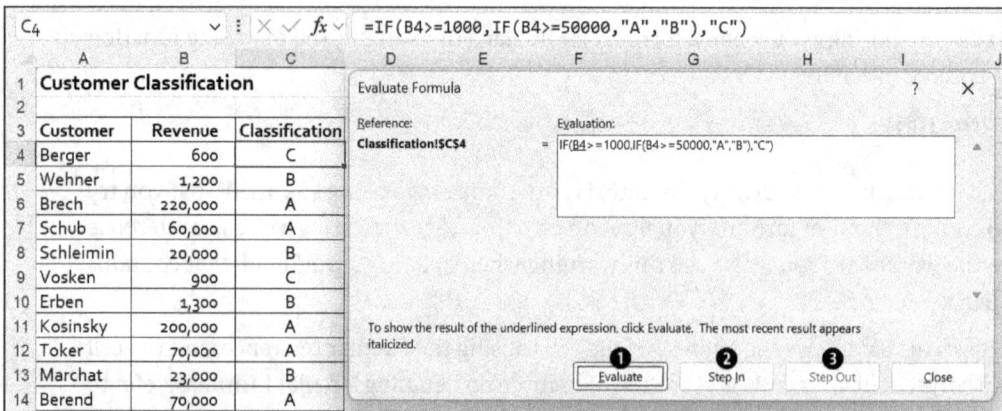

Figure 3.60 Step-by-Step Evaluation of Nested Function

3.6.12 Documenting Formulas

As a useful bonus, especially for documentation or training, Excel includes a full editor for mathematical equations. You can access it via **Insert • Symbols • Equation**, which opens a dedicated **Equation** tab. Along with numerous mathematical symbols in the **Structures** group, it offers several palettes with formula templates to make editing complex formulas easier. When you select a formula or part of one, the **Equation** tab appears to help you customize its appearance. Excel displays these formulas using text boxes.

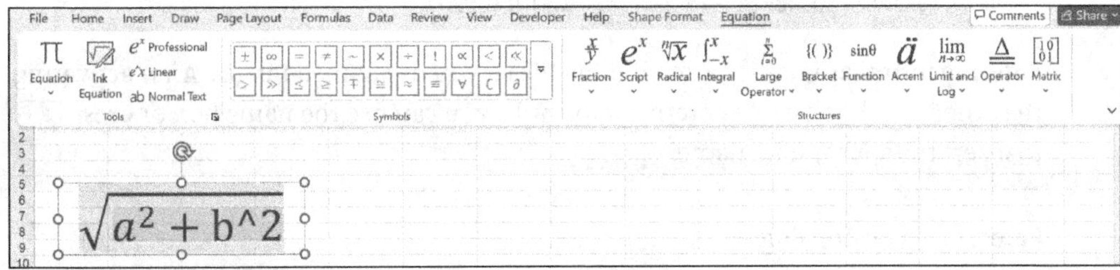

Figure 3.61 Formula as Graphic Object

3.7 Referencing Tables with Formulas

A formula can refer not only to cells in the same sheet but also to cells on another sheet in the same workbook. You can also reference values in other workbooks, and we previously introduced these external references while discussing different types of references in Section 3.3.

A reference can be simple and directly transfer the value from one cell to another, like this:

```
=[SOURCE.xlsx]Sheet1!$G$10
```

On the other hand, the value can be used as an operand in a formula or as an argument in a function, like this:

```
=[SOURCE.xlsx]Sheet1!$G$10*SalesTax
```

The formula with the external reference appears in the dependent workbook and points to the source workbook.

3.7.1 External Reference Notation

You can reference cell ranges in another file by prefixing the usual addresses with the file name, sheet name, and column and row references. To distinguish the file name from a sheet name, enclose it in square brackets, and to separate the sheet name from the rest of the reference, insert an exclamation mark. For ranges, the notation is simplified: you only need to enter the file and sheet names once. This is a valid summation formula:

```
=SUM([SOURCE.xlsx]Sheet1!$G$10:$G$20)
```

If the source workbook is in a different folder or on another drive, you must also include the drive letter and folder path:

```
=SUM('D:\EXTAB\[SOURCE.xlsx]Sheet1!'$G$10:$G$20)
```

Entering formulas is easier when linking files if you use named ranges. After activating the other workbook while entering a formula, you can use the name box or press F3 to insert the appropriate name.

Referencing by Copying

You can also create simple external references by copying through the clipboard. To do this, switch to the source workbook, select the cell or range to use as source data, and copy it to the clipboard. Then, switch to the dependent workbook, select the cell or range where you want the source data to appear, and use the **Paste Link** command from the **Insert** button menu.

3.7.2 Using External References

The need to use external references in Excel has decreased, thanks to the ability to work with large workbooks. In many cases, tables that reference each other through formulas can be combined within a single workbook, and you should prefer this method because it's easier to verify the accuracy of formulas within one workbook.

Still, there are situations where referencing values from other workbooks in formulas is useful and necessary. This is especially true when certain constants are used across multiple, different workbooks. Of course, you could copy these values into each workbook individually—but if these values change, you need to ensure they're updated simultaneously in all workbooks. Using an external reference makes this much easier.

Example: Exchange Rates

Consider a table that tracks daily data for various exchange rates. Several workbooks refer to this table to perform price calculations based on these rates, and here, the workbook with the exchange rates acts as the source workbook while the other workbooks that link to it externally are dependent workbooks.

The formula results depend on value changes in the source workbook. While it's convenient to have both linked files open on the desktop when entering references, it's not required. The source workbook doesn't need to be open for you to establish the connection. Figure 3.62 shows a small excerpt from the exchange rate table and an example of a price calculation based on that source data. For instance, if the dollar rate changes, you only need to update the exchange rate table.

F1 =[EX24_HB_Chap03_Course_table.xlsx]Course_table!C4

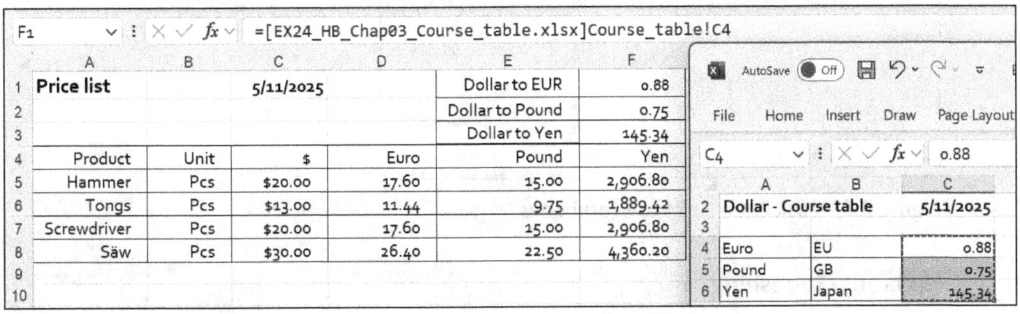

	A	B	C	D	E	F
1	**Price list**		5/11/2025		Dollar to EUR	0.88
2					Dollar to Pound	0.75
3					Dollar to Yen	145.34
4	Product	Unit	$	Euro	Pound	Yen
5	Hammer	Pcs	$20.00	17.60	15.00	2,906.80
6	Tongs	Pcs	$13.00	11.44	9.75	1,889.42
7	Screwdriver	Pcs	$20.00	17.60	15.00	2,906.80
8	Säw	Pcs	$30.00	26.40	22.50	4,360.20
9						
10						

C4 0.88

	A	B	C
2	**Dollar - Course table**		5/11/2025
3			
4	Euro	EU	0.88
5	Pound	GB	0.75
6	Yen	Japan	145.34

Figure 3.62 Linking Two Tables

It's best to create the table with the rates first and save it with its final name. Then, you can create references to this source file in the other workbooks. If you don't follow this order and try to save a dependent workbook before saving the source workbook, you'll be asked if you want to save the table with references to unsaved files.

This is not recommended. Excel will enter the temporary default name *Workbook(s)* into the formula, and since these default names are usually replaced when a table is saved, the reference will be lost because a file with that default name will no longer exist. Then, when you open the dependent file later, you'll see a dialog box titled **File Not Found**.

This solution looks even better when you use a data range in the workbook with exchange rates that pulls data from a table on a website providing current rates. This method is explained in Chapter 19, Section 19.4, using stock prices as an example.

Repairing Links

Links between files (now called *links*) are not foolproof. If a formula in the dependent file can't display values, then the **Workbook Links** task pane—which is found under **Data • Queries & Connections • Workbook Links**—can help. It lets you fix broken links so the affected formulas use correct references again and pull data from the source file. The task pane initially shows all links currently used in the workbook.

To correct a link, select it and choose the **Change source** option from the menu under the three dots. When you select the correct file from the list or use **Search**, the connection is restored, and the formula reference is updated. The menu also lets you break the link or open the source file. The icons at the top let you update or break all links in the workbook.

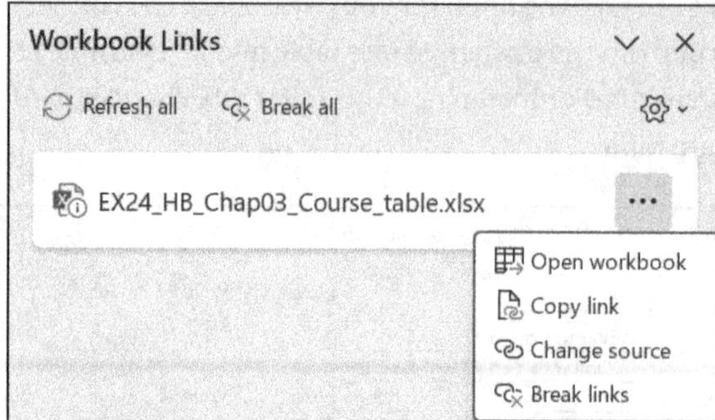

Figure 3.63 Task Pane Links to Workbooks

Updating/Refreshing

You can set more precise update options via the gear icon. Once you save a dependent workbook with linked formulas, it remains connected to the source workbook.

If a value in the source workbook changes a formula in the dependent workbook references, then the dependent workbook's value usually updates as well. If the dependent workbook isn't open at the time, then the update occurs only when you reopen it. When you open only the dependent table, a prompt appears asking if you want to update changes, provided the **Ask to refresh** setting is enabled. You should usually answer this question with **Update**, as this lets the dependent workbook incorporate changes made to the source workbook.

Always refresh skips this prompt. If you prefer manual updates only, select the **Don't refresh** option. However, this means you or other users won't be notified when the source data changes.

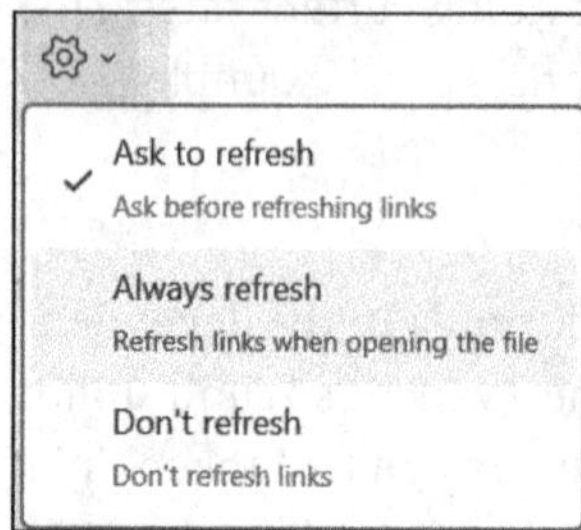

Figure 3.64 Update Options

Links Are Fragile

While these technical features are useful, you should carefully consider the organizational impacts of such links. For example, if someone renames a source workbook, formulas in dependent workbooks will lose their references. This issue can also arise if a file is moved to a different folder or drive. In that case, you can follow the steps described and redirect the link to the other file.

Be careful when copying files with external references to another computer. If you copy only the dependent table, then you cannot restore the reference and you should respond to the update prompt with **No**. The table will then display the recently used values.

3.8 Impact of Removing Cells

If you delete the cell itself instead of just clearing its content, then any formula referencing that cell loses its link and shows the `#REF!` error. You will need to fix the formula manually. If a cell referenced by an array formula is deleted, `#N/A` appears. The situation changes if the deleted cell is part of a range that supplies values to a formula argument. For example, consider a simple sum formula like `=SUM(B2:B12)`. If any cell is deleted, the formula will adjust correctly every time.

Deleting a cell with a formula only causes errors if another formula refers to it. In that case, the `#REF!` error appears. Cells within a result array cannot be deleted individually.

3.9 Recalculation Control

Normally, Excel automatically recalculates all formulas affected by new entries or changes to values, formulas, or name definitions. It does not recalculate formulas that are unaffected by changes. Calculations run in the background; if you're entering data or running certain commands, calculation pauses temporarily and then resumes.

3.9.1 Calculation Options

All options that affect Excel's calculation method are found on the Formulas page, which is accessible via the **File • Options** menu. Under **Workbook Calculation**, the main choice is between **Automatic** and **Manual**, and once you make this setting, it stays active until you change it. Although Excel limits automatic recalculation to cells affected by data changes, turning it off can help with large workbooks by preventing recalculation after every entry. The **Partial** option is designed to exclude large data tables and Python formulas from automatic calculation, since these tables and formulas can demand more processing power.

You activate manual recalculation by selecting the **Manual** option. The **Recalculate workbook before saving** box is checked by default, and usually, there's no need to change this because it guarantees that the file will be fully recalculated whenever it's saved. However, if you're working on a large workbook and save it often, it might be more convenient to uncheck the box.

Calculation options

Workbook Calculation
Automatic
Partial
Manual
Recalculate workbook before saving
Enable iterative calculation
Maximum Iterations: 100
Maximum Change: 0.001

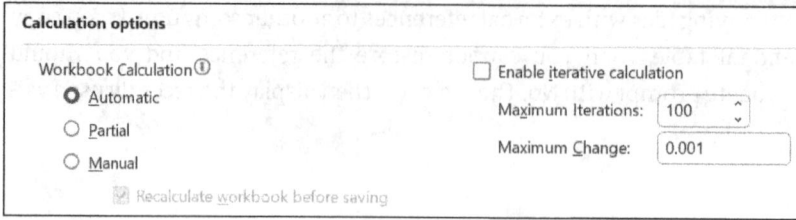

Figure 3.65 Choosing Calculation Options

To switch quickly between calculation modes, use the **Calculation Options** menu in the **Formulas** group under **Calculation.**

These settings always apply to all open workbooks. While Excel saves the chosen option with each workbook, it uses the setting from the last opened workbook. To ensure certain workbooks are calculated manually, open them separately.

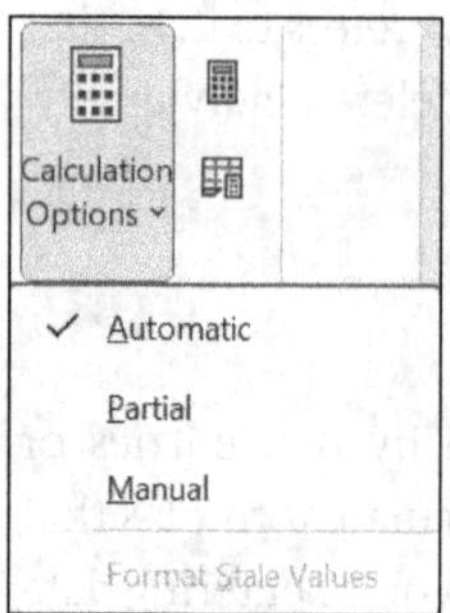

Figure 3.66 Calculation Options on Formulas Tab

If you enable **Manual** calculation along with the **Format outdated values** option, then until they are recalculated, values that would change during recalculation will be struck through to mark them as outdated.

6	Mar	~~14400~~	17280	18	~~50400~~
7	Total	~~36400~~			~~127400~~

Stale Value
Help on this Error
Calculate Now (F9)
Switch to Automatic Calculation
Disable Format Stale Cells

Figure 3.67 Strikethroughs Indicating Outdated Values

With manual recalculation enabled, you can recalculate all open workbooks by pressing [F9] or clicking the **Recalculate** button in the **Formulas • Calculation** group. To recalculate only the active sheet, use the **Calculate Sheet** button or press [Shift]+[F9].

When you enter a value that's related to a formula in the workbook, the status bar prompts you to **Calculate**. Clicking or tapping this button also triggers recalculation.

Refreshing Calculations

Before printing a table or displaying a chart, always recalculate the sheet to avoid printing or showing incorrect results.

3.9.2 Managing Iterative Calculations

Iterative calculations are repeated calculations that involve circular references. For example, Excel calls a reference circular when a formula in cell C7 includes cell C7 as an operand or argument. In this case, the formula refers to itself, causing the calculation to loop.

Such circular references can occur accidentally if you select the wrong cell. In that case, they are unwanted. However, you can also use them intentionally to solve certain calculation problems, such as equations where only approximate values can be found. The assumption is that repeating the calculation gradually approaches a value that no longer changes significantly upon recalculation. Figure 3.68 shows a simple example of this from cost accounting.

D13 =D6+0.05*F13

	A	B	C	D	E	F	G
1	**Iterative calculation: example cost accounting**						
2							
3			Preliminary cost center A		Preliminary cost center B		
4			Levy		Levy		
5			of Cost center B		of Cost center A		
6	**Primary costs**			2000		4000	
7	Levy 1 of A				20	400	
8		B	5	200			
9	Levy 2 of A				20	40	
10		B	5	200			
11	Levy 3 of A				20	4	
12		B	5	2			
13	**Primary and secondary costs**			2222		4444	
14							
15				=D6+(5%*F13)		=F6+(20%*D13)	
16							
17	Input cost center B assumes 20% of the costs of preliminary cost center A						
18	Input cost center A takes over 5 % each from input cost center B						

Reference_correction | Price_list | Circle | Row_colum_totals | Levy

Ready Circular References: D13

Figure 3.68 Iterative Allocation Calculation

In operational cost center accounting, different preliminary cost centers can mutually offset each other. Cost center A takes a percentage of cost center B's expenses, while cost center B takes a percentage of cost center A's costs. This is known as *reciprocal allocation*, and Figure 3.68 shows a step-by-step example of how to calculate it. The tedious process of allocating costs repeatedly until nothing remains can be replaced by two formulas using circular references.

The formula in D13 is this:

```
=D6+(5%*F13)
```

And the formula in F13 is this:

```
=F6+(20%*D13)
```

These correspond to the following equations:

```
D13=D6+(5%*(F6+(20%*D13)))
F13=F6+(20%*(D6+(5%*F13)))
```

Note that D13 and F13 appear on both sides of the equal sign. As a result, the formulas in D13 and F13 essentially refer to themselves because D13 refers to F13, which in turn refers back to D13, and the same is true for F13.

3.9.3 The Number of Iterations and Minimum Deviation

If you allow circular references in a table, you can control how Excel handles them. In the **Excel Options** dialog, which you can access via **File • Options • Formulas**, you can check the **Enable iterative calculation** box under **Calculation options** to permit iterative calculations.

You can also set the maximum number of iterations here. Excel displays the default value under **Maximum Iterations** as 100, which means the calculation stops after the hundredth iteration at the latest and shows the final result. The status bar indicates how many iterations have been completed.

The second setting that controls iterative calculation is **Maximum Change**, which defaults to 0.001. This means the calculation stops repeating when the difference between results is less than the maximum change. This is important when calculating an approximate value because the correct value depends on the dimension of the expected result. For example, if you expect values with six decimal places, 0.001 is too coarse.

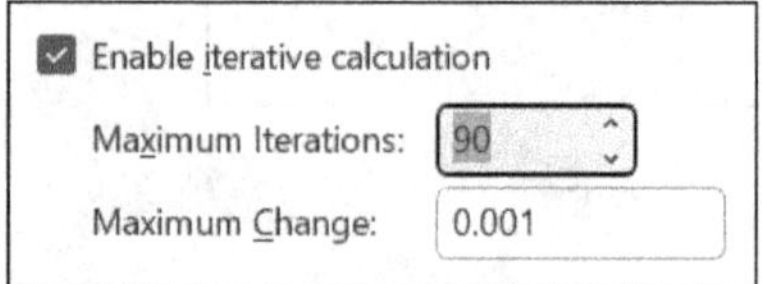

Figure 3.69 Options for Iterative Calculation

3.9.4 Multithreading

In the **Excel Options** dialog, under **Advanced** in the **Formulas** group, you can enable the multithreaded calculation option. For complex calculations, such as large PivotTables or data models, parallel processing can significantly improve speed, and if your system has multiple processors, you can specify how many to use for this feature. Excel automatically detects the number of physical or logical processors and, by default, selects **Use all processors on this computer**.

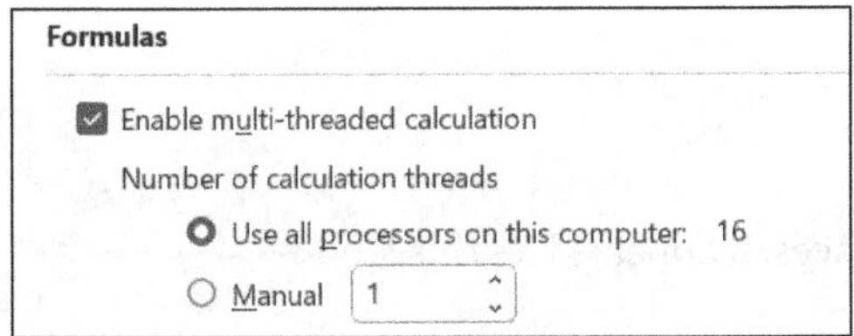

Figure 3.70 Options for Parallel Processing

3.9.5 Workbook Options

While the options we've discussed so far apply to the entire work session, you can set calculation options for individual workbooks in the **Excel Options** dialog under **Advanced**. Options you select under **When calculating this workbook** are saved with the workbook.

If **Update links to other documents** is checked, references to data from other Windows applications, like tables from a Word document, update during recalculation. Otherwise, Excel uses the most recent values that were provided by the external application.

The **Set precision as displayed** option is especially useful for practical work, but it's usually not selected. Why? When it's enabled, Excel replaces the internally stored values in the workbook with the displayed values. What does this mean? If this option is not enabled, then if a formula calculates a result with seven decimal places but you format it to show only two, Excel still uses the full seven decimal places in subsequent calculations. However, if the **Set precision as displayed** option is enabled, then it replaces the stored values with the formatted ones, so the value truly has only two decimal places—which means you'll lose some data. Because this change is significant, Excel asks for confirmation with the **Data will be permanently lost** message. Note that this setting always applies to the entire workbook.

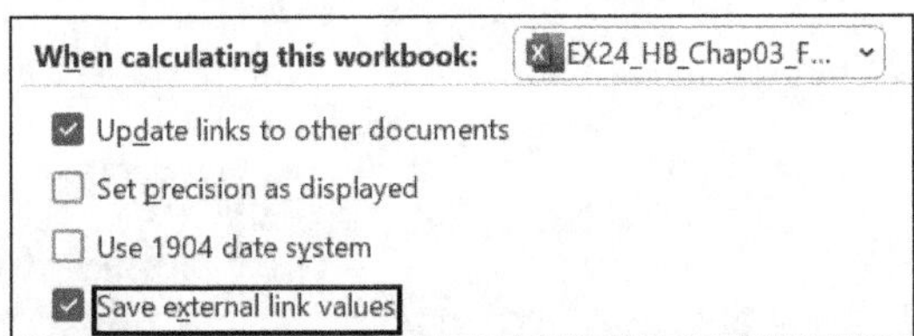

Figure 3.71 Options for Current Workbook

Excel's calculation precision is generally limited to 15 digits, which can be distributed between the integer and decimal parts in any way. Any digit entered beyond the 15th place is displayed and stored as zero. For example, say you enter this value into a cell:

12345678901234567890

The formula bar will show how Excel stores it:

12345678901234500000

In the standard format, it appears like this:

1.234567E+19,

And in number format without decimal places, it appears like this:

12345678901234500000

This doesn't indicate how large a number can be. The limit here is 9,99999999999999+307 for positive numbers and -9,99999999999999+307 for negative numbers, which is almost exactly equivalent to a 1 or -1 followed by 308 zeros.

If you select the **Use 1904 date system** option, the date serial number used for calculations will start at 0 for 1/1/1904 instead of 1 for 1/1/1900. This date system is used on Apple Macintosh computers. Apple avoided the issue that 1900 was not a leap year, and calculations from 1900 ignored this, causing the date 2/29/1900 to be incorrectly accepted. (As mentioned earlier, this no longer applies in the new Strict Open XML date format.)

Users who often exchange workbooks between Mac and PC can use the alternative date system. Note that changing the date system for existing date values in a workbook shifts them by 1,462 days, so you'll need to correct them by adding or subtracting this amount. The easiest way is to use **Paste • Paste Special • Add** or **Subtract**.

If you leave the **Save external link values** box checked, Excel stores copies of values from other workbooks that are linked to the active workbook in a cache to speed up formula calculations. If the linked source data range is very large, this can use a lot of storage space, so you can choose to disable this behavior in Excel if you wish.

Chapter 4
Designing Worksheets

The demands on worksheet design, beyond ensuring data and formula accuracy, vary widely. Correctly formatting worksheets that are used internally is less critical than correctly formatting a customer quote or management report. In any case, you should balance the effort you invest in the design with the importance of the purposes for which you intend to use the spreadsheet. You also need to consider how the results will be presented. For example, the font sizes of text on a spreadsheet that's printed out on paper will be different from the font sizes of text in a presentation that's projected on a wall. Also, publishing a quote online requires different considerations, like accessibility.

Let's see how a plain table can be transformed with some formatting.

	A	B	C	D	E	F
1	Novamedia: Quarterly Results 2024					
2						
3	Category	1st Quarter	2nd Quarter	3rd Quarter	4th Quarter	Total
4	DVD drives	120000	123000	140000	150000	533000
5	Sound cards	90000	95000	92000	90500	367500
6	Scanner	145000	149000	155000	165000	614000
7	Video cards	134000	120000	136000	140000	530000
8		489000	487000	523000	545500	2044500

Figure 4.1 Table Before Formatting

	A	B	C	D	E	F
1	**Novamedia: Quarterly Results 2024**					
2						
3	Category	1st Quarter	2nd Quarter	3rd Quarter	4th Quarter	Total
4	DVD drives	$120,000	$123,000	$140,000	$150,000	$533,000
5	Sound cards	$90,000	$95,000	$92,000	$90,500	$367,500
6	Scanner	$145,000	$149,000	$155,000	$165,000	$614,000
7	Video cards	$134,000	$120,000	$136,000	$140,000	$530,000
8		$489,000	$487,000	$523,000	$545,500	$2,044,500

Figure 4.2 Same Table After Quick Formatting Changes

Of course, taste plays a role and opinions often differ, but professional design aims to support a table's content and present information clearly and neatly. The layout should clarify what belongs together and what should be viewed separately, key information should be emphasized, and striking results should stand out to avoid being overlooked.

Your worksheets always make a better impression when they have a consistent, recognizable design—and you can achieve this with styles and themes, which this chapter covers in detail. Other options are the solutions that we introduced in Chapter 1.

4.1 Formats for Cells and Cell Ranges

Besides column width and row height, Excel's wide range of design options mainly applies to cells and cell ranges. The format of a cell includes all properties and are independent of its content. This raises the following questions:

- How should values be displayed?
- How should the cell content be aligned?
- Which font should be used?
- Should the cell be highlighted with lines or borders?
- What color and pattern should be used for the cell background?
- Should the cell be protected from changes?

4.2 Formatting Tools

The basic tools for formatting cell ranges are grouped on the **Home** tab under **Font**, **Alignment**, **Number**, **Styles**, and **Cells**.

The first three groups also have dialog box launchers in the lower right corner that you can click or tap to open different tabs of the **Format Cells** dialog. You can also access this dialog from the context menu of any cell range by using the same command. Each tab in the dialog addresses one of the questions in the previous list.

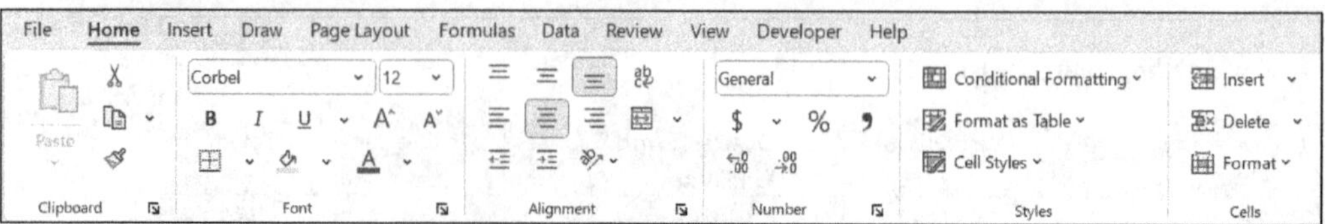

Figure 4.3 Basic Formatting Tools on Home Tab

The context menu for cell ranges also offers a selection of the most commonly used formatting icons (see Figure 4.4).

On a touchscreen, you can access the context menu by briefly tapping the selected range again with your finger, as described earlier. The horizontal menu shows icons like **Font** and **Fill** that let you select colors, and other commands appear only when you open the vertical menu by clicking the arrow at the end (see Figure 4.5).

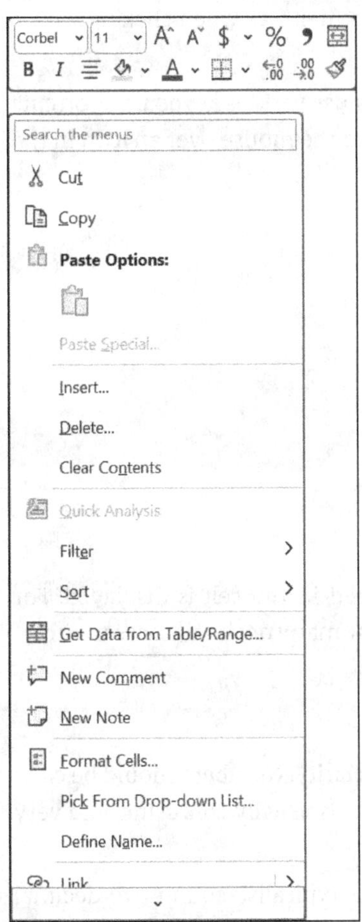

Figure 4.4 Advanced Context Menu for Cell Ranges

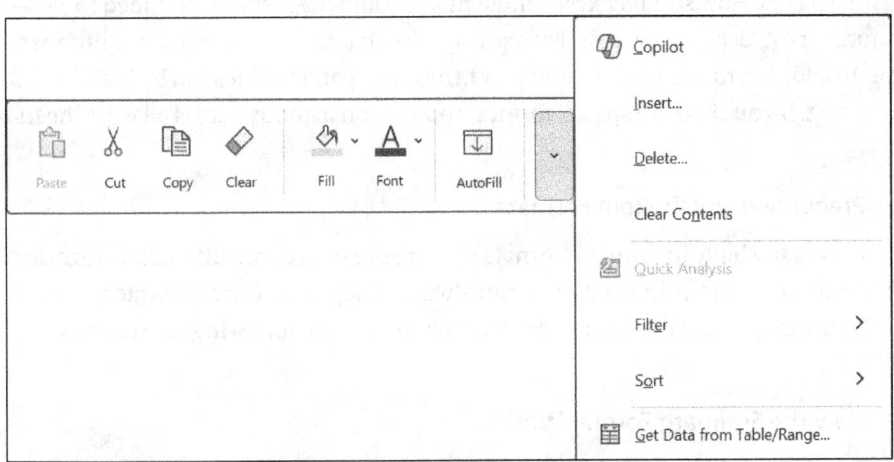

Figure 4.5 Context Menu for Cell on Touchscreen

Tapping the cell range with a pen brings up the same menu as when clicking with a mouse.

If you prefer not to use a mouse, finger, or pen, you can use various keyboard shortcuts to format cells. These shortcuts appear when you hover the mouse over an icon in the ribbon or hold the pen above it.

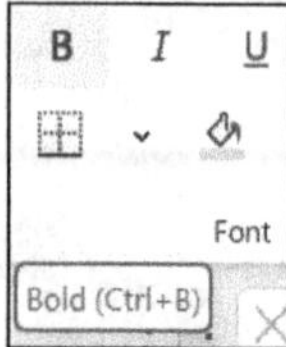

Figure 4.6 Displaying Keyboard Shortcut for Formatting

4.3 Choosing the Number Format

The *number format* determines how the value entered in the cell is displayed. For numeric values, this mainly concerns whether and how many decimal places to show.

4.3.1 Clarity and Accuracy

Two key requirements apply here, and the first one is clarity. Numbers should be clear and easy to read. A column with varying decimal places is usually confusing, and very large numbers are hard to read without digit grouping.

The second requirement is accuracy. Invoice amounts typically require two decimal places, but in a sales report, you can omit decimal places because working with thousands is enough. However, a formula may return results with an unpredictable number of decimal places. How should Excel handle this? Should the result be rounded to a specific number of decimal places? Formatting also controls how negative numbers appear. In addition to minus signs and parentheses, negative values can be highlighted with red text. If you choose a special format, apply it consistently across all worksheets.

4.3.2 Predefined and Custom Formats

Excel offers many built-in number formats that are organized into different categories. If a quote doesn't meet your needs, you can always add custom formats, which you can define by using specific code characters, as explained in the following subsections.

4.3.3 How the Standard Format Works

Unless you change cell formats in a new worksheet, Excel applies a standard format that displays numbers differently depending on the context, so the number format isn't fixed. This format can display up to eleven digits, and leading zeros before the decimal (except the one immediately before it) and trailing zeros after the decimal are omitted. For example, the input `01.30` displays as `1.3`.

If the column width is too narrow for a value, Excel first removes decimal places and rounds automatically. If that doesn't work or the number has more than eleven digits, then Excel displays it in scientific notation. For example, `6.3333E+33` shows the mantissa with the required decimal places and the exponent with at least two digits. The result of `3 * 10^12` appears as `3E+12`.

4.3.4 Input Format Determines Output Format

We've already mentioned a unique feature of this general format: How you enter data can automatically change the standard format to another format. The following table shows some examples.

Input	Output	Format Used
$12000	$ 12.000	Currency with dollar symbol
12,000.33	12,000.33	Number
12.25 %	12.25%	Percent
133e2 or 133E+02	1.33E+04	Scientific
3/7/18	3/7/2018	Date
3/1	1-Mar	Date
0 1/3	1/3	Fraction
6:30	06:30	Time

You can set the cell format by how you enter the data, and fixing the format through data entry is often a practical method because it removes the need for special formatting commands. For example, you can define the format of the first cell in a column by entering a dollar amount with two decimal places and then apply that format to the entire column. This works especially quickly when you use the **Home • Clipboard** group. Use the **Copy Format** button—the one with the paintbrush—and with the first cell selected, click or tap **Copy Format** and then select the corresponding column header. For the most common formats in the commercial sector, this method is entirely sufficient.

4.3.5 Format Icons

Excel offers a format palette and buttons for the most common number formats in the **Home • Number** group. The palette's dropdown list displays the current cell's format and includes many formats from various categories that you can quickly apply to a selected cell range with a click or tap.

On a touchscreen, swipe up in the palette with your finger or stylus if options at the bottom are hidden. These buttons also appear in the mini toolbar of the context menu for a cell range.

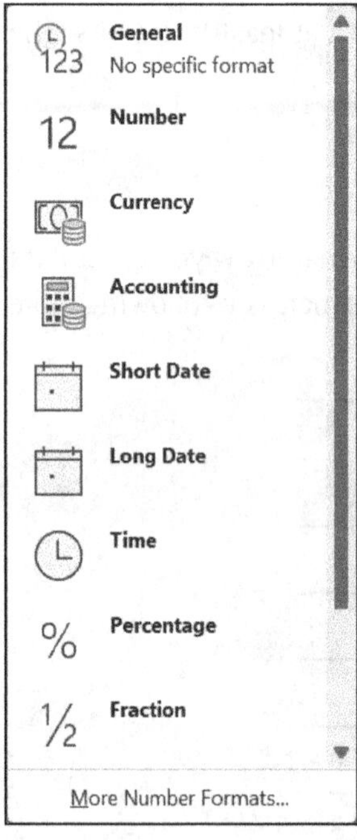

Figure 4.7 Palette of Common Number Formats

Some keyboard shortcuts also apply to number formats, as shown in the following table.

Symbols	Keyboard Shortcuts	Formats
		Accounting number format with two decimal places.
	Ctrl+Shift+$	Currency number format with two decimal places.
	Ctrl+Shift+%	Percent format.
	Ctrl+Shift+!	Number format with comma separators.
		Add a decimal place.
		Remove a decimal place.

You can click the currency symbol's arrow button to opens a menu you can use to quickly switch between different currency formats.

4.3.6 Defining a Specific Number Format

You can get more control over formatting through the **Format Cells** dialog box, which you can also open with `Ctrl`+`1`. It doesn't matter if you format the cells before or after entering data.

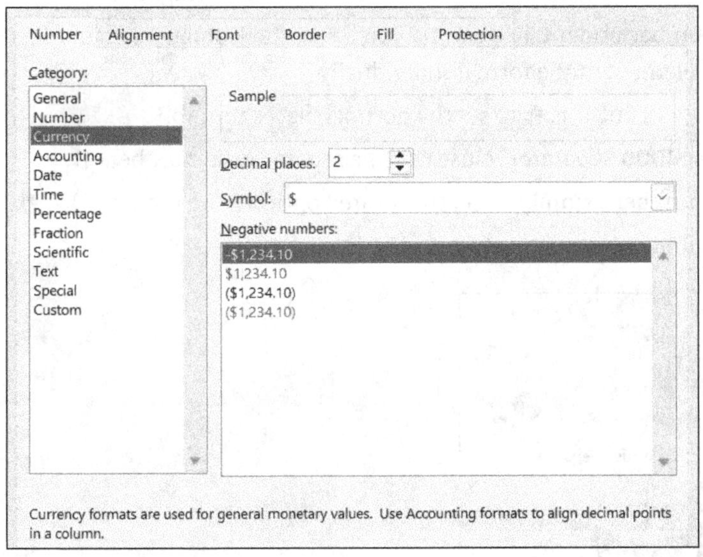

Figure 4.8 Format Cells Dialog Box with Number Tab

[+]

Getting Rid of Incorrect Formats

Say you mistyped a number and entered "12/12" with a slash instead of "12,12" with a comma. Instead of showing as **12,12**, it shows as **12-Dec**. Then, if you enter the number "12,12" correctly in the same cell, the cell will display **12-Jan** and the formula bar will show the corresponding date for the year **1900**.

The problem here is that your first entry changed the cell's standard format to a date format, so the second entry tries to interpret the input as a date but only recognizes the 12 before the decimal. However, if you select the cell and go to **Home • Editing • Clear • Clear Formats** or reapply the number format by using a format icon, then Excel will display **12.12** correctly again.

It's often best to format a cell range before entering data into it so the values appear exactly as you want from the start. You can also set a main format for the entire worksheet first and then format any differing areas separately. To do this, use the **Select All** box.

Number Formatting for a Cell Range

If you want to display numbers with comma separators but no decimal places in a table of statistical data, follow these steps:

1. Select the cell range. To format multiple sheets the same way, group the sheets first and then select the ranges on the active sheet.
2. Open the **Format Cells** dialog box.
3. On the **Number** tab, Excel shows the current format of the active cell unless you've selected a range with mixed formats. In that case, no format is highlighted.
4. Start by selecting **Number** under **Category**. As you'll see, the **Number** category lets you customize each element of the format individually.
5. In this case, use the small buttons ❶ to set the decimal places to zero.
6. Check the box for **Use 1000 Separator** to use comma separators in numbers ❷.
7. To display negative numbers, simply select the desired option from the small list ❸.
8. If the format shown under **Sample** meets your needs, click **OK**.

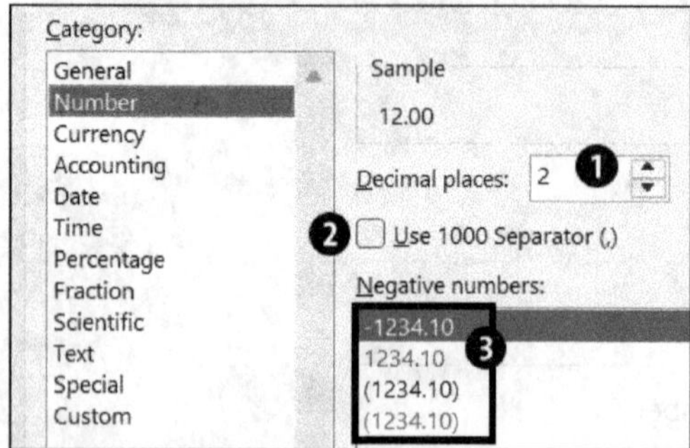

[»]

Resetting to the Standard Format

To reset a format to the standard format, select **Standard** under **Category** in the dialog box. Clearing a cell's format restores the standard format—not just for numbers but also for other formatting like font and color.

4.3.7 International Currency Formats

Excel makes it easy to work with multiple currencies in a single worksheet. Under the **Currency** and **Accounting** categories, you'll find number formats that include currency symbols and that also use comma separators. The currency symbol list offers a wide selection of symbols, and it lets you display euros in one column and US dollars or British pounds in another without any hassle (see Figure 4.9).

Number formats that are designed for accounting data entry, like balance sheets and journals, are available under **Accounting**. In these formats, zero values appear as a dash, and numbers align so the decimal point stays in the same place when decimal places match—regardless of whether currency symbols appear. The minus sign for negative values is left-aligned. In Figure 4.10, you can see how some of these formats work.

	A	B	C	D
1	International price list			5/11/2025
2				
3		Dollar	Euro	Pound
4	Product 1	$54.00	€ 60.48	£45.36
5	Product 2	$81.00	€ 90.72	£68.04
6	Product 3	$108.00	€ 120.96	£90.72
7	Product 4	$135.00	€ 151.20	£113.40
8	Product 5	$162.00	€ 181.44	£136.08
9	Product 6	$189.00	€ 211.68	£158.76

Figure 4.9 Table Showing Different Currency Formats

	A	B	C	D
1	Formats for numbers			
2				
3	Category	Input	Output	Codes
4	Number	12000.25	12,000.25	#,##0.00
5		-2390	-2,390.00	
6		0	0.00	
7				
8	Currancy	12000.25	$12,000.00	$#,##0.00;-$#,##0.00
9		-2390	-$2,390.00	
10		0	$0.00	
11				
12	Accounting	12000.25	$ 12,000.00	_($* #,##0.00_);_($* (#,##0.00);_($* "-"??_);_(@_)
13		-2390	$ (2,390.00)	
14		0	$ -	

Figure 4.10 Various Number Formats and Their Format Codes

For percentages, you can set the number of decimal places in the **Percentage** category.

Under **Fraction**, you'll find many format options. If you apply the **Up to one digit** format to a cell range before entering data, you can type "1/4" without Excel converting it to a date. That means you won't need to enter "0 1/4" thereafter. In the other examples, the number in the cell is rounded up or down to show the count of quarters, tenths, and so on.

Under **Scientific**, you can use the exponential format and freely choose the number of decimal places in the mantissa. This format shows the number as a product of a value and a power of ten, and it works especially well for very large or very small numbers, which are common in technical and scientific fields.

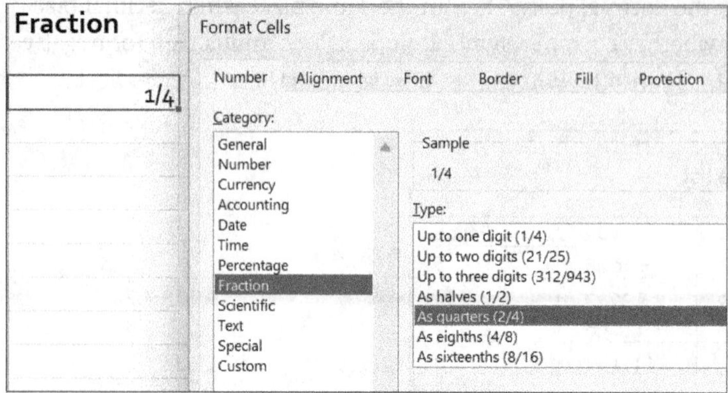

Figure 4.11 Fraction Formats

4.3.8 Date and Time Formats

In addition to number formats, Excel offers many options for displaying dates and times. You can access international formats from various countries by selecting the appropriate setting in the **Locale (location)** dropdown list. The long date format, which shows the day name before the date, is very useful and simplifies the creation of schedules. Figure 4.12 displays the built-in formats. Excel's default output format depends on the **Region** settings in **System Settings**. For example, if you choose the short date format there and select the four-digit year, a two-digit year will automatically display as four digits.

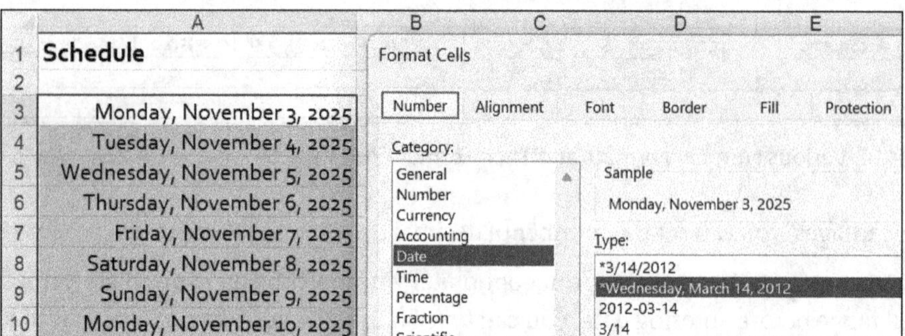

Figure 4.12 Particularly Useful Date Format that Includes Weekdays

If this format feels too unwieldy, you can use a custom format instead, as we explain next. The `TTT TT.MMM` format shows the date with abbreviations for the weekday and month (for example, Mo 24.Sep).

4.3.9 Text and Special Formats

One special format is the **Text** format. When you apply this format to an empty cell range, Excel treats any numbers you enter afterward as text strings and aligns them to the left. This is useful for entering codes, customer numbers, phone numbers, and similar data.

There's also a helpful set of special formats for ZIP codes, ISBNs, Social Security numbers, and more. Formats for specific countries are available here, and you can select them via **Locale (location)**.

4.3.10 Custom Formats

If the built-in number formats don't meet your needs, use the **Format Cells** dialog box to create your own. A format type can have up to four parts that are separated from one another by semicolons:

```
Positive format; negative format; zero format; text format
```

If a number format has only three parts, it means there's no specific text format. If the cell contains text, it displays normally as text. If a number format has two parts, then the first applies to positive and zero values and the second applies to negative values. If there's only one part, then it applies to positive, negative, and zero values. You can create these custom formats using an existing format or manually. For example, if you want six-digit document numbers with leading zeros, you can do the following:

1. Select the cell range.
2. In the **Format Cells** dialog box, under **Category**, select **Custom** ❶.
3. The list box will display format patterns you can use as a starting point for your own format, so choose the entry with the zero.
4. Add five more zeros in the **Type** ❷ input field and then click **OK**.

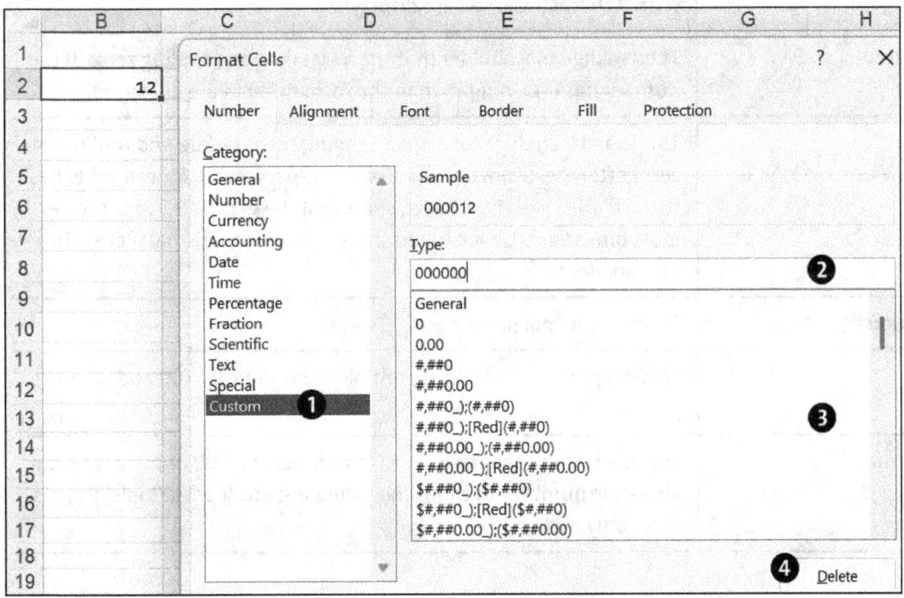

To apply the defined format to another cell, repeat steps 1 and 2 and then pick the new format from the list ❸. Excel always places the new format at the end of the list. You can delete a custom format anytime by selecting it from the list and clicking the **Delete** ❹ button.

Custom formats are saved with the workbook where you created them, and you have different options for using them in other workbooks. You can apply a format to another open workbook by using the **Format Painter** brush icon, or you can share formats through styles, as explained starting in Section 4.8.2.

4.3.11 Format Codes

Excel uses specific code characters to display number formats, which are listed in the following table. Placeholders for digits control how many decimal places appear, and Excel supports up to 15 decimal places. These code characters also control how leading and trailing zeros are handled. Signs can appear before or after the number, and Excel usually shows only negative signs. To display a plus sign, you must include it explicitly in the format pattern.

Format Codes	Meanings
Standard	Data is displayed in standard format.
No entry	Data is hidden. Use this format to hide data like repeating characters (e.g., ;;;).
#	This is a digit placeholder. Leading zeros before and trailing zeros after the decimal point are not shown, and Excel rounds to the specified decimal places. If there are more digits before the decimal point than # characters, they are displayed.
0 (zero)	This is a digit placeholder that shows leading and trailing zeros. If the number has fewer digits than the format, zeros are displayed.
?	This is a placeholder for a digit. Leading zeros before and trailing zeros after the decimal point display as spaces, and if the number has fewer digits than the format, spaces fill the gap. This keeps the decimal point aligned, even when showing varying decimal places. This also applies to fractions.
Period	This is a decimal point.
%	This is a percent sign. Excel multiplies the value by 100 and adds the percent sign.
Comma	This is a comma separator used in numbers of 1,00 or greater. It can also scale numbers to thousands, millions, etc. (For example, #,, displays only millions.)

Format Codes	Meanings
¢, £, and ¥	These are currency symbols you enter by holding NumLock and typing ANSI codes Alt+0162, Alt+0163, and Alt+0165, respectively.
E–, E+, e–, and e+	This is scientific notation format, and the number of zeros or # characters to the right sets the number of digits in the exponent. E– or e– shows only the minus sign, while E+ or e+ shows both signs of the exponent.
€, $, -, +, /, (,), {, }, <, >, ^, ´, = !, &, ~, :, and a space	Each one of these displays the specified character. To display other characters, enclose them in double quotes or precede them with a backslash (\). You can also use a hyphen (-) as a separator, as in ##-###-##.
* and **	The * code fills the cell to the left with the following character, and the ** can be used as a fill character when filling out checks.
_ (underscore)	This inserts a space the width of the next character, and it can be used, for example, to align positive numbers exactly under negative numbers shown in parentheses.
"*Text*"	This displays whatever text is inside the quotation marks. It lets you place the unit directly after a number, so for example, you can use #0.00 "sqm".
@	This is a placeholder for a string of any length
M or MM	This displays the month number without or with leading zeros (e.g., 3 or 03).
MMM	This displays the abbreviated month name (Jan, Feb, etc.).
MMMM	This displays the full month name.
MMMMM	This displays the month name as a single letter.
D or DD	This displays the day as a number without or with leading zeros (e.g., 9 or 09).
DDD	This displays and abbreviated day name (e.g., Sat, Sun).
DDDD	This displays the full day name
YY or YYYY	This displays a two-digit or four-digit year number (e.g., 93 or 1993).
h or hh	This displays hours without or with leading zeros (e.g., 3 or 03), and it displays the hour in 12-hour format if the format includes AM or PM.
m or mm	This displays minutes without or with leading zeros (e.g., 3 or 03). It must follow h or hh; otherwise, it will be treated as a month number.

Format Codes	Meanings
s or ss	This displays seconds without or with leading zeros (e.g., 3 or 03).
[ ]	This supports time formats exceeding 24 hours or 60 minutes/seconds. You enclose the leftmost part of the format in square brackets (e.g., [H]:mm:ss). This format lets you sum multiday periods in hours. Numbers entered without format characters are treated as multiples of 24 hours (e.g., Three equals 72:00:00).
AM /am / A /a PM /pm /P /p	This displays time in 12-hour format: AM, am, A, or a for times before noon (ante meridiem) and PM, pm, P, or p for times after noon (post meridiem). If the time format is not specified, then time displays in 24-hour format.
[Color]	The characters after this code appear in the specified color. Possible entries include BLACK, BLUE, CYAN, GREEN, MAGENTA, RED, WHITE, YELLOW, and COLOR n (where n is a number between 0 and 56 that represents a color from Excel's palette).
[Condition Value]	You can use this for conditional formatting. Condition represents one of the following operators: <, >, =, >=, <=, or <>. Value represents any number you enter. So for example, [>1000] applies the format that comes it if the cell's value is greater than 1,000.

The following figures demonstrate various formats and their effects.

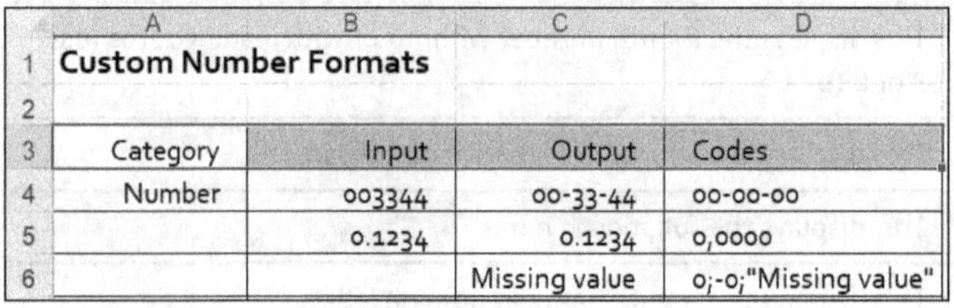

	A	B	C	D
1	**Custom Number Formats**			
2				
3	Category	Input	Output	Codes
4	Number	003344	00-33-44	00-00-00
5		0.1234	0.1234	0,0000
6			Missing value	0;-0;"Missing value"

Figure 4.13 Custom Number Formats

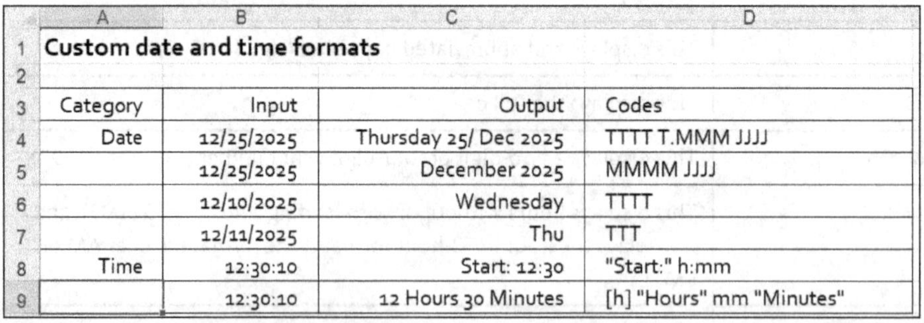

	A	B	C	D
1	**Custom date and time formats**			
2				
3	Category	Input	Output	Codes
4	Date	12/25/2025	Thursday 25/ Dec 2025	TTTT T.MMM JJJJ
5		12/25/2025	December 2025	MMMM JJJJ
6		12/10/2025	Wednesday	TTTT
7		12/11/2025	Thu	TTT
8	Time	12:30:10	Start: 12:30	"Start:" h:mm
9		12:30:10	12 Hours 30 Minutes	[h] "Hours" mm "Minutes"

Figure 4.14 Custom Date and Time Formats

	A	B	C	D
1	**Conditional formats**			
2				
3	Category	Input	Output	Codes
4	Date	1/8/2025	Deadline exceeded 08/01/25	[>31.12.25] "Deadline exceeded" TT.MM.JJ
5				
6	Number	1200	1200 qm	0 "qm"
7		1000	Start value: 1000	"Start value" 0
8		900	Invalid value	[>1000] "Invalid value"; 0

Figure 4.15 Conditional Formats and Formats with Additional Text

[+]

Troubleshooting Time Calculations

If you want to add different time values and enter them as hours and minutes, the total you get may not be what you expect. Excel usually interprets entries like 15:30 as a time of day (i.e., 3:30 PM), not as elapsed time. So, adding 15:00 to 13:00 results in 4:00 as a time of day. That is 13 hours after the first value, if you interpret the values as a time of day; but you want the result to be 28:00, which would be the value if you interpret them as elapsed time. What should you do? To convince Excel to understand your time values as elapsed time and display their sums correctly, you can apply a special format. In the **Format Cells** dialog box, under the **Custom** category, enter the format "[hh]:mm." In our example, this displays 28:00, showing the elapsed time in total hours and minutes.

4.3.12 The Problem with Zeros

In tables filled with data over longer periods, zero values often become an issue. Should formulas display zero values, show other characters, or show nothing at all? You have several options.

The first option is to hide zero values for the entire workbook. To do this, go to **File • Options • Advanced • Display options for this worksheet**, find the **Show a zero in cells that have zero value** checkbox, and uncheck it to hide zero values.

The second option is to define a format that hides zero values or replaces them with a dash (e.g., `0.00;-0.00;;` or `0.00;-0.00;"-"`). This method lets you create more precise solutions by applying formats to specific formulas. For formulas where it's important to know if the result is zero, such as test values, you can choose a different format.

The third option is to enter the `=NA()` function in cells where values are missing. The cell will then show `#NA`, and formulas referencing these cells will also return `#NA` to indicate missing values.

You can also control the display of zero values by using formulas. For example, this formula shows the sum in cell F3 only if it's not zero:

```
=IF(SUM(A3:E3) <> 0,SUM(A3:E3),"")
```

4.3.13 Currency Formats

As mentioned earlier, you can format numbers so the currency symbol automatically appears before or after the amount. You can set which symbol appears as the default currency symbol in the regional options of **System Settings**, and you can access the **Currency** tab by clicking the **More Settings** button.

You can set the dollar as the default currency by selecting the $ symbol from the **Currency symbol** dropdown list, or you can change it to any other currency, like the euro, yen, or pound.

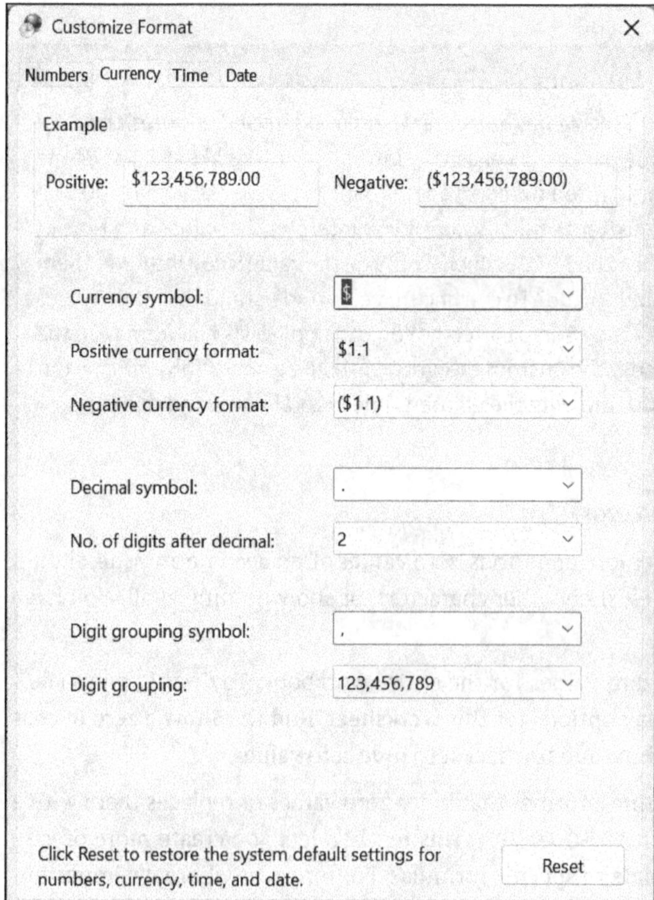

Figure 4.16 Setting Default Currency Symbol

4.3.14 Years

The year 2000 posed a problem for computers because we were used to dropping the century digits from year numbers. The safest approach going forward is to break this habit and always enter the full four-digit year, but this isn't without challenges either. You may need to widen columns to fully display the date.

Excel helps you switch to four-digit years by offering date formats with the full year in the **Format Cells** dialog box. One format shows the date as month number, day of month, and full year—as in 9/7/2021. Others use the full or abbreviated month name, such as Sep 7, 2021.

Interpreting Incomplete Year Numbers

Excel offers two methods to help you properly handle incomplete year numbers. When you enter a date in a cell without specifying the century, Excel will, by default, read an entry like "10/12/07" as October 12, 2007. This applies to all years from 00 to 29. If you enter a number between 30 and 99, Excel will assume that it refers to a date during the 1900s and will prefix the year with 19. For example, it interprets an entry of "10/12/43" as October 12, 1943.

	A	B	C
1	Input Output		
2			
3	10/12/07	-->	10/12/2007
4	10/12/19	-->	10/12/2019
5	10/12/34	-->	10/12/1934
6	10/12/89	-->	10/12/1989

Figure 4.17 Examples of Outputs for Incomplete Years

Excel automatically shows the full year in the formula bar when you select the cell. If you reformat a cell to display the year as four digits, you can verify the result. Excel always stores the full year internally.

If you need a date from 1904, such as your (great-)grandmother's birthdate, you can enter the full four-digit year even if the cell is formatted for two-digit years.

You can also adjust the time window described previously if needed. To do this, in **Control Panel** under **Region**, click the **Additional Settings** button to find a **Date** tab with an option in the **Calendar** group (see Figure 4.18). For example, you might set the range from 1940 to 2039, and then, Excel would interpret an entry of "10/12/33" as a date in the twenty-first century. However, you should only make such a time window shift for good reasons. For instance, if you need to accurately process birthdates from the early twentieth century, resetting the time window can help Excel interpret an entry of "15" as 1915 instead of 2015. It's especially important that all devices in a company use the same time window.

On this tab, you can set two-digit years to automatically display as four digits by default. However, you can still shorten them later using a custom format. To do this, choose a **Short date**, which is a format like **MM/DD/YYYY**.

If you eventually want to find cells in a workbook that use two-digit years, you can use the **Error Checking** feature described in Section 3.6. Enable the **Cells with two-digit years** rule under **File • Options • Formulas** in the **Error Checking Rules**.

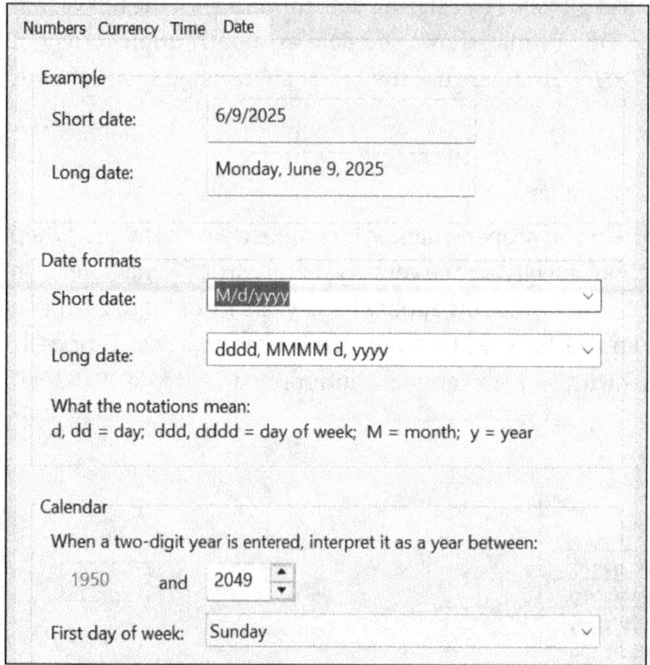

Figure 4.18 Shifting Time Windows in Exceptional Cases

Special Features of the Date Function

Excel handles dates created by a date function differently from dates you directly enter into cells. When you're using `=DATE(Year, Month, Day)` with a two-digit year, Excel doesn't apply the usual time window rule; instead, it always adds 1900. This applies to all year entries from 0 to 1899. For example, you can enter the date 12/31/2000 as "=DATE(100, 12, 31)."

You can also enter dates from 2000 onward by using a four-digit year. Excel processes all four-digit years from 1900 to 9999 directly, and when you're using the 1904 date system, the same rules apply to numbers from 4 to 1899 and from 1904 to 9999. You may find this behavior unusual at first, but it provides greater flexibility for date calculations over extended periods. Keep in mind that many Excel functions require a date as an argument. If the argument is stored as a date in a cell, the windowing rule applies. If the argument is created using the date function, the behavior described here applies.

4.4 Font Style and Alignment

How you apply fonts and align cell contents within cells greatly affects the professional appearance of your worksheets. This involves ensuring that data in table columns is easy to read and creating clear, attractive headings and labels.

4.4.1 Choosing the Right Fonts

When discussing a cell's font, three properties matter: the font family, font size, and font style (the last of which can be a combination of attributes like boldface and italics).

Font

The number of fonts available on computers today is nearly overwhelming, and your font choice should at least suit the document's content. Calligraphic fonts like Zurich-Calligraphic make birthday party invitations look nice—an easy task in Excel—but look out of place in a profit-and-loss statement. Also, mixing too many fonts on the same page can make documents look unprofessional. It's best to keep it down to one or two fonts, especially since you can vary font sizes and styles.

Fonts fall into two main categories:

- Serif fonts
- Sans-serif fonts

Serifs are the small strokes at the ends of letters that were originally created by Romans using chisels to inscribe letters in stone. If you look at a book or magazine, you'll see that most paragraph text—like in this book—is set in a serif font such as Times New Roman. On the other hand, headings often use sans-serif fonts like Calibri, Corbel, or Arial.

Serif fonts are easier to read in longer texts because they guide the eye in recognizing letters. For example, a capital *I* and a lowercase letter *l* are easier to tell apart in serif fonts. However, sans-serif headings tend to look clearer and more open, and sans-serif fonts are also usually better to use for columns of numbers.

Aptos is the new sans serif font in Office
Times New Roman is a well-known serif font.

Figure 4.19 Text without and with Serifs

Font Size

Font size is typically measured in points. A font size of 8 points is quite small, and for number columns, a font size of 10 or 11 points usually works best. When projecting tables on a wall, it's best to use 12- or 14-point font. You can also use 12-point font for labeling rows and columns, while table headers or chart titles should use 14-point font or larger. A well-designed heading always draws attention.

Font Weight

You can freely combine font attributes like **Bold**, **Italic**, and **Underline**, but using them sparingly usually creates a more polished look. You should never use highlights without a specific purpose in mind, and when you do apply them, do so consistently. For example, if you double-underline totals, do so consistently throughout the entire workbook. Styles can help with this and are explained further in the following sections.

Changing the Default Font

For a long time, Excel used 11-point Calibri as the body text font and Calibri Light for headings. However, Microsoft recently introduced a new default font that was designed by Steve Matteson and named after the California city of Aptos.

You can change this via **File • Options • General** at **Use this font as the default font.** Open the dropdown list and select your preferred font and then set the font size under **Font size.** This change will take effect only after you restart Excel.

This book primarily uses the Corbel font, which enhances the readability and clarity of numbers.

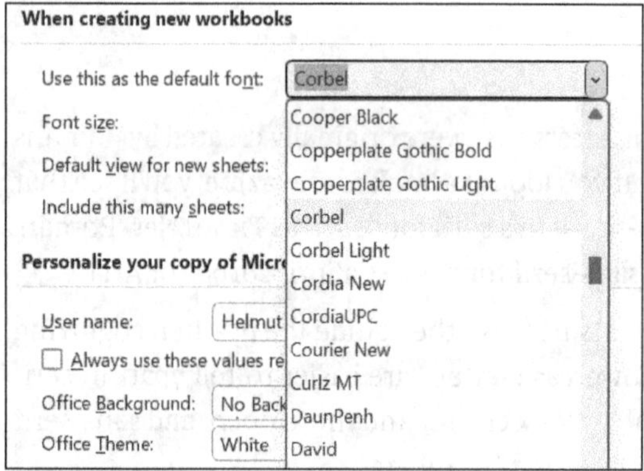

Figure 4.20 Choosing Default Font

The Font Group

To choose a font for a cell range, select the cells and use the buttons and list boxes in the **Home** tab's **Font** group. These tools are grouped together, and the font and font size list boxes always show the current settings of the active cell.

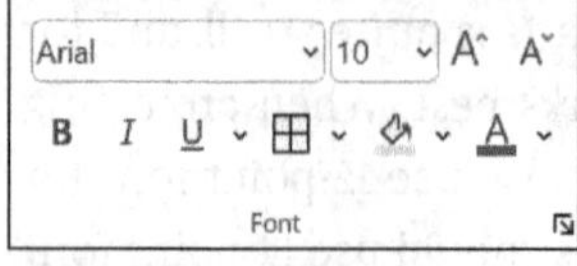

Figure 4.21 Font Group

When you open the font list via the small arrow button, the font names immediately appear in their actual fonts. The cell you select will also temporarily adopt the font of any font name on the list that you hover over with your mouse. This lets you easily find the font you want by trying it out without changing any text, and you can click or tap to apply the font you want to your selection.

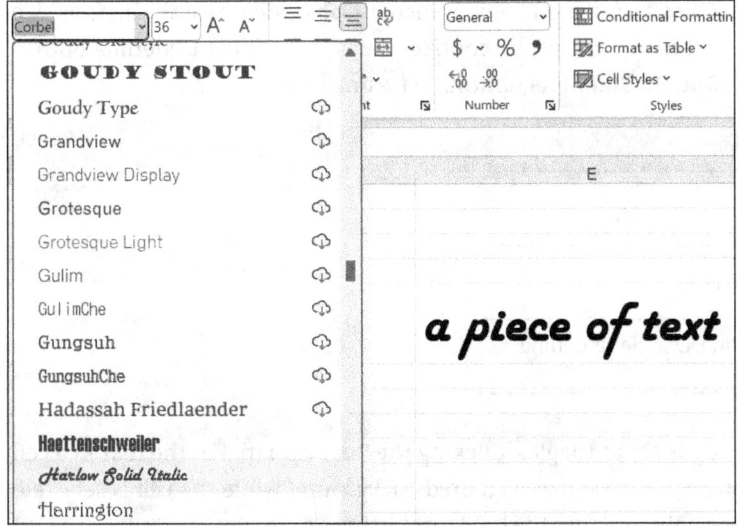

Figure 4.22 Temporary Font Displays

In **Touch input** mode, to make selection easier, options are spaced further apart. You can also drag or swipe up or down to reveal fonts that are currently off-screen, and if you know the name of the font you want, you can usually type the first one or two letters of the name to move the font to the top of the list. Use the handle at the bottom to shorten or extend the list.

The temporary font preview, which you activate by hovering with the mouse or pen before finalizing with a click or tap, is usually enabled as part of the **Live preview** feature. It helps you find the perfect font type, font size, and many other formatting features, but you can disable it in the **Excel Options** dialog under **General** if you need to. However, **Live preview** isn't available in **Touch input** mode.

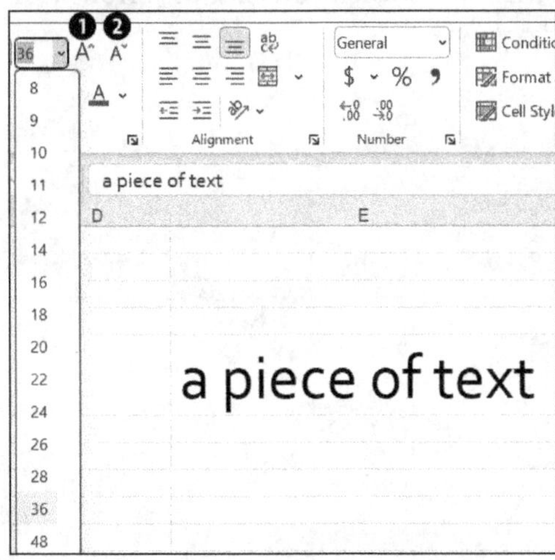

Figure 4.23 Font Size Dropdown List

Alternatively, you can use the two buttons to increase ❶ or decrease ❷ the font size stepwise. The other buttons in this group control font style, and the **Underline** button offers **Underline** and **Double Underline** options via a small arrow.

Symbols	Functions
F	Bold
K	Italic
U ▾	Underline, Double Underline

These font attribute icons act as toggles. Clicking the **Bold** icon makes the text in a cell boldface, which shows up onscreen as a colored background while the cell is selected. Clicking the **Bold** icon again removes the bold formatting. Some font attributes can also be applied with keyboard shortcuts, and Excel sometimes offers two options for this, as described in the following table. These shortcuts also act as toggles.

Keyboard Shortcuts	Effects
Ctrl+2 or Ctrl+B	These toggle **Bold** on or off.
Ctrl+3 or Ctrl+I	These toggle **Italic** on or off.
Ctrl+4 or Ctrl+U	These toggle **Underline** on or off.
Ctrl+5	This toggles **Strikethrough** on and off.

Font Color

The **Font Color** icon links to a color palette that opens when you click the arrow with your mouse.

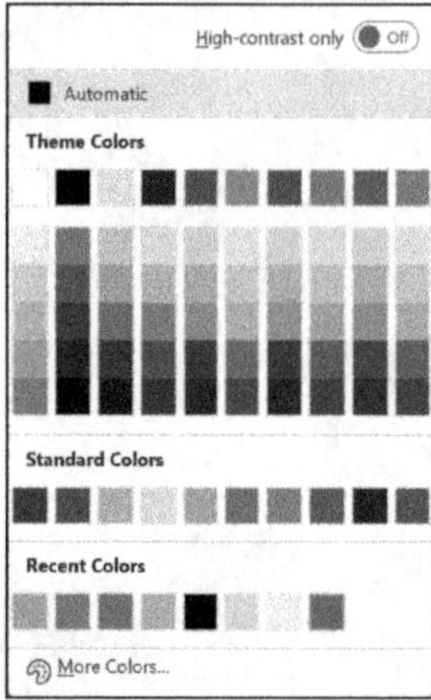

Figure 4.24 Font Color Palette

To assign a color to just one range, click directly on the desired color pattern. The font in the selected range will change color, and the button itself will update to that color. So, when you select another cell range, just click the left part of the button.

In **Touch input** mode, the palette expands to make the color patterns easy to tap. Tap the button to open the palette and select a color by tapping the desired color pattern.

The **Fill Color** button offers a similar palette, which we'll cover in more detail in the following sections. We'll also discuss the border lines button located here later.

The Font Tab

In addition to the tools in the **Font** group, you can open the **Font** tab in the **Format Cells** dialog box by using the group's dialog box launcher. There, you can enter font properties or select options from the list boxes. You can even enter fractional values like 10.5 for the **Size**. When you choose the font properties for selected cells under **Font**, **Font style**, and **Size**, a sample appears in the **Preview**.

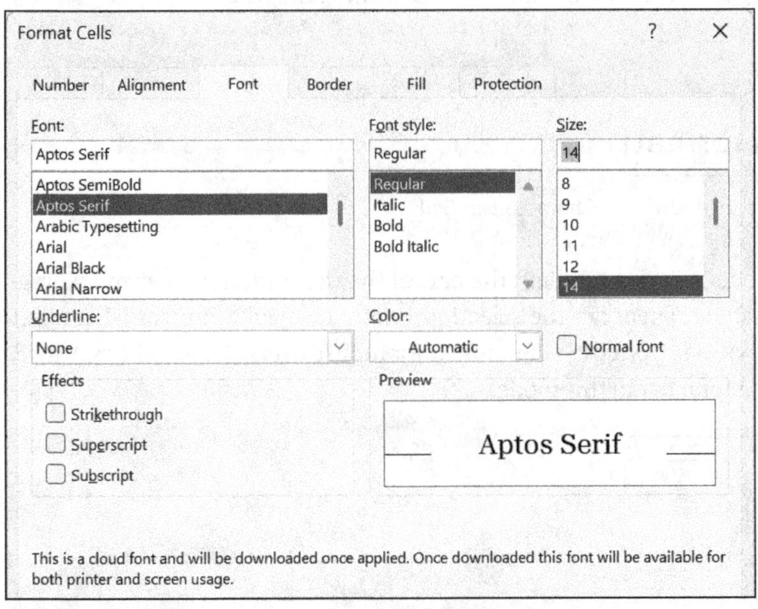

Figure 4.25 Font Tab in Format Cells Dialog Box

> **Highlighting Suggested Changes**
>
> The **Strikethrough** option is especially useful when several people edit worksheets. This lets you suggest striking through certain text or values without deleting the data right away. If your suggestion is rejected, you can remove the strikethrough formatting.

In the **Underline** field, you can choose from various underline styles. To change the text color, click the arrow under **Color** to open the color pattern dropdown list. Under **Effects**, choose whether the text is strikethrough, superscript, or subscript. Finally, you can check the box next to **Normal font** to reset the selected range to the current default font.

You can open the **Format Cells** dialog box from the context menu of a cell selection via **Format Cells**, but it's often simpler to use the buttons on the mini toolbar in the context menu, which also provides key font tools on the left. To set a main font for the entire worksheet, select the whole sheet by using the **Select All** box.

On a touchscreen, you open the context menu—which is arranged slightly differently—by tapping a selected range again. Then, you can use the **Fill** and **Font Color** buttons to open the palettes we described earlier.

Formatting Individual Characters in a Cell

You can change the font not only for an entire cell but also for individual characters or words in it. For example, to highlight the first letter of a heading, select that character directly in the cell or formula bar and then use the font icons on the mini toolbar or the **Format Cells** dialog box. You can handle chart labels the same way.

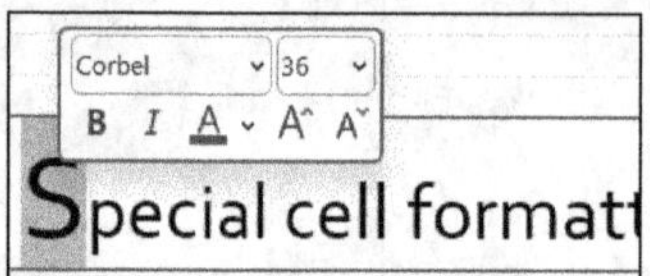

Figure 4.26 Formatting Individual Characters in Cell

In **Touch input** mode, you can highlight the part of the cell content you want by dragging the round handle. Then, tap the selection briefly to open a horizontal context menu with some options from the mini toolbar for mouse users. For all other settings, it's best to use the menu bar in this mode.

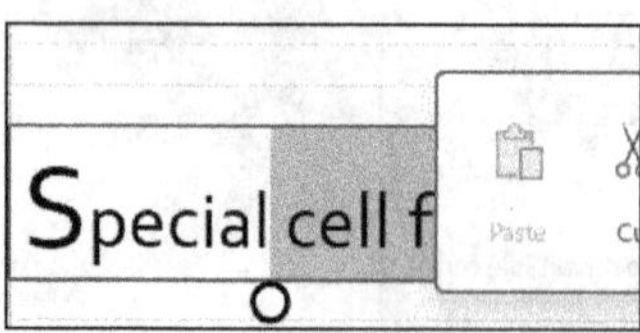

Figure 4.27 Formatting Characters in Cell on Touchscreen

[»]

Watch Out for Pitfalls in Group Editing

Say you selected multiple sheets to which you want to apply a uniform font, and then, you spotted a small error in the heading of the first sheet and fixed it. Then, to your dismay, you find out that all sheets share the same heading because you forgot to turn off group mode after formatting the sheet group. Group mode appears in the title bar, and to exit group mode, you can click or tap a single tab outside the group. If all sheets are grouped, you can open the group's context menu and select **Ungroup Sheets**.

4.4.2 Aligning Labels and Cell Values

By default, Excel aligns text to the left edge of the cell, aligns and numbers to the right, and centers logical values like TRUE and FALSE as well as error values. Therefore, the alignment of a formula's result depends on the data type it returns.

Often, the default alignment isn't ideal. For example, when a text column is to the right of a number column, the text appears too close to the numbers. If the column heading is left-aligned but the numbers are right-aligned, the table can look disorganized, as shown in Figure 4.28.

	A	B
1	**Value**	**Test person**
2	233	Tester 1
3	344	Tester 2
4	123	Tester 3
5	235	Tester 4

Figure 4.28 Columns That Need Their Cell Alignment Changed

Adjusting the alignment helps make the table much clearer. To change the alignment of a cell range, select the cells you want to modify and then use the **Home** tab's **Alignment** group to select the icon with the alignment option you want.

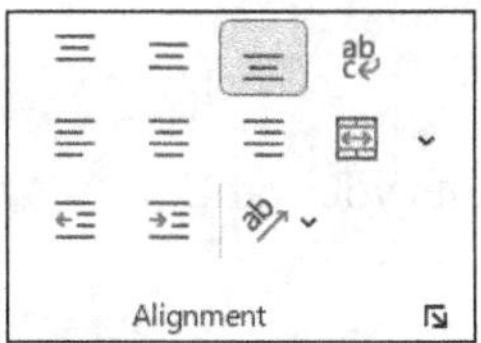

Figure 4.29 Alignment Group

The top toolbar has buttons for vertically aligning cell contents, and next to it is an **Alignment** icon with its own menu. At the end of the row, you'll find a **Wrap Text** icon, which lets you enter multiline labels in a cell.

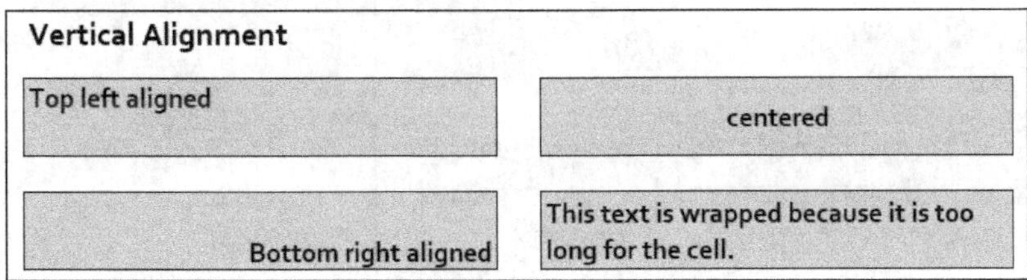

Figure 4.30 Examples of Vertical Alignment

The bottom row in the **Alignment** group features the most commonly used icons for horizontal alignment. You can select **Align Left**, **Center**, or **Align Right**. Note that if text is longer than the column width, it will overlap adjacent empty cells: right-aligned text overlaps cells to the left, left-aligned text overlaps cells to the right, and centered text overlaps cells on both sides. Next to it are two icons you can use to gradually decrease or increase indents. We'll cover them shortly.

The last button in this row opens a menu with commands for merging cells. The first option is **Merge & Center**, and we'll show you an example of it in the next section.

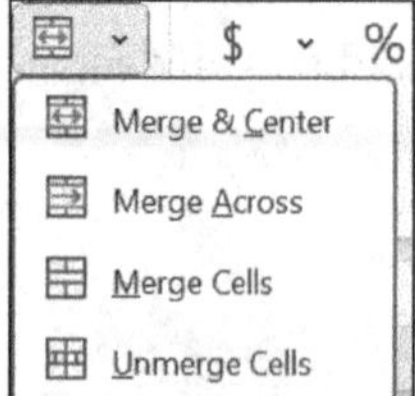

Figure 4.31 Merge & Center Button Menu

4.4.3 Centering Headings Across Multiple Columns

When table headers span many columns, you'll often want to center the heading over the table. To do this, follow these steps:

1. Type the title in the leftmost column.
2. Select a range from that cell to the last used column in the table.
3. On the **Home** tab, under **Alignment**, select **Merge & Center**.

The cells in the selection must be empty or you won't get the results you want.

The Alignment Tab

The **Alignment** tab in the **Format Cells** dialog box, which is accessible via the dialog box launcher, offers even more options. This dialog box provides separate settings under **Text alignment** for horizontal and vertical placement of cell contents (see Figure 4.32).

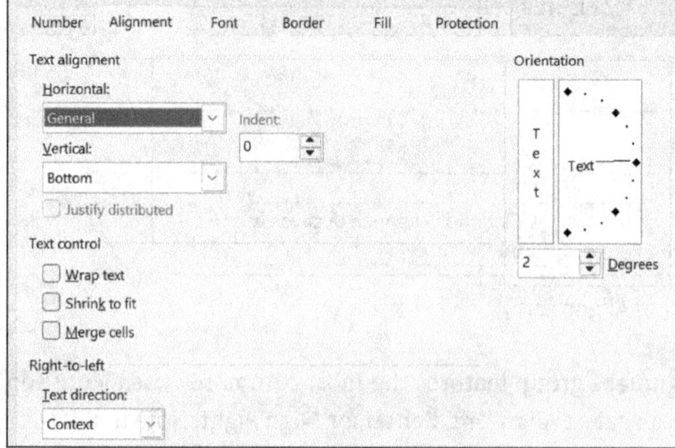

Figure 4.32 Alignment Tab in Format Cells Dialog

Under **Horizontal**, you can select **Center, Left,** or **Right** in addition to the default setting, with an indent option for the latter two.

The **Justify** option aligns text evenly on both sides.

The **Fill** option repeats the entered string to completely fill the cell. For example, it can fill cells with *** (see Figure 4.33).

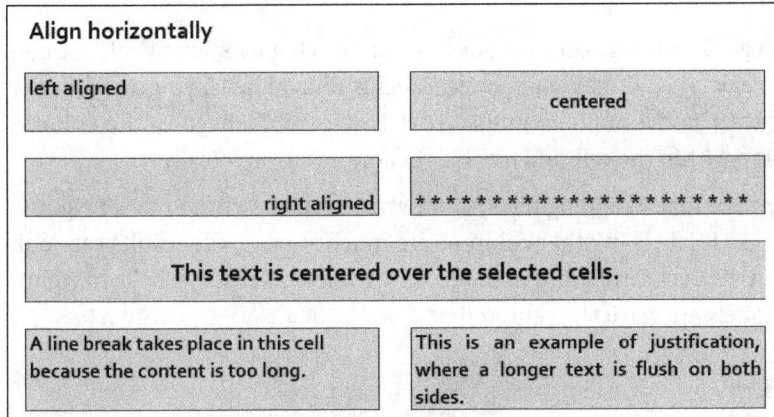

Align horizontally

left aligned

centered

right aligned

****

This text is centered over the selected cells.

A line break takes place in this cell because the content is too long.

This is an example of justification, where a longer text is flush on both sides.

Figure 4.33 Examples of Horizontal Alignment

The **Center Across Selection** option centers the content of the first cell in a selection—such as three cells—without merging them. If you enter something in a neighboring cell afterward, the space between the entries is shared. Since Excel 2002, you've been able to change text direction for international use, such as with Arabic fonts.

Indents to Group Labels

It's often helpful to organize row labels by category, and indenting text clearly shows subordinate categories. The example in Figure 4.34 lists different cost types.

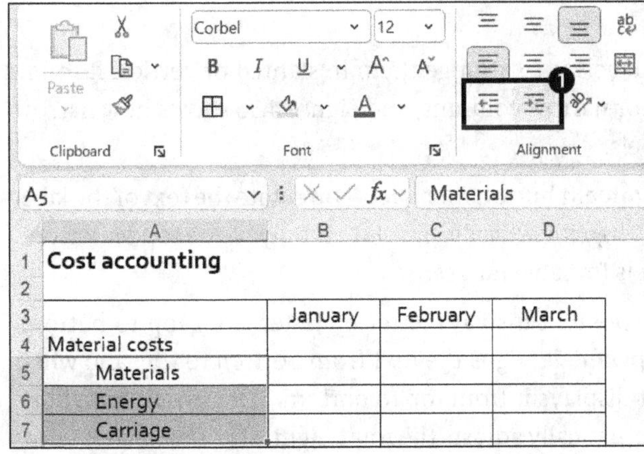

Figure 4.34 Arranging Row Labels by Using Indents

Costs are divided into two main groups—material and personnel—and each item in these groups is indented. On the **Alignment** tab in Figure 4.34, **Left** is selected for text alignment and the **Indent** is set to 3. You can also use the two indent icons ❶ in the **Alignment** group instead of the dialog box, and each click or tap on an icon adjusts the indent by one level so you can easily find the best setting.

Longer Labels

Often, the data in a column (e.g., certain codes) is just one character long while the column label is more detailed. If having a longer label is unavoidable, you can use text wrapping along with horizontal alignment. Text that extends beyond the column boundary will wrap within the cell, and row height adjusts automatically.

However, this type of text wrapping has the drawback that the entire row's height is always adjusted—so if labels in other columns are much shorter, the results can look uneven. Changing the vertical alignment of text in the relevant cells can help. In Figure 4.35, the shorter labels are vertically centered.

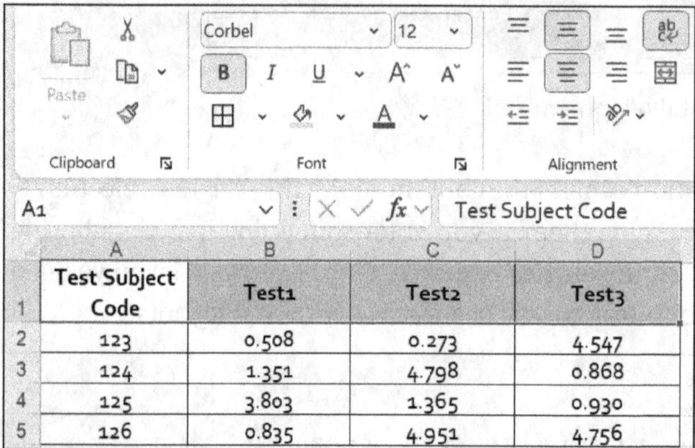

Paste
Corbel
12
Clipboard
Font
Alignment
A1
Test Subject Code
A
B
C
D
Test Subject Code
Test1
Test2
Test3
123
0.508
0.273
4.547
124
1.351
4.798
0.868
125
3.803
1.365
0.930
126
0.835
4.951
4.756

Figure 4.35 Combining Text Wrapping with Vertical Alignment

Slanted and Vertical Labels

Another way to handle longer labels is to arrange them in slanted or vertical fashion. Slanted alignment works well for narrow columns, and Figure 4.36 shows how narrow columns can display longer labels.

When you select a label, the **Alignment** button menu lets you rotate the text of the label clockwise or counterclockwise, align it vertically, or rotate it fully to the top or bottom. You can also create vertical labels for table ranges.

When you select **Vertical Text**, the text displays letter by letter from top to bottom, when you select **Rotate Text Upward**, it aligns the text from bottom to top, and when you select **Rotate Text Down**, it displays it from top to bottom. When you rotate horizontal text vertically, Excel automatically adjusts the row height to fit the longest entry while the column width initially stays the same (see Figure 4.37).

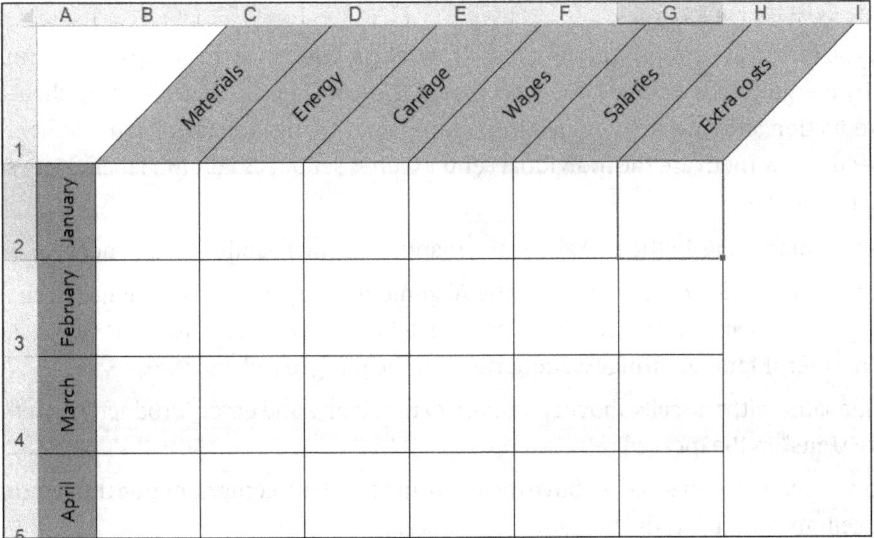

Figure 4.36 Table with Vertical and Slanted Labels

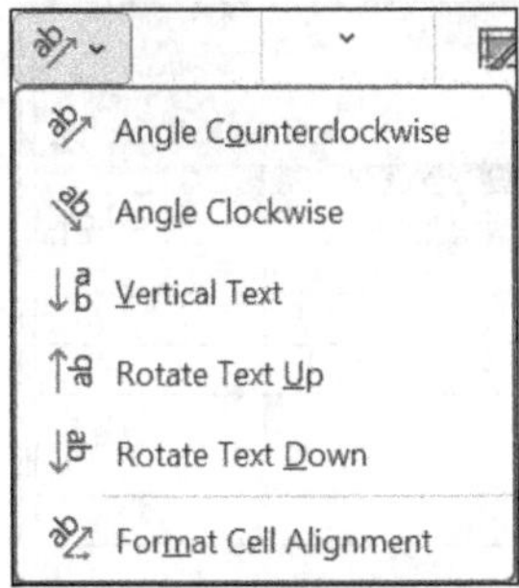

Figure 4.37 Options for Alignment Button

You can also fine-tune these settings in the **Format Cells** dialog box. On the **Alignment** tab, under **Orientation**, choose the angle you want. Then, drag the red dot in the semicircle with your mouse, finger, or pen; click or tap anywhere in the semicircle; or enter a value under **Degrees**.

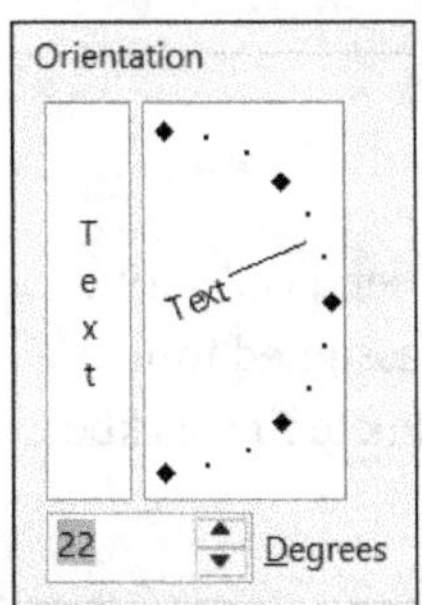

Figure 4.38 Setting Slant

Merging Cells for Longer Text

Just as you can create groups with indents in row labels, you can form groups in column labels by merging cells. Figure 4.39 shows part of the labels for a cost accounting sheet. The production sites are first divided into manufacturing headquarters and auxiliary units, and below them are the individual center names set out as column labels. Here's how to merge cells:

1. Select as many cells in the row above the manufacturing headquarters as needed.
2. Click the **Merge & Center** button in the **Alignment** group, which will merge all the cells you've selected into a single cell with an address that comes from the first cell. Then, enter "Manufacturing Headquarters" in the merged cell.
3. Do the same with the cells above the auxiliary unit labels and enter "Production Auxiliary Units" in the merged cell.
4. One row above, select all cells above the main and auxiliary centers, merge them into one cell, and enter "Production sites" in that cell.

It's convenient to place a table title or longer text in a table into a merged cell.

	A	B	C	D	E	F	G
1	Production sites						
2	Manufacturing Headquarters				Production Auxiliary Units		
3	A	B	C	D	Payroll office	Repair shop	Work office
4							
5							
6							
7							
8							
9							
10							

Figure 4.39 Table Titles and Text in Merged Cells

Merging cells also makes designing tables and forms much easier. It gives you greater flexibility in table layout. Merged cells keep all the properties of a regular cell, they're referenced by a single cell address, and they can be copied, moved, and more.

To unmerge cells, select the **Unmerge Cells** option from the **Merge & Center** button menu.

Automatic Font Size Adjustment

Excel also lets you fit labels or values neatly within a cell. You can set a cell's font to shrink automatically if the content expands, without changing the assigned font size. To do this, on the **Alignment** tab, under **Text Control**, uncheck the **Shrink to fit** checkbox.

Then, if you enter content into one of these formatted cells and it doesn't fit, Excel reduces the font size so all data is visible.

However, this cell feature has its limits. When the content gets too large, the font becomes unreadable. At that point, you'll need to take other steps.

4.5 Borders and Patterns

Lines and borders, along with colors and patterns for cell backgrounds, are effective tools for keeping your worksheets organized. For example, lines help separate total rows from the columns being summed and distinguish labels from data entry areas, and borders help clearly group different sections within a large worksheet. You can also use colors and background patterns to highlight important information, such as especially notable or critical values.

If you want to hide gridlines in the window or for printing, use these tools to organize the table's layout clearly and create an attractive overall impression. This is especially important for frequently used forms like invoices, delivery notes, and order forms.

A cluttered worksheet not only leaves a poor impression and may frustrate the recipient, but it can also make the recipient miss key information. Not everyone has time to review a table from the first to the last row.

4.5.1 Border Line Palette

If you select a cell or range to which you want to add lines or borders, you have several options.

Mouse users can use the **Border** icon in the **Home • Font** group, which works similarly to the **Font Color** icon described previously. You can also choose from twelve preset border styles (see Figure 4.40). Note that except for the **All Borders** option, which applies to every cell in the selected range, all other options only affect the outer border of the entire range.

To separate one range from others, you can, for example, draw a thick border around it. This is shown in Figure 4.41, where thick borders separate the product column on the left from the currency columns to the right and also separate the currency types row at the top from the rows below it.

To remove all borders from a selected range, choose **No Border**. Clicking the **More Borders** option opens the **Borders** tab in the **Format Cells** dialog box. In **Touch input** mode, the borders palette is spaced out to make selection easier, so tap to select the desired pattern.

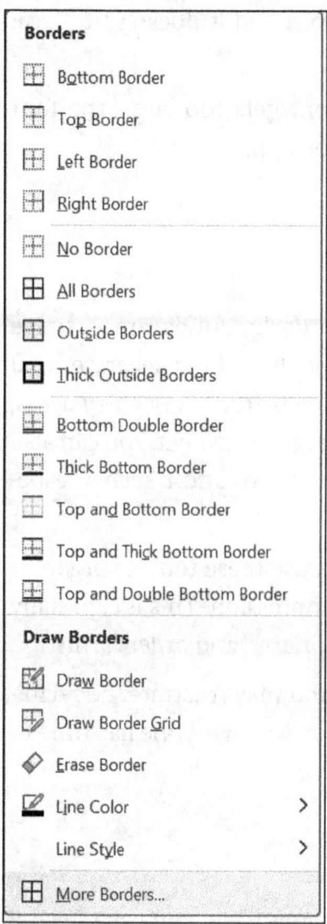

Figure 4.40 Extensive Borders Palette

	A	B	C	D
1	**Border**			5/11/2025
2				
3		Dollar	Euro	Pound
4	Product 1	$54.00	€ 60.48	£45.36
5	Product 2	$81.00	€ 90.72	£68.04
6	Product 3	$108.00	€ 120.96	£90.72
7	Product 4	$135.00	€ 151.20	£113.40
8	Product 5	$162.00	€ 181.44	£136.08
9	Product 6	$189.00	€ 211.68	£158.76

Figure 4.41 Cell Range with Thick Borders Separating Ranges

The Borders Tab

Applying borders and lines via the **Format Cells** dialog box is less convenient since you can only set options for the currently selected range. However, the **Borders** tab offers more options, including diagonal lines to mark a range for later deletion. Here's how you use it to apply borders:

1. Select the cell or cell range you want to frame or add lines to.
2. Choose a preset on the **Borders** tab, which you can then customize. For example, click or tap **Outline** to add a border around the entire selected cell range, use **Inside** to add a grid of lines within the selected range, or select **None** to remove all existing borders. You can also use **Outline** and **Inside** borders simultaneously.
3. The **Border** box displays the current border preview. You can click or tap any line in the box to remove it or add missing lines by clicking inside the box or on one of the icons around it (see the following figure).
4. To use different line styles and colors, click a line pattern in the right panel and then choose a color pattern under **Color**. All lines you add afterward will use the selected style and color until you choose otherwise.
5. You can also freely combine these options. For example, you can add a thin bottom border to every cell in the range while separating columns with bold lines.

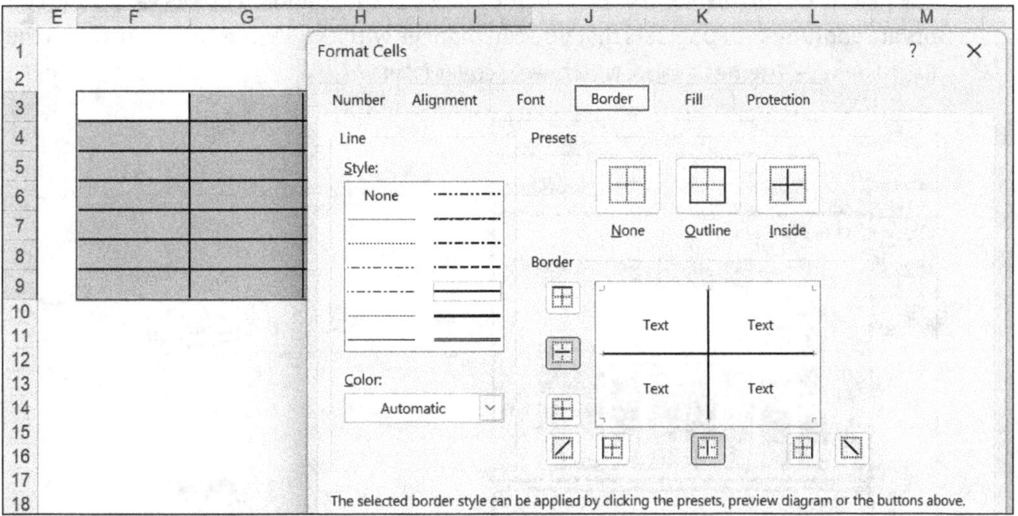

4.5.2 Drawing Borders

Instead of selecting cells first and then adding borders, you can use one of the two drawing tools in the **Borders Palette** at the bottom, where you can preset line style and color by using dropdown lists. The two drawing tools are **Draw Border** and **Draw Grid Border**. The first draws a border around the area you outline with the drawing tool, and the second applies a full cell grid to the selected range. You finish drawing by clicking the **Border** icon again.

To help you make corrections, the palette includes an eraser icon that activates the delete function. You can easily remove misplaced borders or grids with a mouse. Choosing another option in the palette disables the delete function. Border drawing isn't available in **Touch input** mode.

Euro
50,00 €
51,00 €
52,00 €
53,00 €
54,00 €
55,00 €

Figure 4.42 Framing Price Table Using Pen Tool for Borders

4.5.3 Colors and Fill Patterns

To set the cell background color, use the **Fill Color** button in the **Home • Font** group. Clicking or tapping the small arrow opens a palette with color options that are similar to those in the font color palette discussed earlier. This palette is also enlarged in **Touch input** mode.

The palette distinguishes between **Theme Colors** ❶ and **Standard Colors** ❷. Theme colors are coordinated color sets that you can manage with the **Theme Colors** button in the **Page Layout • Themes** group, which we'll cover later.

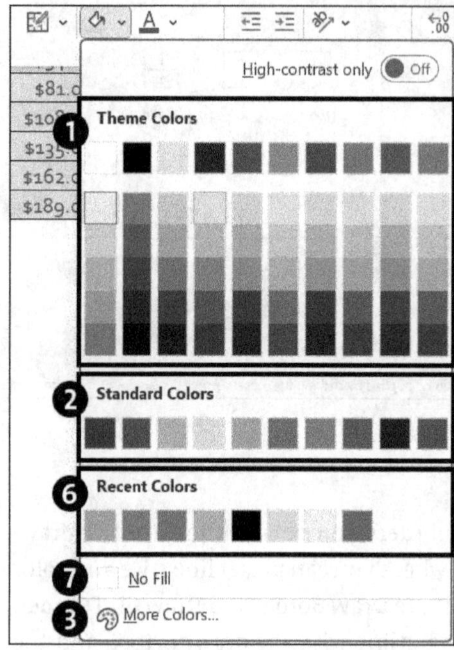

Figure 4.43 Palette for Fill Color Icon

Standard Colors are Excel's preset colors, which are often used for chart fills when users don't assign custom colors. You can click the **More Colors** ❸ button to open a dialog with two tabs, and on the **Standard** tab, you can replace the default standard colors if needed. You can replace the current color with one you select by clicking or tapping the color hexagon.

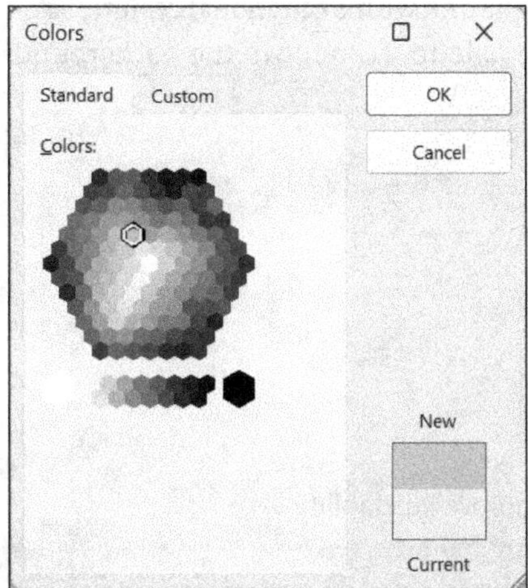

Figure 4.44 Replacing a Standard Color

If this selection isn't enough, you can mix a custom color from the primary colors on the **Custom** tab. In the large color field, you can either click precisely on a spot or drag with your mouse, finger, or stylus until the small preview window below under **New** (❹ in Figure 4.45) shows the selected color. The same applies to the narrow brightness bar ❺, which displays various blends of white or black with the color chosen in the left field.

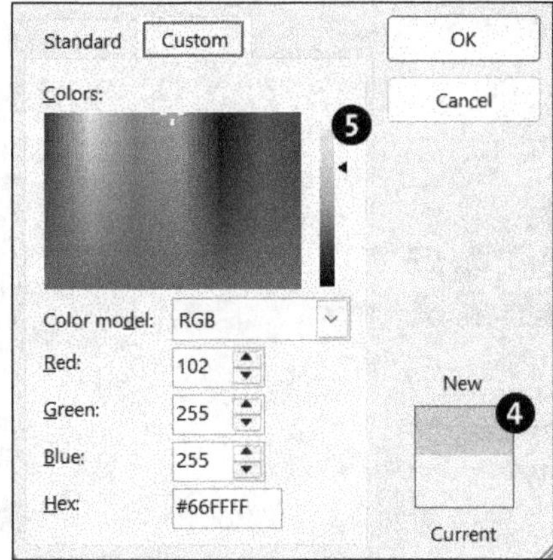

Figure 4.45 Mixing Custom Colors

Instead of dragging, clicking, or tapping in the color fields, you can enter the numeric values for the two color systems Excel supports. The HSL color system uses three factors: **Hue**, **Saturation**, and **Lightness.** The RGB system mixes the colors **Red**, **Green**, and **Blue**. You can use both systems independently, and you can also enter the hex code for the color, as defined in HTML.

Your custom colors appear in the palette under the **Recent Colors** section (refer back to Figure 4.43 ❻), making them easy to access.

If you want to make columns stand out or highlight one with exceptional results, simply select the column and pick a color from the palette. To remove the background color, choose **No Fill** (refer back to Figure 4.43 ❼).

	A	B	C	D
3		Dollar	Euro	Pound
4	Product 1	$54.00	€ 60.48	£45.36
5	Product 2	$81.00	€ 90.72	£68.04
6	Product 3	$108.00	€ 120.96	£90.72
7	Product 4	$135.00	€ 151.20	£113.40
8	Product 5	$162.00	€ 181.44	£136.08
9	Product 6	$189.00	€ 211.68	£158.76

Figure 4.46 Using Different Colors for Columns to Improve Readability

The Fill Tab

You can also color cells through the **Fill** tab in the **Format Cells** dialog. You can select the background color from the same palette used by the **Fill Color** button.

The **Pattern Format** list also lets you apply shading to selected cells, layered over the background. Using **Pattern Color**, you can choose a specific color for the shading. For example, you can overlay a line pattern on a solid background to highlight a header row.

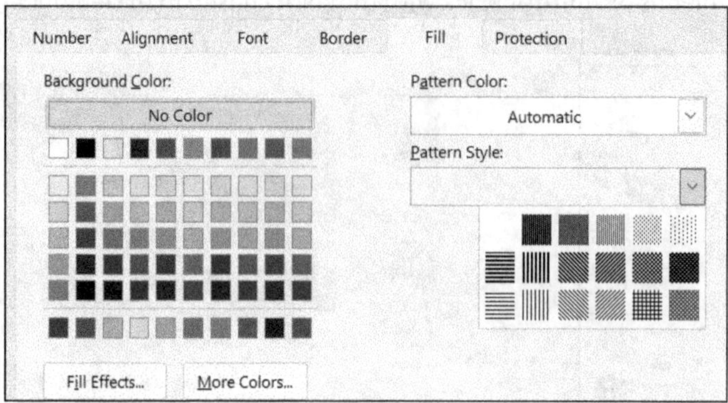

Figure 4.47 Fill Tab

Less Is More

Colors in a worksheet work much like fonts, and the ease of coloring cells can tempt you to overuse them. More than three or four colors on one sheet create a cluttered look, confuse viewers, and quickly tire their eyes. Also, when using such color patterns, it's important to consider how the area will look when it's printed. So at the start, experiment a bit to find the combinations that really work—and remember that data needs to be easy to read.

4.5.4 Using Colors as an Organizational Tool

Using colors and patterns can be effective for many purposes, even without data. For example, for vacation planning in a department or small company, these patterns can quickly create clear overviews. Figure 4.48 provides an example of this. To fill out such a plan, place a color pattern in a cell. To enter a multiday vacation, click that color cell, click the **Copy Format** button, and drag the brush over the relevant days.

Holiday schedule 2025

Name	1-Jan	2-Jan	3-Jan	4-Jan	5-Jan	6-Jan	7-Jan	8-Jan	9-Jan	10-Jan
Klaris										
Toka										
Border										
Ernest										
Berwang										
Deront										
Dumas										
Heming										

Figure 4.48 Vacation Planning Using Cell Background Colors

Colors work well in lists you want to process because they help you mark what's already done.

Controlling with colors

Nr	Image check
Chapter 1	
Chapter 2	
Chapter 3	
Chapter 4	
Chapter 5	
Chapter 6	

Figure 4.49 Marking Completed Items with Background Color

4.5.5 Checkboxes

Instead of marking completed items with background colors, you can now use a familiar Excel feature: cells with checkboxes. To add these organizational tools in a column to the left or right of a task column or across multiple columns, select the column, go to the **Insert** tab, and in the **Controls** group, select the **Checkbox** option. Thereafter, when a task is completed, the box will show a checkmark in its box (see Figure 4.50).

You can also arrange checkboxes and tasks in a single row if you prefer. You create boxes and checkmarks by using formats, and the cell should contain one of two Boolean values: TRUE (checked) or FALSE (unchecked). Deleting the format does not remove the Boolean values.

	A	B	C
1	**Pending tasks**		
2			
3	☑	Text review	
4	☑	Checking the images	
5	☐	Checking web links	
6	☐	Control of indices	
7	☐	Checking the examples	
8			

Figure 4.50 Completion Control with Checkboxes

To test whether the checkbox in cell A3 is checked, use a function starting with `=IF(A3=TRUE...`. Adding cells A3 through A7 individually gives 2, while `=SUM(A3:A7)` returns zero. Combining multiple tests and evaluating them is useful, as shown in Figure 4.51.

	A	B	C	D	E	F
10	**Completed**	**Nr**	**Image check**	**Examples**	**Index**	**###**
11	☐	Chapter1	☐	☐	☐	☐
12	☑	Chapter2	☑	☑	☑	☑
13	☐	Chapter3	☐	☐	☐	☐
14	☐	Chapter4	☐	☐	☑	☑
15	☐	Chapter5	☐	☐	☐	☐
16	☐	Chapter6	☐	☐	☐	☐

Figure 4.51 Example of Combined Controls

4.5.6 Image Backgrounds

Using **Page Layout • Page Setup • Background** lets you place images as a background behind the data in a worksheet. This command opens the **Insert Pictures** page, which offers various image sources.

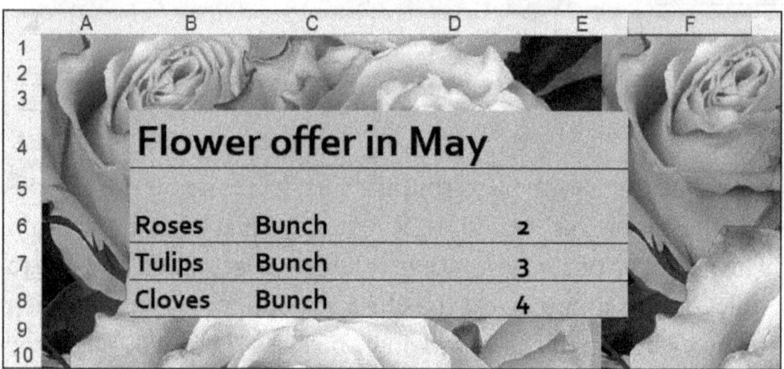

Figure 4.52 Image Background for Worksheet

Clicking the **Browse** button under **From a File** opens a dialog box where you can select a graphic file, and all common file types are supported. Excel tiles the entire worksheet with the selected image, and it lets you use paper textures, decorative patterns, or photos as a background. To remove the background, select **Delete Background** in the **Page Setup** group.

Selecting the **Bing Image Search** option opens the online image dialog, while **OneDrive – Personal** lets you access photos stored on OneDrive.

4.6 Enhancing Sheets with Themes

Even documents focused on numbers don't have to look dull, and most people prefer reading a well-designed worksheet. To assist you in improving the look of your documents, Microsoft introduced themes. A *theme* combines coordinated colors, fonts, and effects. This concept applies across multiple applications. You can apply a theme you like consistently across your Word documents, PowerPoint presentations, and Excel worksheets to create a unified look for all your files.

Many built-in themes are available. The default Office theme appears first when you click the **Page Layout** tab and select the **Themes** button in the **Themes** group. The standard theme has been updated with new fonts and colors, especially to enhance readability. We've already introduced you to the new default font, Aptos.

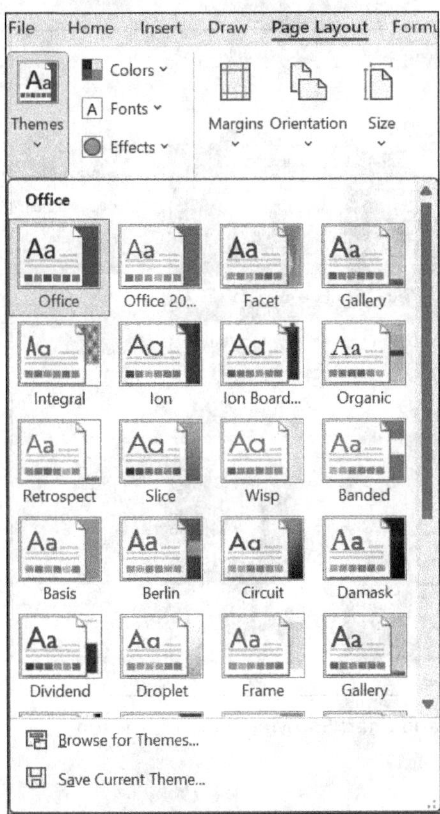

Figure 4.53 Theme Palette

Themes affect the following elements in Excel:

- Cells and cell ranges (These can use theme colors instead of standard colors for background or font color. When you change the theme, the colors update accordingly.)
- Table styles
- Charts and sparklines
- PivotTables and slicers
- Graphic objects
- SmartArt and WordArt objects
- Colors for sheet tabs

Assigning a new theme updates the font, colors, and graphic effects for SmartArt objects all at once.

4.6.1 Applying a Different Theme

By hovering over a theme pattern, you can preview its effect on the current worksheet if the worksheet contains affected elements, like a chart. This requires the **Enable Live Preview** option to be enabled under **Options • General.**

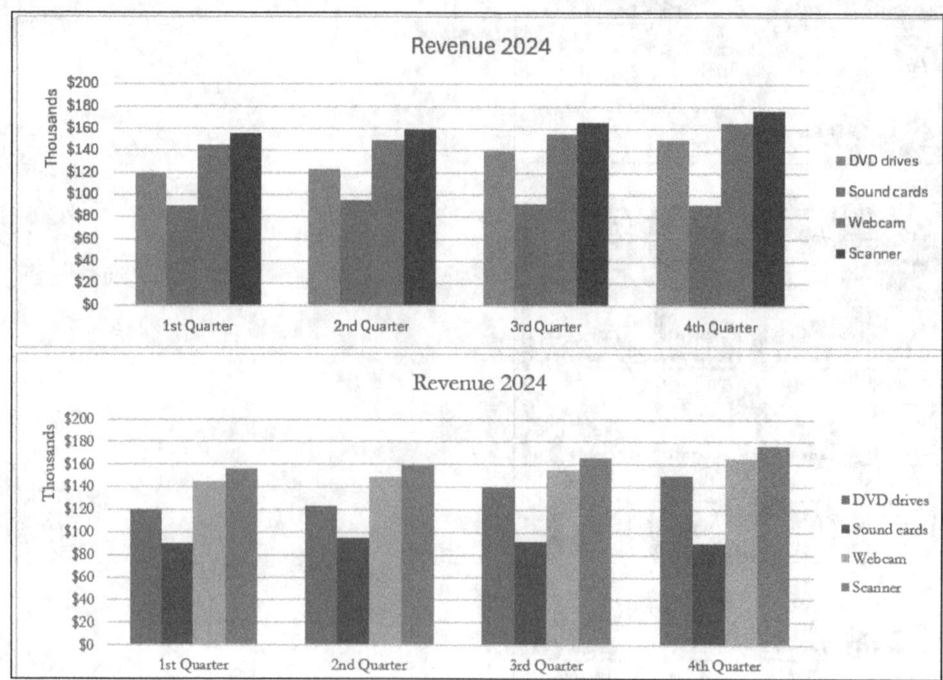

Figure 4.54 Chart Showing Office Theme (on Top) and Chart Showing View Theme (on Bottom)

You can click a theme option to apply it immediately. In **Touch input** mode, just tap the design; live preview isn't available here. Applying a theme always affects the entire workbook, and if you apply a theme later, you might need to adjust the page layout because some fonts may have changed.

4.6.2 Customize Themes

In the **Themes** group, you'll find three buttons you can use to change individual aspects of a selected theme. For example, if you like the fonts in the **Organic** theme but not the colors, you can take the colors from another theme. You can also take other properties from different themes. Follow these steps:

1. Select the **Organic** theme by using the **Themes** button.
2. Click the **Colors** button to open the palette with color combinations from different themes.
3. Choose the color scheme you want by clicking or tapping it in the palette. Alternatively, use the **Customize Colors** option to create your own color schemes.

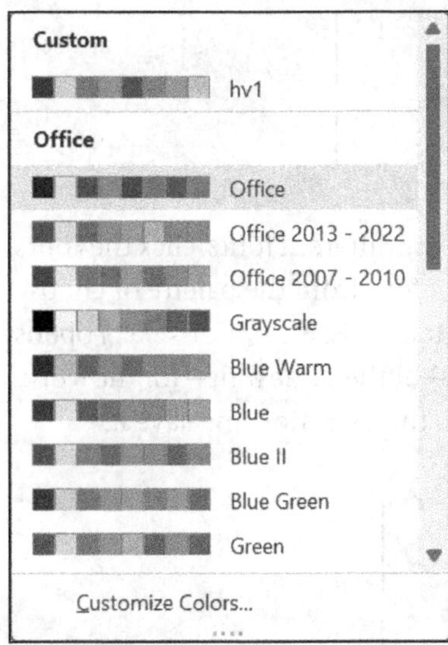

4. Open the individual color palettes by using the buttons under **Theme Colors**, then select the colors you want. The dialog box has four buttons for text and background colors, and the colors you choose appear in the **Fill Color** and **Text Color** palettes. You can use the **Accent 1** through **Accent 6** buttons to control chart colors and the last two buttons to manage the display of hyperlinks. Under **Sample**, you can instantly see how your color choices look. Finally, enter a name for the color selection you've created under **Name** and click **Save** to confirm.

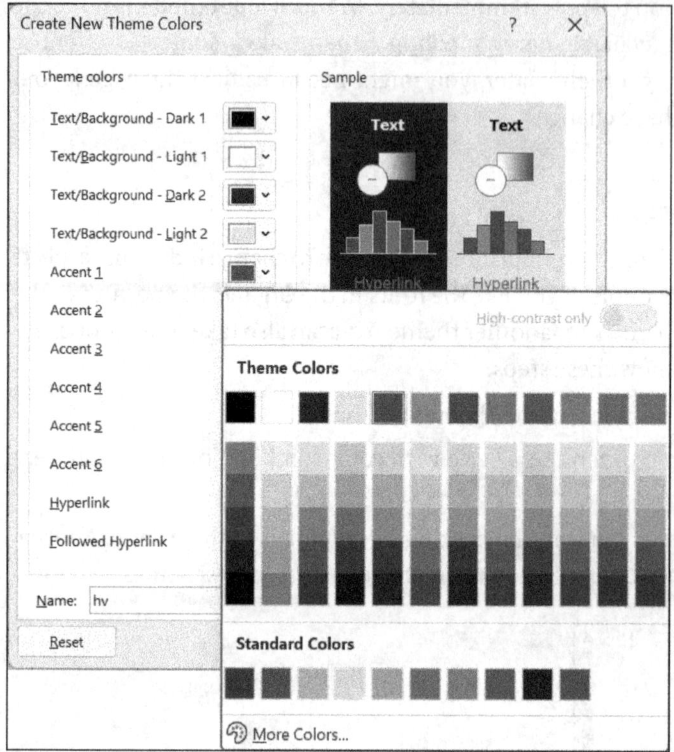

5. If you like the colors of a specific theme but want to use different fonts, click the **Fonts** button in the **Themes** group. You can either select a font from the palette or choose the **Customize Fonts** option. If you choose the **Customize Fonts** option, Excel opens a small dialog where you can assign different fonts—or the same font—for the worksheet's headings and data. Finally, assign a name to this selection and save it.

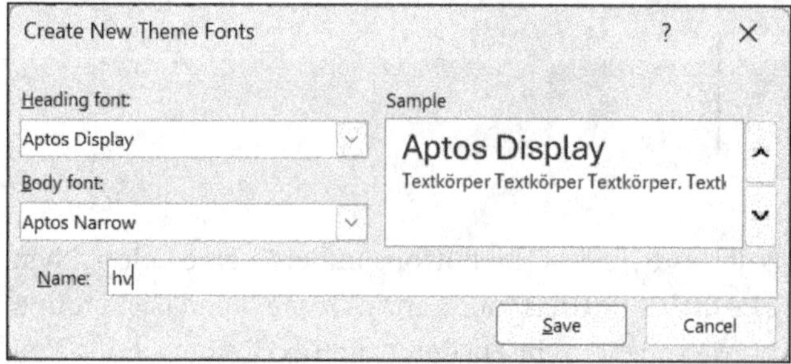

The **Effects** button in the **Themes** group lets you choose graphic effects for SmartArt objects. You can also apply effects from another theme to a specific design, but custom adjustments are not available here.

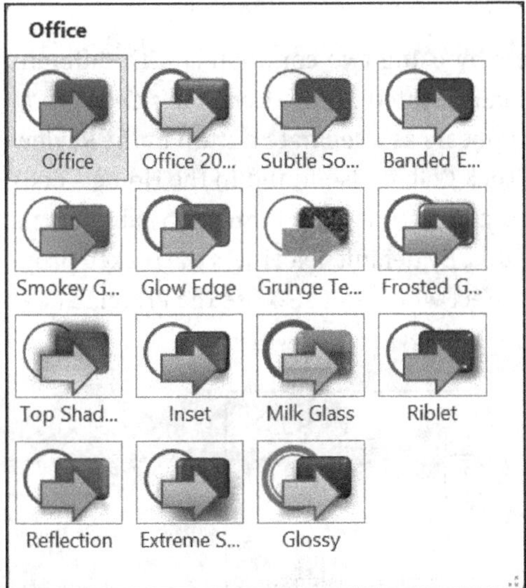

Figure 4.55 Range of Graphic Effects

Besides saving your own color or font choices, you can also save a custom theme as a design file by using the **Save Current Theme** command in the **Themes** palette. Excel will automatically suggest the *Document Themes* folder, which you can create as a subfolder in the user-specific templates folder, as we'll explain in Section 4.8. The name of this theme appears in the palette under the **Custom** category.

4.7 Protecting Sheets and Workbooks

In Excel, cell locking is treated as a cell format property, so we'll cover this topic here. This also means that when you copy or move cells, this property transfers with them. Normally, all labels, numbers, and formulas in an open workbook can be changed at any time. Changes can be intentional—authorized users can make them for legitimate reasons, or unauthorized users can make them to manipulate values. But you can also make changes unintentionally. For example, you could overwrite a complex formula with a number if you were to select the wrong cell.

To help you avoid making unintentional changes, Excel will warn you about cells containing formulas that aren't locked against changes, but only if the **Unlocked cells containing formulas** rule is enabled under **File • Options • Formulas** in the **Error Checking Rules** section. Once you've finalized a worksheet's structure, tested its formulas, and set certain data—like final values for a year or month—you should lock those cells to prevent further changes. You can apply this to an entire workbook, whole sheets, or selected ranges within worksheets.

4.7.1 Allowing or Preventing Changes

You can protect individual Excel worksheets by using two commands with different scopes. The broader command, **Protect Sheet**, is located in the **Home** tab's **Cells** group under the **Format** button menu, and it also appears as a separate button in the **Review** tab's **Protect** group. The second command, **Lock Cell**, is also found in the **Home • Cells** group under the **Format** button menu. You can also use the **Protection** tab and the **Locked** option in the **Format Cells** dialog for this. By default, Excel sets the **Locked** property for all cells, but this feature only takes effect when you also use the **Protect Sheet** command.

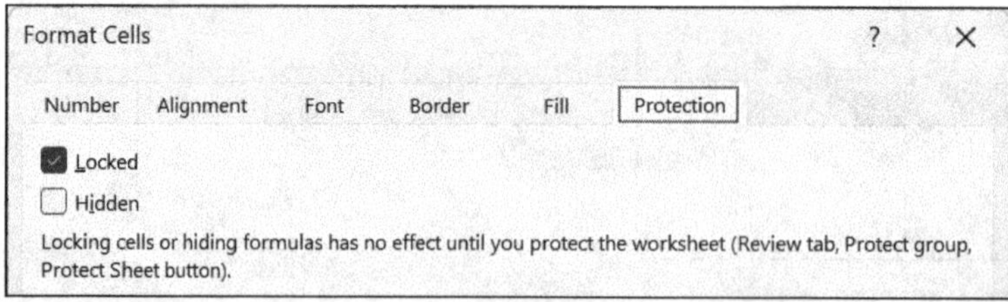

Figure 4.56 Protection Tab in Format Cells Dialog Box

You can also use the **Protection** tab to prevent formulas from appearing in the formula bar. Check **Hidden** to enable this functionality. If you want to make it harder for unauthorized users to copy your complex formulas, this option can help. However, some ranges require calculation verification—for example, in accounting—so you may not want to use this option in all cases.

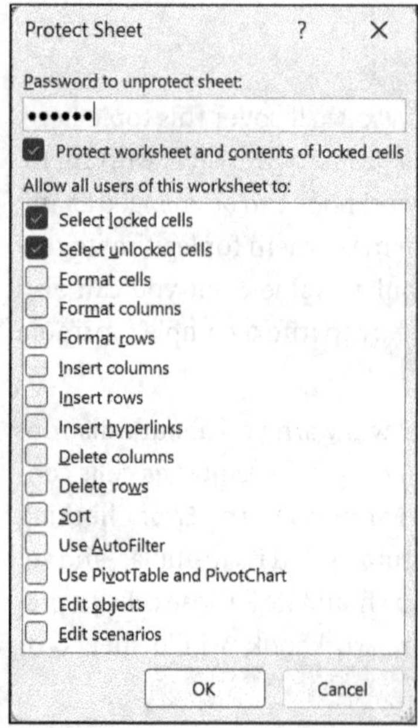

Figure 4.57 Protect Sheet Dialog Box

This option only works after you lock the sheet with the **Protect Sheet** command. To enable protection, keep the **Protect worksheet and contents of locked cells** option selected in the **Protect Sheet** dialog box. To prevent others from easily bypassing this protection by using the **Unprotect Sheet** command at the same location, set a password in the field at the top.

You can adjust the protection's strictness by selecting or clearing the listed options. For example, to allow the table to be reformatted, check **Format cells**. In this way, you can precisely tailor the protection settings to your needs. Commands for disabled options will be hidden on this sheet.

To protect the entire workbook, go to the **Review • Protect** group and select **Protect Workbook** to open the **Protect Structure and Windows** dialog box. Then, select **Structure** to prevent sheets from being moved, inserted, or deleted in the workbook. The **Windows** option is disabled here because each application window displays only one workbook, which fills the entire window.

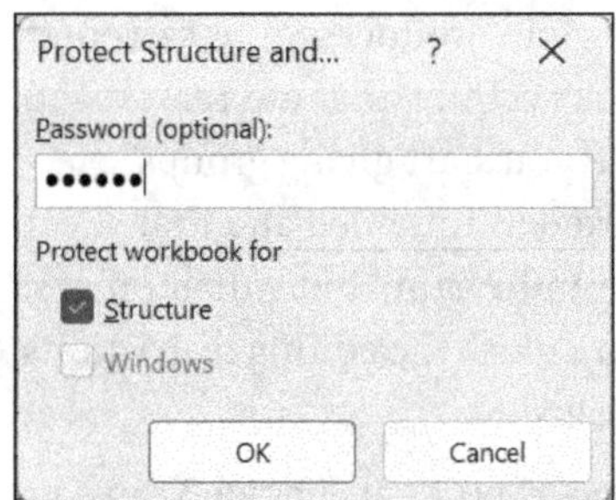

Figure 4.58 Protect Workbook Dialog Box

4.7.2 Unlocking Input Ranges

It's useful to put a full protection lock on entire sheets once you've set their final content. For example, statistics on last year's results usually remain unchanged, so it makes sense to put a full protection lock on sheets containing such data.

However, if data still needs to be entered in certain worksheet ranges, you can protect only the formulas, labels, and values that shouldn't be changed—like constants or reference values from the previous year or month. To do this, first remove the default protection from the cells that should remain editable, and then refer back to the example in Chapter 2. In the quarterly report, it's best to unlock the cells with change values and lock everything else—labels and formulas. Follow these steps to do it:

1. Select all cells that can be edited, which in this case are B4:E7.
2. Right-click and choose **Format Cells** from the context menu, and on the **Protection** tab, uncheck **Locked**. To visually distinguish unlocked ranges, you might apply a different background or text color to the selection.
3. Go to the **Review** tab and select **Protect • Protect Sheet** to lock the rest of the worksheet.

	A	B	C	D	E
1	Novamedia: Quarterly Results 2024				
2					
3	Category	1st Quarter	2nd Quarter	3rd Quarter	4th Quarter
4	DVD drives	$120,000	$123,000		
5	Sound cards	$90,000	$95,000		
6	Scanner	$145,000	$149,000		
7	Video cards	$134,000	$120,000		
8		$489,000	$487,000		

Format Cells
Number Alignment Fon
Locked
Hidden

When entering data in the unlocked range, always use the Tab key so the cell pointer stays within the unlocked range. This is especially useful for creating input forms.

4.7.3 Selective Range Protection

Excel lets you grant specific users permission to edit certain ranges within a workbook. In our example, only the person who's responsible for certain product groups can enter values. To give different groups access to the workbook sections that are relevant to them, go to **Review • Protect • Allow Edit Ranges**. Before you start giving groups selective access, make sure that the specified ranges are locked as intended and that sheet protection is not enabled yet. Then, click the **New** button to name individual ranges, select them, and assign a password to them so users can unlock the editing restrictions with it. Then, share the password with the authorized users.

Alternatively, you can use the **Permissions** button to select users or groups who can edit selected ranges without a password.

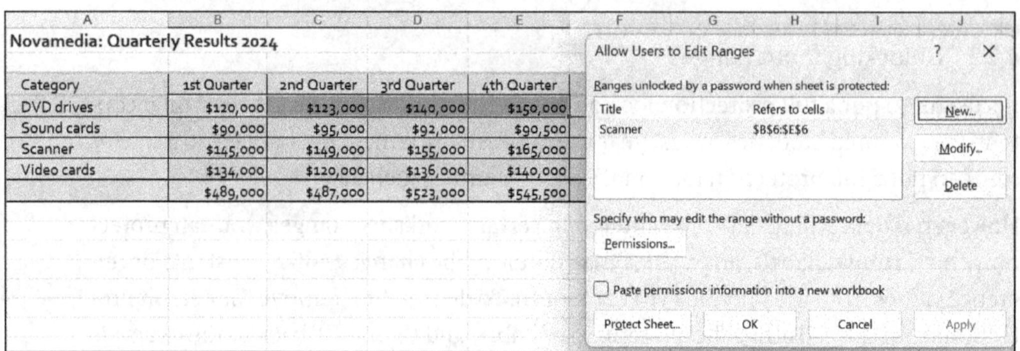

A	B	C	D	E
Novamedia: Quarterly Results 2024				
Category	1st Quarter	2nd Quarter	3rd Quarter	4th Quarter
DVD drives	$120,000	$123,000	$140,000	$150,000
Sound cards	$90,000	$95,000	$92,000	$90,500
Scanner	$145,000	$149,000	$155,000	$165,000
Video cards	$134,000	$120,000	$136,000	$140,000
	$489,000	$487,000	$523,000	$545,500

Figure 4.59 Setting User Permissions for Individual Ranges

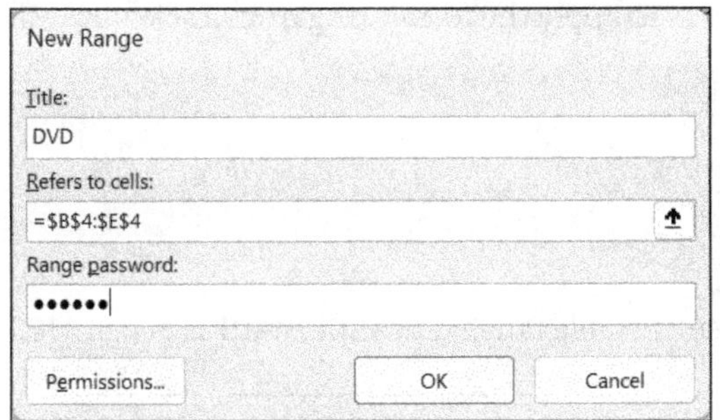

Figure 4.60 Protecting Ranges with Passwords

4.8 Consistent Design Using Styles

Perfecting a tasteful worksheet design takes time. Often, the best formatting design emerges gradually, and to avoid starting over each time, you can reuse formats you've already created. This ranges from simply copying formats to using styles, both of which this section will cover.

4.8.1 Copying Formats

When you're using the **Copy** and **Fill** functions, Excel usually transfers both the cell content and its format to the target ranges. To transfer a cell's format to other cells without its content, use the **Formatting** button under **Home • Paste** in **Other Paste Options**. You can also find this option by clicking the **Paste Options** button that appears at the end of the target range after you paste.

The **Format Painter** button in the **Home • Clipboard** group is also very useful, and we covered it briefly in the section on number formats. Clicking the button changes the cell pointer into a paintbrush, and all cells you drag over with the paintbrush will be formatted like the active cell. Double-click the button to format multiple cell ranges until you click it again, and note that all formatting properties currently applied to the source cell will be transferred to the other cells.

You can also use the **Format Painter** button to copy formats in **Touch input** mode. To do this, tap the paintbrush icon and then tap the cell where you want to apply the format. The only thing missing here is the paintbrush-shaped mouse pointer. You also can't extend the target range by dragging, but if you tap a column or row header instead of a cell, then the format will apply to the entire column or row.

You can also use the **Format Painter** button to apply the formatting from a selected cell range to another range, as follows:

1. Select cells A3 through F8.
2. Click or tap the paintbrush button.
3. Click or tap cell A10.

Figure 4.61 shows the formatting of the entire table range repeating, starting at cell A10.

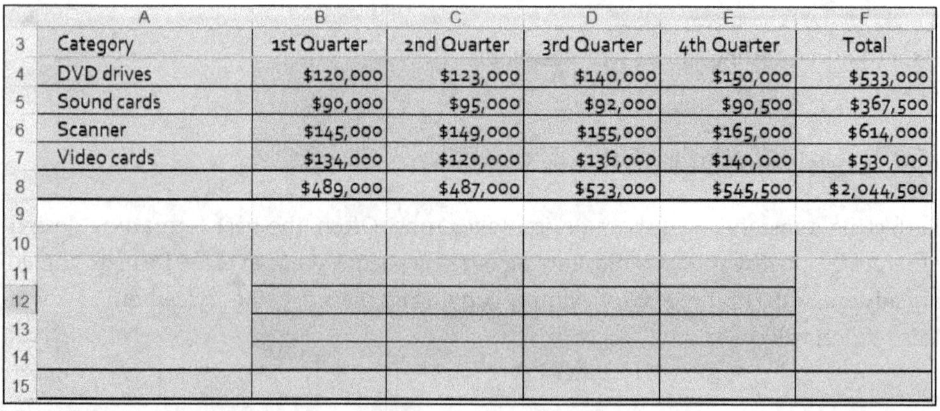

	A	B	C	D	E	F
3	Category	1st Quarter	2nd Quarter	3rd Quarter	4th Quarter	Total
4	DVD drives	$120,000	$123,000	$140,000	$150,000	$533,000
5	Sound cards	$90,000	$95,000	$92,000	$90,500	$367,500
6	Scanner	$145,000	$149,000	$155,000	$165,000	$614,000
7	Video cards	$134,000	$120,000	$136,000	$140,000	$530,000
8		$489,000	$487,000	$523,000	$545,500	$2,044,500
9						
10						
11						
12						
13						
14						
15						

Figure 4.61 Applying Formatting from Filled Table Range

4.8.2 Reusing Styles

Styles provide an efficient way to reuse specific format patterns. Applying all formatting options to individual cells or ranges can be time-consuming, but you can use cell style templates to combine multiple formatting properties and apply them all at once. Excel provides a wide range of cell formats for various purposes, and they're accessible via the **Home** tab • **Styles** group and the **Cell Styles** button, which opens the menu. To apply a style to a selected cell range, simply click or tap the desired template. You can also hover over the style to display a live preview in the selected range. (If it doesn't display, **Live Preview** might be disabled under **File • Options • General.**)

If Excel's built-in styles don't meet your needs, you can create your own. For example, to format a total cell with a slightly larger bold font and a specific background color, you'd normally perform three separate steps—applying different formatting icons or switching between tabs.

If a special format is applied to a sum cell and fixed for a single cell, you can easily create a custom style for the workbook. You can also use these styles in other workbooks.

Figure 4.62 Assigning Built-In Cell Styles

Defining with a Sample Cell

The easiest way to define a style for a sum cell, for example, is to use a sample cell. Follow these steps to do it:

1. Manually apply all the format properties that the style requires to the sample cell.
2. Select the sample cell, go to **Home • Styles • Cell Styles**, and at the bottom of the gallery, click **New Cell Style** ❶.
3. Enter the name *highlighted_total* as the **Style name** of the style ❷. Then click **OK**, and Excel will associate the active cell's formatting with this name thereafter.

The new style will appear in the style palette under **Custom,** and you can apply it to a selected cell range with a click or tap, just like any other style. As the dialog box shows, a style can combine multiple formatting features so you can update a cell range's design with all these elements at once. Styles also help you maintain a consistent look across your workbooks. If you don't want to include a formatting feature in the style, deselect it first.

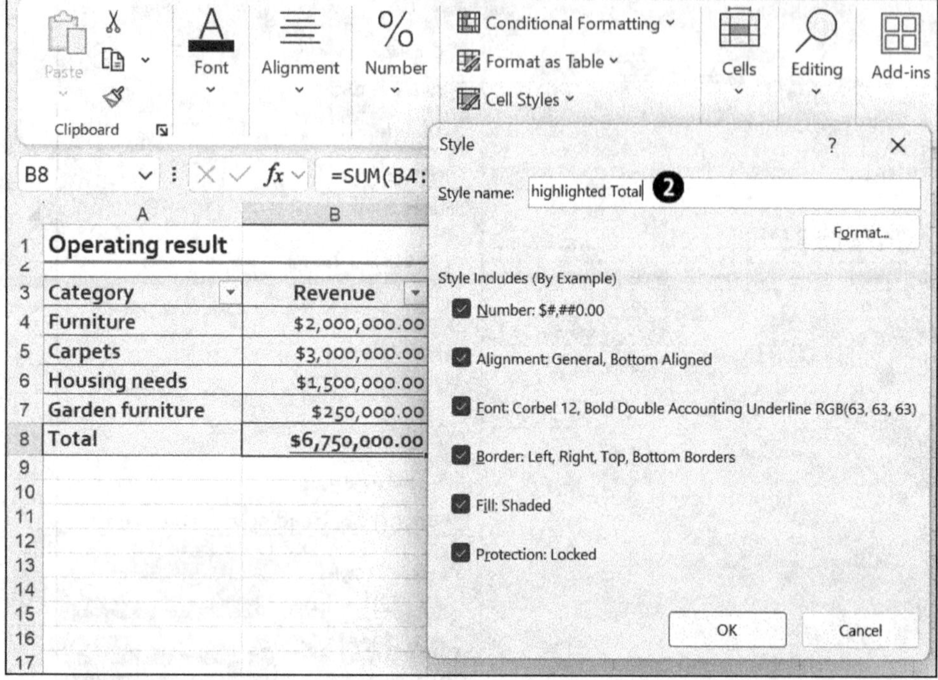

Editing Existing Styles

Suppose that you want to change the **Standard** styles because you don't like the font and that you want the table to have a gray pattern. To edit an existing style, right-click the corresponding style sample in the palette and choose **Modify** from the context menu. In **Touch input** mode, tap the pattern until the context menu appears.

Then, the **Style** dialog box will open and display the name of the **Standard** style. Click the **Format** button to open the **Format Cells** dialog, where you can choose the desired settings. If you haven't changed the font or pattern of any cells in this workbook before, the entire workbook will show a gray cell background, and all existing data will appear in the new font. Only cells where you've already changed the pattern color will keep that color, and if you want these cells to have the gray color too, reapply the **Standard** style from the palette.

Changing an existing style updates all cells assigned to that style—but cells with manual formatting changes won't be affected, and the manual formatting will remain unchanged. For example, if you set a different font—like **Segoe UI Semibold**—in columns A through Z and then add a gray pattern to the **Standard** style, those columns will turn gray but keep the semibold font.

Overlaying Styles

You can easily apply multiple format templates to a single cell, and this happens, for example, whenever you choose a predefined number style instead of the default standard format. Since these templates only define the number format, that format is the only thing that changes. Other attributes—font, borders, and so on—stay the same. A new style only changes the attributes it defines.

Deleting Styles

If you no longer need a style, you can delete it to keep your list tidy. Right-click the style sample and select **Delete** from the context menu. Usually, it's not a good idea to delete styles that are still in use in the worksheet. If you do so by mistake, cells assigned to that style will revert to the standard format.

4.8.3 Importing Styles into Other Workbooks

Changes you make to styles or new styles you create typically apply only to the current workbook, but the need for specific design styles often goes beyond a single file. So, once you find a combination of fonts, colors, and patterns you like, you'll want to reuse it. There are two main ways to do this:

1. Importing styles from another workbook (which we cover in this section).
2. Inheriting styles by using templates (which we cover in Section 4.8.4).

The first method involves importing styles from an existing workbook into the current one. Both workbooks must be open, and in the workbook receiving the styles, you go to **Home • Styles • Cell Styles** and then select **Merge Styles** from the menu. Then, in the dialog box, choose the workbook that contains the styles you want to import.

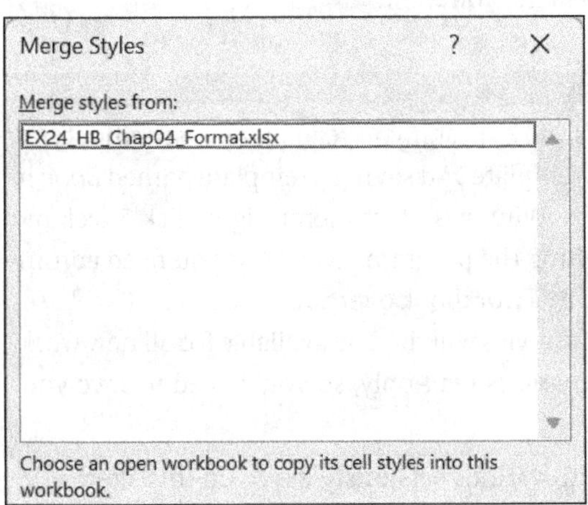

Figure 4.63 Merge Styles Dialog Box

After you confirm, Excel checks for matching style names between the two workbooks. If the selected workbook only has styles that aren't present in the active workbook, then the merge is simple. These formats are imported into the active workbook and can be used as usual.

Typically, both workbooks include some styles with the same name, like the Standard style. That's why Excel prompts you first to confirm whether to merge styles with identical names. If you select **Yes**, the active workbook's styles are replaced by those from the other workbook. If you select **No**, your existing definitions for styles with the same name stay intact, and only the other styles are imported. This workbook-to-workbook merging method is most useful when you're applying very specific formatting that you use only in certain workbooks.

4.8.4 Defining Styles Without a Sample Cell

To define or modify a style without relying on sample cells, use the **Style** dialog box as follows:

1. Enter the name of the style.
2. If you enter a new name, the format attributes of the currently active cell will appear under **Style Includes**. Check the format dimensions you want to include in the new style; leave the others unchecked or remove the checks from them.
3. Use the **Format** button to select the specific format attributes, and on the various tabs, choose the combination of formats to include in the style. After you confirm by clicking **OK**, you can review your selection again in the initial dialog box.
4. After you click **OK**, the new template appears in the **Cell Styles** palette.

To create a style based on an existing one, use the **Duplicate** command in the context menu of the style sample. The **Style** dialog then adds a number to the original name, and you can just replace it with the new name you want.

The Convenience of Automatic Templates

The second method of applying formats works best if you regularly use specific styles. To use it, start by creating an automatic template and saving a template named *Book* in the *XLSTART* startup folder. Then, in the Windows Start menu, right-click Excel and select **Run as administrator** when launching the program. (Note that you need administrator rights to do this.) The settings in this workbook override the defaults Excel uses when opening a new workbook, and these styles will then be available for all new worksheets. However, Excel opens this workbook as read-only, so you'll need to save your work under a different name.

You can also use different templates with various sets of styles. To do this, create the necessary styles in templates and save them in the *Templates* folder, as explained in Chapter 1.

4.9 Table Styles

If you don't want to format individual parts of your worksheet, Excel can handle much of the formatting for you. The program offers automatic formatting for tables that have a fairly regular structure. This applies only to font style, cell backgrounds, and borders; number formatting is not included. Table styles are helpful because most tables share a similar basic layout, and Excel provides several ready-made table styles for these.

As shown in Figure 4.64, many tables have a header made up of one or more label rows and one or more label columns to the left of the main data. When data is summarized by calculations, a row with the results often appears at the bottom, and sometimes, a column appears to the right of the data block, such as a total sum. Occasionally, a row or column includes more complex calculations referencing values in the upper or lower rows or the left or right columns.

Two common table structures

Table with column and row evaluation

Name	2023	2024	Total
Karol	30000	32000	62000
Bernd	35000	34000	69000
Gangus	32000	34000	66000
Total	97000	100000	197000

Data list with heading line

Name	Phone	City
Karol	+49 0221367889	Cologne
Jess	+44 113 2233465	Leeds
Gangus	+49 030 76 56 78	Berlin
Kern	+1 617 878989	Boston

Figure 4.64 Typical Table Structures

Another common table type is data lists. These tables have a single heading row at the top, followed by multiple data records (as shown in the right pane of Figure 4.64). A summary row may appear at the bottom.

4.9.1 Applying a Table Style

To apply a table style to an empty cell range as a guide for future data entry, follow these steps:

1. Select the cell range designated for the table.
2. On the **Home** tab, in the **Styles** group, click the **Format as Table** button to open the table styles gallery.
3. Click or tap the table style you want to use.

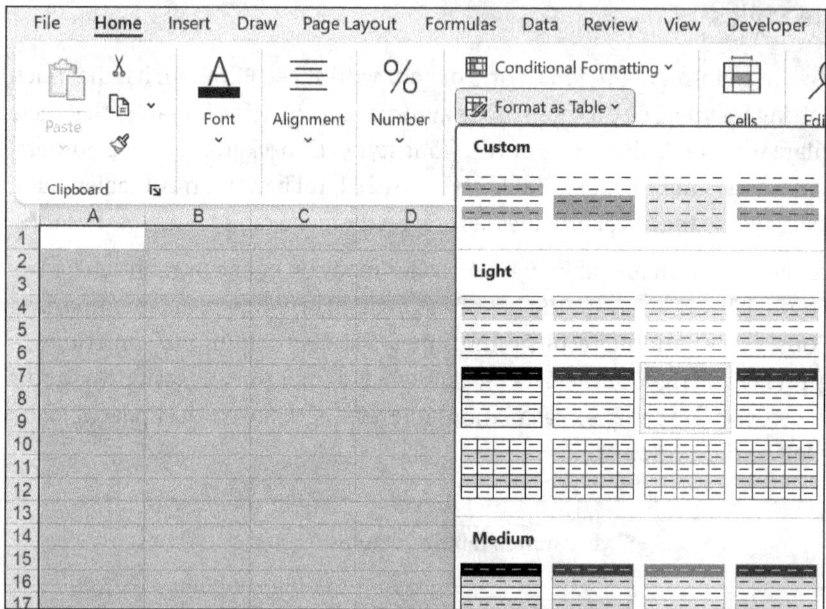

4. The program requires confirmation of the selected range, so in the small dialog, also check the **My table has headers** option.

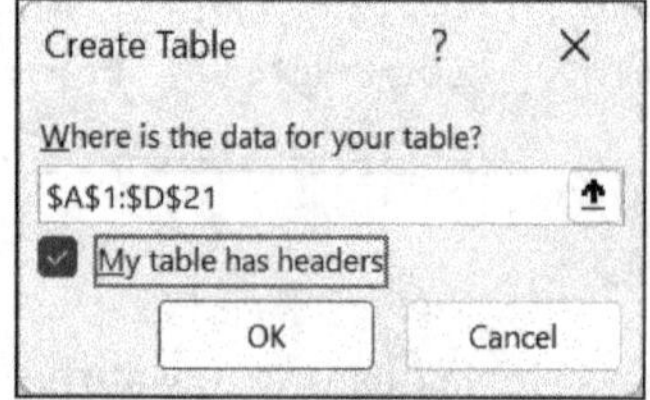

5. The style will be applied to the empty range, which you can then fill with content. The first row will show consecutively numbered placeholder labels for the columns.

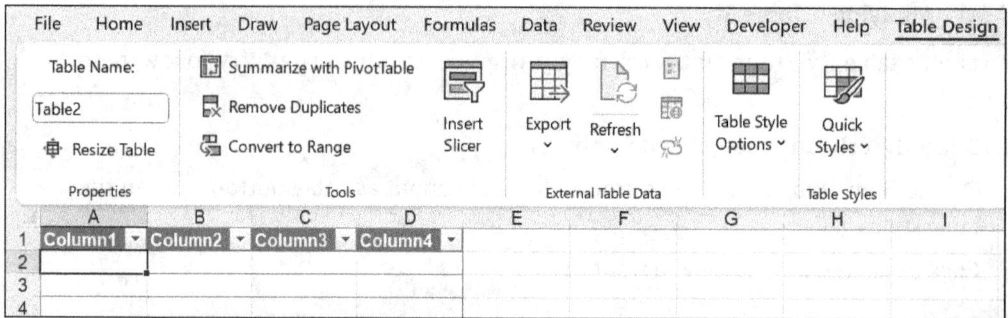

Applying a table style converts the cell range into a table range, and the **Table Design** tab appears automatically.

4.9.2 Applying a Table Style

To assign a table style to a cell range that already contains data, you follow a similar process. To apply a predefined table style to our previous example of a profit calculation table, follow these steps:

1. Select any cell within the table.
2. Pick your preferred style from the table style palette.
3. Excel will try to guess the correct table range and highlight it with a dashed border. You can adjust the range if you need to in the **Format As Table** dialog box, and you can also check a box to specify whether the range includes a header row. Once you confirm, the style you choose will apply to the range.

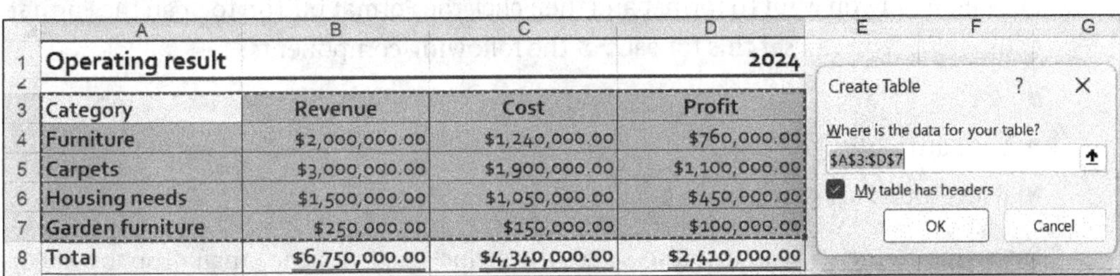

4. On the **Table Design** tab that appears, you can further customize the formatting. For example, to highlight the first column as a label column, check the **First Column** option in the **Table Style Options** group, and to emphasize the profit column, select **Last Column**. Checking the **Banded Rows** option formats rows in alternating pairs, and you can apply the same formatting to columns.
5. If you don't like the format changes from the options you choose, open the palette again by using the **Quick Styles** button. It offers many styles based on your current selections. If there's enough space, a sample strip appears instead of the button, and you can expand it by using the **More** button at the bottom right.

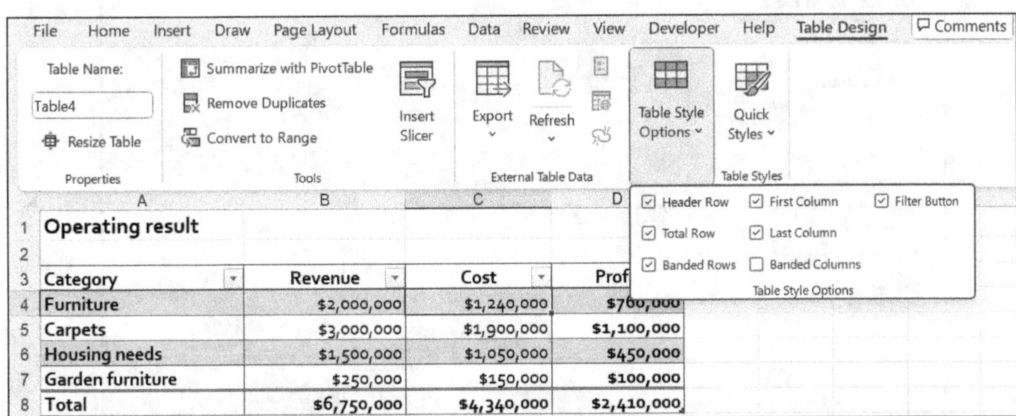

You can edit the automatically formatted table like any other if you want to change specific formatting. For example, depending on the text length in the first row, Excel might make some columns too wide or too narrow.

4.9.3 Designing a Table Style

Although the **Table Style** palette offers many variations, the suggested formats might not match your preferences. That's why the palette includes the **New Table Style** option at the bottom, which you can click to open the **New Table Style** dialog. Start by entering a nice name for the table style; this name will appear when you hover over the sample.

Below that, the list box shows the various table elements you can format. Select each table element you want to format and then click the **Format** button to open the **Format Cells** dialog. You can set this for each of the following components:

- Border style
- Font attributes
- Background and pattern

For stripe elements, you can also select the stripe size from the small dropdown. For example, to format two rows identically, choose **First Row Stripe** and **Second Row Stripe** with **Stripes Size 2** for each, then assign different background colors to the two table elements using the **Format** button.

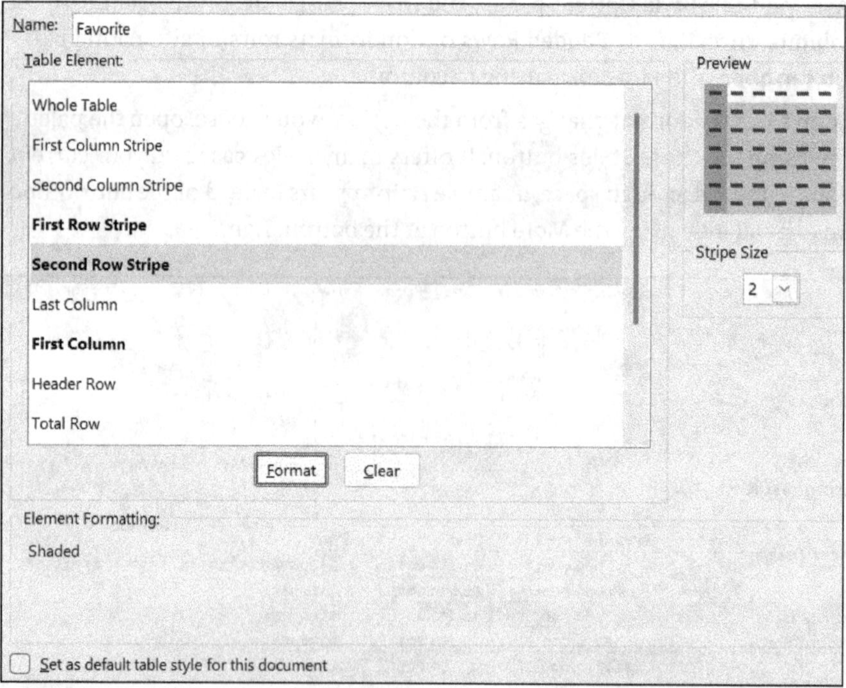

Figure 4.65 Creating Custom Table Styles

The last option in the dialog lets you set the style as the default for the table style if you want to. After you complete the definition, a format sample appears in the style palette under the **Custom** category.

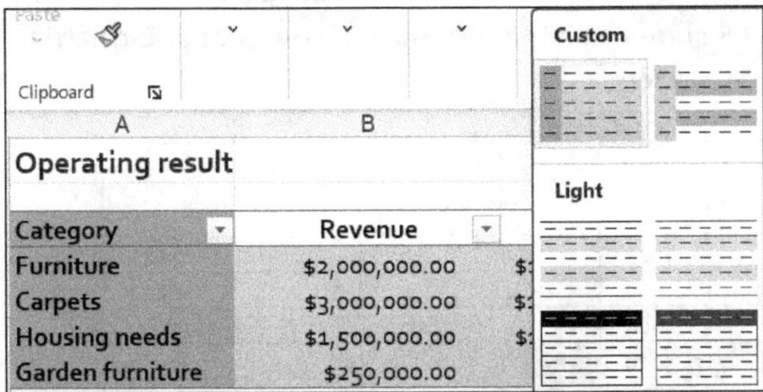

Figure 4.66 Custom Table Styles in Palette

Instead of creating custom styles from scratch, you can duplicate built-in styles and modify them as needed. The context menu of format samples has the modify option for built-in styles disabled, but it allows modifications to custom styles. The duplicate option is enabled for any table style.

The context menu also lets you choose whether to clear existing formats or keep them when applying a format to a table range.

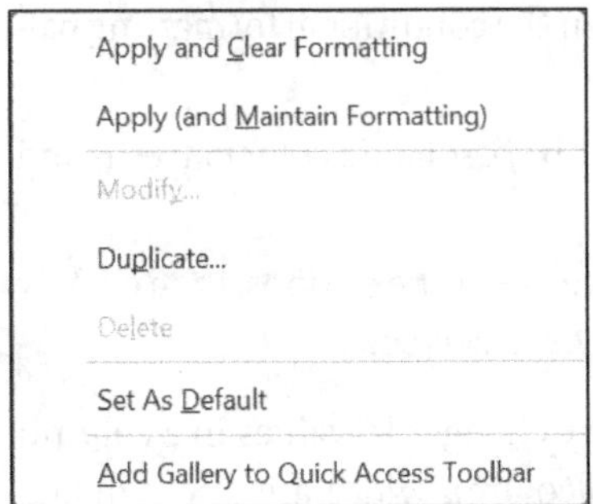

Figure 4.67 Context Menu for Table Style

4.9.4 Deleting Table Styles

To delete a table style completely, place the cell pointer inside the table, select **Table Design • Table Style** in the palette, and choose **Delete**. Even after you delete the style, the worksheet range remains an Excel table and isn't converted back to a normal cell range. This only happens when you select **Table Design • Tools • Convert to Range,** which keeps the current format but removes the special features that are available for table ranges. You'll also notice that all table-specific options will disappear from the context menu of the normal cell range. For more on working with table ranges, see Chapter 16.

4.10 Data Analysis with Conditional Formatting

One especially useful feature is conditional formatting. The **Conditional Formatting** button in the **Home • Styles** group offers various palettes you can use to highlight cells based on their values. Figure 4.68 shows a simple example of conditional formatting using data bars, color scales, and icon sets.

	A	B	C	D
1		Bars	Gamut	Symbols
2	Category	2022	2023	2024
3	DVD drives	120000	132000	108000
4	Sound cards	90000	99000	144000
5	Scanner	145000	159500	160800
6	Video cards	134000	147400	276000

Figure 4.68 Some Examples of Conditional Formatting

4.10.1 Data Bars

Data bars represent each value in a column with a colored bar that fills the column width proportionally. They provide a practical way to visualize values, like bar charts do. The process of creating data bars is very simple:

1. Select the range of relevant values.
2. Go to **Home • Styles • Conditional Formatting** to open the conditional formatting palette.
3. Click or tap the **Data Bars** option to open a palette with six solid fill color patterns and six gradient fill options.
4. Preview the result live by hovering your mouse or pen over the options, then click or tap to apply the sample you want to the cell range you selected.

The columns instantly format in the chosen color. Any changes to values in a column automatically update the color bars, adjustments to column width are also handled automatically, and you can easily change the background color of a cell range at any time. Applying a table style later also works with conditional formatting, which uses rules to control how formats are applied. With data bars, the rule is mostly preset, but you can adjust it if you need to via the **More Rules** ❶ option in the sample submenu.

The first list box shows different rule types, and the lower section provides controls you can use to customize the rules.

The simplest method is to format all cells in a range based on their values. You can adjust the preset distribution of bar lengths across the cell range by using the **Type** ❷ (see Figure 4.69) and **Value** ❸ fields under **Minimum** and **Maximum**. For example, instead of setting the shortest bar to **Type: Percent** and **Value:** 0, you can set it to 20. This shortens the bars and effectively changes the chart's scaling.

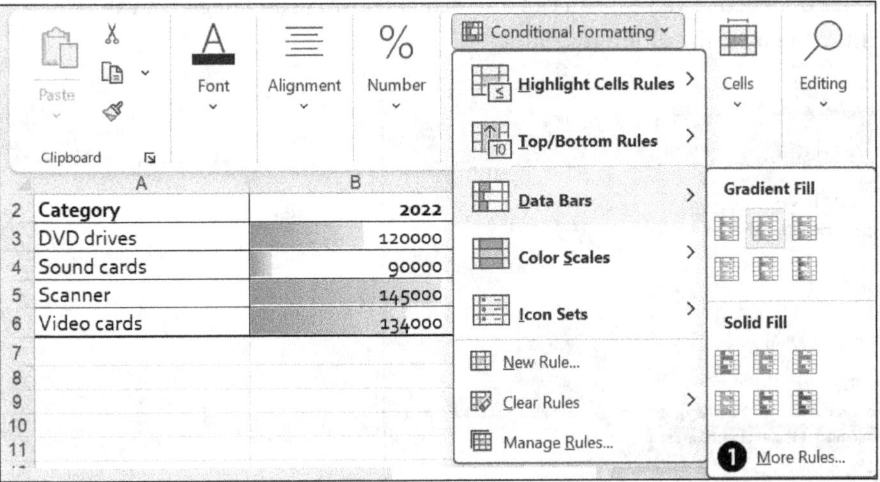

The **Color** dropdown list lets you pick any color, so you're not limited to the preset color patterns. You can color the bar borders separately, and you can also completely hide the values in the column by selecting the **Show Bar Only** ❹ option. You can also click the **Negative Value and Axis** ❺ button to opens a dialog where you can assign a separate color to bars representing negative values.

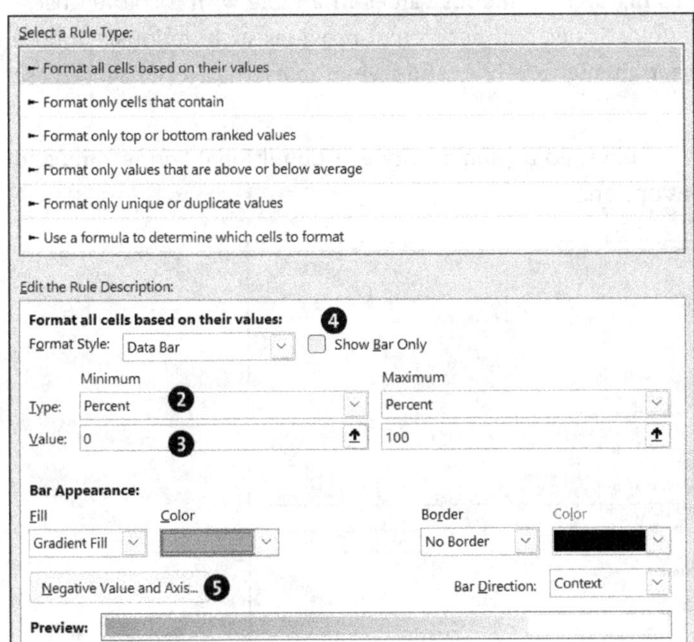

Figure 4.69 Rule Definition Dialog

Then, in **Axis settings** in the dialog, you can, for example, set the zero point to the center of the cell. By selecting the **Automatic** ❻ option, you can position the zero point based on the size of negative values. You can also display positive and negative values in the same direction by using different colors.

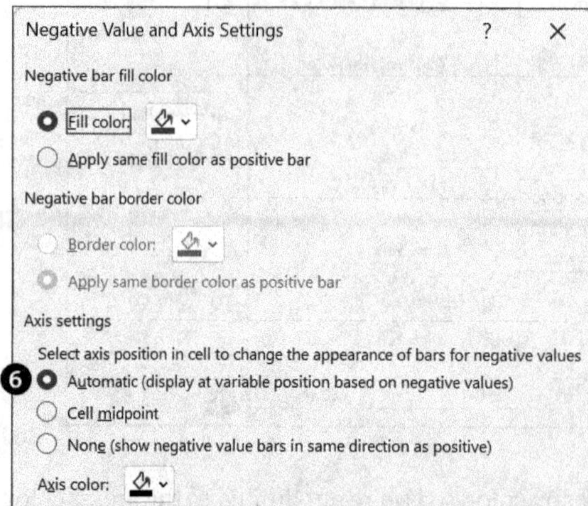

Figure 4.70 Negative Value Options

4.10.2 Color Scales

Sometimes, it's helpful to highlight standout values in a table with a colored background. The conditional **Color Scales** number format provides such options. Suppose you want to quickly spot certain high and low values when analyzing a test. Follow these steps:

1. Select the range of test values, go to **Home • Styles • Conditional Formatting,** and choose the **Color Scales** option.

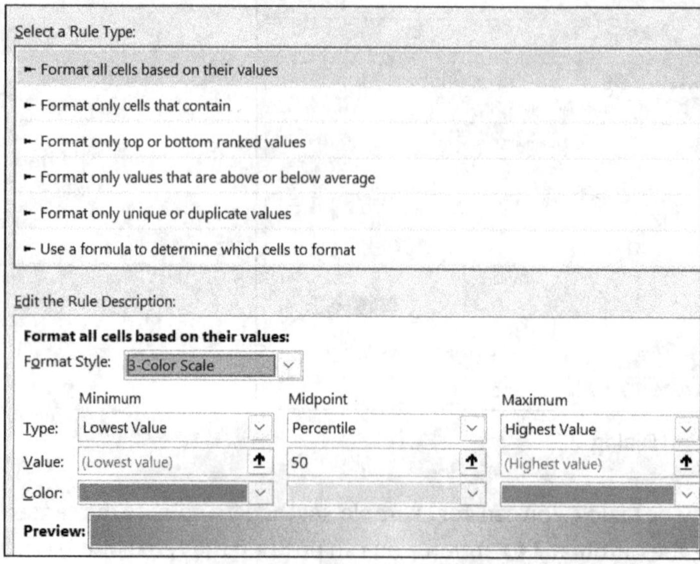

2. Instead of a preset color pattern, select **More Rules**.
3. In the **Rule Type** list, select the first rule type.
4. Under **Format Style**, pick **3-Color Scale** to display controls for **Minimum**, **Midpoint**, and **Maximum**.
5. For example, under **Type**, choose **Percent** to show percentage distribution. Enter the appropriate values in the three fields.
6. You can change the preset colors by using the palettes in the three list boxes as needed.

Each conditional format appears only while the condition in its rule is met. When the value changes, the condition is re-evaluated and the color changes, as in the following figure.

	A	B
1	**Test results**	
2		
3		Score
4	Test person 1	93.43
5	Test person 2	2.28
6	Test person 3	43.85
7	Test person 4	91.65
8	Test person 5	178.81
9	Test person 6	142.48
10	Test person 7	181.59
11	Test person 8	21.22
12	Test person 9	2.11
13	Test person 10	116.55
14	Test person 11	100.52

4.10.3 Icon Sets

It's often important for critical values in a table to stand out immediately. One way to do this is by using icons with conditional formatting. The following figure lists 8-hour average ozone levels, and values exceeding the alarm threshold of 240 $\mu g/m^3$ and the information threshold of 180 $\mu g/m^3$ are highlighted. One solution is to use traffic light symbols, and to create them, follow these steps:

1. Select the range of averages.
2. Go to **Home • Styles • Conditional Formatting**, choose **Icon Sets** from the **Format Style** dropdown list, and select **More Rules**.
3. Under **Icon Style**, pick the 3-light traffic signal and click the **Reverse Icon Order** button, since high values should be marked as critical this time.
4. Enter 240 as the threshold for the red light and 180 for the yellow light.

3		8-hour average
4	10/10/2024	140.00
5	10/11/2024	160.00
6	10/12/2024	130.00
7	10/13/2024	180.00
8	10/14/2024	260.00
9	10/15/2024	270.00
10	10/16/2024	200.00
11	10/17/2024	170.00
12	10/18/2024	140.00
13	10/19/2024	130.00
14	10/20/2024	120.00
15		

Format all cells based on their values:

Format Style: Icon Sets — Reverse Icon Order

Icon Style: — Show Icon Only

Display each icon according to these rules:

Icon		Value	Type
when value is	>=	240	Number
when < 240 and	>=	180	Number
when < 180			

OK Cancel

The **Icon Sets** menu offers four groups of icons. Under **Directional**, you'll find symbols that highlight changes compared to reference values in the selected cell range. Under **Shapes**, you'll find basic forms like circles, triangles, and quadrilaterals. **Indicators** help you distinguish critical values from accepted ones, and **Ratings** offer several ways to clearly visualize value differences.

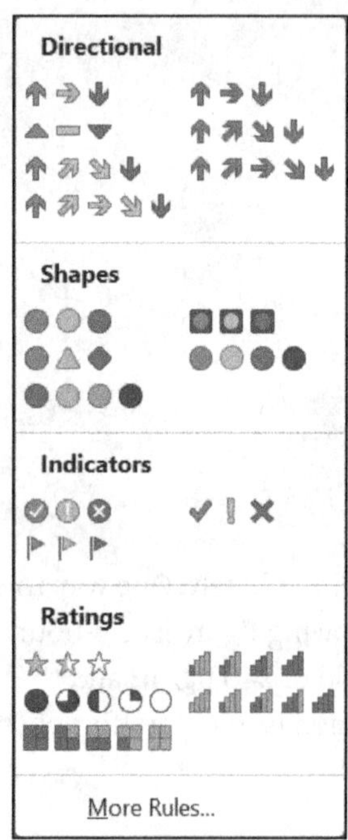

Figure 4.71 Range of Symbol Sets

4.10.4 Simple Comparison Rules

The conditional formatting palette also offers options for simpler solutions. For example, to highlight values above a certain limit, choose the **Highlight Cell Rules** option in the palette for highlighting cells and choose **Greater Than**. Then, in the small dialog box, just enter the threshold and select the formatting you want.

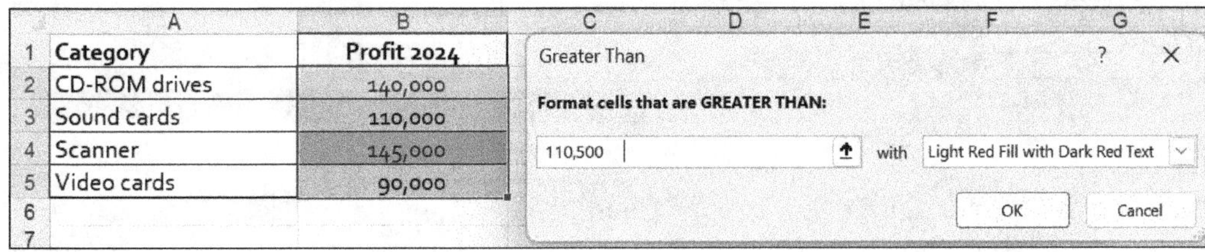

Figure 4.72 Highlighting Values Above Limit

To quickly find duplicate values in a column, simply choose the **Duplicate Values** option. The **Top/Bottom Rules** palette offers various criteria, like **Top 10%** or **Below Average**. You can usually confirm the small dialog box here unless you want a different highlight color.

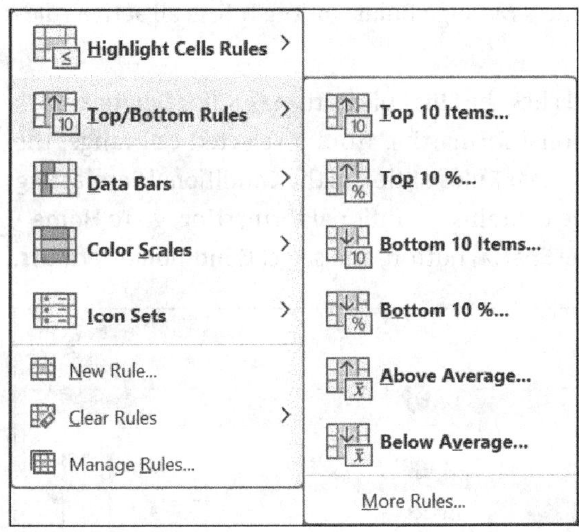

Figure 4.73 Rule Options for Conditional Formatting

4.10.5 More Complex Rules

Use the **New Rule** option (❶ in Figure 4.75) to create more complex rules that use different rule types than before. For example, if you don't want formatting based on fixed values, you can use logical formulas in the rules. To highlight a profit value in a table when it exceeds last year's value, select the two columns with the values to compare and then choose **New Formatting Rule** in the dialog. Under **Select a Rule Type**, choose the **Use a formula to determine which cells to format** row. Then, in the first text box, enter either the logical formula or a reference to one. In our example, we've entered "=C3>B3" here. The formatting isn't predefined, so set it by using the **Format** button (see Figure 4.74).

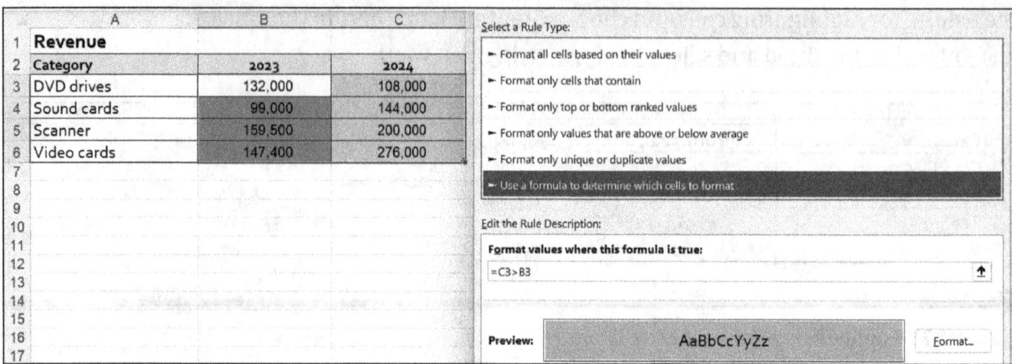

Figure 4.74 Example of Rule Using Logical Formula

4.10.6 The Rule Manager

Since a workbook can accumulate many conditional formats, Excel offers a handy tool under the **Conditional Formatting** menu's **Manage Rules** option. It lists all active rules and lets you edit them as needed.

To edit a rule, select it from the list and click the **Edit Rule** button ❷. Click **Delete Rule** ❸ to remove a rule. To remove conditional formatting from a selected cell range, the entire sheet, or a selected table, use the **Clear Rules** option in the **Conditional Formatting** pane. If you're unsure which cell range contains conditional formatting, go to **Home • Editing • Find & Select • Go To,** click the **Special** button, and select **Conditional Formats.**

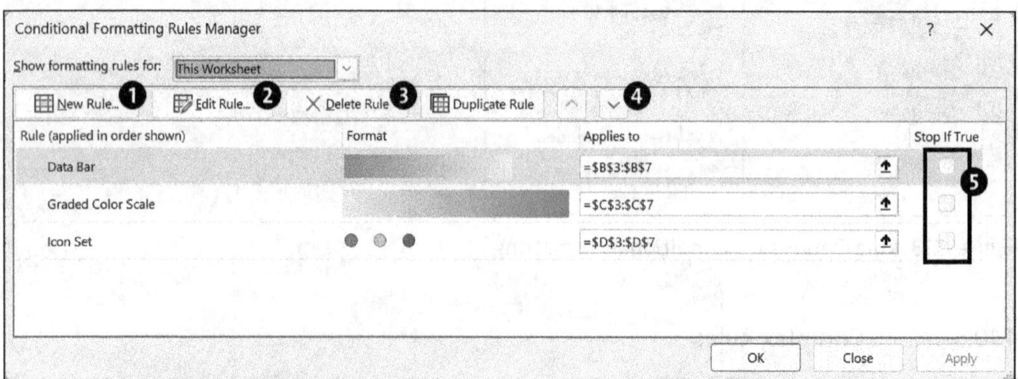

Figure 4.75 Rule Manager Dialog

Overlaying Conditional Formats

You can define multiple rules for a cell range. Formatting a cell range as a data list and showing icons for specific values in that range don't conflict with each other; you can easily combine both formatting features. However, sometimes two rules for a range conflict. For example, if Rule 1 sets a cell's font to red but Rule 2 sets it to green, then the rule priority determines which format applies.

	A	B	C	D
1	Revenue			
2				
3	Category	2022	2023	2024
4	DVD drives	120000	132000	108000
5	Sound cards	90000	99000	144000
6	Scanner	145000	159500	200000
7	Video cards	134000	147400	276000

Figure 4.76 Combining Two Conditional Formats

The priority follows the order in which the rules were created. You can see this order in the rule manager dialog as a list of rules. If you need to deprioritize a rule, select it in the dialog and move it down with the small arrow button (❹ in Figure 4.75).

You can also disable rules without deleting them. Just check the **Stop If True** ❺ option. This is useful, for example, when you're converting the workbook to an older format that doesn't support multiple rules applied simultaneously.

Transferring Conditional Formats

Like other formats, you can copy conditional formats to other cells by using the **Format Painter** and dragging the fill handle. Alternatively, on a touchscreen, you can do it by dragging the **AutoFill** icon.

4.10.7 Quick Formatting with the Quick Analysis Tool

Excel offers a simple way to apply six common formats by using the Quick Analysis tool introduced in Chapter 1. After you select the range to format, click the **Quick Analysis** button at the bottom.

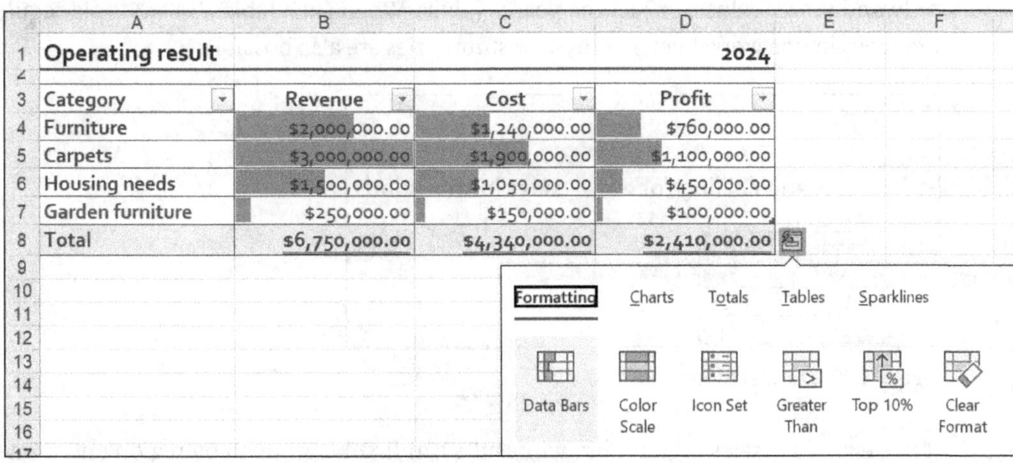

	A	B	C	D
1	Operating result			2024
2				
3	Category	Revenue	Cost	Profit
4	Furniture	$2,000,000.00	$1,240,000.00	$760,000.00
5	Carpets	$3,000,000.00	$1,900,000.00	$1,100,000.00
6	Housing needs	$1,500,000.00	$1,050,000.00	$450,000.00
7	Garden furniture	$250,000.00	$150,000.00	$100,000.00
8	Total	$6,750,000.00	$4,340,000.00	$2,410,000.00

Figure 4.77 Applying Conditional Formats with Quick Analysis Tool

The **Formatting** tab is already selected at the top, and six conditional format icons appear below it. You can select any format and apply it with just one click or tap, except for the **Greater Than** option, which requires you to enter a comparison value as well. The last button lets you delete conditional formatting in the currently selected range.

4.11 Improving Clarity with Outline Levels

Inspired by the outlines used in writing, Excel offers an outlining feature for tables. What does it do? Tables often group information in certain rows or columns that detail adjacent rows or columns. A simple example is a series of twelve-month columns with values summarized in a year column. This table has a two-level hierarchy: the month level and the year level. The outlining feature lets you hide the individual month columns when you want to, showing only the yearly totals. To do this, simply click the button with the minus sign.

Branch billing for 2024 - sales in thousands

Branch	Jan	Feb	Mar	Apr	May	Jun	Jul	Aug	Sep	Oct	Nov	Dec	Total
A	120	120	110	120	120	120	140	120	120	130	130	120	1470
B	120	130	120	120	130	120	120	120	110	110	120	120	1440
C	120	90	120	50	120	90	120	120	120	120	120	120	1310

Figure 4.78 Example with One Outline Level

You can easily expand the example by adding columns for quarterly results. In that case you would use three levels. Users can choose to view or print all levels, only the quarterly and yearly values, or just the yearly values. When such tables list multiple retail branches in one worksheet, row-by-row summaries are also possible.

Branch billing for 2024 - sales in thousands

Branch	Total
A	1470
B	1440
C	1310

Figure 4.79 Compact View

If there are four sales regions, branch results can first be grouped by region and then shown as a total. Figure 4.80 shows the full table.

Branch billing for 2024 - sales in thousands of Dollars

		C	D	E	F	G	H	I	J	K	L	M	N	O	P	Q	R	S
2		Jan	Feb	Mar	Q1	Apr	May	Jun	Q2	Jul	Aug	Sep	Q3	Oct	Nov	Dec	Q4	Total
3	A	22	42	1	65	81	98	86	265	74	68	12	154	83	28	76	187	671
4	B	99	53	41	193	63	26	68	157	25	22	25	72	90	21	72	183	605
5	C	34	69	86	189	59	99	34	192	86	46	55	187	56	13	14	83	651
6	North	155	164	128	447	203	223	188	614	185	136	92	413	229	62	162	453	1927
7	D	47	56	16	119	20	57	91	168	69	44	71	184	76	47	80	203	674
8	E	42	74	95	211	90	25	97	212	77	78	40	195	35	38	13	86	704
9	F	56	82	27	165	63	70	51	184	98	18	23	139	68	60	71	199	687
10	East	145	212	138	495	173	153	240	566	244	139	133	516	179	145	164	488	2065
11	G	57	42	47	146	4	57	52	113	75	18	41	134	90	17	81	188	581
12	H	22	28	22	72	60	34	78	172	11	89	91	191	13	48	9	70	505
13	I	54	36	19	109	70	99	49	218	8	19	70	97	60	49	31	140	564
14	South	133	106	88	327	134	190	179	503	94	126	203	423	163	113	121	397	1650
15	J	33	27	18	78	45	26		71	49	87	75	211	31	99	21	151	511
16	K	73	95	48	216	26	90	41	157	71	11	84	166	11	84	34	129	668
17	L	50	61	54	165	56	72	56	184	7	21	88	116	34	67	80	181	646
18	West	156	183	120	459	127	188	167	482	127	119	247	493	76	250	135	461	1895
19	Total	589	665	474	1728	637	754	774	2165	650	520	675	1845	647	570	582	1799	7537

Figure 4.80 Complete Table with All Outline Levels

To achieve a clear result at every outline level, you need to carefully distribute labels so it's always clear what the data refers to and which data are combined in a formula. It's best to review each level separately.

4.11.1 Managing the Outlining Feature

Excel offers both automatic and manual outlining, and you can use automatic outlining if the table structure allows it. If the worksheet contains only the outlineable table, then the cell pointer's position doesn't matter. If you want to outline only a specific range, first select the entire range of outlineable data. Then, on the **Data** tab in the **Outline** group tab, select **Auto Outline** from the menu of the **Group** button.

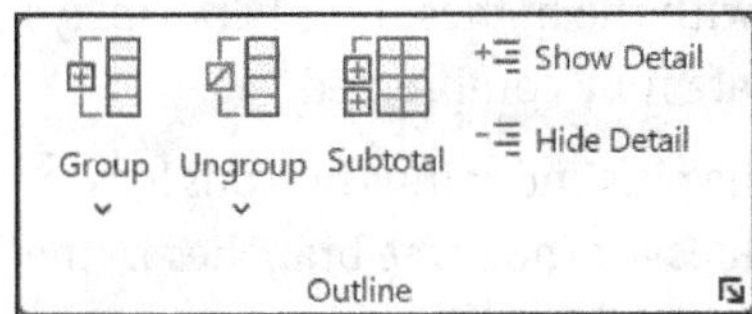

Figure 4.81 Outline Group on Data Tab

You can control how Excel performs automatic outlining through the **Settings** dialog box, which is accessible via the dialog box launcher in the **Outline** group. This is only necessary if you want to change the default checked settings.

Under **Direction**, the default checked option is **Summary rows below detail**. This matches a table layout where column totals appear below the number columns. If the sum appears above the number column, then the box must be unchecked. The default is also to place main columns to the right of detail data because formulas like cross totals usually appear to the right of the summed values. If you want the formula column to appear to the left of the value columns, uncheck the box.

Checking the **Automatic styles** box tells Excel to apply predefined styles to different outline levels. **ColumnLevel_1** is the style for the next higher level above the actual data, and this format applies to entire columns or rows. If you leave the box unchecked, you can assign styles later. To do this, select either the entire outlined range or a specific outline level, reopen the **Settings** dialog, and choose **Apply Styles**. In both cases, the special styles Excel uses will remain available for further editing in the **Cell Styles** palette. The outline applies your chosen settings when you click the **Create** button, and if an outline already exists, you'll be prompted to confirm whether you want to modify it.

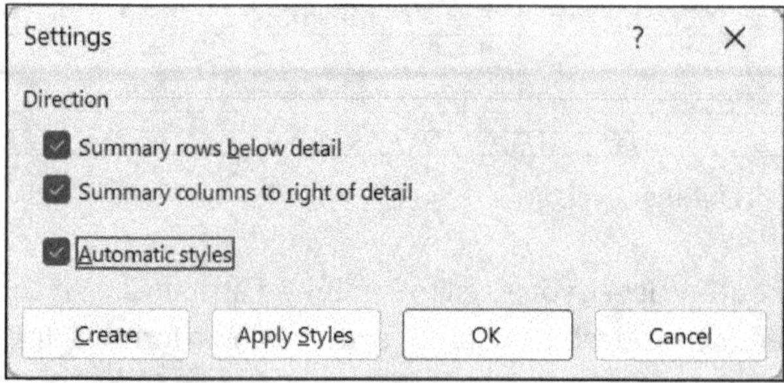

Figure 4.82 Outlining Feature Settings

Showing or Hiding Levels

Unless the **Show outline symbols if an outline is applied** box is unchecked under **File • Options • Advanced in Display options for this worksheet**, you can identify the outline result by the special outline symbols that appear above or to the left of the table.

Use these small buttons to switch between outline levels with a click or tap. These small number icons represent different row or column levels, and the number of these icons shows how many levels are in use. Clicking the icon with the highest number displays all levels, while clicking the one labeled 1 shows the data fully compressed.

Number icons always show or hide entire levels, but the plus and minus buttons let you expand or collapse parts of a level. In Figure 4.83, values for the three branches in the North region are shown, while only total values appear for the other regions.

	B	C	D	E	F	G	H	I
1		**Branch billing for 2024 - sales in thousands of Dollars**						
2		Jan	Feb	Mar	Q1	Apr	May	Jun
3	A	22	42	1	65	81	98	86
4	B	99	53	41	193	63	26	68
5	C	34	69	86	189	59	99	34
6	North	155	164	128	447	203	223	188
10	East	145	212	138	495	173	153	240
14	South	133	106	88	327	134	190	179
18	West	156	183	120	459	127	188	167
19	Total	589	665	474	1728	637	754	774

Figure 4.83 Detail Data for One Region Only

Restrictions to Keep in Mind

Automatic outlining applies only under these conditions:

- The table contains formulas referencing cells in the same column or row.
- References are consistent and point in only one direction.

If you try to automatically outline a table with only text and constants but no formulas, the **Auto Outline** option will be disabled. In that case, you can outline manually.

Condition 2 on the list is crucial. Automatic outlining won't work if a sum formula refers to both a cell on the left and a cell on the right. In that case, you'll see the error message **Cannot create an outline**.

Analyzing Outlined Tables

A key advantage of outlining is that each level can serve as the base for further processing. This is especially useful for creating charts, such as the following one showing quarterly results.

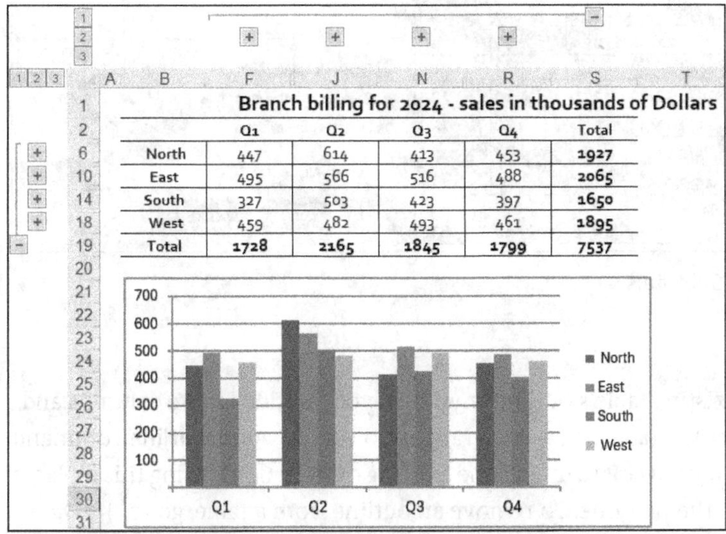

Branch billing for 2024 - sales in thousands of Dollars

	Q1	Q2	Q3	Q4	Total
North	447	614	413	453	1927
East	495	566	516	488	2065
South	327	503	423	397	1650
West	459	482	493	461	1895
Total	1728	2165	1845	1799	7537

Figure 4.84 Chart of Quarterly Results

Excel supports up to eight outline levels in both directions—by columns and rows. However, you can apply an outline only once per table, so you can't have multiple subtables with separate outlines in the same worksheet.

4.11.2 Creating Subtotals

In the previous examples, higher-level groups were created using sum formulas. Excel also offers a **Subtotal** button in the **Outline** group that you can click to automatically create subtotals in a list and link them to the automatic outline. Clicking this button opens a dialog where you can choose the summary type.

In Figure 4.85, multiple values are entered for each of three days and need to be summarized. To make this happen, simply place the cell cursor anywhere in the table and select **Subtotal**. Then, specify which column contains the grouping criterion—here, it's the date column. Next, choose the summary function in the second list box and the columns to summarize in the third. Then, confirm the dialog and Excel will create the subtotals and insert an automatic outline.

	A	B	C	D
1	Branch billing for 2024 in thousands			
2				
3	Date	A	B	C
4	1/22/2024	120	120	120
5	1/22/2024	120	130	90
6	1/22/2024	110	120	120
7	1/22/2024	120	120	50
8	22.01.2024 Result	470	490	380
9	1/23/2024	120	130	120
10	1/23/2024	120	120	90
11	1/23/2024	140	120	120
12	1/23/2024	120	120	120
13	23.01.2024 Result	500	490	450
14	1/25/2024	120	110	120
15	1/25/2024	130	110	120
16	1/25/2024	130	120	120
17	1/25/2024	120	120	120
18	25.01.2024 Result	500	460	480
19	Total	1470	1440	1310

Figure 4.85 Creating Subtotals

Adjusting Outlines

You can adjust an existing table's outline anytime if you've added more columns and/or rows. If you select the table's new total range and run the **Auto Outline** command again, Excel will ask if you want to update the existing outline. Confirming this replaces the old outline with the new one. To remove an outline from a table, go to the **Data • Outline** group and choose **Clear Outline** from the **Ungroup** menu.

4.11.3 Manual Outlining

You use manual outlining when automatic outlining conditions aren't met. For example, imagine a worksheet with several short tables stacked vertically. Longer texts between the tables explain their content, and these texts should be hideable when that's necessary. Here, the tables form the parent level and the texts form the child level. Because multiselection isn't allowed, you must select each text range individually and use the **Data • Outline • Group** command each time to apply it.

	A	B	C
1	**Sales Planning**		
2		Number	Visits per year
3	Customers Class A		
4	District 1	120	6
5	District 2	110	6
6	District 3	90	6
7	If it is possible to recruit 3 more representatives, the number of		
8	visits will be increased to 8.		
9			
10	Customers Class B		
11	District 1	100	4
12	District 2	95	4
13	District 3	120	4
14	The number of visits has been reduced because sales		
15	of this customer class could not be increased.		
16			

Figure 4.86 Example of Manual Outline

In a small dialog box, choose whether to group rows or columns in the selected range if full rows or columns aren't selected. Select the **Data • Outline • Ungroup** command to remove groupings from the selected range, and click or tap the Level 1 symbol to hide text as needed, leaving only the tables visible.

Copying Outlines

When you copy an outlined table to another file, you can include the outline if you want. Just click the **Select All** box in the upper-left corner before using **Copy**. In the worksheet where you want to paste the outlined table, you must also check the **Select All** box. Then, when you use **Paste**, the table and its outline will be inserted.

If you only want to copy the data, select the appropriate range in the table and the outline won't be copied to the new table.

4.12 Data Entry Using Controls

One way to design your tables is by using controls like those commonly found in dialog boxes. This applies not only to buttons that run macros, as covered in Chapter 22, Section 22.2. Using list boxes, scroll bars, and spin buttons can also help you simplify data entry and reduce errors in worksheets.

4.12.1 Selecting Data with a Combo Box

You can use a list box in a worksheet to select a specific item from a list. For clarity's sake, we'll use a simple list of month names here. To link this list to a combo box, first ensure that the **Developer** tab is visible. To do this, go to **File • Options • Customize Ribbon**, and in the window on the right, under **Main Tabs**, check the box for the **Developer** tab.

1. On the **Developer** tab, click the **Insert** button in the **Controls** group, and from the menu, select the combo box icon ❶ under **Form Controls**.

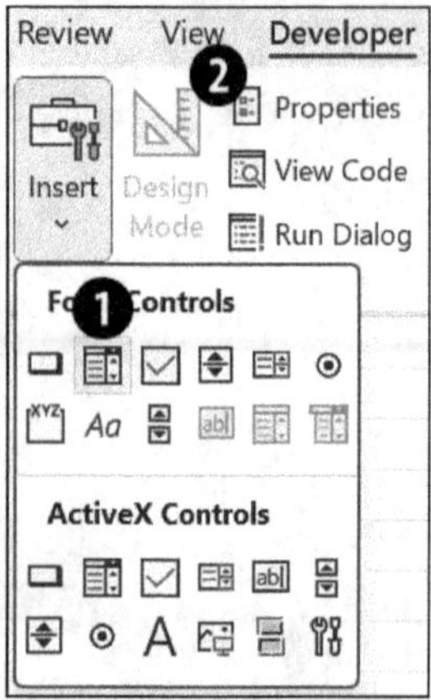

2. Draw a rectangle for the combo box on the worksheet.
3. Next, in the **Controls** group, click the **Properties** button ❷ and make sure the item is selected.
4. On the **Control** tab of the dialog, enter the range containing the month names as the **Input range** ❸.
5. Under **Cell link** ❹, enter the address of the cell where the selected list value should appear. This cell will only show the number of the selected position.

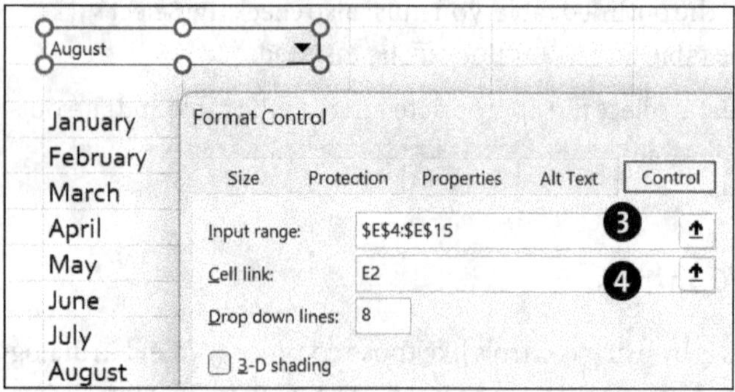

6. With the following simple trick, you can easily convert this number into a month name for another cell. In the illustration, the output range is a cell behind the combo box, so it will remain hidden.
7. In the cell where the selected month name should appear—which is A3 here—you can use the following function to do this (the values in the angle brackets are placeholders):

```
=INDEX(<array>, <row>, <column>)
```

For `array`, enter the list range with the month names; for `row`, use the address of the output link cell; and for `column`, enter 1, since the month column is a single column. The function therefore looks like this:

```
=INDEX(E4:E15,E2,1)
```

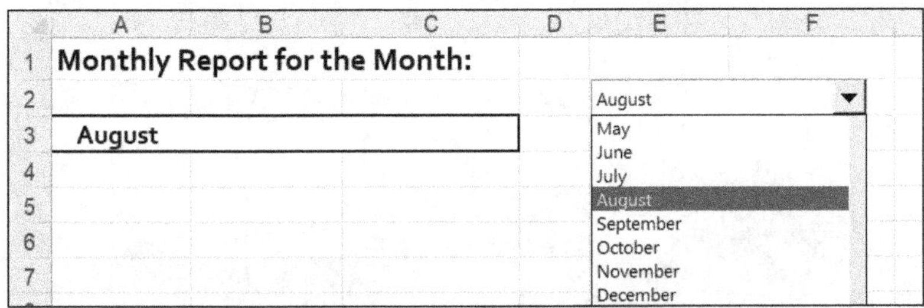

4.12.2 The Scroll Bar and Spin Button

The **Scroll Bar** and **Spin Button** controls let you output numbers directly, with customizable ranges and step sizes.

You do not need an input list range for these two elements, but you do need a **Cell link**. You can assign the value selected with these elements directly to a cell, and Figure 4.87 demonstrates this with a present value calculation in which different values for the monthly payment are to be tested. You can enter the **Minimum value** and **Maximum value** under **Properties** in the **Controls** group, where you can also enter the **Incremental change**.

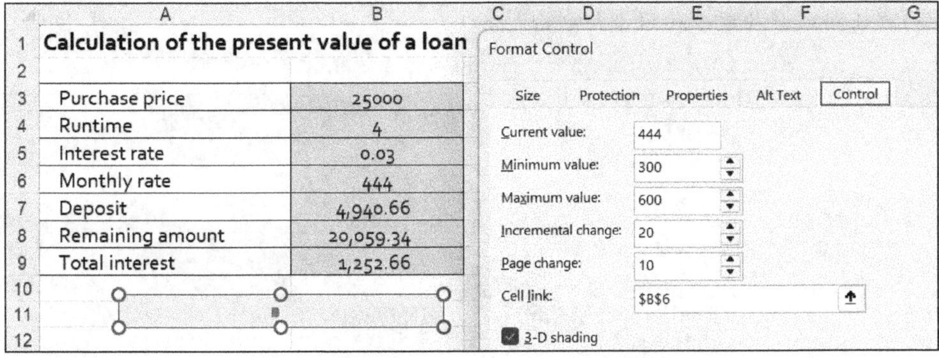

Figure 4.87 Calculating Present Value Using Scroll Bar

Useful applications include simple what-if analyses, charts where you adjust values with the **Spin Button** or **Scroll Bar**, and more.

Chapter 5
Analysis and Forecasting

Once you've built a calculation model with data and formulas, you can reuse it repeatedly. For example, you can update a weekly report model with new data each time by just changing the week number. Excel's formulas calculate the new data, but using formulas isn't the only way to have Excel compute and analyze your data. The program also provides advanced tools you can use to analyze existing data and models. The following sections present these options through practical examples.

5.1 Calculations Without Formulas

First, we'll briefly cover a simple way to perform calculations by using existing data and without entering formulas. One method focuses on multiplying a column of numbers by one or more factors, and the other involves adding, subtracting, or otherwise manipulating entire cell ranges.

5.1.1 Multiplying a Price Column by a Percentage

Imagine a table with a column listing the sales prices of several items. You decide to reduce all sales prices by 3% at the start of the month, and in this case, you don't need to re-enter the sales prices, nor do you need to create formulas for this (although you could). Instead, you can do the following:

1. Enter the price reduction factor into an empty cell. In this example, you want to reduce the price to 97% of the original, so you enter 97% or 0.97.
2. Copy the cell's value using **Home • Clipboard • Copy.**
3. Select the column with the original sales prices.
4. Go to **Home • Clipboard • Paste • Paste Special.** Under **Paste,** select **Values,** and under **Operation,** choose **Multiply.** Then click **OK.**

Excel will multiply each previous sales price by 0.97 and display the result as a value in the cell, and you can then delete the cell containing the percentage. Instead of multiplying, you can also choose to **Add, Subtract,** or **Divide.**

Instead of recalculating all values in the target range with the same number, as in this example, you can work with different values. In this case, either the copied range and the target range must be exactly the same size, or the source range must fit multiple times into the target range. For example, if the source range has two cells in one column and the target range has 20 cells, then the multiplication will alternate between the first and second cells of the source range. On the other hand, if the target range contains 21 cells, you'll get a message saying the target range doesn't fit.

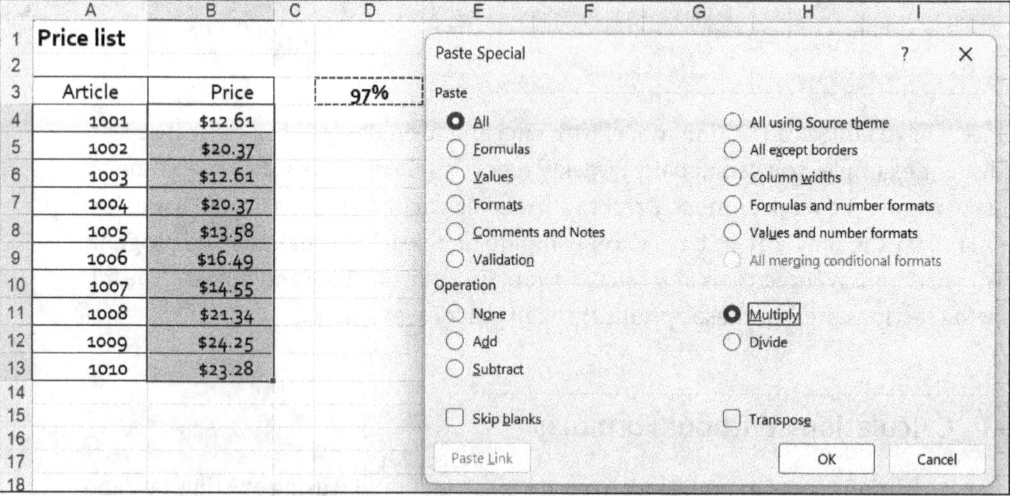

5.1.2 Combining Ranges

You'll usually employ this function to combine two equally sized ranges. For example, to combine two columns of numbers without using formulas, add column B to column A and then delete column B. The function can also work with data from different worksheets or workbooks.

5.2 Consolidating Results

The **Consolidate** function offers you a practical way to combine data from different ranges. Although this function has lost some relevance due to the calculation options enabled by the three-dimensional structure of workbooks, it still provides useful ways to merge and analyze data from multiple sources.

Excel offers two methods of consolidating data: based on the data's position in the worksheets or based on their category.

5.2.1 Consolidating by Position

In the first method, the source ranges must be in the exact same location on all sheets, and in the second, the source ranges can be located anywhere in the source files.

Imagine you receive workbooks from three plants at the end of each month, with monthly data organized on different sheets. Each workbook includes a consistently structured table showing production across various product groups, and you want to combine these three individual results into a single summary in a new workbook.

Of course, you could combine the different sheets containing these ranges into one workbook, but that might not be desirable. For example, those sheets may include other data that aren't relevant.

	A	B	C	D	E
1	**Monthly bulletin**	**Factory I**		**Month: Jan 2024**	
2					
3	Category	Units	Revenue	Costs	Profit
4	Washing machines	8,000	9,600,000	6,720,000	2,880,000
5	Dryer	6,000	5,400,000	3,780,000	1,620,000
6	Fridges	12,000	7,200,000	5,040,000	2,160,000
7	Chest freezers	7,000	6,300,000	4,410,000	1,890,000

Figure 5.1 Results Sheet of Plant

Consolidating Events: Setting the Target Range

The first step involved in consolidating the results from the three plants is to set a target range where the data will be combined. While the source ranges and target range can be in the same workbook, this example assumes they are not. Follow these steps:

1. Open a new workbook and select a worksheet. It's a good habit to name the sheet by double-clicking or double-tapping the sheet tab.
2. Label the sheet accordingly by copying the label from one of the source ranges to the exact same spot here and adjusting it as needed. Since we're first describing the consolidation method based on data position, make sure the cell addresses align with their labels. For example, the number of washing machines sold should always appear in cell B4.
3. Select the range where the consolidated data will appear. Simply select the first cell of the range, which is B4 here.
4. On the **Data** tab, go to the **Data Tools** group and click the **Consolidate** icon ❶.

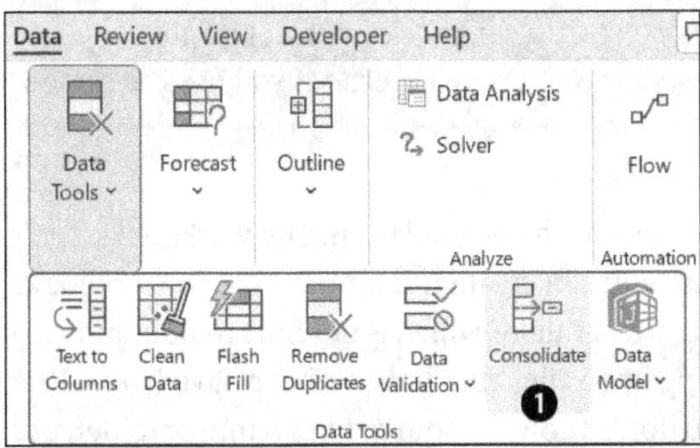

5. The **Function** ❷ field defaults to **Sum** because the program assumes that values from the ranges will be added. The list also offers other options for analyzing data, including basic statistics like standard deviation and variance. In this example, **Sum** is the right choice.
6. Excel now requires you to make an entry in the **Reference** ❸ field. The detail needed in the lookup depends on where the source data is located. If the source data is on the same worksheet as the target range, simple cell references suffice, but if it's on a different worksheet in the same workbook, then you'll need to include the sheet name before the cell address. When referencing other workbooks, you must specify the file name, and if the workbook isn't in the default folder, you should include the full file path.

   In this example, the reference points to another workbook, so you should enclose the file name in square brackets and place the sheet name followed by an exclamation mark before the range address.
7. `[EX24_HB_Chap05_Factory_A.xlsx]Production!$B$4:$E$7` is a correct reference. After entering or selecting the reference in the open workbook, click the **Add** button ❹ and the reference you entered will appear in the **All references** list. Proceed similarly with the other references.
8. Finish by clicking **OK**, and Excel will calculate the total values for all three plants and output them as values in the target range.

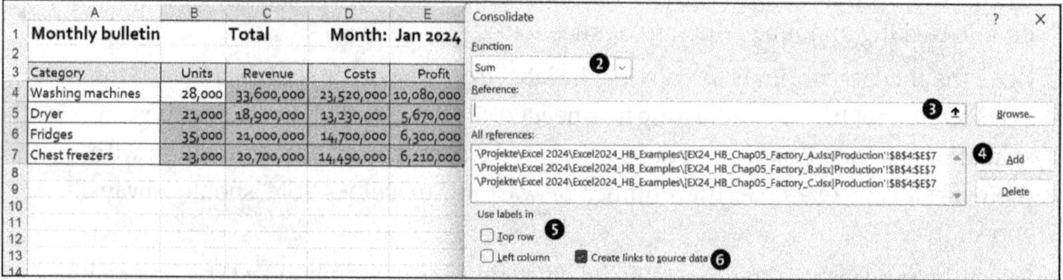

Use Range Names for Better Results

Using range names is more elegant than entering range references. Since the data in this example comes from different workbooks, it doesn't matter if you use the same range name each time.

You can specify up to 255 source ranges in such a consolidation. The workbooks don't need to be open, but it's helpful because you can create references to open workbooks by selecting them—instead of typing them manually, which is not only tedious but also more prone to errors. Avoid referencing open files that haven't been saved yet, since their final file names aren't fixed. If you don't know the source file's name or folder, use the **Browse** button to open the **Browse** dialog box and locate your data.

5.2.2 Consolidating by Category

If source data isn't always in the same spot on a sheet, or if not all source ranges include every item, use the second method: consolidation by category. For example, with this method, Excel won't have a problem if one plant enters results outside the A3 to E7 range, say in A104 to E108. It also won't mind if a plant doesn't produce refrigerators and the corresponding row is missing. This method doesn't require manually entering labels in the destination range because Excel pulls them from the source ranges and compiles them accordingly. Follow these steps to use this method:

1. Select the top-left cell of the destination range, including the label row.
2. In the **Data** group, click **Data Tools** and then select the **Consolidate** icon.
3. Under **Reference**, enter each source range with labels included, clicking **Add** after each one.
4. To have Excel use labels, specify under **Use labels in** (❺ in the previous figure) where the labels come from. In this case, you can check both boxes.
5. Click **OK** and Excel will copy the labels to the specified location in the target range and consolidate the data there.

Editing a Consolidation

Consolidation data is saved with the workbook containing the target range. You can refresh the summary anytime if the source data changes, and you can also modify the source ranges at any time if needed. Here's how you do it:

1. Select the target range and reopen the **Consolidate** dialog box.
2. To add more source ranges, enter the new range under **Reference**, click **Add**, and then click **OK** to recalculate.
3. To exclude a source range from the calculation, select it under **All References** and click **Delete.**
4. To modify only the selected reference, update the address under **Reference.**

Creating Links to Source Ranges

Excel typically consolidates data by using the current values from the source ranges, calculates the result, and displays it in the target range. This completes the process, and if the values in the source ranges change later, you need to repeat the consolidation with the same settings in the dialog box to refresh the results.

To have the destination range update automatically when any source range changes, check the **Create links to source data** box (❻ in the previous figure). After this, the data in the workbook will update automatically whenever you open it, and in the target range, you'll see external references to individual cells in the source ranges instead of constant values.

Excel adds new rows or columns and formulas for each source range to summarize the data based on the function chosen in the dialog box. The table is automatically grouped and initially shown in its most compact form, and you can use the grouping symbols to reveal all the detailed data in the target range.

	A	B	C	D	E
1	**Monthly bulletin**		**Total**	**Month:**	**Jan 2024**
2					
3	Category	Units	Revenue	Costs	Profit
4		8,000	9,600,000	6,720,000	2,880,000
5		10,000	12,000,000	8,400,000	3,600,000
6		10,000	12,000,000	8,400,000	3,600,000
7	Washing machines	28,000	33,600,000	23,520,000	10,080,000
8		6,000	5,400,000	3,780,000	1,620,000
9		6,000	5,400,000	3,780,000	1,620,000
10		9,000	8,100,000	5,670,000	2,430,000
11	Dryer	21,000	18,900,000	13,230,000	5,670,000
15	Fridges	35,000	21,000,000	14,700,000	6,300,000
19	Chest freezers	23,000	20,700,000	14,490,000	6,210,000

Figure 5.2 Linked Consolidation

5.3 Add-Ins for Statistical Data Analysis

For statistical analysis of table data, you can use not only the many statistical table functions introduced in Chapter 15 but also a comprehensive analysis tool add-in that simplifies statistical evaluation of your data. If you have installed the Analysis ToolPak and added it through **File • Options • Add-Ins** to Excel's list of add-ins, an additional **Analysis** group with the **Data Analysis** button will appear on the **Data** tab.

Before using any of the options in this command, you should arrange the data to be analyzed in a worksheet so the analysis function you select can process it. How data must be arranged for a specific function depends on that function, and covering each function in detail is beyond the scope of this book.

However, here's one example: creating a histogram. This type of chart is commonly used to show frequency distributions.

5.3.1 A Histogram Showing the Distribution of Deviations

A common use for a histogram is to display the frequency distribution of deviations from a set standard. For example, imagine a table listing the percentage deviation from normal weight for a sample of 500 people:

1. The function requires an input range in the worksheet containing only numeric data, except for the column or row labels. The data can be arranged in columns or rows.

2. You can also create a class range in addition to the input range. You can enter boundary values here to define the limits of each class, and you should list boundary values in ascending order. When one or more boundary values are present, Excel counts the cases that fall into each class, and it separately summarizes values below the lowest boundary and above the highest boundary. If you don't use a class range, Excel automatically divides the data range into equal-width intervals.
3. After completing these preparations, you can use the **Data Analysis** command.
4. Select the analysis function to apply to the data in the worksheet, which in this case is **Histogram**.
5. In the **Histogram** dialog, enter the reference for the range to analyze under **Input Range.** Include the cell containing the label.
6. Under **Bin Range,** select the range with the boundary values, including the cell with the label.
7. Make sure the **Labels** checkbox is checked to include labels in the analysis. If it's unchecked, default labels will be generated.
8. Specify the upper-left cell of the range where you want the analysis results to appear under **Output Range.** If the range isn't empty, you'll get a warning before the data is inserted. You can also choose the **New Worksheet Ply** option to send the output to a new sheet, and the data will start at cell A1 on that sheet. You can enter the sheet name directly in the input field, and you can also select **New Workbook** as the output destination.
9. You can check **Pareto** to sort the results by descending frequency. Otherwise, the function uses the ascending frequency sequence. You can also choose **Cumulative Percentage,** which will make the function add an extra column with cumulative frequencies and insert a corresponding curve into the histogram. To plot a chart directly from the results, check **Chart Output.**

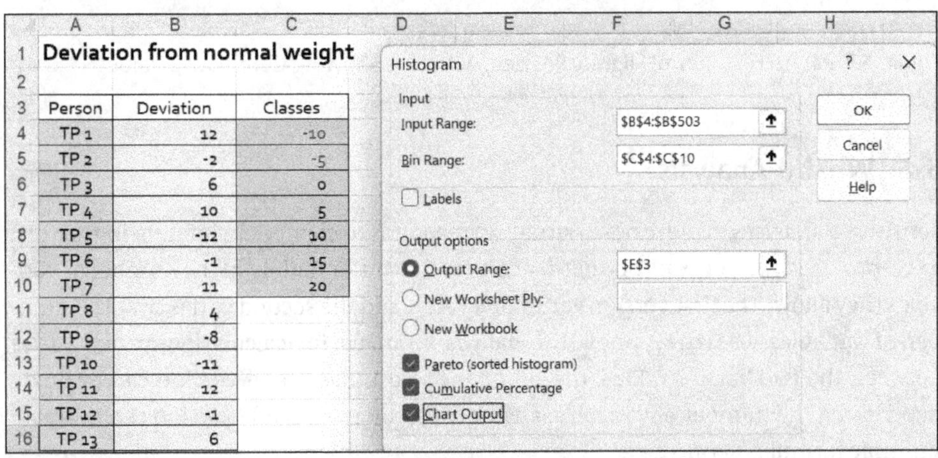

10. After you confirm, Excel creates a frequency table and chart.

Bin	Frequency	Cumulative %	Bin	Frequency	Cumulative %
-10	72	14.40%	10	89	17.80%
-5	80	30.40%	15	87	35.20%
0	82	46.80%	0	82	51.60%
5	75	61.80%	-5	80	67.60%
10	89	79.60%	5	75	82.60%
15	87	97.00%	-10	72	97.00%
20	15	100.00%	20	15	100.00%
More	0	100.00%	More	0	100.00%

11. The chart will look more like a histogram if you set the gaps between columns to zero. To do this, click the chart and then right-click any column. Choose **Format Data Series** and then **Series Options**. On a touchscreen, tap and hold a column until the context menu appears, tap the arrow at the end, and select **Format Data Series**.

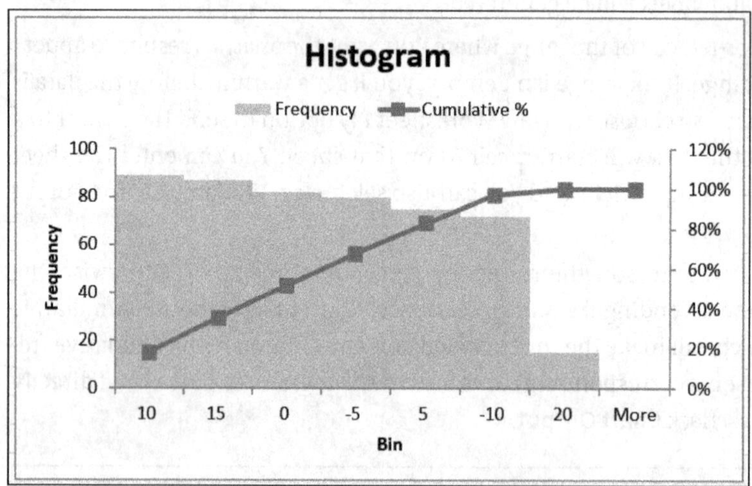

Figure 5.3 Pareto Histogram, Slightly Revised, with Line Showing Cumulative Values

5.4 What If Analysis

Sensitivity analyses explore how a situation responds to changes in influencing factors. Excel provides two versions of what-if analyses that differ mainly in the number of variables they allow: The first version varies one value, and the second varies two. The number of variables also affects how the data is arranged for calculation, which varies between the two versions. Plus, the number of formulas that a version can evaluate depends on the number of variables, and only the single-variable version can evaluate multiple formulas at once.

5.4.1 One-Variable Data Tables

Here's a simple example. In Figure 5.4, you'll find a small application using a present-value function. The question might be this: If the interest rate is 1.3%, how much must be deposited on 1-Jan to receive an amount after three years that's equal to three annual deposits of $40,000?

	A	B	C
1	**Calculating the present value**		
2			
3	Deposits	$40,000.00	
4	Periods	3	
5	Interest rate	1.30%	
6	Present value		-$116,946.31
7	Other interest rates	0.75%	
8		1.00%	
9		1.15%	
10		1.40%	

Figure 5.4 Preparing Present-Value Calculation as Batch Operation

The formula with the PV() (present value) function in cell C6 doesn't use a fixed interest rate but refers to cell B5, where 1.3% is initially entered. Follow these steps:

1. To observe the effect of different interest rates, enter a series of rate variations in a free column (which is column B in Figure 5.4). The **Other interest rates** label shows these values, and you can enter the alternative values one below the other in a single column. The order isn't important, but sorting them in ascending or descending order makes the results clearer.
2. Prepare the formula that will use these alternative values, which in this example is the present value function. Enter this formula in the row directly above the first row of alternative value, which must be shifted at least one column to the right because the program will place the newly calculated values directly below this formula. In Figure 5.4, cell C6 is the correct place for the formula.
3. Select the range to use for the data table (earlier versions called this a *multioperation*). In Figure 5.4, this range is shaded for clarity. The range is defined by the column with alternative values and the row containing the formula or formulas the command will recalculate. Unlike a normal formula, this operation's result doesn't appear in the cell with the formulas but in a continuous series of cells.
4. From the **Forecast** group on the **Data** tab, select the **Data Table** command from the **What-If Analysis** button menu. This opens the dialog box, where you can enter the cell address for the interest rate—which in this case is B5. This is the input cell where you can enter different interest rates one after another to recalculate the present value formula each time. Since the alternative interest rates are arranged in a column, enter the address "$B$5" in the **Column input cell** field. If the alternative interest

rates were arranged in a row—say, from D5 to L5—you would enter the cell address in the **Row input cell** field.

	A	B	C	D
1	**Calculating the present value**			
2			Data Table	? ×
3	Deposits	$40,000.00		
4	Periods	3	Row input cell:	
5	Interest rate	1.30%	Column input cell:	$B$5

5. When you confirm the entries in the dialog box, Excel calculates the present value formula for all alternative interest rates and shows the results in the column below the formula.

You can read the table like this: An alternative interest rate of 1.4% yields a present value of $116,716, while 1.7% yields $116,032, and so on. The original formula in cell C6 and the initially assumed interest rate of 1.3% in cell B5 remain unchanged.

If you look at cell C7, you'll see an array formula with the `TABLE()` function in the formula bar. The cells that show the results of the multioperation form a result matrix, so the special rules described in Chapter 3, Section 3.5 apply. You cannot change or delete individual cells within this range, but you can freely modify the other interest rates you've entered. If you've set the other arguments of the present value function with cell references, you can now test different values for the number of periods or regular payments, and the entire multiple operation table will recalculate each time.

	A	B	C
1	**Calculating the present value**		
2			
3	Deposits	$40,000.00	
4	Periods	3	
5	Interest rate	1.30%	
6	Present value		-$116,946.31
7	Other interest rates	0.75%	-$118,222.25
8		1.00%	-$117,639.41
9		1.15%	-$117,292.00
10		1.40%	-$116,716.79
11		1.50%	-$116,488.02
12		1.70%	-$116,032.72
13		1.88%	-$115,636.77
14		2.05%	-$115,243.08
15		2.23%	-$114,851.62
16			

Figure 5.5 Calculated Alternatives

Evaluating Multiple Formulas

When working with a single variable, you can observe how multiple formulas behave simultaneously. For example, you can calculate the present value assuming payments are due at the beginning of the period. You just need to add the argument for the due date to the formula, and in this case, you select the range from B7 to D16 for the multiple operation.

3	Deposits	$40,000.00		
4	Periods	3		
5	Interest rate	1.30%		
6	Present value		-$116,946.31	-$118,466.61
7	Other interest rates	0.75%	-$118,222.25	-$119,108.92
8		1.00%	-$117,639.41	-$118,815.80
9		1.15%	-$117,292.00	-$118,640.86

Figure 5.6 Present-Value Calculation for Payments in Advance and in Arrears

5.4.2 Multiple Operations with Two Variables

The second version of the multioperation accepts two variables, or two input cells, but evaluates only one formula at a time. The table layout differs slightly from the first version. Enter the formula in the previously empty top-left cell of the selected table range.

Then, you enter the alternative values for the first variable in the left column of the table range, as usual, and you enter the values for the second variable in the first row, starting with the cell to the right of the formula cell. So, this version requires the values for the second variable to be where the formulas were in the first version.

Example: Rate Calculation

Figure 5.7 shows the layout of this specific table range, and once again, we'll use a financial function as the example. A rate calculation uses three arguments: a principal amount (such as a loan), an interest rate, and the number of periods. The loan and interest rate will be varied. The values for the first variable (the loan) are in column B, while the values for the second variable (the interest rate) are in row 6. The reference to the rate formula is in the top-left cell of the table range, which is cell B6.

	A	B	C	D	E	F	G
1	**Rate calculation**						
2							
3	Interest rate	1.10%					
4	Loan	9000					
5	Periods	4					
6	Rate	-$2,312.21	1.25%	1.40%	1.55%	1.70%	1.85%
7		$10,000.00	-$2,578.61	-$2,588.11	-$2,597.62	-$2,607.15	-$2,616.68
8		$12,000.00	-$3,094.33	-$3,105.73	-$3,117.14	-$3,128.57	-$3,140.02
9		$13,000.00	-$3,352.19	-$3,364.54	-$3,376.91	-$3,389.29	-$3,401.69
10		$15,000.00	-$3,867.92	-$3,882.16	-$3,896.43	-$3,910.72	-$3,925.03

Data Table ? ×
Row input cell: $B$3
Column input cell: $B$4
OK Cancel

Figure 5.7 Payment Calculation with Two Variables

First, select the table range from B6 to G10, and then, use the **Data Table** command. This time, fill in both input fields, and for the **Row input cell**, enter the address of the cell with the interest rate, which is B3. For the **Column input cell**, enter B4, which contains the loan amount.

This command creates a table of values you can read like an *xy* chart. To find the payment for a $13,000 loan at a 1.55% interest rate, locate the row starting with $13,000 and then move right to the column headed by 1.55%, which contains the corresponding payment.

Convert to Values

After the array operation returns its results, it's often helpful to convert the array formulas in the result range into constant values. To do this, select the entire result array—using [Ctrl]+[/] or **Home • Editing • Find & Select • Go To Special • Current Array**—copy the data, and then use **Paste Special** with the **Values** option to paste over the selection. Excel will automatically select the entire array range, but if you want to keep the original formula(s), select only the cells with the array formulas!

5.5 Planning Scenarios

The multioperations covered in the last section deliver useful results as long as one or two values can change—but in many cases, that's not enough. Excel offers a more powerful tool for what-if analysis called the Scenario Manager, and it supports many more variables—up to 32.

5.5.1 What Scenarios Are For

In Excel, a *scenario* is like a simulation that estimates how different factors affect a situation. Typically, you use scenarios to explore solutions by comparing different outcomes.

They often start with a question you want to answer using one or more scenarios. For example, how would sales likely increase if two new branches, each with four salespeople and 300 square feet of sales space, were opened—assuming sales per salesperson and per square foot remain similar to the current branches? Wouldn't a larger branch with six salespeople and 500 square feet of sales space generate more? And how would storage costs for the ten key raw materials change if the order cycle and quantity were adjusted? These are simple examples of questions for scenarios.

In an Excel model, running a scenario means running the model with a specific set of variable values. Scenario A uses value set 1, Scenario B uses value set 2, and so on. You could create these scenarios by entering a specific group of values into an existing model, saving it as *Plan1*, then entering a new group of values for the same cells and saving it as *Plan2*, and so forth. However, while this method works, comparing the results is difficult and cumbersome. A better approach is to create several uniform tables in a workbook and enter different values in each—but even then, comparing the alternatives remains tedious.

This is where the Scenario Manager helps. It lets you easily manage a worksheet model with multiple variants and compare the effects of different assumptions in a summary report. However, using the function is not very helpful if the worksheet model only has a label column and a column with values and formulas. In that case, it's better to list planning alternatives in the adjacent columns, which gives you an immediate overview of the different calculation options. In other words, the Scenario Manager becomes more useful as the worksheet model behind the scenarios grows more complex.

5.5.2 Planning Alternatives for the Advertising Budget

We'll demonstrate how the Scenario Manager works with a simple example so you can easily follow the illustrations in the figures. This example focuses on planning advertising budgets, and the table in Figure 5.8 summarizes the costs of various advertising media used for different products.

To calculate the advertising cost per unit sold, you need to compare planned sales figures for products to costs. To do this, start by building the worksheet model completely with all formulas. You can use test data or enter values for the first scenario right away, or you can start by entering just the formulas.

	A	B	C	D
1	**Advertising budget plan**			
2				
3	Media	Product 1	Product 2	Product 3
4	Newspapers	$35,000.00	$37,000.00	$22,000.00
5	Internet	$40,000.00	$33,000.00	$27,000.00
6	Television	$100,000.00	$90,000.00	$120,000.00
7	Poster	$12,000.00	$13,000.00	$11,000.00
8	Direct mail	$10,000.00	$12,000.00	$12,000.00
9	Total	$197,000.00	$185,000.00	$192,000.00
10	Planned sales	35,000	30,000	33,000
11	Advertising costs per sales unit	$5.63	$6.17	$5.82

Figure 5.8 Planning Advertising Budget

Which Factors Change?

Once you've built the model, you need to decide which values can change across scenarios and which cannot. This depends entirely on the situation the model represents. Often, some values in a model are fixed from the start and cannot be changed, while for other values, there's room for adjustment within a certain range.

In this example, you can set the number of ads in daily newspapers higher or lower. Planned sales assumptions may vary between optimistic and pessimistic figures, and you should also test a value in the middle.

Variable Cells and Result Cells

In Excel terminology, cells containing values to be changed are called *variable cells*. These values are independent, meaning they don't depend on other values. Variable cells usually don't contain formulas, and if they do, Excel displays a warning. When displaying the scenario, the formula is replaced by its result and removed.

The next issue is which data the model's built-in formulas affect and which data they don't. Cells influenced by the adjustable cells' values are the scenario's possible result cells—which are the values you monitor. In our example, we enter varying amounts for daily newspaper ads and different sales estimates into the adjustable cells. The result cells of interest are those that show the total advertising budget for a product and the advertising cost per sales unit.

Strongly Recommended: Using Range Names

After making these preparations, you can open the Scenario Manager. While it's possible to proceed without them, using names for the cells you want to change in a scenario and for the result cells is highly recommended for two reasons:

1. Excel uses these names to create a simple data entry form for inputting values into the adjustable cells. Without names, the dialog box only shows abstract cell addresses.
2. The defined names are automatically included in reports that compare results across different scenarios. This is what makes the reports readable.

As for the names themselves, keep in mind that although you can use longer names, Excel only displays the first 16 characters of each name in the data entry dialog boxes. The name should be unique within the first 16 characters, even if it's longer. Take the time to assign names to the relevant cells because it makes working with scenarios much easier and clearer.

5.5.3 Defining a Scenario

After completing these preparations, you can begin defining your scenarios. Follow these steps:

1. On the **Data** tab in the **Forecast** group, select **Scenario Manager** from the **What-If Analysis** button menu. The first dialog box will tell you that no scenario has been defined for this worksheet yet, so click the **Add** button.
2. Excel manages scenarios by assigning names, so the program first requires you to enter a name for each scenario. It's best to choose a name that clearly describes the scenario since that makes it easier to locate the data associated with that name later.

For example, you could name the first scenario *Optimistic Scenario* if you plan to enter corresponding sales figures. Generic names like *Version_1* and *Version_2* aren't prohibited, but they're impractical since you'll probably forget what assumptions they represent, especially as the number of scenarios increases.

3. The second input field asks for the references or names of the cells whose values need to be updated for each scenario. Excel first shows the address or name of the currently selected cell, so, as is usually best, you should select the first editable cell or range before running the command. You can also make a multiselection, or you can enter the references or names one at a time under **Editable Cells** and separate them with semicolons. If the cells are adjacent, enter the range reference. You can also select the references directly in the worksheet. To do this, from the second reference onward, hold down the `Ctrl` key as you enter each one, and Excel will add the semicolon automatically. References to other sheets are not allowed here. You can also insert names using `F3`.
4. Under **Comment**, your name and the date the scenario was created or modified appear automatically. To add a comment, click inside the text box and type your notes.
5. Under **Protection**, you can check the box labeled **Prevent changes**. This prevents changes to the scenario, as long as the file is fully protected afterward via **Review • Protect • Protect Sheet**. You can also manage each scenario differently—for example, you can protect the most likely variant from changes while allowing edits to other variants for different numbers.
6. If you check the **Hide** box, the current scenario won't appear in the list in the first dialog box when the file is fully protected.

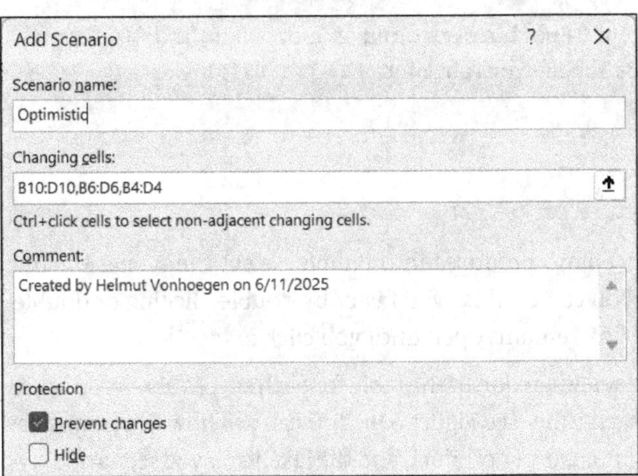

7. After you confirm the dialog box by clicking **OK**, another dialog box will open with input fields for the editable cells you specified. If the cells were named earlier, their names will appear before the corresponding input fields. If you entered values for the

first scenario when creating the table, those values will show here and you can accept them easily.

8. To add more scenarios, click **Add**; otherwise, click **OK**. In the first step, Excel saves the entered values and reopens the **Add Scenario** dialog box, where you can enter the name for the next scenario. In our example, a name like *Pessimistic* would work. You don't need to change the references for adjustable cells if the second scenario uses the same values as the first.

   You can also create scenarios on a worksheet with different adjustable cells.

9. Add a comment to the new scenario if you need to and confirm the dialog box, and the form for adjustable cells will appear again. After you enter the last scenario, Excel will return you to the first dialog box, where you'll see the names of all defined scenarios.

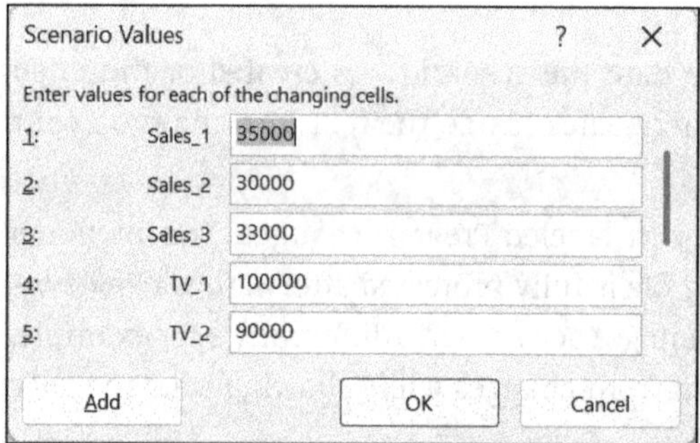

[»]

Using Formulas Instead of Fixed Amounts

You can enter not only constant values but also formulas in the input fields. For example, if the value 22000 appears initially, you can enter =22000 * 1.05 to increase the value by 5%.

Viewing Individual Scenarios

To see the results a specific scenario produces in the table, select the scenario name from the list and click **Show**. You can do this even faster by double-clicking or double-tapping the name. The dialog box remains open until you click or tap **Close**.

The **Scenario Manager** creates a new version of the table for each set of values you enter, and all formulas recalculate each time (see Figure 5.9). You can see how different sales assumptions impact advertising costs per product, and the last displayed scenario stays visible in the table after you close the dialog box.

To switch quickly between scenarios, add the **Scenario** command to the Quick Access Toolbar. Thereafter, clicking the button will reveal a list of scenario names (see Figure 5.10).

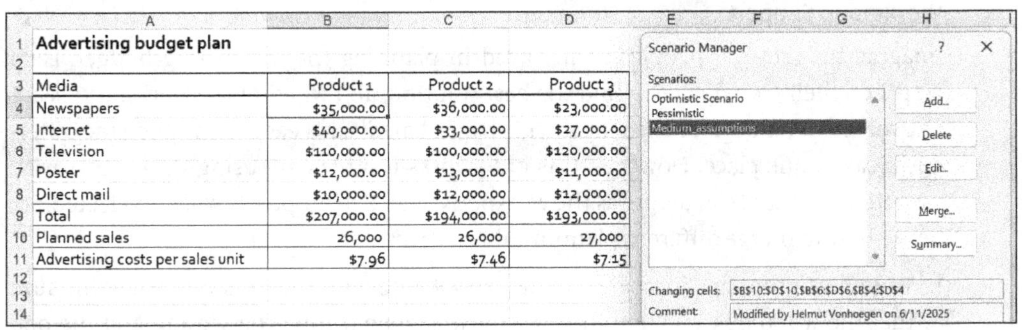

Figure 5.9 Displaying a Specific Scenario

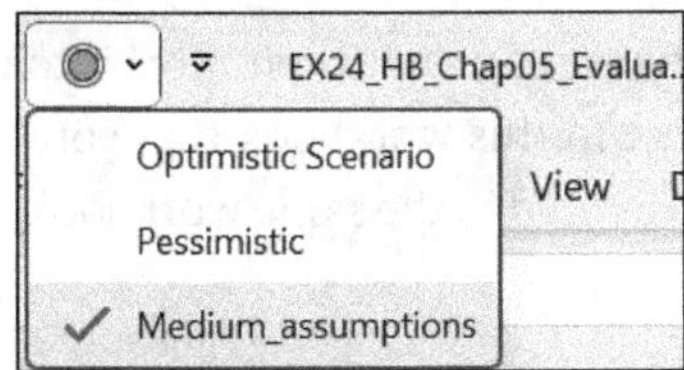

Figure 5.10 Selecting Scenarios from Quick Access Toolbar

5.5.4 Editing Scenarios

Once you've defined scenarios, you can edit them anytime unless their changes are locked. For example, if daily newspaper ad prices change, you don't need to create a new scenario. You only need to update the scenario affected by this change as follows:

1. In the **Scenario Manager** dialog box, select the appropriate scenario name and click **Edit**.
2. If you need to, you can also change the scenario name, adjustable cell references, and comments in the next dialog box. If you don't need to make changes, simply click **OK** to close the dialog box.
3. To enter different values for the adjustable cells, input them into the next dialog box and click **OK**.

Excel automatically logs the date and the name of the person who made each change. This creates a journal of all scenarios for a worksheet.

Deleting Scenarios

If you want to delete individual scenarios (for example, because the alternative they represent is unacceptable), select the name in the dialog and click **Delete**. This lets you gradually narrow the alternatives to those that truly matter to you. If you delete all scenarios from a worksheet, remember that the sheet will keep the values from the last displayed scenario. Also, these delete actions cannot be undone.

Workgroup Support

Imagine that several people are involved in planning the advertising budget. Each works on their own device to find the best solution, and whether the workstations are networked or data is still exchanged via physical media, it's easy to combine the different ideas in one place. However, this only works if others are using worksheets with essentially the same structure as the workbook where all proposals will be collected. So, here's how to merge different planning alternatives:

1. To import proposals from another workbook into your model, start by making sure that the workbook you're importing from is open during the session and the one you're importing to is the active workbook.
2. In the **Scenario Manager** dialog box, click the **Merge** button.
3. Under **Book**, select the other workbook that contains the scenario data to import into the active workbook. Under **Sheet**, select the worksheet in this workbook that contains the values. You can also merge scenarios that are defined in the same workbook but on different sheets.

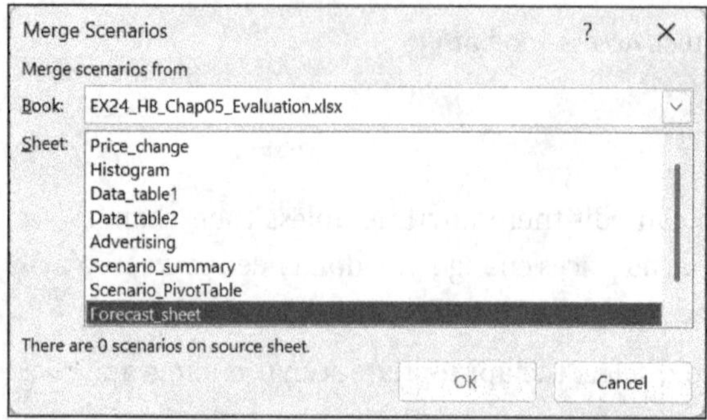

If merging causes conflicts with scenario names, Excel appends the creator's name, the date, or a number to the merged scenario name to differentiate them.

5.5.5 Summary Reports

Excel displays only one version of the data you compiled for a calculation model on the worksheet. To compare results from different scenarios, you can generate summary reports.

Use the **Summary** button in the **Scenario Manager** dialog box to do this. Then, in the small **Scenario Summary** dialog box, you first choose the type of report. Choosing **Scenario summary** means creating a standard table with summarized results. Excel inserts a worksheet for this table into the active workbook, placing it before the sheet with the data model. The table is automatically outlined, and the outline symbols appear.

Alternatively, you can choose **Scenario PivotTable report** to display the same data directly in a PivotTable. This creates an interactive table on a new sheet before the active worksheet, allowing you to rearrange and reorganize the report data as PivotTables let you do (see Chapter 18). Excel assigns default names to the sheet tabs for both report types.

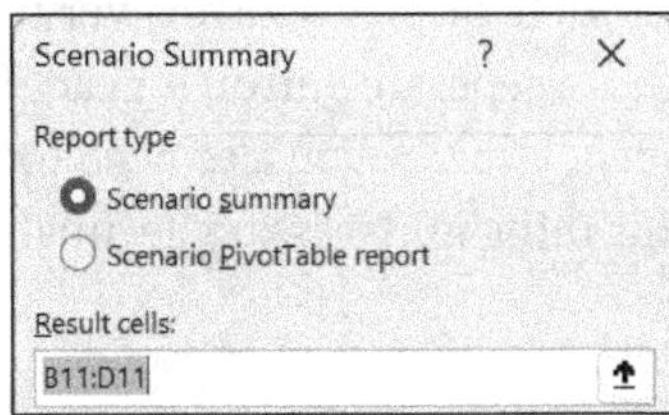

Figure 5.11 Scenario Report Dialog Box

Scenario Comparison

The main purpose of the scenario report is to show how different assumptions in the model impact the calculated results. You can specify which results appear in the report in the **Result Cells** field. To use more than one result cell, hold `Ctrl` and select them with the mouse or use the semicolon. You can also leave the result cells blank in the scenario summary, and in that case, only the values of the changing cells will be compiled. (For the PivotTable report, you must specify the result cells; otherwise, you'll get an error message.)

The next two figures show examples of the two report types. As you can see, Excel uses the previously assigned cell names so you can instantly understand the meaning of each value.

Excel also inserts the author's name and the creation or modification date into the header of the summary report. This is very helpful when multiple people are involved in the project. In Figure 5.12, the data is hidden using the outline symbol. Of course, this table can be edited like any other. For example, the column widths have been adjusted here.

Scenario Summary	Current Values:	Optimistic Scenario	Pessimistic	Medium_assumptions
Changing Cells:				
Sales_1	26,000	35,000	22,000	26,000
Sales_2	26,000	30,000	21,000	26,000
Sales_3	27,000	33,000	22,000	27,000
TV_1	$110,000.00	$100,000.00	$120,000.00	$110,000.00
TV_2	$100,000.00	$90,000.00	$100,000.00	$100,000.00
TV_3	$120,000.00	$120,000.00	$130,000.00	$120,000.00
Press_1	$35,000.00	$35,000.00	$35,000.00	$35,000.00
Press_2	$36,000.00	$37,000.00	$40,000.00	$36,000.00
Press_3	$23,000.00	$22,000.00	$25,000.00	$23,000.00
Result Cells:				
Costs_P1	$7.96	$5.63	$9.86	$7.96
Costs_P2	$7.46	$6.17	$9.43	$7.46
Costs_P3	$7.15	$5.82	$9.32	$7.15

Notes: Current Values column represents values of changing cells at time Scenario Summary Report was created. Changing cells for each scenario are highlighted in gray.

Figure 5.12 Scenario Report

PivotTable from Scenarios

The second report type, the PivotTable, consolidates plan data from different authors onto separate pages. If you choose a PivotTable, use the arrow in the first row to choose whose suggestions are shown. For example, you can choose **All** to display data for everyone involved in the planning at once. The PivotTable creates a separate button for each set of editable cells, and you can freely change the field labels. Usually, you also need to format the table slightly to get a clean report, such as by trimming unnecessary decimal places.

	A	B	C	D
1	Author	Helmut Vonhoegen		
2				
3	Scenarios	Costs_P1	Costs_P2	Costs_P3
4	Medium_assumptions	7.96	7.46	7.15
5	Optimistic Scenario	5.63	6.17	5.82
6	Pessimistic	9.86	9.43	9.32

Figure 5.13 Slightly Updated PivotTable Report

5.6 Forecasting Based on Existing Data

Since the release of Excel 2016, a group of forecast functions have been available under statistical functions to expand the capabilities of the earlier `FORECAST()` function. The `FORECAST.ETS()` function provides estimates for future periods based on known values from past periods. It uses a version of the ETS algorithm that applies exponential smoothing, and the predicted values are seen as a continuation of the past data series. The values are mapped to a timeline, with the target date extending that timeline. A key new feature is the ability to factor in seasonal cycles, but this function only works with timelines that have equal intervals, such as one hour, one day, one week, one month, or one year. Instead of time or date values, you can also use a numeric index.

The function involves at least three arguments:

- `Target_Date`
  This specifies the data point to predict, which can be a date/time or a numeric value. Values are the past data points used as the basis for the prediction, and you can provide this argument as an array or a range.
- `Timeline`
  This is either an array or a range of numeric data. Intervals must be equal, and the axis cannot include zero values, but the values don't need to be sorted because the function sorts them internally.
- `Seasonality`
  Other arguments are optional, and unlike a simple trend calculation, this argument lets you account for seasonal cycles or patterns such as sales trends or visitor counts.

In addition to letting you manually enter functions into this group, Excel provides a **Forecast Sheet** button on the **Data** tab within the **Forecast** group, which you can use to create a forecast sheet based on a table of existing values. In Figure 5.14, we present a simple table with sales results for three weeks. You can see that the values follow a weekly pattern, with Sunday's value always the highest and Monday's the lowest.

To create a forecast sheet for the upcoming weeks based on this table, first select the table data, which in this case is A2:B22.

	A	B
1	Day	Revenue
2	Mon, Nov 04, 2024	2,010
3	Tue, Nov 05, 2024	5,000
4	Wed, Nov 06, 2024	4,700
5	Thu, Nov 07, 2024	4,500
6	Fri, Nov 08, 2024	5,500
7	Sat, Nov 09, 2024	6,000
8	Sun, Nov 10, 2024	7,000
9	Mon, Nov 11, 2024	2,000
10	Tue, Nov 12, 2024	5,188
11	Wed, Nov 13, 2024	4,804
12	Thu, Nov 14, 2024	4,659
13	Fri, Nov 15, 2024	5,675
14	Sat, Nov 16, 2024	6,053
15	Sun, Nov 17, 2024	7,009
16	Mon, Nov 18, 2024	2,085
17	Tue, Nov 19, 2024	5,323
18	Wed, Nov 20, 2024	4,820
19	Thu, Nov 21, 2024	4,747

Figure 5.14 Historical Data for Forecast Sheet

Then go to **Data • Forecast • Forecast Sheet**, and in the dialog box, click or tap the small triangle next to **Options** ❶ to access all available settings.

At the top of the dialog, you'll see a chart based on the preset settings. It updates immediately when you make changes under **Options**, and you can use the two chart icons at the top right ❷ to choose between a line or column chart. Under **Forecast End** ❸ and **Forecast Start** ❹, the calendar controls let you set the forecast period precisely. Typically, it's best to set the forecast start at the end of the existing data so the function can analyze the full historical timeline. You can also select an earlier period to compare actual data with forecasted data.

If you leave the **Confidence Interval** option checked, Excel will generate confidence interval values alongside the estimates and display them in the chart. This provides insight into the forecast's reliability. Then, use the spinner control ❺ to set the confidence level. The default is *0.95*, which creates a confidence interval with a 95% confidence level. This means 95% of possible values at the target date will fall within the confidence interval range.

Under **Seasonality** ❻, you can choose between **Detect Automatically** and **Set Manually**. In the first case, Excel tries to identify the trend pattern automatically. The spinner lets you set the number of elements in a seasonal pattern manually, and here, we select "7" for a weekly pattern. If the time axis shows months, use "12" for an annual season or "4" for quarterly. The maximum value is 8760, the number of hours in a year.

Checking the **Include forecast statistics** ❼ option adds a small table of key statistics to the forecast sheet, and it shows the smoothing parameters used and several error metrics. Under **Timeline Range** ❽, you can see the time axis with its units, and under **Values Range** ❾, the data range is linked to the time axis, which you can adjust if needed. The dropdown list for **Fill Missing Points Using** ❿ lets you choose between **Zeros** or **Interpolation**, and the list for **Aggregate Duplicates Using** ⓫ lets you decide whether to sum or average duplicate values for the same time. Then, click the **Create** button to add the forecast sheet as a new worksheet in the workbook.

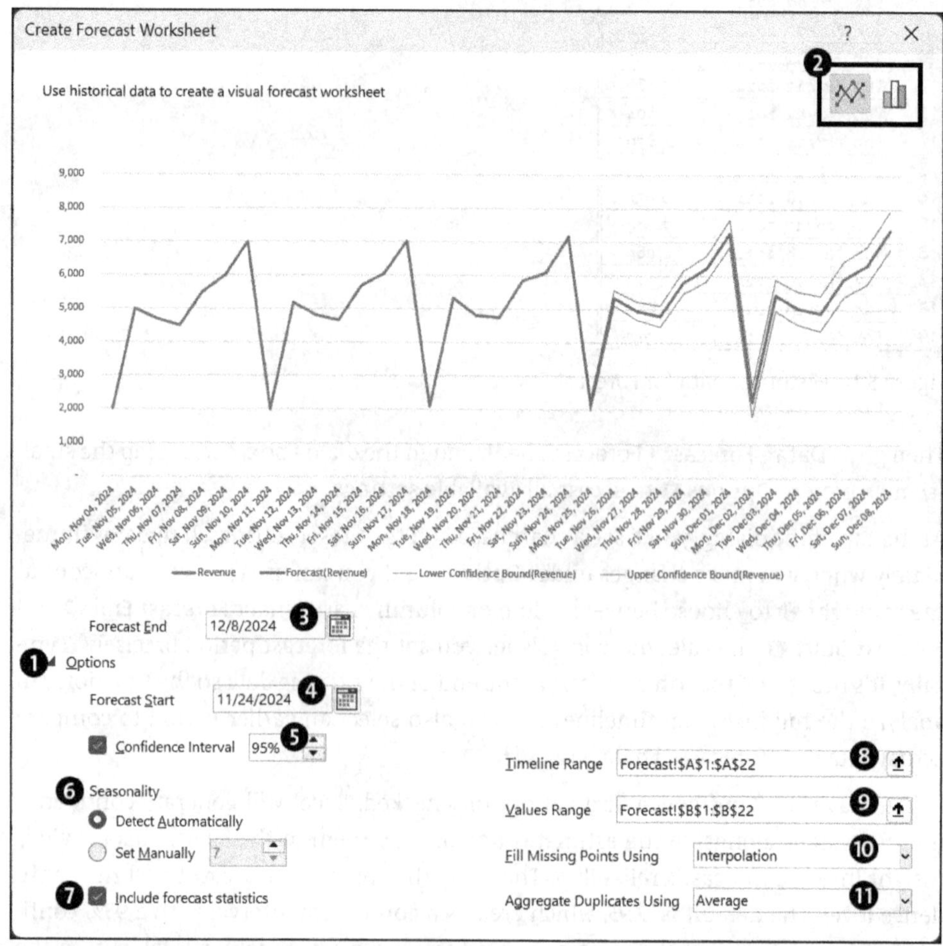

Figure 5.15 Forecast Sheet Dialog

Figure 5.16 displays the diagnostic sheet based on the data and options you previously selected. The new sheet repeats the evaluated data in columns A and B, and forecast data is stored in column C using the `FORECAST.ETS()` function. If the **Confidence Interval** option is enabled, formulas appear in columns D and E that calculate the lower and upper confidence interval limits. The `FORECAST.ETS.CONFINT()` function subtracts this value from the estimate in column C for the lower limit and adds it for the upper limit. From the start of the forecast, the chart displays a line for estimated values and two lines showing the confidence interval. For more details on these forecast functions, see Chapter 15.

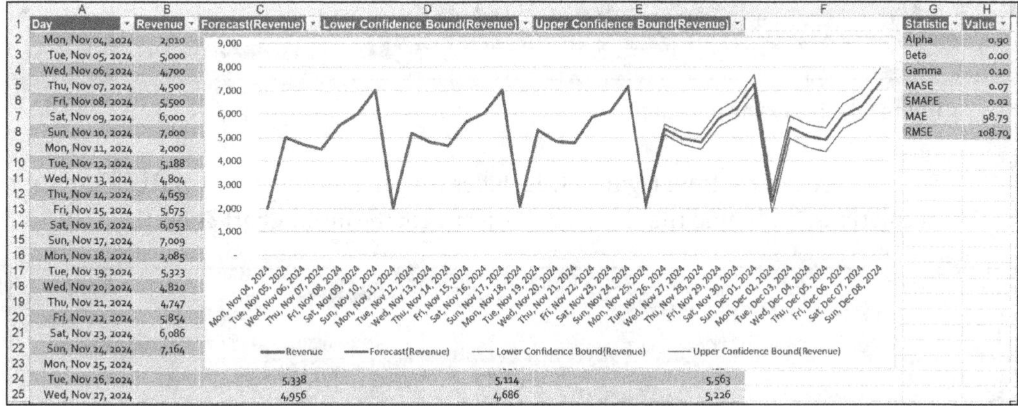

Figure 5.16 Forecast Sheet with Chart and Statistical Summary

5.7 Automatic Data Analysis

Once you've filled a worksheet with data, you might wonder if you can analyze, summarize, or visualize it. This is why Excel offers its **Analyze Data** feature, which you can access by clicking the button with the same name ❶ in the **Analysis** group on the **Home** tab (see Figure 5.17). Excel will open a task pane that immediately displays several analysis templates using the current data.

If the sheet contains a table, selecting any cell inside it will generate suggestions for the entire table. You can also choose to select specific columns to get targeted analysis suggestions, and you can start this selection by using the search box at the top of the task pane ❷. A list of column names with checkboxes will then appear. You can also click **Update** to get suggestions for options based on your selection.

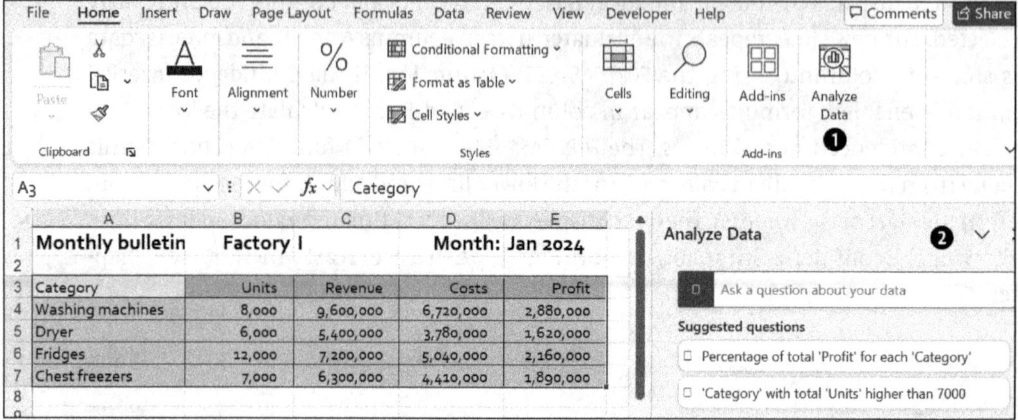

Figure 5.17 Task Pane for Automatic Data Analysis

The simple example from Figure 5.17 includes suggestions for a chart and column sum calculations. If you find the chart helpful, just click the **Insert Chart** ❸ button.

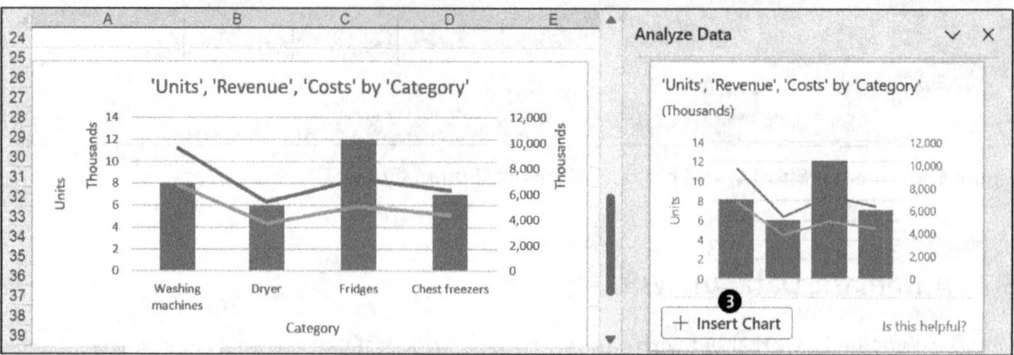

Figure 5.18 Chart Based on Selected Cell Ranges

Chapter 6
Optimization

With Goal Seek and Solver, Excel offers two tools that help you find specific values. With Goal Seek, you specify the result a formula in a cell should produce and Excel calculates the value a specific precedent cell must have to achieve that target.

Solver offers many more options. You have a cell for a target value, but you can specify more than just a fixed target—you can also set the target to reach a maximum or minimum. You can also create models that calculate things like the conditions for achieving optimal profit and models aimed at minimizing waste. Typically, these models include constraints that must be met while pursuing the goal, such as keeping certain values within a specific range.

6.1 Goal Seek

Excel's Goal Seek feature mathematically solves an equation with one unknown. Consider the task of calculating the interest earned on a certain amount over a specific period, given the amount, interest rate, and duration. You solve this problem using the following formula:

```
Interest = Amount * InterestRate * Years
```

The `Interest` value is the unknown in this equation (the other values are known), and without Excel, you'd solve this problem with a calculator or by hand. You can also use the same formula if the question changes. For example, if you know the interest amount and the principal over a certain number of years, and if you need to find the interest rate, then you must rearrange the equation to solve for the interest rate:

```
InterestRate = Interest / (Amount * Years)
```

Working with simple equations like this is not difficult, but working with more complex problems can become time-consuming and error prone. You could also work this way within Excel, but two problems would arise. First, you'd have to give up Excel's built-in table functions since they can't be easily rearranged, and second, you'd need to rearrange the equation manually (which requires knowing the math) because Excel doesn't provide a way to rearrange equations. However, you can avoid all this effort by using Goal Seek.

6.1.1 Determining the Maximum Loan Amount

With Goal Seek, Excel uses built-in functions but no longer calculates the result from given values. Instead, it sets the result and determines what one of the other values must be to reach the desired outcome.

Here's a typical example from financial math. Suppose you want to take out a larger loan and pay it off over 20 years. You have $1,000 available each month for payments (principal + interest), and the bank's loan interest rate is 4.7%. You can solve this problem by using the spreadsheet's present value function `PV()`:

```
= PV(4.7%/12,20*12,-1000)
```

The function returns a value of $155,401.10, meaning you can afford to take out a loan of this amount. However, you then realize that this loan isn't enough—you need $170,000. Since the bank won't negotiate the interest rate, you have only two options: extend the term or increase the annual payment.

To demonstrate goal seeking options, Figure 6.1 shows a table where the calculation has been copied twice to explore both alternatives and the calculation headings have been updated accordingly. To find the monthly payment for the $170,000 loan, select the result cell (D9 in the example) and then start the calculation by using **Data • Forecast • What-If Analysis • Goal Seek.**

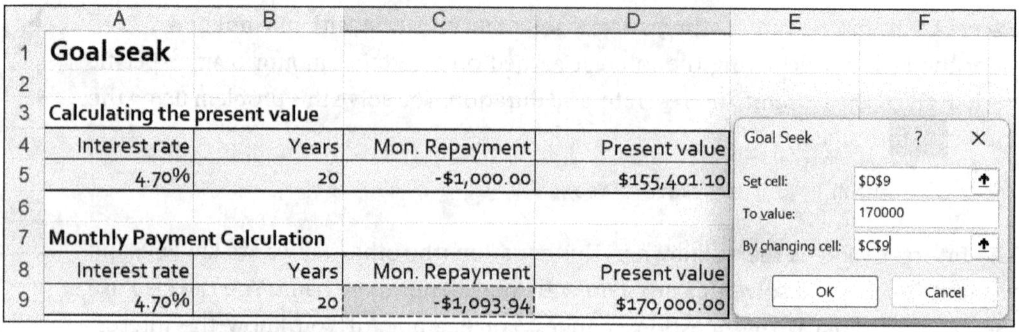

	A	B	C	D
1	**Goal seak**			
2				
3	**Calculating the present value**			
4	Interest rate	Years	Mon. Repayment	Present value
5	4.70%	20	-$1,000.00	$155,401.10
6				
7	**Monthly Payment Calculation**			
8	Interest rate	Years	Mon. Repayment	Present value
9	4.70%	20	-$1,093.94	$170,000.00

Goal Seek ? ×
Set cell: $D$9
To value: 170000
By changing cell: $C$9
OK Cancel

Figure 6.1 Goal Seek for Loan Problem

The **Set cell** is already entered correctly, with the **To value** at *170000*. The **By changing cell** is the one with the monthly payment (in this example, C9). Clicking **OK** opens a dialog box where Excel confirms it has solved the problem, and clicking **OK** again enters the target value and the calculated payment into the worksheet.

You can use the same method to find out how to finance the loan by extending the term. The target value is *170000* again, but this time, the **By changing cell** is the one with the number of years (in this example, B13).

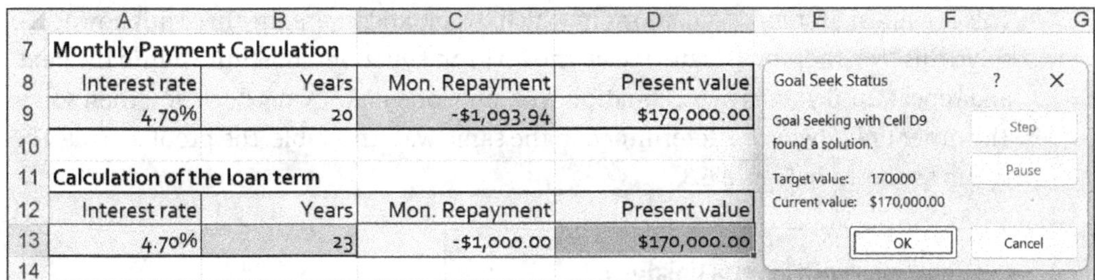

Figure 6.2 Solution Notification

Trying to finance a $300,000 loan with a longer repayment period triggers a message that ... **may not have found a solution**. This happens because the loan isn't feasible that way; with a loan balance of $300,000, the monthly interest alone exceeds $1,000, leaving nothing for principal repayment.

6.2 Finding Solutions with Solver

Goal Seek solved a very simple problem. From the table's perspective, a cell contained a formula whose result was the target value—and this formula could reference many other cells, which themselves depended on others, and so on. During Goal Seek, the value of one of these predecessor cells—which are called *adjustable cells*—was changed to reach the desired target value.

6.2.1 Advanced Solver Options

Solver works in much the same way, but with far more extensive options. You can still set a target value—but now, it can be a fixed value, a maximum, or a minimum. Instead of a single adjustable cell, you can specify multiple cells. Solver is an Excel add-in, so it's only available if you've added it to your add-ins list. To add it, go to **File • Options • Add-Ins**, and under **Manage**, select **Excel Add-ins** from the dropdown. Then, click **Go** to open the **Add-Ins** dialog and enable the **Solver** option.

6.2.2 How Solver Works

Mathematically, Solver tackles problems that can be expressed as equations or systems of equations with multiple unknowns, including inequalities. Here's a simple example that may be familiar from math class at school:

```
(3 × ux) + (4 × uy)+ (2 × uz) = 57
(2 × ux) - (4 × uy) + (3 × uz) = 18
  ux + (8 × uy) - (3 × uz) = 15
```

This is a linear system of equations in which *ux*, *uy*, and *uz* are the three unknowns. To solve this by hand, you isolate one unknown in an equation, substitute it into the next, and repeat until you have an equation with only one unknown, which you then solve. The other unknowns are determined in the same way. In a table, the problem could be set up as shown in Figure 6.3.

	A	B	C	D	E
1	**Equations with several unknowns**				
2					
3	ux	uy	uz		
4	3	6	12		
5					
6	Equation 1:	3*ux+4*uy+2*uz		=	57
7	Equation 2:	2*ux-4*uy+3*uz		=	18
8	Equation 3:	ux+8*uy-3*uz		=	15

Figure 6.3 Problem for Solver

Using Solver

To solve this task with Solver, follow these steps:

1. Assign a cell to each unknown and enter any initial value. Use the unknowns' names as labels and cell names via **Formulas • Define Names • Create from Selection.**
2. Enter the three equations into cells E6 through E8. The results won't match the original equations yet, since the unknowns initially have arbitrary values. For clarity, the equations are also displayed as text in column B in the following figure.
3. Place the cursor in the cell with the first equation (E6), and open Solver via **Data • Analysis • Solver**. The **Solver Parameters** dialog box will then open.
4. Under **Set Objective** ❶ in the dialog, specify the target cell, which is $E$6 in this case.
5. Under **To** ❷, select **Value Of** and enter "57" in the field. This is the value the first equation in cell E6 must equal when the correct values for the three unknowns are found.
6. In the **By Changing Variable Cells** ❸ field, select $A$4:$C$4 as the variable cells—which are the three cells containing the unknown values (currently set to arbitrary values).
7. The **Subject to the Constraints** field lists the additional conditions that must also be met. Click the **Add** button ❹ to enter the other two constraint equations: `$E$7 = 18` and `$E$8 = 15`.
8. Under **Select a Solving Method** ❺, choose **Simplex-LP**, which is the module for linear Solver problems, since this is a linear system. Then, click the **Solve** button ❻ to start Solver, and Excel will adjust the variable cell values until it finds a result that satisfies the conditions. The values for the three unknowns are correctly determined as 3, 6, and 12.

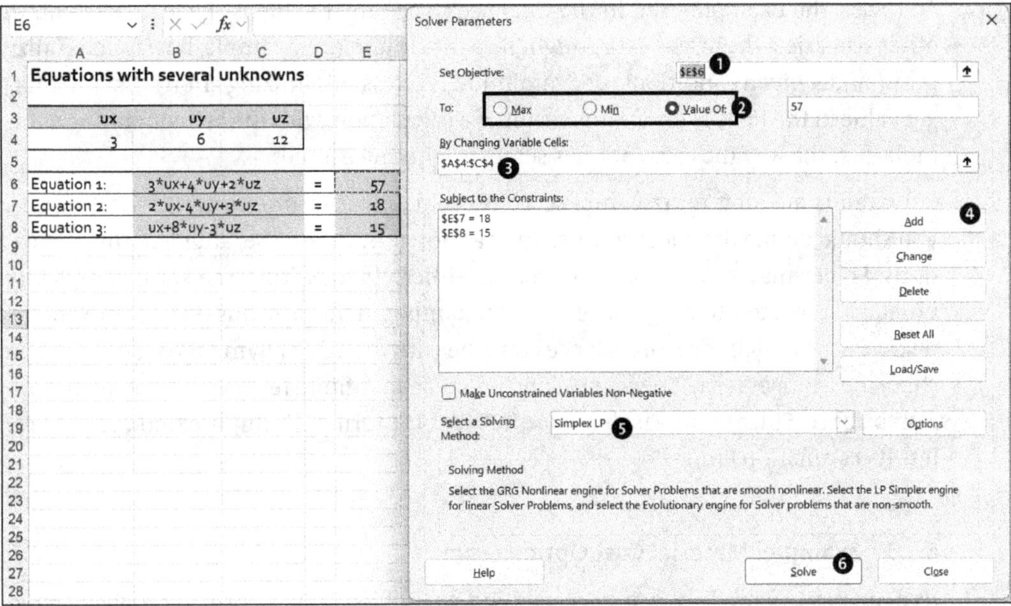

9. In the **Solver Results** dialog, Solver confirms it has found a solution and displays it on the worksheet. You then select **Keep Solver Solution** to apply this solution or choose **Restore Original Values**.

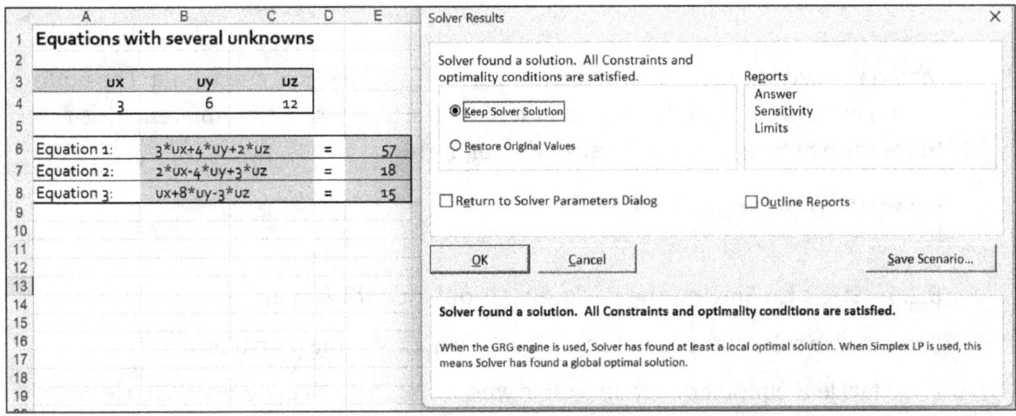

This tool allows you to solve many problems computationally that would otherwise require extensive mathematical effort.

Defining the Target Value and Constraints

The main challenge when using Solver is to formulate and prepare a model that Solver can process to find a solution. The simplest step is defining the target value, which means deciding what the final outcome should be.

To revisit the example used in the Goal Seek, the target value is the loan amount. In other scenarios, the target value might be a maximum (for example, how to maximize profit under given conditions) or a minimum (such as minimizing effort). Since the target value in the table is always the result of a calculation, with the cell containing a formula, just the way the calculation is set up might cause issues.

Constraints are a bit more complex. To prevent Solver from returning trivial results, you should define the constraints as fully as possible. Otherwise, after lengthy calculations, Solver may produce a result that is either obvious from the start, completely unrealistic, or likely to trigger an error. For example, in the previous Goal Seek example, if you ask how high the loan balance could be with monthly payments of $1,000 when choosing a longer term, Solver will conclude that an infinite term yields the maximum loan balance. Since *infinity* isn't a value in Excel, the term will simply extend until Excel hits its calculation limit.

6.2.3 Example: Material Cost Optimization

As a practical example of using Solver, here's a problem you'd normally solve by hand with differential calculus. It's a packaging problem. The goal is to produce cans with a specified volume (e.g., 1,000 ml) while minimizing material use and costs. The volume of a can is calculated by multiplying the base area (a circle) by the can's height, and the base area is calculated as the area of a circle. The formula is this:

```
Vol = r^2 * PI * h
```

Materials used to make this can include the bottom, the side, and the lid. The bottom and lid are both circles (`r^2 * PI`), and the side area is the can's circumference (`2 * r * PI`) times the height. So, the total area of the can is this:

```
Area = 2 * r^2 * PI + 2 * r * PI * h
```

6.2.4 Steps for Solving the Packaging Problem with Solver

You can also solve this problem by using Solver after some preparation:

1. To tackle the problem, set up a table where you first calculate material use and can volume by using arbitrary numbers.
2. Enter the appropriate formulas into the cells for material use and volume (the cell addresses depend on the table layout):

   ```
   C6:  =2 * A6 ^ 2 * PI() + 2 * A6 * PI() * B6
   C7:  =A6 ^ 2 * PI() * B6
   ```

   Using arbitrary values for the can's radius and height, the formulas initially yield arbitrary results for material use and volume. If the can's dimensions are in cm, material use is in cm^2 and volume in ml.

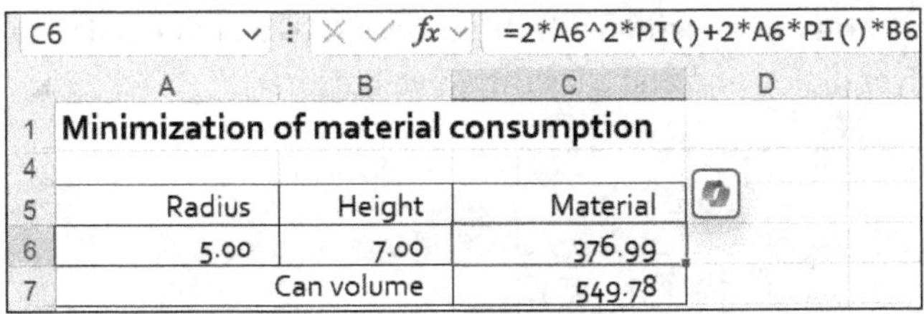

C6
=2*A6^2*PI()+2*A6*PI()*B6
Minimization of material consumption
Radius
Height
Material
5.00
7.00
376.99
Can volume
549.78

3. After these preparations and before opening Solver, place the cell pointer in the target value cell (in this example, the material consumption cell: C6) to save an entry. Then, open Solver via **Data • Analysis • Solver.** If Solver was previously used in the current worksheet, it will still hold the old entries. This is helpful when continuing an unfinished problem, like changing the target value or adding constraints. For a new problem, select **Reset All** after opening Solver to clear old entries completely.
4. If the cell address for **Set Objective** is already selected when you open Solver, you can simply accept it. In the **To** field, select **Min** (for minimum) in this example, since the goal is to find the solution with the least material use.
5. The adjustable variable here are the two cells containing the radius and height. In this example, the variable cells are arranged in a contiguous range. This doesn't have to be the case. If the variable cells aren't in a contiguous range, you can separate the individual cells or ranges with semicolons. The easiest way to select them is by highlighting them in the worksheet while holding down the Ctrl key to add more cells or ranges.

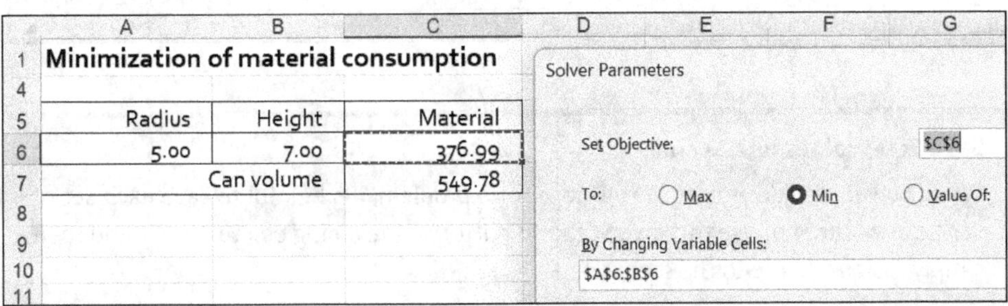

6. This example has only one key constraint: The can's volume must be 1,000 ml. Without this constraint, Solver would produce negative values for the material requirement. Clicking **Add** opens the constraints dialog box, where you can enter the cell reference for the can volume (ideally by selecting it in the table). Then, under **Constraint**, enter "1000" as the value. From the comparison operator dropdown list, select = since the volume must exactly equal this value. If you need to add more constraints—which isn't the case here—click **Add** again to save the current one and define the next one.

7. After confirming the constraints by clicking with **OK**, choose **GRG Nonlinear** as the solving method in the Solver dialog box and click the **Solve** button.

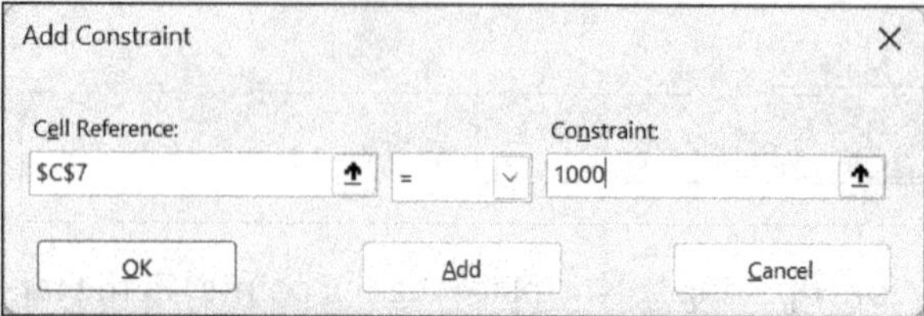

8. After a brief calculation, Solver provides the desired result in the **Solver Results** dialog box, with intermediate results possibly appearing in the status bar. The result shows a radius of about 5.67 and a height of about 9.90, which closely matches the differential calculus calculation (the height should be twice the radius).
9. Click **OK** to insert the result into the worksheet or **Cancel** to reject it.

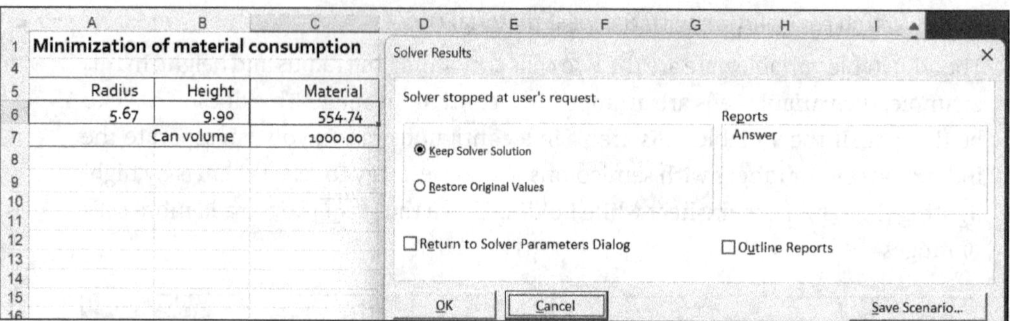

[+]

Save Your Values as a Scenario

When Solver provides multiple solutions for a problem, it's helpful to save each set of values as a scenario. Use the **Save Scenario** button in the dialog box to do this and then simply assign a name to the scenario.

Notes on Solution Methods

As mentioned earlier, Solver offers several solution methods in the **Solver Parameters** dialog box. Here's a brief overview of their differences.

- **Simplex LP**
  This is the method for linear optimization. This is the right choice if all relationships are linear, meaning adjustable cells are only multiplied or divided by constants (not by each other) and no powers, roots, or functions like sines or logarithms are used. Use this method only if you are certain that the relationships are strictly linear.

- **GRG Nonlinear**
  This is the default solution method. It supports continuous nonlinear optimizations and allows formulas for the target value that include powers, roots, or functions like sines and logarithms. The adjustable cells can be multiplied, divided, or raised to powers of each other.
- **Evolutionary**
  This is a solution method for noncontinuous optimizations, based on biological evolution and models describing genetic changes. Solver generates repeated solution proposals that gradually get closer to the desired target values.

Additional Options and Settings

The method described so far works well for problems that don't require complex calculations. However, you might not get result you want if the calculation time is too short or the accuracy isn't sufficient. In these cases, you can adjust Solver's settings by clicking the **Options** button in the **Solver Parameters** dialog box, which opens another dialog box where you can extensively control how Solver works.

To understand how these settings affect Solver's operation, you need at least a basic understanding of its process. Solver simply tests which values the adjustable cells need to produce the desired result in the target cell. It changes the values in the adjustable cells step by step, checks if the target cell's value moves closer to the goal, and verifies that all constraints are satisfied. This cycle repeats until it finds a solution, the preset time runs out, or the maximum number of iterations is reached.

The **Options** dialog has three tabs: two dedicated to the **GRG Nonlinear** and **Evolutionary** methods and one **All Methods** tab with general settings. The first general setting is **Constraint Precision**, which defines the precision at which Solver treats a constraint as satisfied. The higher the precision (that is, the smaller the value), the longer Solver will take.

Enable **Use Automatic Scaling** when your calculation includes both very large and very small numbers. Solver scales numbers internally to a comparable range and converts the results back afterward, which can sometimes improve accuracy. Enabling **Show Iteration Results** lets Solver display the result after each attempt in the status bar and asks whether to continue searching for a solution. This is helpful for complex, lengthy calculations because it lets you monitor Solver's progress.

If you need to enforce strict integer constraints, disable **Ignore Integer Constraints** and set **Integer Optimality** to 0%. However, requiring integer values for adjustable cells can significantly increase calculation time. Increasing the tolerance—which is set by default to 1%—can reduce the time spent, though it means accepting larger deviations.

The lower option group controls the solution limits that influence Solver's performance. Under **Max Time (Seconds)**, you can set a time limit for the total time in seconds that Solver has to solve a problem. You can set the number of solution attempts under **Iterations**, with a default of 100.

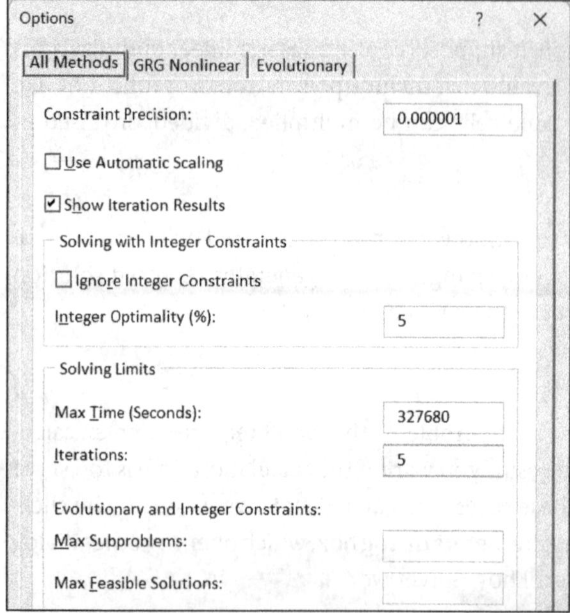

Figure 6.4 Solver Options

Often, you can improve Solver's results without changing any settings. You can do this by testing different starting values before running Solver to find the most favorable ones.

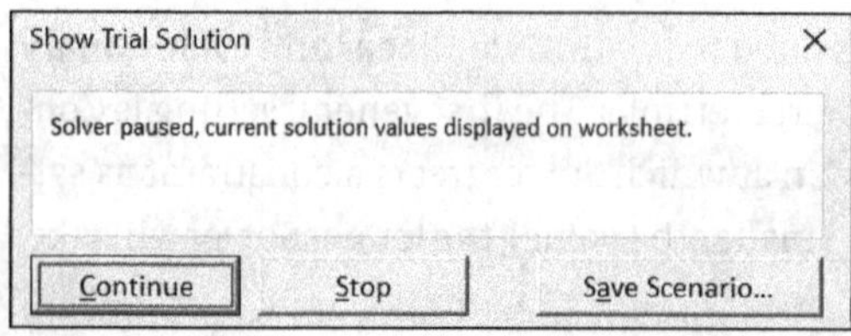

Figure 6.5 Showing Intermediate Results

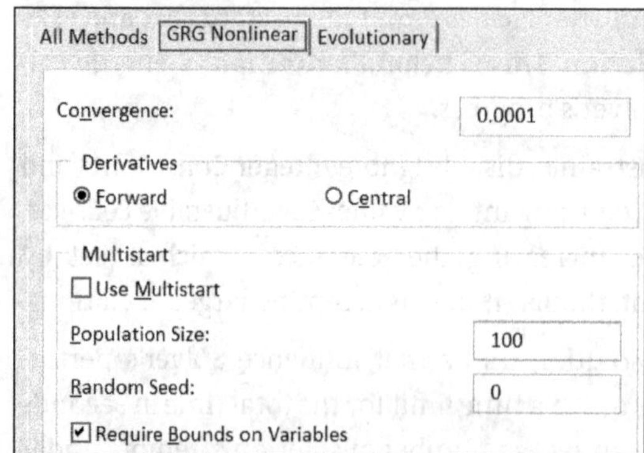

Figure 6.6 Special Options for GRG Nonlinear Method

For more information on how the Solver add-in from Frontline Solvers works, visit *www.solver.com/excel-solver-online-help*.

6.2.5 Evaluating Results and Reports

After you complete your work with Solver, a message will appear confirming whether a solution was found. You can also document Solver's process with several reports. To do this, select the reports you want under **Reports**. The answer report summarizes the results in a clear table, and it displays both the starting values and the values found by Solver.

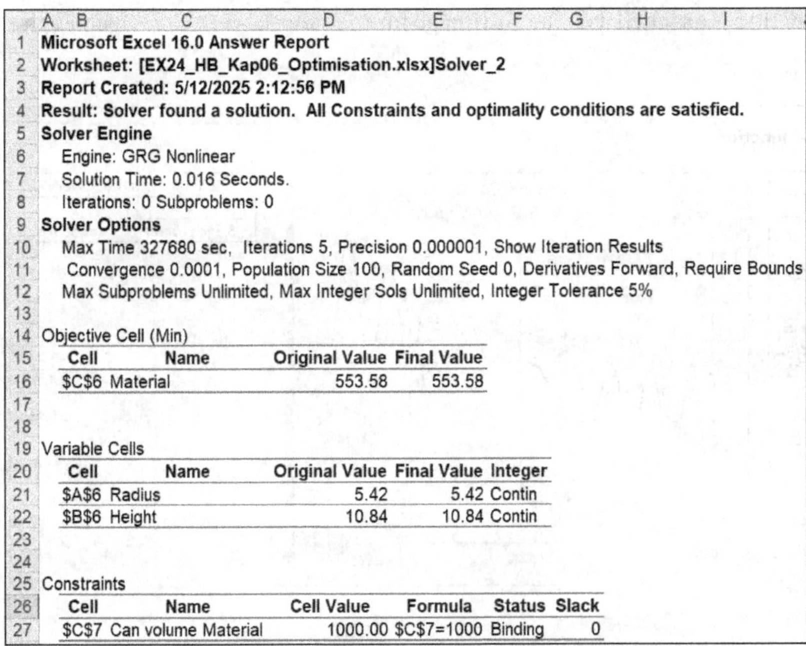

Microsoft Excel 16.0 Answer Report
Worksheet: [EX24_HB_Kap06_Optimisation.xlsx]Solver_2
Report Created: 5/12/2025 2:12:56 PM
Result: Solver found a solution. All Constraints and optimality conditions are satisfied.
Solver Engine
Engine: GRG Nonlinear
Solution Time: 0.016 Seconds.
Iterations: 0 Subproblems: 0
Solver Options
Max Time 327680 sec, Iterations 5, Precision 0.000001, Show Iteration Results
Convergence 0.0001, Population Size 100, Random Seed 0, Derivatives Forward, Require Bounds
Max Subproblems Unlimited, Max Integer Sols Unlimited, Integer Tolerance 5%

Objective Cell (Min)

Cell	Name	Original Value	Final Value
$C$6	Material	553.58	553.58

Variable Cells

Cell	Name	Original Value	Final Value	Integer
$A$6	Radius	5.42	5.42	Contin
$B$6	Height	10.84	10.84	Contin

Constraints

Cell	Name	Cell Value	Formula	Status	Slack
$C$7	Can volume Material	1000.00	$C$7=1000	Binding	0

Figure 6.7 Answer Report

The other two reports, the sensitivity report and the limit report, provide information about the method Solver used. These reports are mainly useful when the results aren't entirely satisfactory because they can serve as a starting point for alternative problem formulations. The evolutionary algorithm (EA) method also includes a population report that provides statistical values for the variables. If the linear Simplex LP method is applied to a problem that isn't linear, the linearity report it offers will help identify which formulas are nonlinear.

6.2.6 Additional Notes

The previous example produces a usable result right away. That's expected of a book example, but in practice, it's not always the case. Getting a result may be delayed by minor issues like errors in problem formulation, but after a few attempts, you can resolve those difficulties.

However, it becomes more challenging when the task itself causes difficulties. This happens when multiple solutions of varying usefulness exist or when the mathematical relationships prevent Solver from handling the task effectively. Since listing all possible variants here is impossible, we'll demonstrate some options using a mathematical example. We'll start with this mathematical function, whose results will be clear from the start:

```
y=x+x*SIN(3*x)
```

When this function is graphed, you get the image for the range x = -10 to x = 10 in Figure 6.8.

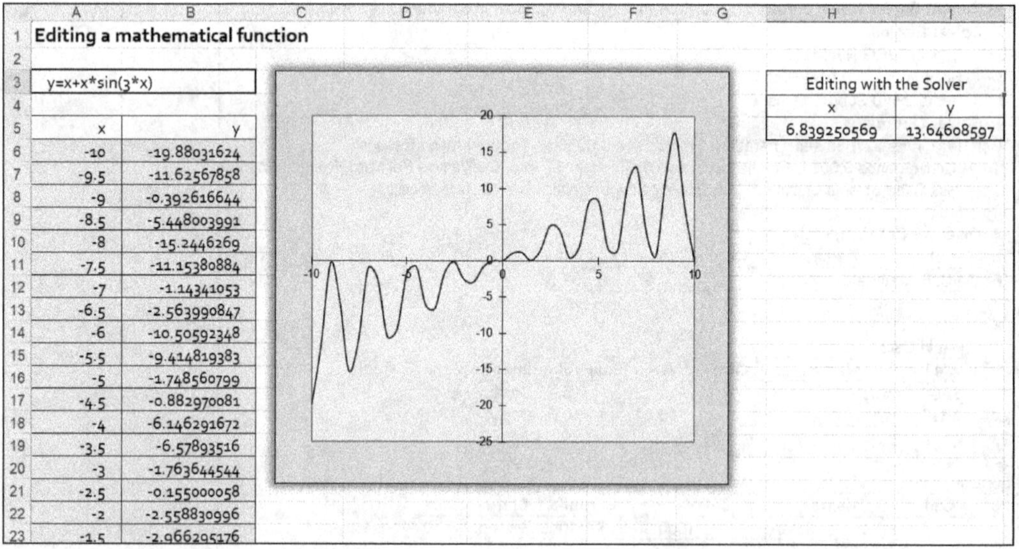

x	y
-10	-19.88031624
-9.5	-11.62567858
-9	-0.392616644
-8.5	-5.448003991
-8	-15.2446269
-7.5	-11.15380884
-7	-1.14341053
-6.5	-2.563990847
-6	-10.50592348
-5.5	-9.414819383
-5	-1.748560799
-4.5	-0.882970081
-4	-6.146291672
-3.5	-6.57893516
-3	-1.763644544
-2.5	-0.155000058
-2	-2.558830996
-1.5	-2.966295176

Figure 6.8 Periodic Function as Solver Task

Similar starting points can occur in many problems, where *similar* doesn't necessarily mean periodic but rather a problem without a perfectly regular pattern. Suppose you want to use Solver to find the X value where Y reaches its maximum.

You're already familiar with the setup. Enter any X value (e.g., 1) into one cell (the adjustable cell) and enter the formula to calculate Y into another. After you run Solver, it returns an X value of 0.79 and a Y value of 1.34. A quick look at the graph clarifies how Solver worked. It first found that y decreases as x increases and increases as x decreases. Solver continues searching in this direction and finds that at x = 0.79, a maximum value is reached, which decreases again as x gets smaller. This means Solver has solved the problem.

If you start the search at 7 instead, you get a completely different result: x is now 6.84 and y is 13.64. Here too, Solver searched until it found the next maximum and then stopped.

Choosing Suitable Starting Values Helps Solver

For problems like these, you have two options: start with a value that's nearly optimal or set Solver's search range as a constraint. Unfortunately, Solver can't search for all solutions at once when multiple solutions exist.

It often helps to begin by selecting a graphical representation, as shown in this section's example, which already offers clues for the solution. XY charts work well here, with the target value as the Y value and the values it depends on (the adjustable cells) as the X value(s). When the target depends on multiple values, it's best to use several charts, each of which varies only one input value within acceptable limits.

Chapter 7
Presenting Data Graphically

The best way to highlight key numerical data is by turning it into a graphical presentation like a chart, which lets you grasp the message of the numbers at a glance. Excel has always been well equipped for this task, and since Excel 2016, many new chart types have been added. 7

This chapter and the next cover the key tools for graphical analysis. This chapter explains how to design freeform graphical objects to enhance your documents. These include organizational charts and images that you can insert directly into a workbook from digital devices.

7.1 Graphical Analysis with Charts

Creating a chart from a table in Excel is very easy. Simply select the data range in the table, and with a few clicks or taps, Excel generates a chart. However, keep in mind that not every chart type is equally suitable for any given purpose. Charts that display numerical values consist of several elements, which we'll briefly introduce first. Not all elements need to be included in every chart.

7.1.1 Chart Elements

Most charts use the Cartesian coordinate system as their external frame of reference. The two lines are called *axes*: the horizontal one is the *x-axis*, and the vertical one is the *y-axis*. These axes serve as scales for displaying values, with the x-axis usually representing the independent variable and the y-axis representing the dependent variable. Excel generally follows these conventions, although the x-axis is also called the *category axis* (or just category) and the y-axis is also called the *size* or *value axis* (see Figure 7.1).

This naming hints at where Excel departs from the conventions: in *bar charts* (horizontal bars representing data size), the category axis (the x-axis) is vertical and the value axis (the y-axis) is horizontal. In 3D charts, a third axis is added to the x- and y-axes, and it's commonly called the *z-axis* or the *depth* or *series axis* (see Figure 7.2).

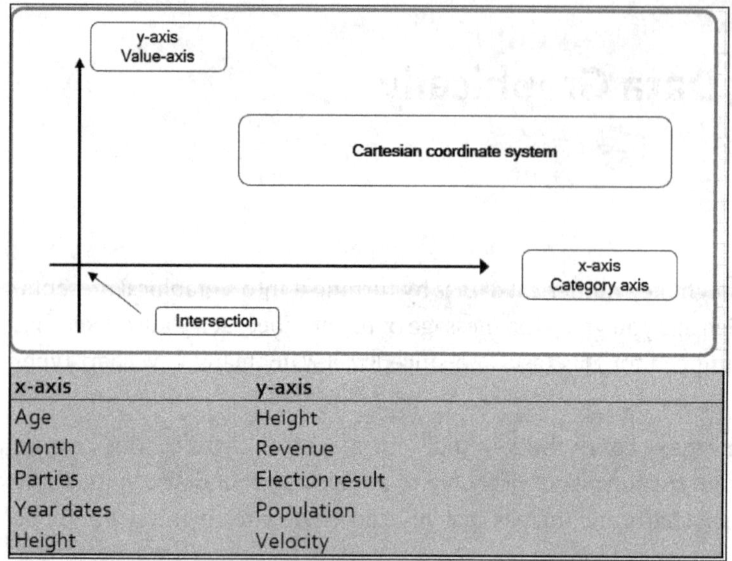

x-axis	y-axis
Age	Height
Month	Revenue
Parties	Election result
Year dates	Population
Height	Velocity

Figure 7.1 Examples of Labeling X- and Y-Axes

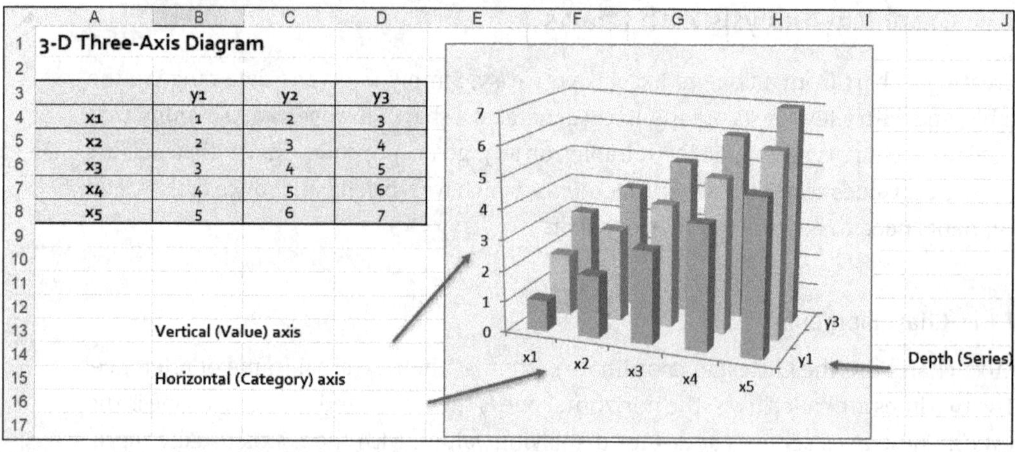

	y1	y2	y3
x1	1	2	3
x2	2	3	4
x3	3	4	5
x4	4	5	6
x5	5	6	7

Figure 7.2 Three Axes in 3D Chart

Axis Scaling

Axis scaling is usually shown by dividing the axes with ticks and subticks and then labeling them. This only makes sense if the axes are actually numerical. For example, in the previous 3D column chart, only the y-axis is divided numerically. Most Excel chart types have only one truly numerical axis.

7.1.2 Nonrectangular Coordinate Systems

Besides rectangular coordinate systems, two others are also used. One appears in *pie charts* and *doughnut charts* (layered pie charts). Instead of a category axis, these use a circle, which, unlike an axis, has no starting point. Instead of a value axis, the angle around the circle represents data size, so different data values correspond to different angles.

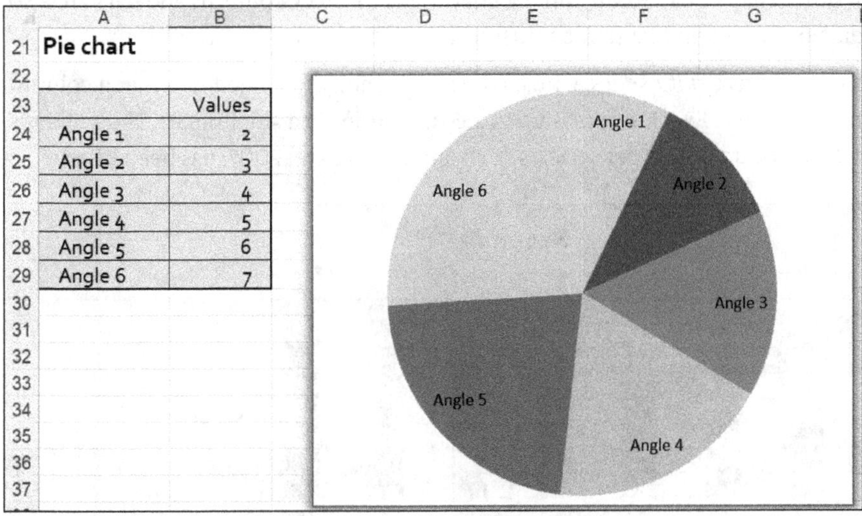

Figure 7.3 Pie Chart with Angles Representing Value Axis

Another coordinate system uses *radar charts*. Here, the categories are arranged in a circular pattern and each category has its own value axis where data points are marked.

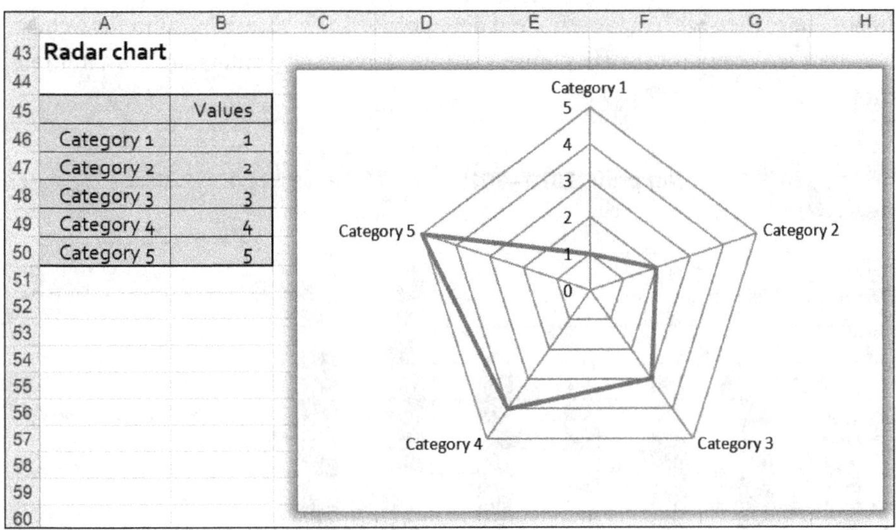

Figure 7.4 Radar Chart with Value Axis for Each Category

7.1.3 Data Series and Data Points

The coordinate system provides the frame of reference for graphically displaying table data. Data is usually shown as *data points*, which can appear as columns, bars, areas, lines, or other shapes, depending on the chart type. In Excel, columns and bars are also called data points—don't let that confuse you.

Related data forms a *data series*, which consists of values arranged in either a column or a row of the table. Excel supports up to 255 data series in a single chart. The previous limit of 32,000 data points per series, which applied to Excel 2007, has been lifted.

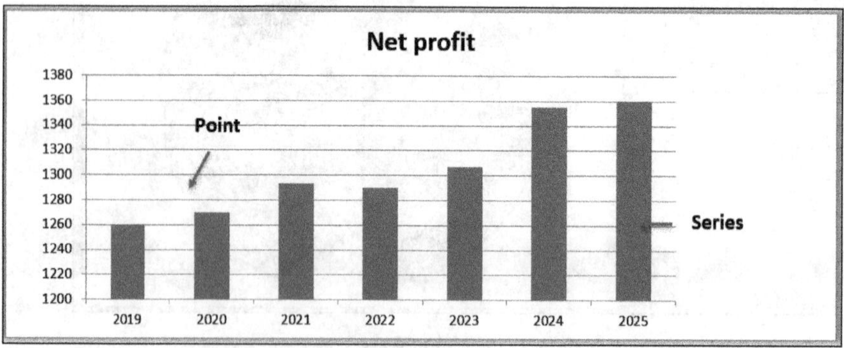

Figure 7.5 Data Series and Data Points in Chart

7.1.4 Additional Chart Elements

A coordinate system with data series and data points is the bare minimum for a chart. Such a chart isn't very informative unless the viewer already understands what it represents.

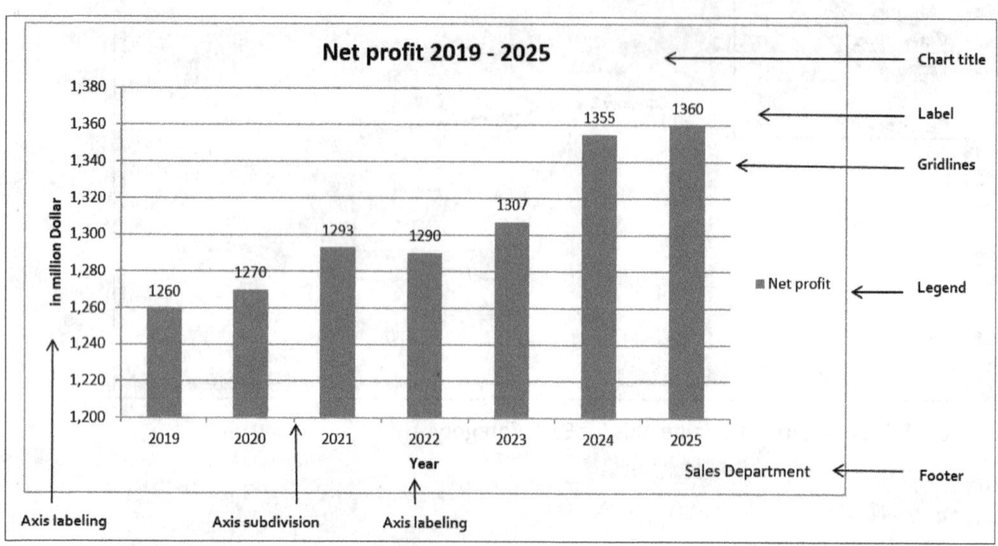

Figure 7.6 Additional Elements of Chart

To make a chart truly meaningful, labels are essential to clarify what each column or bar represents. Common elements include the following:

- **Chart title**
  This is a brief, clear description of what the chart shows.
- **Axis labels**
  These explain what each axis represents and what its units are.
- **Axis subdivisions**
  These help make axis values easier to read.
- **Gridlines**
  These are lines across the plot area that help you read values more precisely.
- **Legend**
  This specifies what each graphical element of the chart (bars, columns, lines, pie segments, etc.) represents.
- **Footers**
  These provide information about the source of the material (in source citations) and include additional explanations as needed.

Taking the time to create clear and informative labels is especially helpful for those who don't have direct access to the original chart data.

7.1.5 Chart Area and Plot Area

A chart's appearance in Excel depends on how the program arranges its elements and divides the available space. This happens automatically at first, but you can adjust nearly everything afterward. To allocate space, Excel places the chart's graphical elements—like columns, bars, pies, or bubbles—within the plot area that is inside a special chart. You can then select and resize or move the chart area by dragging its border or handles.

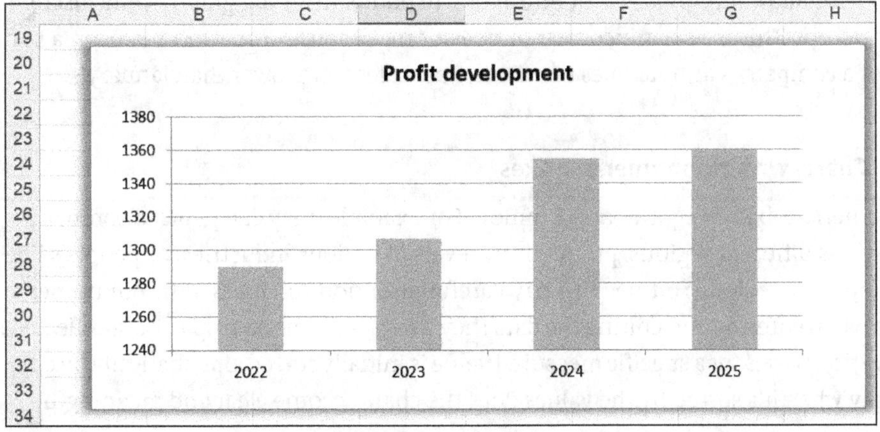

Figure 7.7 Chart with Selected Chart Area

Clicking or tapping the plot area inside the chart area displays handles that let you resize it by dragging. Labels usually appear outside the plot area, axis labels stay linked to the plot area, and elements like the legend and titles can be moved independently.

7.2 Chart Types in Excel

Creating an effective chart from numerical data requires considering which chart type works best for each kind of data. We cover this topic in this section, which reviews Excel's chart types. Here are some initial thoughts to help you get oriented. In daily use and in the media, two main chart types dominate: charts with rectangular coordinate systems—meaning two-axis charts—and proportion charts, which are usually pie charts.

7.2.1 Charts with Rectangular Coordinate Systems

Most charts—bar, column, line, and area—fall into the category of rectangular coordinate system charts. Nearly all data can be displayed using one of Excel's built-in chart types, though this often makes choosing one more difficult. They can be categorized by whether the x-axis is divided numerically or not.

7.2.2 Discrete or Continuous Subdivisions

A numerical subdivision can be temporal (years, months, etc.), but many other types also appear, especially in the natural sciences. Additionally, numerical subdivisions can be discrete or continuous. For example, when graphing annual sales, you use discrete values, such as years. In contrast, when you're showing a beam's deflection based on its length, you need to use continuous values on the x-axis, which measures length.

Even if the transitions seem smooth—years are divisible, and deflection might be measured in 10-cm increments, which are discrete values—there's a key difference: Continuous values allow interpolation of intermediate points from the graph, while discrete values do not. This is either incorrect or doesn't make sense. You can't tell from a bar chart of a company's annual sales that December sales were lower than January's.

7.2.3 Charts with Nonnumeric X-Axes

Nonnumeric x-axis categories are common. For example, they can depict a company's sales across different regions, productivity levels in various industries, or people's life expectancy by region. You need to pay careful attention to charts with nonnumeric x-axes when entering and editing the data they use. Here's one example: Data collected from various cities for a specific measured value is initially sorted alphabetically by city, and only when it's sorted by the values does the chart become clear and meaningful.

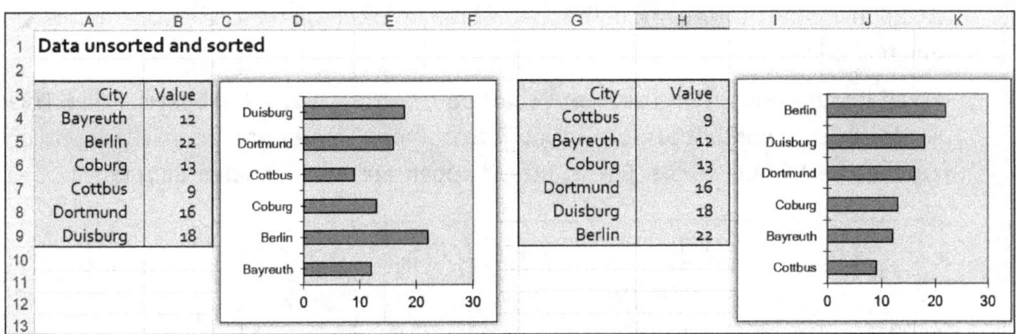

Figure 7.8 Sorting Data to Create Clearer, More Readable Charts

7.2.4 Charts Without Rectangular Coordinate Systems

The most common charts without rectangular coordinate systems are pie charts, which are often called proportional charts. They get this name because a full circle is divided proportionally based on the values, which makes pie charts ideal for data where parts sum up to a whole. The most familiar example is showing party shares of vote percentages in elections. Doughnut charts offer a good alternative for comparing multiple data series because they display several proportional charts in concentric rings.

7.3 From the Table to the Chart

For the first example of creating a chart, we'll use a quarterly sales report for various product groups. A column chart will compare the quarterly results of each group, and the total annual sales per group will be excluded from the chart because including them would distort the view.

7.3.1 Creating a Column Chart: First Attempt

To create a chart from the sales table, start by selecting the data range you want to analyze. Given the table's layout, the range from A3 to E7 works well. It includes the quarterly sales figures, the column with product group names for the x-axis labels, and the row with quarter labels to use as the legend.

	A	B	C	D	E	F
1	Novamedia: Quarterly Results 2024					
2						
3	Product group	1st Quarter	2nd Quarter	3rd Quarter	4th Quarter	Total
4	DVD drives	$120,000	$123,000	$140,000	$150,000	$533,000
5	Sound cards	$90,000	$95,000	$92,000	$90,500	$367,500
6	Webcam	$145,000	$149,000	$155,000	$165,000	$614,000
7	Scanner	$156,000	$159,000	$166,000	$176,000	$657,000
8		$511,000	$526,000	$553,000	$581,500	$2,171,500

Figure 7.9 Selecting the Range to Be Charted

Avoid including the title row in the selected range. Excel will try to use that row for legend text or labels.

You'll find the basic tools to create a chart on the **Insert** tab, in the **Charts** group. This section offers icons for the most common chart types. To create a column chart, click or tap the **Insert Column or Bar Chart** button to open a palette of related chart types.

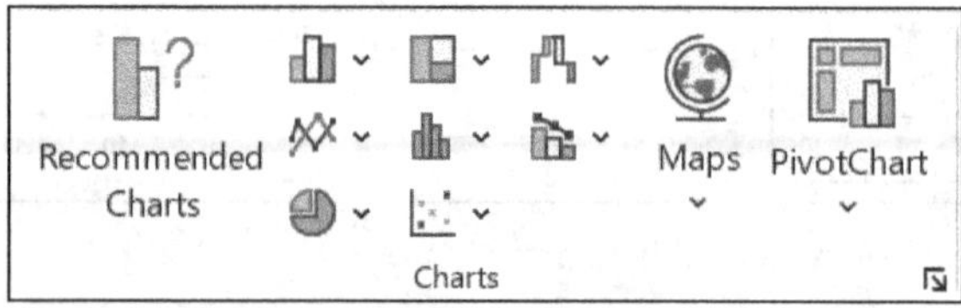

Figure 7.10 Chart Tools in Charts Group

Choosing the Chart Type

The palette includes various column and bar chart types, separated into 2D and 3D options. The last option, **More Column Charts,** opens the **Insert Chart** dialog box, which shows all chart types at once. (This dialog also appears if you use the group's dialog box launcher.)

In this example, start by selecting a simple 2D column chart—the first one in the palette. You can easily change the chart type later.

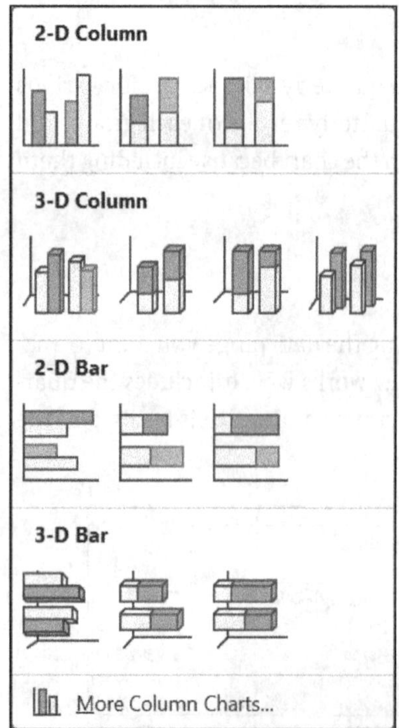

Figure 7.11 Column and Bar Chart Palette

Once you select the desired style, Excel immediately draws a chart matching the table in the center of the application window. If you don't like the size, you can resize the chart by using the border handles. Figure 7.12 shows the default chart created by Excel. The width was slightly adjusted so the x-axis labels don't need to be angled.

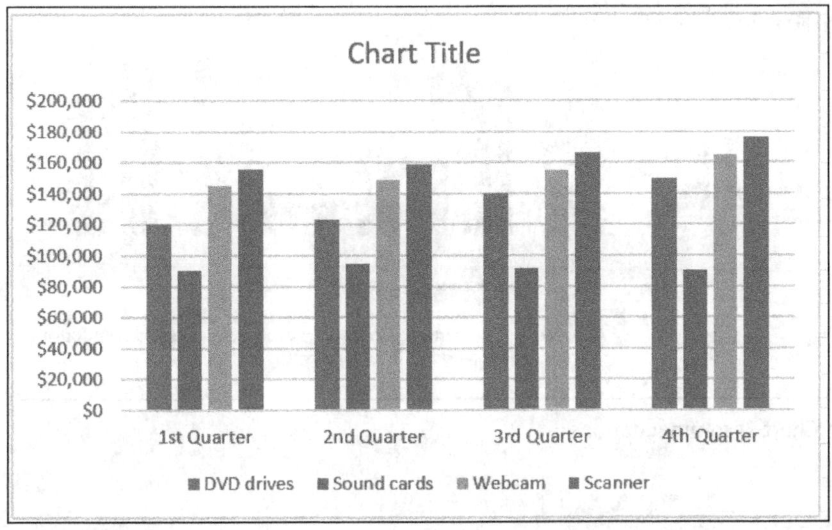

Figure 7.12 Default Chart from Selected Data

Default Axis Assignment

If the selected table has more rows than columns, Excel treats the first column as the x-axis labels and the adjacent columns as data series. Otherwise, as in this example, Excel suggests using the first row for the labels.

For the quarterly report, you can accept the default setting. In this example, the data is arranged so that quarterly sales are listed vertically in one column, while sales for each product group run horizontally in a row. If you want the chart to show and compare sales by product group over the year, then this layout is the best choice. It assigns each product group its own color or pattern, making the data easy to distinguish visually.

7.3.2 Recommended Charts

If you're not familiar with the different chart types, let Excel recommend one for you. After selecting the data as described, go to **Insert • Charts** and click the **Recommended Charts** button. In the dialog, you'll find a series of chart thumbnails on the tab with the same name that are already linked to the selected data.

Clicking or tapping a template shows how the chart will look in the right pane and offers tips about that chart type. The first suggested template usually fits the data best, but the others also work well. When you confirm a template, Excel creates the corresponding chart.

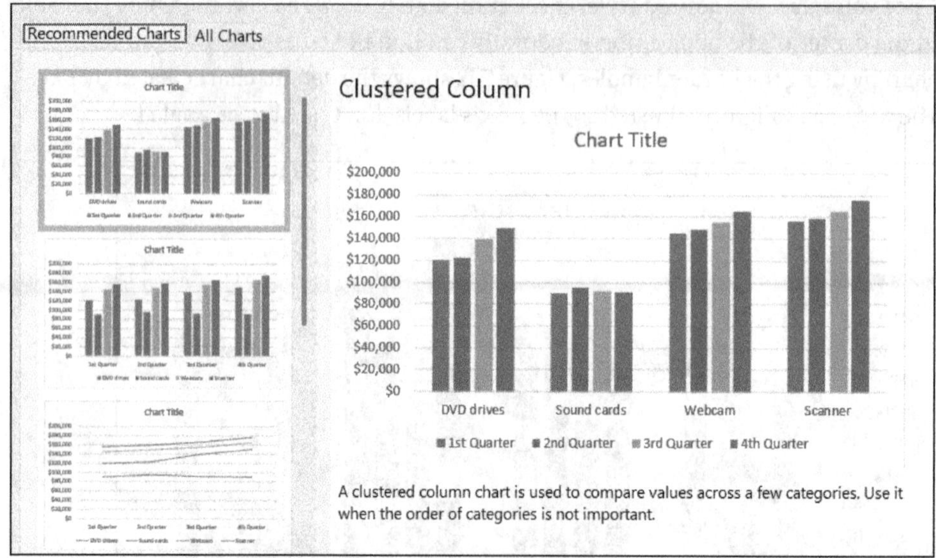

Figure 7.13 Chart Recommendations

7.3.3 The Quick Chart

If you haven't disabled Quick Analysis options via **Options • General**, you can select your data range and then click the **Quick Analysis** or press Ctrl+Q to create a chart instantly. You can select **Charts** in the top row of the palette to display five commonly used chart types as buttons, and you can hover your mouse or use the arrow keys to preview different charts based on your data. Click or tap to create the selected chart, and for other types, use the **More Charts** button to open the **Insert Chart** dialog.

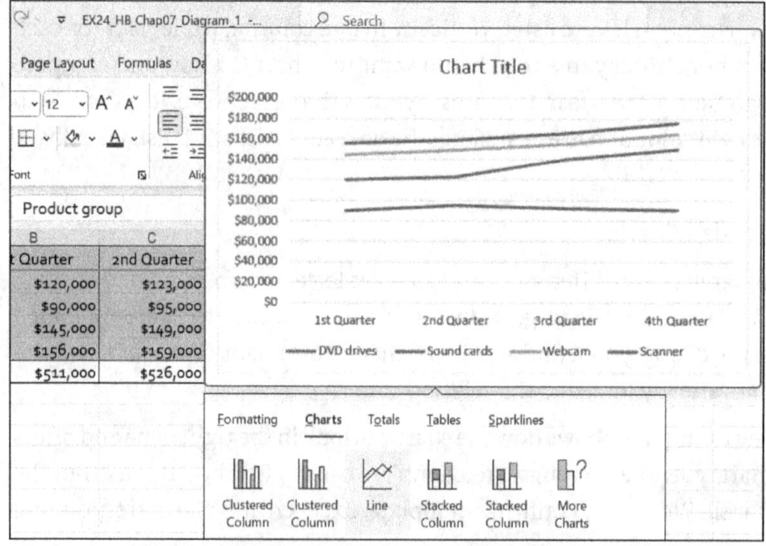

Figure 7.14 Creating a Chart by Using the Quick Analysis Tool

7.3.4 Overview of Chart Design Tools

When a chart is selected, two tabs appear for detailed editing. The **Chart Design** tab includes five groups that handle setting chart data, choosing the chart type, adding necessary labels, scaling the axes properly, optionally adding trendlines or error bars, assigning chart layouts and styles as needed, and moving the chart to another worksheet or a separate sheet if required.

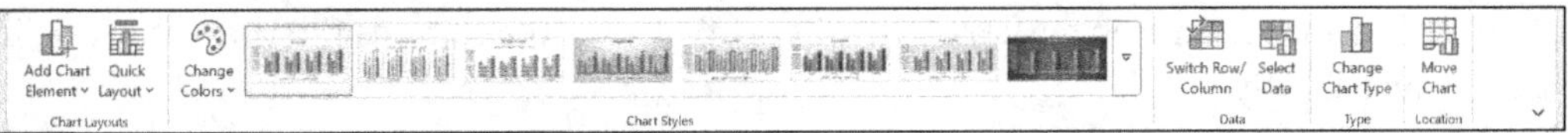

Figure 7.15 Groups on Chart Design Tab

On the **Format** tab, you'll find seven command groups. The first group, **Current Selection**, lets you pick individual chart components from a dropdown list to apply specific formatting. The **Insert Shapes** group includes options for selecting specific shapes like borders for the legend or title, applying related effects, inserting WordArt styles, adding alternative text, and arranging and sizing graphic elements in the chart. The **Shape Styles**, **WordArt Styles**, and **Size** groups have small buttons in the lower right corner, known as dialog box launchers, which open the corresponding task panes.

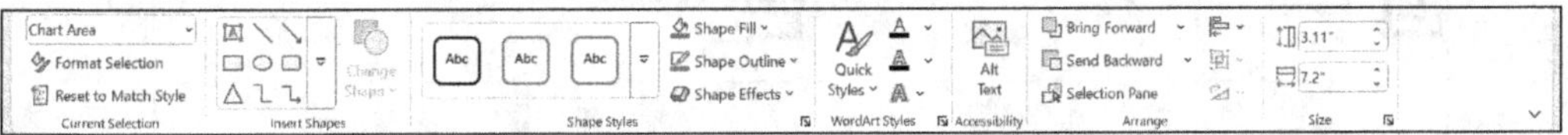

Figure 7.16 Groups on Format Tab

Fixing the Table Range

A common mistake when creating charts is selecting the wrong table range—one that's missing a column or row or includes too much data. If the initial chart output shows that the worksheet range doesn't fit the chart's purpose, then you can fix it on the **Chart Design** tab, which allows you to make corrections. The **Data** group includes the **Select Data** button, which you can click to open the **Select Data Source** dialog box.

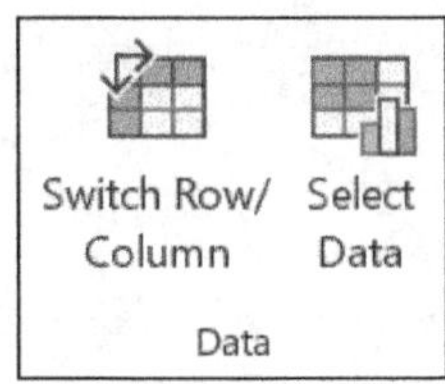

Figure 7.17 Data Group

The first input field displays the current chart data range using absolute references. You can correct the range here if you have made a mistake or need to expand or shrink it. Just select the correct range. Then, if the dialog box blocks your view, click or tap the **Collapse Dialog** icon ❶.

To select noncontiguous ranges, either press Ctrl or Shift+F8 as usual to add more selections or enter the addresses directly after a semicolon.

Instead of adjusting the entire chart data range, you can remove or add individual series within the selected data range as needed. To exclude scanner data from the chart, select the appropriate entry under **Legend Entries** and click the **Remove** ❷ button. To add a series, click the **Add** ❸ button and then select the range in the table that includes the name cell and the value cells. To temporarily hide specific rows or categories from the chart, uncheck the boxes next to the items ❹. This is the same as filtering, which we'll cover shortly.

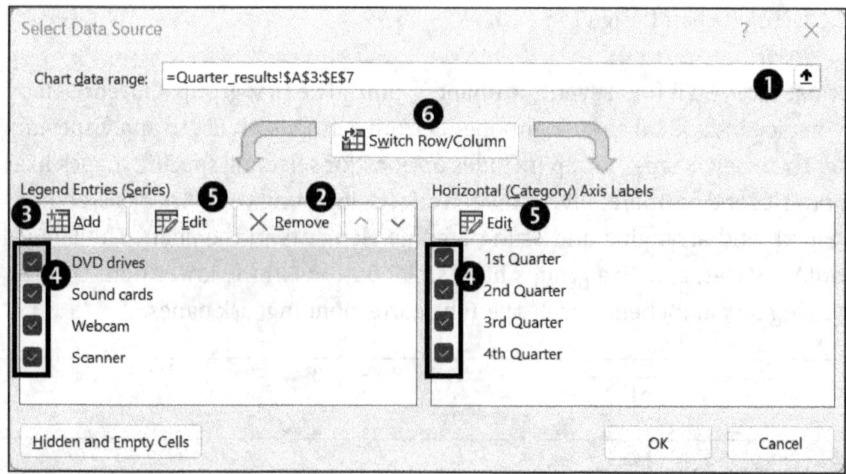

Figure 7.18 Editing Data Source

The **Edit** button ❺ lets you adjust the range references for both legend entries and categories.

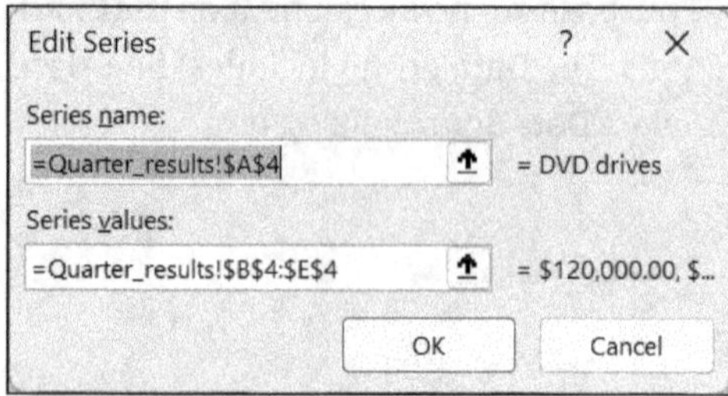

Figure 7.19 Editing Data Series

The Switch Row/Column button ❻ is also available directly in the **Data** group. It swaps the table ranges that are assigned to the two axes. In this example, the chart's x-axis would be divided by product groups.

7.3.5 Chart Filter

When you select the chart, you can click the button with the filter icon to open a control with two tabs. Under **Values**, you can select from the data series and categories currently in the table. You can hover over an item to highlight it in the chart, and you can deselect it to temporarily remove it from the chart when you confirm with **Apply**. Unlike the permanent change to the chart's data source described earlier, this only temporarily hides specific data series or categories. You can click or tap outside the control or press Esc to close it.

The second page, **Names**, lets you hide data series or category labels using the **(None)** option. These labels are then replaced with placeholder names or numbers.

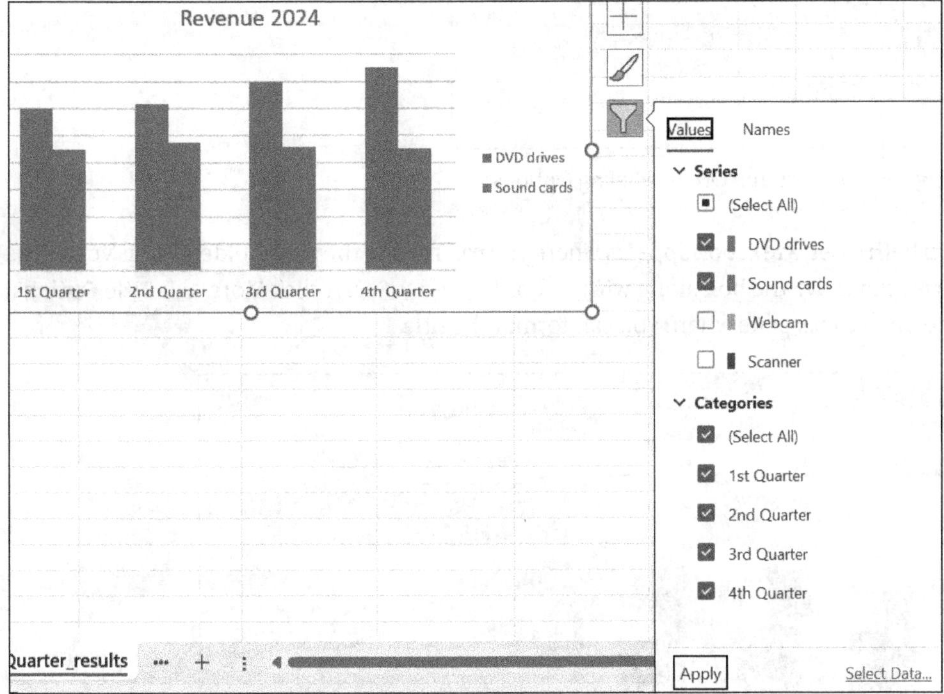

Figure 7.20 Example of Applying Filter

7.3.6 Titles, Legends, and Other Options

The automatically generated chart can benefit from some additions, especially in labeling. On the **Chart Design** tab, find the **Add Chart Element** button in the **Chart Layouts** group and click it to open a menu with many options for labels and other chart enhancements.

Next, under **Chart Title**, choose whether to center the title inside the plot area above the chart—with the **Centered Overlay** option—or place it above the plot area by using **Above Chart**.

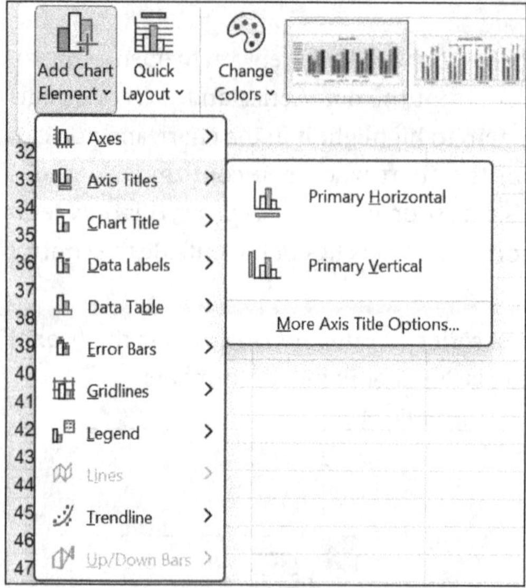

Figure 7.21 Titles and Other Labeling Options

In both cases, a text box appears where you can replace the placeholder title as you wish. You can move the box along with its border, and you can click **More Axis Title Options** to open a task pane where you can format the title.

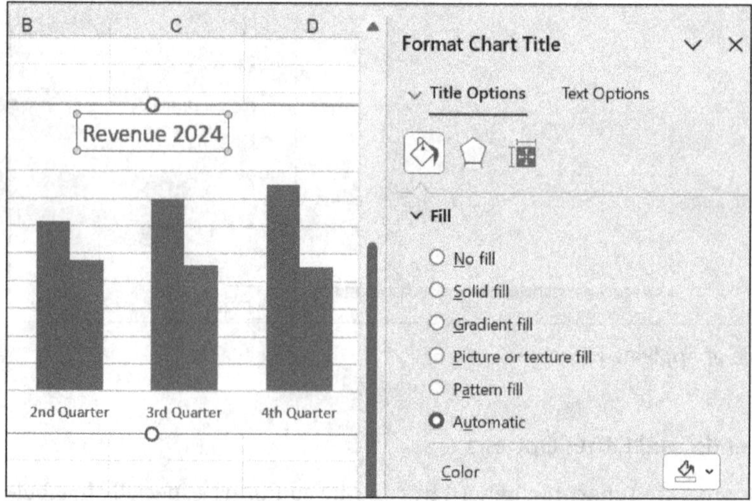

Figure 7.22 Selected Chart Title and Its Task Pane

While you've already determined the source of the legend texts by selecting the chart data range, you can click the **Legend** option from the menu in Figure 7.21 to open a palette of positions where you can place the legend. Legends are always necessary when—like in this case—multiple data series are present. By the way, you can't change the text of a legend directly in the chart itself but only in the cell from which it is taken.

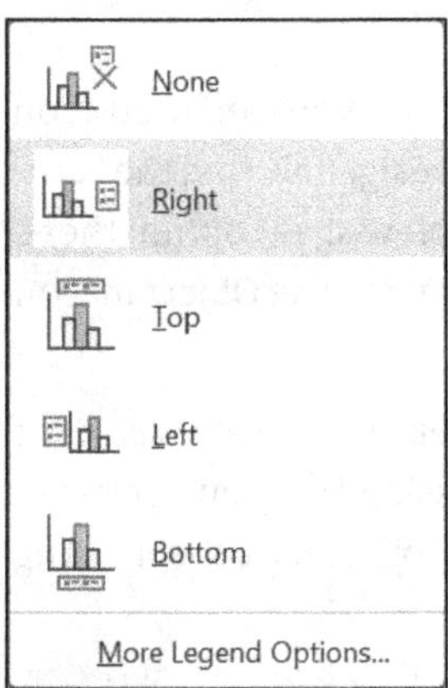

Figure 7.23 Legend Options

In 3D charts, a data series label can also serve as an axis label. It's best to choose either a legend or an axis label, since using both usually clutters the chart. For long data series labels, legends are generally more effective.

Excel also offers a quick way to add missing chart elements here. Click or tap next to the selected chart on the button with the plus sign, and from the list of chart elements, select the ones you want. If additional options are available, open the menu by clicking the small triangle next to the name. All options include a live preview so you can see their effects before applying them. With a mouse, just a click is enough. With a pen, hover over the option. With your finger, tap the option and the list will stay open so you can change your choice if you don't like the chart's result; you can tap outside the list to confirm your current selection.

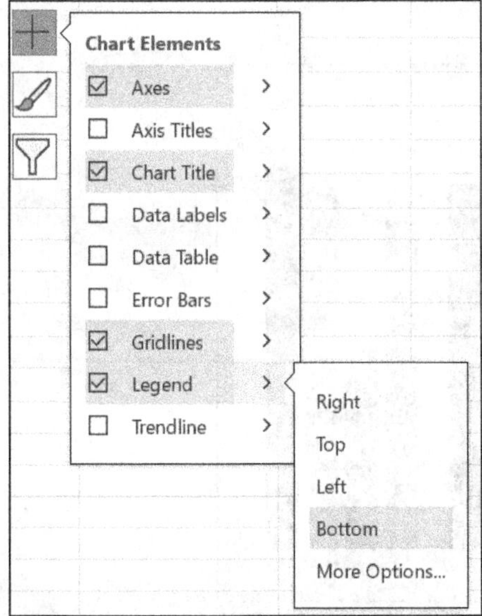

Figure 7.24 Chart Elements Palette

7.3.7 Setting the Chart Location

Instead of placing a chart on the same worksheet as its source table, you can insert it on its own sheet or another worksheet. To do this, go to the **Chart Design** tab, find the **Location** group, and click the **Move Chart** button. If you choose the **New sheet** option there, give the chart sheet a fitting name and replace the default. If you choose **Object in**, you can select the worksheet from the dropdown menu.

If you choose a separate chart sheet, the chart fills the workbook window at its current size. You can create a chart on its own sheet even faster by pressing F11 on a selected range of table data. The selected data will immediately appear on a chart sheet, using the chart type that's set as the default.

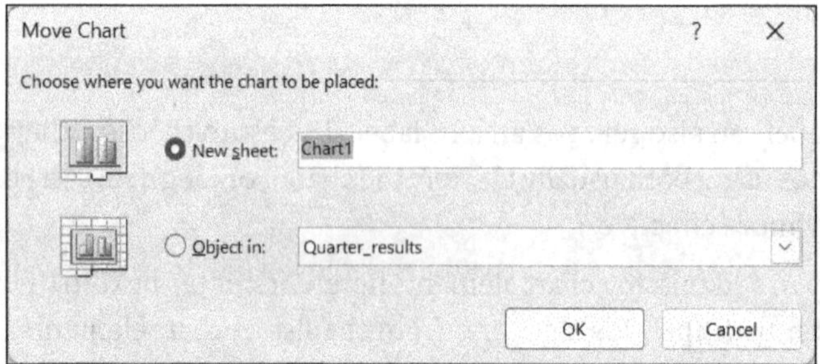

Figure 7.25 Deciding Where to Place Chart

If you choose a worksheet as the chart location, the chart initially appears centered in the current window, and you can easily drag it to the spot you want by its border. To resize the chart, simply drag the handles along its edges. These handles appear when you select the chart by clicking or tapping it, and you can drag the corner handles to resize proportionally.

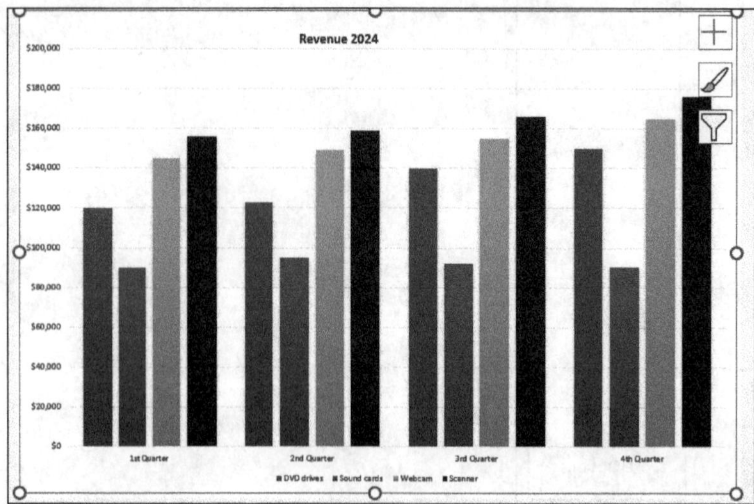

Figure 7.26 Column Chart on Separate Chart Sheet

7.4 Linking the Table and Chart

When you select a chart that's embedded in the worksheet, the related table data is highlighted with colored borders. Click or tap a data series bar to display its corresponding data in the table.

Hover over a data point to see which series, point, and value are selected. With pen input, simply hovering the tip is enough, and this keeps the connection between the table and chart clear at all times. Whenever the table data changes, the chart updates immediately.

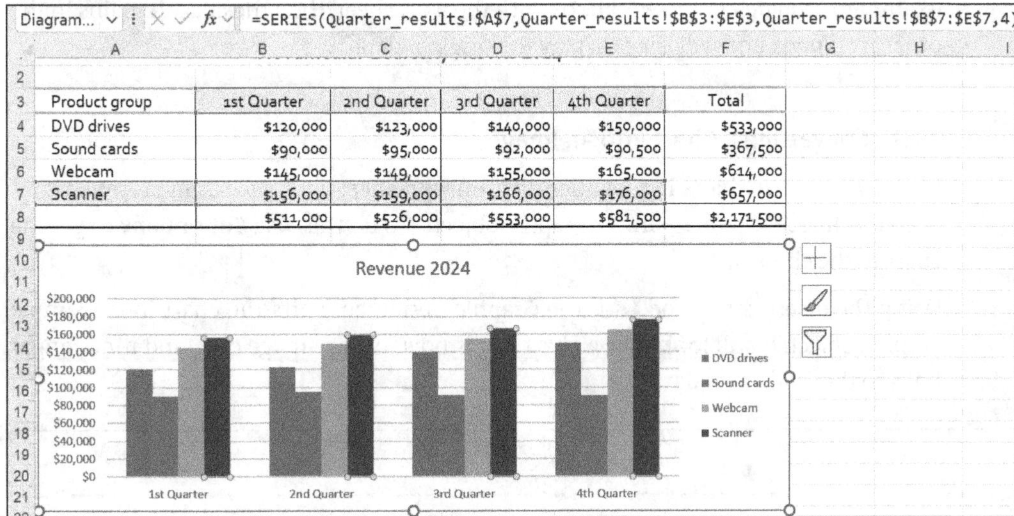

Product group	1st Quarter	2nd Quarter	3rd Quarter	4th Quarter	Total
DVD drives	$120,000	$123,000	$140,000	$150,000	$533,000
Sound cards	$90,000	$95,000	$92,000	$90,500	$367,500
Webcam	$145,000	$149,000	$155,000	$165,000	$614,000
Scanner	$156,000	$159,000	$166,000	$176,000	$657,000
	$511,000	$526,000	$553,000	$581,500	$2,171,500

Figure 7.27 Selected Data Series and Highlighted Table Data

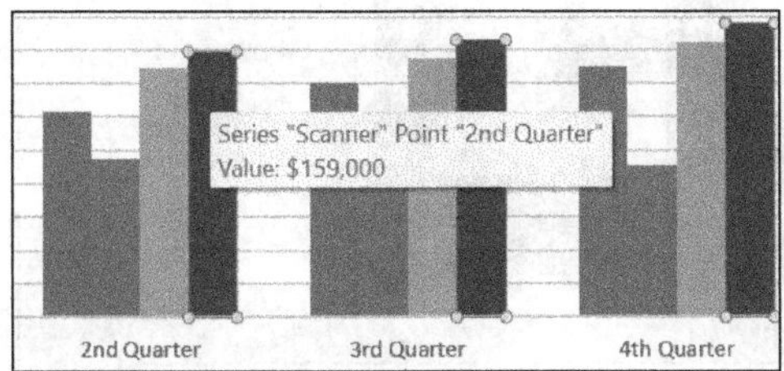

Figure 7.28 Selected Data Series Showing Data Point's Value

A chart's basic connection with its table is that it visually represents the table's numbers. You create this link by using `SERIES()` functions, which work like other Excel functions. When you select a data series in a chart, the related `SERIES()` function appears in the formula bar, and if you like tinkering, you can even make changes there. The function works like this: The first three arguments are the absolute addresses of the worksheet ranges, and the last argument is a number that controls the row order:

```
=SERIES(RowNames,ColumnNames,Values,PlotOrder)
```

It gets a bit more complicated when the worksheet and chart are in different workbooks, but that's entirely possible. Just select the data in the appropriate workbook by using the **Select Data Source** dialog, and if the chart has its own sheet, you can move or copy the entire sheet to another workbook. When the chart and worksheet are separated this way, the link is made through external references, as you've already seen. As long as both workbooks are open, changes in the worksheet immediately update the chart, but if the chart file isn't open, changes in the worksheet only take effect when you reopen the workbook containing the chart.

You can also break the link between the two workbooks without losing the chart. For example, if you delete the workbook with the worksheet, then the chart remains in the other workbook and will then display the original data.

7.4.1 Converting a Chart into an Image

If you want to use charts independently from worksheet data, such as in presentations, you can "freeze" them as simple graphic objects. To do this, use **Cut** or **Copy** to get the chart to the clipboard.

Using the **Insert** command with the **Graphic** option converts the chart into a simple graphic object. It still looks like a chart but is now just an image of it, and it no longer links to the worksheet data.

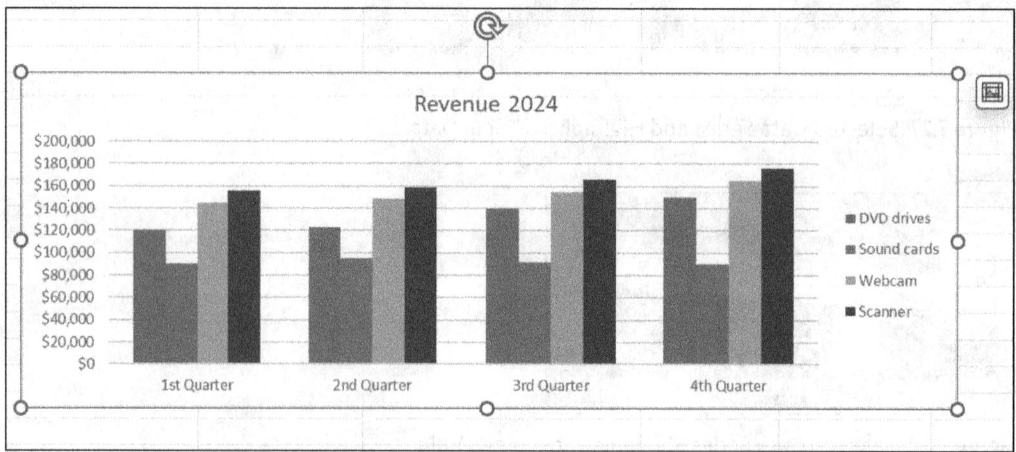

Figure 7.29 Inserting Chart Graphic

7.5 Optimizing Charts

Charts created by the previously explained methods often need improvement. The headings or axis labels may be too large or too small, and the automatic proportions Excel sets might not fit your purpose. Fortunately, Excel offers many options to customize your chart however you like. To edit the chart, just click or tap on it once. This also works for individual elements, and you don't need to select the entire chart first to access its parts.

7.5.1 Formatting the Current Selection

As mentioned, the **Format** tab includes the **Current Selection** group when you select a chart or one of its elements. The dropdown always lists the currently selected chart objects, clicking the chart area displays **Chart Area**, and clicking the plot area shows **Plot Area** as the object name.

You can also pick the object to edit from the list. For example, to format the **Series "Sound cards"** differently, open the dropdown and select it. This is the best method to use when elements are small and hard to click, like narrow columns in charts with many data series. Alternatively, you can use the sliders at the bottom of the window to zoom in temporarily.

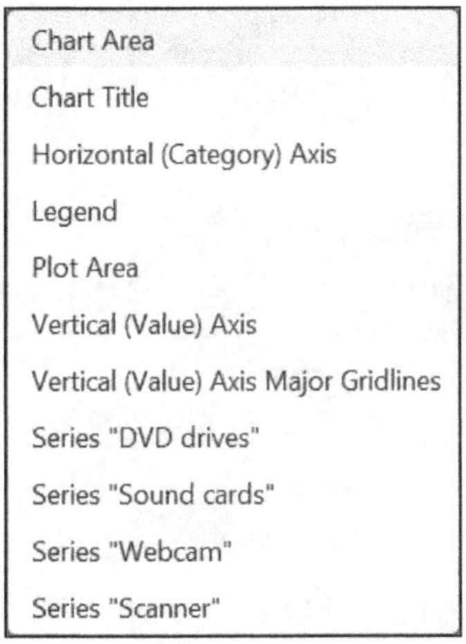

Figure 7.30 Object Selection via Dropdown

Special commands for formatting the currently selected object appear immediately when you choose the **Format Selection** option below the dropdown. All chart formatting task panes have a consistent design. They usually offer multiple pages through links ❶ under the heading. Below the links, icons ❷ group related option subcategories. Options for a selected icon appear below that, and these are grouped under different headings ❸, which also serve as buttons to expand or collapse details.

To open a task pane, double-click or double-tap the item you want to format. Most settings you choose in the task pane update the chart immediately, even before you close the task pane with **Close** ❹. Selecting a different element in the chart instantly shows its options in the task pane.

You can keep the task pane open until the chart is fully formatted. For example, start by formatting the plot area, then the axes one by one, and finally the legend. The task pane updates automatically with each selection. You can also select elements using the menu that opens when you click the small triangle next to the first link in the task pane.

You can also pull the task pane completely out of the application window. The small arrow next to the heading lets you choose **Move** ❺. When enabled, the pane follows your mouse anywhere on the screen; click or tap to place it. You can also drag the pane by holding down the mouse button. On a touchscreen, the pane follows your finger or stylus. Double-click or double-tap to snap it back to the right or left side of the application window.

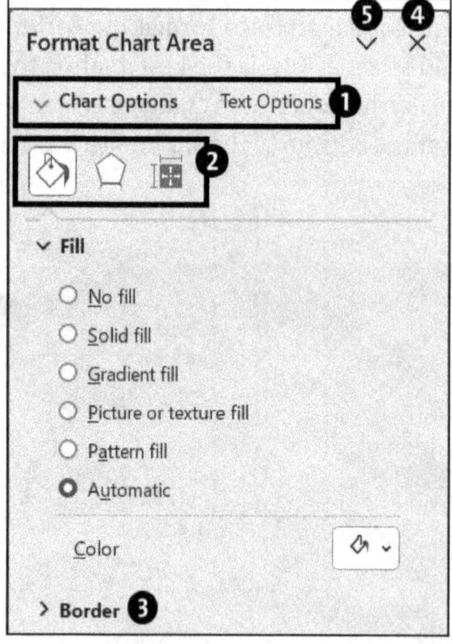

Figure 7.31 Task Pane for Formatting the Chart Area

7.5.2 Context Menus

Context menus are available for the entire chart and all its individual elements. Besides formatting options, these menus offer commands related to the selected object. Like with cell range context menus, a mini toolbar appears at the top.

You'll also find a dropdown list of chart objects, which let you quickly change fonts for the legend or chart title and assign different colors to elements like the plot area of a 2D chart or the floor and walls of a 3D chart.

On the touchscreen, context menus for chart elements open when you press and hold your finger on an element. The layout is similar to the menus for cell ranges: Some common commands are arranged horizontally, while others appear vertically when you tap the arrow button.

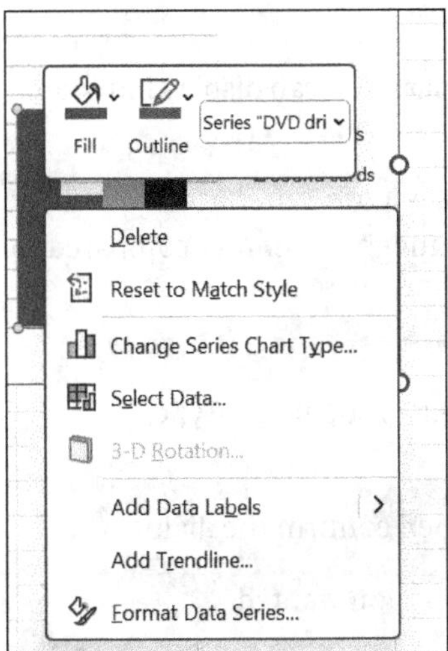

Figure 7.32 Context Menu for Data Series

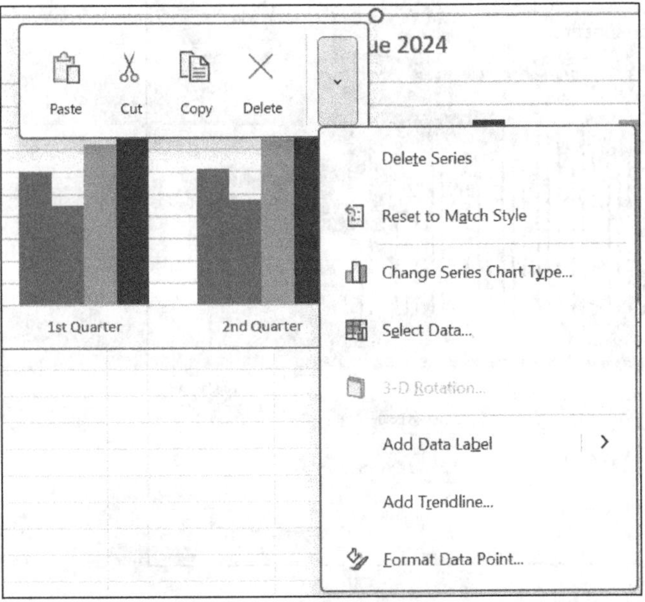

Figure 7.33 Context Menu for Data Point on Touchscreen

Editing Directly with Your Mouse, Finger, or Pen

You can make many changes simply by dragging with your mouse, finger, or pen. To move a title or legend, just click or tap it and drag the border to the desired spot. There are special options for designing 3D charts. We'll cover this in more detail in the following sections.

7.5.3 Combining Chart Types

In addition to changing the overall chart type for your chart, you can display individual data series differently (though only on 2D charts).

For example, one of the many chart types you can create is a column chart from a table showing warehouse inventory changes. It's helpful to highlight inventory compared to inflows and outflows, and you can do that as follows:

1. Select the chart with a click or tap.
2. Click the **Change Chart Type** button in the **Chart Design** group under **Type**.
3. Select the **Combo** chart type.
4. For the **Stock** data series, choose **Line with Markers**, then confirm the dialog.

Now the inventory data series is displayed as a line, just as you wanted.

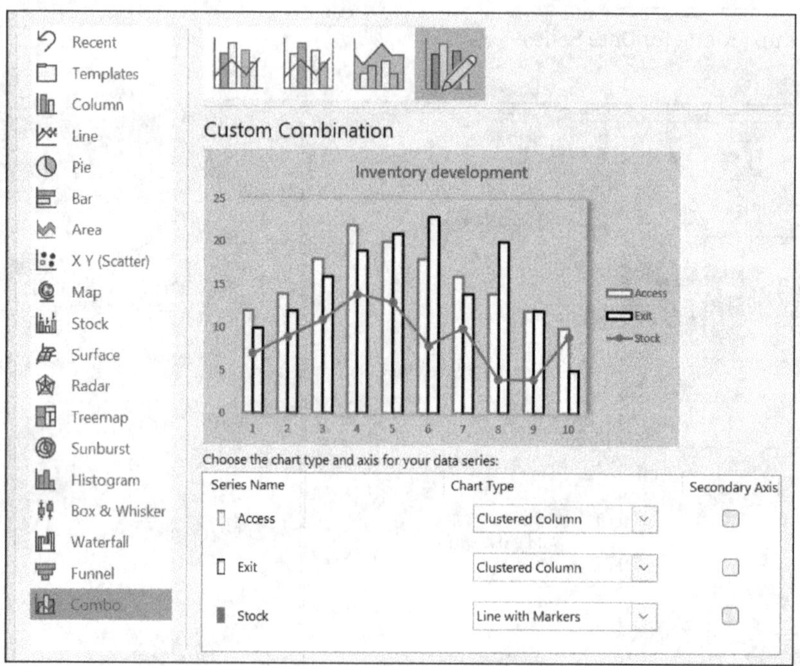

In a line chart, you can customize the markers for each data point in many ways:

1. In the example, select the stock line by double-clicking or double-tapping it.
2. Use the **Fill and Line** ❶ icon and then click the **Marker** ❷ link below.
3. Under **Marker Options**, you'll find a dropdown to customize the markers when you enable the **Built-in** option.
4. Pick a marker symbol that fits and then choose the size you want below.

To help the markers stand out against the bars, you can add colored outlines to the bars using the **Shape Styles** palette.

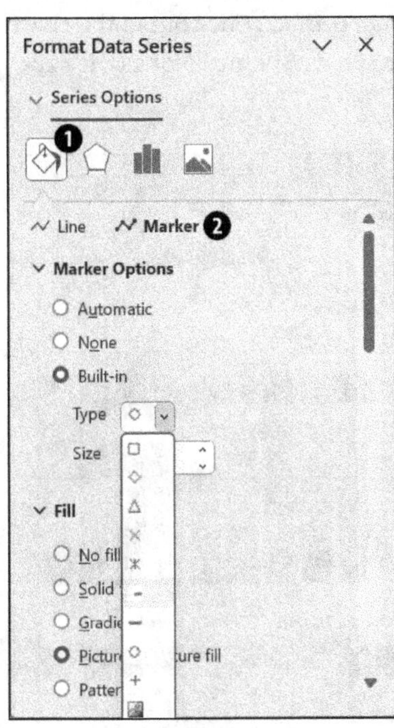

7.5.4 Improving Shapes

Sometimes, a few shape adjustments can enhance how data appears in a chart.

1. When you select the plot area in your chart, you can apply effects from the **Shape Effects** palette, like a shadow that subtly lifts the area. These effects are also available in the **Shape Styles** group.

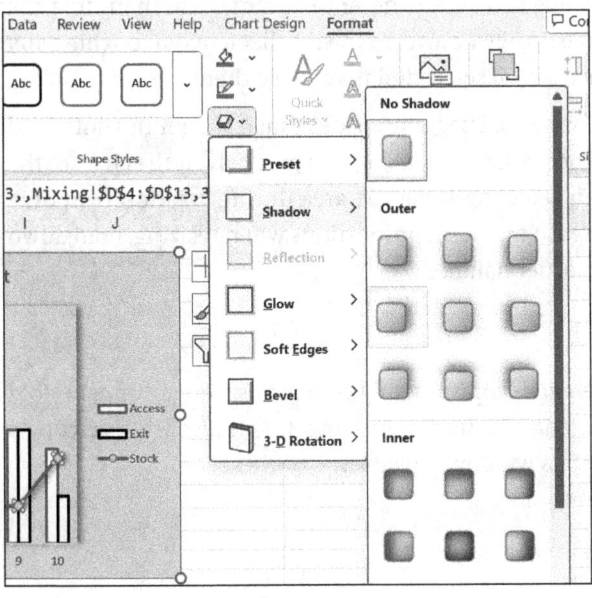

2. You can also highlight the entire chart area with a shape effect. Click or tap the chart border, then use one of the **Shape Styles • Fill Effects** to set a theme color as the background.

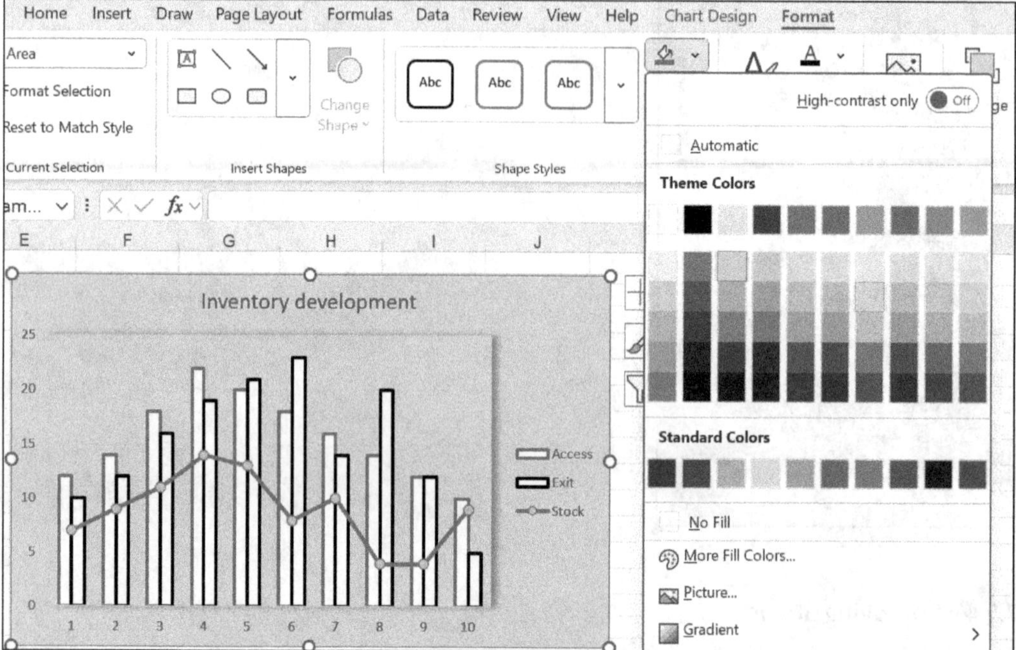

7.5.5 Resizing and Repositioning the Chart

If you're unhappy with the chart's size or position, you can easily adjust it. Click or tap an empty spot along the chart's edge to select the entire chart area, then drag the selected chart by its border to move it anywhere in your worksheet. Avoid touching any elements inside the chart area, as this will select and may move them.

Resize the chart by dragging the handles: Dragging a side adjusts height or width, and dragging a corner adjusts both. When resizing, you can also precisely fit the chart to the rows and columns, just like when selecting the chart area during creation. You only need to press `Alt` while dragging. To maintain proportions when resizing, hold down the `Shift` key while dragging a corner handle.

Copying Charts

When you select a chart, you can create a copy by using your mouse, finger, or pen: Hold `Ctrl` while dragging to copy the chart instead of moving it. This also lets you copy a chart to another open workbook if its window is visible.

The same techniques apply within the chart. You can resize the plot area just like the overall chart area, and you can also drag labels to exactly where you want them. For example, it often makes sense to drag the legend directly into the plot area, and you can then enlarge the drawing area itself.

Pulling Out Pie Slices

In pie and doughnut charts, you can pull out all or individual slices or segments. To select a single slice, click or tap to select all and then click or tap again on the slice you want.

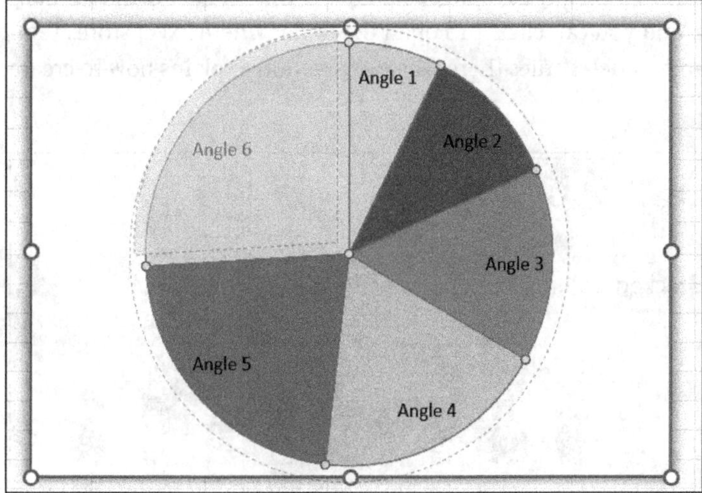

Figure 7.34 Pulling Out Slice of Pie Chart

7.6 Designing Charts

Excel offers so many chart design options that it's helpful to have a general overview of the menu groups, dialog boxes, and task areas available.

7.6.1 Changing the Chart Type

The most significant change you can make to a chart is choosing a different chart type. or subtype. To change the type, select the chart, go to the **Chart Design** tab, and then click the **Change Chart Type** button in the **Type** group. This command is also available in the chart's context menu and some of its components.

The **Change Chart Type** dialog box that opens this way is the same as the **Insert Chart** dialog box mentioned earlier, except for the title. On the **All Charts** tab in the left pane, it shows the different base types. The right pane displays sample subtypes, with the first subtype selected by default for each base type. You'll see a preview of the chart with the selected data, and one option will use column values for the data series while the other will use row values. The preview enlarges when you hover the mouse or hold the pen over it, and you can choose the option you prefer and confirm the dialog. Double-clicking or double-tapping the preview works the same way.

The **Recent** ❶ button appears at the top and to the left of the base types, and you can click it to display the types you've used before. Below it, you'll find a folder button labeled **Templates** ❷, and if you click it and you've created custom templates, they'll appear in the right window. To help you manage templates, this dialog includes a **Manage Templates** button, and you can click it to open the folder where Excel stores templates so you can rename or delete files there. The next section explains how to create templates.

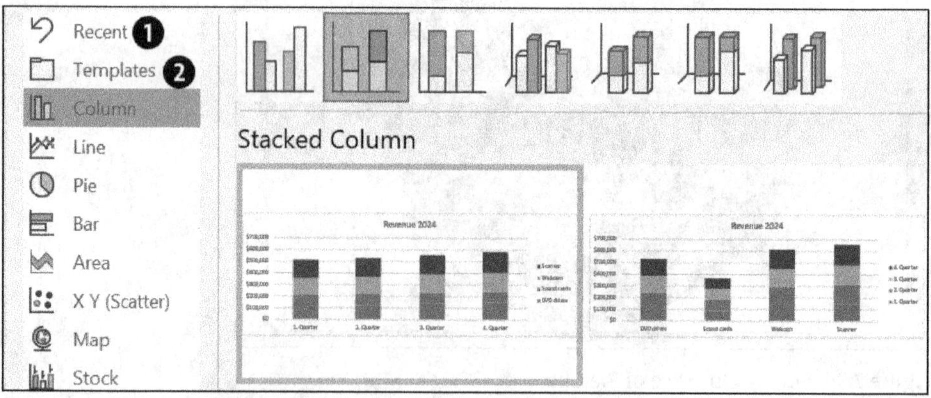

Figure 7.35 Change Chart Type Dialog Box

7.6.2 Chart Layouts and Chart Styles

Compared with earlier versions, the selection of subtypes has been streamlined: the previous pyramid and cone types are no longer listed separately. However, you can assign the corresponding shape property to any 3D column or 3D bar later, through the **Format Data Series** pane under **Series Options** and the **Column shape** settings.

Excel's chart types and subtypes offer much more than you might expect. After you select a chart type, Excel provides a set of quick layouts tailored to that type along with a wide variety of chart styles. Quick layouts control the number and arrangement of chart elements—like the legend's position and whether gridlines and data labels appear. Chart styles focus on styling these elements, including colors, column and line styles, and the fill of the plot and chart areas.

The **Quick Layout** palette appears on the **Chart Design** tab in the **Chart Layouts** group when the chart is selected. Clicking or tapping a pattern applies that layout to the current chart and rearranges its elements. Figure 7.37 shows the second layout as an example.

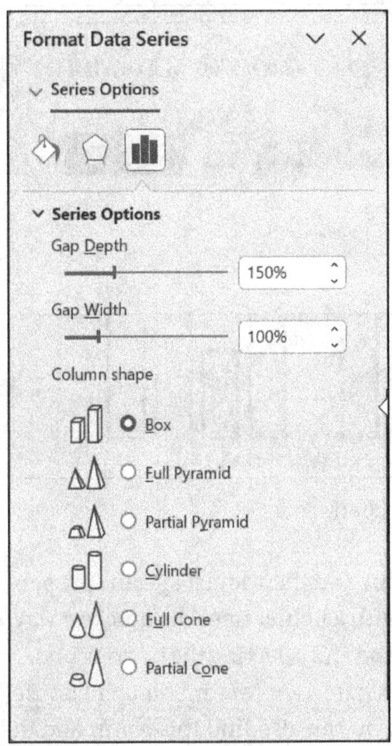

Figure 7.36 Alternative Column Shapes

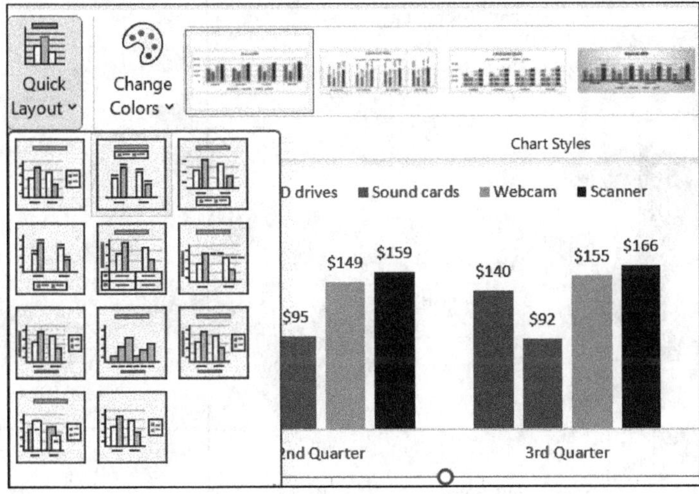

Figure 7.37 Column Chart with Layout 2

The chart styles palette on the **Chart Design** tab can be browsed via the scroll bar or fully expanded. Clicking or tapping a sample applies that style to the selected chart. Only samples for the current color scheme are shown. Use the **Change Colors** button to select other schemes, which will display different samples.

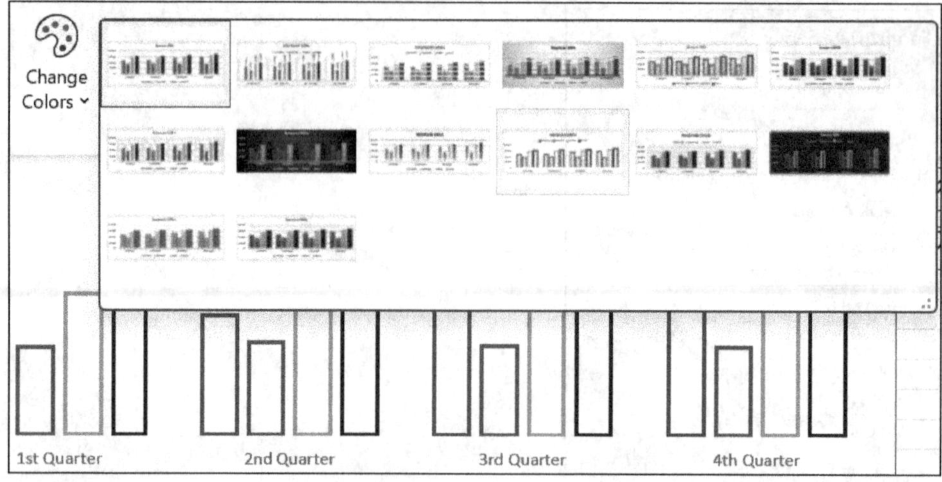

Figure 7.38 Chart Format Template Palette for Column Chart

Often, it's helpful to apply a predefined style to a chart and then adjust specific properties, like replacing a dark fill color in the chart area with a lighter one. The quickest way to do this is by going to the **Format Chart Area** task pane and selecting the options under **Fill**. If you don't like the adjustment, just open the chart's context menu and choose **Reset to Match Style**. This will undo your changes. You can also find this command in the **Format • Current Selection** group.

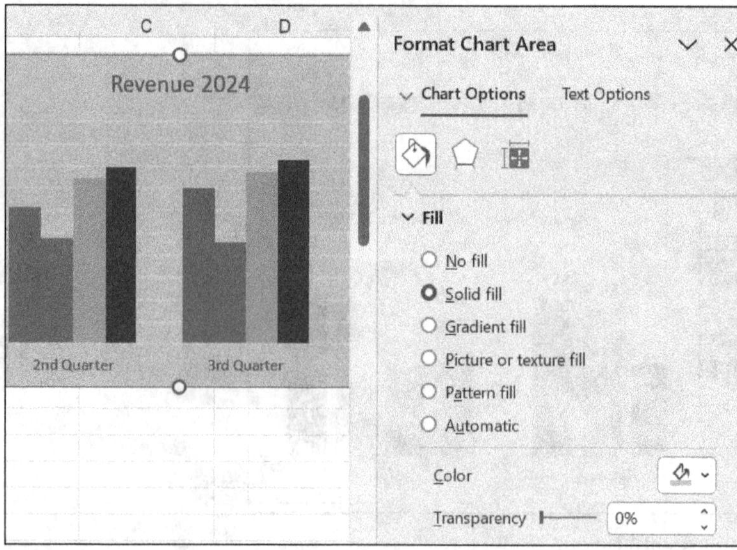

Figure 7.39 Detailed Changes to Chart Format Template

Besides the samples in the **Chart Styles** group, Excel offers the second of three buttons to the right of a selected chart to give you even faster access to chart styles.

Under **Style**, small preview images match the data and the current chart. These types are easy to apply with a click or tap. To change the chart's color scheme, use **Color** and select the row with the combination you prefer.

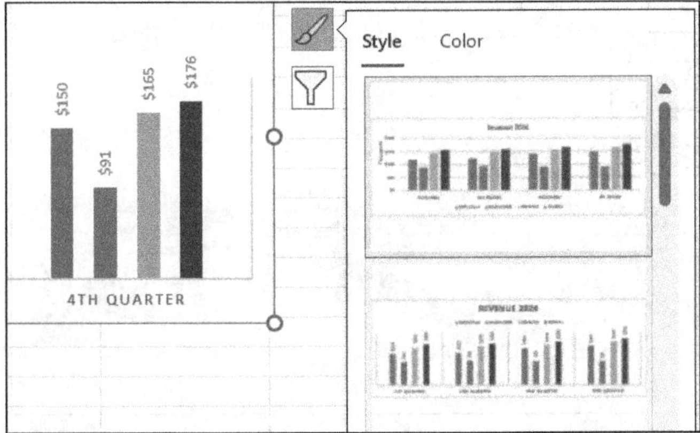

Figure 7.40 Selecting Style with Chart Style Button

7.6.3 Custom Templates

If you want to reuse this color-modified version of the built-in chart style later, you save it as a custom template. To do this, select **Save as Template** from the context menu of the finished chart. Then, in the dialog box, enter a suitable name for the chart template file where Excel saves the selected formatting. The file uses the **.crtx* extension, and by default, Excel offers the user's *...Templates/Charts* folder. The custom template then appears under **Templates** in the **Insert Chart** or **Change Chart Type** dialog.

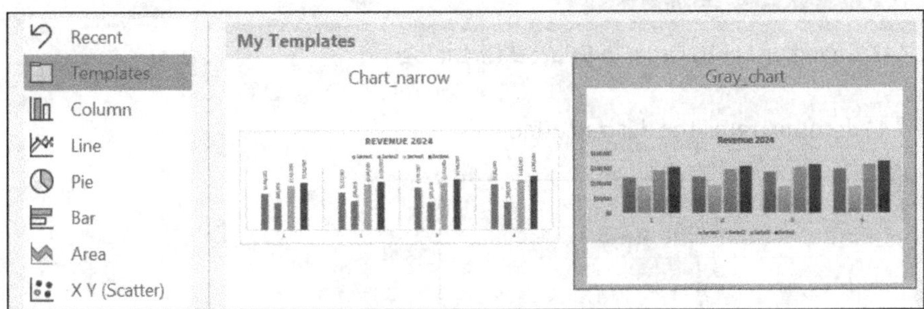

Figure 7.41 Custom Template in Dialog Box

7.6.4 Arranging Data Series

Often, the column arrangement in a table doesn't yield the best results in a chart based on that range. For example, if a column with smaller values sits between or after columns with larger values, it usually looks better to display the bars with large values first and the smaller ones afterward, or vice versa. In 3D charts, data series often disappear completely or partially behind others due to their size. You can fix this by moving an invisible series to the front. Figure 7.42 shows an example of this.

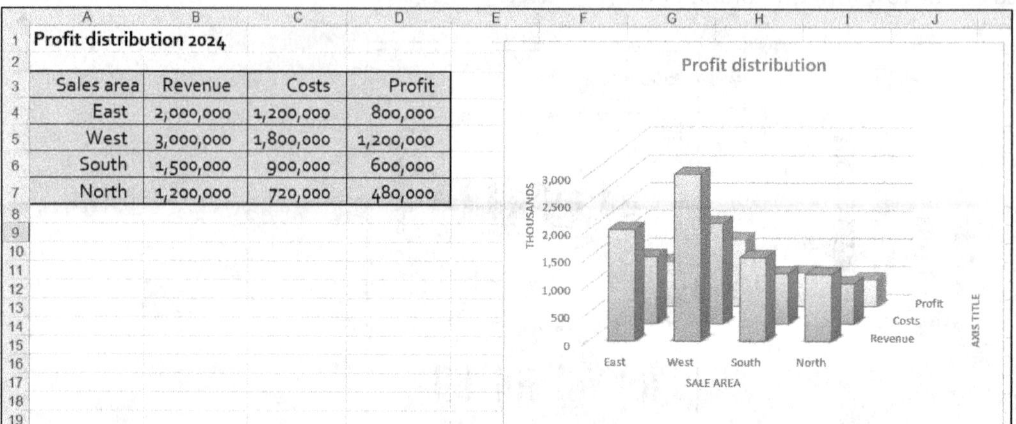

Profit distribution 2024

Sales area	Revenue	Costs	Profit
East	2,000,000	1,200,000	800,000
West	3,000,000	1,800,000	1,200,000
South	1,500,000	900,000	600,000
North	1,200,000	720,000	480,000

Figure 7.42 Poor Column Arrangement

You can rearrange data series differently from in the table by using the **Select Data Source** dialog, which opens when you click the **Data • Select Data** buttons on the **Chart Design** tab. Select the data series you want to move and then use one of the two arrow buttons ❶.

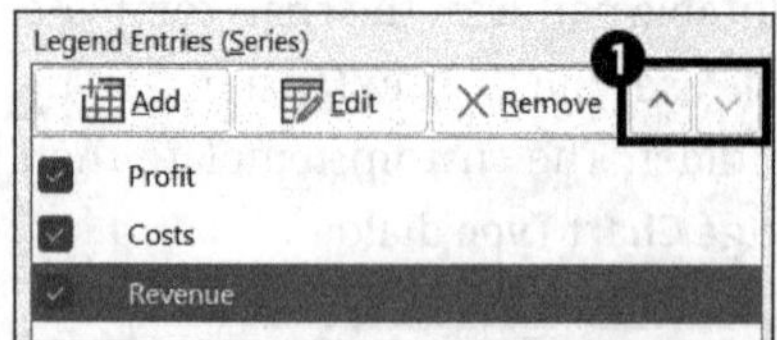

Figure 7.43 Changing Series Order in Select Data Dialog

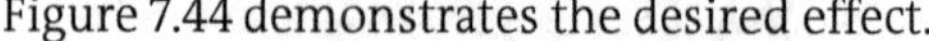

Figure 7.44 demonstrates the desired effect.

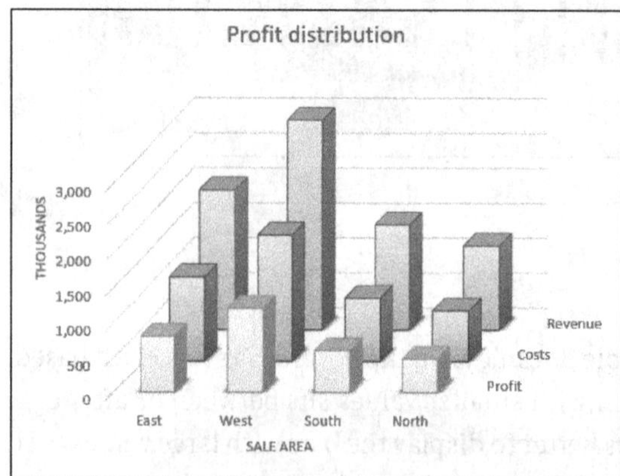

Figure 7.44 Improved Row Arrangement

Adjusting the Viewing Angle

In 3D charts, you'll often need to adjust the viewing angle to make all data visible. You can do this by selecting the plot area and using the 3D rotation settings in the **Format Chart Area** task pane under **Chart Options.**

Here, you can enter angle values to rotate each chart axis. Changing the **X** value rotates you around the chart: at 0 you face the front, and at 360 you've rotated once counterclockwise. By changing the **Y** value from –90 to +90 degrees, you can move around the chart almost like in a semicircle, from bottom to top. At 0 degrees, you're level with the chart. Depending on the chart type, you can also adjust the **Perspective**. Setting a value near zero creates an isometric view, with no spatial distortion, while setting a value of 100 applies the maximum spatial foreshortening. Changes appear immediately in the chart, making them easy to review.

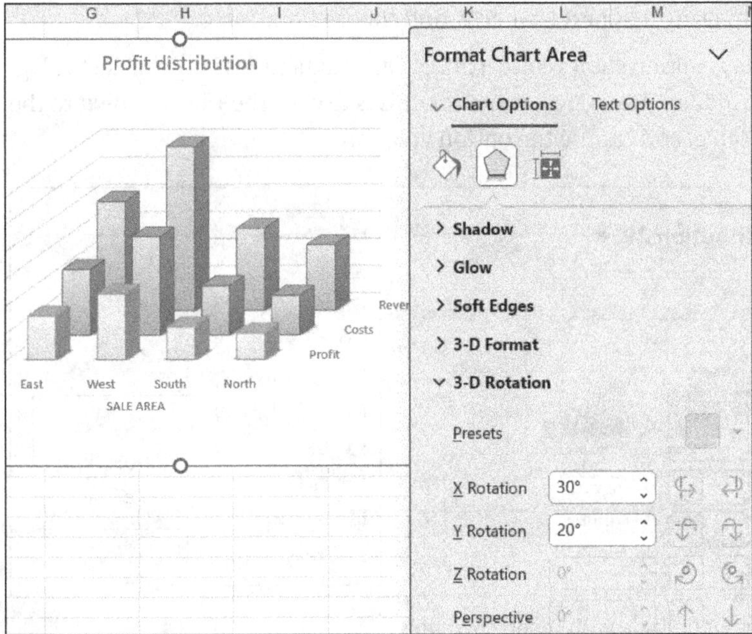

Figure 7.45 Adjusting Viewing Angle

Format Options for Data Series

On the **Data Series Options** page in the **Format Data Series** task pane, you can adjust specific values under **Series Options** via the chart icon, depending on the chart type. For all types of column or bar charts, you can adjust element overlap or spacing here. For pie charts, set the angle of the first segment, and for doughnut charts, adjust the size of the inner ring.

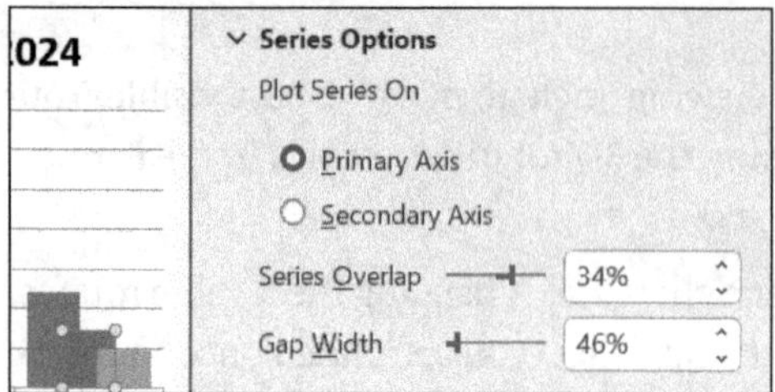

Figure 7.46 Adjusting Column Overlap

Including the Data Table with the Chart

It's often helpful to attach the underlying data table to the chart, especially when sharing a chart on a separate sheet. A predefined layout mentioned earlier handles this automatically. Alternatively, you can use the **Chart Design** tab in the **Chart Layouts** group, click the **+** button to bring up **Chart Elements**, and choose **Data Table**.

The **With Legend Keys** option shows color icons for each data series in the table; the legend can usually be hidden. The fastest way to do this is to click the **+** button next to the chart, select **Data Table,** and choose the option you want.

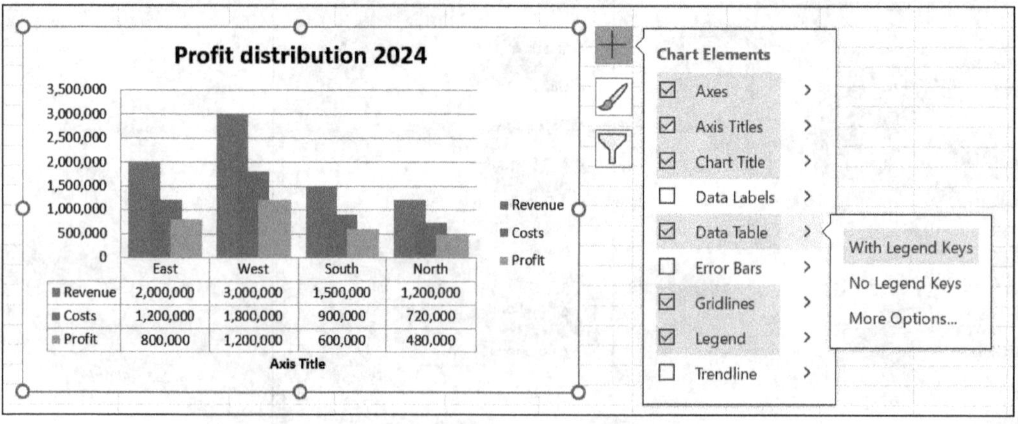

Figure 7.47 Inserting Data Table with Legend Symbols

Changing Fonts in the Chart

When you select the chart area, you can reset the font for the entire chart, along with the previously mentioned pattern and border options. To do this, select the **Font** option from the chart's context menu. On the **Character Spacing** tab, you can reduce character spacing if label readability is an issue. When you change the font for the entire chart, it's usually best to adjust the font size for titles and axis labels separately to make them stand out.

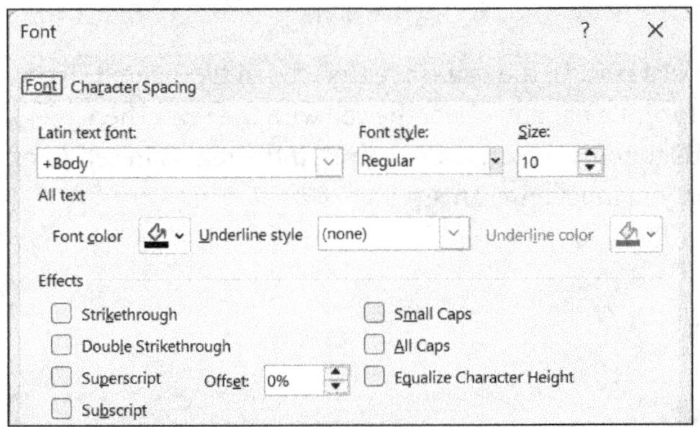

Figure 7.48 Changing Font for Entire Chart

7.6.5 Inserting and Formatting Labels

If a label, such as an axis title, is missing, you can use the options under the **Add Chart Element** button on the **Chart Design** tab. Alternatively, you can click the **+** button next to the chart, which lets you add multiple elements at once before hiding the options by selecting any cell.

These labels are inserted in preset locations with default text and font settings, but you can easily customize them. Simply select the text element by double-clicking or tapping it, and once you select it, start typing right away. To finish, click or tap outside the text box or choose **End Text Editing** from the context menu.

The **Add Chart Element** menu offers extra options like **Lines** and **Up/Down Bars**, which are useful for line and area charts.

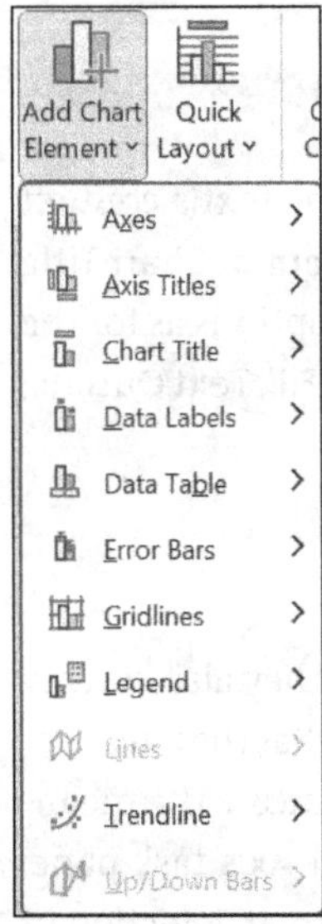

Figure 7.49 Buttons for Labels on Chart Design Tab

Variable Chart Titles

You can also enter a reference to a label in the worksheet instead of typing text. To do this, type an equal sign in the formula bar and select the cell with the text. Then, press [Enter], and the text will be added to the chart. For example, if the text is in cell A1 of the *Data_table* worksheet, then you must enter this:

```
='Data_table'!$A$1
```

Don't enter =A1. That won't work.

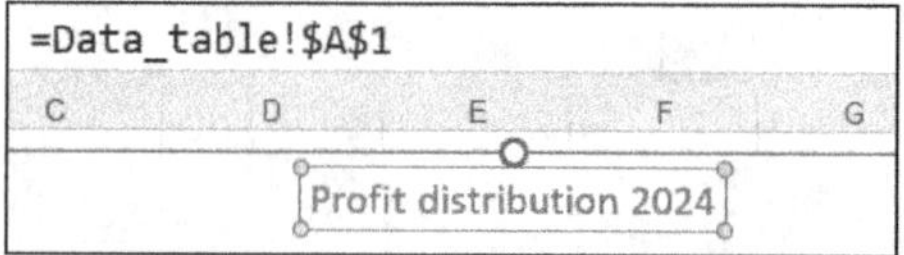

Figure 7.50 Using Cell's Content as Title

Though this method may seem a bit complex, it lets you create dynamic labels. For instance, if you generate a monthly report chart regularly, you can automatically include the month name in the chart title.

Adding a Chart Subtitle for Professional Presentations

Unfortunately, the default text options still lack a chart subtitle (often called a footnote), which is essential for professional presentations. This is typically where you include source information, sometimes including notes placed below the chart. You can insert a custom text box by using the **Text Box** button, which is found in the **Insert • Text** group.

Fill and Borders for Labels

Each label has a multipage task pane, which is easily accessible via the text's context menu—for example, you can use **Format Axis Title** for an axis title or **Format Chart Title** for a chart title. Under **Title Options • Fill,** you'll find most of the same options as for the chart or plot area borders. Under **Text Options**, you can customize **Text Fill, Text Outline, Shadow,** and **3-D Format** to style the labels as you like.

7.6.6 Axis Formatting

The axes of a chart are not just its vertical and horizontal boundaries; they also convey key information. Along with titles that clearly and concisely describe what the axes represent, they include a scale that reflects proportions and labels that indicate the meaning of each tick mark. You can adjust all these settings in the **Format Axis** task pane, which has multiple tabs. Access it by selecting the axis and clicking **Format Selection** in the **Current Selection** group, using the axis's context menu, or simply double-clicking or double-tapping the axis.

The **Alignment** group is key for axis labels. Select **Axis Options** and then the **Size & Properties** icon. You can use the same slanted text method from tables in charts, and this often helps with longer labels. Excel first tries to display labels horizontally, and if they overlap, it shows them slanted or vertically.

You can set a value under **Custom angle** to slant the label, and you can see the effect immediately while the chart stays visible. The **Text direction** dropdown list also offers two angles with which you can align text vertically.

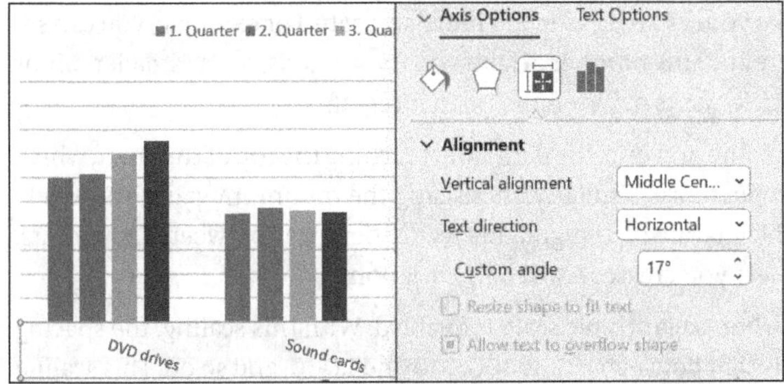

Figure 7.51 Setting Slanted Headings

In the **Axis Options** group, which shows the **Axis Options** icon, you'll find the **Automatically select based on data** option. As Figure 7.52 shows, this option also works with multilevel hierarchies; here, the year and quarter columns are simply included in the data range the chart uses.

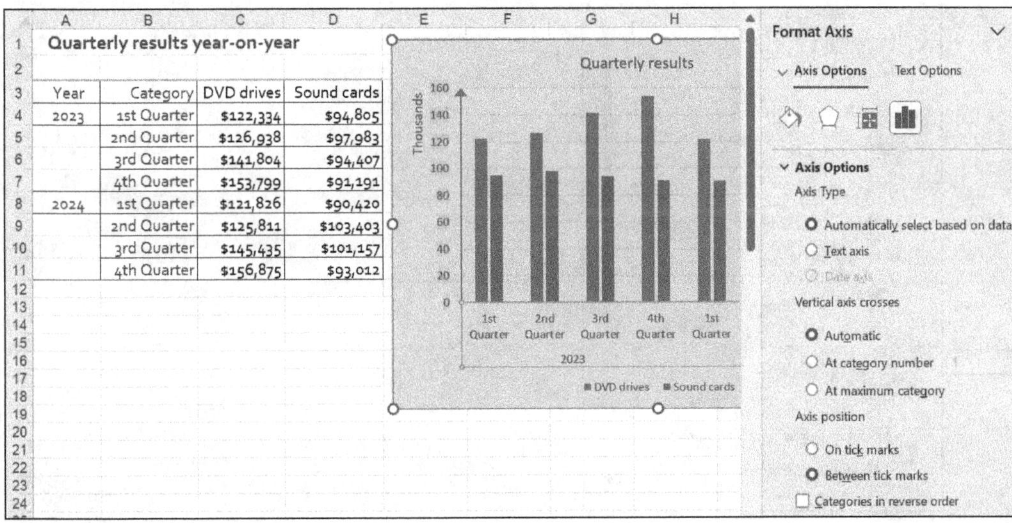

Figure 7.52 Axis Options Page

If the axis entries are date or time values, Excel defaults to a **Date axis** and automatically fills in missing points to keep spacing even. To prevent this, you can choose the **Text axis** option instead. You can also select options for positioning major and minor tick marks under **Tick Marks**. The **Categories in reverse order** option lets you flip the entry sequence, which can be helpful for bar charts.

Axis Scaling

For numeric axes, the **Axis Options** page mainly provides settings you can use to scale the axis. If it's a value axis (both axes in xy charts or just one in others), this page lets you control the number of tick marks and labels individually.

Proper scaling can make your chart clearer and more accurate. For example, you can set larger major intervals, reduce the number of labels on the size axis, or use smaller minor intervals to add more tick marks between the main intervals.

Raising the maximum value prevents the data from reaching the top of the chart, which usually improves its appearance. Similarly, increasing the minimum value effectively trims the bottom of the chart, which highlights size differences between data points. Adjusting axis scaling lets you create more chart variations.

This is especially true when **Logarithmic scale** is enabled. With this scaling, the spacing between 1 and 10 equals that between 10 and 100, 100 and 1,000, and so on. This scaling is common in scientific charts.

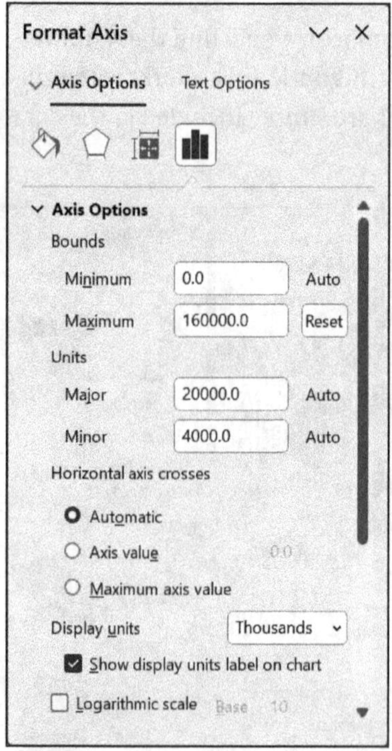

Figure 7.53 Scaling Axis

Axis Lines

The **Format Axis** pane offers many design options for customizing axes on the **Fill & Line** and **Effects** tabs. We won't cover all of them here since most are self-explanatory. For example, you can display the value axis as an arrow line, like in Figure 7.54. To do this, choose the arrowhead shape under **End Arrow type** and the size under **End Arrow size** in the **Line** group.

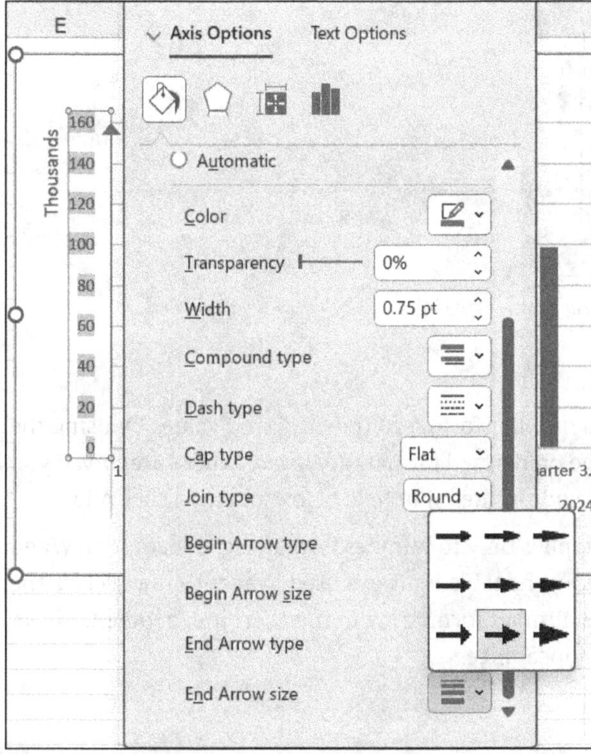

Figure 7.54 Displaying Value Axis as Arrow

Choosing the Unit of Measurement

If your table contains very large numbers, you can "shrink" the values shown in the axis labels. To do this, use the **Axis Options** icon, select a display unit from the **Axis Options** dropdown list under **Display units,** and choose the appropriate option.

For example, if you want to show only thousands or millions, select the corresponding option here. You can also check the box to show the label, and you should make sure the numbers are formatted correctly, as explained in the next section. Also, turn off the currency format here; otherwise, $200,000 might display as just $200.

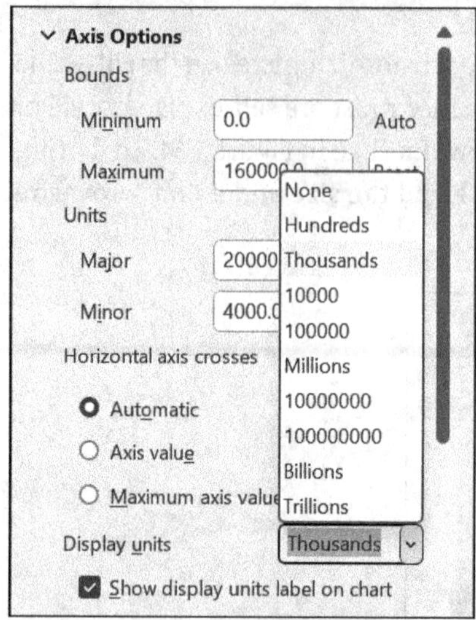

Figure 7.55 Choosing Display Unit on Numeric Axis

Number Formats in the Chart

For a value axis, you can set the number format of the displayed values by using the options in the **Number** group. The predefined formats are listed under **Category**. If you need a variation, edit the format code in the input field below and then click **Add**.

The **Linked to source** checkbox applies only to number formatting, not content. When it's checked, the data's format from the table is always used; when it's unchecked, the format isn't. You can also define number formatting in the chart independently from the table to improve the chart's appearance.

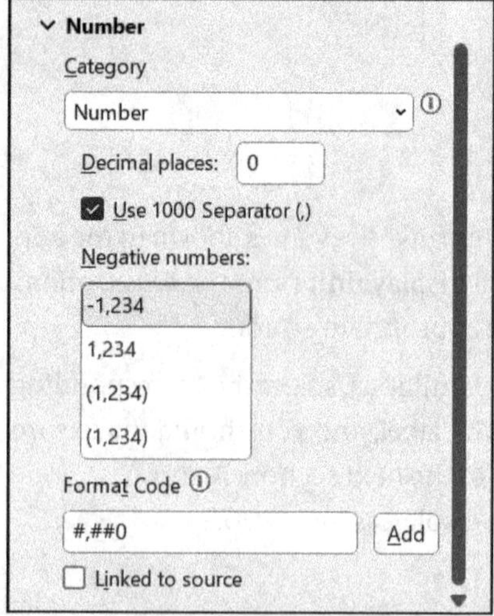

Figure 7.56 Formatting Numbers on Axes

7.6.7 Improving Readability with Gridlines

Gridlines are directly linked to the axes, where the tick marks on one or both axes appear as solid lines on the chart background. This type of display makes the chart look a bit busier but allows for the more precise reading of values from the graphical representation. All these settings are directly linked to the scaling options. The graph of the function y = x ^ 2 shown in Figure 7.57 is suitable only for illustrative purposes, without gridlines.

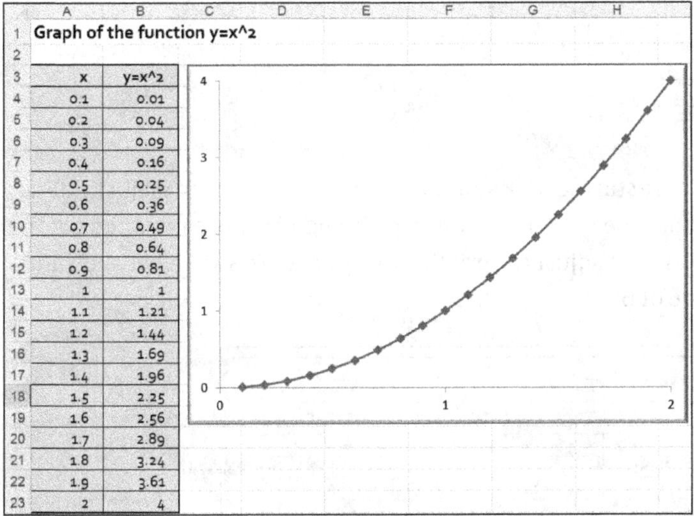

Figure 7.57 xy Chart of Function Without Gridlines

However, you can easily format this chart to look like it was drawn on graph paper:

1. After creating the chart, go to the **Chart Design** register, then in the **Chart Layouts** group, click the **Add Chart Element** button and select **Gridlines.** It opens a palette where you can choose major and minor gridlines for both the horizontal and vertical axes. You can find the same option by clicking the **+** button next to the chart and selecting the gridlines in the submenu.

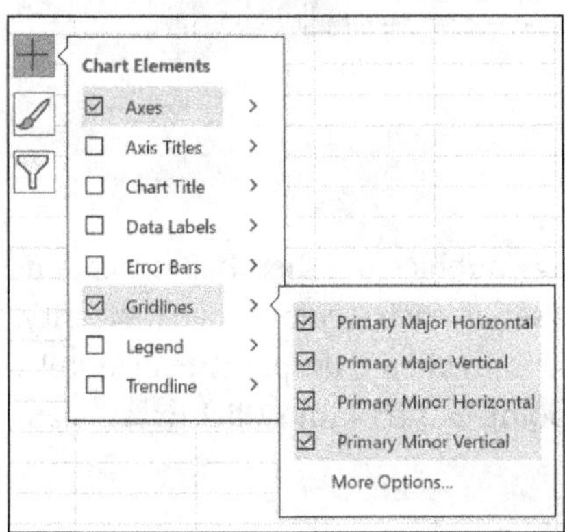

In the **Format Axis** task pane, the y-axis **Maximum** is set to 4 and the x-axis **Maximum** is set to 2. The **Major** tick marks on both axes are set to 1, and the **Minor** tick marks are set to 0.1.

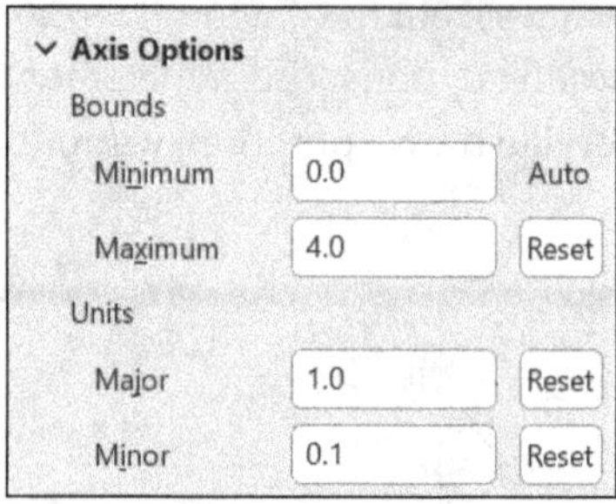

2. If the default gridlines are still too dark and clutter the chart, open the major gridlines formatting pane and use the **Fill & Line** icon to pick a lighter line color or adjust the **Weight** setting. You can also adjust the gridlines as needed. To switch between lines, just click or tap on one of them.

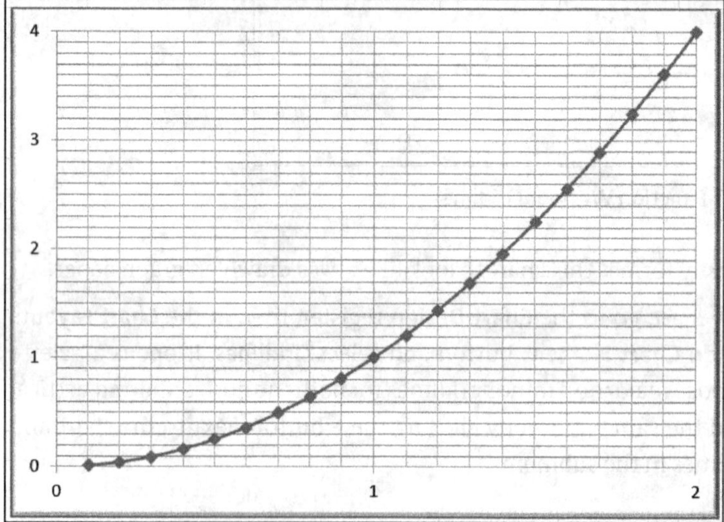

Figure 7.58 Chart on Graph Paper

7.6.8 Formatting Data Series and Data Points

Data series appear as bars, columns, areas, lines, rings, bubbles, or other shapes, depending on the chart type. When there are multiple data series, each is color coded for clarity. Excel lets you format each data series, and even individual data points, separately. Select a data series by clicking or tapping any data point, like a column. Click or tap again to select a single data point.

[+]

Selecting Objects by Using the Keyboard

You can use the keyboard to select chart objects. This is especially handy for charts with many small elements. You use the arrow keys to switch between data series and data points. Use the [↑] and [↓] keys to switch between data series and the [→] and [←] keys to move between individual data points within a series.

Instead of selecting chart elements directly, you can use the dropdown list in the **Current Selection** group or the mini toolbar to open a context menu. After selecting a data series, you can select the **Format Selection** command or its context menu option to open the **Format Data Series** task pane.

Then, via **Fill**, you can change the coloring of the data series. There, you can explore the available options by clicking the **Fill Effects** button in the **Format • Shape Styles** group. For example, use the **Gradient fill** option ❶ to create multistep gradients where you can adjust the colors, direction, and angles. Click or tap the displayed color bar or use the **Gradient stops** button ❷ to insert movable stop markers. For each selected stop marker, you can set the color individually by using the controls below ❸.

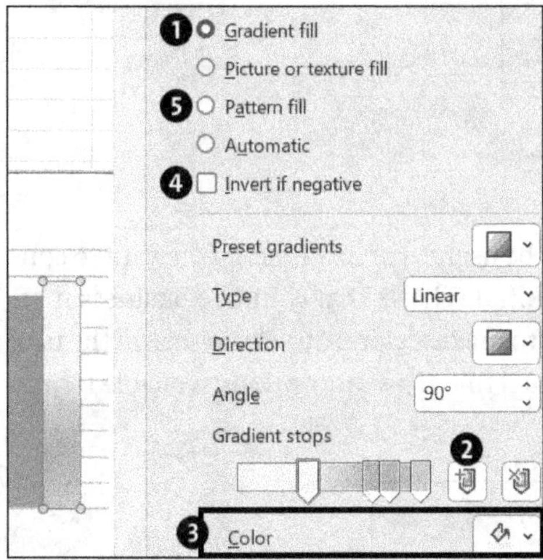

Figure 7.59 Formatting Data Series with Gradients

You can also select a special option for negative values in the chart: **Invert if negative** ❹. This swaps the foreground and background colors for negative values to highlight them.

For line, grid, and XY charts, the **Line** tab offers many options for line color and thickness; under **Marker**, you can choose the shape and color of data points. You can also format individual data series with unique color patterns to consistently assign the same color to a specific product group across different charts.

[+]

If You Don't Plan to Print in Color

It's not very helpful if charts look good on-screen but the columns are barely distinguishable in black-and-white print. In such cases, you can assign different patterns to the data series by using the **Fill** tab and the **Pattern fill** option ❺. You can choose both **Foreground** and **Background** colors.

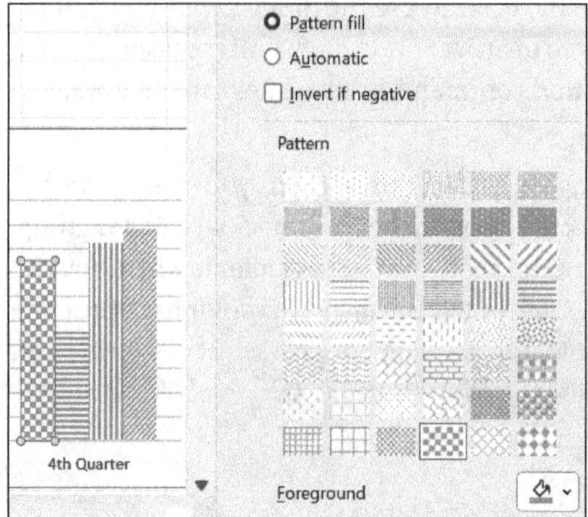

Figure 7.60 Palette for Pattern Fill

Highlight Individual Values

Formatting a single data point separately can be useful. For example, if you want to consistently highlight your own company in multiple charts that compare business metrics among different firms, you can use color to make it stand out. To do this, in the task pane, select **Format Data Point.** Under **Fill,** you'll find the same options available for the data series.

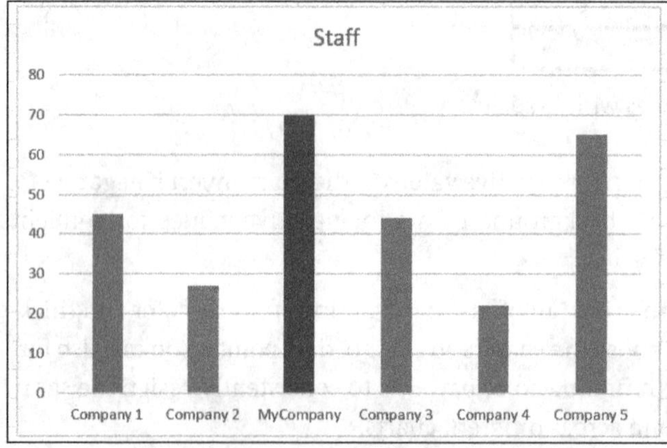

Figure 7.61 Highlighting Data Point

Effects

For data points and series, the **Effects** icon offers many options, including **Shadow, Glow, Soft Edges,** and **3D Format**. For example, the **Glow** settings can create a glowing effect behind a bar, while the **Soft Edges** settings smooth the edges of the objects.

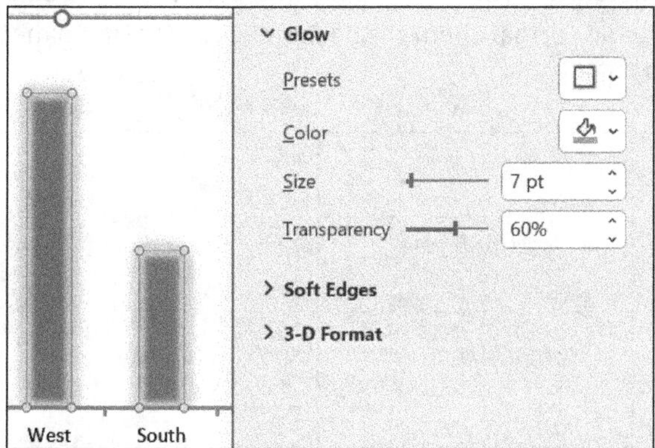

Figure 7.62 Glowing Bar

7.6.9 Data Labels

In pie charts, values are often shown directly on or near each slice. These elements are called *data labels*. They're available in many chart types, but they often overlap, so they need to be placed a bit farther from the data they describe. The solution is usually to use connector lines linking the data labels to their data points.

You can quickly add data labels by using the **+** button. When you check **Data Labels** on the menu, different options appear under the small triangle, depending on the chart type. You'll also find **More Options** here, and it opens the detailed **Format Data Labels** pane. The four icons let you access groups for **Fill & Border**, **Effects**, **Size & Alignment**, and special **Label Options**, including number formatting.

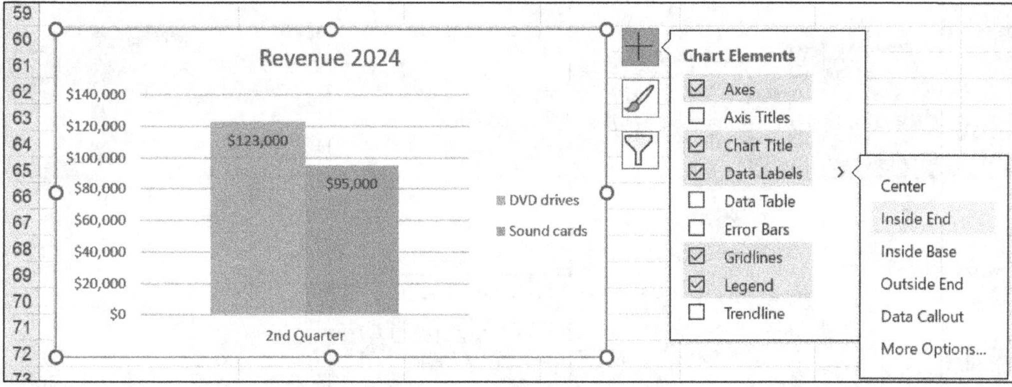

Figure 7.63 Inserting Data Labels

Under **Label Options**, you first choose the content of the labels, and besides the raw number, you can also add the category name or data series name. If **Show Leader Lines** are enabled, they connect the label to the data point when you move the label from its default position.

When you drag labels to a new location, the leader lines adjust automatically. You can select these leader lines in the chart and format them in the **Format Leader Line** pane under **Line**, using any of the available line styles.

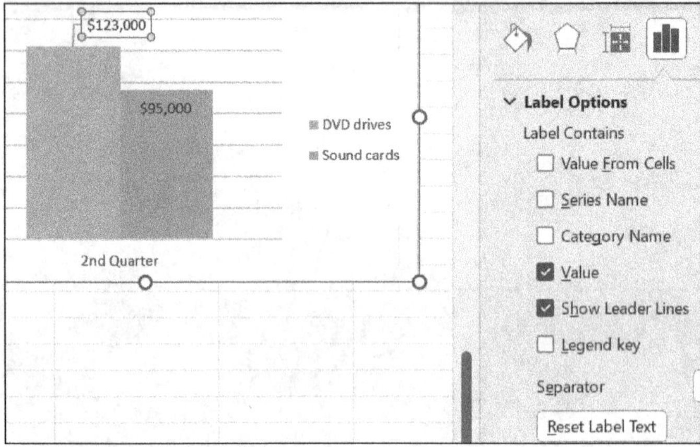

Figure 7.64 Label Options

You can replace the label's shape with others from the shapes gallery, which Excel offers under **Insert • Illustrations • Shapes**. To do this, open the context menu on a label shape, select **Change Data Label Shapes,** and pick the shape you want from the gallery. Excel will update the shape for all labels in the chart.

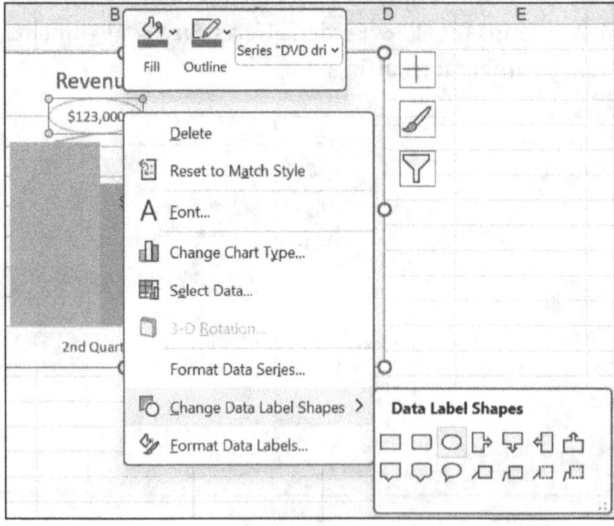

Figure 7.65 Changing Label Shape

You can also add an extra label to a specific data point to highlight or comment on it. Double-click the label so only it is selected, and then, from the context menu, choose **Insert Data Label Field**. Select one of the available options, and if you choose **Cell**, you can select a cell containing the label text. This text will be added to the existing label.

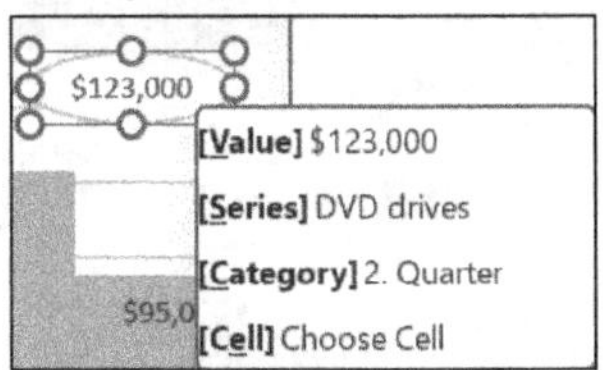

Figure 7.66 Expanding Single Label

Fortunately, Excel preserves the effort you put into designing labels, even if you decide to change the chart type again. The program tries to retain your existing settings as much as possible when switching to a new chart type.

7.6.10 Trend Calculation

For data that suggests a mathematical relationship, Excel can add a trend calculation to a chart. In the chart shown in Figure 7.67, it's clear that the data follows a trend.

Once the chart is active and you've selected a data series, open the **Trendline** palette via **Chart Design • Chart Layouts • Add Chart Element**. Choose the appropriate option, and Excel adds the corresponding line to the chart; selecting **None** removes any existing trendline.

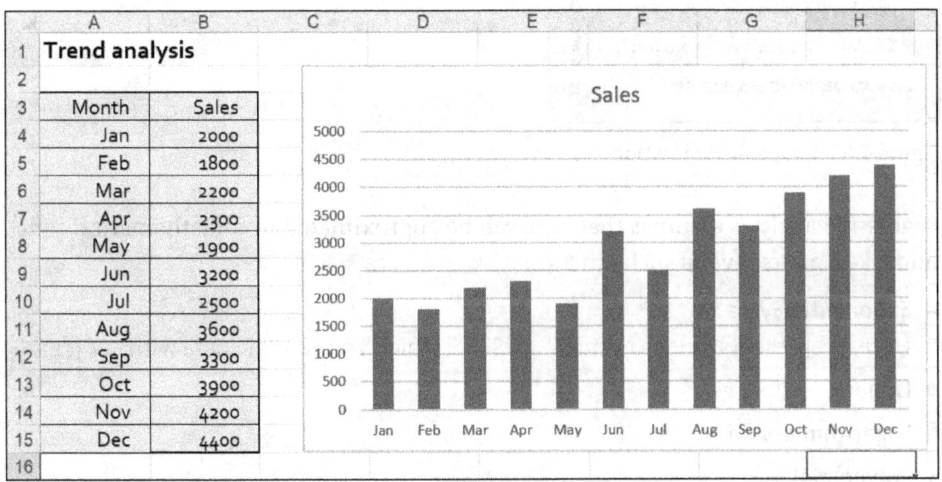

Figure 7.67 Using Table and Chart as Starting Point for Trend Analysis

You get access to more options by opening the **Format Trendline** pane through **More Trendline Options**, which you can also use later to modify a selected trendline. In addition to the formatting options under **Line**, you'll find six regression types under **Trendline Options**.

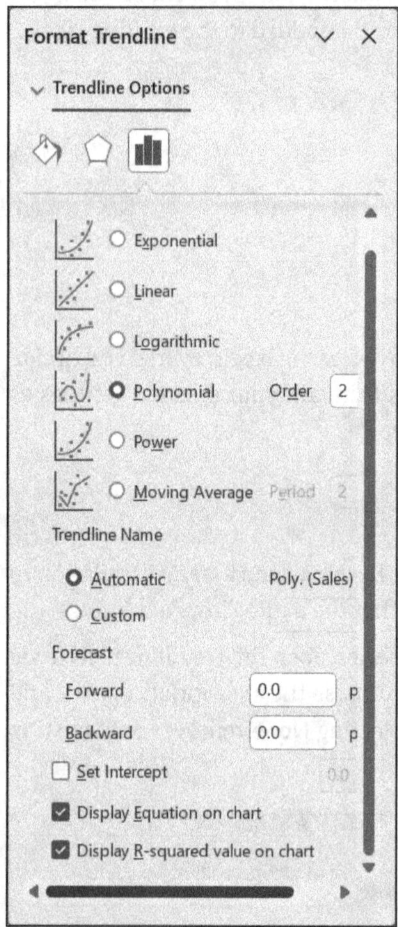

Figure 7.68 Choosing Regression Type

Regression analysis assumes that data can be approximated by a mathematical function. Excel offers several such functions:

- **Exponential:** `y = a * e ^(b * x)`
  Here, e represents Euler's number 2.71828..., which you can calculate with `=EXP(1)`.
- **Linear:** `y = m * x + b`
- **Logarithmic:** `y = a * ln(x) + b`
- **Polynomial:** `y = a * x ^ 2 + b * x + c`
  With this option, you can set the starting power of x under **Order:** 2 for `x ^ 2` and 3 for `x ^ 3`, up to a maximum of 6. Increasing the **Order** value generally improves the trendline fit but reduces its mathematical significance.
- **Power:** `y = a * x^b`

- **Moving Average**
  The trend value is calculated by averaging adjacent values. You can adjust the number of adjacent values used via the **Period** setting, where higher numbers result in a more accurate fit.

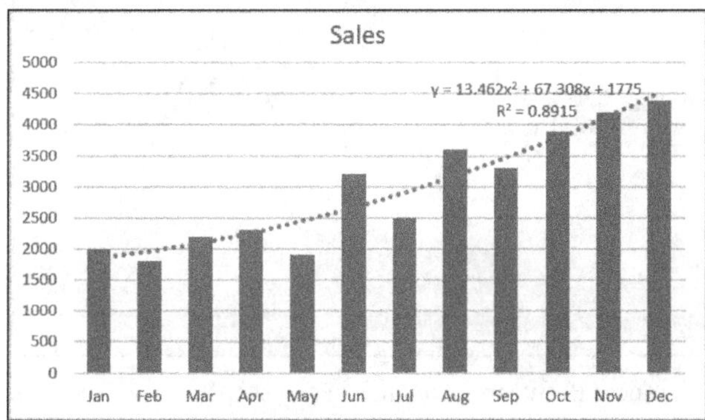

Figure 7.69 Regression—Third-Degree Polynomial

You can also assign a custom name to the trendline in the task pane under **Custom.** You can specify how many data points Excel should consider, either forward or backward, and you can also display the equation that Excel uses and the coefficient of determination on the chart. You can find information about the coefficient of determination in the RSQ(), LINEST() and LOGEST() statistical table functions. This lets you perform complex statistical analyses without using any statistical functions yourself.

7.6.11 Drop Lines, High/Low Lines, and Up/Down Bars

Line charts let you add elements that are unavailable in other chart types: drop lines, range lines, and error bars. Drop lines are also available in area charts. For example, consider a table with two columns showing monthly profits from last year and this year. When the line chart is selected, you can add vertical lines connecting each data point to the x-axis by going to **Chart Design • Add Chart Element • Lines • Drop Lines.** These lines make it easier to read the values.

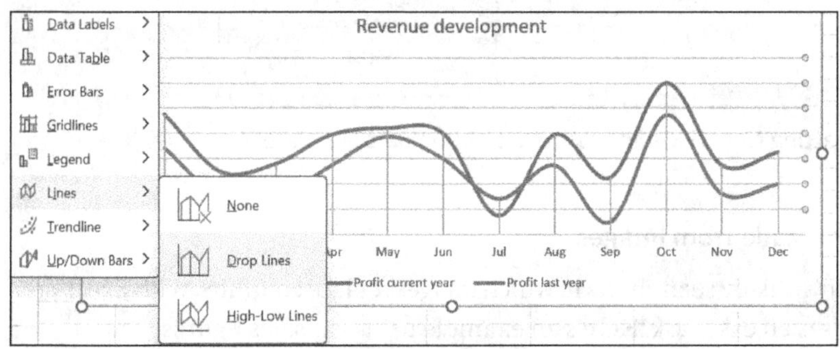

Figure 7.70 Adding Reference Lines to Line Chart

Alternatively, you can use high/low lines here, which show the differences in monthly results, as Figure 7.71 illustrates.

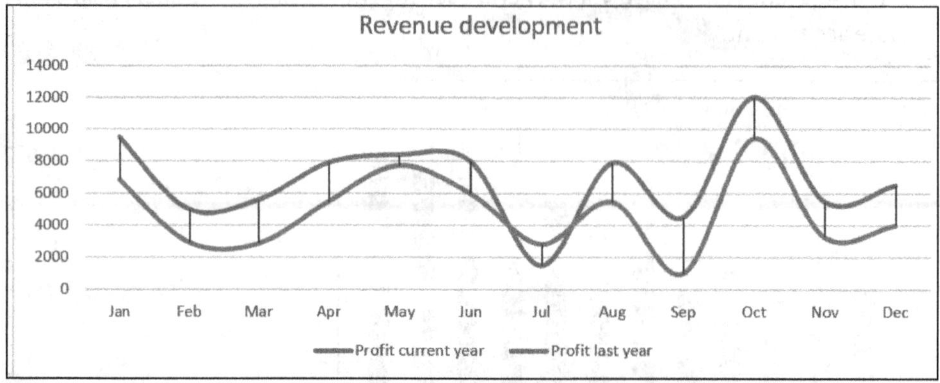

Figure 7.71 Range Lines Display Differences Between Two Values Each Month

Instead of range lines, you can also add bars to the chart to highlight negative or positive deviations. Go to **Chart Design • Add Chart Element • Up/Down Bars • Up/Down Bars**. This adds small, color-coded bars at each data point, showing the difference between the second data column's values (the current year's profits) and the first column's values (last year's profits).

The default colors are black and white, but you can easily change them in the task pane under **Down Bar Options** and **Up Bar Options**.

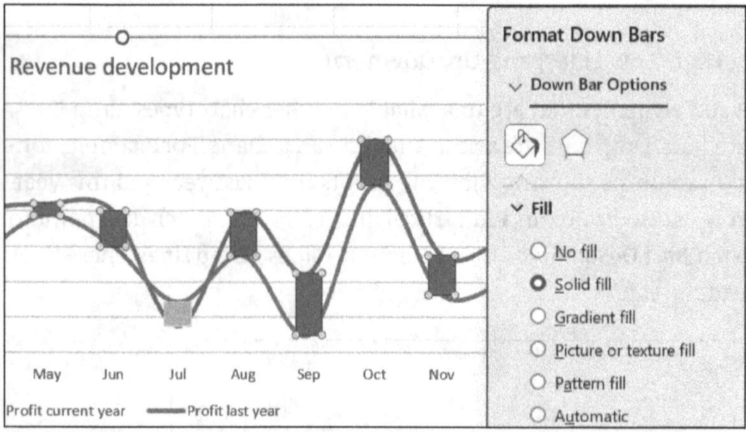

Figure 7.72 Formatting Negative and Positive Deviation Bars Separately

7.6.12 Columns Made from Images

In the media, you've likely seen charts in which images replace columns. With Excel, creating such charts is an easy task. Here's an example using car sales figures:

1. Start by creating a column chart from the table and formatting it with all needed labels and details.
2. Usually, it's best to reduce the **Gap Width** between columns (or bars) under **Format Data Series** in **Series Options** to make the elements wider.
3. In the **Fill** group, select the **Picture or texture fill** option. This activates the feature. Click the **Insert** button, select the image file you want, and the image will appear in the chart.
4. You now have several options for how to display the image in the column.
   - **Stretch** resizes the image to fill the column.
   - **Stack** repeats the image until the column is filled, keeping the original size.
   - **Stack and Scale with** lets you set, under **Units/Picture**, how often the image repeats per unit. If necessary, the image will be resized accordingly.

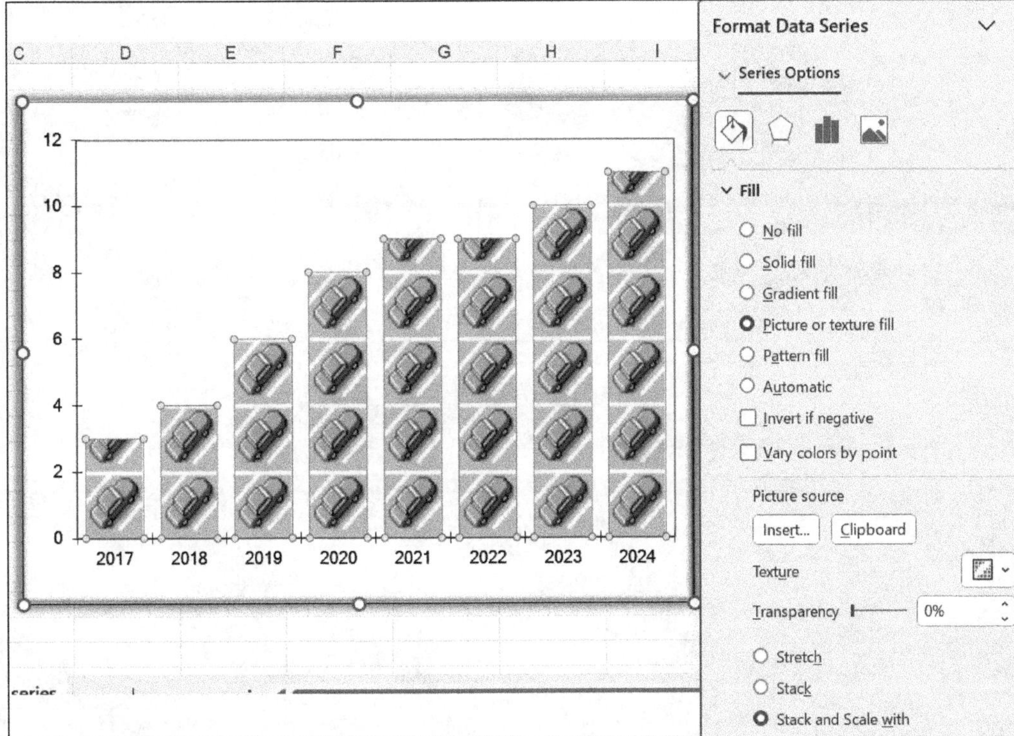

Figure 7.73 Chart with Inserted Graphic

Chapter 8
Using Charts Effectively

When converting a table into a chart, you need to decide which chart type and subtype best fit your purpose. This chapter will help you make that choice. Using examples, it offers an overview of the many options and shows which chart types suit different kinds of data best. It also highlights specific design features of various chart types.

8.1 Criteria for Choosing a Chart Type

Here are some key points to consider when selecting a chart:

- A chart should be simpler than the table it represents. Overloading it with unnecessary, ornate details defeats this purpose.
- The chart should accurately reflect the data. Avoid scaling tricks that exaggerate or obscure differences, although in rare cases, you'll need to clearly disclose them.
- Within an overall presentation, use the same chart types for identical data to maintain clarity.

Unfortunately, these rules are often ignored. It starts with overusing 3D effects and ends with many charts being so distorted in scale that they only show that something is increasing or decreasing.

8.2 Standard Charts

Column and bar charts are the two most common chart types in the media. Pie charts are used to show how a given amount of something is distributed among people or other entities. For example, pie charts that show how corporate budgets or taxpayer dollars are allocated show "how the cake is cut." Line charts are mainly used to analyze large amounts of data.

8.2.1 Column Charts

Column charts are widely used because each column is seen as a distinct, independent element. Few people try to infer values between two adjacent columns based on their height—they represent only the exact data from the underlying table. Since the horizontal axis is traditionally used as a time axis, it naturally makes it easy to use column charts show developments in a broad sense.

However, the impact of a column chart on viewers largely depends on the scaling, as the following example clearly demonstrates. This chart shows the debt progression of a (fictional) country over time. Using a logarithmic scale, the debt increase appears far less dramatic.

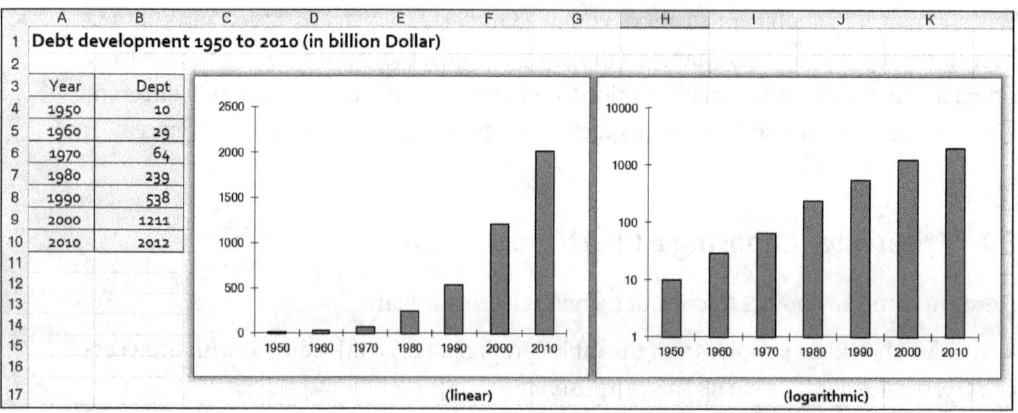

Figure 8.1 Debt Progression, Shown with More and Less Dramatic Differences

Column Options (Side by Side or Stacked)

For multiple data series, you can display columns either side by side or stacked. These are called stacked or clustered columns. There are seven subtypes of such columns (see Figure 8.2). Subtype 1, *clustered columns*, has each column in a group in a single color. Subtype 2, *stacked columns*, has the columns stacked on top of each other when multiple data series are present. This is useful when the total shown actually matters, like the total debt across different debt types.

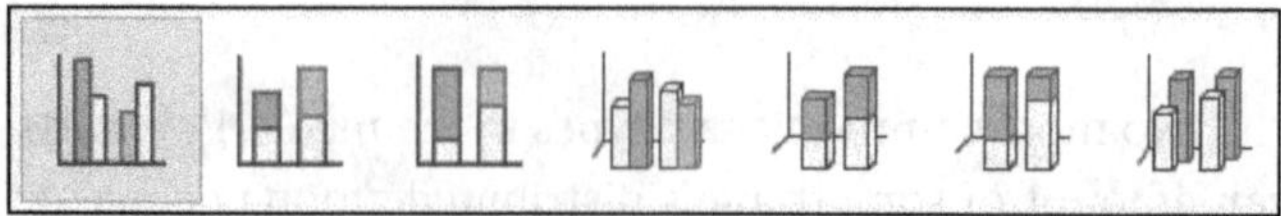

Figure 8.2 Subtypes of Column Charts

Subtype 3 also stacks columns but converts the data into percentage shares. This chart type is helpful when you want to show proportional breakdowns—which are normally displayed as separate pie charts—across multiple time periods, regions, or similar categories. For example, you can use the results of German federal elections. This data would require twelve charts if displayed as pie charts; as a ring chart, it would be completely confusing.

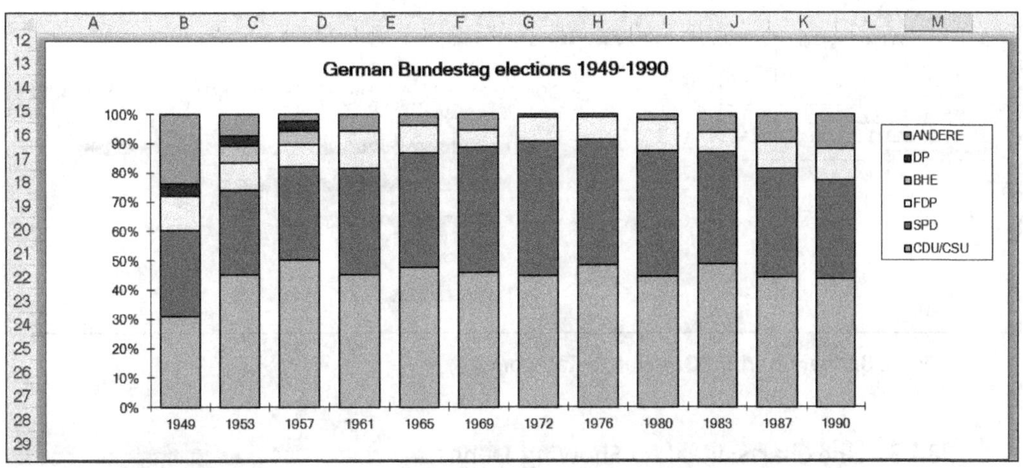

Figure 8.3 German Election Results—Column Charts Show Proportions of Votes

Subtypes 4 through 6 match subtypes 1 through 3, except that they show individual columns in 3D. Only subtype 7 is a true 3D chart that uses three axes.

8.2.2 Bar Charts: Ideal for Long Category Labels

Bar charts are structured like column charts, except on the category axis. In bar charts, the vertical axis and the size axis are arranged horizontally. Essentially, any data that can be shown in a column chart can also be displayed in a bar chart, but you should generally avoid using a bar chart when the category axis represents a time sequence. This conflicts with the common association of the horizontal axis with time progression.

On the other hand, bar charts are ideal when the categories represent qualitative features. While you could use a column chart in these cases, there's often a strong reason to choose a bar chart: Qualitative category labels tend to be long and don't fit well in a column chart. The bar chart subtypes are almost identical to those of column charts, except the category axis is vertical and the value axis is horizontal (see Figure 8.4).

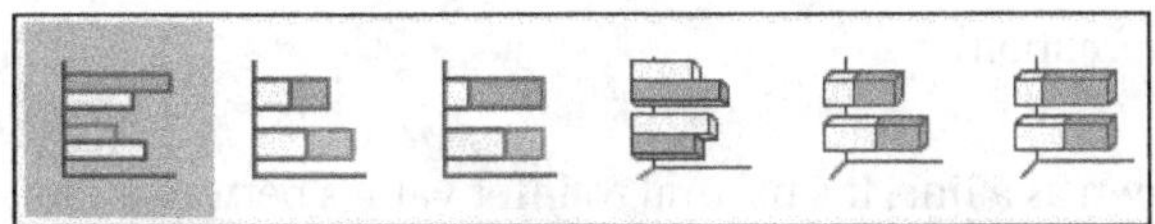

Figure 8.4 Bar Chart Subtypes

Figure 8.5 is a typical example of a bar chart. It shows the increase in nominal wages in Germany in 2014 compared 2013, broken down by position in the professional/managerial hierarchy. As with all qualitative categories, it's best to order them by a clear criterion, and in this case, ordering by rank in the professional/managerial hierarchy works well. If the categories don't suggest a natural order, sort them by size.

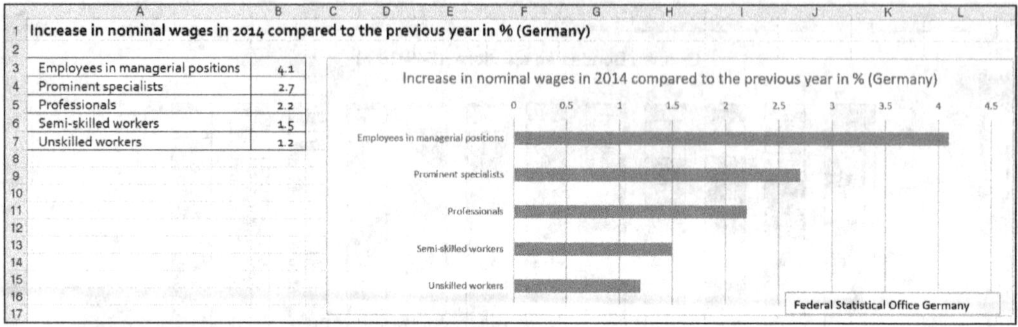

Increase in nominal wages in 2014 compared to the previous year in % (Germany)

Employees in managerial positions	4.1
Prominent specialists	2.7
Professionals	2.2
Semi-skilled workers	1.5
Unskilled workers	1.2

Figure 8.5 Bar Chart for Qualitative Categories

8.2.3 Line Charts: Ideal for Showing Trends

Charts that display data as a line within a rectangular coordinate system are the original form of charts. Line charts have the major advantage of fitting in many more data points than do chart types that use space-consuming elements like bars or columns for each point. The downside of line charts is that individual data points don't stand out as clearly as they do in bar or column charts. For example, here's a chart showing energy consumption trends among different energy sources.

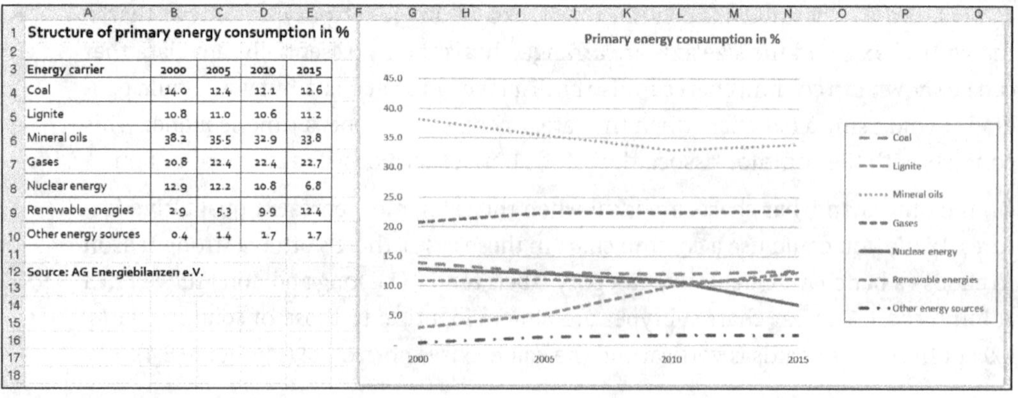

Structure of primary energy consumption in %

Energy carrier	2000	2005	2010	2015
Coal	14.0	12.4	12.1	12.6
Lignite	10.8	11.0	10.6	11.2
Mineral oils	38.2	35.5	32.9	33.8
Gases	20.8	22.4	22.4	22.7
Nuclear energy	12.9	12.2	10.8	6.8
Renewable energies	2.9	5.3	9.9	12.4
Other energy sources	0.4	1.4	1.7	1.7

Source: AG Energiebilanzen e.V.

Figure 8.6 Line Chart of Energy Consumption

In all chart types where data is shown as a line, it's natural to infer values between data points (interpolate), meaning data that isn't actually in the source table. So, line charts should only be used when this inference is appropriate and justified. Also, line charts work best when the (horizontal) category axis represents time or another quantitative progression. If the category axis contains purely qualitative data, a line chart is not suitable.

Line Chart Subtypes

Subtype 1 shows lines without data points. Subtypes 2 and 3 are stacked or 100% stacked lines, like stacked columns. Subtypes 4 to 6 correspond to types 1 to 3 but also display data points. Subtype 7 displays bands instead of lines, creating a 3D effect.

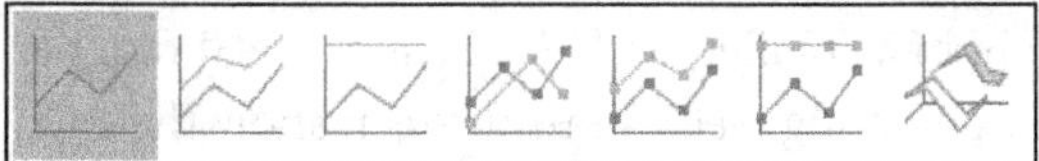

Figure 8.7 Line Chart Variations

8.2.4 Pie Charts: For Showing Proportions

Pie or doughnut charts are the standard charts for data that show the composition of a whole. They don't require a coordinate grid, and they differ from previously mentioned charts by showing only proportions. The total area of the circle always represents 100%, and the area of each sector reflects the percentage share of that sector's component. The chart in Figure 8.8 displays the percentage shares of separately collected recyclable materials.

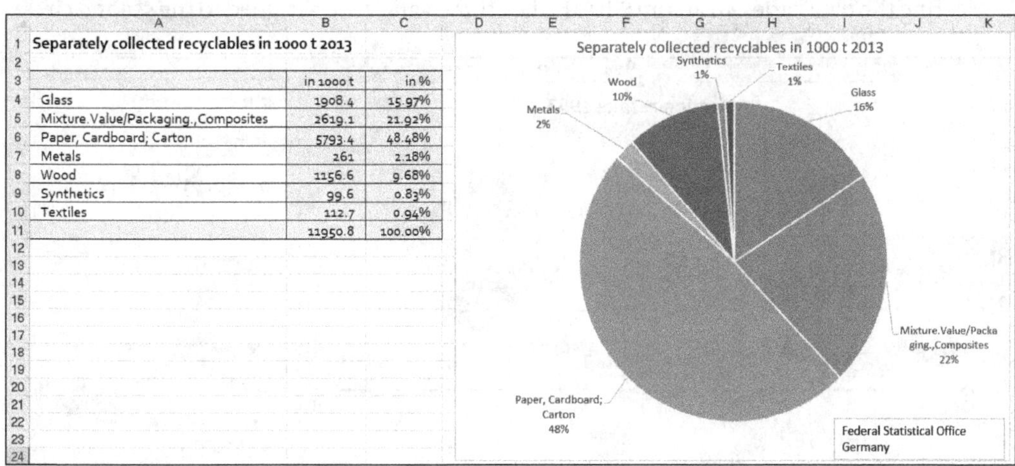

Separately collected recyclables in 1000 t 2013

	in 1000 t	in %
Glass	1908.4	15.97%
Mixture.Value/Packaging.,Composites	2619.1	21.92%
Paper, Cardboard; Carton	5793.4	48.48%
Metals	261	2.18%
Wood	1156.6	9.68%
Synthetics	99.6	0.83%
Textiles	112.7	0.94%
	11950.8	100.00%

Figure 8.8 Displaying Proportions in Pie Chart

When proportions can't be meaningfully ordered, as in stacked bar or column charts, a pie chart is the most effective way to visualize the data. As in all charts, too many categories in a bar chart reduce its readability. If you need to show multiple data series, you can place several pie charts side by side. An alternative is the doughnut chart (which we describe later).

Pie Chart Subtypes

There are five subtypes of pie charts (see Figure 8.9). In subtypes 1 and 2, the pie segments stay together, while in subtypes 3 and 4, they are separated. Subtype 2 adds only a 3D effect, while subtypes 3 and 4 are meant for cases where, alongside some larger segments, several smaller segments need to be shown. These small parts can be combined into one segment or displayed separately in a smaller circle (as in subtype 3) or a bar block. This way, for example, a pie chart depicting election results can show the total share of votes won by small parties while also revealing the distribution within that group.

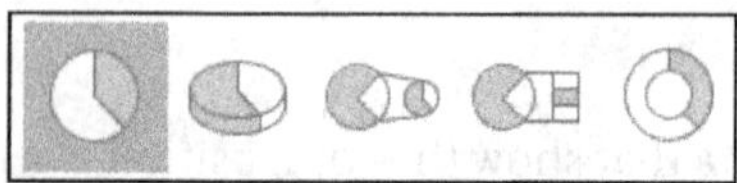

Figure 8.9 Pie Chart Subtypes

On the **Series Options** page, you can choose which parts appear in the second circle of subtype 3. To do this, select **Split Series By**, choose **Position**, and enter the number under **Values in second plot**. Alternatively, you can split by a percentage. Set this when formatting the data series. In Figure 8.10, the last three values are assigned to the second circle.

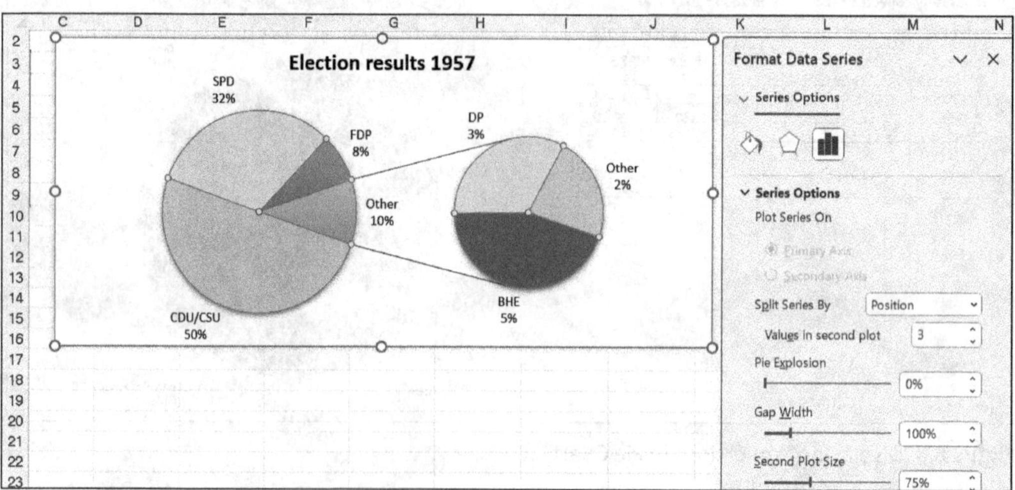

Figure 8.10 Example of Subtype 3

All values below the selected percentage are placed in the second circle. You can adjust the size of this second circle smoothly by using the slider. Subtype 5 is a doughnut chart, which was a separate category in older versions.

Labels in Pie Charts

Whether a pie chart is effective largely depends on the type of data labels, and you can set them in the task pane under **Data Labels.** Several formatting options are available under **Label Options.**

In the top group, you define the label's content. You can choose to display the **Value From Cells**, data **Series Name**, **Category Name**, or **Percentage** for each pie segment. After setting a data label, you can also add a **Legend key**—a small color pattern—next to it. You can also show leader lines that connect the label to its pie segment, which helps clarify small segments. Under **Label Position**, you can specify where the label should appear.

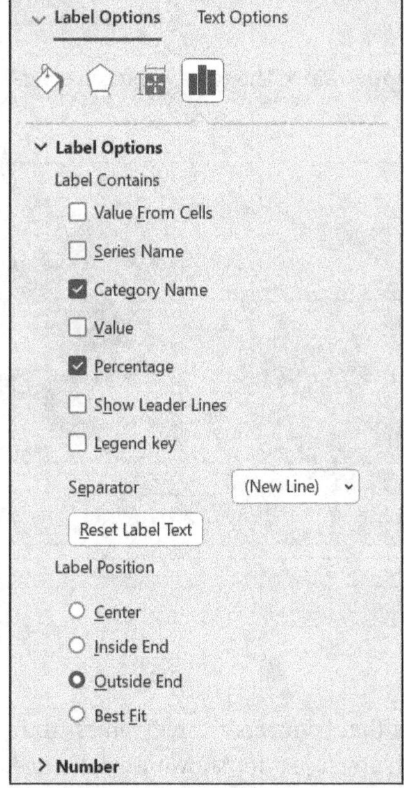

Figure 8.11 Options for Data Labels in Pie Charts

8.3 Value Differentiation with Area and Range Charts

Area and range charts are effective for illustrating changes in share distributions or value differences.

8.3.1 Area Charts

An area chart closely resembles a line chart, except that the spaces between lines are filled with color. This chart type is found under the **Area** category. Subtype 1 overlaps areas, subtype 2 stacks them, and subtype 3 stacks areas that always total 100%. Subtype 4 is a true 3D area chart that requires three axes, while subtypes 5 and 6 simply add a 3D effect. When the category axis shows a time or spatial sequence, an area chart works well for visualization.

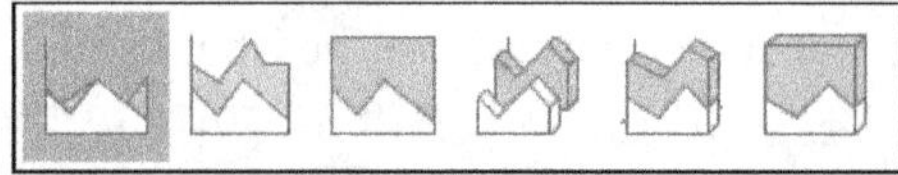

Figure 8.12 Area Chart Subtypes

Figure 8.13 shows soil composition at different depths. Since the table values are percentages, the areas always total 100%.

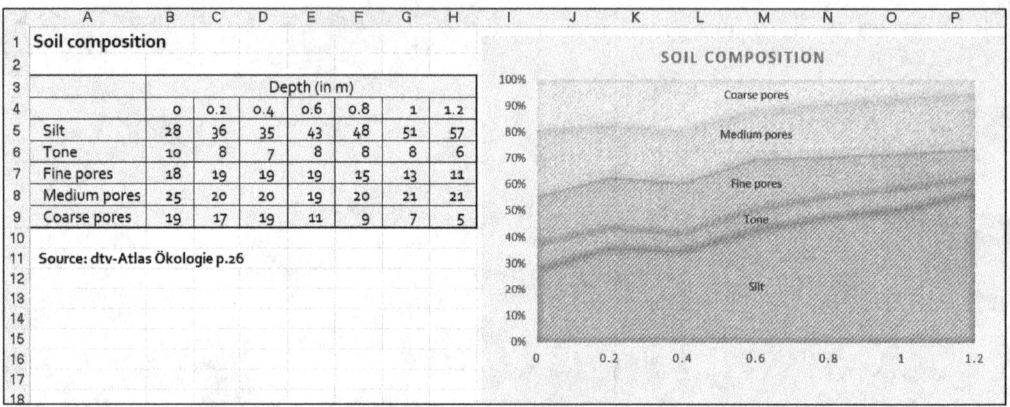

Soil composition

	Depth (in m)						
	0	0.2	0.4	0.6	0.8	1	1.2
Silt	28	36	35	43	48	51	57
Tone	10	8	7	8	8	8	6
Fine pores	18	19	19	19	15	13	11
Medium pores	25	20	20	19	20	21	21
Coarse pores	19	17	19	11	9	7	5

Source: dtv-Atlas Ökologie p.26

Figure 8.13 Visualizing Soil Composition with Area Chart

8.3.2 Range Charts: Not Just for Stocks

Under the name **Stocks**, Excel offers a chart type that connects related values from multiple data series. These are range charts, which aren't just for showing stock price trends. The following example of a week-long temperature recording demonstrates this, with daily maximum, minimum, and average temperatures.

However, this chart type is commonly used for stock charts. In this case, these range charts display three to five of the following values:

- Opening price, meaning the stock price at market open
- Highest value, which is the highest price reached within a given time period (e.g., a day)

- Lowest value
- Closing price, meaning the price at market close
- Volume, meaning the number of shares traded

Make sure the columns are arranged in the order that matches the selected type.

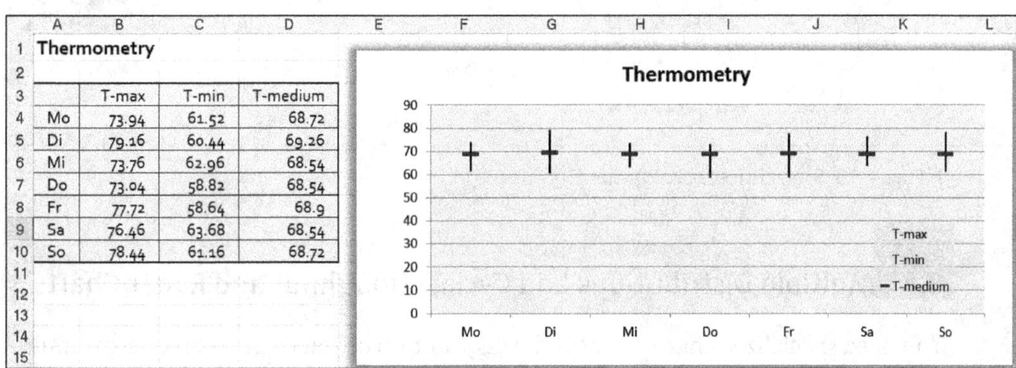

Thermometry

	T-max	T-min	T-medium
Mo	73.94	61.52	68.72
Di	79.16	60.44	69.26
Mi	73.76	62.96	68.54
Do	73.04	58.82	68.54
Fr	77.72	58.64	68.9
Sa	76.46	63.68	68.54
So	78.44	61.16	68.72

Figure 8.14 Range Chart for Temperature Fluctuations

Subtypes for Three to Five Data Series

Range chart subtypes differ depending on whether they require three, four, or five data series. Subtype 1 shows the high, low, and closing prices, and subtype 2 adds the opening price. In subtype 3, the first data series shows trading volume instead of the opening price. Subtype 4—**Volume-Open-High-Low-Closing**—requires five data series.

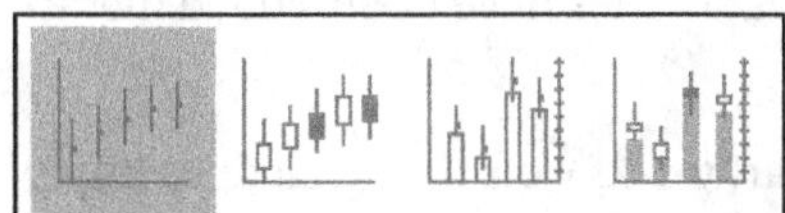

Figure 8.15 Range Chart Subtypes

Data series must be entered in the order specified by the type name. In subtypes 2 and 4, the opening and closing values are combined into a rectangle, which appears white if the first value is lower than the second (indicating a price increase for stocks). If the closing value is lower, the rectangle is filled in with a dark color. High and low values appear as vertical lines above and below the rectangles, clearly showing all four values.

One drawback of range charts is that individual data points can be hard to see. You can select the error bars in the chart and use the task pane to format the **High-Low Lines**, and you can adjust their thickness and color under **Line**. You can also select the deviation bars in the chart and format them differently in the task pane under **Up-Bars** and the **Low-Bars**.

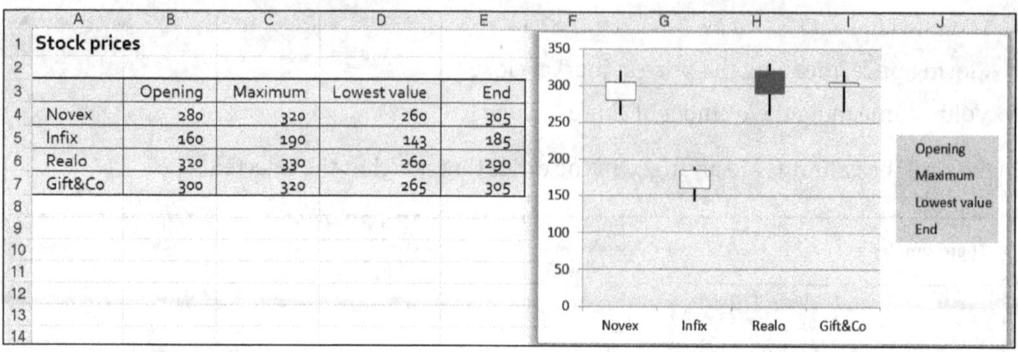

Stock prices

	Opening	Maximum	Lowest value	End
Novex	280	320	260	305
Infix	160	190	143	185
Realo	320	330	260	290
Gift&Co	300	320	265	305

Figure 8.16 Stock Price Development

8.4 Multiple Distributions and Cycles: Doughnut and Radar Charts

You need specialized charts to show the distribution of values in a whole across multiple cases, not just one. The same applies to displaying values that change in a specific sequence.

8.4.1 Doughnut Charts for Comparing Data Groups

A doughnut chart combines several pie charts, nesting them inside each other rather than placing them side by side. For multiple data groups shown as proportions, Excel offers several options: stacked columns, bars, and stacked area charts. Using multiple pie charts is also a good approach, especially when the data series have no intended hierarchy.

If there aren't too many data groups, a doughnut chart works well, though it does imply a hierarchy. When arranging circles concentrically, make sure any chronological order is clear, with earlier data groups inside and later or more important ones outside. In the legend, this chart type provides a label for each ring segment. A ring chart isn't a separate chart type but a subtype of the pie chart. The example in Figure 8.17 displays election results from three German Bundestag elections in one chart. The most recent result appears in the outermost ring.

Often, it's helpful to reduce the size of the doughnut hole on the **Data Series Options** page under **Format Data Series**. The size ranges from a minimum of 10% to a maximum of 90%, and shrinking the hole frees up more space to display data. You can also rotate the ring to any position by adjusting the angle of the first segment. A value greater than 0 for **Ring explosion** adds spacing between the rings.

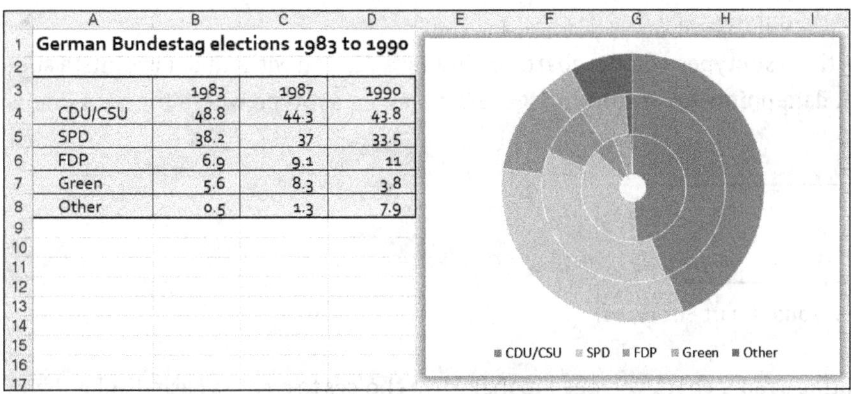

German Bundestag elections 1983 to 1990
1983
1987
1990
CDU/CSU 48.8 44.3 43.8
SPD 38.2 37 33.5
FDP 6.9 9.1 11
Green 5.6 8.3 3.8
Other 0.5 1.3 7.9
CDU/CSU SPD FDP Green Other

Figure 8.17 Ring Chart Combining Multiple Pie Charts

8.4.2 Radar Charts for Cycles

Radar charts most closely resemble polar coordinate charts, but with three key differences:

- The angle isn't set directly; it's determined by the number of values.
- The direction or order follows a clockwise rotation (counterclockwise in polar coordinates).
- The size axes can include negative values.

Radar charts are ideal for displaying cyclical processes or data. Examples include brightness by angle, temperature or precipitation throughout a day or year, and a plant's growth rate over an annual cycle. The (fictional) example in Figure 8.18 shows the average daily temperature in a country over the twelve months of the year.

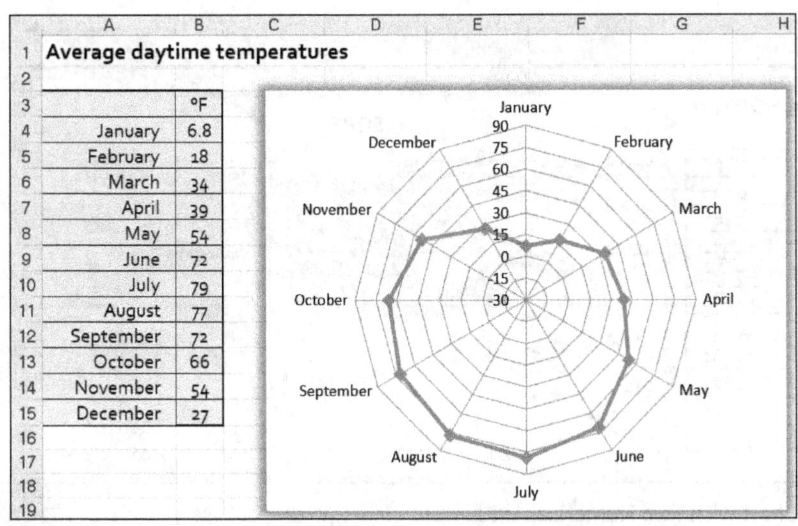

Figure 8.18 Radar Chart for Displaying Temperature Cycles

Radar Chart Subtypes

There are three subtypes of radar charts, including a version without data points, a subtype with data points (as shown in Figure 8.19), and a subtype where the data line is filled.

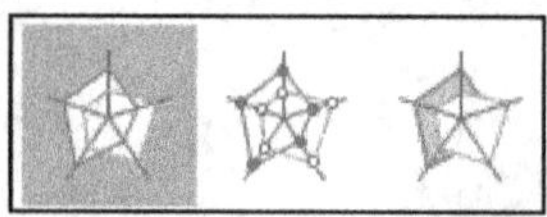

Figure 8.19 Radar Chart Subtypes

When editing a radar chart, you need to make two choices to create a clear display. First, choose data lines and points that are easy to see. Second, use the standard axis formatting options:

- Careful scaling ensures all values are easy to read.
- Selecting appropriate intervals prevents axis labels from overlapping the data.

This chart type can handle multiple data series, but with more than two or three, the chart becomes cluttered.

8.5 Value Relationships: Scatter and Bubble Charts

You'll want to use scatter charts or xy charts whenever you analyze values across two numerical axes. You'll find many examples of xy charts in math books (such as function graphs) and scientific publications (such as charts showing measurement data). Here's a demonstration graph of the function:

```
y=SQRT(x)
```

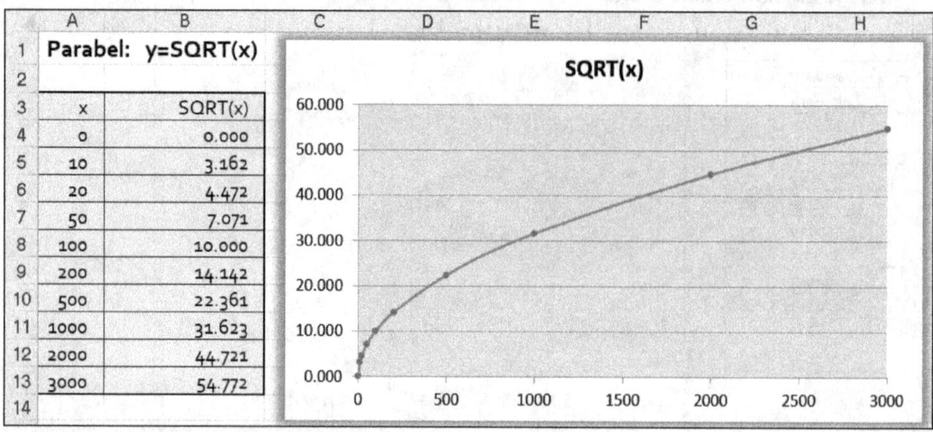

Parabel: y=SQRT(x)

x	SQRT(x)
0	0.000
10	3.162
20	4.472
50	7.071
100	10.000
200	14.142
500	22.361
1000	31.623
2000	44.721
3000	54.772

Figure 8.20 xy Chart with Two Numerical Axes

This example keeps the data points to show that the spacing between X values varies. This is necessary to prevent the curve from becoming too inaccurate in the range of small X values. To create the same curve with a line chart, you'd need to enter many X values into a table at equal intervals (0, 2, 4, ..., 800). The xy chart only resembles a line chart on the surface; the second numerical axis is the key difference.

8.5.1 A Rather Limited Selection of Subtypes

Although the **X Y (Scatter)** chart—with its two numerical axes—is essential for displaying scientific data, Excel still offers surprisingly few subtypes of it. 8

Subtype 1 shows only the data points. Subtype 2 adds a connecting curve based on interpolated values, and subtype 3 also draws a curve from interpolated values but omits the data points. These two subtypes are particularly useful for function graphs because you don't need to generate countless values to get a clear representation. In subtypes 4 and 5, the data points are simply connected by straight lines—and one shows the data points while the other doesn't.

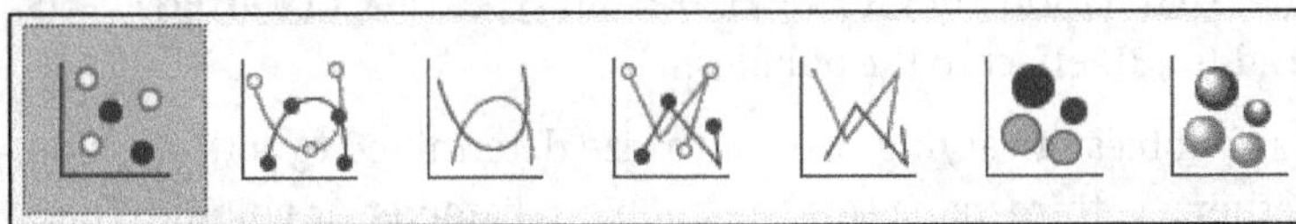

Figure 8.21 Subtypes of xy Chart

8.5.2 Editing a Chart Type

To make up for the limited variety of chart types, it's especially helpful to create your own chart variants and save them as templates. Pay special attention to subtypes that include only lines (curved or straight solid ones):

- Charts with gridlines on one or both axes
- Charts with logarithmic scaling on one or both axes, with or without gridlines
- Charts with two identical-looking data series for functions with positive and negative values

To display the SQRT(X) graph correctly, you need to create two data series. When you select subtype 3, the curve for negative values appears in a different color. To create a uniform curve across the entire graph, manually apply the same line format to both data series by using the **Fill & Line** button for a consistent look.

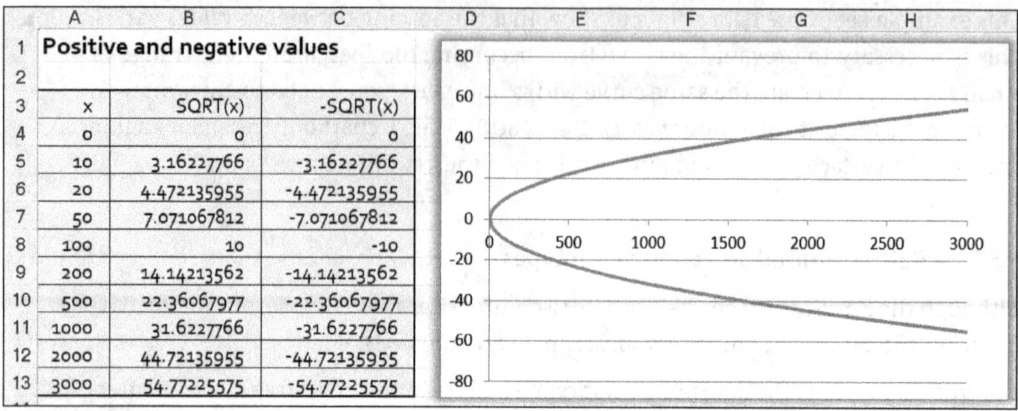

Positive and negative values

x	SQRT(x)	-SQRT(x)
0	0	0
10	3.16227766	-3.16227766
20	4.472135955	-4.472135955
50	7.071067812	-7.071067812
100	10	-10
200	14.14213562	-14.14213562
500	22.36067977	-22.36067977
1000	31.6227766	-31.6227766
2000	44.72135955	-44.72135955
3000	54.77225575	-54.77225575

Figure 8.22 Function Graph with Positive and Negative Values

8.5.3 Bubble Charts as a Type of XY Chart

Bubble charts, like xy charts, can display multiple numerical data series. In this case, there are three, and for that reason, they are offered as subtypes of **X Y (Scatter)** charts. The second subtype adds a 3D effect to the bubbles.

Data points appear as bubbles of varying sizes, with size determined by a third value. You can choose whether the third value sets the bubble's diameter or volume. These charts work best with a relatively small number of data points; otherwise, the message becomes unclear. For example, the following chart shows selected products with prices on the x-axis and sales on the y-axis. At the same time, the bubble size should represent the market share of the products.

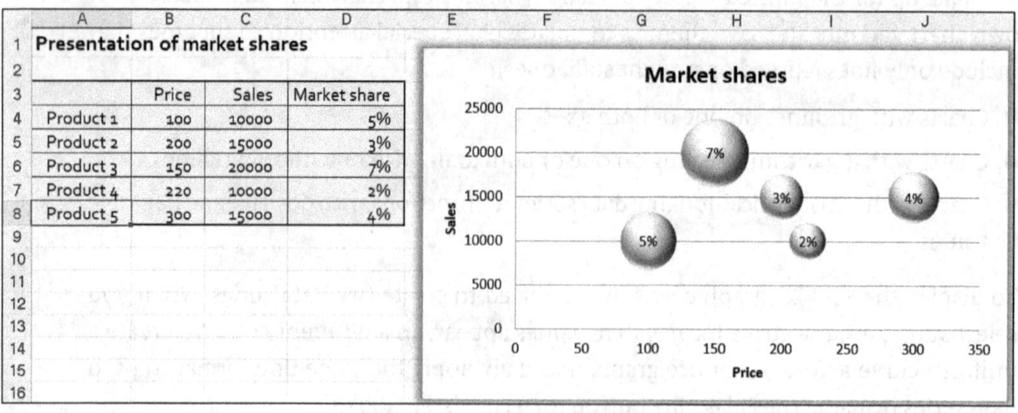

Presentation of market shares

	Price	Sales	Market share
Product 1	100	10000	5%
Product 2	200	15000	3%
Product 3	150	20000	7%
Product 4	220	10000	2%
Product 5	300	15000	4%

Figure 8.23 Market Shares in Bubble Chart with Centered Data Labels

To create the chart, select the three value columns. In the task pane,in **Format Data Series** under **Series Options,** you can choose whether the size in the table should represent the bubble area or the bubble diameter. You can also choose whether to display bubbles for negative values. For this chart type, it's often helpful to add a centered data label using the **+** button. Then, in the task pane, in **Format Data Labels** under **Label Options,** you can select **Bubble Size** as the label to display the corresponding percentage. This will make the chart's message much clearer.

8.6 Combo Charts

Often, it's helpful to combine multiple chart types in one chart to show relationships among very different data. Here's an example of combining column and line charts. Figure 8.24 shows revenues and expenses, with an additional highlighted data series for profit (revenues minus expenses) on a separate vertical axis on the right.

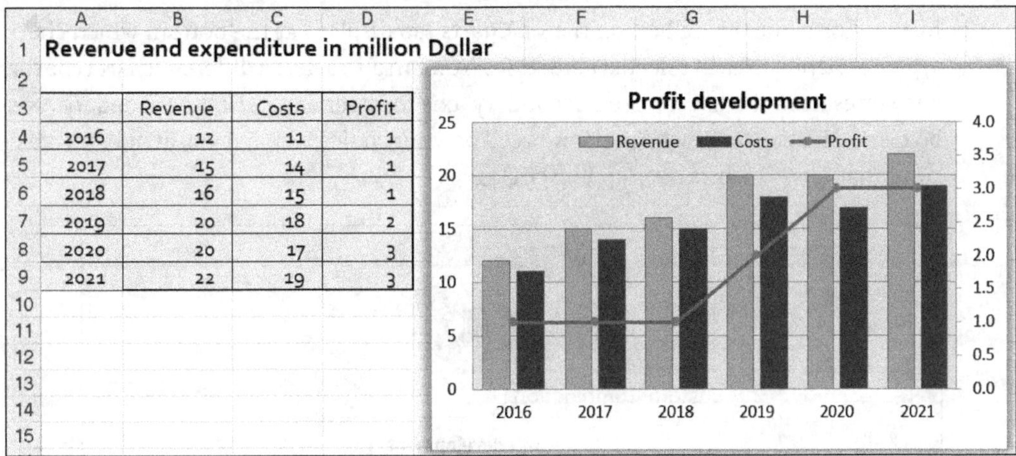

Revenue and expenditure in million Dollar

	Revenue	Costs	Profit
2016	12	11	1
2017	15	14	1
2018	16	15	1
2019	20	18	2
2020	20	17	3
2021	22	19	3

Figure 8.24 Combo Chart Highlighting Specific Data Series

8.6.1 Balancing Differences in Scale

A common challenge is comparing data of very different types or at different scales in one chart. For example, comparing manager salaries (in hundreds of thousands of dollars per year) with profits (in millions of dollars per year) isn't effective with simple chart types like a column chart, which would show a mix of very large and very small columns in this situation.

Sometimes, only one data series in a chart needs to be highlighted, even if it could be grouped with the others. One way to solve this is by combining multiple chart types in a single chart. There are many ways to combine line, column, area, bubble, and range charts.

8.6.2 Combination Types

Excel 2013 introduced **Combo** as a separate chart type, and its first three subtypes are common combinations. Subtype 1 combines clustered columns and lines, subtype 2 combines clustered columns and lines using a secondary value axis, and subtype 3 combines stacked areas and grouped columns. Subtype 4 lets you create custom combinations.

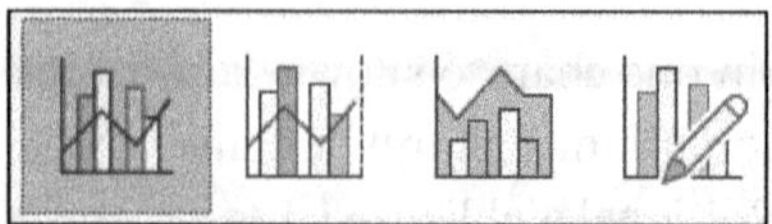

Figure 8.25 Combo Type Variants

In every case, the dialog box on the **All Charts** tab displays at the bottom which chart type is assigned to each selected data series. You can also select a different chart type for any series from the list. To use a secondary axis for a series, check the **Secondary Axis** box. Excel instantly shows a preview based on your choice, helping you avoid unusable combinations—such as overlapping circles.

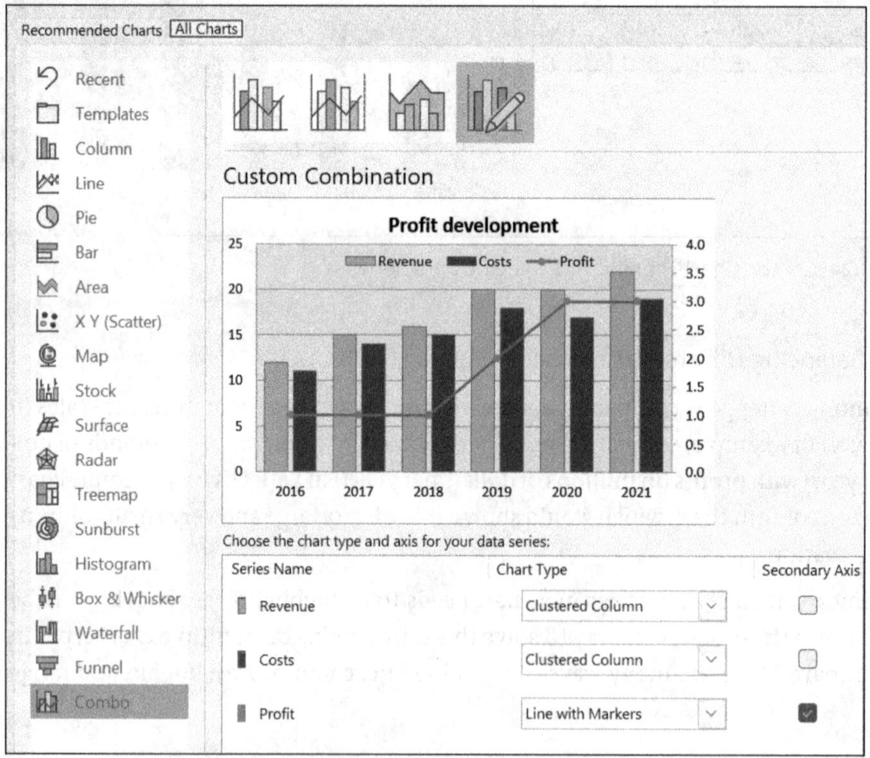

Figure 8.26 Choosing Chart Types in a Combo Chart

If you frequently need special combo charts, it's best to create them as custom types for easy reuse. After fully formatting a chart, select it and choose **Save as Template** from the context menu, then give it a clear name that identifies the chart type.

8.7 3D Effects and True 3D Charts

Turning a 2D chart into one with 3D effects can add visual interest but often causes unwanted side effects. For example, the pie chart in Figure 8.27 displays the values 1, 2, 4, and 8, each of which is twice as large as the one before, and the angles of each slice double accordingly. It's relatively easy to see the angles doubling in the pie chart on the left, but in the 3D view on the right, it's harder because of the different viewing angle.

8

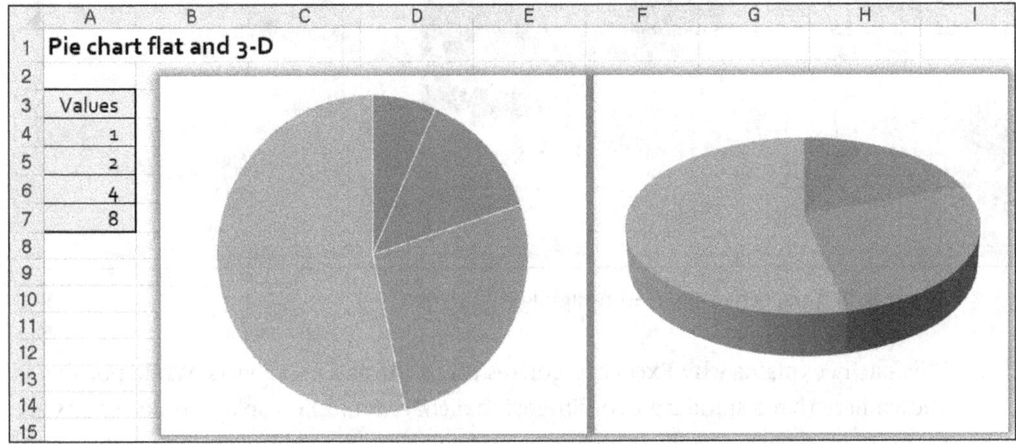

Figure 8.27 Size Ratios with and without 3D Effects

8.7.1 Real and Pseudo 3D Charts

Unfortunately, Excel gets a bit confused when it comes to the term *3D*. A truly three-dimensional chart has three dimensions: length, width, and height—or mathematically, an x-, y-, and z-axis. This only applies to some chart types that Excel labels as "3D charts." All others have 3D effects applied purely for visual appeal.

8.7.2 The Viewing Angle Is Key

All 3D charts, whether real or pseudo, allow you to freely choose the viewing angle. When you select the chart area, you can access the **3D Rotation** option from the context menu. As mentioned earlier, in the **Format Chart Area** task pane, under the **Effects** button on the **3D Rotation** page, you can adjust the axes settings to move the chart and set the viewpoint almost anywhere. On the same page, under **Depth (% of base)**, you can adjust the preset value to change the 3D depth effect.

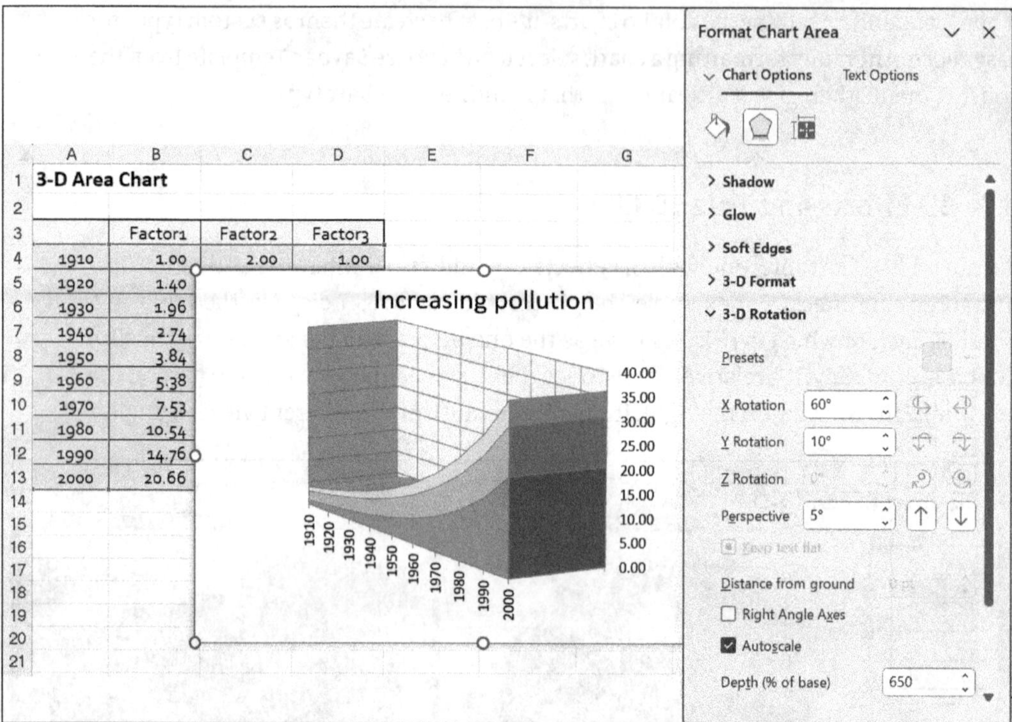

Figure 8.28 Perspective and Depth Highlight the Trend

This partly explains why Excel categorizes these charts as 3D charts. While nothing is shown here that a standard two-dimensional chart couldn't display, some aspects are easier to emphasize. To illustrate a concerning trend like rising environmental pollution, a 3D area chart with the right perspective and depth is very effective.

True 3D Charts

When data involves more than two dimensions, a true 3D chart is necessary for visualization. The example in Figure 8.29 shows life expectancy trends in different world regions and time periods. In the table range, the first row lists the time periods and the first column lists the regions. The values for the three axes are arranged in rows. Follow these steps to create a 3D chart that's shows these values:

1. Select table range A4:F8 and then choose the three-dimensional subtype of the area chart.
2. By default, Excel uses the specified time periods as labels for the x-axis and the regions as labels for the depth axes.
3. Disable the legend because it's redundant with the matching axis labels.
4. To improve readability, use the **+** button and select **Gridlines** to enable **Depth Major** gridlines, **Primary Major Horizontal** gridlines, and **Primary Major Vertical** gridlines.

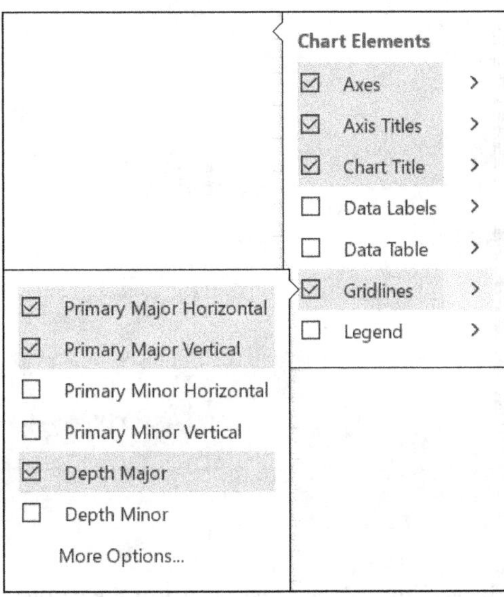

5. Create a title for the entire chart that clearly explains what it shows.

Any other 3D chart type would work here as well. Figure 8.29 also highlights that in a 3D area chart, one of the two dimensions of the chart range is clearly emphasized: All data in a series (in this example, the regions) is grouped into a single color-consistent and visually consistent block. You can do the same with a 3D line chart.

If you want to avoid this effect, a 3D column chart is a better choice because you can assign the same color to all data series.

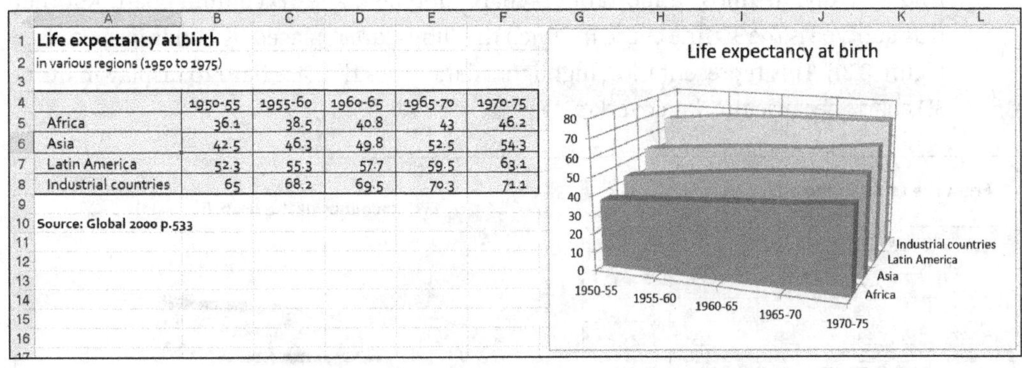
Life expectancy at birth
in various regions (1950 to 1975)

	1950-55	1955-60	1960-65	1965-70	1970-75
Africa	36.1	38.5	40.8	43	46.2
Asia	42.5	46.3	49.8	52.5	54.3
Latin America	52.3	55.3	57.7	59.5	63.1
Industrial countries	65	68.2	69.5	70.3	71.1

Source: Global 2000 p.533

Figure 8.29 Life Expectancy by Region and Period, Displayed with 3D Area Chart

8.7.3 Charts with Three Axes

Excel offers true 3D formats for column, area, and line charts, featuring three axes. You can swap the depth (series) axis and the category axis when creating or editing the chart. Each series' data is grouped by color.

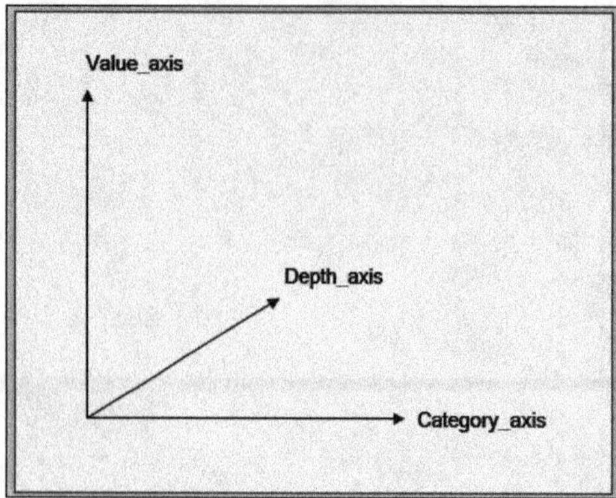

Figure 8.30 Axes in 3D Charts

8.7.4 Examples of 3D Chart Uses

Certain data sets lend themselves well to three-dimensional visualization. Examples include the following:

- Average annual income broken down by the north-south divide over the past ten years
- Administrative effort in companies by company size over recent years
- Changes in air pollution over the years, broken down by region

8.7.5 The True 3D Subtypes

Undoubtedly, the most important 3D chart subtype is the 3D column chart. Another true subtype is the 3D line chart, in which the "lines" are replaced by bands in space (see Figure 8.31). This representation highlights data series that are hard to display in other 3D charts because the data overlap.

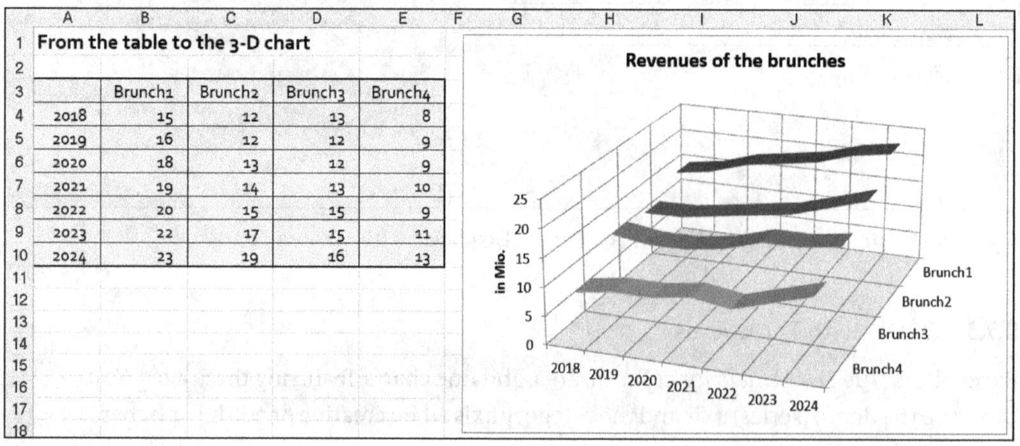

From the table to the 3-D chart

	Brunch1	Brunch2	Brunch3	Brunch4
2018	15	12	13	8
2019	16	12	12	9
2020	18	13	12	9
2021	19	14	13	10
2022	20	15	15	9
2023	22	17	15	11
2024	23	19	16	13

Figure 8.31 3D Bands in Space

8.7.6 A 3D Chart with Equally Weighted Series and Categories

As mentioned, preset 3D chart types emphasize certain dimensions more than others through coloring. You can change this by using a custom chart type for a 3D column chart that adheres to these requirements:

- The columns must have a square cross-section.
- The spacing between data series and categories must be equal.
- All columns must share the same color.

While the first two requirements are easy to meet, the last one is more challenging because the color of each column group can't be set globally, only individually. To format a 3D column chart this way, follow these steps:

1. Create a table with at least as many data series as you might need later. A table with ten to twenty columns should be sufficient, and two entries per column are enough. You can easily create the values by using the fill handle or the **AutoFill** feature.

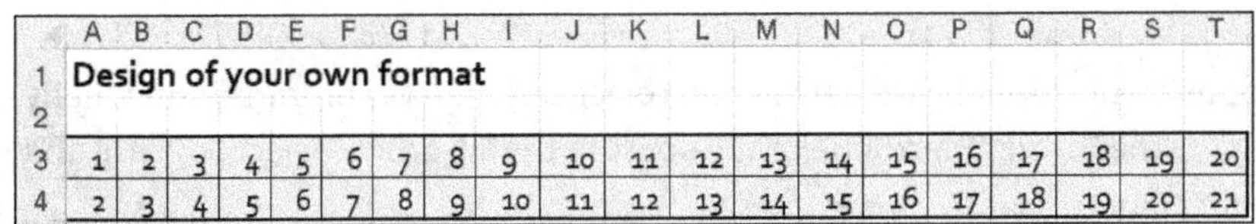

	A	B	C	D	E	F	G	H	I	J	K	L	M	N	O	P	Q	R	S	T
1	Design of your own format																			
2																				
3	1	2	3	4	5	6	7	8	9	10	11	12	13	14	15	16	17	18	19	20
4	2	3	4	5	6	7	8	9	10	11	12	13	14	15	16	17	18	19	20	21

2. Using this table, insert a 3D column chart via **Insert • Charts • Insert Column or Bar Chart.** Use the **+** button menu to delete the **Legend** selection to remove the default legend.
3. In the same menu, click **Axes** and uncheck **Depth** axis to hide its labels.
4. Under **Chart Design • Data**, click the **Switch Row/Column** button to arrange data series by columns.
5. Select each data series in turn, and in the **Format Data Series** pane under **Series Options**, set both the **Gap Depth** and **Gap Width** to exactly 100%. Gap width is the distance between two columns in a data series, while gap depth affects the cross-section of each column.

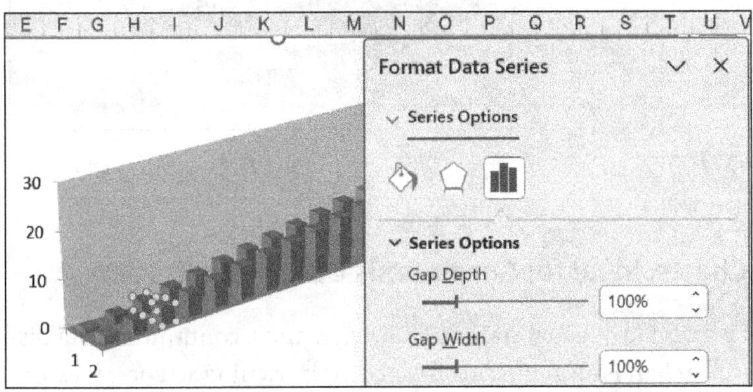

6. From the menu on the small triangle next to the heading above the icons, select **Floor**, **Back Wall**, and **Side Wall** one by one, then assign each a preset fill color using the **Line & Fill** button.

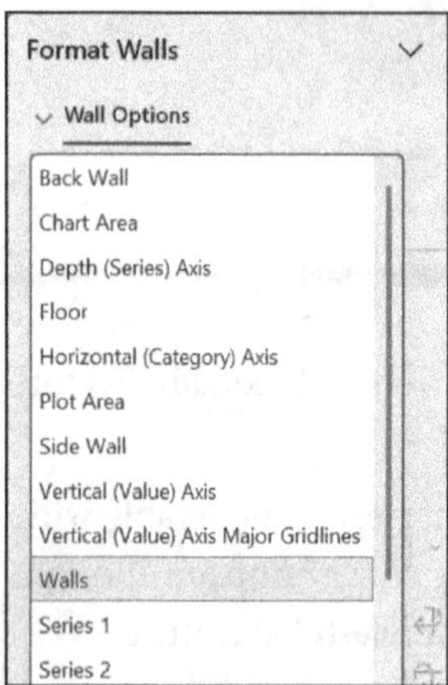

7. Assign a consistent color to each data series. Select each pair of columns individually while keeping the **Format Data Series** pane open. In the **Fill** group, select **Solid Fill** and choose your color.
8. Finally, save the chart as a custom chart type via the chart's context menu by selecting **Save as Template**.

To apply the new format, select your data and pick the custom chart type under **Templates** in the **Insert Chart** dialog. Figure 8.32 shows a small example of the resulting chart.

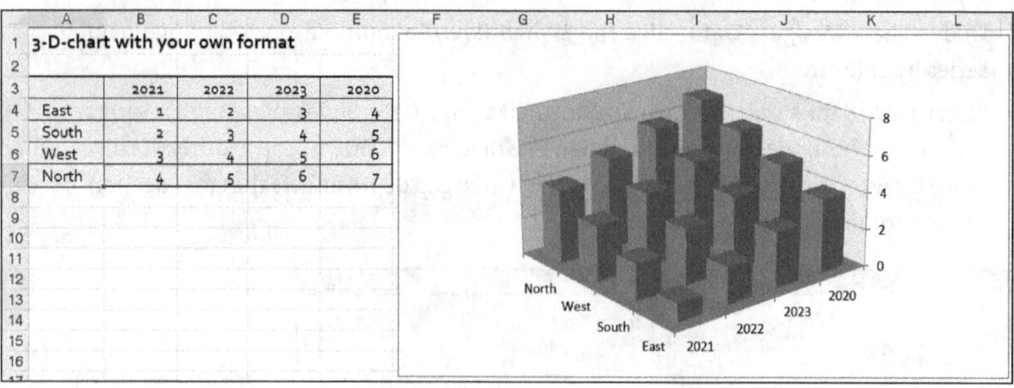

3-D-chart with your own format
2021 2022 2023 2020
East 1 2 3 4
South 2 3 4 5
West 3 4 5 6
North 4 5 6 7
North
West
South
East
2021
2022
2023
2020
0
2
4
6
8

Figure 8.32 Chart in New Format

8.8 3D Surface Charts: Ideal for Continuous Data Visualization

Whenever the x- and y-axes are of equal status in a 3D view and a continuous data display is needed, the surface chart is an almost ideal tool. This intriguing 3D chart is essentially a line chart that's extended by one axis, and it closely matches what mathematics calls a 3D chart. We'll use a graph of a well-known mathematical function as an example.

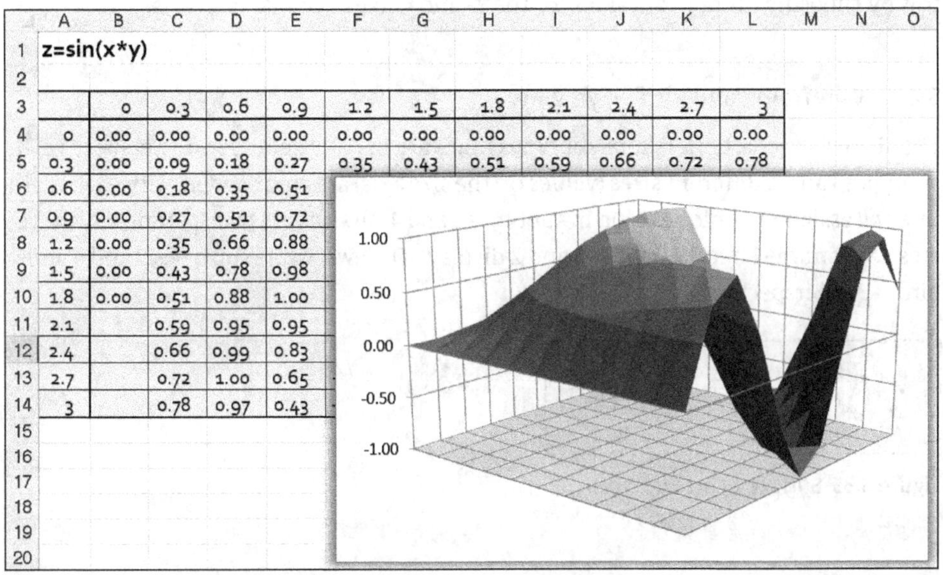

Figure 8.33 Graph of Three-Dimensional Function

This chart type, which creates mountains and valleys, is especially useful for showing geographical features. Here's a small example, though it requires some editing. For the vertical axis under **Axis Options**, set the base to intersect not at 0 but at the **Axis Value**. At the same time, enter "700" as the lowest axis value under **Minimum**. Also, hide all walls by selecting the **No Fill** option.

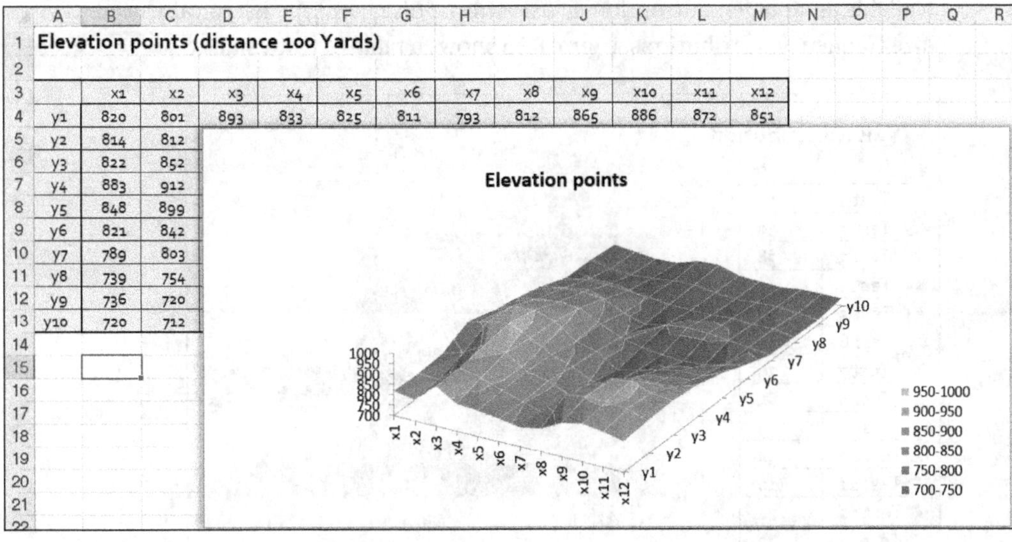

Figure 8.34 Landscape as Surface Chart

Unfortunately, the 3D surface chart, like the line chart, has only one numerical axis (the z- or size axis), while the other two axes are nonnumerical. You can usually work around this by choosing numerical values for the x- and y-axes at equal intervals.

8.8.1 Wireframe and Bird's-Eye View

The 3D surface chart offers four subtypes that vary in color scheme and perspective. In subtypes 1 and 3, different sizes (values on the size axis) are color-coded, with each main interval assigned a unique color. In subtypes 2 and 4, all sizes are transparent, which creates a wireframe effect. Subtypes 1 and 2 display 3D views, while subtypes 3 and 4 show bird's-eye perspectives.

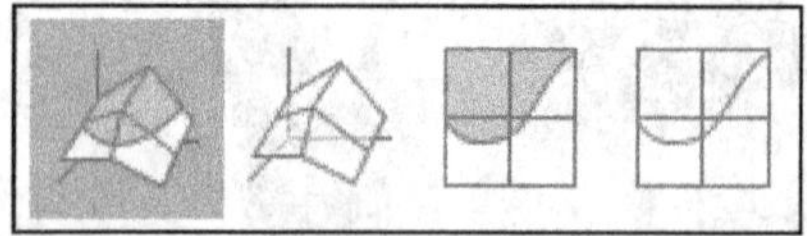

Figure 8.35 Surface Chart Subtypes

8.9 Additional Chart Types

Since the release of Excel 2016, several chart types have been introduced. This section presents them with brief examples.

8.9.1 Statistical Charts

A chart type you previously created as a column chart through special settings is now available as a readymade option: a *histogram*. This chart type is typically used to visualize frequency distributions. Figure 8.36 shows a table of test values.

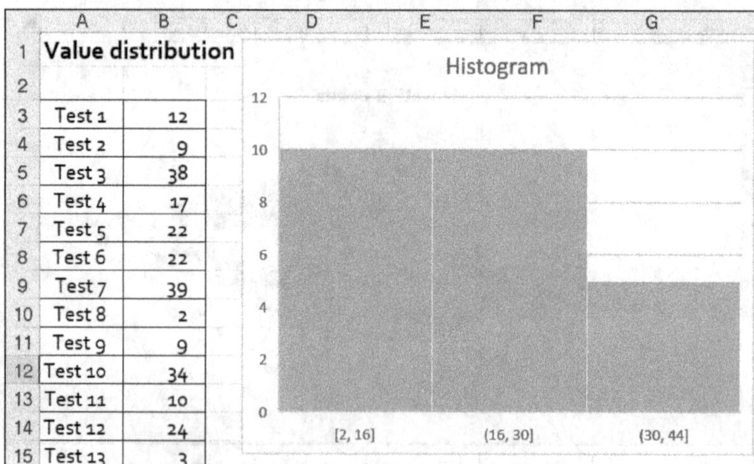

	A	B
1	**Value distribution**	
2		
3	Test 1	12
4	Test 2	9
5	Test 3	38
6	Test 4	17
7	Test 5	22
8	Test 6	22
9	Test 7	39
10	Test 8	2
11	Test 9	9
12	Test 10	34
13	Test 11	10
14	Test 12	24
15	Test 13	3

Figure 8.36 Histogram of Frequency Distribution of Test Values

If you assign the **Histogram** chart type to this table (using the **Insert Statistic Chart** button in the **Charts** group or the **Insert Chart** dialog), Excel automatically groups the values; in Figure 8.36, there are three classes labeled on the x-axis.

To change the breakdown, double-click or tap the axis, and then, under **Axis Options • Bins**, select a different **Number of bins**. You can also manually set the lowest and highest groups under **Underflow bin** and **Overflow bin**. Excel calls classes in a histogram chart bins.

A subtype of this chart is the *Pareto chart*, which adds a curve to the histogram to show cumulative frequency, with a second value axis displaying percentages.

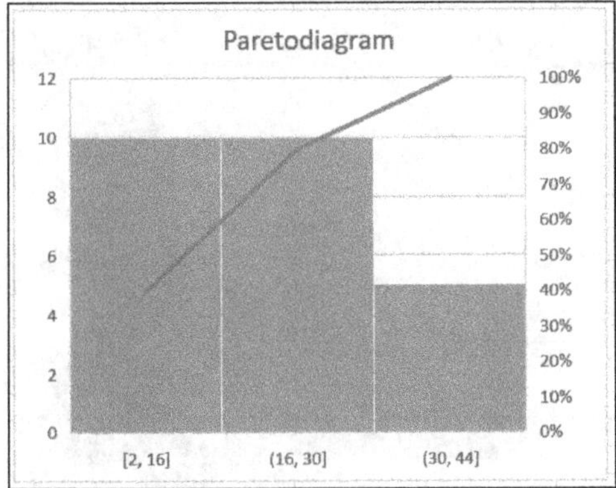

Figure 8.37 Example of Pareto Chart

The second chart type that's available via the **Insert Statistic Chart** button is the **Box and Whisker** chart, which is also suitable for displaying frequency distributions. In Figure 8.38, participant numbers for courses on various topics are listed consecutively for two conference locations.

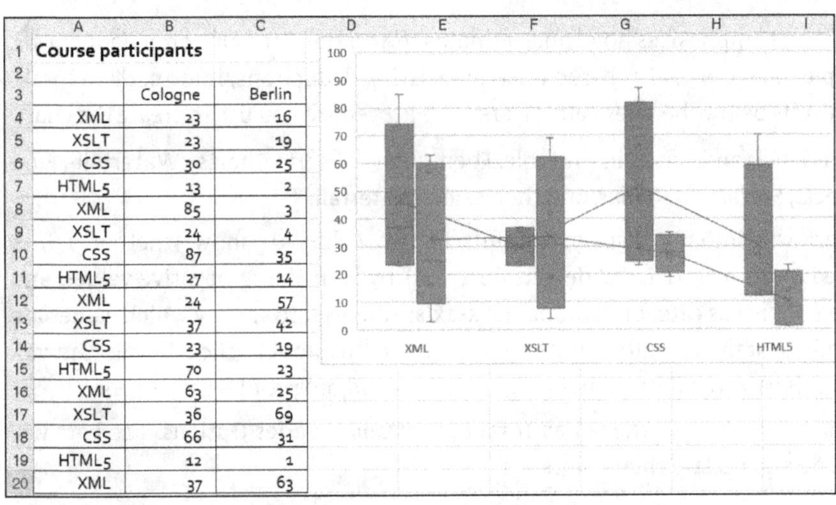

	A	B	C
1	**Course participants**		
2			
3		Cologne	Berlin
4	XML	23	16
5	XSLT	23	19
6	CSS	30	25
7	HTML5	13	2
8	XML	85	3
9	XSLT	24	4
10	CSS	87	35
11	HTML5	27	14
12	XML	24	57
13	XSLT	37	42
14	CSS	23	19
15	HTML5	70	23
16	XML	63	25
17	XSLT	36	69
18	CSS	66	31
19	HTML5	12	1
20	XML	37	63

Figure 8.38 Box Chart Showing Distribution of Course Participants

The box chart displays a bar for each course and location, showing the range of the most common values, along with a T-line that marks the lowest and highest values, which represent outliers. A mean value line is also included.

For this chart type, choosing the right series options is key to how the values are displayed. You can also show points for individual values inside the bars by selecting **Show inner points**.

You can also hide outlier points, mean markers, and the average line. Under **Quartile Calculation,** choose whether to calculate the median, including or excluding outliers.

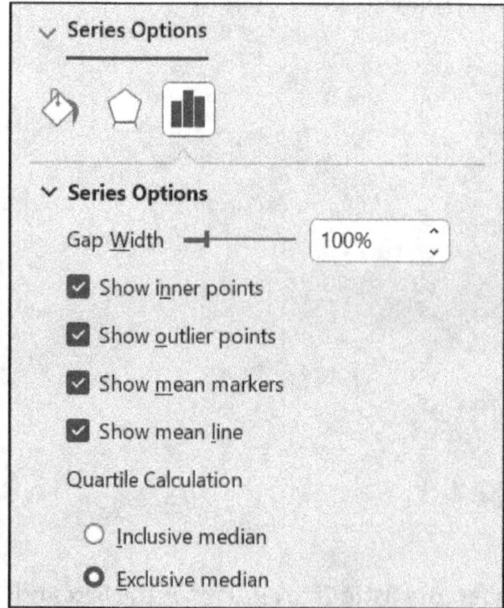

Figure 8.39 Series Options for Box Chart

8.9.2 Waterfall Charts

A waterfall chart emphasizes how much remains from a starting value after deductions. The example in Figure 8.40 is based on a table listing deductions, intermediate totals, and final results below the sales value. Here's the procedure you use to create the chart:

1. Select the labels and values in the table. Then, choose **Insert • Charts • Waterfall-, Funnel-, Stock-, Surface- or Radar**, and then select **Waterfall**.
2. The chart will initially display columns in two colors to show whether values increased or decreased. Excel detects decreases by identifying negative values, and the labels appear as categories along the x-axis. You need to specify which values are intermediate or final results in the task pane. To do this, double-click the relevant values—in this case, the data points for gross profit and net profit—one by one.
3. In the task pane, enable the **Set as Total** option under **Series Options.** The bars will recalculate and update their colors.

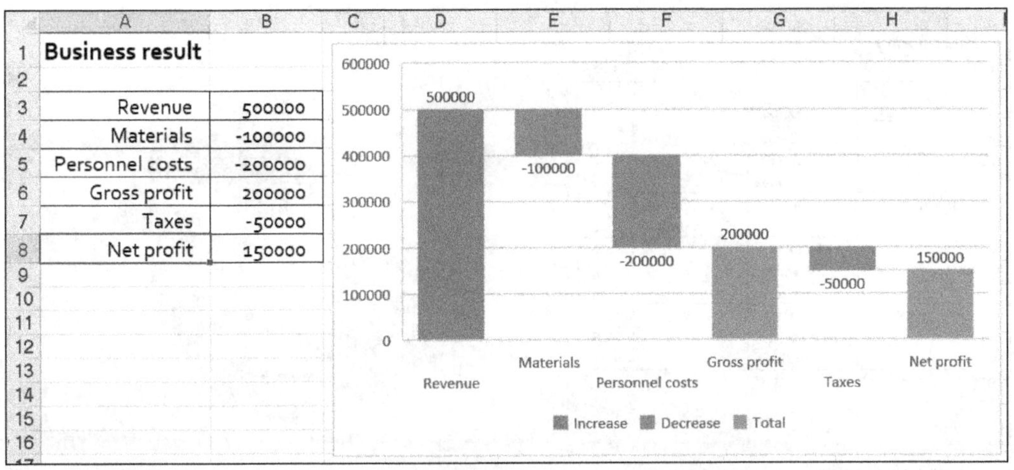

Figure 8.40 Waterfall Chart Showing Business Results

8.9.3 Sunbursts

You can find the **Sunburst** and **Treemap** chart types under **Insert • Charts** by clicking the **Insert Hierarchy Chart** button.

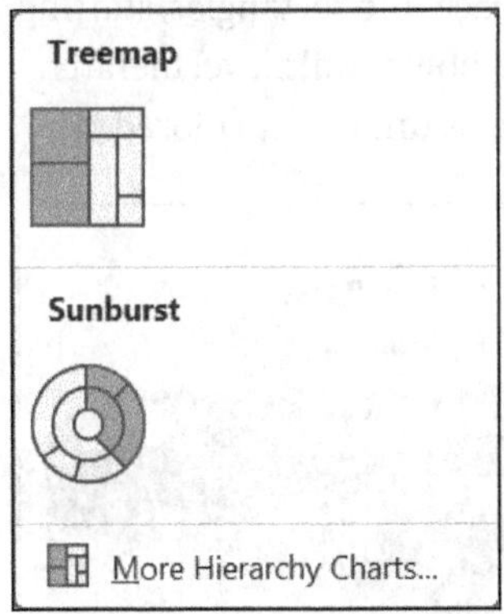

Figure 8.41 Two Chart Types for Hierarchical Data

The sunburst chart resembles a doughnut chart but displays hierarchies across multiple levels, which allows some levels to be skipped. In Figure 8.42, the musical genre is the first level and musicians and composers are the second. The third level applies only to one composer, so cells B9, C4 through C7, and C10 through C15 remain empty. The fourth level shows the number of albums, which determines the segment sizes. The first level also acts as the legend, which is marked by different colors.

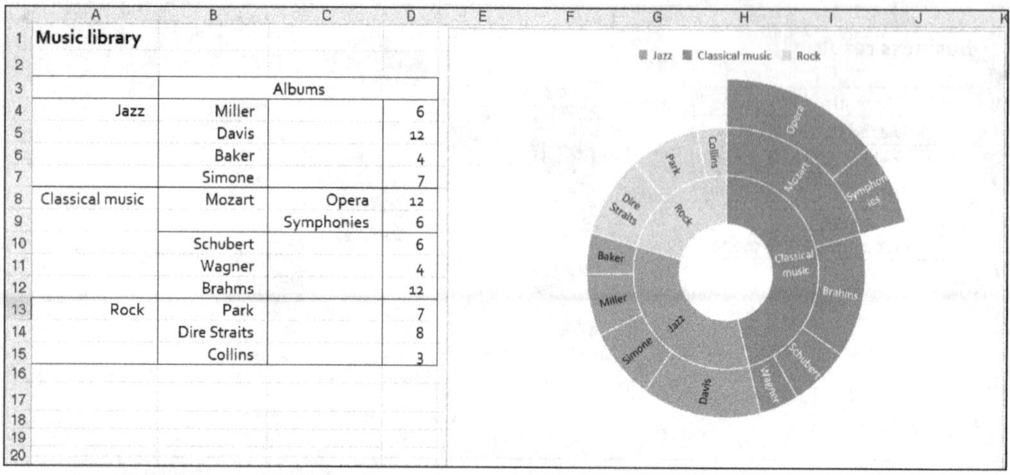

Music library
Albums
Jazz
Miller 6
Davis 12
Baker 4
Simone 7
Classical music
Mozart
Opera 12
Symphonies 6
Schubert 6
Wagner 4
Brahms 12
Rock
Park 7
Dire Straits 8
Collins 3
Jazz
Classical music
Rock

Figure 8.42 Chart Displaying Number of Albums in Different Musical Genres

8.9.4 Treemaps

A treemap is a simple type of hierarchical chart you can use to analyze a table that has multiple columns or rows: one or more for labels and one for values. This chart type doesn't use a coordinate system. It automatically divides space into rectangles, starting on the left with the largest values. Figure 8.43 demonstrates how a multilevel hierarchy can be analyzed on a treemap. In this case, the country level is uniformly colored.

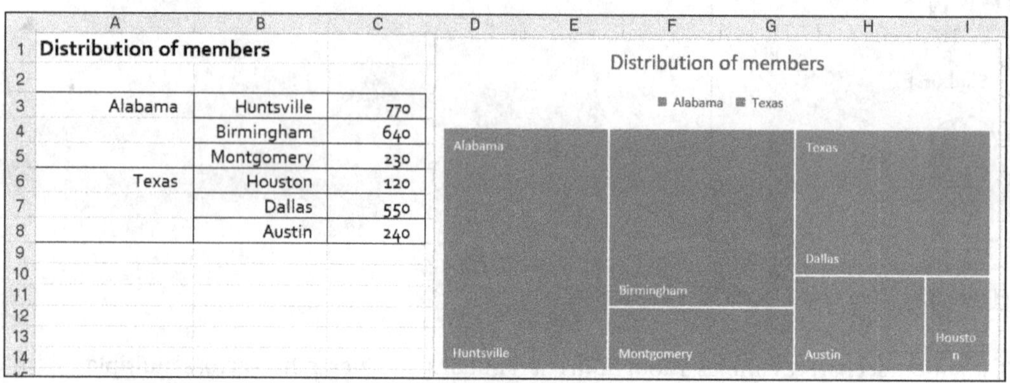

Figure 8.43 Treemap Showing Member Distribution

8.9.5 Funnel Charts

A funnel chart does a good job of illustrating how a quantity is consumed over time. Figure 8.44 shows how an annual budget is spent over twelve months.

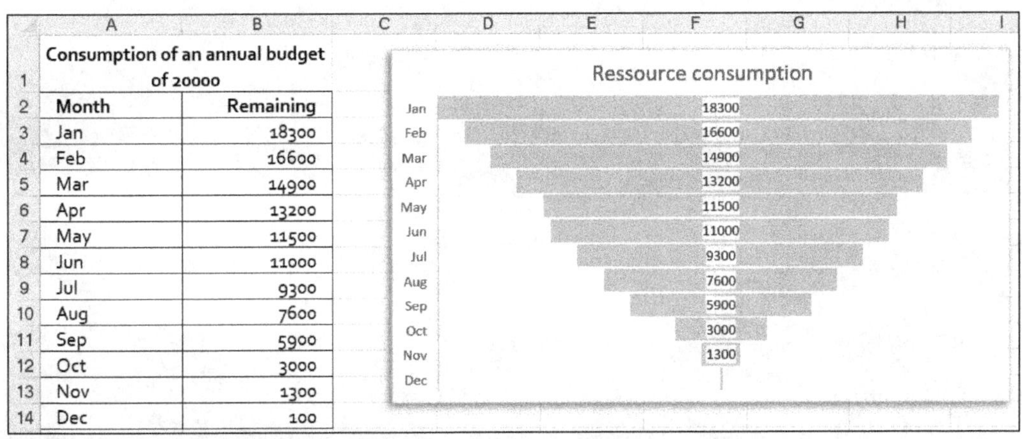

Consumption of an annual budget of 20000	
Month	Remaining
Jan	18300
Feb	16600
Mar	14900
Apr	13200
May	11500
Jun	11000
Jul	9300
Aug	7600
Sep	5900
Oct	3000
Nov	1300
Dec	100

Figure 8.44 Funnel Chart Displaying Budget Outflow Throughout the Year

8.9.6 Map Charts

For data linked to geographic locations, you can use two-dimensional maps. The base table must include at least one column with geographic data—like country names, regions, or postal codes—so that you can analyze at least one column with related values.

Country	Branches
Germany	77
Italy	64
France	23
Spain	40

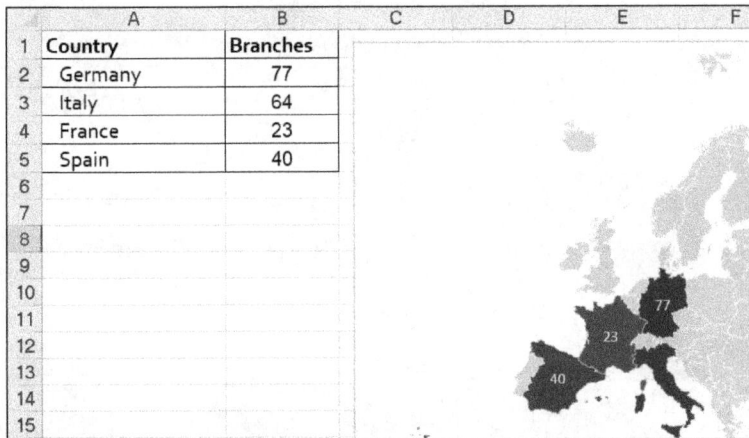

Figure 8.45 Country Chart

Use the **Format Data Series** pane to set the map area and series color.

Chapter 9
Data Visualization with Sparklines

There are several ways to visualize data. One option besides using charts is to use *sparklines*, which are the "sparkling" lines also known as *number images* or *word graphics*. Both of the latter terms hint at some uncertainty about what sparklines really are—they're something like charts, but not quite. Edward Tufte, a Yale professor, is credited with creating sparklines, which he called "small, intense, simple datawords." He designed them to visualize data within a sentence without expanding the line's space.

9.1 Features and Uses

Sparklines offer a compact way to present data. They rely on keeping the text and data that provide context closely connected to the data visualization. These *datawords* are especially useful when numbers show sequences or trends. A simple example is a row like the one shown in Figure 9.1.

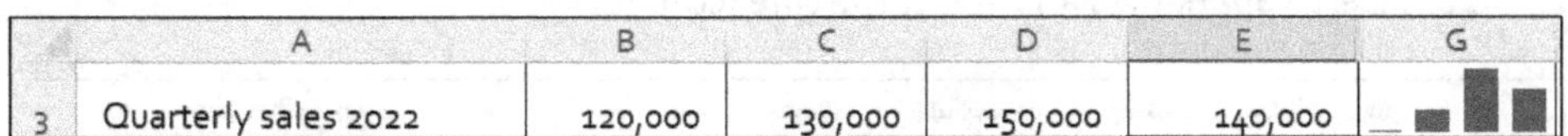

	A	B	C	D	E	G
3	Quarterly sales 2022	120,000	130,000	150,000	140,000	

Figure 9.1 Data Row with Sparkline

The first cell states the topic, the next four cells contain the data, and the sparkline visualizes it. You can rearrange these elements, but usually, the label, numbers, and graphic should stay close together. If the numbers change, the chart updates immediately. When printing, sparklines are treated like cell contents and will always print.

Unlike charts, sparklines aren't graphic objects; they're part of the cell's content, even if they don't appear in the formula bar. The sparkline serves as the cell's background, so you can also enter text (such as a label for the chart) in the cell. You can also place another background beneath the sparkline.

Since the sparkline is linked to the cell, any change to the cell's height or width instantly adjusts the sparkline. You can also use a merged cell as the space for a sparkline. When you merge multiple cells using the **Merge & Center** button, a sparkline in the first cell will stretch across the entire range. If you move a cell containing a sparkline, any relative cell references will update accordingly. You can also extend a range of sparklines using the fill handle, and you can enter data for these cells later if needed.

What are sparklines used for? Common uses include data that show trends, seasonal changes, or distributions over time, such as the following:

- Price trends
- Price developments
- Sales, volume, or profit trends
- Temperature data over time
- Political distributions

[+]

Copying into Word

To copy a sparkline from Excel into a Word document, paste it as a graphic.

9.2 Inserting Sparklines

Here's a typical example: A table showing fictional price values over the course of a week. The first column lists the securities, and the following columns show daily prices. To insert sparklines, follow these steps:

1. First, select the cell range where the sparklines will appear; in this example, it's G4:G9.
2. Go to **Insert • Sparklines** and select the **Line** icon.

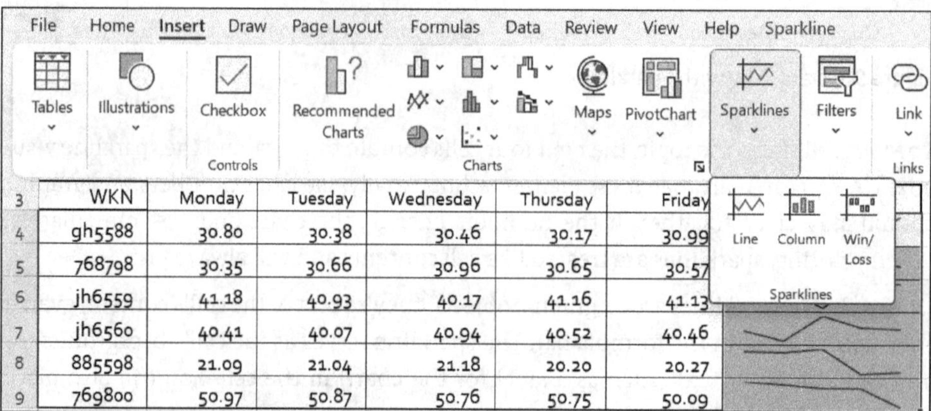

3. In the **Create Sparklines** dialog, enter the price data range under **Data Range** to use for the sparklines that should be visualized. Then, drag across the range B4 to F9 or select the previously named range by pressing F3.
4. Since the position range for the sparklines is already set by the selection in step 1, you can close the dialog.

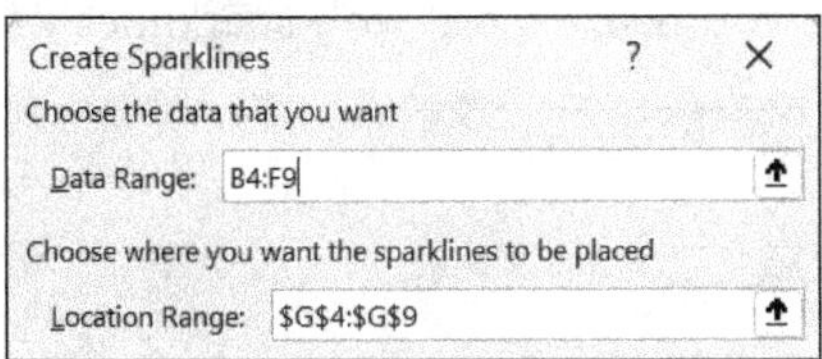

5. Excel then shows the **Sparkline** tab on the ribbon. In the display group ❶, you'll find options to highlight specific points: **High Point**, **Low Point**, **First Point**, **Last Point**, and **Negative Points**. In this case, it's best to highlight at least the high and low points. Selecting the **Data Points** option highlights every data point.
6. Depending on your choice under **Display**, the **Style** group will offer different design samples ❷. Click or tap to apply the style.
7. In addition to the formats in a chosen style, you can pick custom colors from the palette by using the **Sparkline Color** ❸ and **Marker Color** ❹ buttons. For data points, you set colors separately for each point type, with the current selection shown in that type's menu. On touchscreens, the palettes are enlarged to make selecting patterns easier.

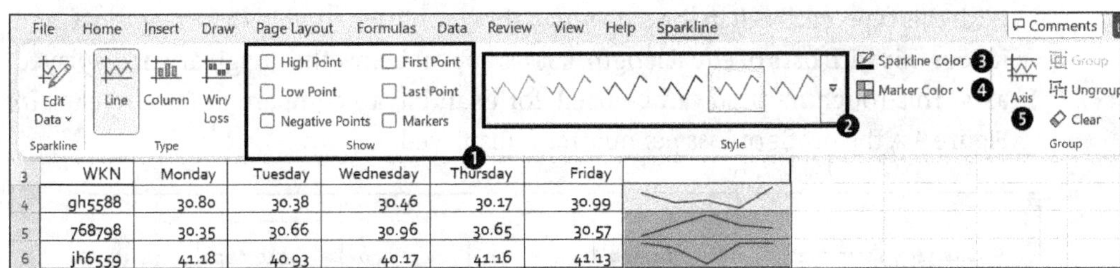

8. In the **Group** group, a menu is available for the **Axis** ❺ button that offers options to customize the axes. You can adjust the minimum and maximum values of the vertical axis here to change the scale and better highlight differences. If negative values appear, you can show the horizontal axis to clearly indicate the transition below zero.

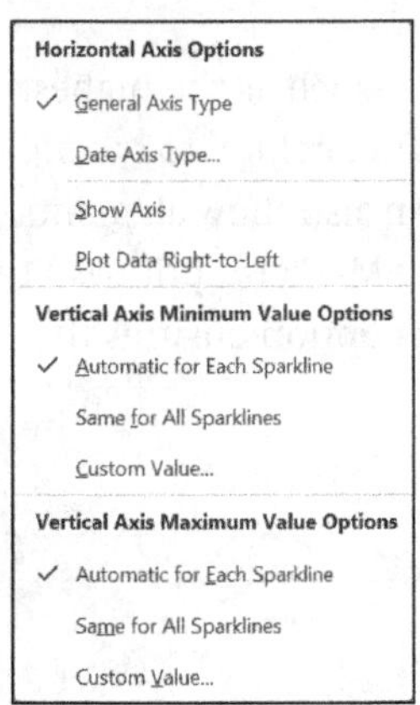

Adjusting the column width and row height lets you resize a cell with sparklines to make the chart lines as clear as possible.

9.3 Display Options

Excel offers three types of sparklines, each of which has many format templates and customization options:

- **Line**
  This type is a standard line chart that's based on the discrete values in the specified data range that are embedded in a cell. It lacks details—axes are usually hidden, and there are no legends or titles. This type can clearly depict even large data ranges, especially when you want to see the overall trend or highlight key points.
- **Column**
  With this type, a simple bar chart without labels is created and negative values appear below the baseline. You can also display this axis as a simple line if you need to, and you can do that via the **Axis** menu in the **Sparkline • Group** section.
- **Win/Loss**
  With this type, each cell in the data range shows only whether it's a gain or a loss, and it does this with bars of equal length. Losses appear below the imaginary or displayed axis. This rough method can be used, for example, to compare different years. In Figure 9.2, the profit or loss per quarter is displayed.

	A	B	C	D	E	G
3	Quarterly sales 2022	120,000	130,000	150,000	140,000	
4	Quarterly sales 2023	125,000	120,000	115,000	-50,000	
5	Quarterly sales 2024	140,000	130,000	120,000	129,000	

Figure 9.2 Marking as Profit or Loss

9.3.1 Highlighting Points

In the **Sparkline • Show** group, you can highlight specific points—such as the highest and lowest points or the start and end points—with colors. It's often helpful to specifically mark points with negative values. With the **Line** type, you can also show all points on the line by using the **Markers** option. With the **Column** type, the **Markers** option isn't available, and that makes sense in this case. The **Negative Points** option ensures that negative bars appear in a different color.

9.4 Editing Sparklines

Once you've added sparklines to a worksheet, you can change almost everything about them afterward. This includes the type, color assignments, axis handling, grouping, and data range. As long as the default group mode is enabled, selecting one cell in a sparkline group selects the entire group. Changing settings like the style automatically applies to the whole group.

9.4.1 Changing the Type

If the data supports it, you can switch between the three types anytime. To do this, select a cell within a group or a single cell with a sparkline, and then, under **Sparkline • Type**, choose the new type.

9.4.2 Assigning Colors

On the **Sparkline** tab, the **Style** group offers 36 styles that are accessible via the small scroll bar or the button at the bottom right. In **Touch input** mode, an arrow appears on the right to fully expand the palette.

Styles vary by color combinations. You can also separately set the color of the line or bars and the colors of individual data points, and any selection you make will override the corresponding setting in the style.

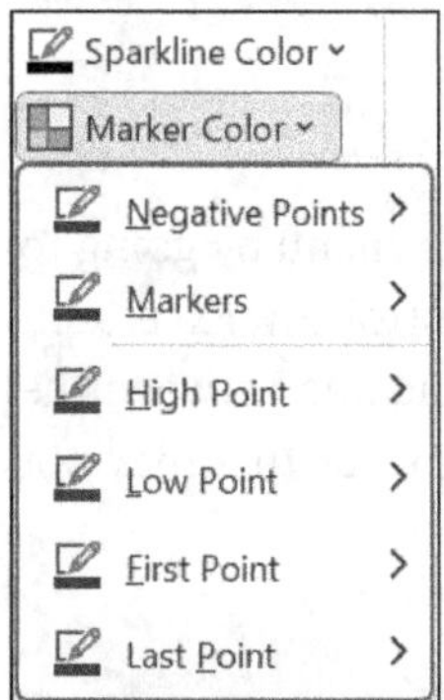

Figure 9.3 Options for Coloring Lines and Data Points

9.4.3 Axis Settings

The menu for the **Axis** button in the **Sparkline • Group** lets you show the axis and set custom minimum and maximum values for the vertical axis. This lets you scale the axis to make differences between sparkline values easier to see. For example, if all values are high and vary slightly, then you can set a specific minimum value to trims the bottom of the bars and thus highlight the differences.

If both values are set to **Automatic for each sparkline**, then scaling adjusts each cell relative to its own values. The bars for the highest values in different sparklines are all the same length, even though the absolute values can vary greatly. The **Exact for all sparklines** setting ensures that the bar lengths accurately reflect the different values across sparklines.

Besides the default **General** axis type, there is also the **Date** axis type setting. In the dialog, you can specify the cell range containing the date values. This option is useful when data is ordered by dates that don't occur at regular intervals. Figure 9.4 shows values from three consecutive days and two values each from a week later, and the sparkline displays values while accounting for time proportions.

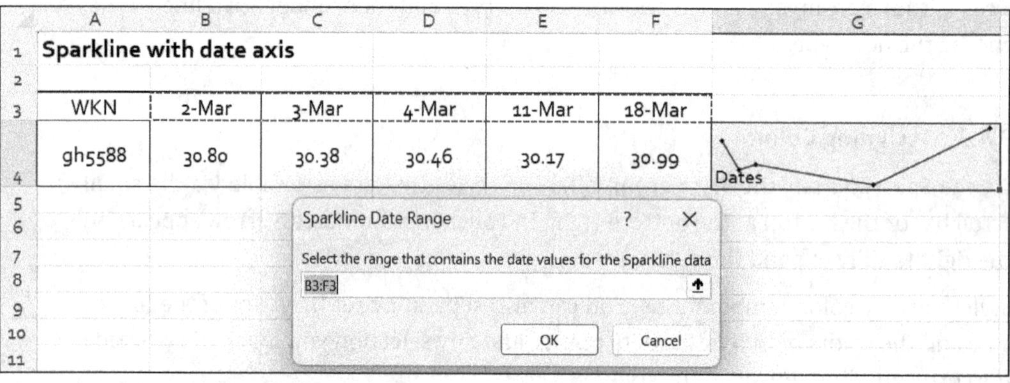

Figure 9.4 Sparkline with Date Axis

9.4.4 Handling Empty Cells

If your data range contains empty cells, you can open the **Edit Data** menu by going to **Sparkline • Sparkline** and use the **Hidden & Empty Cells** option to control how Excel handles them. You can leave a gap, treat the cell as zero, or connect the line between the previous and next values—although this assumes an intermediate value that may not accurately reflect reality.

9.4.5 Group or Individual Handling

Sparklines are usually used in groups, so when you specify a cell range in the **Create Sparklines** dialog under **Location Range**, Excel automatically groups that range. Clicking any cell in this range selects the entire range, and subsequent commands (such as changing colors or styles via the **Sparkline** tab) typically affect all sparklines in the range. Axis settings in the **Group** section apply to all items in the group.

9.4.6 Ungrouping Cells

Alternatively, you can work with individual sparkline cells. Follow these steps do to that:

1. To ungroup a single cell or a set of cells, specifically select them by clicking or using [Ctrl]+click or [Shift]+click. On a touchscreen, tap each cell individually. To ungroup the entire group, select the whole range.
2. Follow the **Sparkline • Group • Ungroup** path.
3. Edit each sparkline cell independently.

Note that Excel keeps unselected cells in a group as a remainder group during this action. If you need to regroup isolated cells later, use the **Group** command in the same group.

9.4.7 Clearing Sparklines

When you want to clear sparklines, there are two commands available in the **Group** group via the delete icon. Use **Clear Selected Sparklines** to clear previously selected cells, and the remainder of the group will stay intact. To clear an entire group, select any cell within it and choose **Clear Selected Sparkline Groups**. As the name implies, you can clear multiple groups at once, so for example, to clear sparklines in three columns, you select three cells in a single row within those groups.

9.4.8 Editing Data Sources

If you need to change the source data range later, you can do so for a single sparkline or an entire group. To edit a group, select any sparkline cell within it and go to **Sparkline • Edit Data • Edit Group Location & Data** to open the dialog, where you can modify both the data range and the location range. In this way, you can expand or move both ranges.

To change the data range for individual sparklines, select them, go to **Sparkline • Edit Data • Edit Single Sparkline's Data**, and in the dialog, enter the new source range. All commands from the previous sections are also available in the sparklines context menu under **Sparklines**.

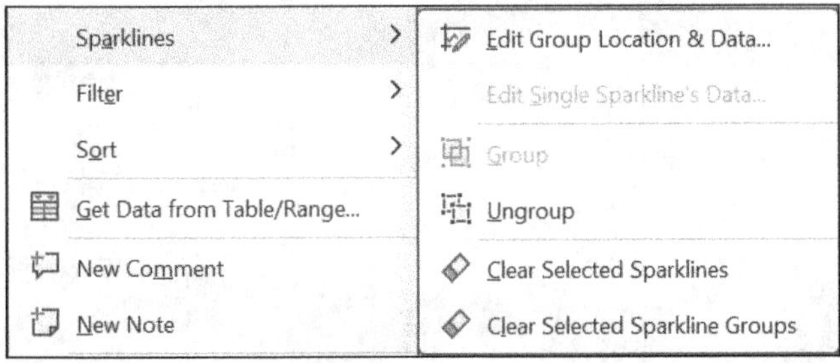

Figure 9.5 Context Menu for Sparkline Group

Chapter 10
Refining Worksheets with Graphics

To support organizational tasks, simulations, and especially presentations, you'll often want to pair charts with diagrams, sketches, logos, images, and similar visuals. You can also use graphic objects as triggers for macros or hyperlinks. If you don't want to use a separate program, Excel provides easy-to-use yet powerful tools for this. With a touch-screen, you can draw handwritten notes or sketches on the worksheet or mark spots using your finger or stylus.

10.1 Overview of the Graphic Tools

Excel offers several ways to create graphic objects beyond charts and sparklines. These tools are found on the **Insert** tab, within the **Illustrations** and **Text** groups.

❶ You can insert finished graphics from existing files by using the **Pictures** button and then edit them with tools on the **Picture Format** tab once you select the object. The **Stock Images** option offers a themed catalog of images on various topics. You can also import graphics and images from the web by using the **Online Pictures** option, which opens a dialog where you can select image sources like **Bing** and **OneDrive**.

❷ To create your own drawings, such as flowcharts, use the **Shapes** palette, which is a comprehensive collection of graphic elements. When you select an object, the **Shape Format** tab becomes available for editing.

❸ Under **Icons**, you'll find a comprehensive dialog box with numerous symbols sorted by category. These symbols come from an SVG library. The **Graphics Format** tab will open for editing.

❹ Click the **3D Models** icon to insert your own 3D illustrations or online models into the workbook. You can freely rotate the models to view them from any angle, and a dedicated **3D Model** tab will open for editing.

❺ Use the **SmartArt** button to create organizational charts and other schematic drawings via a special dialog box that offers an extensive SmartArt catalog. The **SmartArt Design** and **Format** tabs will appear for detailed editing.

❻ The **Screenshot** button lets you insert an area of the screen that you select into the worksheet.

❼ Use the **Text • WordArt** button to easily add decorative text as headings or logos in a workbook.

❽ You can also embed graphics or videos in Excel or link the workbook to those objects by using the **Text • Object** button. You'll learn more about that in Chapter 21.

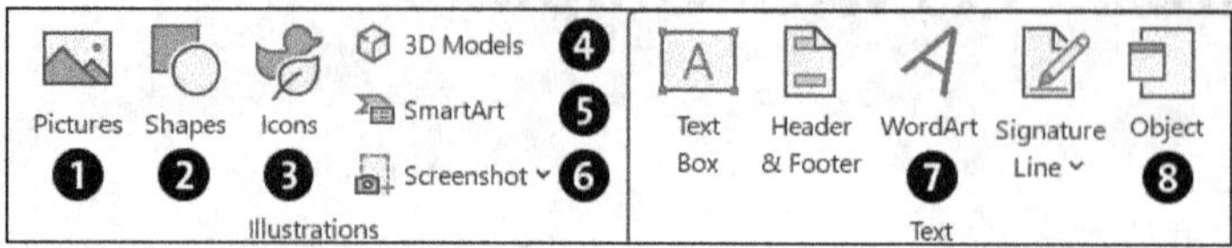

Figure 10.1 Illustrations and Text Groups on Insert Tab

10.2 Drawing Preset and Freeform Shapes

Under **Insert • Illustrations • Shapes**, you'll find a wide range of drawing tools. You can use the handle at the bottom right to change the height and width of any shape, and the first group shows your most recently used shapes for quick access.

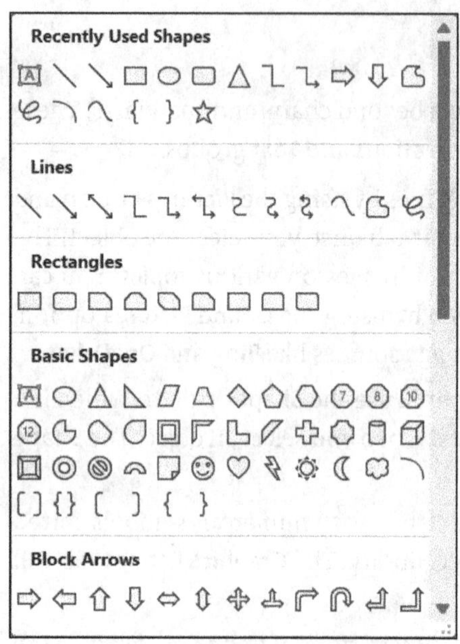

Figure 10.2 Shapes Palette

10.2.1 Drawing a Simple Shape

To draw a rectangle on the worksheet, follow these steps:

1. Click a rectangle in the **Shapes** button group. The cell pointer will switch to the drawing cursor, and the palette will close. You can also select the icon by using the arrow keys and confirm by pressing Enter instead of using the mouse.
2. Position the cross-shaped mouse pointer where you want one corner of the rectangle to be.

3. Hold down the mouse button and drag toward the opposite corner until the rectangle is the size you want it to be. Release the mouse button to finish drawing.

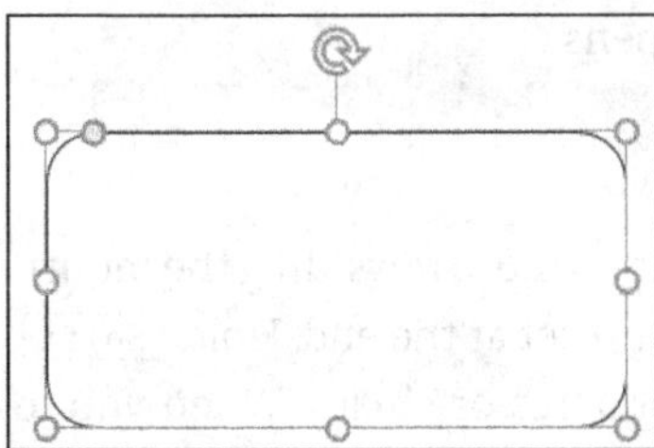

The object will display handles and stay selected for further editing.

[+]

Use Tools Several Times

Normally, a drawing tool turns off when you release the mouse or click any cell. To draw several arrows in a row, right-click the arrow icon and choose **Lock Drawing Mode.** It will stay active until you turn it off by clicking the arrow button again.

Drawing on the Touchscreen

On the touchscreen, finger mode works a bit differently because you can't drag directly with your finger. Follow these steps to use it:

1. Tap one of the rectangles in the **Rectangles** group on the **Shapes** button palette.
2. Excel will insert an object of preset size in the center of the screen.
3. Drag the object to where you want it.
4. Use the round handles to resize it.

The object will stay selected for further editing. You can immediately start labeling, coloring, or adding effects.

If you hold your finger on the object briefly, a horizontal context menu will appear with the **Fill** and **Edit Text** buttons. You can find additional commands by clicking the arrow at the end.

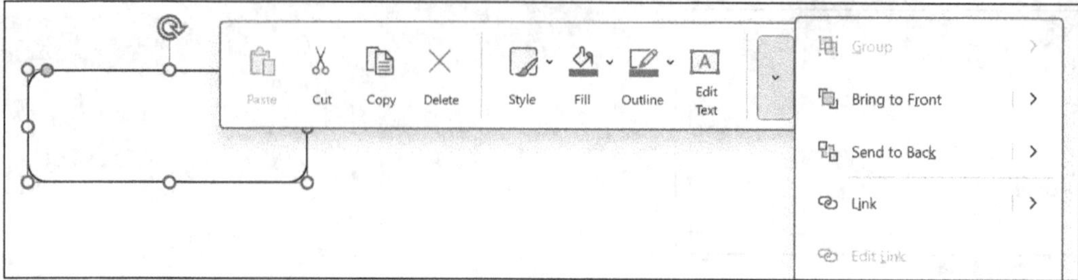

Figure 10.3 Round Handles and Context Menu on Touchscreen

If you use a pen, you can draw the rectangle just like with a mouse. Touching with the pen tip opens the same context menu as a right-click opens.

Circles, Ellipses, Lines, Arrows, and Squares

The process for drawing a circle or ellipse is similar. For lines and arrows, drag the mouse from the start point to the end point. The arrowhead is preset at the end. Hold `Shift` while drawing a rectangle or ellipse to create perfect squares or circles. When you're drawing lines, you can hold the `Shift` key to restrict angles to multiples of 15 degrees.

You can hold `Ctrl` while drawing with the **Rectangle** or **Ellipse** tools to draw the shape from its center, and you can hold both `Shift` and `Ctrl` to draw a square or circle from the center.

Instead of dragging to draw a rectangle, you can click the spot that sets the upper-left corner first. Excel will then create a preset shape for the selected object type; for example, the rectangle tool will draw a square and the ellipse tool will draw a circle. You can adjust the size later, during editing.

If you want the cursor to select only graphic objects and not cells, activate the option on the **Home** tab in the **Editing** group under **Find & Select** by choosing **Objects**. This activates the setting, and it stays active until you turn it off explicitly. This is especially helpful when editing many objects.

10.2.2 Freeform Lines

For freehand drawings, in the **Shapes** palette under **Lines**, you'll find (among others) three object types we'll take a closer look here: **Curve**, **Freeform: Shape**, and **Freeform: Scribble**. All three are special types of Bézier curves that let you easily create precise curves and arcs. Each curve consists of one or more segments, each of which has a start and end point. The end point of one segment is the start point of the next, and the connection between points is either a straight or a curved line segment. The curve's shape depends on the position of control points located at each curve point.

Think of the control points as magnets that shape the curve like a metal band stretched between the start and end points.

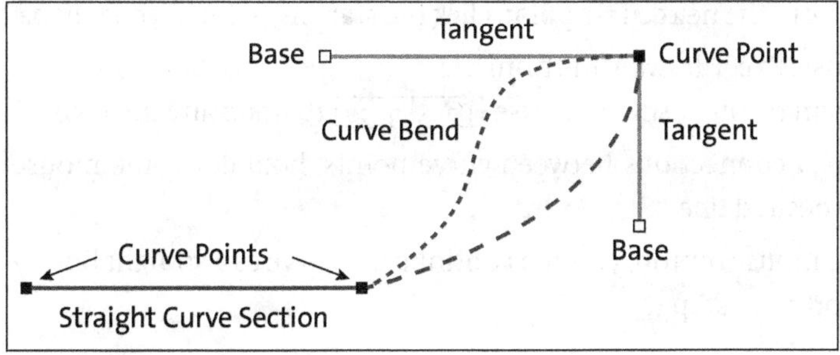

Figure 10.4 Structure of Bézier Curve

Drawing a Curve

Before you begin drawing, it's usually best to turn off the gridlines on your worksheet. On the **View** tab, choose the appropriate option in the **Show** group and then follow these steps:

1. Activate the **Curve** tool under **Insert • Illustrations • Shapes,** which is located in the **Lines** group. Click the starting point of the curve with the crosshair cursor.
2. Set the vertex for the first arc by clicking that spot.
3. Add all other vertices to shape your curve. Excel automatically smooths the connections between points.
4. Double-click to set the last curve point. Click the starting point to close the shape.

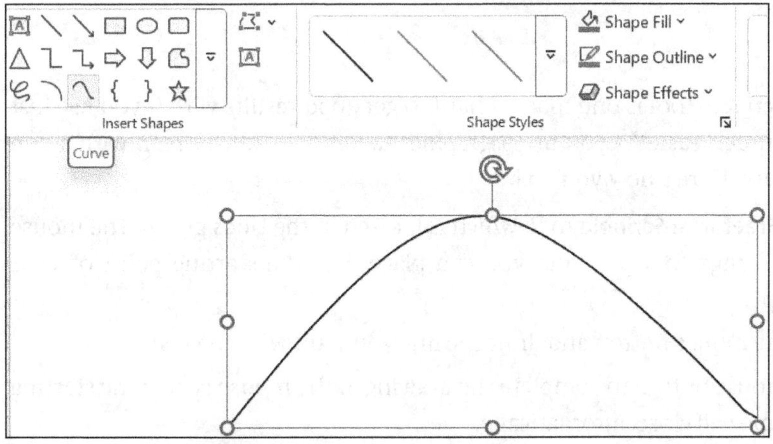

Draw a Freeform Shape

Use the **Freeform: Shape** tool to draw any polygon shape as follows:

1. Select the **Freeform: Shape** tool in the **Lines** group. Position the crosshair at the starting point of your freeform shape.
2. Set the last curve point by double-clicking or by closing the shape when you click the starting point.

3. To create a straight line to the next curve point, click the starting point of your shape.
4. As you move the mouse, Excel draws a line from the starting point to the cursor. Click to fix the line and begin the next segment. Use [Backspace] to undo any mistakes.
5. To create freely shaped connections between curve points, hold down the mouse button and draw the desired line.
6. When you release the mouse button, you can continue the curve as a straight line by using the click method from step 3.

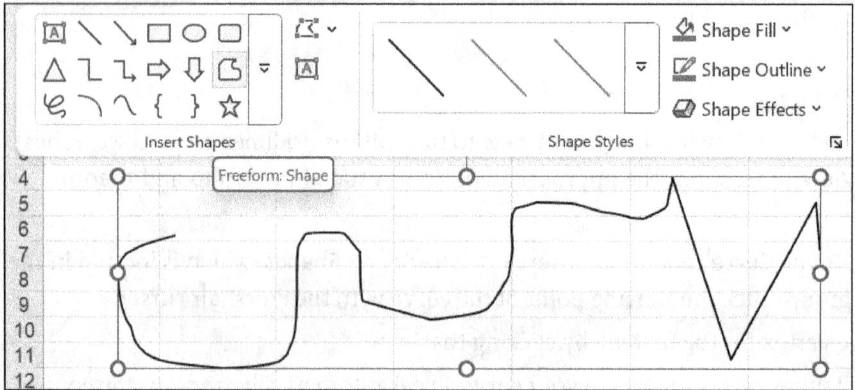

After you draw the first element, the shapes palette will become available in the **Insert Shapes** group on the **Shape Format** tab.

Scribble

The **Freeform: Scribble** tool is one that it's hard to get good results with if you use just the mouse. It's much easier if you use a graphics tablet or a touchscreen with a pen instead of a mouse. Here's how you do it:

1. Activate the **Freeform: Scribble** tool, which is located in the **Lines** group. The mouse pointer will change to a pen, and you can place it at the starting point of your freeform shape.
2. Hold down the mouse button and draw the line you want with the pen.
3. Release the mouse button to complete the drawing. If the mouse is over the starting point, the shape will close automatically.

On a touchscreen, you can easily use the **Curve**, **Freeform Shape**, and **Scribble** tools with your finger or, better yet, a pen. Tapping works like a mouse click.

10.2.3 Creating a Flowchart

Shapes are especially useful for building drawings from premade elements. A common example is a flowchart for planning an IT solution, which you can create as follows:

1. Add the individual elements of the diagram by using the symbols in the **Shapes** palette under **Flowchart.**

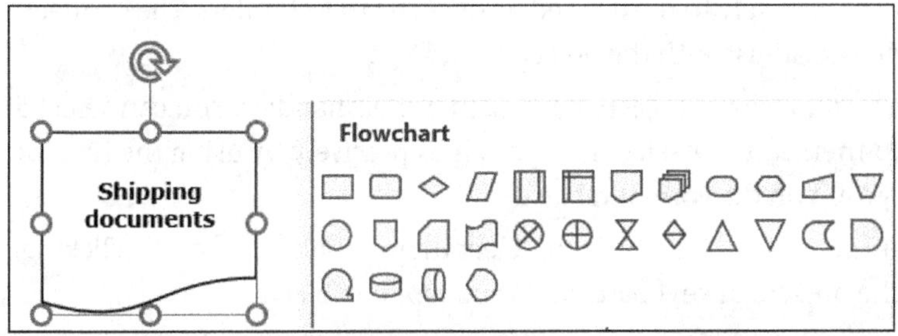

2. To label an element, choose **Edit Text** from the context menu and then type your text. When you highlight it, the mini toolbar will appear and let you change the font, size, and style. Click or tap outside the object to finish or select **Exit Editing Text** from the context menu.
3. Draw a rectangle around all objects to select them (make sure **Select Objects** is enabled). To center all text, click the **Center** button under **Home • Alignment**.
4. To insert connector lines between objects, choose the **Double Arrow Line** or **Connector: Elbow Double-Arrow** icon from the **Lines** group.
5. Hover over the object where the connector line should start until the small connection points appear.
6. Click the desired connection point and drag to the object where the connector line should end. All connection points will appear again. Click a connection point to draw the connector line and then drag the handles at the ends of the line to the connection points by using your finger.

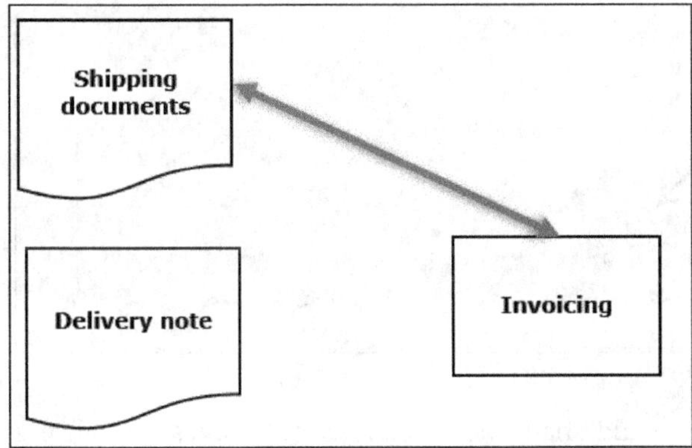

Fine-Tuning the Flowchart

After you create the flowchart, you can make a few finishing touches to improve it:

1. Drag selected objects precisely to the desired position by their borders. The connecting lines will move and adjust with the object.
2. You can also resize and reshape objects by dragging their handles. You can select a graphic object and then enter or select its size values precisely by using the two list boxes in the **Size** group on the **Format** tab.
3. To align objects at the same height or flush, select them. Hold `Ctrl` while clicking, or tap `Ctrl` on the on-screen keyboard, and then tap the objects.
4. On the **Shape Format** tab in the **Arrange** group, select the options under **Align**.
5. To change outline and fill attributes for objects or connectors, select the object and pick the attributes from the style palette in the **Theme Styles** group. For example, the following image shows how you select thicker lines. If the preset styles aren't enough, you can create custom options through the palettes for **Shape Fill**, **Shape Outline**, or **Shape Effects** settings.

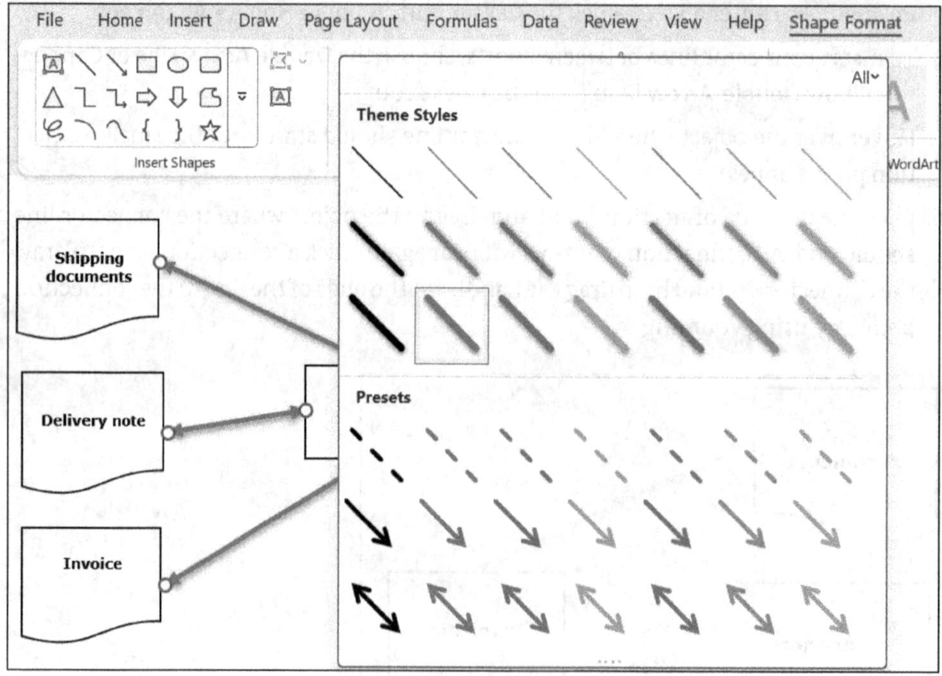

6. Use the **Shape Effects** button to add shadows, reflections, soft edges, or lighting effects to selected objects to give them a professional look.

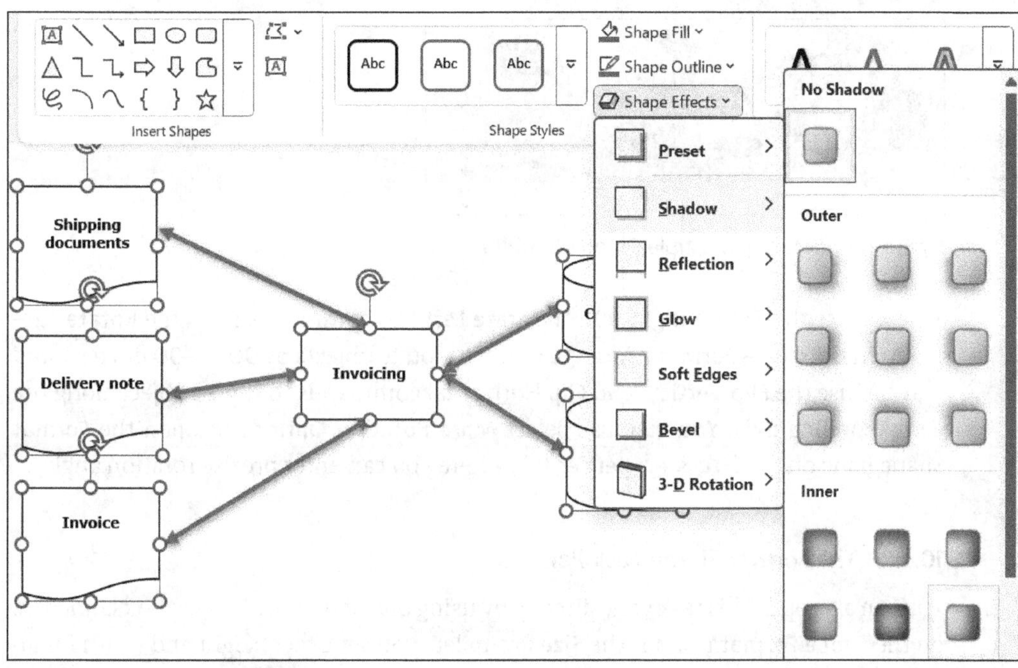

Figure 10.5 Shadows Subtly Emphasize Individual Objects

10.3 Fine-Tuning Graphic Objects

After inserting objects into a worksheet, you often need to make some adjustments to get the desired result. For example, an object you insert by simply clicking or tapping usually isn't the right size.

10.3.1 Adjusting Object Size, Proportion, and Rotation

Two-dimensional objects display eight resize handles and a rotation handle, while lines have two resize handles. You can hold `Shift` while dragging a corner handle to resize the object proportionally, handles on the sides let you stretch or compress the object, and you can pressing `Ctrl` to move the opposite handle or side accordingly.

You can drag the rotation point to rotate the object around its center, and you can also select multiple objects to rotate them all at once. Hold the `Shift` key to rotate the object in 15-degree increments, and if you press `Ctrl`, the rotation will pivot around the selection point diagonally opposite the one you've chosen instead of the object's center.

10

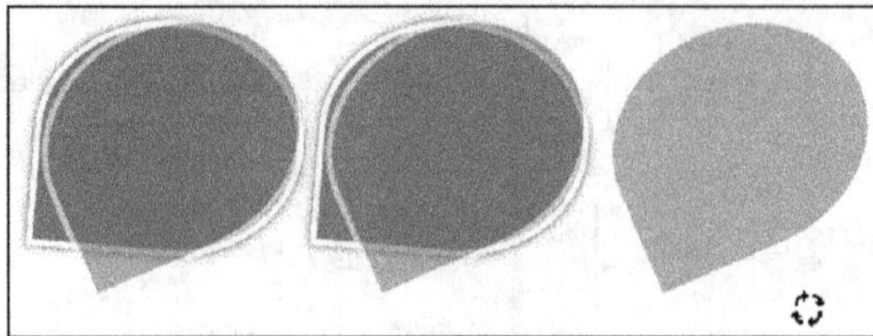

Figure 10.6 Rotating Multiple Objects Simultaneously

You can use the **Rotate Right 90°** or **Rotate Left 90°** commands from the **Rotate** submenu in the **Shape Format • Arrange** group to rotate objects by 90 or –90 degrees, and you can use the **Flip Vertical** and **Flip Horizontal** commands to flip an object along the corresponding axis. You can also select **More Rotation Options** to open the **Format Shape** pane on the **Size & Properties** tab, where you can enter precise rotation angles.

10.3.2 The Format Shape Task Pane

You can also open this task pane directly by using one of the three dialog box launchers on the **Shape Format** ribbon. The **Size** group lets you set exact height and width for an object, and under **Scale Height** or **Scale Width**, you can choose percentage values based on the current size.

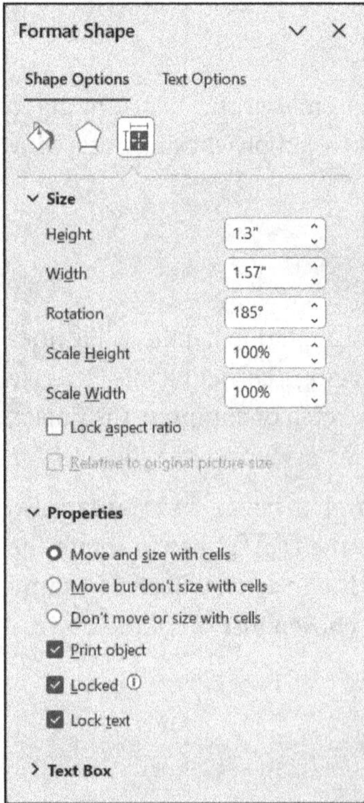

Figure 10.7 Size Group in Format Shape Task Pane

The **Properties** group also lets you lock the object to prevent changes, as described for cell ranges. If the **Locked** option is turned off, then the object will remain editable even after file protection is enabled. For text boxes, you can also use **Lock text**, but both will only take effect after the entire sheet is protected.

This also controls how a graphic object interacts with the cells on its worksheet. The default setting is **Move and size with cells**, which means that when cells are moved, the object within the cell range will move with them, and if cell sizes change, the object will resize with them. **Move but don't size with cells** means that resizing cells won't affect the object, but when cells move, the object will move with them. If you choose the **Don't move or size with cells** option, the object will remain completely unaffected by the worksheet.

You can also choose whether an object will print when you print the worksheet. This choice matters for objects that you use only within the worksheet, like text boxes for instructions or macro triggers.

Alternative Text for Graphic Objects

For all graphic objects, Excel always offers the **Edit Alternative Text** option in the context menu. You can also find the **Alt Text** icon under **Shape Format • Accessibility** in the menu bar. Both open the **Alt Text** task pane, and in the text box, you can enter text that will display or be read aloud instead of the graphic to help people with visual impairments understand the object's content.

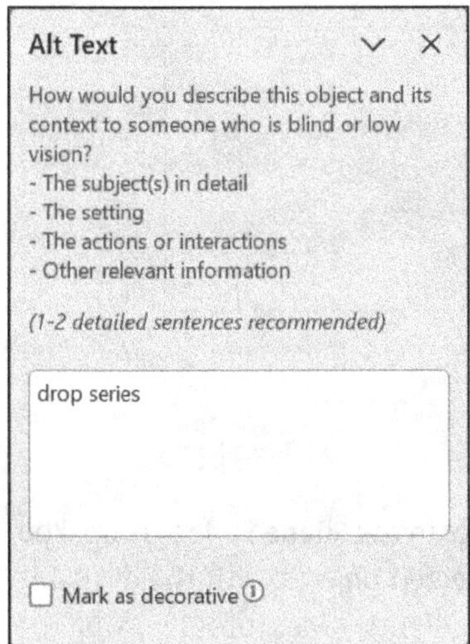

Figure 10.8 Example of Alternative Text for Graphic

10.3.3 Moving and Copying Objects

To move an object, place the pointer over a visible part of it (the outline or fill). The mouse pointer will change to a four-headed arrow at its tip when over the object, and you can drag the object to the desired location while holding down the left mouse button or drag it with your finger or pen. You can hold `Ctrl` while dragging to copy the object.

10.3.4 Object Attributes

Each shape object has properties beyond its outline, including line color, line style, and fill. You can adjust these by using the various dropdown lists in the **Shape Styles** group. You can also use the **Format Shape** task pane, which you can open from the object's context menu by selecting the same command.

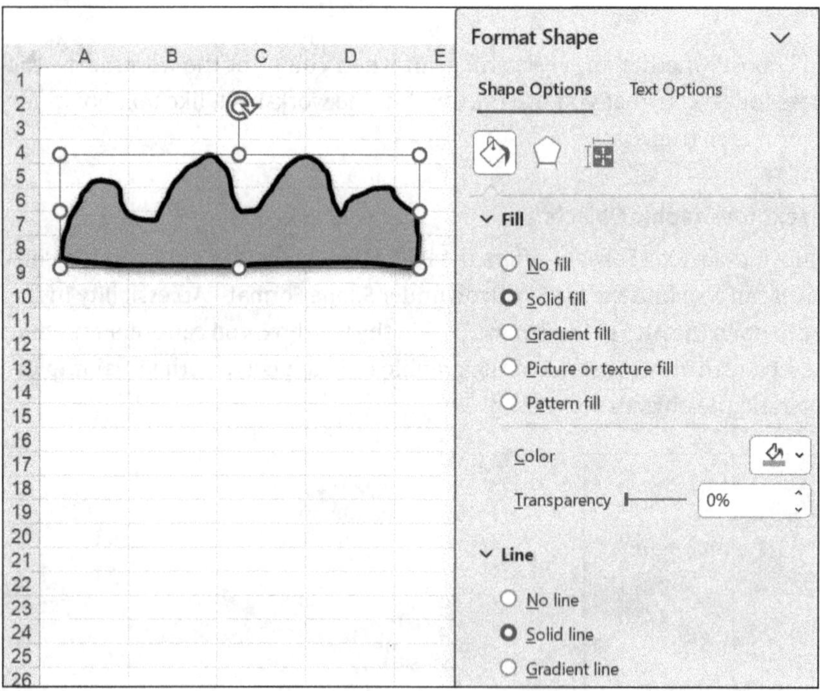

Figure 10.9 Fill and Line Option Groups for Graphic Shape

10.3.5 Outline and Fill

For quick outline and fill styling, use the samples palette in the **Shape Styles** group. You can click or tap to apply the style you want to the selected objects, or if the available styles don't meet your needs, you can assign colors, gradients, or textures by using the **Shape Fill** palette in the **Shape Styles** group (see Figure 10.10).

If the available colors aren't enough, you can use the **More Fill Colors** option to open the **Colors** dialog, where you can precisely mix custom colors by specifying red, green, and blue values. If you know the hexadecimal color code used in HTML, you can enter it directly in the **Hex** field (see Figure 10.11).

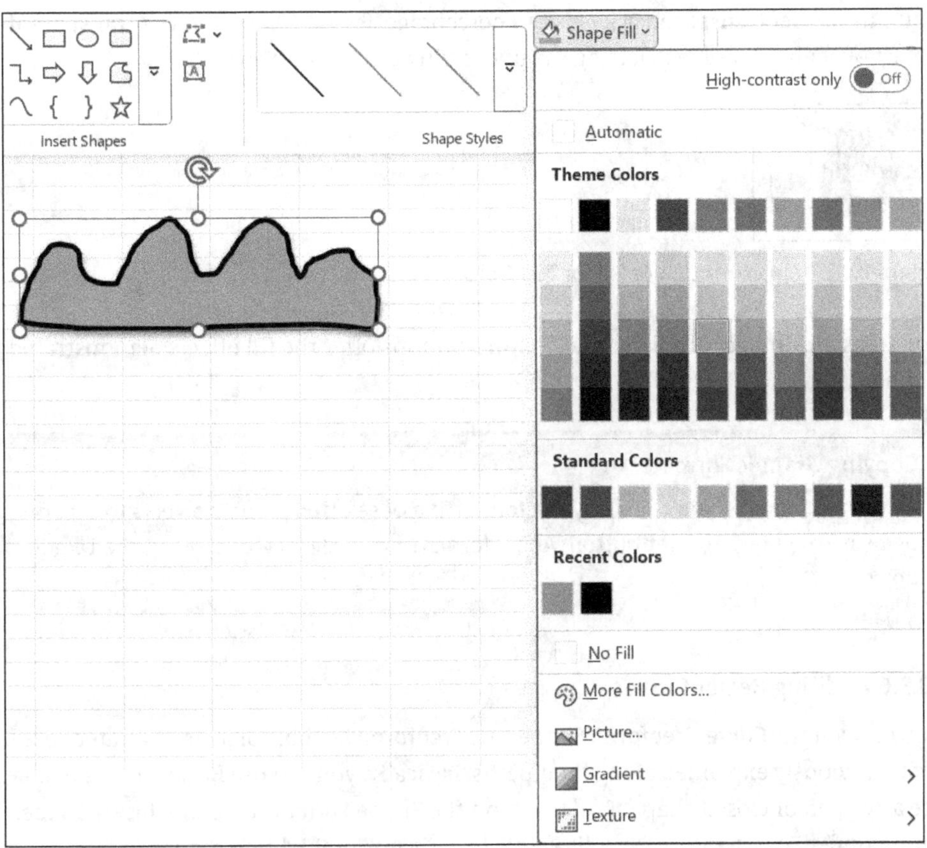

Figure 10.10 Design Fills for Selected Objects

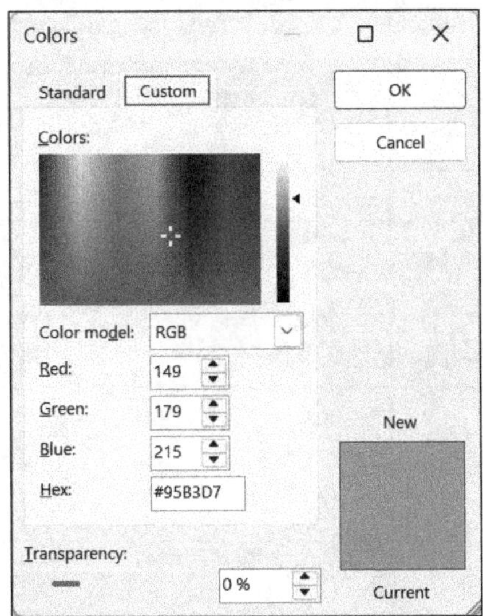

Figure 10.11 Custom Color Selection

You can also set transparency levels as a percentage. Figure 10.12 displays a transparent object over another object, letting the underlying color show through.

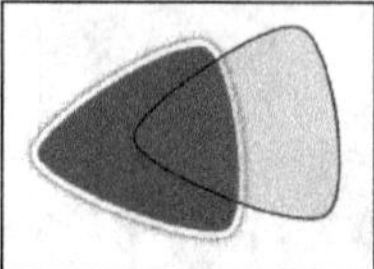

Figure 10.12 Transparent Color Fill for Top Object

For areas, the **Image**, **Gradient**, and **Texture** options offer the fill effects demonstrated in Chapter 8.

Copying Graphic Formats

As with cell ranges, you can apply the formatting of selected graphic objects to others by using the **Copy Format** button. Apply formatting by clicking or tapping the target object.

10.3.6 Editing Bézier Curves

You can edit the **Curve**, **Freeform: Shape**, and **Freeform: Scribble** objects in greater detail than previously explained along their paths. Basically, you can use Bézier curves to create any open or closed shape. To do this, on the **Shape Format** tab in the **Insert Shapes** group under **Edit Shape**, select **Edit Points** to activate **Point Editing** mode.

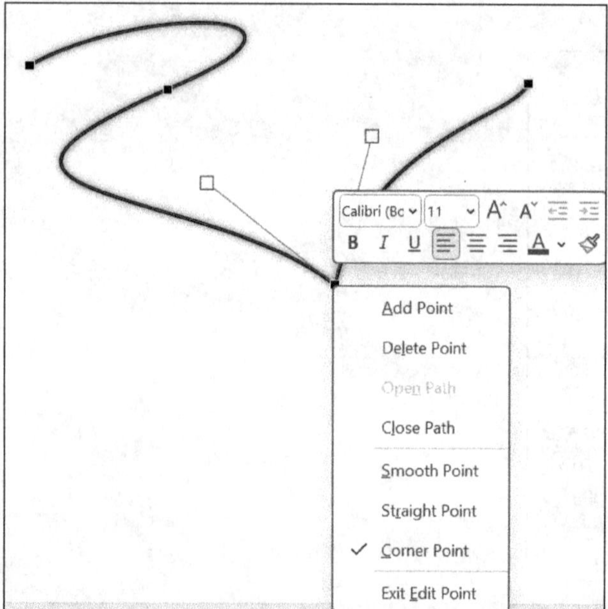

Figure 10.13 Context Menu for Visible Curve Points

All points of the selected Bézier curve will then become visible and editable. During point editing, point editing commands appear only in the context menu, which you open by right-clicking a point or curve segment or by pressing and holding with a finger or stylus.

You perform point editing by following these steps:

1. Select the curve or freeform shape to edit, then choose **Edit Points**. The Bézier curve will show all points that define its shape.
2. You can drag each point to a new position by using a mouse, finger, or stylus. This changes the curvature of the lines connecting to neighboring points.
3. Drag from the middle curve segment with a mouse, finger, or stylus to insert a new point and adjust the curve.
4. To delete an unwanted point, choose **Delete Point** from its context menu.

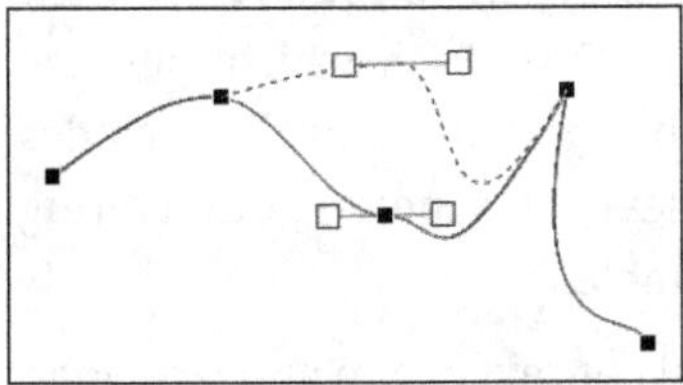

Different Point Shapes

Control points determine the path of the line connecting two points on a curve, and moving control points changes the curve's shape. Control points and their connecting lines appear only when a point is selected by clicking or tapping. Different point types depend on the control points' positions relative to each other and the point, and you can convert any point into another by using the context menu.

At the **Smooth Point**, the control points are symmetrically arranged along a tangent through the point. Both control points are equally distant from the point, and moving one control point shifts the other in the opposite direction. The curvature is the same on both sides of an even point. By contrast, the **Straight Point** option allows control points to be at different distances from the point.

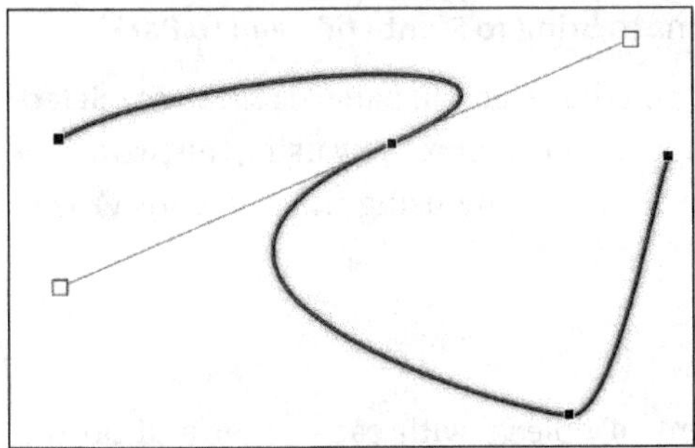

Figure 10.14 Smooth Point

At a **Corner Point**, the two control points are completely independent and can form any angle, thus allowing sharp points.

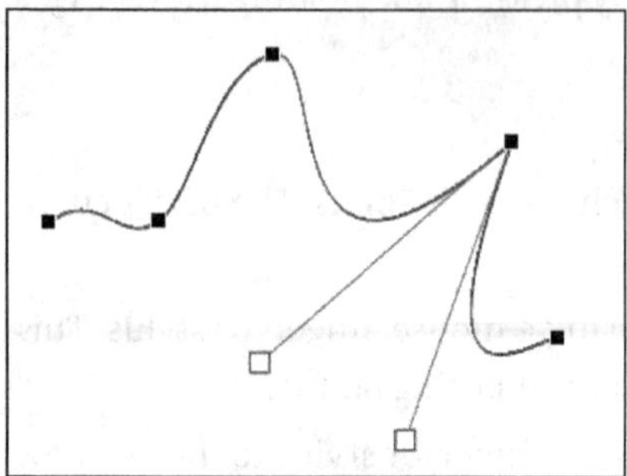

Figure 10.15 Corner Point with Tangents

You can easily turn an open curve into a closed one by using the **Close Path** context menu option. Likewise, you can reopen a closed curve with **Open Path**, which splits the point where the curve was closed during construction. The context menu also includes the **Exit Point Editing** command, which you can use to leave the special point editing mode. This also happens when you click on a cell in the table.

On the touchscreen in **Touch input** mode, the horizontal context menu appears when you hold your finger on the period briefly. You'll only see the period commands after tapping the arrow button at the end.

10.3.7 Techniques for Complex Drawings

Complex graphics are created by combining multiple objects and then aligning them side by side or stacking them as needed. All objects are arranged in different layers relative to one another: one object sits above another, which in turn sits below a third.

Order

The hierarchy is initially set by the order in which objects are created, and you can freely change it by using the options in the **Arrange** group on the **Shape Format** tab. You can move selected objects forward or backward one level by using the **Bring Forward** or **Send Backward** buttons, which also include options to **Bring to Front** and **Send to Back**.

For more extensive regrouping, it's best to open the Selection pane via **Arrange • Selection Pane**. Here, you can reorder selected shapes in the hierarchy by using the two arrow buttons ❶, and you can also temporarily hide objects by using the eye icons ❷ (see Figure 10.16).

Alignment

Complex graphics require precise alignment of objects with each other and on the worksheet, and the **Arrange** group offers tools for this under **Align**. Various options will appear when you select at least two objects, and you can use the **Distribute Horizontally** and **Distribute Vertically** commands to arrange three or more objects with equal spacing horizontally or vertically (see Figure 10.17).

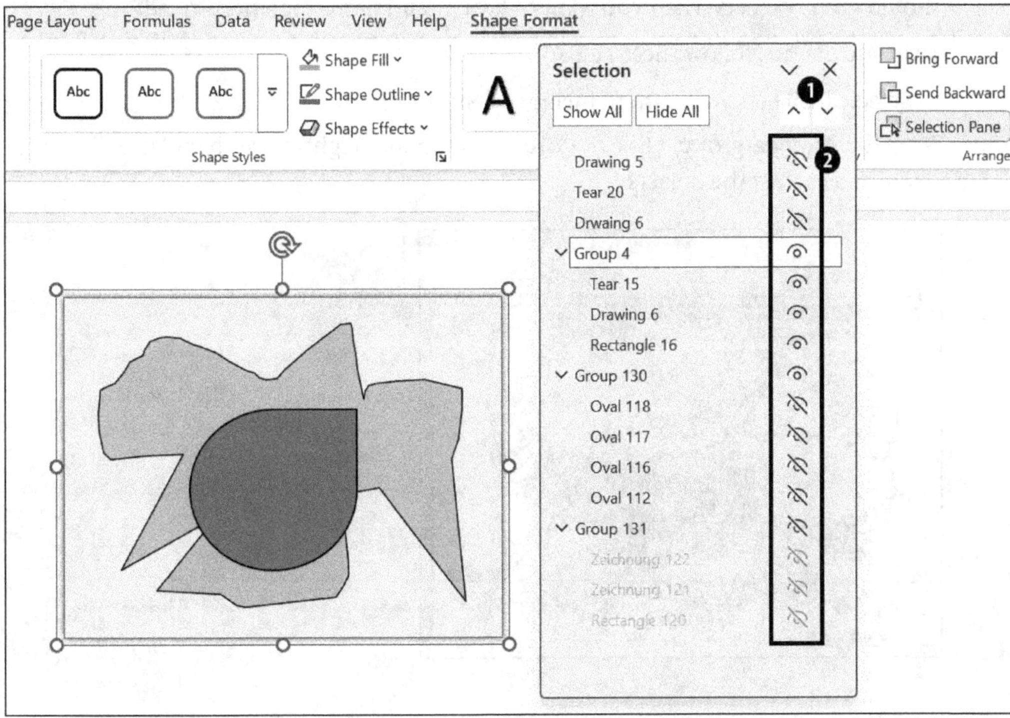

Figure 10.16 Layering Order

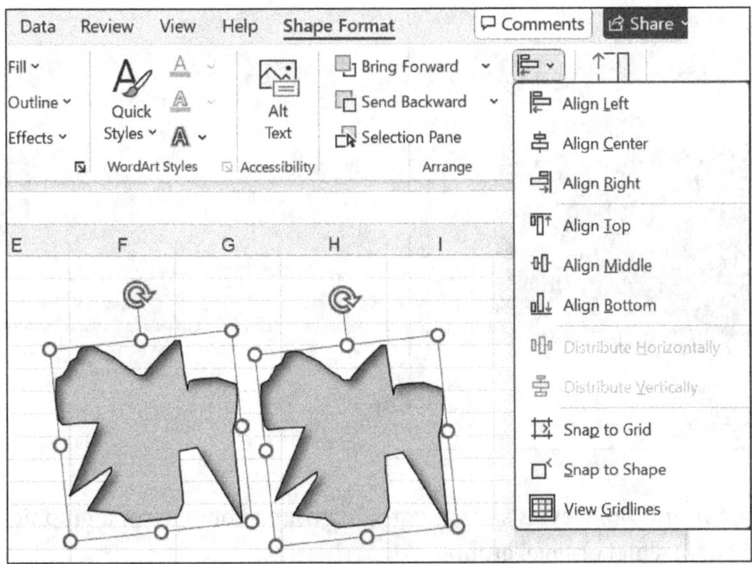

Figure 10.17 Aligning Objects

Grouping Objects

If a drawing contains multiple objects, you can group them together. Excel will then treat the group as a single object and allow you to move or resize it all at once. You should group objects when you want to keep their relative positions fixed.

To group objects, follow these steps:

1. Select all objects you want to include in the group.
2. In the **Arrange** group, click the **Group** command. Eight resize handles will appear around each of the objects.

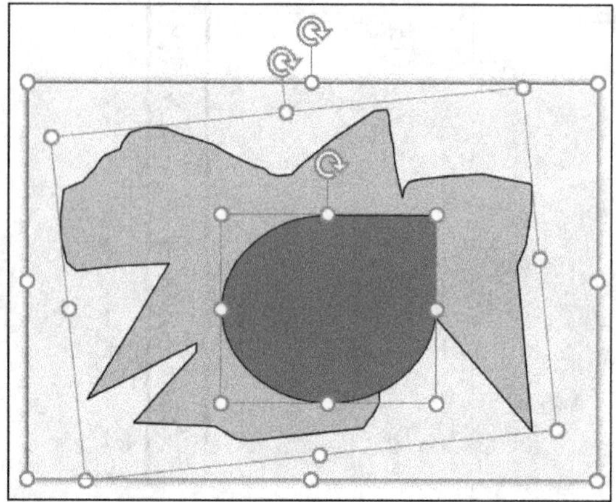

3. If you want to edit a single object within the group later, choose **Ungroup** from the **Group** menu.

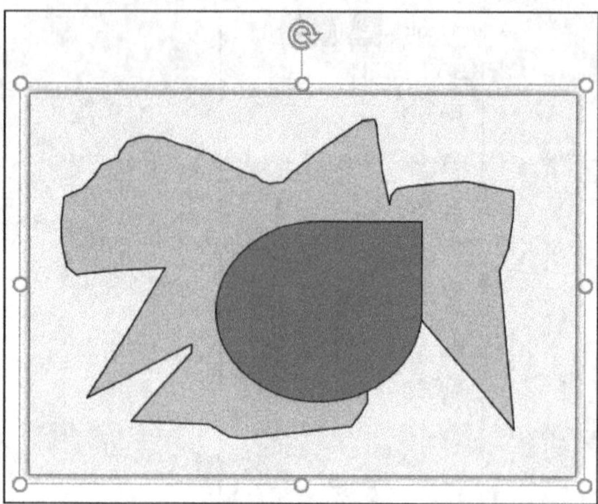

You can apply **Group** in multiple levels so that you can group subranges of a graphic later and include them in a higher-level group.

10.3.8 Shape Effects

You can add shadows, reflections, glows, soft edges, bevels, and 3D effects to most graphic objects. You can access all these options through the **Shape Effects** button we mentioned earlier in the **Shape Styles** group and the **Format Shape** or **Format Graphic** task panes. Here's an example of how to do this:

1. Select an object, open the **Shape Effects** button menu, and choose the **Shadow** palette.
2. Click or tap to pick a shadow style.

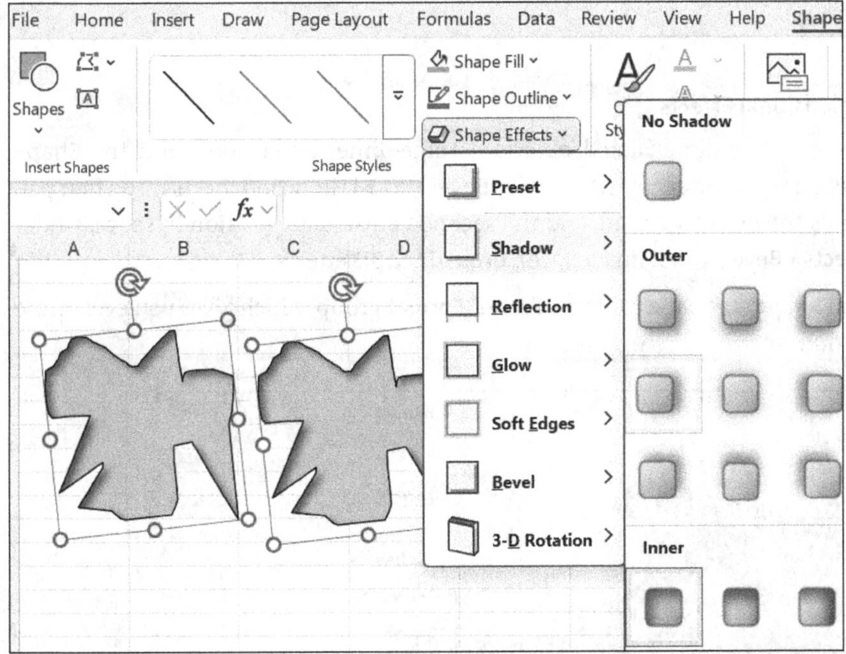

3. If the preset styles aren't enough, open **Shadow Options** to access the **Format Shape** task pane, go to the **Effects** page and the **Shadow** group, and use the small sliders or spin boxes there to fine-tune all settings.
4. Under **Angle**, move the shadow up, down, right, or left to get the effect you want.
5. You can change the shadow's color by using the **Color** palette.
6. You can also freely adjust the shadow's **Transparency** there.

The **Format Shape** pane lets you select any objects while it's open. Instead of a shadow, you can apply a glow effect that surrounds the object with an aura. You'll find many glow options under **Glow** and **Soft Edges**, and you can pick any color for these effects by using **Color**. The **Soft Edges** patterns let you soften the shape's outline to different degrees, and similar effects are available under **Reflection**.

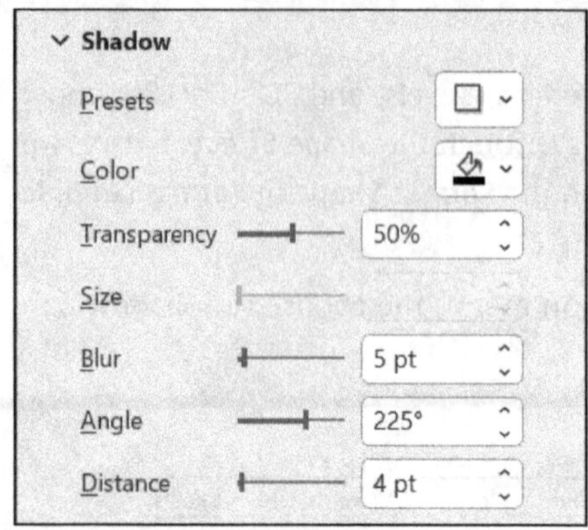

Three-Dimensional Effects

3D effects give two-dimensional objects a three-dimensional look, and the **Shape Effects** button menu offers palettes under **Bevel** and **3-D Rotation** for this. The first palette controls the object's design, and the second controls its position in space. Under **Shape Effects • Bevel**, you'll find a dozen predefined 3D effects.

The **Format Shape** task pane contains the **3-D Format** group, which gives you even more design options.

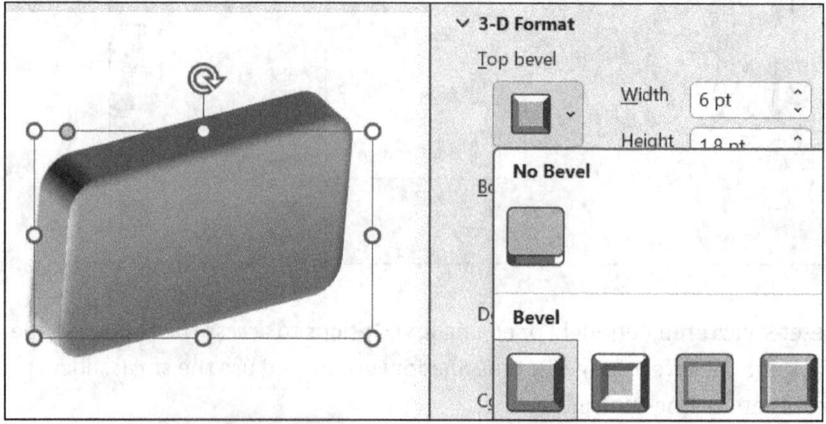

Figure 10.18 Example of Beveling Rectangle

In addition to values that define the bevel type, you can choose specific surface effects under **Material** and **Lighting** (see Figure 10.19).

Under **Shape Effects • 3D Rotation,** you'll find many patterns for positioning objects in space. You can select either a perspective or parallel projection (see Figure 10.20).

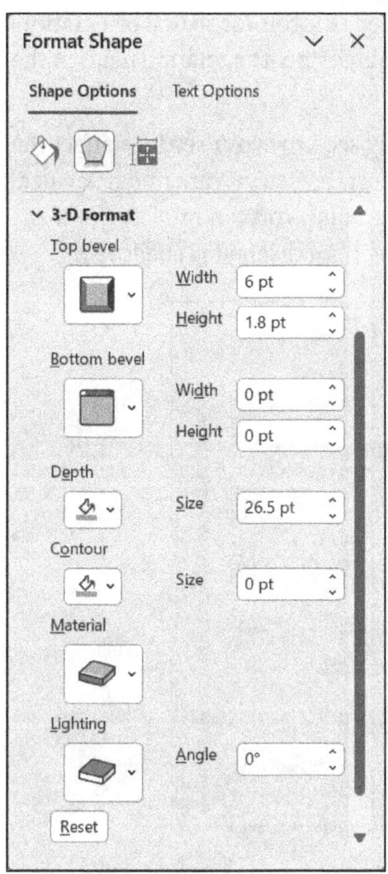

Figure 10.19 The 3D Format Options Group

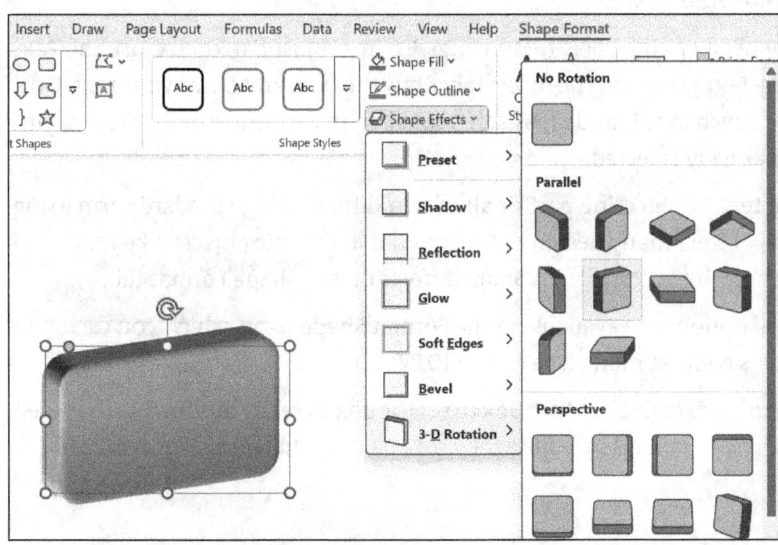

Figure 10.20 Selecting Object's 3D Representation

In the task pane under **3D Rotation**, you'll find detailed controls for spatial orientation, and you can adjust the values for the three axes by using either the rotation fields or the small arrow buttons.

When you enable the live preview, you can instantly see how your settings affect the object. If you select a perspective pattern, you can adjust the field of view under **Perspective** to be narrower or wider. Under **Distance from ground**, you can raise or lower the object relative to a virtual base position, which is useful for displaying shadows.

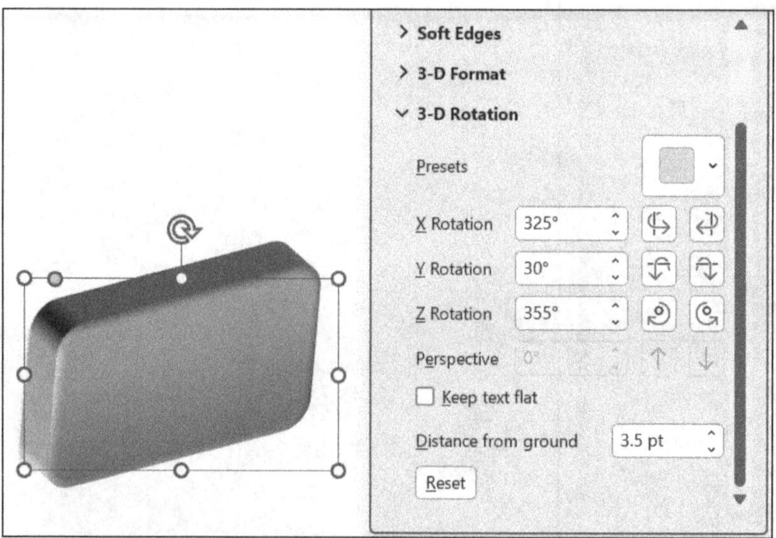

Figure 10.21 Options for Rotation in Space

10.3.9 Freely Movable Text Boxes

You'll frequently need to add labels to drawings. To insert just a text box, use the **Text Box** button in the **Text** group on the **Insert** tab, draw the box, and enter your text. On a touchscreen, use **Touch input** mode to insert a text box in the center of the screen with placeholder text already selected.

You can edit the text by selecting part or all of it and then using standard formatting tools. The text box offers many design options found in graphic objects, like reflection or 3D rotation. These are accessible via **Shape Effects** on the **Shape Format** tab.

Additional special options are available in the **Format Shape** pane, which you can open from the text box's context menu (see Figure 10.22).

Under **Text Options**, you can select **Text Box** to set the box to resize automatically, based on the text. You can also precisely adjust the box's borders there (see Figure 10.23).

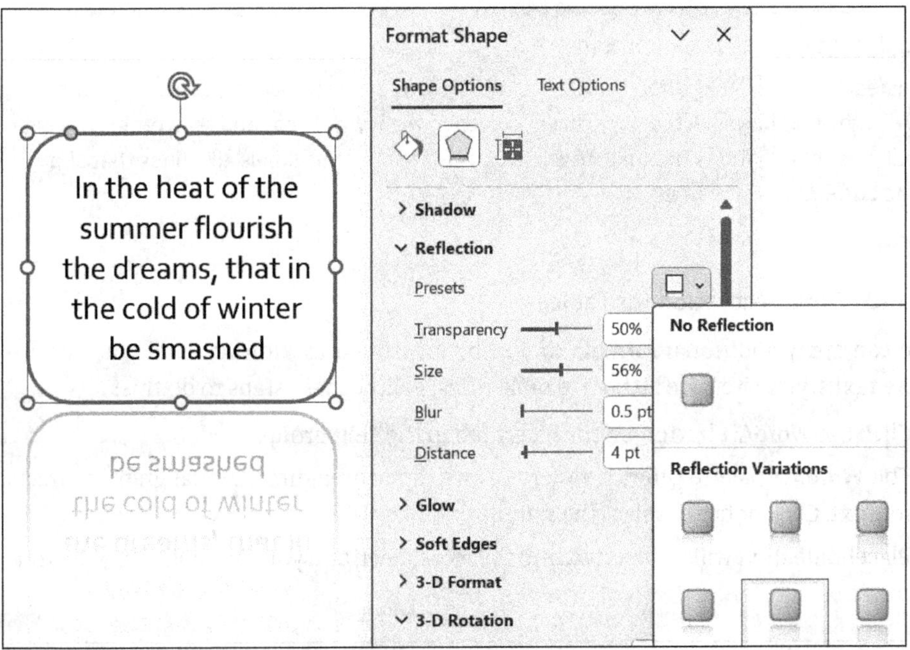

Figure 10.22 Mirrored Text Box

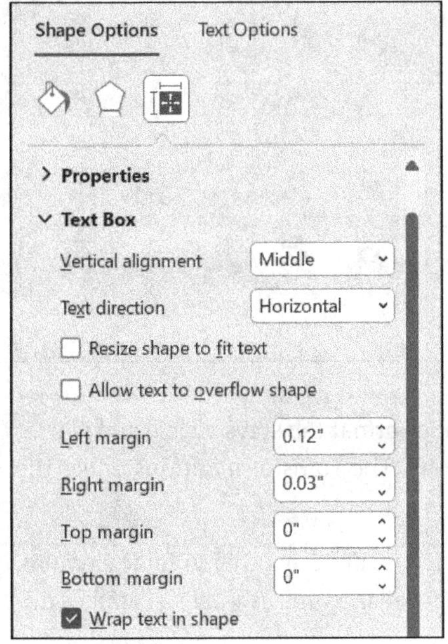

Figure 10.23 Text Box Options

[+]

Link Text Boxes to Cell Contents

Instead of typing text directly into a text box, you can reference the contents of a cell. With the text box selected, type an equal sign in the formula bar and then enter the cell address that contains the text. This lets you create dynamic labels, like ones that show the current month name.

10.3.10 Text Decoration for Tables

You can apply additional formats to text by creating it as WordArt, and you can use these text styles to create attractive table titles. Follow these steps to do this:

1. Click the **WordArt** button on the **Insert** tab in the **Text** group.
2. The **WordArt** palette offers a variety of two- and three-dimensional graphic effects for text. Click or tap to select the sample you like.
3. Placeholder text will be inserted into the worksheet, and you can replace it with your own text.

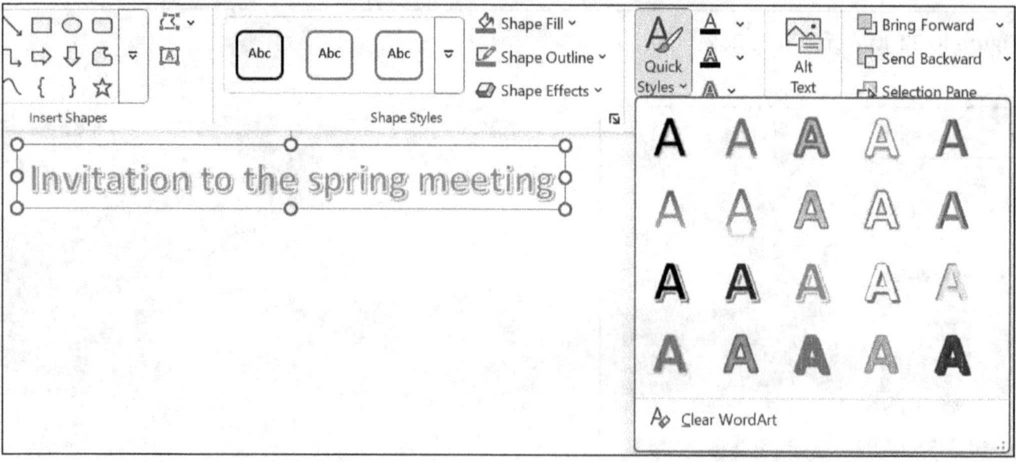

4. While the WordArt object is selected, the **Shape Format** tab stays visible and will let you apply styles from the **WordArt Styles** group. The **Transform** options under the **Text Effects** menu are especially useful here.

You can move, resize, and rotate WordArt objects just like any other graphic element, and you can change the effects of a WordArt object at any time as well. Just apply a different effect.

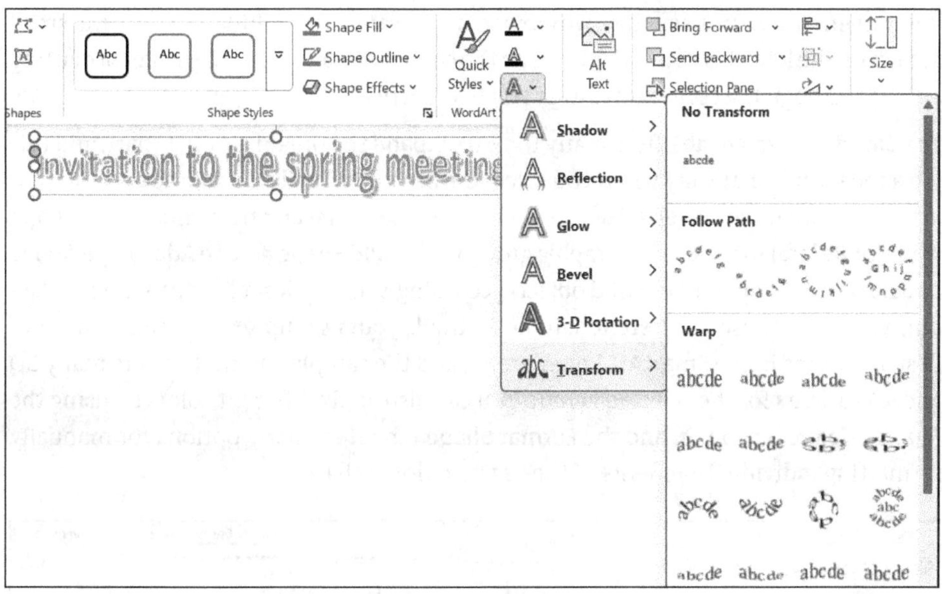

10.4 Creating Organizational Charts in a Hurry

Are you planning to reorganize a department or company? Excel offers a handy tool for this. Click the **SmartArt** button in the **Insert • Illustrations** group to open a dialog where you can choose from various basic shapes to visually represent specific relationships. If you select an organizational chart under **Hierarchy**, Excel creates a layout template where you can enter your company's data.

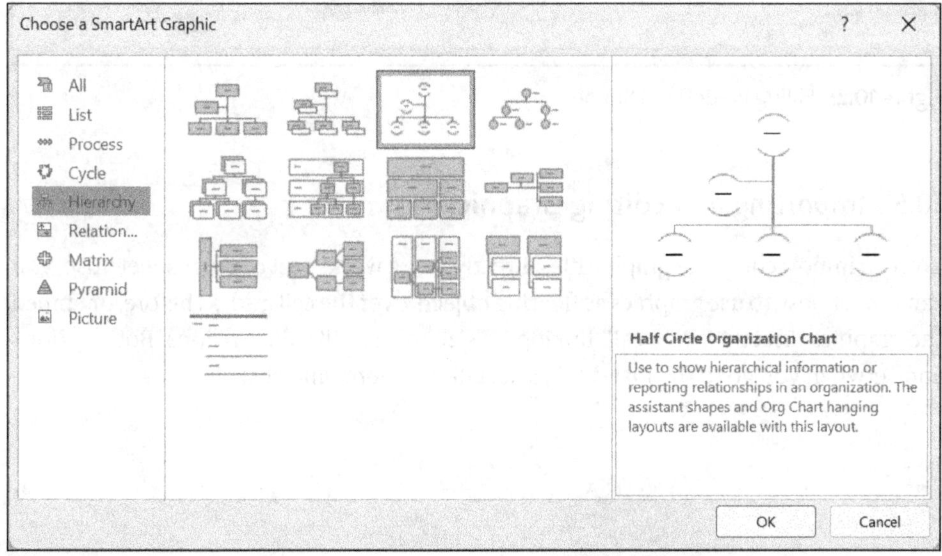

Figure 10.24 Choosing the Type of Diagram

You can enter text in a separate window that you can show or hide by using the small arrows on the left. This simplifies labeling the chart. You can access the usual formatting options through the context menu of any text entry.

The **SmartArt Design** tab offers many tools to expand the organizational chart, and this tab appears automatically when you select the organizational chart. To add an element at a specific hierarchy level, select the adjacent element first and then choose **Add Shape After** (or **Before**) in the **Create Graphic** group under **Add Shape** ❶. Use **Add Shape Above** or **Below** to insert parent or child objects accordingly. If you don't like the scheme's layout, you can choose a different sample in the **Layouts** group ❷. You can find more design options in the **SmartArt Styles** group, and the sample palette ❸ offers many 2D and 3D variants for the selected layout. You can also apply different colors by using the **Change Colors** button ❹, and the **Format Shape** tab offers many options for manually formatting individual elements of the organizational chart.

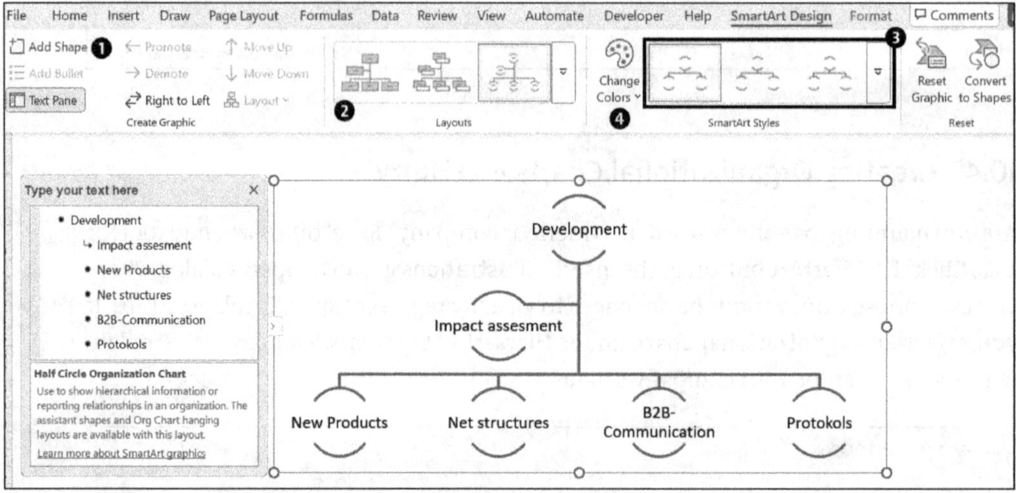

Figure 10.25 Filling Predefined Shape

10.5 Importing and Editing Graphics

You can import complete graphic files directly into a worksheet or chart sheet. Also, you can now choose to use graphics as floating objects over the cell grid, as before, or embed the graphics directly into a cell, linking its size to the cell's dimensions. Both options appear when you go to the **Insert** tab, select **Illustrations**, and click **Pictures**.

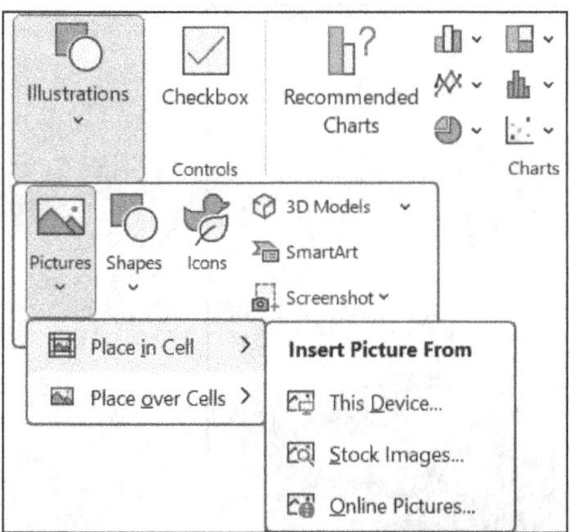

Figure 10.26 Selecting Purpose of Importing Image

10.5.1 Inserting Pictures into Cells

Inserting small pictures inside cells of a column is useful when creating lists of products, locations, buildings, or participants. Here's a simple example of how to do it:

1. Select the cell in the worksheet where you want to insert the image.
2. On the **Insert** tab, under **Illustrations**, select **Pictures • Insert into Cell • This Device**, which opens the **Insert Picture** dialog. Then, select the file type to display files of that type. If you select a file, Excel shows a thumbnail—if the corresponding view is enabled—so you can easily confirm that it's the right file.
3. Click **Insert** to place the picture into the active cell.

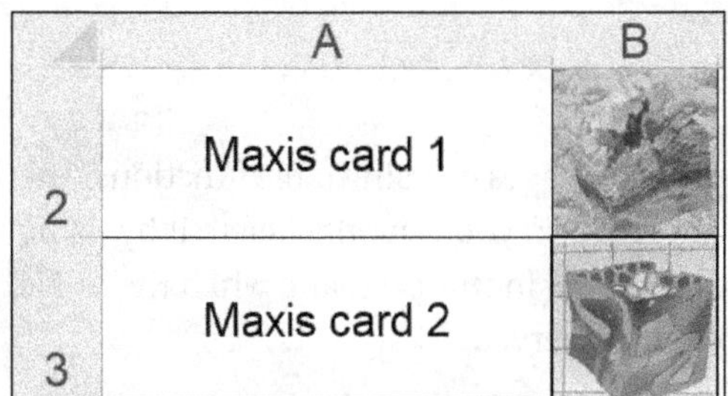

Figure 10.27 Picture Inside Cell

Autogenerated alternative text will appear in the formula bar, and you can resize the picture by adjusting the row height and column width. A button will also appear near the top right of the picture, and it will let you convert the cell image into a floating picture above the grid. A button will also appear on the floating image, and you can click it to return the image to the cell's frame.

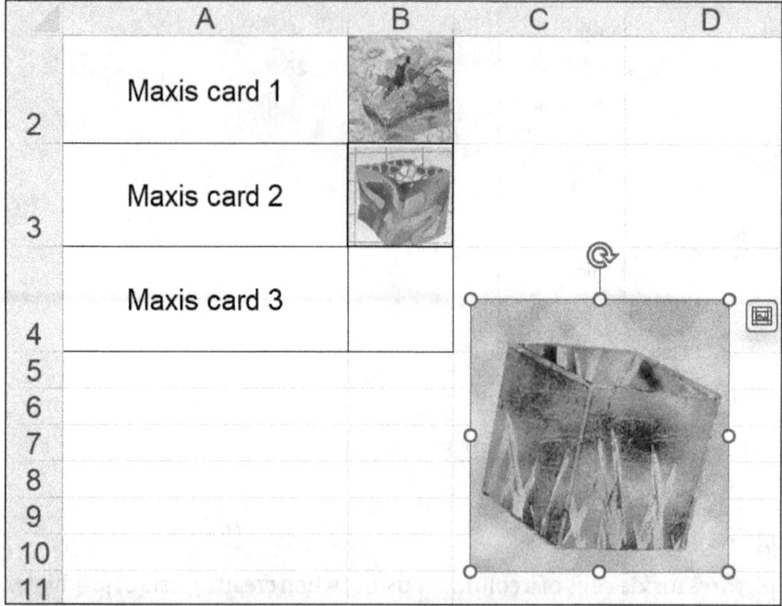

Figure 10.28 Floating Image

The context menu for the image in the cell also offers options to show a larger preview, create a lookup to the cell, or edit the alternative text.

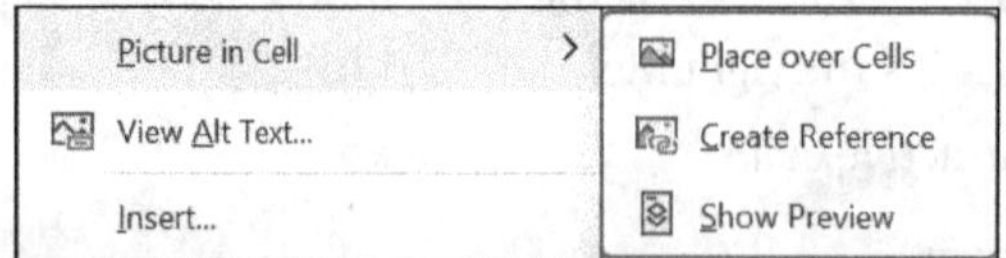

Figure 10.29 Context Menu for the Cell Image

Section 10.5.3 also explains how to create such cell images by using table functions. The image in the cell is the cell's content, not its format, so you can only delete it by using **Clear Contents**. The `=ISBLANK()` formula returns `FALSE` in this case, and when the cell is copied or moved, any changes to its size are not preserved.

10.5.2 Inserting Images over Cells

A simple way to enhance your worksheets is by importing images from a file on your computer or from the web. However, Excel may fail to load an image even if the file type is supported, and that often happens if the file wasn't saved correctly or the file type doesn't match the actual format. Issues can also occur if the image was created with older software versions that use outdated file type variants. In these cases, it's best to resave the file in the original program or convert it to a different format if needed. Follow these steps to do that:

1. To insert an image file, go to the **Insert** tab and then, under **Illustrations**, select **Pictures • Insert Over Cells • This Device**. This opens the **Insert Picture** dialog box.
2. From the **Insert** button menu, you can either insert the image directly into the Excel file or link to the image file. After you confirm the dialog box, the selected image appears in your worksheet.
3. You can move and resize the image as usual. To perform further editing, use the tools on the **Picture Format** tab, which we explain in the next section. This tab isn't available for images in cells, but you can easily switch between the two image states.

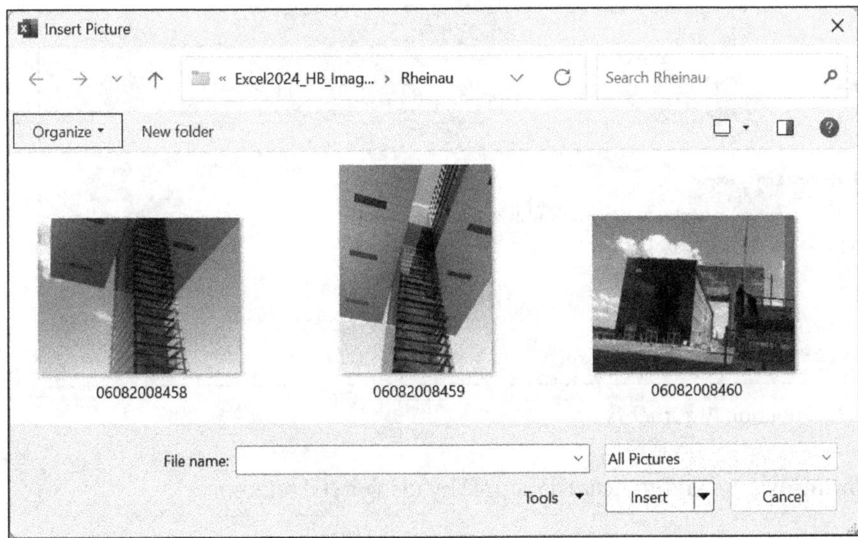

In the graphic's context menu, you'll find a **Change Image** option if the wrong image was inserted by mistake, and you can select it to reopen the **Insert Picture** dialog. Excel now supports the newer image format WebP, which is designed specifically for displaying images on the web. WebP supports lossy and lossless compression as well as transparency, and in either mode, it's about 25% smaller than JPEG or PNG. However, not all platforms and applications support WebP yet.

10.5.3 Inserting Images by Using a Function

A new feature lets you insert images into cells by using the `IMAGE()` lookup function. The source is the image file's URL path using the HTTPS protocol. The function supports BMP, JPG/JPEG, GIF, TIFF, PNG, ICO, and WebP file formats.

Optional parameters include alternative text describing the image and dimension settings. For **Sizing**, you can use the following values:

- **0**: The image will be fitted into the cell while maintaining its aspect ratio.
- **1**: The image will fill the cell, ignoring the aspect ratio.

- **2**: The image will keep its original size, which may extend beyond the cell boundaries.
- **3**: The image size will be set with the **Height** and **Width** arguments.

In the formula, you can specify the **Height** and **Width** in pixels. If you enter only one value, the aspect ratio will be preserved.

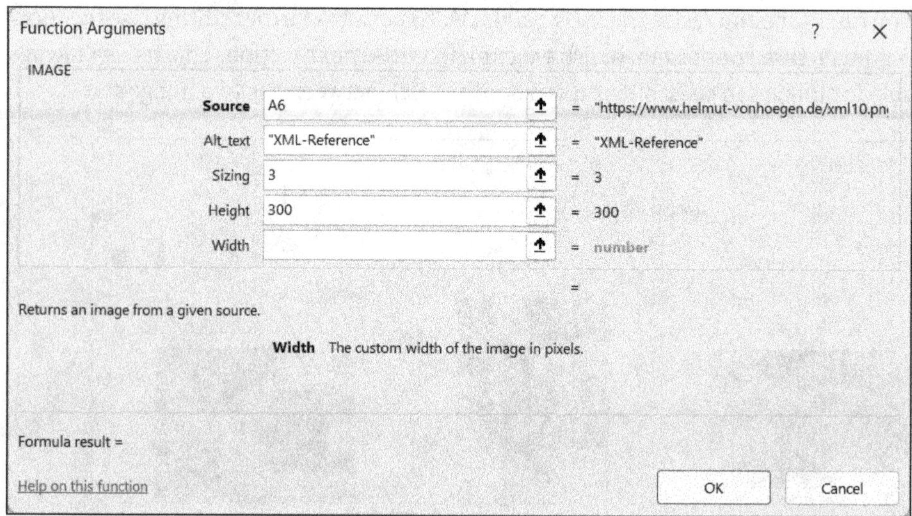

Figure 10.30 Function Arguments

The following example references the source by using a cell reference:

```
=IMAGE(A6,"XML-Reverenz",3,300)
```

This function's advantage is that it supports conditional image display. For example, an `IF()` function can show image A in one case and image B in another. However, the image may not display if the network resource is inaccessible.

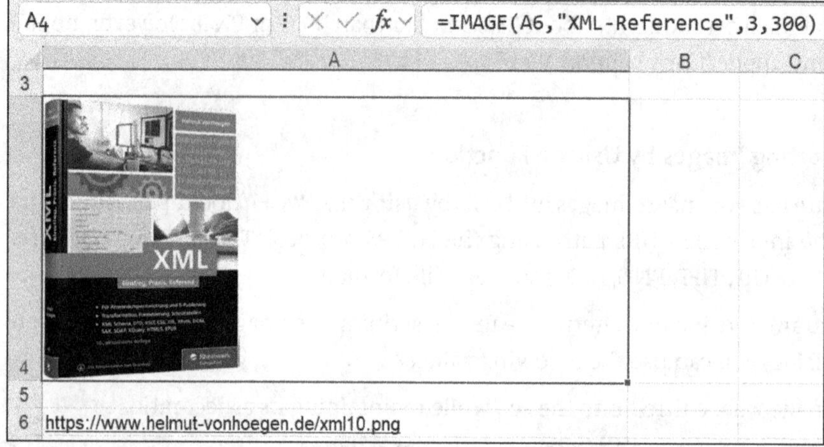

Figure 10.31 Importing Image into Cell

10.5.4 Editing Images Directly

To help you avoid having to switch to a separate image or graphic editor to make minor changes, Excel provides tools on the **Picture Format** tab. Basic image editing functions are available, and you can easily select effects like chalk sketches or watercolors from the sample palette under **Artistic Effects.** In the first group, **Adjust**, you'll find a **Remove Background** button ❶ that opens its own **Background Removal** tab.

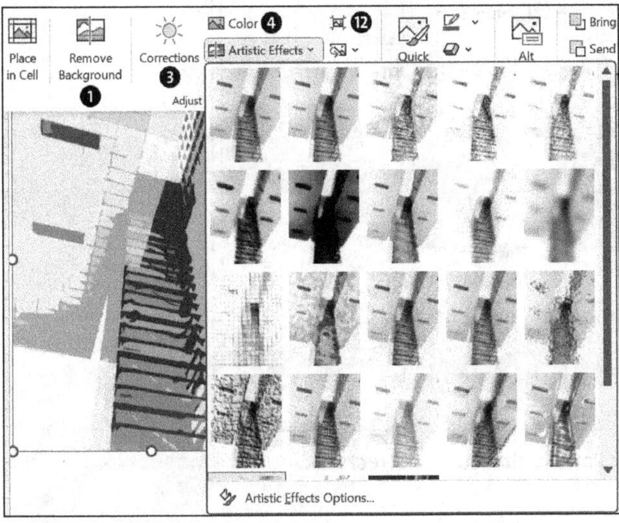

Figure 10.32 Picture Tools on Picture Format Tab

Use the buttons in the **Refine** group ❷ to drag markers and specify which areas to remove or keep.

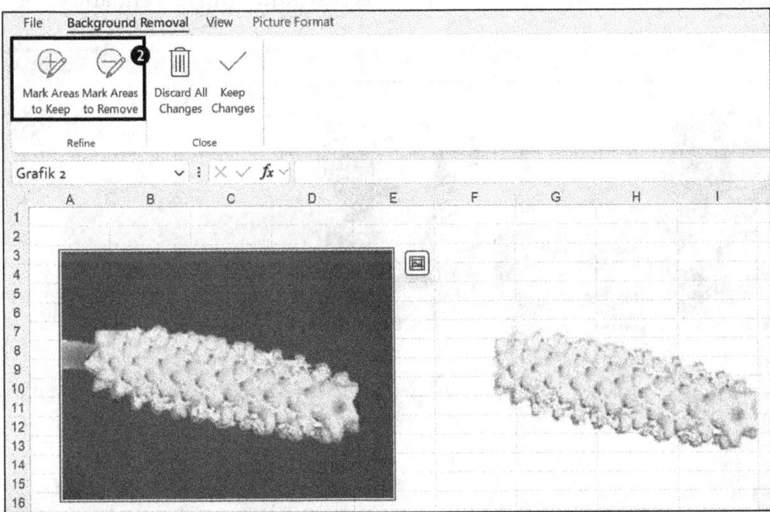

Figure 10.33 On the Left Is the Suggested Area to Remove, and on the Right is the Cropped Result with a Transparent Background

The **Corrections** ❸ button offers a palette of graduated options to adjust brightness and contrast. Selecting an option immediately previews the effects on the image.

If you need a different setting, use **Picture Correction Options** to open the **Format Picture** task pane, which lets you make continuous adjustments on the **Picture** tab under **Picture Corrections.**

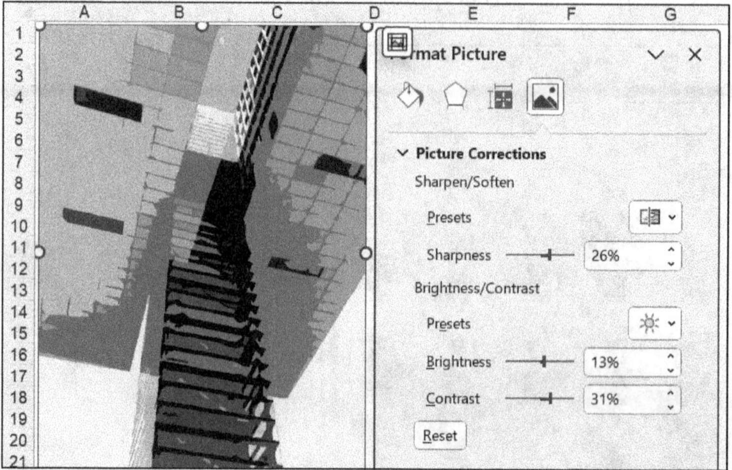

Figure 10.34 Format Picture Task Pane with Picture Correction Options

The **Color** button in the **Picture Format • Adjust** group (refer back to Figure 10.32 ❹) changes the picture's color mode and offers many light and dark variants as samples. The sample named **Washout** creates a watermark-like effect.

Click or tap a color and then use the **Set Transparent Color** ❺ option to make that color transparent. This lets you, for example, make a solid background transparent and effectively isolate the image content.

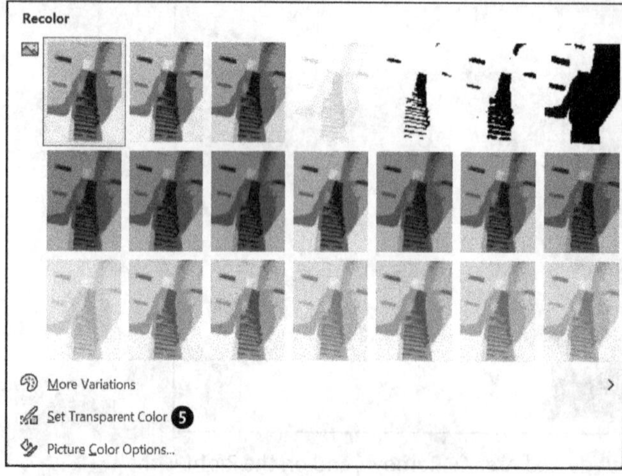

Figure 10.35 Color Correction Palette

The **Transparency** ❻ button opens a palette ❼ with various transparency levels for the selected image. Click or tap to apply a level. The **Picture Transparency Options** button opens the **Picture Transparency** group in the **Picture Format** task pane.

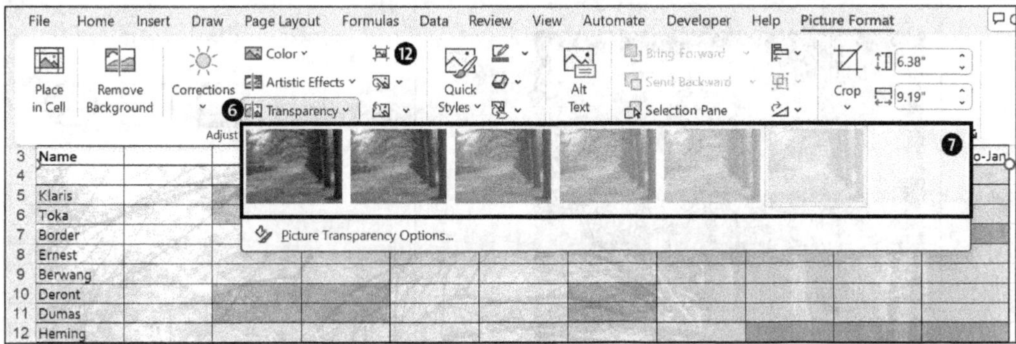

Figure 10.36 Table with Transparent Image

If you only need part of the image, use the **Crop** tool in the **Size** group, which you can use to resize the border around the image with the eight handles. The button's menu also includes the **Crop to Shape** option, which you can use to trim the image to one of the shapes in the **Shapes** palette. The **Size** group also has list boxes ❽ for setting the image size precisely.

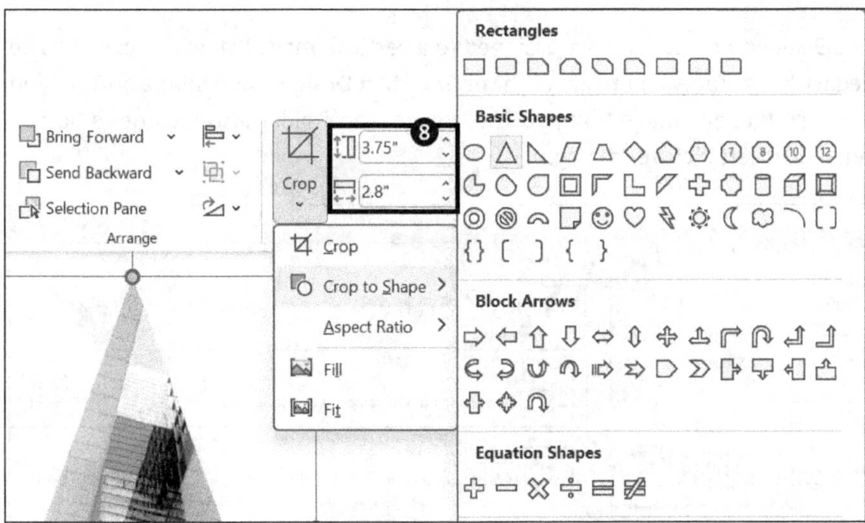

Figure 10.37 Cropping Tool

Under **Picture Styles,** you'll find a variety of useful display styles and frames for the image. You can change the border color by using the **Picture Border** palette ❾.

The **Picture Layout** button ❿ lets you open the **SmartArt Layouts** palette for more options. This is how you can insert a picture into one of these layouts.

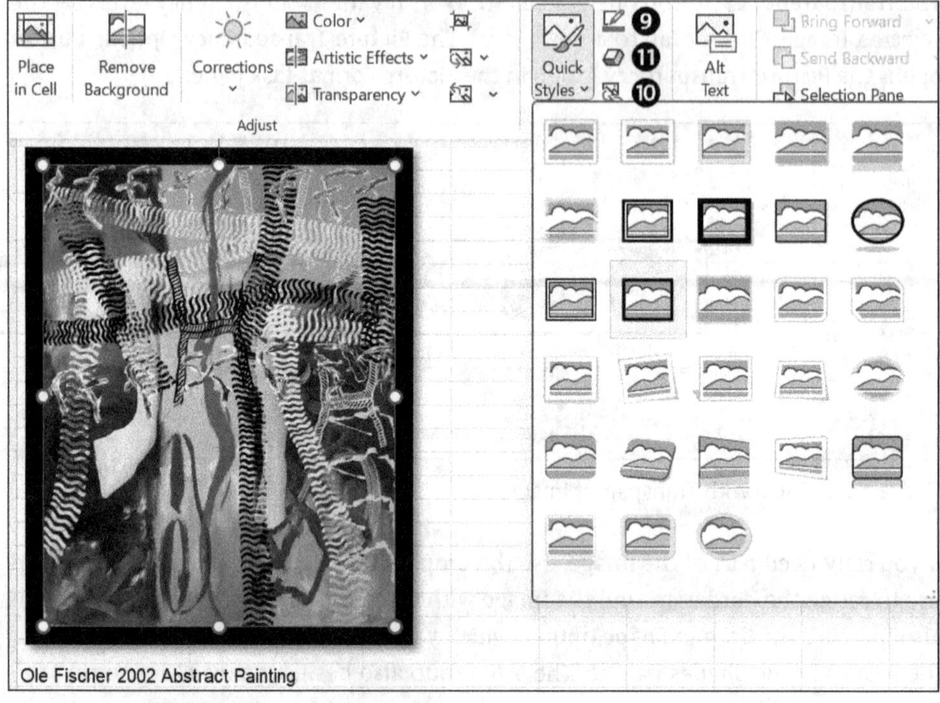

Figure 10.38 Using Image Format Templates

Figure 10.39 shows how images are assigned to a vertical image list, which can then be enhanced with text for each image. With the **SmartArt Design • Add Shape** button, you can easily insert a new image-text pair. The image placeholder also includes a button that opens the **Insert Pictures** dialog.

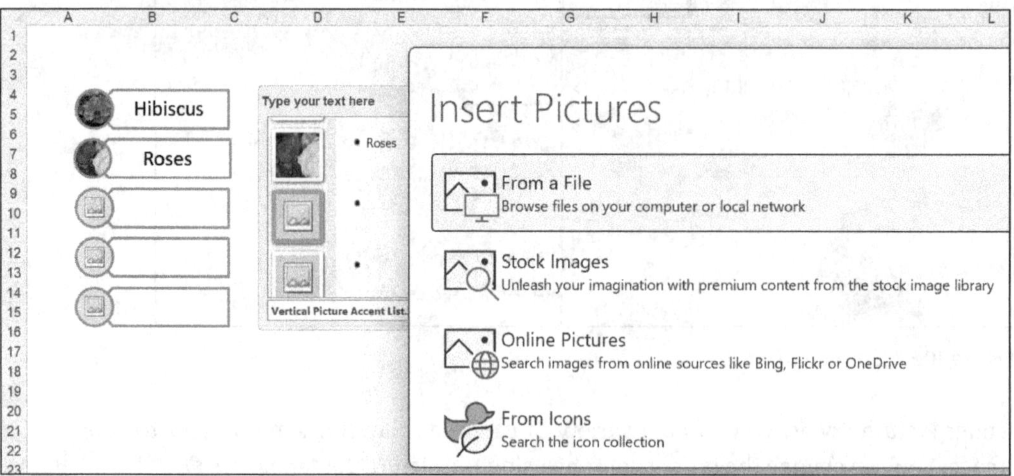

Figure 10.39 Template for Illustrated Product Lists

The options under **Picture Format • Picture Effects** match those we described earlier for graphic objects under **Shape Effects**. To reduce file size, you can compress graphics in Excel. The **Compress Pictures** button (⓬ in Figure 10.36) in the **Picture Format • Adjust** group lets you set different resolutions under **Resolution** for printing, screen display, or email.

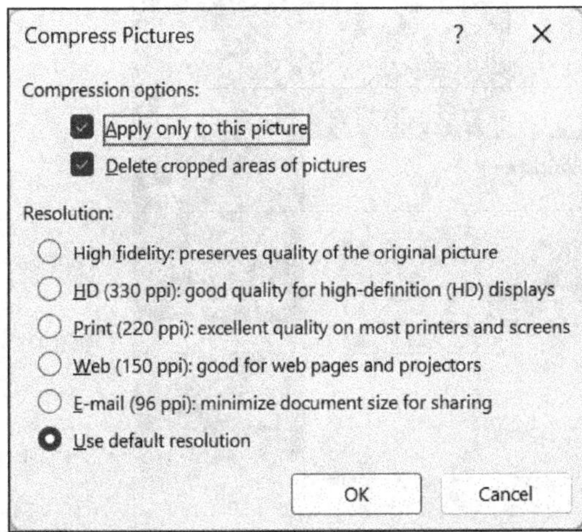

Figure 10.40 Save Space

If you make a mistake while editing a picture, you can click the **Reset Picture** button under **Adjust** to restore the original image. You can also choose to reset only the picture size here. To replace the picture in that picture's spot, use the **Change Picture** button in the same group.

10.5.5 Inserting Screenshots

A handy feature in the **Insert • Illustrations** group is the **Screenshot** button. If other windows, like your browser, are open, they appear as image sources. Clicking or tapping inserts the image at the current spot in the worksheet.

If the image is a webpage screenshot, you'll be asked whether to add the page's URL link. If you agree, clicking or tapping the screenshot in the worksheet will open the page in your browser. The link also shows when you hover the mouse over the inserted image. You can disable this prompt in **Options • Advanced** by selecting **Do not automatically hyperlink screenshot.**

You can select the **Screen Clipping** option to temporarily hide the Excel window, which lets you capture any part of the screen by drawing a rectangle there. When you release the mouse, the clipping will appear as an image object in the worksheet, and you can then treat it like any other image.

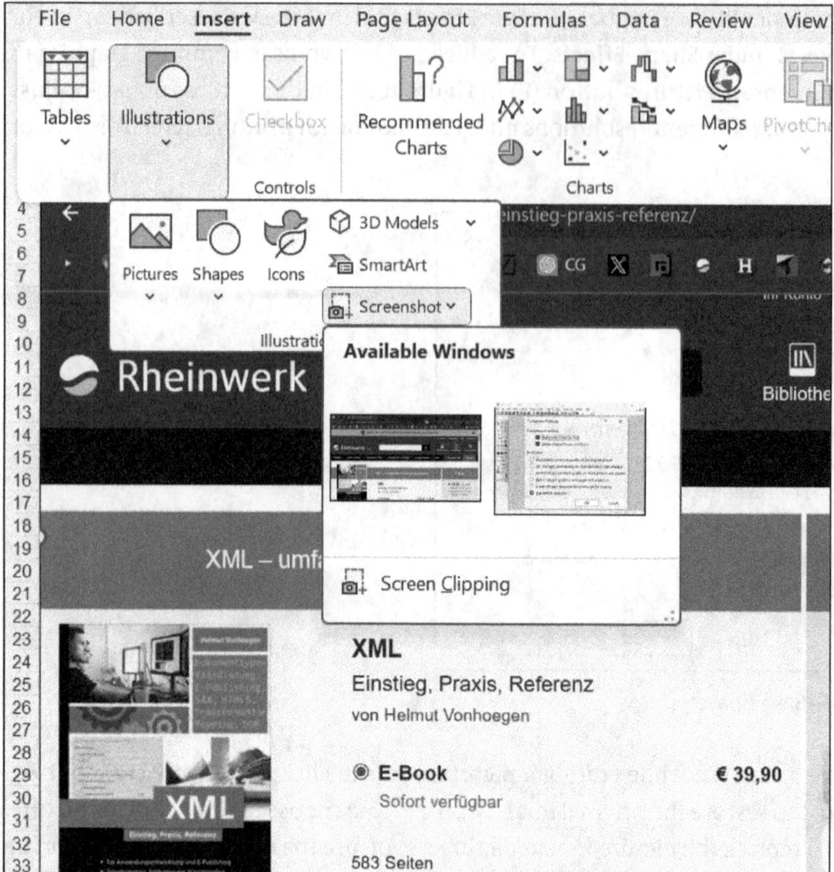

Figure 10.41 Using Screenshot Button Menu to Import Image from Website

10.6 Using Icons

Icons are typically part of a symbol system that's designed for language-independent communication. The dialog opened by the **Illustrations • Icons** button offers a wide selection of categories at the top, each of which has related symbols.

10.6.1 Using the Icons Library

Choose the icon you want and insert it into the worksheet at the mouse pointer location by using **Insert**. The icon will be added as a graphic object, and to edit it, use the **Graphics Format** tab.

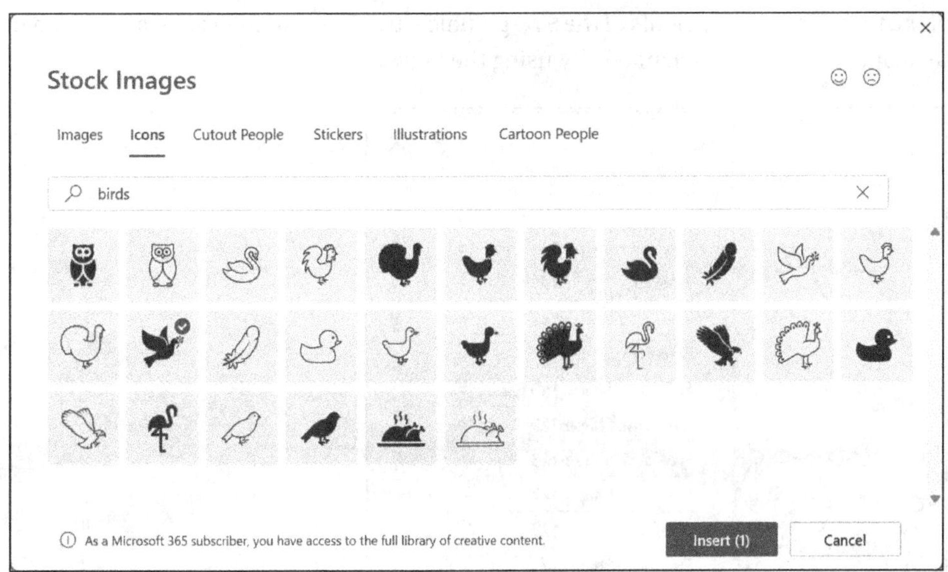

Figure 10.42 Dialog for Icons Button

10.6.2 Inserting Custom Vector Graphics

The icons we've described are in the open XML Scalable Vector Graphics (SVG) format. Excel also lets you insert your own graphics in this format, and to do it, you set the file type to **Scalable Vector Graphics** in the **Insert Pictures** dialog.

The **Graphics Format** tab is also available here. Notably, you can break the graphic into its parts by using the **Convert to Shape** button in the **Change** group.

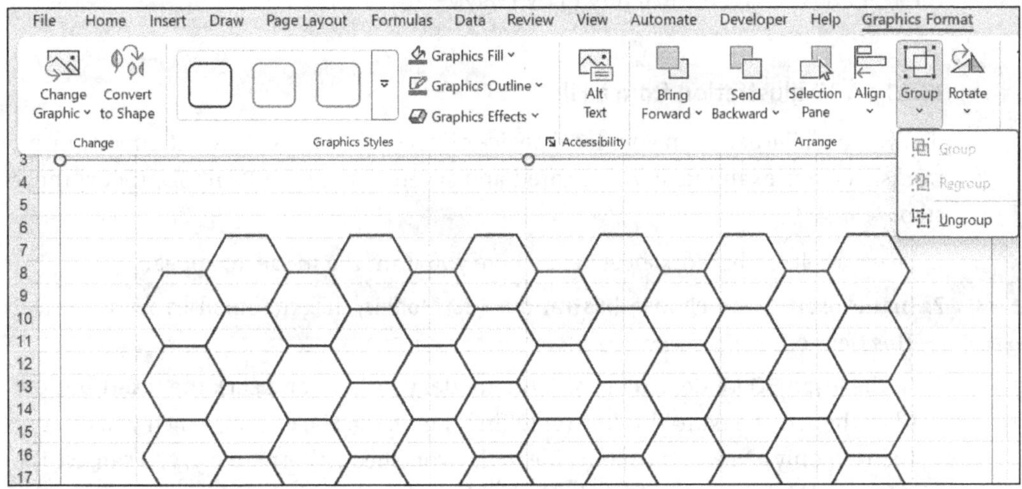

Figure 10.43 SVG Graphic in Worksheet

Next, the individual elements of the SVG graphic—two layers in this case—are available as shapes you can edit separately by using the drawing tools.

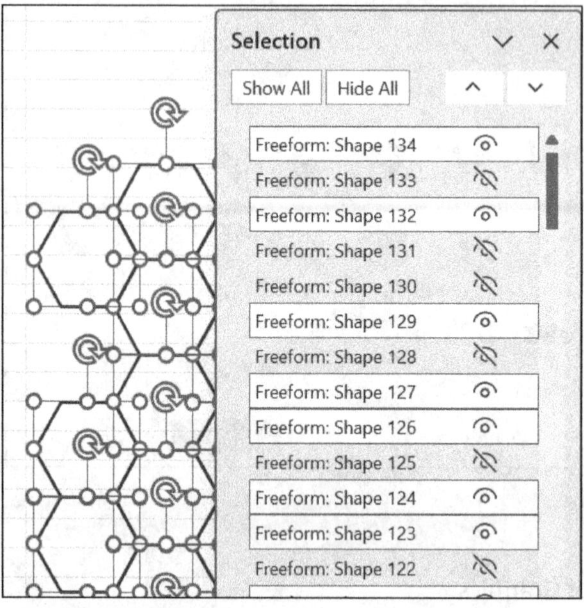

Figure 10.44 SVG Graphic Split into Two Layers for Editing

10.7 3D Models and 3D Maps

To show an object from multiple angles, it's helpful to use 3D illustrations that you can rotate. Excel offers this capability in two ways.

10.7.1 A 3D Illustration from a File

If a 3D model file exists on your local device or network, you can insert it directly into a worksheet. For example, if you created and saved a model in Paint 3D, follow these steps:

1. Select a cell on the worksheet near where you want the model to appear.
2. In the **Insert** group, choose **Illustrations • 3D Models**, click the small arrow, and select **This Device**.
3. In the **Insert 3D Model** dialog, pick the model you want and click the **Insert** button. Click the arrow next to this button to find the **Link to File** option, which you can use to insert only a link to the model file in the worksheet. Alternatively, you can use the **Insert and Link** option to insert both the file and the link so that changes to the external file will update in the worksheet.

4. The model appears on the worksheet with the usual handles, and a button in the center lets you rotate the model in any direction.

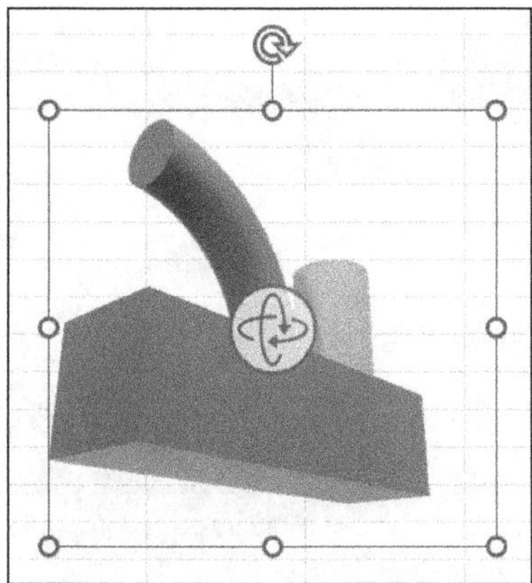

5. Right-click the frame to open a context menu with the **Views** button, which offers different positions for the current object. Simply click a preset to position the object accordingly.
6. Use the **Pan & Zoom** button to display small arrows you can use to zoom the object in or out. You can also move the object to a different spot within the frame.

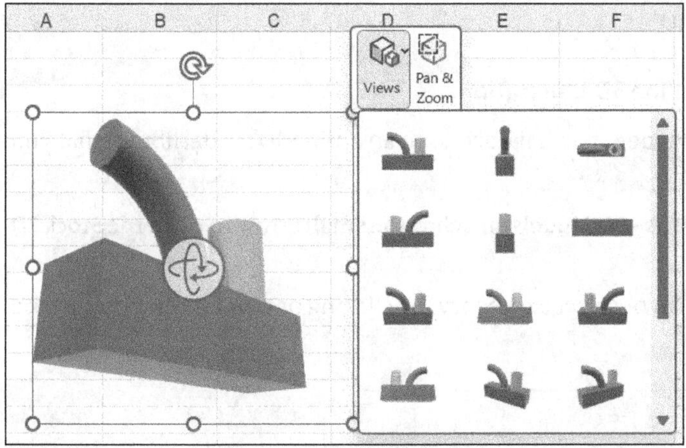

7. In the object's context menu, click **Format 3D Model** to open the task pane, where you can precisely adjust rotation around all three axes in the **3D Model** group under **Model Rotation**.

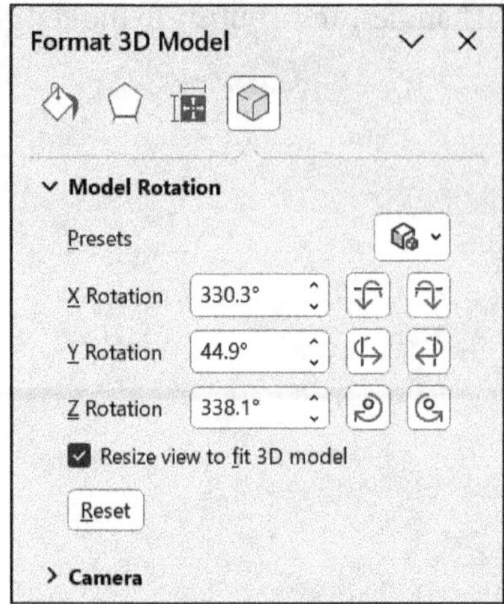

When you select a 3D object, the **3D Model** tab displays several groups for further editing. If you get lost while editing—which is easy to do with 3D—the **Reset 3D Model** button can help.

Figure 10.45 3D Models Tab

10.7.2 Importing an Online 3D Illustration

Besides locally saved 3D models, you have access to an online 3D model library that you can use as follows:

1. Go to **Insert • Illustrations • 3D Models** and click the small arrow to open the **Stock 3D Models**.
2. Either type a search term or select a category (by clicking or tapping) and then pick a suitable object.

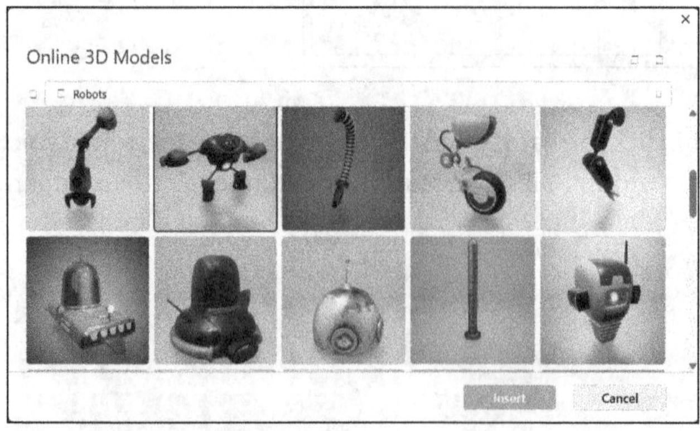

3. Using **Insert** places the selected object near the mouse pointer on the worksheet. Further handling is the same as with locally imported objects.

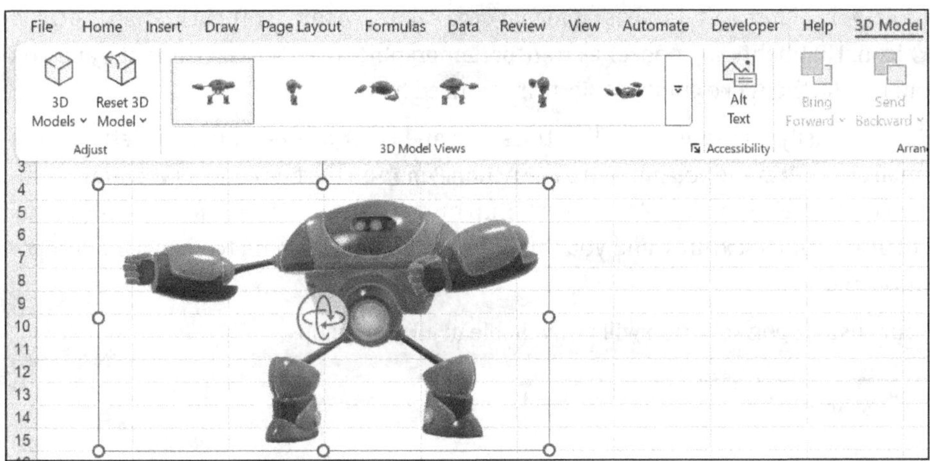

Figure 10.46 3D Model of Video Drone

10.8 Freehand Drawing

The **Draw** tab ❶ is especially useful when working with a touchscreen and pen. Touch input is possible but less precise.

Even without a touchscreen, you can use this tab to highlight specific values in a table with your mouse. You can show or hide the tab via **File • Options • Customize Ribbon**.

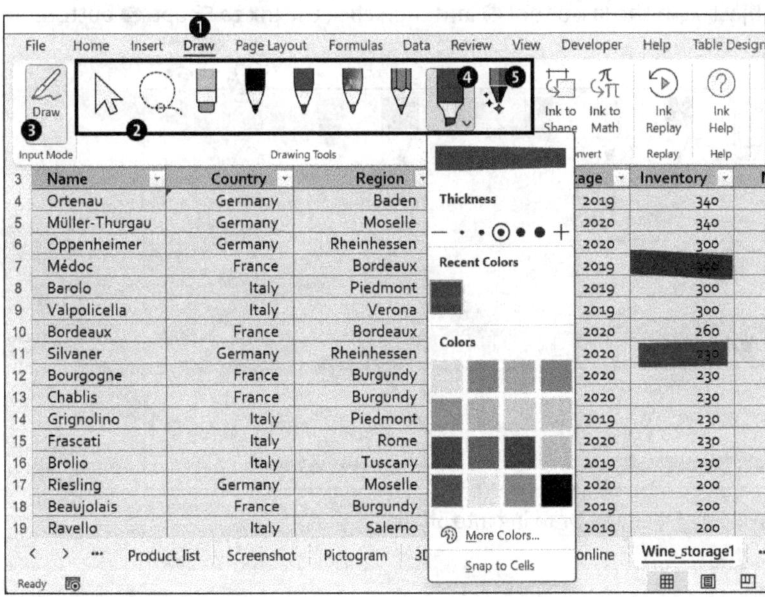

Name	Country	Region	...age	Inventory
Ortenau	Germany	Baden	2019	340
Müller-Thurgau	Germany	Moselle	2020	340
Oppenheimer	Germany	Rheinhessen	2020	300
Médoc	France	Bordeaux	2019	[illegible]
Barolo	Italy	Piedmont	2019	300
Valpolicella	Italy	Verona	2019	300
Bordeaux	France	Bordeaux	2020	260
Silvaner	Germany	Rheinhessen	2020	230
Bourgogne	France	Burgundy	2020	230
Chablis	France	Burgundy	2020	230
Grignolino	Italy	Piedmont	2019	230
Frascati	Italy	Rome	2020	230
Brolio	Italy	Tuscany	2019	230
Riesling	Germany	Moselle	2020	200
Beaujolais	France	Burgundy	2019	200
Ravello	Italy	Salerno	2019	200

Figure 10.47 Draw Tab Provides Pens and Markers in Customizable Colors and Sizes

10.8.1 Highlights and Handwritten Comments

To highlight specific values in a table, just click or tap one of the markers under **Drawing Tools** (❷ in Figure 10.47) and use it to mark the values. To turn it off, click or tap the **Draw** ❸ icon. Highlights will appear as transparent graphics over the cell contents, and you can handle them like graphic objects if you need to.

You can easily adjust all pens' line thickness and color by opening the palette via the small arrow ❹ at the bottom right. The **Action Pen** ❺ on the far right lets you draw numbers that Excel automatically converts into numeric values. Pens offer not only colors but also various textures, and you can use a pen's context menu to change its composition and order.

Your custom pen selection will be available in all Office apps.

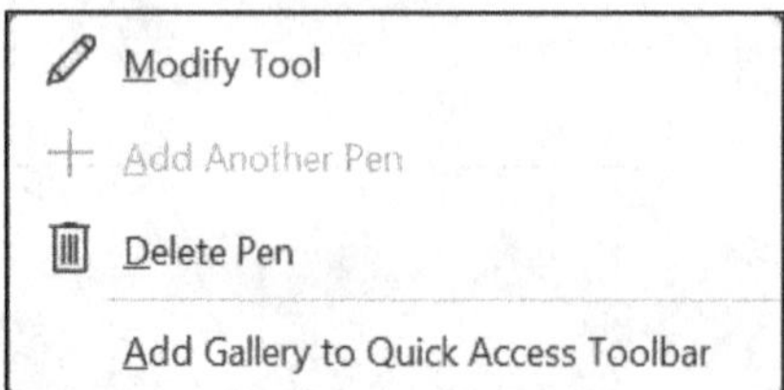

Figure 10.48 Context Menu for Pen

10.8.2 Creating Sketches

Besides highlighting or handwriting notes on table data, you can use pens to create quick sketches. Excel can even turn a rough line into a perfect triangle or circle. To do this, circle the object with the lasso tool ❶ and then click the **Ink to Shape** ❷ button.

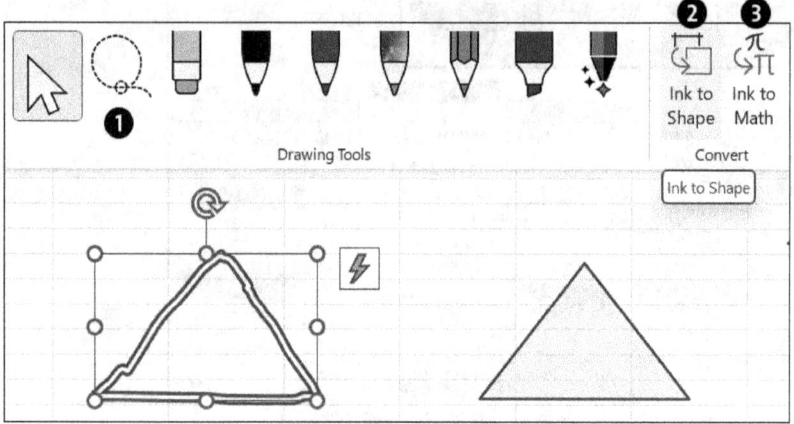

Figure 10.49 Converting Freehand Drawing into Shape

10.8.3 Inserting Mathematical Equations

It works the same way with math. To insert a mathematical equation into a sheet to document a calculation, click **Ink to Math** (see Figure 10.49 ❸) and draw the formula by hand after clicking or tapping **Write** in the dialog box. Your handwriting will instantly be converted into math in the header. If there are errors, you can circle them by using **Select and Correct** and usually fix them by choosing from the suggested list. Click **Insert** to add the formula to the worksheet as a graphic object.

To remove a hand-drawn element, use the **Erase** icon ❹ and drag the small eraser over the element. This doesn't apply to converted shapes or equations, which you can delete like regular graphic objects. To select multiple items at once, click the **Lasso Selection** ❺ icon and draw a lasso around the items.

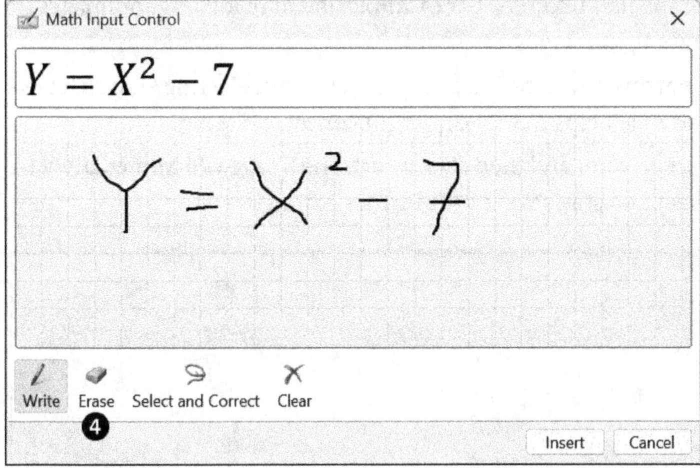

Figure 10.50 Entering a Formula by Hand

You can also use the **Ink Replay** ❻ feature, which plays your freehand drawings in the order you created them.

On the **Review** tab, use **Ink • Hide Ink** to hide all freehand elements when you don't need them. They won't be deleted, though, and you can click the same button to show them again.

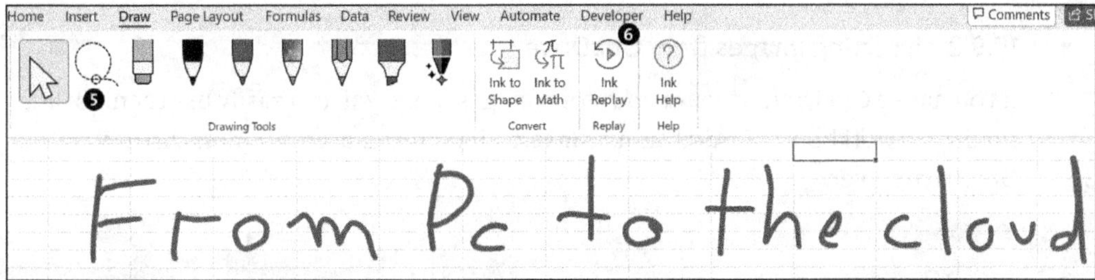

Figure 10.51 Playing Back Freehand Drawings

10.9 Finding Images on the Web

When you open the Online Pictures feature in Excel, it shows images from **Bing** by default. If you're signed in to **OneDrive**, your personal OneDrive account will also appear as an additional source.

10.9.1 Inserting a Photo into a Worksheet

If you need an illustration on a specific topic, try searching **Bing** first to see if it has something suitable. Here's how to insert an image:

1. Select the cell where you want to insert the image.
2. Go to **Insert • Illustrations • Pictures • Insert Over Cells • Online Pictures**.
3. Type a search term in the first text box. For example, if you're looking for images of roses, you'll find results right away.
4. Use the **Filter** icon to narrow down the image type, such as by selecting **Photo**. Choosing **Creative Commons only** helps reduce copyright concerns.
5. Click or tap the image you want and then click **Insert**. The image will immediately be copied to the worksheet and stay selected for further editing.

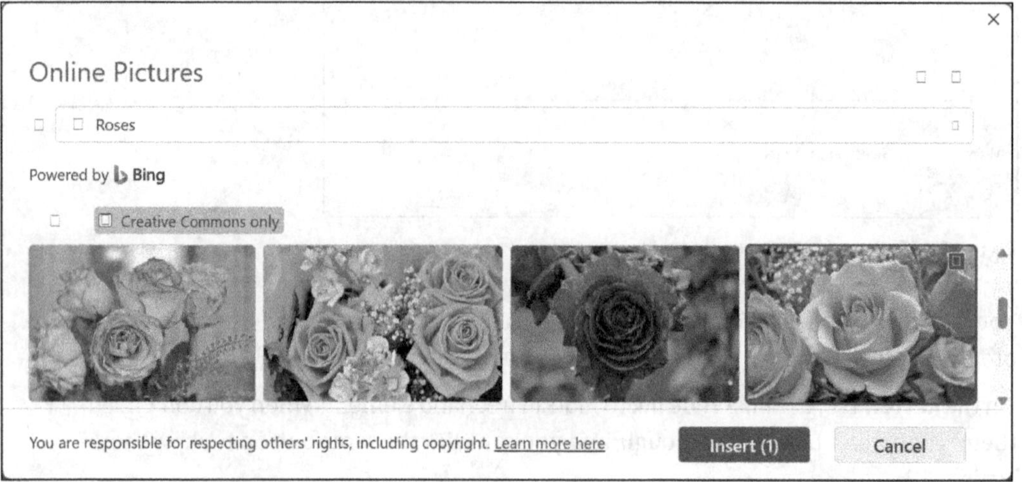

10.9.2 Inserting Images from OneDrive

If you have a OneDrive account and store images there, you can easily use them to illustrate your workbooks. To do this, use **Insert • Pictures** to access folders directly on the OneDrive server.

10.10 Table Snapshots

We conclude this chapter with a special method for creating graphic objects. It creates snapshots from previously selected table ranges and treats them as graphic objects. Each object stays linked to the table range through a formula. Here's how to use this method:

1. Select the worksheet range you want to capture.
2. Copy the selection to the clipboard.
3. Select the top-left cell in the sheet where you want to place the picture.
4. In the **Home** tab, under the **Clipboard** group, click the **Paste** button menu and choose **Other Paste Options • Linked Picture**. All options available for graphic objects will then be accessible. 10

When you select the new object, Excel shows a formula with the worksheet range in the formula bar. This creates a link. A handy use of this feature is to copy small excerpts from your key daily workbooks onto a single sheet. Double-clicking the excerpt opens the related workbook. These table snapshots can also consolidate key data from multiple tables onto one sheet, making it easy to review critical values at a glance.

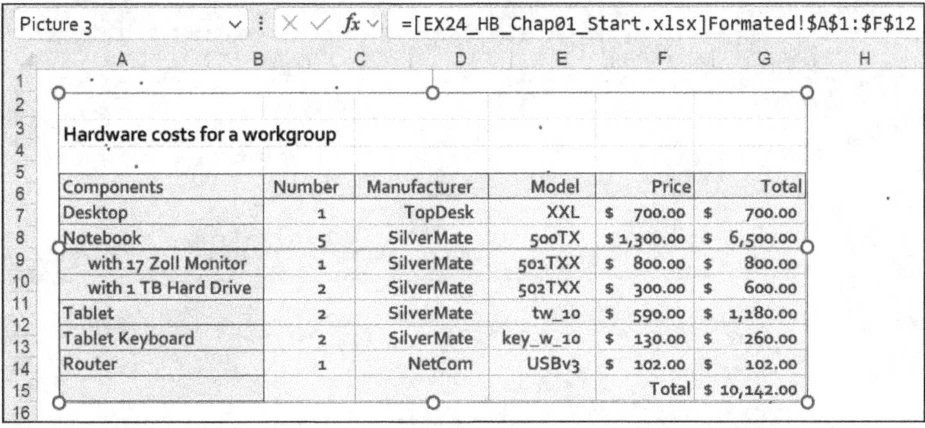

Components	Number	Manufacturer	Model	Price	Total
Desktop	1	TopDesk	XXL	$ 700.00	$ 700.00
Notebook	5	SilverMate	500TX	$ 1,300.00	$ 6,500.00
with 17 Zoll Monitor	1	SilverMate	501TXX	$ 800.00	$ 800.00
with 1 TB Hard Drive	2	SilverMate	502TXX	$ 300.00	$ 600.00
Tablet	2	SilverMate	tw_10	$ 590.00	$ 1,180.00
Tablet Keyboard	2	SilverMate	key_w_10	$ 130.00	$ 260.00
Router	1	NetCom	USBv3	$ 102.00	$ 102.00
				Total	$ 10,142.00

Figure 10.52 Snapshot of a Table with a Linked Formula

10.10 Table Snapshots

Chapter 11
Preparing Documents for Publishing

If a workbook is intended for more than personal use, you can take several steps to prepare the document for distribution. For example, you can link specific metadata to the document to ensure proper filing and quick retrieval. This was already discussed in Chapter 1, Section 1.8.

Alternatively, you may want to remove certain data or metadata before sharing the document. For sensitive information, saving it encrypted can help you prevent unauthorized access. Another layer of protection is a digital signature, which ensures the document hasn't been altered or damaged during sharing.

When several people contribute to a document, it's helpful for them to clearly mark its completion and notify everyone involved. If you plan to share data with users of older Excel versions, check the workbook's compatibility with those versions before distributing it. All the measures we've briefly mentioned here are available in Excel under **File • Info**.

11.1 Document Inspection

Before sharing documents, it's often wise to review key elements that affect their future use. When working with a workbook, you often store information that isn't intended for the final version of the document. You might test formulas on a sheet that you plan to hide later, and you may also use data in a column that you then hide. Comments and notes can also contain content that isn't meant or suitable for publication, such as: "Erna messed up again here."

Excel provides a Document Inspector that can check the current workbook for such elements, and you can access it via **File • Info** and the **Check for Issues** button. The menu for this button offers **Inspect Document** as the first option, and before opening the dialog, Excel automatically checks whether the workbook has unsaved changes and displays a message if it does. You should then confirm saving by clicking **Yes**.

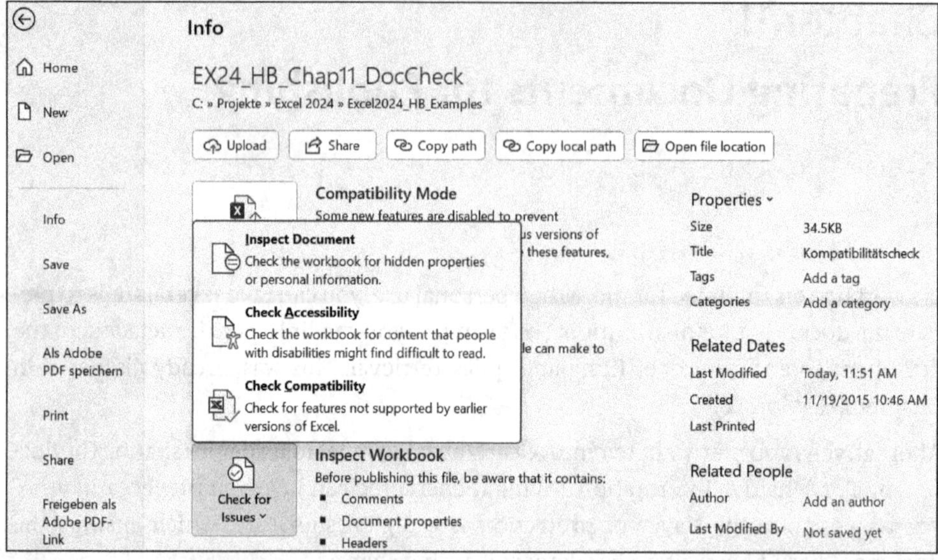

Figure 11.1 Menu for Check for Issues Button

11.1.1 Document Inspector

In the **Document Inspector** dialog, you can select or deselect a range of predefined check points. Select the items you want to check, such as hidden rows and columns, hidden worksheets, outdated comments, or objects that are formatted as invisible. You can also check for hidden metadata and personal information, headers and footers, and custom XML data. Use the **Inspect** button to start the process.

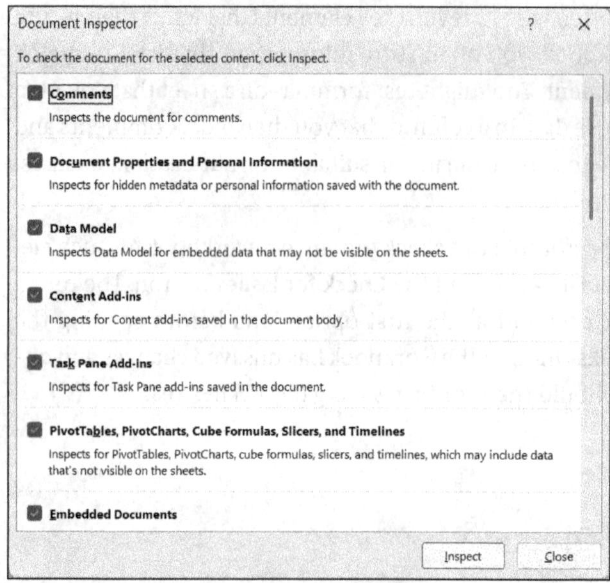

Figure 11.2 The Document Inspector Dialog

If any selected items are found, a general notification will appear and a button to remove each item will be provided.

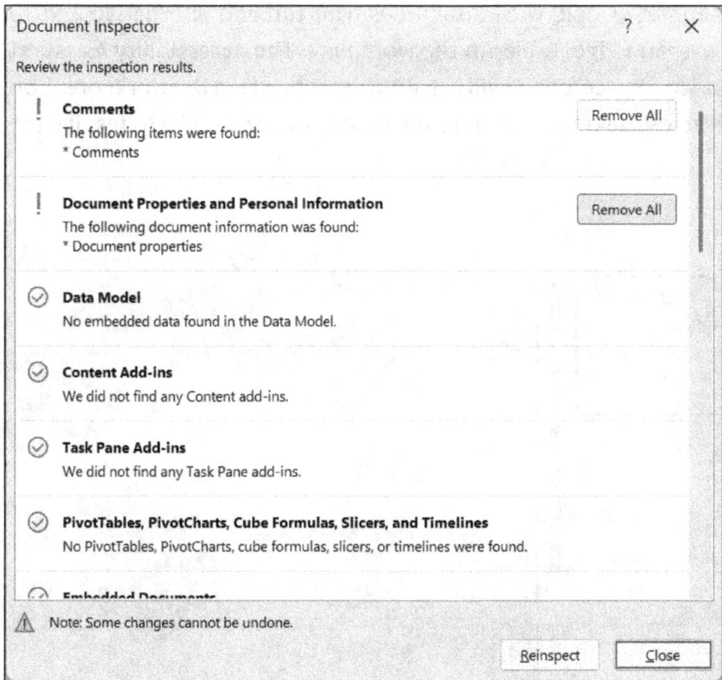

Figure 11.3 Document Inspector Messages

When document properties and personal information are removed, a notice appears under **Inspect Workbook** indicating that the file now has a setting to automatically remove this data when saving. To let you adjust this setting for future saves, an **Allow this information to be saved in your file** link is also provided. Alternatively, you can immediately verify the removal of metadata in the right section of the page under **Properties**.

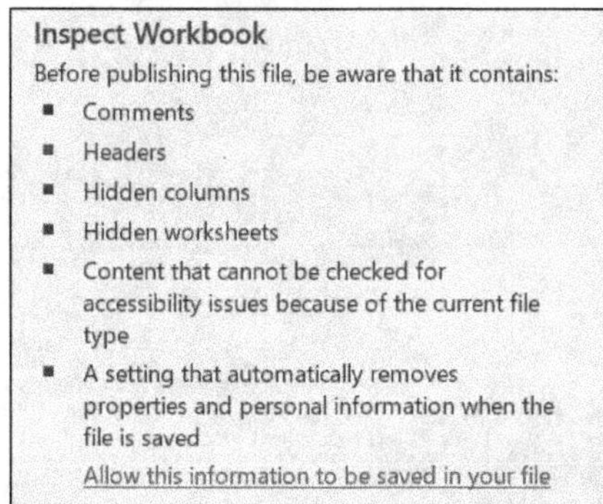

Figure 11.4 File Status Notice

11.1.2 Accessibility

The second option on the **Check for Issues** button menu lets you check the document for accessibility. To support people with disabilities, you can add alternative text to graphical objects, charts, and PivotTables in the workbook. The **Accessibility Assistant** pane will appear in the workspace, and it will list all critical objects in the workbook. You can clicking the small arrow next to an item to reveal steps you can take to resolve the issue.

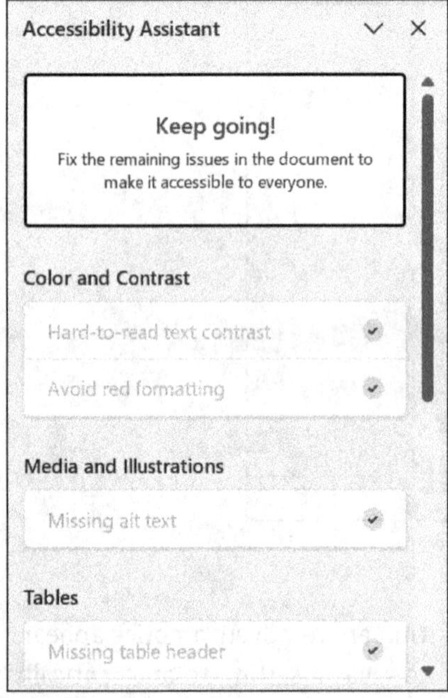

Figure 11.5 Accessibility Pane

You can also run this check within the workbook by going to the **Review** tab, clicking the button in the **Accessibility** group, and selecting **Check Accessibility**.

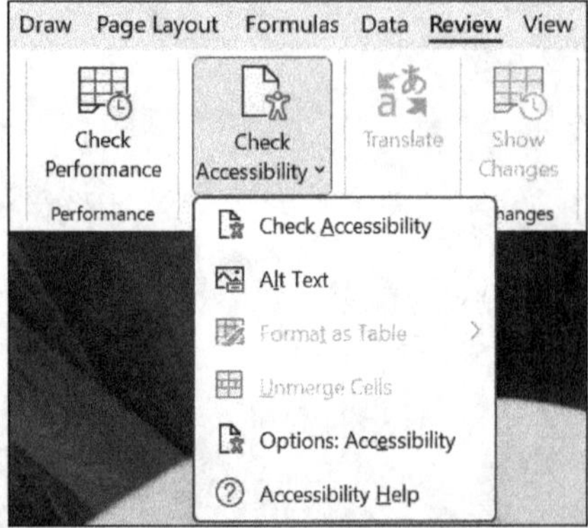

Figure 11.6 Accessibility Check Menu

The menu leads to **Options: Accessibility**, where you can enable preferences to ensure accessibility.

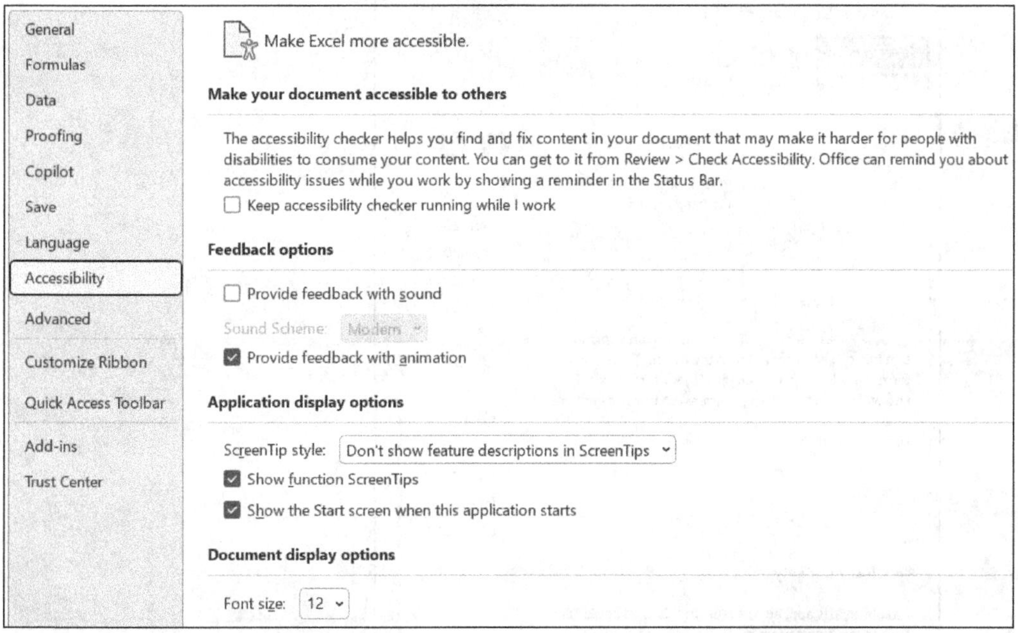

Figure 11.7 Options for Improving Accessibility

11.1.3 Checking for Compatibility

If you plan to share workbooks in an older file format, check for compatibility issues with the formatting and features used before saving. To do this, go to **File • Info • Check for Issues • Check Compatibility**.

In the dialog, use the **Select versions to show** button ❶ to select the target format. All detected issues will be listed and prioritized. For example, if the workbook contains any newer table functions, you'll receive corresponding alerts.

The **Find** ❷ link takes you directly to the relevant location, while the **Help** ❸ link offers guidance on how to resolve the issue. A common example is sparklines, which older versions don't support.

Less severe compatibility issues, such as when table formats are used that older versions don't provide, fall under the **Minor loss of fidelity** group. In these cases, when you save the workbook, formats will be replaced with similar ones that are compatible.

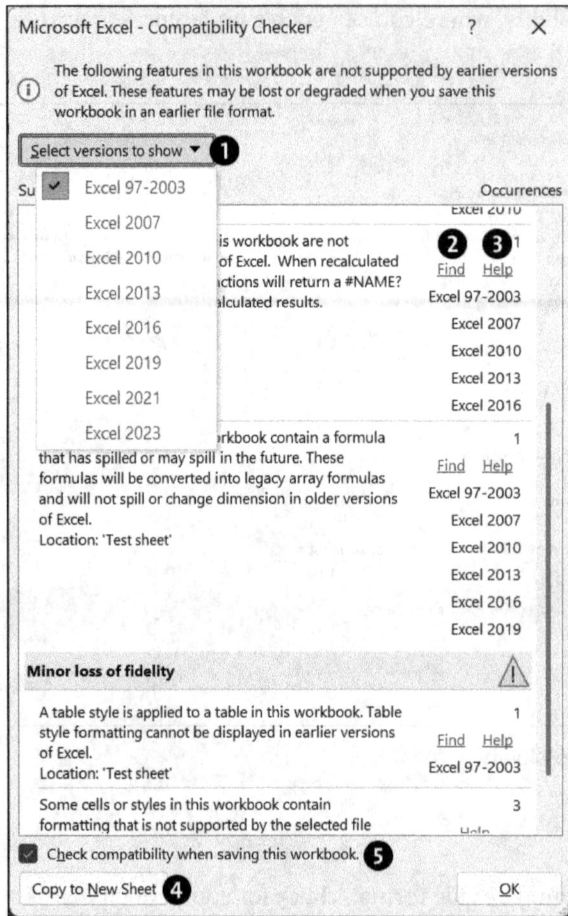

Figure 11.8 Compatibility Check Results

11.1.4 Capacity Issues

A specific problem occurs when working with very large tables that exceed the capacity of very old Excel versions. For example, Excel 2003 allowed only 256 columns and 65,536 rows per sheet, and if you save a larger file in that older version, the extra rows and columns won't be saved.

If the compatibility check detects cell ranges that exceed the old row or column limits, you can use the **Find** link in the message to highlight these ranges and, if you need to, copy parts of them to another sheet.

Functions like `SUM()` and `AVERAGE()` have allowed up to 255 arguments since Excel 2007, while older versions support only 30 arguments for these functions.

If formulas exceed the limits of the older version and are detected during the check, you can use the **Find** link to locate and modify the affected cells. If the file is saved in the older format, those formulas will return the `#VALUE!` error.

Another limitation of older versions concerns formula size. Excel supports formulas up to 8,192 characters and 1,014 operands, while older versions allowed only 1,024 characters and a maximum of 40 operands. When nesting functions, up to 64 levels are now allowed, compared to just seven in earlier versions.

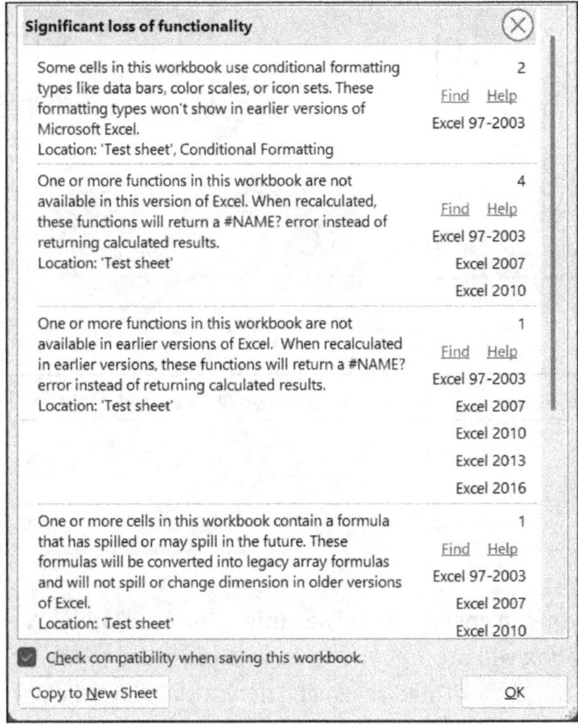

Figure 11.9 Capacity Issue Message

Use the **Copy to New Sheet** button (refer back to Figure 11.8 ❹) to save the check results in a separate worksheet if you need to. To check compatibility every time you save, enable the **Check compatibility when saving this workbook** setting (refer back to Figure 11.8 ❺). When a workbook is opened in the **Excel 97–2003** file format, Excel automatically switches to **Compatibility Mode**.

11.1.5 Finalizing Documents

Once a workbook reaches its final version, you can mark the document as **Final**. To do this, go to **File • Information • Protect Workbook • Mark as Final** (see Figure 11.10).

You must confirm this step before Excel will take any action. It disables any further input, editing commands, and spell check marks in the workbook. All related buttons on the ribbon will be disabled, and the file will also open as read-only.

When a finalized file is opened, a message bar appears below the ribbon stating **MARKED AS FINAL,** along with an **Edit Anyway** button that allows you to remove the editing restrictions if necessary. The same happens if you select **Protect Workbook • Mark as Final** again (see Figure 11.11).

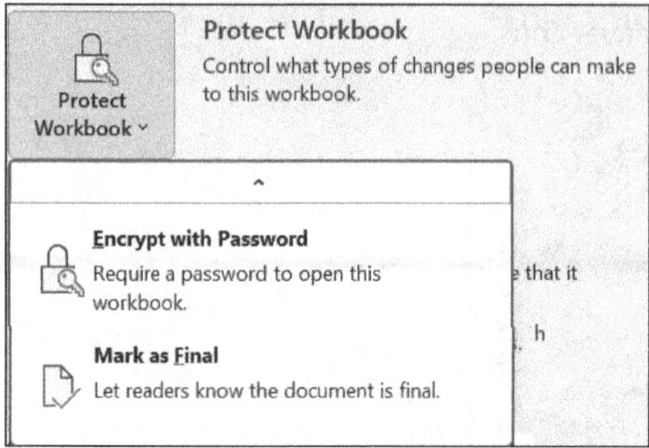

Figure 11.10 Finalizing a File

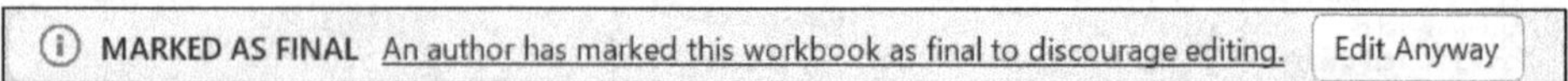

Figure 11.11 Finalized File Notification

11.2 Encrypt Documents

To ensure a workbook's integrity, you can encrypt it via **File • Info • Protect Workbook • Encrypt with Password**. The dialog box will prompt you to enter a password to secure access to the workbook, which must be confirmed as usual. Encryption adds an extra layer of security beyond the simple password protection described in Chapter 1.

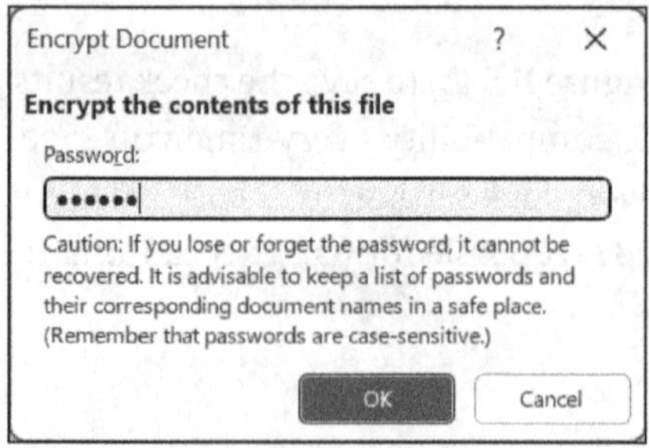

Figure 11.12 Encrypting a Workbook

Encryption also prevents the ZIP package containing the workbook from being opened. To remove encryption, run the encryption command again and confirm the dialog box by leaving the password field blank. In the **Protect Workbook** button menu, you'll also find the **Protect Current Sheet** and **Protect Workbook Structure** options. These correspond to the **Protect Sheet** and **Protect Workbook** options in the **Review • Protect** group, which were covered in Chapter 4.

Chapter 12
Publishing Workbooks

This chapter covers issues related to publishing your tables and charts, whether by printout or email. When you're printing, the main focus is on how to best fit your tables onto the pages. When you're sending by email, features that support collaboration on tables and the review and approval of spreadsheets are especially important. This chapter also explains how to send files as PDF or XPS copies.

12.1 Preparing Worksheets for Printing

The constant storage of data on computers, computer networking, and the vast capabilities of data transmission through global telecommunications have reduced the need to exchange data on paper. However, these same technological advances have also led to a much larger volume of data being processed. At the same time, quality standards for printed data have risen significantly, so what appears on the screen cannot simply be transferred directly onto paper.

When printing worksheets or entire workbooks, you must consider several factors to effectively convert screen displays to printed output:

- The screen displays data in windows that are formatted differently than on paper, and you can freely move the window view.
- On the screen, you can freeze header rows for long tables, while on paper, a header row is usually required on each page.
- The colors in your tables and charts differ from the colors your printer can produce, so if you don't need color printing, consider converting colors to grayscale or using patterns to distinguish chart columns.

When printing, you also usually need to make a series of decisions:

- What should be printed?
- What should the print page layout look like?
- Which printer should be used and with what settings?

Using a multipage item list as an example, we'll show the individual steps and various design options in printing from Excel.

	A	B	C	D	E
1	**Items for home use**				
2					
3	Product group	Article number	Article description	Price	Revenue
4	15	1202	Stor rall white	11.40	684.00
5	15	2803	Stor rall white varnish	21.10	1200.00
6	15	2802	Stor rall black	21.10	814.00
7	15	7121	Stor rall gray	5.00	100.00
8	15	7141	Stor rall beige	5.80	580.00
9	15	7281	Stor rall light green	9.90	1980.00
10	15	5316	DoubleCarrier T1	7.45	298.00

Figure 12.1 Excerpt from Multipage Table

12.1.1 Setting the Print Range

First, determine exactly what needs to be printed. Excel offers several ways to handle this. You can either let Excel decide what to print from each sheet or explicitly set the print area.

Automatic Selection

If you let Excel choose the print area, it will print the entire range on a sheet that contains data, objects, or visible formatting such as borders or cell fills. Specifically, the cell in the lowest row and rightmost column that still contains data, formatting, or objects will determine the number of rows and columns printed.

The automatic selection of data to print can apply to the active sheet, a group of selected sheets, or the entire workbook. To print the complete item list, simply activate the sheet with the list beforehand, since printing the current sheet selection is the default. However, if you let Excel choose the selection, unexpected results can occur. For example, if you once entered a small intermediate calculation in a distant table range and forgot to delete it before printing, then Excel might print an unnecessarily large area. This also applies if you format any cell or range with a background pattern.

Selection by Highlighting

Selecting the sheet is enough to print the entire list, but to print part of the list, you must precisely define the print area. For example, if you want only the first 20 items to appear on paper, you have two options.

The easiest way is to select the range before using the **Print** command via the **File** tab and then choose **Print Selection** from the first menu under **Settings**. You can also make multiple selections or selections for a previously chosen group of sheets, and each such subrange will print on a separate page.

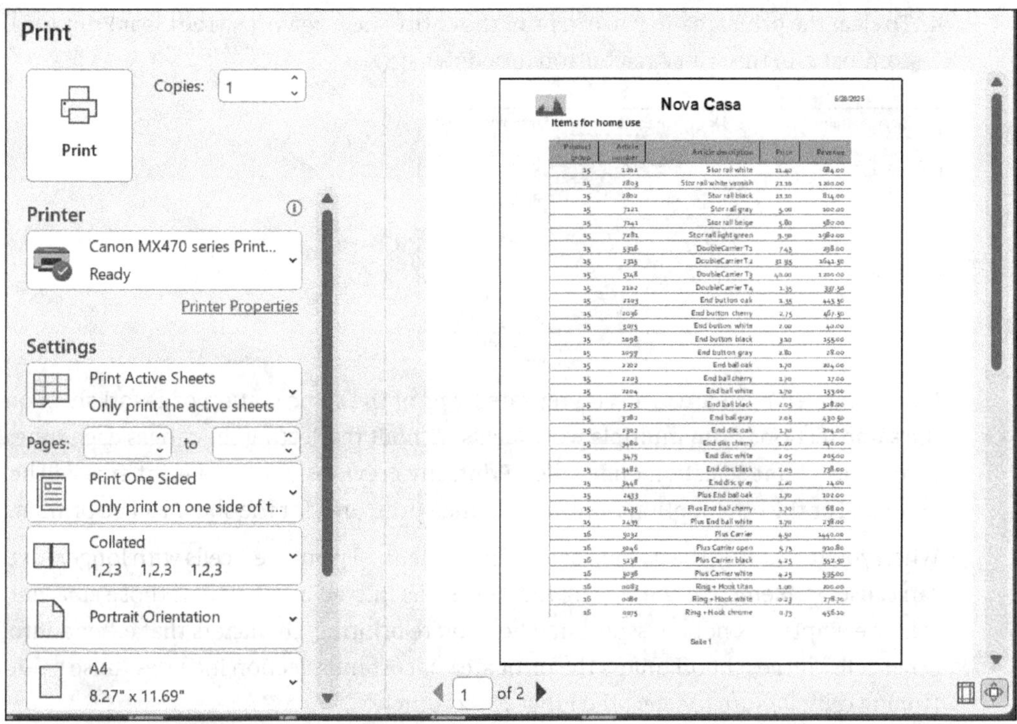

Figure 12.2 Print Preview of Table Section

Defining a Print Area

The second option is to explicitly define a print area for a worksheet. Excel automatically names this range *Print_Area*, so you can refer to it by that name. To set a print area, follow these steps:

1. Select the range you want to print. Selecting a group of columns or rows is sufficient; empty cells will be ignored during printing.
2. On the **Page Layout** tab, in the **Page Setup** group, select **Set Print Area** from the drop-down list of the **Print Area** button. Excel names the selected range *Print_Area*, as shown in the **Name** box, but you can set this name separately for each sheet. The print area, like all ranges, is dynamic—if you insert rows or columns within the range afterward, the range definition will adjust automatically.
3. To print an additional cell range immediately afterward, select it and choose the **Add to Print Area** option when it appears.

   Alternatively, you can set a print area through the **Page Setup** dialog box, which you can open via the dialog launcher in the group with the same name. On the **Sheet** tab, you can enter or select the reference or name for the print area under **Print Area**. In this case, the name *Print_Area* is assigned.

4. To clear the print area so you can print the entire sheet again, use the **Clear Print Area** command in the **Print Area** button dropdown list.

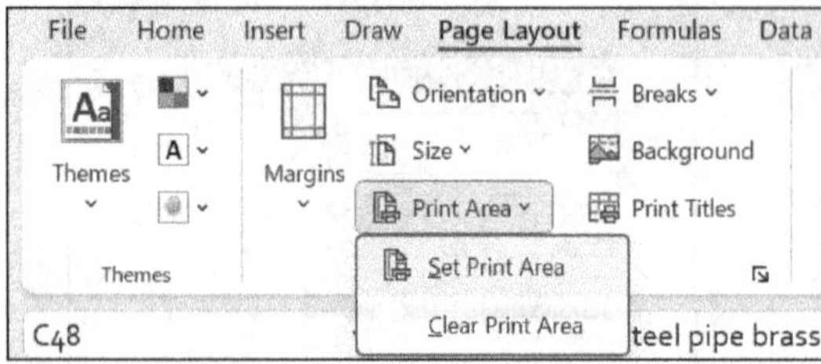

Using a named print area lets you repeatedly print the same section of a worksheet or the same sections from multiple worksheets without reselecting the ranges each time. If you select **Print Selection** under **File • Print**, any previously defined print area will be ignored. For the other options, only the defined print area for the sheet will be printed.

When you're printing the current selection or a defined print area, cells with longer text can cause problems. On the screen, this text spills into adjacent cells. If those adjacent cells are empty, it's not an issue, but when you're printing, characters that extend into other cells will get cut off unless the print area or current selection includes those overlapping cells. Be sure to select the ranges accordingly.

12.1.2 Page Layout

When you're printing a large list, the question is how to best distribute the data across pages. You can leave this entirely to the program, but the results aren't always satisfactory. The question is how the program splits the large table and how you can control this. This mainly involves deciding on the page break, which can be automatic or manual.

Automatic Page Breaks

If, as with the item list, the table range to be printed doesn't fit on one page, the program usually inserts a page break automatically between columns or rows. If one or more columns no longer fit on the page, then those columns are printed on the next pages, and if the maximum number of rows that fit on a page is reached, then a new page starts.

How this automatic break appears depends on the settings you choose when selecting the printer and setting up the page (more on that in the following sections). Automatic page breaks appear as thin lines unless you hide them via **File • Options • Advanced • Options for this worksheet**.

Manual Page Breaks

You can also insert page breaks manually. This is useful when automatic breaks would otherwise separate table columns that logically belong together. A manual page break overrides the automatic one and is also shown as a thin line. To insert one, go to the **Page Layout** tab and the **Page Setup** group, and then, under **Breaks,** select **Insert Page Break**. This command lets you specify the starting column and row for a new page, based on the current cursor position.

For example, to start a new page after row 20, place the cell pointer in cell A21 or select the header of row 21. Figure 12.3 shows a reduced view of how many pages result from these three steps. You can set both horizontal and vertical page breaks simultaneously by placing the cell pointer where the two break lines intersect.

	A	B	C	D	E
6	15	2802	Stor rall black	21.10	814.00
7	15	7121	Stor rall gray	5.00	100.00
8	15	7141	Stor rall beige	5.80	580.00
9	15	7281	Stor rall light green	9.90	1980.00
10	15	5316	DoubleCarrier T1	7.45	298.00
11	15	2315	DoubleCarrier T2	31.95	1641.50
12	15	5148	DoubleCarrier T3	40.00	1200.00
13	15	2102	DoubleCarrier T4	1.35	337.50
14	15	2103	End button oak	1.35	445.50
15	15	2036	End button cherry	2.75	467.50
16	15	3075	End button white	2.00	40.00
17	15	1098	End button black	3.10	155.00
18	15	1099	End button gray	2.80	28.00
19	15	2202	End ball oak	1.70	204.00
20	15	2203	End ball cherry	1.70	17.00
21	15	2204	End ball white	1.70	153.00
22	15	3275	End ball black	2.05	328.00
23	15	3282	End ball gray	2.05	430.50
24	15	2302	End disc oak	1.70	204.00
25	15	3480	End disc cherry	1.20	240.00
26	15	3475	End disc white	2.05	205.00
27	15	3482	End disc black	2.05	738.00
28	15	3448	End disc gray	1.20	24.00
29	15	2433	Plus End ball oak	1.70	102.00
30	15	2435	Plus End ball cherry	1.70	68.00
31	15	2439	Plus End ball white	1.70	238.00
32	16	5032	Plus Carrier	4.50	1440.00
33	16	3046	Plus Carrier open	5.75	910.80
34	16	5238	Plus Carrier black	4.25	552.50

Figure 12.3 Manually Divided Pages

To remove a specific page break, place the cell pointer just to the right of or below the dashed line. If the selected cell touches a page break line, you can use the and **Remove Page Break** command from the **Breaks** menu. To remove all manual page breaks at once, select the **Reset All Page Breaks** command.

Before printing, it's best to check the page preview under **File • Print** or **View • Page Layout** to ensure that the manual page breaks produce the desired result. Later on, you'll learn more about this later, and we'll also explain how to review and adjust the page layout in the **Page Break Preview**.

Fitting More Data onto a Single Page

Once you know what needs to be printed, the next question is how the print pages should be formatted. This mainly concerns how much data can fit on one page. The more data you can neatly fit onto a page, the less paper you use. For example, five hundred printed copies of a document may consist of two or six pages each, depending on whether one column and a few rows spill over from one page onto the next.

There are several ways to control the number of pages in a document. Since the row height in the table depends on the font size, using a smaller font lets you fit more rows onto a page. For example, reducing the standard font size from 10 to 9 points increases the row count from 56 to 60 on an A4 page, and with an 8-point font, you can fit 64 rows onto a page. A smaller font also allows more columns.

Another factor is choosing the optimal column width for all columns, and you can also save space by setting the default margins slightly narrower. A different way to fit more data onto a page is scaling, which we'll explain in a later section.

Page Layout Options

Excel offers a full **Page Layout** tab with five groups for designing page layouts, as well as the **Page Layout** view, which is accessible via the button of the same name in the **View** group under **Workbook Views**.

The dialog launchers in the **Page Setup**, **Scale to Fit**, and **Sheet Options** groups all open different tabs within the **Page Setup** dialog box. Some buttons in the groups provide palettes for quick option selection, while others open tabs in the dialog box. The options that are available in the dialog box depend on the type of sheet you activate. When you select a chart sheet, the **Sheet** tab is replaced by **Chart**. Certain options also depend on the printer you select, since not all printers offer the same features.

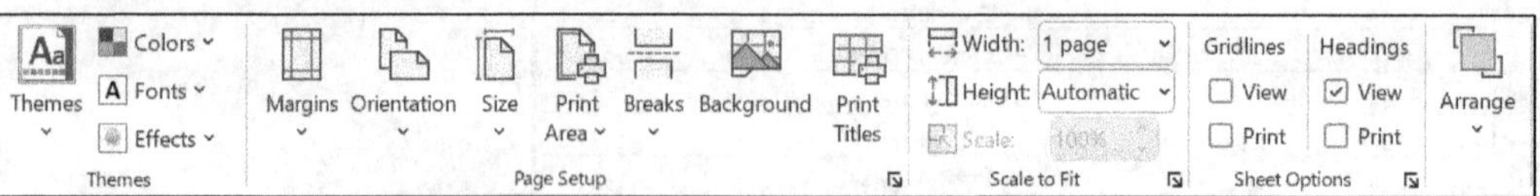

Figure 12.4 Page Layout Tab

Page Layout View

It's best to activate **Page Layout View** before working with the options in the **Page Layout • Page Setup** group, so you can immediately see how specific settings affect the layout. You activate this view by clicking the **Page Layout** button ❶ in the **View • Workbook Views** group. You can access it even faster via the **Page Layout** button in the status bar.

The rulers shown in this view help you precisely position objects on each page. The measurement unit defaults to your computer's regional settings, but you can change it under **File • Options • Advanced • Display • Ruler Units**. This view also lets you check page breaks, especially when combined with the zoom options in the status bar.

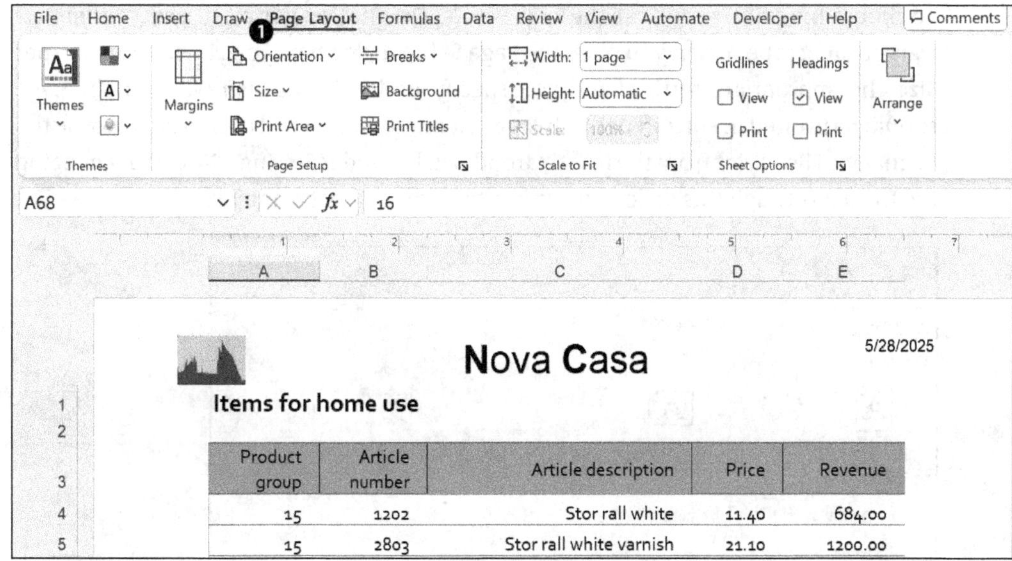

Figure 12.5 Table in Layout View

12.1.3 Choosing the Paper Size and Print Layout

How much data fits on the paper depends largely on the paper size you use and the print layout you apply when printing the data. To select the paper size, go to the **Page Layout** group and click the **Size** button, which opens a menu with many common standard sizes.

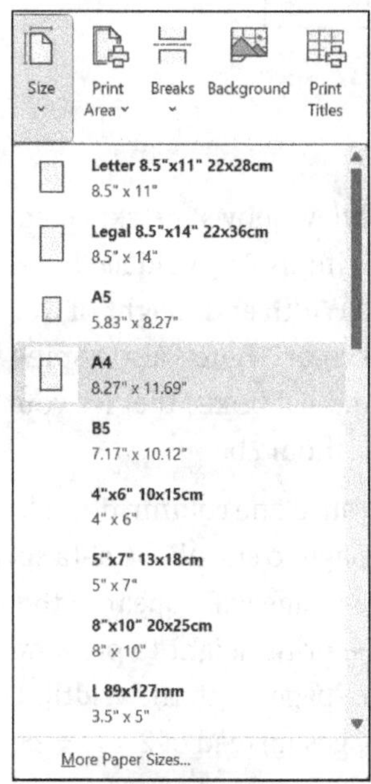

Figure 12.6 Selecting Paper Size

To choose the print layout, use the **Page Setup • Orientation** button to select either **Portrait** or **Landscape**. When you open the **Page Setup** dialog box, you choose under **Paper Size** the appropriate setting, and you specify under **Orientation** whether the page should be printed in **Portrait** or **Landscape** orientation. The third option is to choose the format via **File • Print** from the fourth dropdown list under **Settings**. Note that changing the format through this dialog box only affects the current print job.

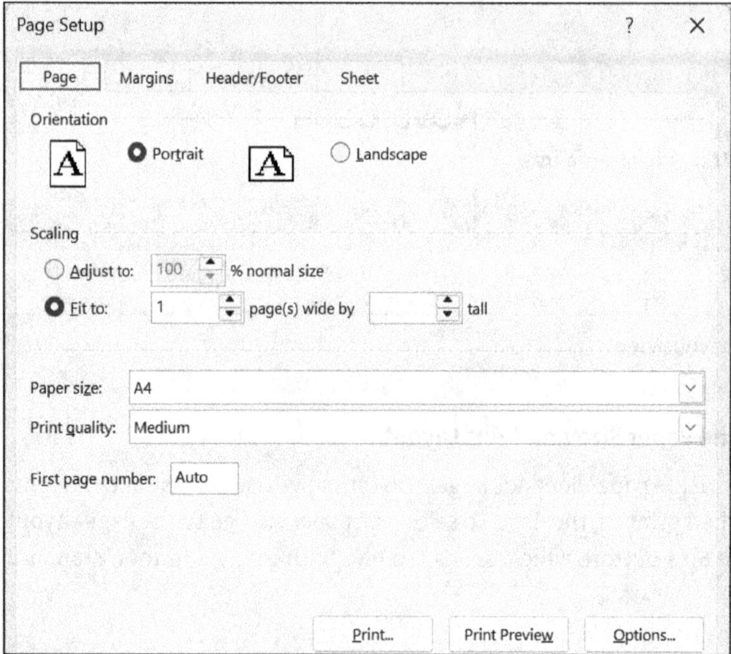

Figure 12.7 Page Setup Dialog Box

Scaling

If the item list is printed in portrait mode, then the print preview shows the expected number of pages, although the last page may only display two items. To avoid having a nearly empty last page, use the two dropdown lists to set the **Width** and **Height** of the page, which you'll find under the **Page Layout** group • **Fit to Format**. There's also a spinner control you can use to set the page scaling percentage and list boxes that let you specify how many pages wide and/or tall the entire print area should be.

For example, if a normal printout would produce 28 pages because one column doesn't fit on each of the first 14 pages, you can set Excel's width to **1 page** to shrink the data so the table is exactly one page wide. After that, the scaling percentage will appear in the now grayed-out spin box. You can set a similar option for the print height to prevent one or two rows at the end of each page from adding an extra page. With the **Width: 1 page** and **Height: 13 pages** setting, you only need 1 × 13 = 13 pages instead of 2 × 14 = 28 pages. That's a significant savings.

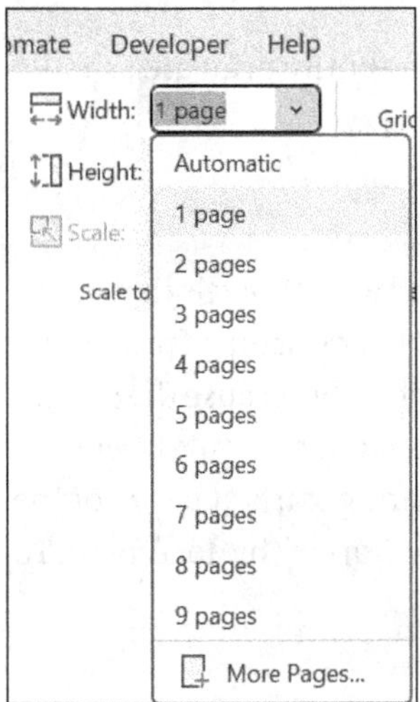

Figure 12.8 Scaling Options

Alternatively, you can use the dial for **Scaling** to directly set a specific reduction or enlargement. In this case, you must set both dropdown menus to **Automatic**. Some preset scaling levels are also available under **File • Print** in the last dropdown menu under **Settings**.

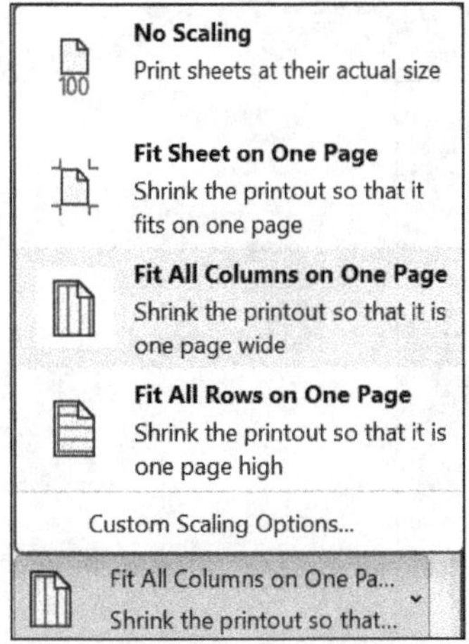

Figure 12.9 Scaling Presets

You can make similar settings in the **Page Setup** dialog box under the **Paper Size** tab in the **Scaling** section:

- **Reduce/Enlarge**
  Select this option to enter a percentage in the adjacent box that sets the scale of reduction or enlargement. The original size is set at 100%, and allowed values range from 10 to 400.

- **Adjust**
  Use this option to specify how many pages wide and tall the entire print area should be.

Additional Page Layout Options

The **Paper Size** tab lets you choose, under **Print Quality,** various quality levels that determine the resolution for printing the page. Dots per inch (DPI) is a measure of print quality, and higher DPI improves quality but slows printing significantly because the printer must place many more dots. Additionally, under **First page number,** you can enter a starting number for the page numbering. **Auto** means numbering starts at page 1 or the corresponding page number if you don't begin printing at the start of the document. To start with a higher number, simply overwrite this entry.

Column Titles and Row Labels

With a long list like the item list, the challenge is how to label columns on the second and following pages. In lists with many more columns that must be printed in two or three sections, row labels are missing from the second section onward. You'll encounter a similar issue if you only print columns five through nine of a table. In that case, the row labels will be completely missing. Before taping the pages together, consider the alternative Excel offers. In all these cases, use the **Page Layout • Page Setup • Print Titles** option, which opens the **Sheet** tab in the dialog box.

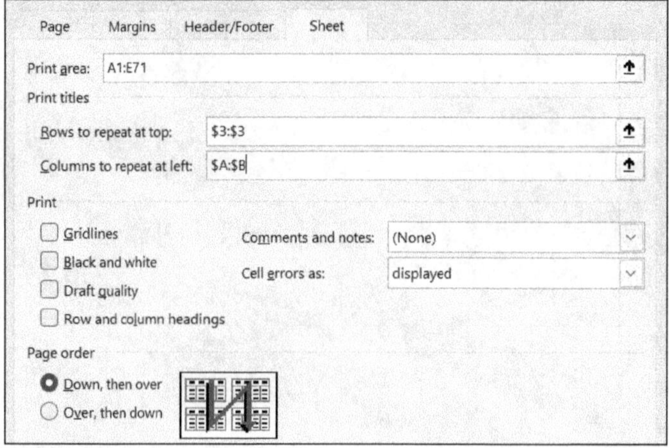

Figure 12.10 Sheet Tab

Two input fields are available. In the **Rows to repeat at top** field, you can specify which rows with column labels should print on every page. If you want to specify only one row, you can just enter the address of a cell in that row, but for multirow labels, you'll need to specify a range of cells. In the item list example, you can select cells A1 to A3 or just cell A3. In the dialog box, you can just specify the row numbers.

For any row labels, enter the corresponding column or columns under **Columns to repeat at left**. The program inserts only the labels it needs. If the print area is, for example, F10:J200, then it will insert only the headers of columns F through J and the row labels for rows 10 through 200. To disable the column titles and row labels, delete the entries in both text fields.

Showing and Hiding Table Elements

Under **Print** on the **Sheet** tab (see Figure 12.10), you can control which worksheet elements are printed and which are not:

- **Gridlines**
  This setting toggles the printing of gridlines on or off, and it's independent of the screen display.
- **Comments and notes**
  This dropdown list lets you choose whether and where to print comments and notes. They can be printed at the end of the sheet or as they appear on the screen.
- **Row and column headings**
  This setting toggles printing of row and column headers on or off.
- **Cell errors as**
  This setting lets you specify, for example, that error values print as blank cells.

You can also toggle printing gridlines and row and column headings under **Page Layout • Sheet Options**.

Print Quality

Two additional settings affect print quality. **Black and white** (see Figure 12.10) enables or disables black-and-white printing for cells and shape objects. This means that in color-formatted tables, all foreground colors except white will print as black and all background colors except black will print as white. This option provides an alternative to grayscale printing on a black-and-white printer, and on color printers, it can speed up printing, such as when printing drafts. If the **Draft quality** box is checked, the table will print without embedded charts, graphics, or gridlines. Logos in the header will also be printed.

Vertical or Horizontal Page Strips

When a print job includes multiple pages, you need to decide the order in which the pages will print. For this, there are two **Page order** options that are illustrated by a small table pattern. By default, Excel prints pages stacked vertically in the worksheet first and then moves to the top of the next strip. This matches the **Down, then over** option. Alternatively, you can choose to print pages side by side first and then continue with the strip below. The best order depends on the table's layout. For example, if you have several tables stacked vertically that are one page long but several pages wide, you should choose the second method.

Printing Chart Sheets

When printing chart sheets, the page order and scaling options described earlier aren't available. On the **Chart** tab, only the **Draft quality** and **Black and white** options are available.

12.1.4 Headers and Footers

A *header* is an extra area that appears on every page that's printed in the print area. You've seen this in books or magazines, where the header often includes the chapter title and page number. The *footer* also appears on every page, just below the content printed from the worksheet. You can set a header and footer for each worksheet, and Excel provides a convenient way to design headers and footers through the ribbon.

Header and Footer Tools

If you edit a table in **Normal** view and click the **Header & Footer** button under **Insert • Text**, Excel automatically switches to **Page Layout** view and opens the **Header & Footer** tab, which includes four groups that are dedicated to customizing headers and footers. Clicking or tapping outside the header or footer closes the tab.

In **Page Layout View**, you can insert header or footer elements directly by using the buttons in the **Header & Footer Elements** group. Three input sections appear above and below the worksheet grid. Click or tap to place the insertion point in one of these ranges and then click or tap a button to insert the corresponding code at that position.

For example, to add page numbering in the center of the footer, click the **Page Number** ❶ button. To also show the total number of pages, type the word "of" after the &[Page] code in the input field and then click the **Number of Pages** ❷ button. The code is &[Page] of &[Pages], and it shows the display you want, which appears immediately in the **Page Layout** view when you select a different input field.

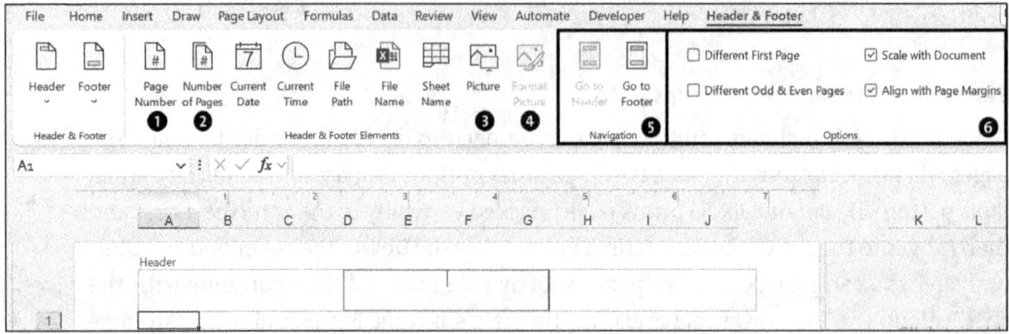

Figure 12.11 Header and Footer Tab

To enlarge the page numbering font, you can highlight the code and format the text either by using the mini toolbar that appears or, like any other text, through the **Home** tab and the **Font** group. Multiline entries are also allowed, and you press `Alt`+`Enter` to insert a new line. You can also edit the codes manually in the input fields, and you can insert some predefined headers or footers by using the menus in the **Header & Footer** group.

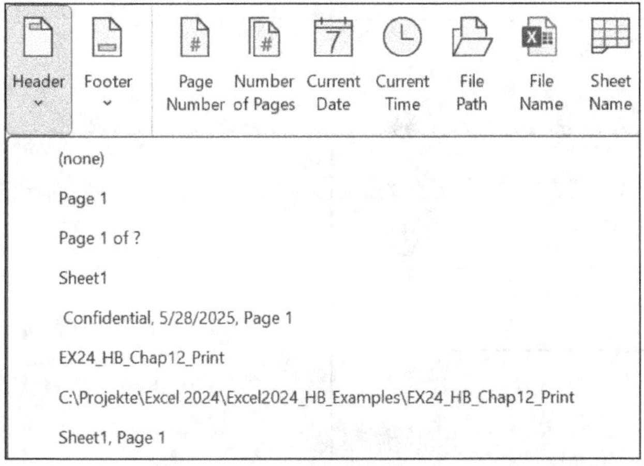

Figure 12.12 Preset Header Menu

Excel also lets you insert graphics, like logos. You can click the **Picture** ❸ button to open the option, and then, you can use **Format Picture** ❹ to adjust the image. The two buttons in the **Navigation** ❺ group (see Figure 12.11) let you toggle between header and footer, and the settings in the **Options** ❻ group are also important. When **Different Odd & Even Pages** is checked, you can select a second version of the header or footer. Checking the **Different First Page** option removes the header and footer from that page, checking the **Scale with Document** option ensures that rows are included when scaling the sheet, and checking the **Align with Page Margins** option ensures that, for example, increasing the page margin also applies to the header and footer.

Figure 12.13 shows an example of the header for the item list and the result in print preview.

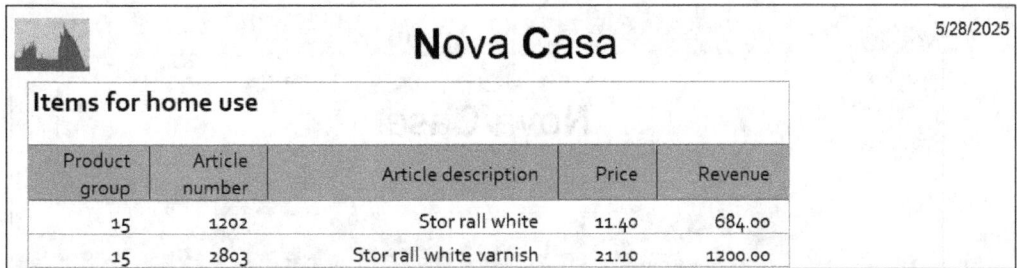

Nova Casa

5/28/2025

Items for home use

Product group	Article number	Article description	Price	Revenue
15	1202	Stor rall white	11.40	684.00
15	2803	Stor rall white varnish	21.10	1200.00

Figure 12.13 Header for Item List

Via Dialog

If you click the arrow in the **Header** ❶ or **Footer** ❷ fields on the **Header/Footer** tab in the **Page Setup** dialog, you'll see several options that are useful in many situations. Whenever you select an option, the display above the **Header** field shows how the row you've selected will look.

Figure 12.14 Header/Footer Tab

If you don't like the options, click the **Custom Header** or **Custom Footer** button to edit the selected presets or completely re-enter the lines with the codes.

Each row has three ranges that you can edit separately. To enter text into a range, click or tap that range. A vertical bar will mark the insertion point. Multiline entries are allowed, and you press `Shift`+`Enter` to insert a new line.

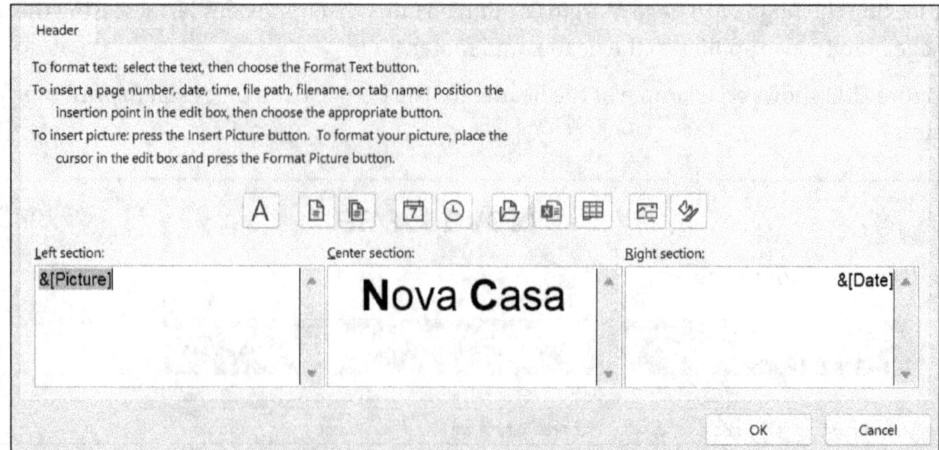

Figure 12.15 Defining Header in Header Dialog Box

[+]

Headers and Footers for Individual Sheets

When printing individual sheets or forms—such as an order form—choose the **(none)** option from the list to ensure that no unnecessary items are included. A single sheet doesn't need a page number, and you should remove any elements you don't require.

Certain elements have symbols that insert corresponding codes into the range (see the following table). The **Format graphic** symbol is only available if a graphic has been inserted.

Symbols	Meanings	Codes
A	Text formatting	–
	Page number	&[Page]
	Total pages	&[Pages]
	Current date	&[Date]
	Current time	&[Time]
	File path	&[Path]
	File name	&[File]
	Sheet name	&[Tab]
	Insert image	&[Picture]
	Format graphic	–

The **A** icon lets you format selected parts of a range. It opens the **Font** dialog box, where you can underline or bold individual words or characters in the header or footer as needed. Note that for multiline entries, you must adjust page margins and the distance from the sheet edge to prevent data overlap.

12.1.5 Setting the Margins

By default, Excel prints a 0.75-inch margin at the top and bottom and a 0.7-inch margin on the left and right of each page. If you use a header, it prints within this margin, and the default margin is about 0.3 inches from the top edge of the page. The same applies to the footer. If this setting isn't suitable for a print job, you can choose from several alternatives via the **Page Margins** button in the **Page Layout • Page Setup** group. This palette is also available under **File • Print** in the **Settings** section.

The **Normal**, **Wide**, and **Narrow** options provide coordinated margin settings for all pages, and you should check whether these options work for many of your print jobs. If you need to use different settings, open **Custom Margins**, go to the **Margins** tab in the **Page Setup** dialog, and set the values you want individually. The most recently defined page margin setting is automatically added to the margin palette as the first option, and that allows you to reuse it anytime. For example, if you set a wider left margin to facilitate hole-punching the table and if you have left-aligned elements in the header or footer, then Excel will adjust the header and footer margins accordingly.

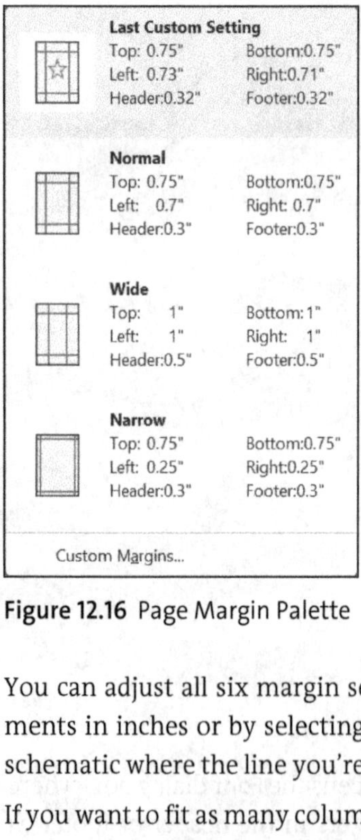

Figure 12.16 Page Margin Palette

You can adjust all six margin settings on the **Page Margins** tab by entering measurements in inches or by selecting them with the small buttons. On the tab, you'll see a schematic where the line you're currently adjusting is always highlighted.

If you want to fit as many columns as possible onto one sheet, you may want to narrow the margins. Keep in mind that such paper documents are usually filed in some type of binder, and a sufficient margin for hole punching will be necessary to prevent data loss when filing. If you want to leave space for comments, choose a wider right margin.

[»]

Avoid Overlap

When setting the values for **Header** and **Footer**, ensure that they match the **Top** and **Bottom** values to prevent overlap.

The **Center on page** option controls the placement of the print area within the margins. By default, the print area aligns directly with the left and top margins. If the printable area doesn't fill the page completely, you can center it horizontally and vertically.

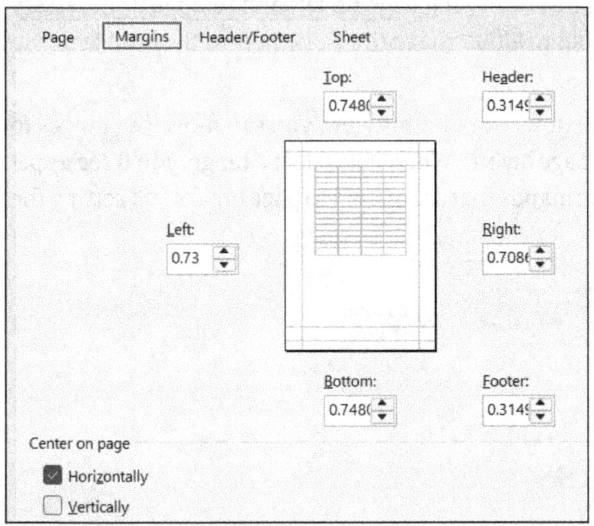

Figure 12.17 Page Margins Tab

12.1.6 Interactive Page Break Preview

Excel provides another tool to help you get the results you want when printing multi-page documents. The **View** tab includes the **Workbook Views** group, which has the **Page Break Preview** button. This view displays the worksheet so you can easily see page breaks and numbers. The worksheet is usually scaled down considerably, and the normally faint page break lines are replaced with thicker ones.

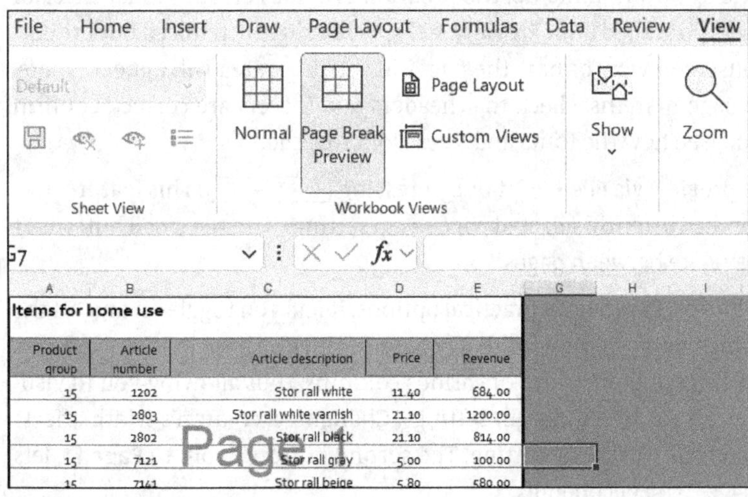

Figure 12.18 Worksheet in Page Break Preview

You can also move these page break lines to adjust the distribution of columns and rows. This lets you quickly fix a page break where one or two rows of a table extend onto the next page. Excel automatically scales each page to fit the selected ranges.

As described previously, you can insert missing page breaks and remove unnecessary ones. The page layout always applies to either the entire worksheet or the print area you select.

Since normal copy and move functions work in this view, you can move cell ranges to another location to improve the page layout. When you select a range, you'll see a special context menu with all the commands that are related to page breaks and setting the print area.

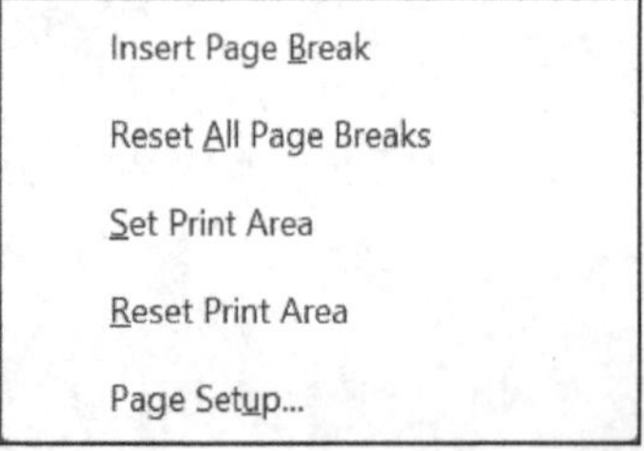

Figure 12.19 Additional Commands in Context Menu of Cell Range in Page Break Preview

To exit **Page Break Preview**, you can use one of the other view buttons in the **View • Workbook Views** group, or more conveniently, the view icons in the status bar.

12.1.7 Reviewing the Layout in Print Preview

Excel offers so many design options that, especially when you're new to the program, it's worth experimenting to find the fonts and sizes that best suit your preferences. To prevent your recycle bin from filling up, the program provides an additional check of the expected print output called **Print Preview**. It displays a slightly reduced but proportionally accurate onscreen view of how the pages in the print area will appear. There, you can verify the page margins, check that headers and footers are correct, confirm page numbering, and see how the fonts and sizes work together.

You can open print preview via **File • Print** or by pressing [Ctrl]+[F2]. This feature uses either the default or the currently selected page setup settings. Use the small buttons at the bottom ❶ to navigate between pages.

The **Show Margins** button ❷ provides practical options. It lets you toggle on and off the margin markers, which you can move with a mouse or finger. These markers indicate different margin settings and the width of various columns, thus allowing you to visually identify and correct incorrect margin settings. Changing a column's width affects both the printout and the workbook setting. The second button, **Zoom to Page** ❸, lets you enlarge the image to inspect details.

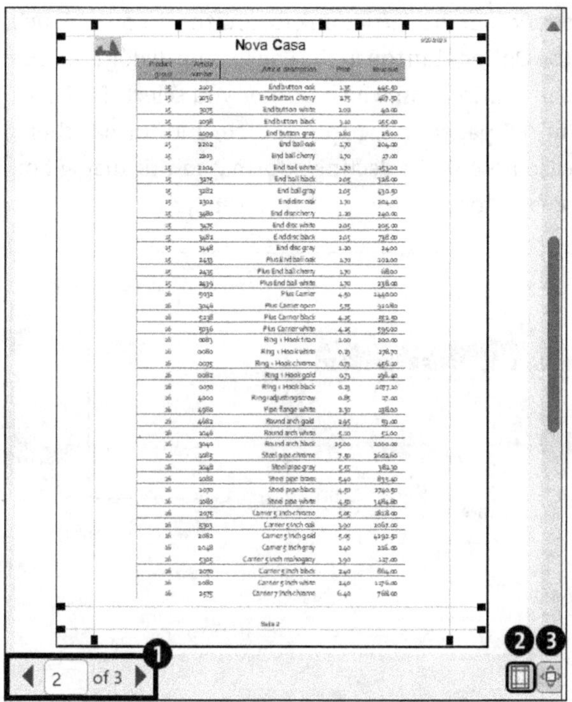

Figure 12.20 Print Preview on File Tab

12.2 Printer Selection and Printer Settings

Depending on how many printers you have installed on Windows, you can choose between different output devices using the **Printer** dropdown list. Choose the printer you want to use. This list shows all printers and other output devices that you have installed on Windows, such as print servers, the Microsoft XPS Document Writer, and fax machines.

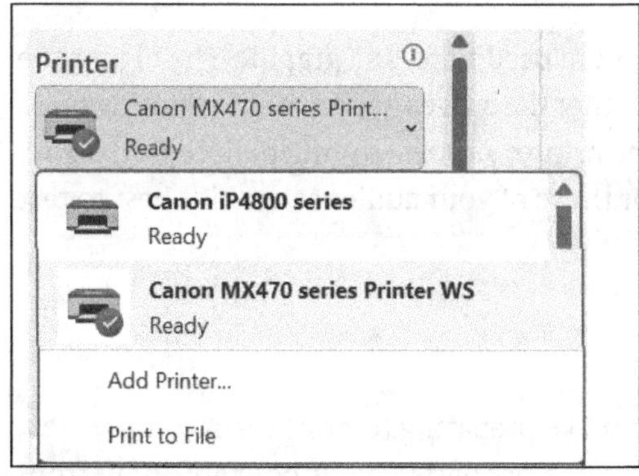

Figure 12.21 Selecting Output Device

To adjust a printer's settings, click the **Printer Properties** link (below the list box), which opens the same dialog box as does the **Options** button in the **Page Setup** dialog box. The appearance of this dialog box depends entirely on which printer you select. Typically, you can change the graphic resolution, paper size, paper source, and format there. Figure 12.22 shows an example. Please note that selecting a format in this dialog box temporarily changes the printer's default format—only for this print job.

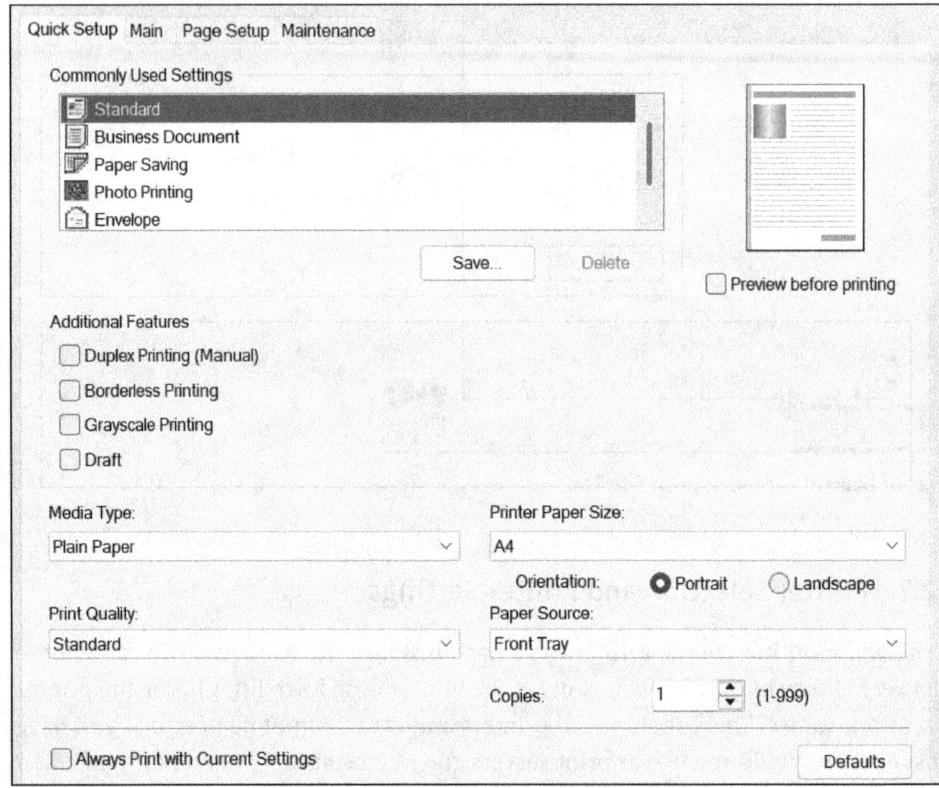

Figure 12.22 Options for Printer Settings

Quality settings determine how many DPIs are used to print a graphic. The higher the DPI count, the finer the image and the smoother the curves or circles. The printing takes a bit longer at a high resolution because the printer simply has much more work to do. Choose based on your quality standards or those of your audience, and it's best to test the different versions to compare them.

12.2.1 Quick Print and Printing Options

The availability of so many settings might make preparing to print seem complicated. However, in most cases, Excel's default settings work perfectly, without any input from you. This especially applies to smaller tables that don't require page breaks. However, for large print jobs, it's worth taking some time beforehand to ensure the first printout is optimal. Excel saves the settings you apply to a workbook or to each individual sheet within it along with the workbook file, and when you open a new workbook, the default settings will be used again.

[+]

Save Print Settings in a Template

If you want to use the same settings in multiple workbooks, you can save them in a template.

12.2.2 Quick Print

To quickly print a small table, go to **File • Print** and click the **Print** button. Excel will print immediately without further prompts, using the current settings for the workbook—either the default settings or those you've set for this workbook. It's also usually helpful to add the **Quick Print** command to the Quick Access Toolbar via the context menu, so you can run it with a single click.

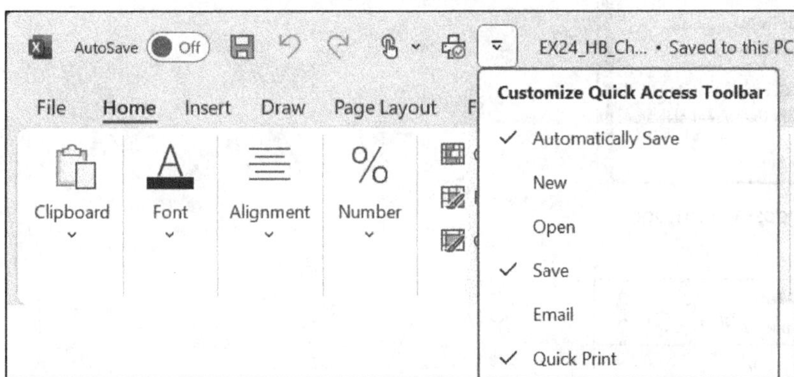

Figure 12.23 Adding Quick Print Option to Access Toolbar

12.2.3 Choosing Printing Options

You can gain more control over printing by selecting the **Print** command on the **File** tab and viewing the options. Under **Settings**, the **Print Active Sheets** option is usually selected by default, so unless you explicitly select multiple sheets, only the active sheet will print. To print the entire workbook, select **Print Entire Workbook**. We've already covered the **Print Selection** option (see Figure 12.24).

If necessary, use the **Ignore Print Area** option to override defined print areas for the current print job. The print areas will remain available. If you want to print multiple copies, you can increase the count by changing the **Copies** setting.

If printing spans multiple pages, you can specify a page range by using the spin boxes under **Settings**. First, enter the page number of the first page under **Pages**, and to print from that page to the end, leave the **to** field blank; otherwise, enter the last page number to print. Below that, choose whether to print the sheets single-sided or double-sided (see Figure 12.25).

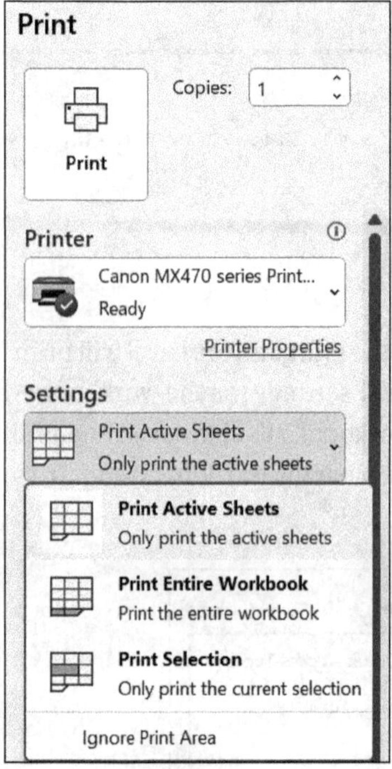

Figure 12.24 Choosing Print Job

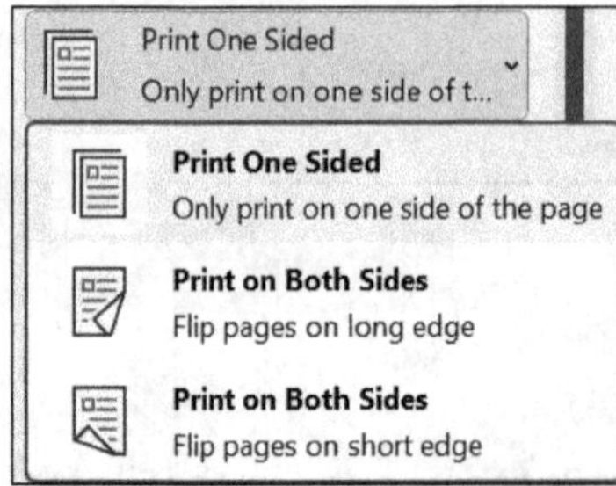

Figure 12.25 Choosing Between Single- and Double-Sided Printing

When you're printing multiple multipage copies, **Settings** also offers the **Collated** and **Uncollated** options. **Collated** means each complete multipage copy prints before the next one starts. **Uncollated** means the specified number of copies of each individual page print before moving on to the next page, and after that, the copies must be manually assembled in the correct order.

The additional buttons under **Settings** let you switch between portrait and landscape layouts, specify paper size, adjust margins, and perform scaling. These settings are also available in the **Page Layout** tab and described there, but the advantage of adjusting these settings here is that the print preview updates immediately and allows you to easily review the results. Once you've selected all settings, you can start printing by clicking the **Print** button.

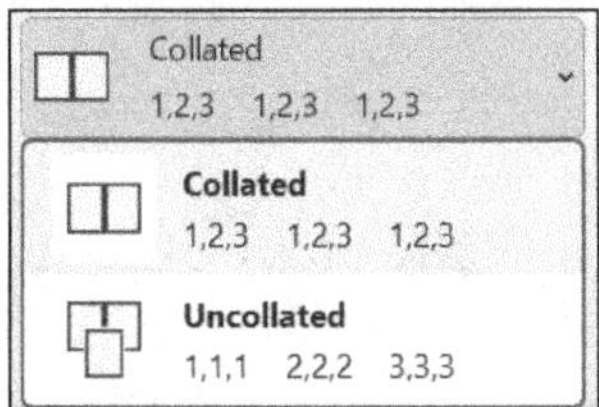

Figure 12.26 Choosing Print Range and Order

12.3 Sending Worksheets by Email

Excel provides two ways to send worksheets by email. One option is to send the entire workbook as an attachment so the recipient can open it in Excel and continue working. On the other hand, if the workbook is for viewing only or the recipient doesn't use Excel, then sending it as a PDF is a good alternative.

12.3.1 Sending a Workbook as an Attachment

To send a workbook as an email attachment, you must have Outlook or another email program installed. The process is straightforward:

1. With the workbook selected, go to the **File** tab and then the **Share** page.
2. Click **Excel Workbook** ❶ to open your default email program with a new message window and include the workbook in the subject line and as an attachment.

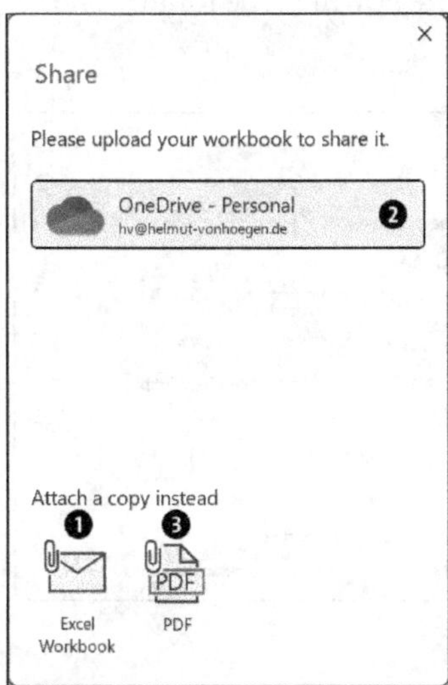

3. Add the address and, if needed, a message, and then send it.
4. We'll discuss the sharing option via a OneDrive link ❷ in the dialog in Chapter 14, Section 14.2.

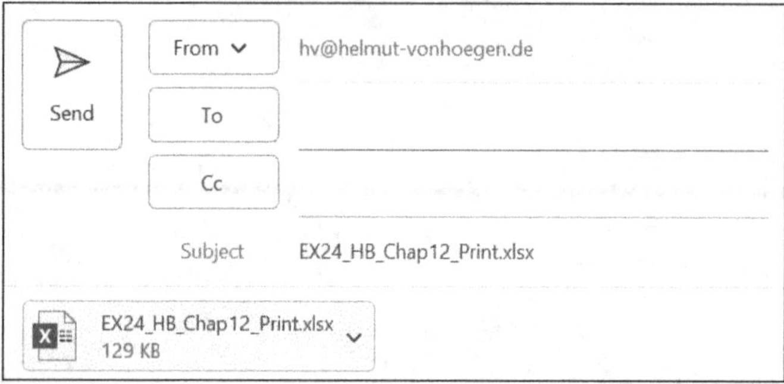

12.3.2 Sending a Workbook as a PDF

The **PDF** ❸ button sends a copy of the workbook as a PDF attachment, using the default email program. The workbook is converted to PDF before you send it, and the attachment can then be opened, for example, in the Edge browser.

12.4 Creating a PDF or XPS Copy

Instead of immediately emailing a copy of a workbook in PDF or XPS format, you can first create such copies locally via **File • Export**—for example, to offer them later for download on a web page. Use the **Create PDF/XPS Document** command, which then displays a **Create PDF/XPS** button. The process differs for the two competing file formats only in that the **Publish as PDF or XPS** dialog defaults to the **PDF** file type for PDF and to **XPS** for XPS.

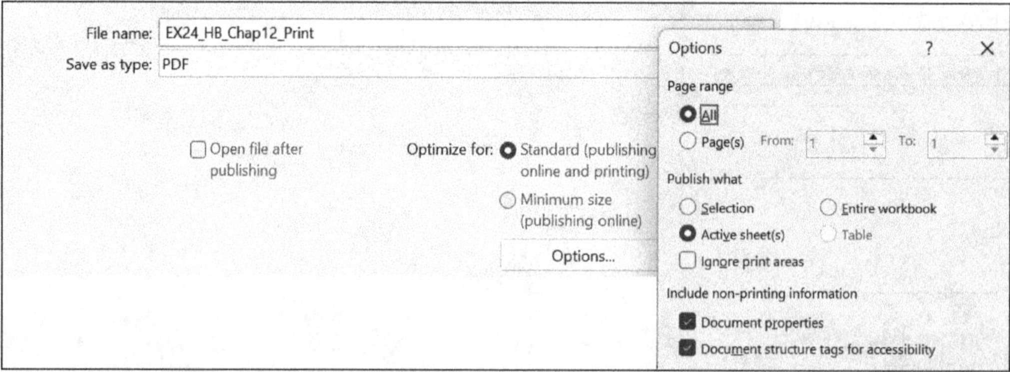

Figure 12.27 PDF Dialog and Its Options

You can also select **Open file after publishing** to automatically open the generated copy in the appropriate reader or viewer. Then, under **Optimize for**, choose either **Standard** or **Minimum size**. The second option is primarily designed for online publishing to save bandwidth. You can also specify exactly what to include in the copy by using the **Options** button. For PDFs, the **PDF/A-compliant** option is also available.

If you select the XPS format, the file will open in the built-in Microsoft XPS Viewer.

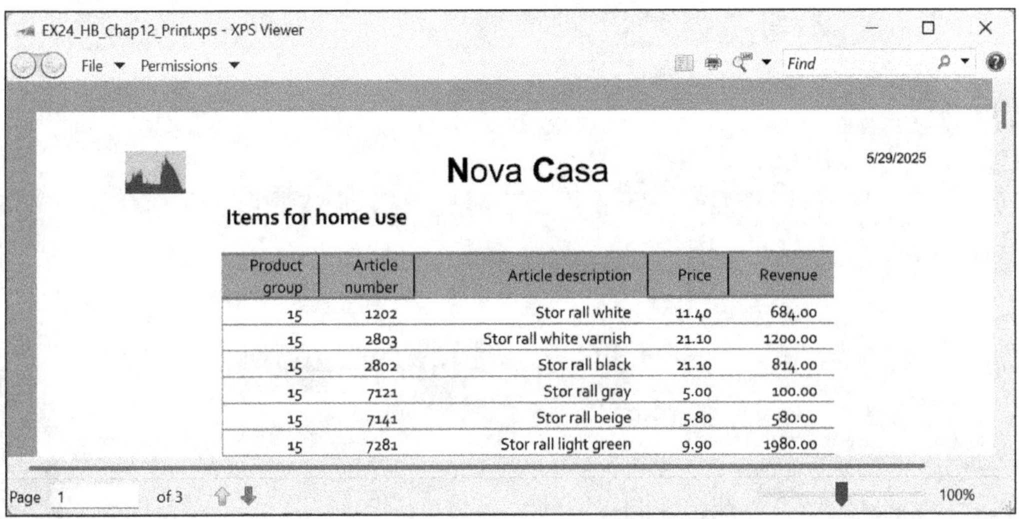

Nova Casa

5/29/2025

Items for home use

Product group	Article number	Article description	Price	Revenue
15	1202	Stor rall white	11.40	684.00
15	2803	Stor rall white varnish	21.10	1200.00
15	2802	Stor rall black	21.10	814.00
15	7121	Stor rall gray	5.00	100.00
15	7141	Stor rall beige	5.80	580.00
15	7281	Stor rall light green	9.90	1980.00

Figure 12.28 XPS Copy Displayed in XPS Viewer

Chapter 13
Excel Data on the Web

The web enabled worldwide access to information while setting standards that are platform- and operating system-independent. After that, applying internet methods to information exchange within smaller networks—such as those of a company, institute, or organization—was a natural step. Nowadays, local networks are increasingly becoming intranets that transform how information is shared and collaboration is organized.

13.1 Integration with the Internet and Intranet

Excel integrates with the internet and intranet in two ways. Excel serves as a tool that provides content for both the company's internal intranet and the global internet. Conversely, Excel can directly access data on both networks. Information from web pages can be imported directly into Excel and further analyzed, provided it is presented in table format. We discuss this topic in Chapter 19.

Excel can also automatically generate web pages from a worksheet or an entire workbook. It creates the necessary HTML source code without any user involvement, and the user doesn't even need to understand what it is. It can both import and generate data in XML format; this is explained in Section 13.5.

The technique of linking documents via hyperlinks applies to all Office documents. From a workbook, such a link lets you quickly view a web page or another Excel file and then return to the current sheet. This increasingly creates a situation where it no longer matters whether a document is stored on your PC, on servers in a company network, or on a web server anywhere in the world. The growing use of cloud services has greatly accelerated this development.

13.2 From Excel to HTML and Back

When an Excel file is converted to HTML, it creates a text file that contains the data from the cells and numerous *markup tags*—which are markers with instructions on where and how to place content on the page. Hence, the name *HyperText Markup Language* (HTML). If a tag contains a style instruction that a particular browser doesn't recognize, the instruction is usually ignored; otherwise, the page content remains viewable. In this respect, the format is quite forgiving compared to XML, which is discussed in Section 13.5.

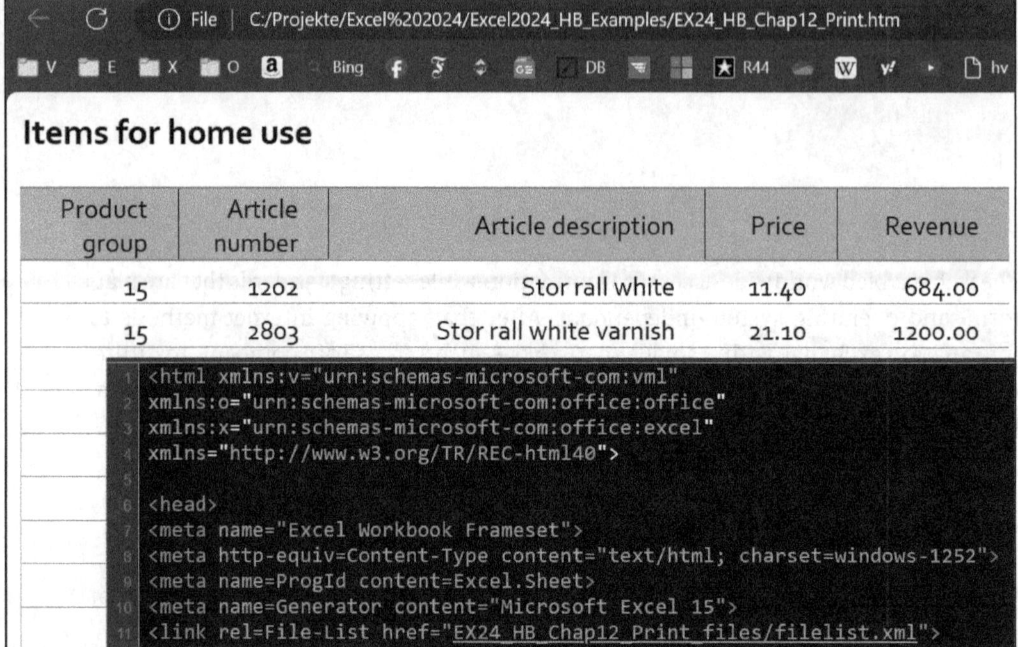

Figure 13.1 Excerpt from Source Code of Table Converted to HTML

To expand design options for web pages—such as fonts or displaying tables and images—a language called Cascading Style Sheets, (CSS) was developed. CSS consists of separate groups of style instructions for displaying different elements on a web page, and these are similar to the styles you can define in Excel. CSS separates formatting from the raw data so it can be changed independently of the content. Excel either embeds these style sheets directly into the generated HTML code or saves the formatting instructions in separate CSS files.

13.2.1 Component Distribution

When you save an Excel table as a web page, Excel creates a set of supporting files alongside the main file in HTML format. Typically, a subfolder named after the main file with the *_files* suffix is created in the main file's folder to store these files.

filelist	5/29/2025 11:10 AM	Microsoft Edge HTM...	1 KB
sheet001	5/29/2025 11:10 AM	Microsoft Edge HTM...	32 KB
sheet002	5/29/2025 11:10 AM	Microsoft Edge HTM...	2 KB
stylesheet	5/29/2025 11:10 AM	CSS Dokument	10 KB
tabstrip	5/29/2025 11:10 AM	Microsoft Edge HTM...	1 KB

Figure 13.2 Folder Containing Supporting Files

This folder always includes a *filelist.xml* file that lists all the files in XML format.

```
<xml xmlns:o="urn:schemas-microsoft-com:office:office">
  <o:MainFile HRef="../EX24_HB_Chap12_Print.htm"/>
  <o:File HRef="stylesheet.css"/>
  <o:File HRef="tabstrip.htm"/>
  <o:File HRef="sheet001.htm"/>
  <o:File HRef="sheet002.htm"/>
  <o:File HRef="filelist.xml"/>
</xml>
```

Figure 13.3 Example of File List in XML

Otherwise, the folder in this example will contain files for the various worksheets: *sheet00x.htm*, a *stylesheet.css* file, and a *tabstrip.htm* file that replicates the sheet tabs. All these files are linked to the main file. If you want to view the source code that Excel generates when converting to HTML, use the **View Source** command from the context menu of the displayed web page, for example in Microsoft Edge. However, some Excel features, like scenarios, can't be converted to HTML. When you save as a web page, you'll receive a warning, and if you proceed, the table with data from the currently selected scenario will be included on the web page. 13

The supporting folder for the web page is specially protected. If you copy or move the main file to another folder, then the supporting folder will be included automatically. Deleting the supporting folder also deletes the main file, and vice versa. If you try to rename the main file, a warning will appear stating that the link to the supporting folder will be lost.

13.2.2 Web Archives

As an alternative to separating files, you can save as a **Single File Web Page** Excel file type to create an MHTML document instead. This format bundles all files into a single web archive, making it especially convenient for transporting to other locations. Microsoft Edge can open these files by automatically switching to Internet Explorer mode.

13.2.3 Web Options

Anyone who publishes web pages should consider who the intended audience is. Not every browser supports all CSS features or certain HTML extensions. To ensure that your website displays correctly on older Microsoft browsers, you can select a target browser. To do this, go to **File • Options**, and then on the **Advanced** page under **General**, click the **Web Options** button ❶ and select the **Browsers** tab. Then, in the first dropdown list, choose the browser type that must be supported. Based on your selection, different options will be preset, but you can still deselect them.

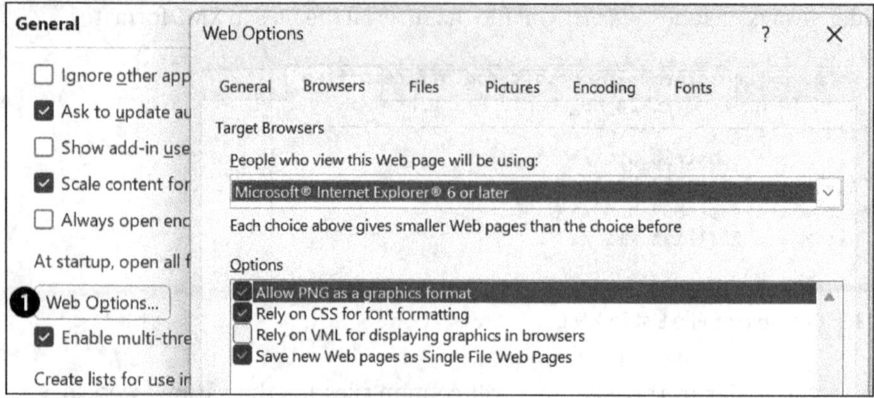

Figure 13.4 Browsers Tab in Web Options Dialog

13.3 Providing Data for the Web

This section will first explain how to publish specific tables, charts, and lists on the web in a way that is easy and straightforward. "Easy" means making data from Excel workbooks available to visitors of a web page as viewable information. This method is especially suitable for information exchange on a company intranet, where the focus is primarily on the data and the web page layout is secondary.

13.3.1 Publishing Excel Data on the Web

If the goal is simply to publish material from Excel to a web page, the first question is: What should be published? Should it be an entire workbook, a single sheet, or individual objects from a workbook—such as a chart, a filtered table, or a selected range of cells? All necessary settings can be configured through the **Save As** dialog box, and the process is largely the same in all cases.

To begin with, here's a simple example: Say you need to publish two price lists on your company's intranet, plus a chart that displays the sales trends of the relevant products. In this case, you can publish the workbook entirely as a web page. Here's how you do it:

1. Place the price lists on two worksheets, the sales figures on a third, and the chart on its own sheet. Rename the sheet tabs to reflect their content.
2. It's usually best to save the workbook as a standard Excel file and then create a copy in HTML format. To do this, use the **Save As** command under the **File** tab, enter an appropriate file name, and select **Web page** in **Save as type**.

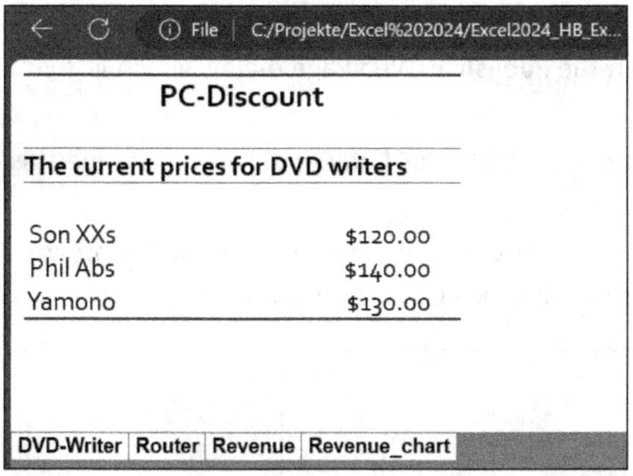

The workbook's structure with its various sheets will be simulated on the web page using a frame layout, with each sheet loading into a separate frame. In the HTML code, script functions handle navigation between sheets by loading each sheet into its corresponding frame.

This example is fairly static. The workbook's data is transferred to the web once, and the process is finished. However, Excel also supports a dynamic solution. For instance, prices for the displayed products may need to be updated. This is a good opportunity to create a link between the workbook and the website so that whenever the workbook changes, the updated version will be sent to the website. Do this as follows:

1. Use the **Save As** dialog and select the **More options** link instead of the **Save** button.
2. Choose to save either the entire workbook ❶ or just a new table range ❷.

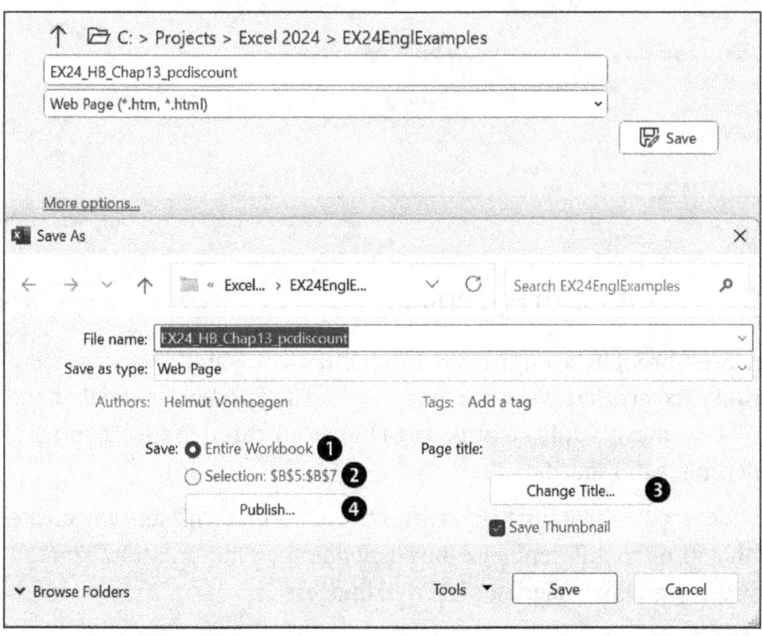

3. Use the **Change Title** button ❸ to set the page title that appears in the browser.
4. Click the **Publish** button ❹ to open the **Publish** as **Web Page** dialog, which lets you review the selected data.
5. To link the workbook to the web page, enable the **AutoRepublish every time this workbook is saved** setting ❺.
6. If the **Open published web page in browser** ❻ option remains selected, the browser will open the new page as soon as you click the **Publish** ❼ button.

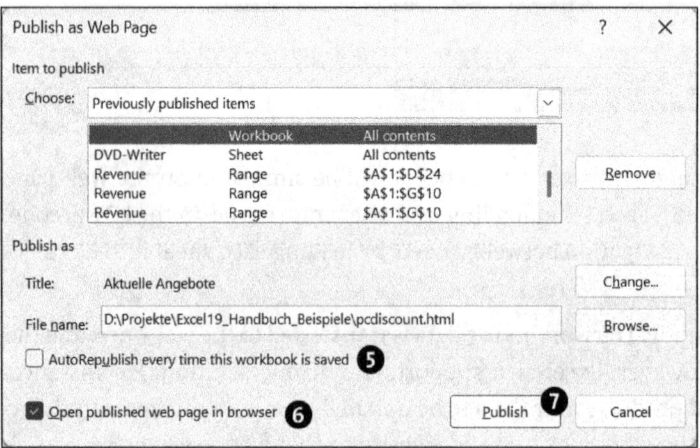

If the workbook is saved again as a web page later, you'll receive a notification. You can then decide whether to apply the current changes in the workbook to the website.

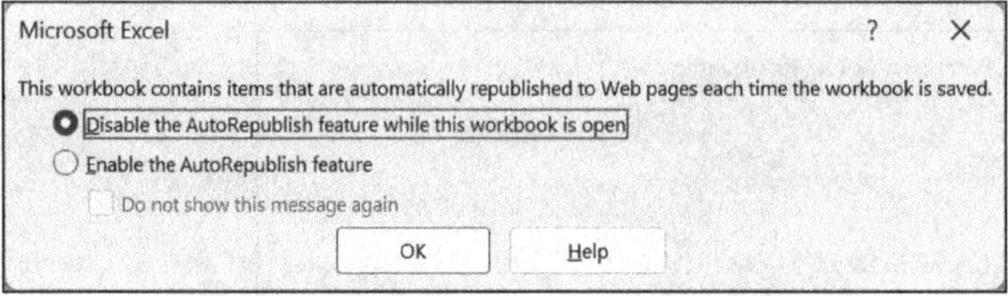

13.4 Linking Documents with Hyperlinks

One feature that's been brought from the web into Office is the ability to link documents with hyperlinks. Excel offers two ways to use a cell as a hyperlink. In the first, the cell's content stays the same; it's only identified as a hyperlink through special formatting—usually underlining and specific colors.

In the second method, a worksheet function creates the hyperlink. This lets you create conditional links that jump to different locations depending on the situation. You can also create links by using graphic objects or inserted images.

13.4.1 Jumping from a Cell

To use a cell as a jump point to a web page, another document, or a specific location in another document or the same workbook, follow these steps:

1. Enter an appropriate note in the relevant cell. For example, if the jump target is a web page for corporate bonds, the cell could say *Corporate bonds*.
2. Keep the cell selected and click the **Link** button in the **Insert • Links** group. Alternatively, you can press [Ctrl]+[K].
3. The dialog box opens with **Link to: Existing File or Web Page** selected, and under **Text to display**, it shows the text that's currently in the cell. Use **ScreenTip** to add an extra note that appears when you hover over the cell with the mouse pointer. If you don't enter anything here, the target address will be displayed.
4. Under **Address**, enter the web page you want or use the small **Browse the Web** button to search for the page directly in the browser. If the address has been used before or multiple pages are currently open in the browser, you can also select it from the list by using the **Browsed Pages** or **Recent Files** buttons. If the page you want is currently visible in the browser, take the address from there by simply clicking or tapping the address field in the dialog.

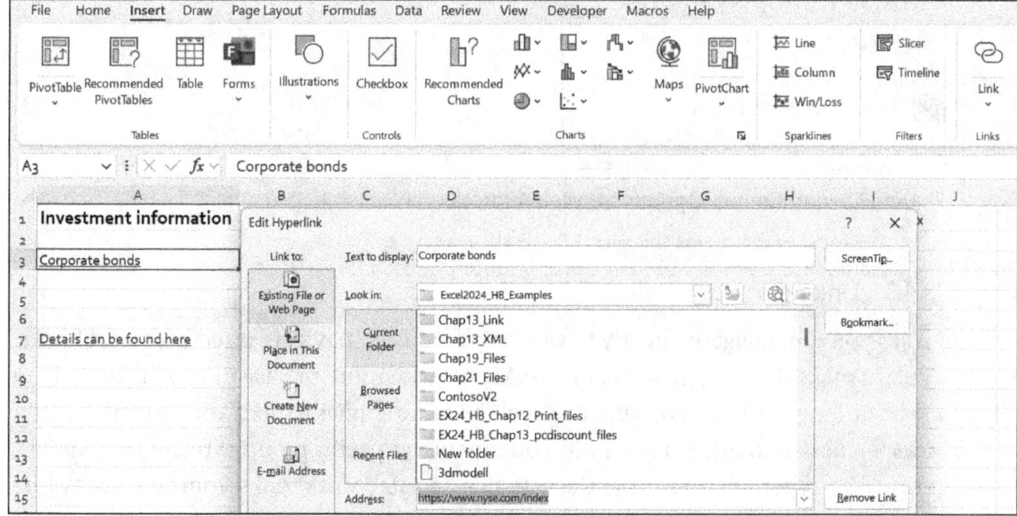

5. When you confirm the entry with **OK**, the cell's content appears formatted with the style currently set for hyperlink cells. Excel uses a special style named **Hyperlink** for this.
6. When you hover the mouse pointer over a hyperlink cell, it turns into a small hand. A click or tap is all it takes to jump to the specified destination. If the internet connection is already established, the browser opens the corresponding page immediately. Otherwise, it tries to connect to the internet and load the page right away.

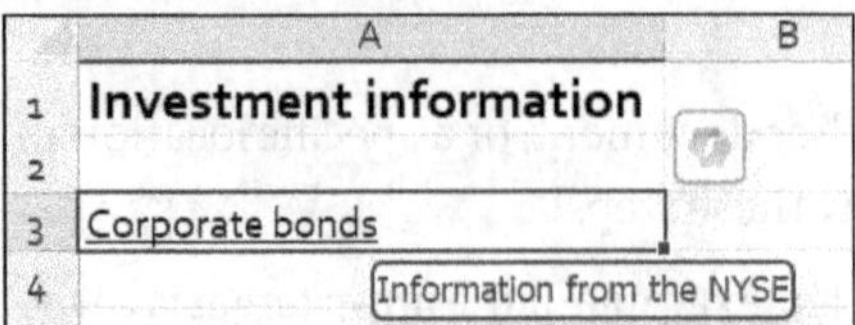

After you use a hyperlink, the link cell's text color changes to show that you've already clicked it.

13.4.2 Automatic Link Creation

When you enter something into a cell that Excel recognizes as a web address or network path, it automatically converts the entry into a hyperlink to that address—unless you've disabled this feature in the **AutoCorrect Options** dialog under **Options • Proofing • Change how Excel corrects and formats as you type**. For such hyperlinks, Excel provides a small button with **AutoCorrect options** that let you remove the hyperlink from the cell if you don't want it. If you want to enter a whole range of addresses as plain text, you can also disable automatic hyperlink creation entirely.

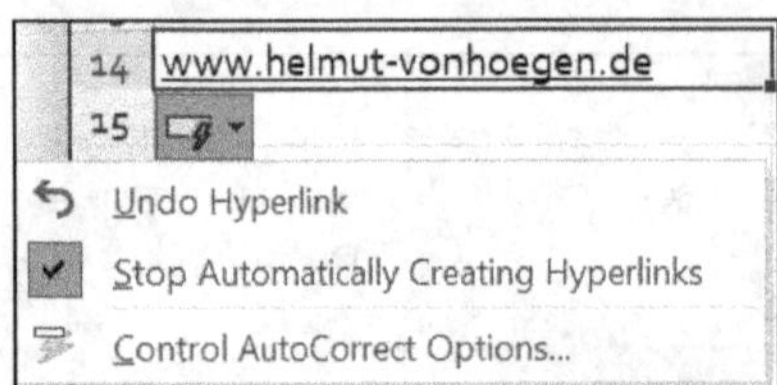

Figure 13.5 AutoCorrect Options for Web Address

13.4.3 Linking to Documents

Just as you can navigate directly to web pages, you can navigate directly to workbooks. You can make the jump to a specific workbook even more precise by linking directly to a specific location, such as a named range. The same applies to text; you can link directly to a specific bookmark. For example, you can turn a specific group of related workbooks into a large hypertext that lets you access all associated workbooks from one place while also allowing you to create unlimited cross-links between them. Once the links are established, you can largely forget the names and paths of all the workbooks since you no longer need to navigate the **Open** dialog box.

13.4.4 A Hyperdocument Composed of Workbooks

Here's a simple example to illustrate: Suppose your daily work mainly involves several workbooks. To let you switch between workbooks with a click, you can create a worksheet containing hyperlinks to all the workbooks—essentially, a table of contents that spans across them. Follow these steps to do it:

1. Create all the workbooks you want to link. If you want to link to specific ranges, name those ranges appropriately.
2. Create a sheet in a separate workbook that lists the titles of the other workbooks.
3. Select each title cell individually and insert hyperlinks to a workbook or specific ranges within the workbooks. In the dialog box, use the **Look in** dropdown list to find and select the target files.

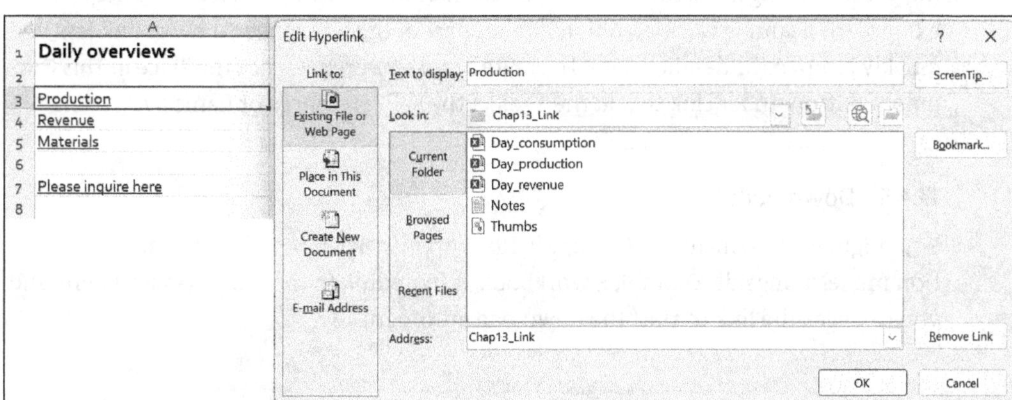

4. To link to a named range, click the **Bookmark** button and select the sheet name under **Cell Reference.** Alternatively, select the range name you want under **Defined Names**. If the names don't appear, click the plus sign.

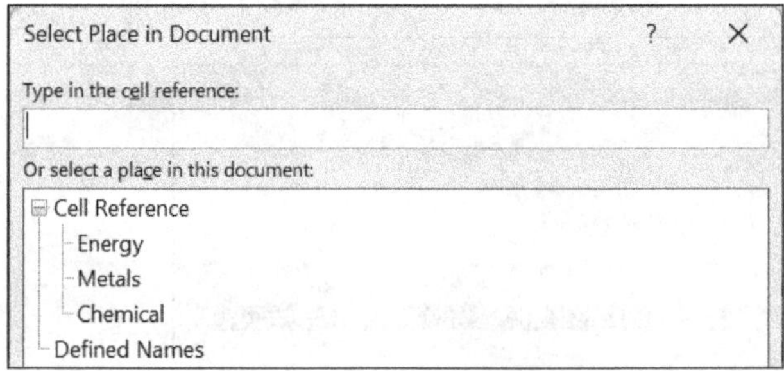

5. After completing all links, save the file again. After that, you'll be able to open all linked files from the table of contents.

To simplify navigation between linked documents, add the **Back** and **Forward** icons to the Quick Access Toolbar. You can find these in the **Excel Options** dialog under **Customize.** Also, select **All Commands** from the **Choose commands from** dropdown list, after which you'll be able to return each time by clicking the **Back** arrow.

[»]

Creating Links to Bookmarks

To create a link to a bookmark in a Word document, manually add it by typing a "#" symbol after the file name and then the bookmark name. You can't find Word bookmarks in Excel by using the **Bookmark** button.

You can also create cross-references between workbooks. Excel automatically tracks the path you take from one jump target to another, and you can always return to the starting point by using the arrow keys. Links within a large worksheet can also be helpful. For example, if a sum is calculated in a hidden part of the worksheet, a hyperlink lets you quickly jump to the detailed data. To create a cross-reference, click the **Place in This Document** button under **Link to** and then select the cell references or names you want.

13.4.5 Downloads

You might want to include a PDF or ZIP file with instructions on how to use the calculation model alongside a complex workbook. A hyperlink to the file works well here, and you can click the link to start the download automatically.

13.4.6 Email Links

When you're sharing a workbook, it can be helpful to add links with your email address. To do this, select the **E-mail Address** button in the dialog and then type "mailto:" followed by the appropriate address in the address field. When the viewer clicks this cell, a new email dialog opens immediately to that address, if an email program is installed.

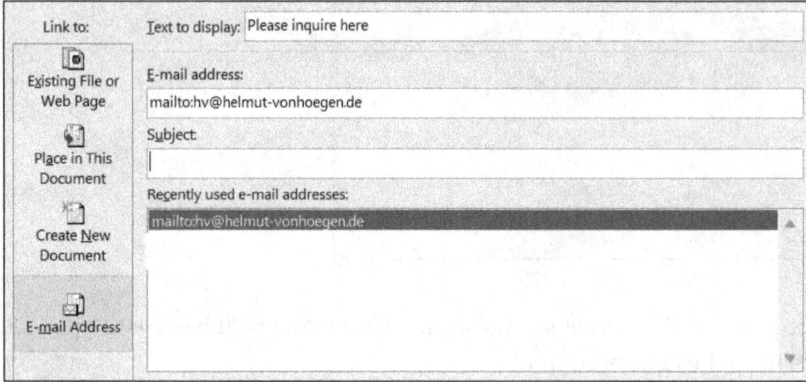

Figure 13.6 Entering Link to Email Address

13.4.7 Hyperlinks Using the Table Function

In Excel, you can create and manage hyperlinks by using a table function, as mentioned earlier. You can insert the `HYPERLINK()` function into a cell in the usual way, and to activate the hyperlink, click or tap the cell, just like with other hyperlinks. The function has two arguments, and the first is the hyperlink address to connect to. You can also specify what appears in the cell, such as descriptive text. This argument is called **Friendly_name**, and if you omit it, the cell simply displays the specified hyperlink address.

You can enter the hyperlink address directly, and if you do, the system will automatically add quotation marks. You can also reference a cell that contains the address as text or returns it as a formula result. If the hyperlink target can't be found for any reason, an error message will appear in the cell. To change the link target, edit the formula as usual.

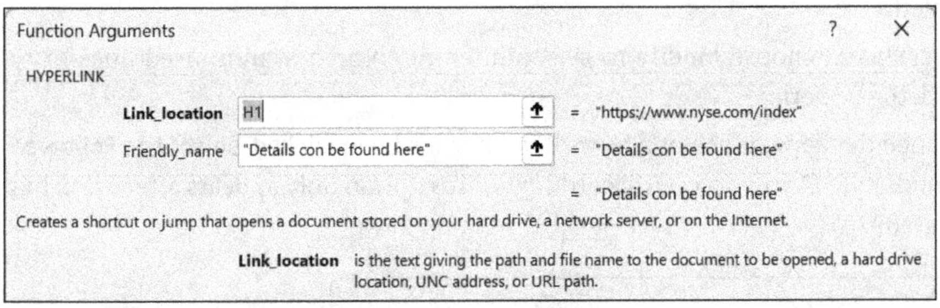

Function Arguments ? X

HYPERLINK

Link_location H1 = "https://www.nyse.com/index"

Friendly_name "Details con be found here" = "Details con be found here"

= "Details con be found here"

Creates a shortcut or jump that opens a document stored on your hard drive, a network server, or on the Internet.

Link_location is the text giving the path and file name to the document to be opened, a hard drive location, UNC address, or URL path.

Figure 13.7 Example of HYPERLINK() Function

The advantage of using the `HYPERLINK()` function is that it supports variable jump targets. For example, you can store a folder name in cell H1 and a file name in cell H2, without typing quotation marks in either cell. Then, you can enter the following hyperlink function in cell B3:

```
=HYPERLINK(H1&H2, "Find details here")
```

In this case, the address is made up of two text parts. What's the advantage? Suppose your workbook contains a series of such hyperlinks, all of which point to files in the same folder. If you reorganize your hard drive or copy the files to another drive, you only need to update the entry in cell H1 to restore the links.

Conditional Links: Sometimes Here, Sometimes There

The introduction to this section already mentioned the possibility of conditional hyperlinks. Assume you prepare quotes sometimes for a German customer and sometimes for a French customer and that the two price lists are stored in separate tables. In the offer table, you can enter a marker in cell K2 to distinguish German offers from French ones. Then, in another cell, you can enter a hyperlink formula that links to the price tables based on this marker. The formula can be as follows:

```
=IF(K2="Fr", HYPERLINK("C:\Office\PLISTEFR.XLSX", "French price list"),
HYPERLINK("C:\Office\PLISTEDE.XLSX","German price list"))
```

The text and the link that will appear will depend on the value in cell K2.

13.4.8 Formatting Hyperlinks

You can change how hyperlinks are displayed in the worksheet for all hyperlinks in the current workbook at once. Excel uses special styles for hyperlinks and visited hyperlinks, and these styles only appear in the style list once the workbook contains hyperlinks or after a hyperlink has been used to jump to another document or location and back. Follow these steps to format hyperlinks:

1. Select a hyperlink cell by using the arrow keys, and then, in the **Home • Styles** group, click the **Cell Styles** button.
2. If the hyperlink hasn't been clicked yet, the **Hyperlink** ❶ style will be highlighted in the palette.
3. Right-click and choose **Modify** to select different colors for all unused links or to remove the underline.
4. To change the appearance of a hyperlink after it's been clicked, select the **Followed Hyperlink** style ❷ and proceed accordingly. (This option only appears after a link has been used.)

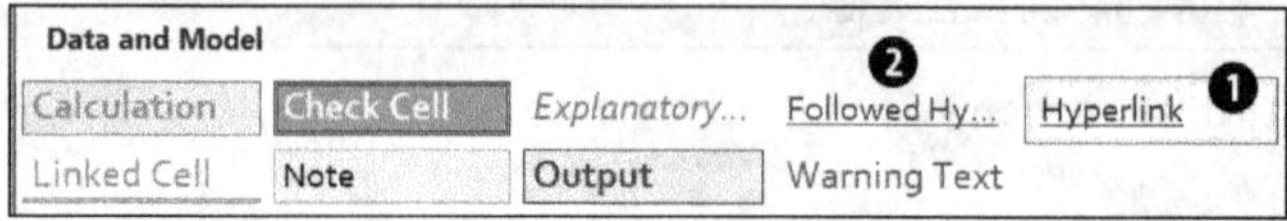

The different hyperlink colors help you remember which pages and documents you've already viewed, so choose colors that are clearly distinguishable from one another.

13.4.9 Hyperlinking from a Graphic Object

Graphic objects in a worksheet can also serve as links to web pages or other documents. Here's a simple example of hyperlinking a graphic object:

1. Using the **Shapes** button in the **Insert • Illustrations** group, draw a block arrow. Then, from the arrow's context menu, select **Edit Text** to label it.
2. Keep the object selected so the handles are visible and then use the **Link** button.
3. Enter the hyperlink target under **Address** or choose it from **Browsed Pages**.

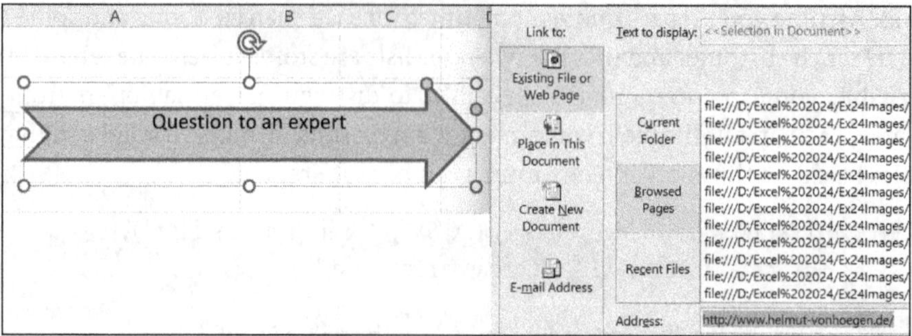

13.4.10 Editing Hyperlinks

Sometimes, you need to update a hyperlink, especially if the file specified as the link target has been moved to another folder or renamed. Automatic updates for such changes would be helpful but are unfortunately not supported. Follow these steps to fix the link:

1. Right-click the cell or graphic object where the links were created. Alternatively, press and hold until the context menu appears and then tap the arrow at the end.
2. Select the **Edit Hyperlink** option.
3. Enter the new target address.

To remove hyperlinks from a workbook, right-click and select **Remove Hyperlink** from the context menu. For cells, this removes the link without deleting the cell content. For graphic objects, it only removes the associated link. However, if you press `Delete` on a selected hyperlink cell or graphic object, both the link and the cell content will be deleted. If you accidentally delete a link, simply click the **Undo** button.

13

13.5 Processing XML Data

As described in Chapter 1, Excel uses XML documents that comply with the Open XML standard as the default format for saving workbooks. Excel also offers specialized tools for creating and processing other XML documents, which can be structured with custom schemas. However, you need to enable these tools in the ribbon before you can use them in Excel. The related commands and buttons are located on the **Developer Tools** tab within the **XML** group. This tab appears when you go to **File • Options**, select **Customize Ribbon,** and check the **Developer Tools** box in the list of main tabs on the right.

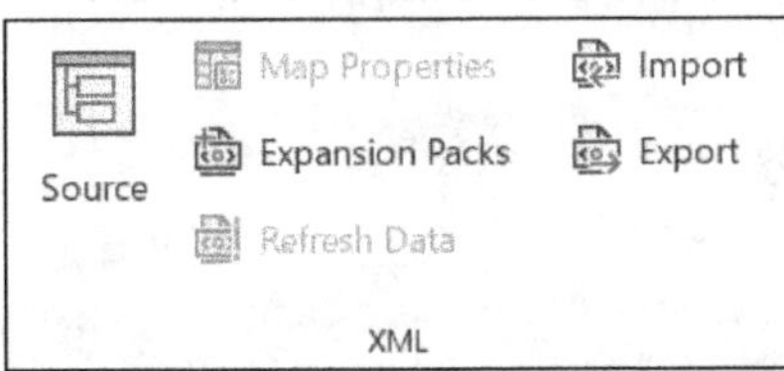

Figure 13.8 XML Group on Developer Tools Tab

13.5.1 Importing XML Data

Excel not only supports data in XML format but also validates such data with corresponding schemas. An *XML schema* is a well-formed XML document that defines the data structure for an entire class of XML documents. Using XML schemas allows companies to enforce mandatory data structures in different business areas through appropriate templates and forms. You can specify a schema or let Excel generate one for you.

When Excel opens an XML file with an associated schema, it automatically uses that schema if it's found at the path that's specified in the file. If the imported data lacks an associated XML schema, Excel tries to generate a suitable schema from the current XML document. However, this method is only somewhat reliable for simple structures because it can only consider the structural elements that are present in a single document. Schemas also typically impose restrictions on an entire class of documents, which can vary widely in detail. For example, if an element allows three possible values, Excel can't infer this from a document that uses only one of those values. Therefore, it's generally best to work with explicitly defined XML schemas, and since they are text documents, you can usually use a simple text editor if you need one.

Excel can open a well-formed XML document directly through the standard **Open** dialog. The following options are initially offered:

- Importing data as an XML table
- Opening the file as a read-only workbook
- Using the data to define a data structure in the **XML Source** task pane

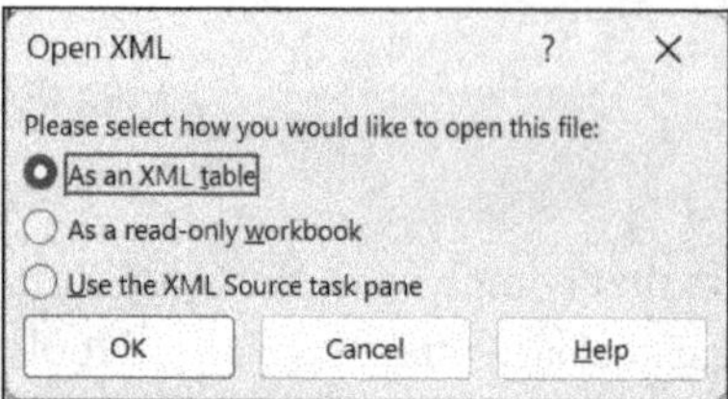

Figure 13.9 Options When Opening XML Document

Importing Data as an XML Table

What happens when you select the first option? The following listing shows a small XML document with order data, structured according to the XML schema specified within it:

```
<?xml version="1.0" encoding="UTF-8"?>
<orderform nr="01000" date="2024-11-01" editedBy="Sylvia Kaily"
  xmlns:xsi="http://www.w3.org/2001/XMLSchema-instance"
  xsi:noNamespaceSchemaLocation="order.xsd">
  <customer>
    <name>Hanna Maier</name>
    <street>Titusstreet 12</street>
    <zip>50678/zip>
    <city>Cologne</city>
  </customer>
  <items>
    <item>
      <itemnr>0045</itemnr>
      <description>Roller blind XBP 312</description>
      <package>pcs</package>
      <quantity>5</quantity>
```

```
      <price>50.00</price>
    </item>
    <item>
      <itemnr>0046</itemnr>
      <description>Roller blind MMX</description>
      <package>pcs</package>
      <quantity>4</quantity>
      <price>40.00</price>
    </item>
  </positions>
</orderform>
```

The schema defines the allowed elements and attributes, specifies their frequency and order, and indicates whether they are required or optional. Here is the schema in this case:

```
<?xml version="1.0" encoding="UTF-8"?>
<xsd:schema xmlns:xsd="http://www.w3.org/2001/XMLSchema"
  elementFormDefault="qualified">
  <xsd:element name="orderform" type="form"/>
  <xsd:complexType name="form">
    <xsd:sequence>
      <xsd:element name="customer" type="customer"/>
      <xsd:element name="items" type="items"/>
    </xsd:sequence>
    <xsd:attribute name="nr" type="xsd:short" use="required"/>
    <xsd:attribute name="date" type="xsd:date" use="required"/>
    <xsd:attribute name="processing" type="xsd:string" use="required"/>
  </xsd:complexType>
  <xsd:complexType name="customer">
    <xsd:sequence>
      <xsd:element name="name" type="xsd:string"/>
      <xsd:element name="street" type="xsd:string"/>
      <xsd:element name="zip" type="xsd:int"/>
      <xsd:element name="city" type="xsd:string"/>
    </xsd:sequence>
  </xsd:complexType>
  <xsd:complexType name="items">
    <xsd:sequence>
      <xsd:element name="item" minOccurs="0" maxOccurs="unbounded">
        <xsd:complexType>
          <xsd:sequence>
            <xsd:element name="itemnr" type="xsd:string"/>
            <xsd:element name="description" type="xsd:string"/>
            <xsd:element name="package" type="pkg"/>
```

13

```
            <xsd:element name="quantity" type="xsd:decimal"/>
            <xsd:element name="price" type="xsd:decimal"/>
          </xsd:sequence>
          </xsd:complexType>
      </xsd:element>
    </xsd:sequence>
  </xsd:complexType>
  <xsd:simpleType name="pkg">
    <xsd:restriction base="xsd:string">
      <xsd:enumeration value="pcs"/>
      <xsd:enumeration value="Lbm"/>
      <xsd:enumeration value="Inch"/>
    </xsd:restriction>
  </xsd:simpleType>
</xsd:schema>
```

If you open the XML document with the **Open XML • As an XML table** option, the worksheet displays a table that flattens the original three element levels into a two-dimensional table.

	A	B	C	D	E	F	G	H	I	J	K	L
1	nr	date	processing	name	street	zip	city	itemnr	description	unit	quantity	price
2	1000	3/1/2026	Sylvia Kaily	Hanna Maier	Titusstreet 12	50678	Cologne	0045	Roller blind XBP 312	Pcs	5	50
3	1000	3/1/2026	Sylvia Kaily	Hanna Maier	Titusstreet 12	50678	Cologne	0046	Rollo blind MMX	Pcs	4	40

Figure 13.10 Table Generated from XML Document

Excel automatically creates a table range for the imported data, as described in detail in Chapter 16, Section 16.5. The first three columns repeat the attribute values of the <orderform> element in each row. The names of the <customer>, <items>, and <item> parent elements do not appear in the table, but the names of the lowest child elements are used as column headers, with their contents listed below.

A link is established between the XML document and the table in the workbook, and you can control how this is handled through the **XML Map Properties** dialog, which is accessible via **Map Properties** in the **Developer • XML** group. There, you can specify whether to validate the data against the XML schema during import or export or to only check for well-formedness. Generally, you should keep the setting that saves the data source definition, or schema, with the file.

In addition to some formatting options, you can specify whether data that's newly imported into the table range should replace existing data or be appended. The second option is needed, for example, to add more order items to the list. If the original XML document behind the list changes, you can use **Developer Tools • XML • Refresh Data** to update the table in Excel. Alternatively, you can use the **XML • Refresh XML Data** context menu command.

If you try to export the data in this state back to XML format by using the command in the **XML** group, Excel will warn you that the data cannot be exported in this form. This occurs because the data structure has been compressed into a two-dimensional table, causing the attribute values for the element `<orderform>` to appear multiple times, as if it were a repeating element. To keep the data exportable, you must store the attribute values in individual cells outside the list, as we explain later in the section on the third method: linking the data source and table.

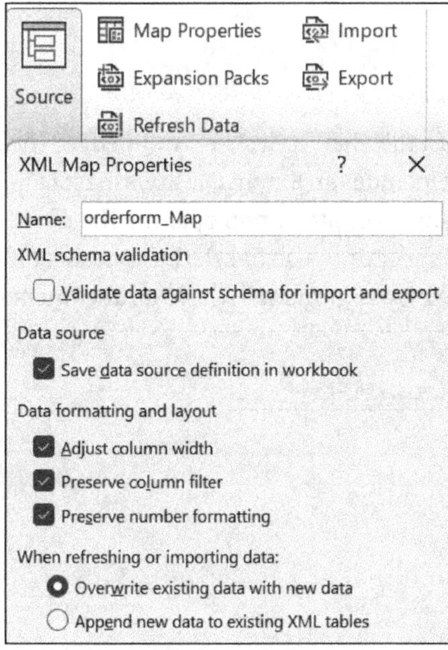

Figure 13.11 XML Mapping Dialog

Instead of opening an existing XML file directly in Excel, you can import XML data into an existing workbook by using the **Developer Tools • XML • Import** command. As with importing data from other sources, you can specify a specific range or worksheet as the destination. Otherwise, the process is the same as opening an XML file as an XML table.

Opening a File as a Read-Only Workbook

If you choose the second option to open an XML file, the document opens as a read-only XML file. This creates a cell range with columns that are automatically labeled with XPath expressions based on the existing element and attribute names. *XPath* is a specialized language for creating expressions that target parts of an XML document. For example, the expression */orderform/customer/name* returns the customer's name, and the expression */@nr* uses the @ symbol to indicate it refers to an attribute.

This method doesn't link back to the original document, and the XML document's hierarchy is also flattened into a two-dimensional table. Write protection prevents data from being written back to the original file, thus preserving its original structure. Figure 13.12 shows how this appears with order data. This version offers some insight into the XML document's structure but has limited practical use.

	A	B	C	D	E	F	G	H	I
1	/orderform								
2	/@date	/@nr	/@nr/#agg	/@processing	/@xsi:noN	/customer/city	/customer/name	/customer/street	/customer/zip
3	3/1/2026	1000	1000	Sylvia Kaily	order.xsd	Cologne	Hanna Maier	Titusstreet 12	50678
4	3/1/2026	1000		Sylvia Kaily	order.xsd	Cologne	Hanna Maier	Titusstreet 12	50678

Figure 13.12 XML Data Imported as Read-Only

Using Stylesheets

Importing XML data as a read-only file is more useful when stylesheets defined for the XML document are applied. If an XML document includes an Extensible Stylesheet Language Transformations (XSLT) stylesheet, you'll be prompted upon opening the document to import the data without using the stylesheet or to use or select a stylesheet if multiple stylesheets are assigned. You can make this assignment in the XML document with a processing instruction like this:

```
<?xml-stylesheet type="text/xsl" href="orderform.xsl"?>
```

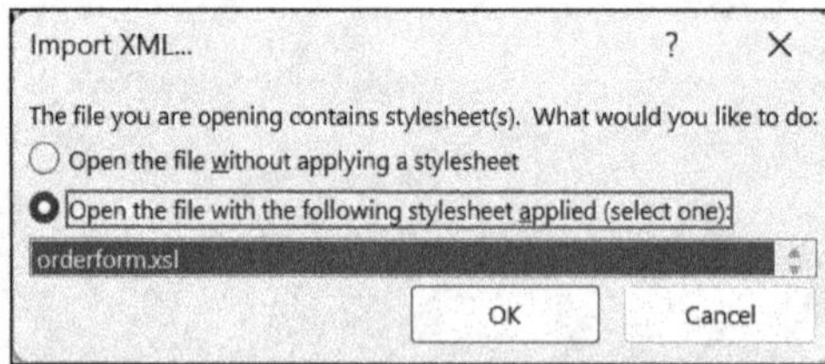

Figure 13.13 Selecting Stylesheet Before Importing XML Data

This stylesheet defines how the data should be output using XSLT and HTML tags. Excel can then directly import the HTML output into its cell structure and thus let you work with the result like a regular worksheet.

Here is a simple example of such a stylesheet that outputs the XML document's data using three `<template>` elements, with the item data imported into a table via a loop.

	A	B	C	D	E
1	**Order**				
2					
3	Hanna Maier				
4	Titusstreet 12				
5	50678 Cologne				
6					
7	**Nr**	**Description**	**Unit**	**Quantity**	**Price**
8	45	Rollo XBP 312	Pcs	5	$50.00
9	46	Rollo MMX	Pcs	4	$40.00

Figure 13.14 Outputting Order Data with an XSLT Stylesheet

Here is the source file without further comments:

```
<?xml version="1.0" encoding="UTF-8"?>
<xsl:stylesheet version="1.0" xmlns:xsl="http://www.w3.org/1999/XSL/Transform">
  <xsl:output method="html" encoding="UTF-8"/>
  <xsl:decimal-format name="dollar" decimal-separator="."
        grouping-separator=","/>
<xsl:template match="/">
    <html>
      <head>
        <title>Order</title>
      </head>
      <body>
        <h3>Order</h3>
         <xsl:apply-templates select="//customer"/>
         <table border="1" cellpadding="5" cellspacing="5">
         <xsl:apply-templates select="//items"/></table>
      </body>
    </html>
  </xsl:template>
<xsl:template match="customer">
      <p><xsl:value-of select="name"/><br />
      <xsl:value-of select="street"/><br />
      <xsl:value-of select="zip"/><xsl:text> </xsl:text>
         <xsl:value-of select="city"/> </p>
   </xsl:template>
<xsl:template match="items">
  <tr>
    <th>Nr</th>
    <th>Description</th>
    <th>Unit</th>
    <th>Quantity</th>
    <th>Price</th>
  </tr>
  <xsl:for-each select="item">
     <tr>
      <td><xsl:value-of select="itemnr"/></td>
      <td><xsl:value-of select="description"/></td>
      <td><xsl:value-of select="unit"/></td>
      <td><xsl:value-of select="quantity"/></td>
      <td><xsl:value-of select="format-number
         (price, '&#36; ##,###.00')"/></td></tr>
  </xsl:for-each>
</xsl:template>
</xsl:stylesheet>
```

13.5.2 Linking the Data Source to the Table

The third option when opening an XML file is to use the **XML Source** task pane. In this case, the program initially imports only the XML document's data structure into the **XML Source** task pane and displays it as a tree that's similar to the folder structure in File Explorer. As mentioned earlier, this structure can be defined by the schema file linked to the XML document or generated by Excel if you confirm the prompt.

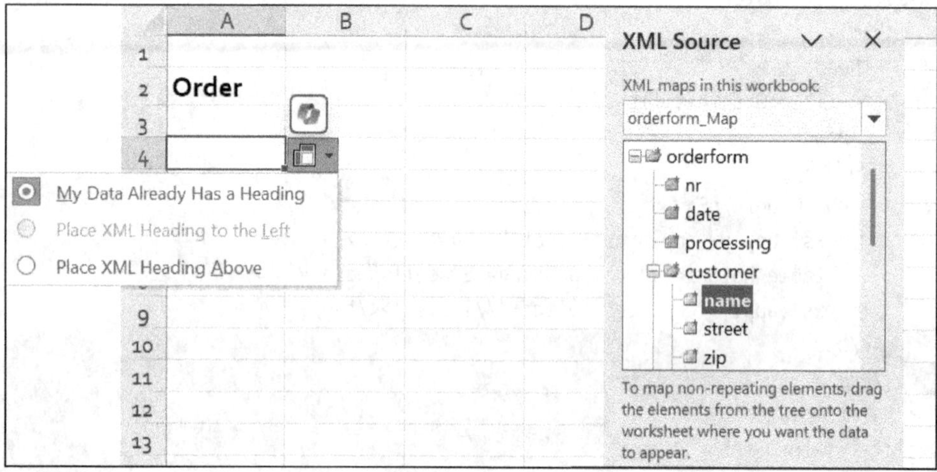

Figure 13.15 XML Source in Task Pane and Menu of Associated Cell Range

If a worksheet requires data from the XML document, you can drag data objects from the task pane to the desired spot in the table by using your mouse or finger. You can choose the order freely. Excel distinguishes between elements that repeat, which are inserted into the table in tabular form, and those that are needed only once, which are placed in individual cells above or beside the table.

The attributes of the `<orderform>` element and the child elements of `<customer>` each appear only once. When you drag the icon from the task pane into the worksheet, a button appears with a menu that lets you use the attribute name as the label or enter a custom label for the field.

For repeating elements, Excel automatically creates a table range while using the element and attribute names as column headers. In this case, you can simply drag the `<item>` symbol into the table range (see Figure 13.16).

Cell ranges that are linked to XML data are outlined with borders that are ignored when printing. Use **Developer Tools • XML • Refresh Data** to import data for the selected cells from the XML file. This command is only enabled when the cell pointer is within a linked cell range. In Figure 13.17, the **Price** column was formatted with a currency style and the column headers were capitalized.

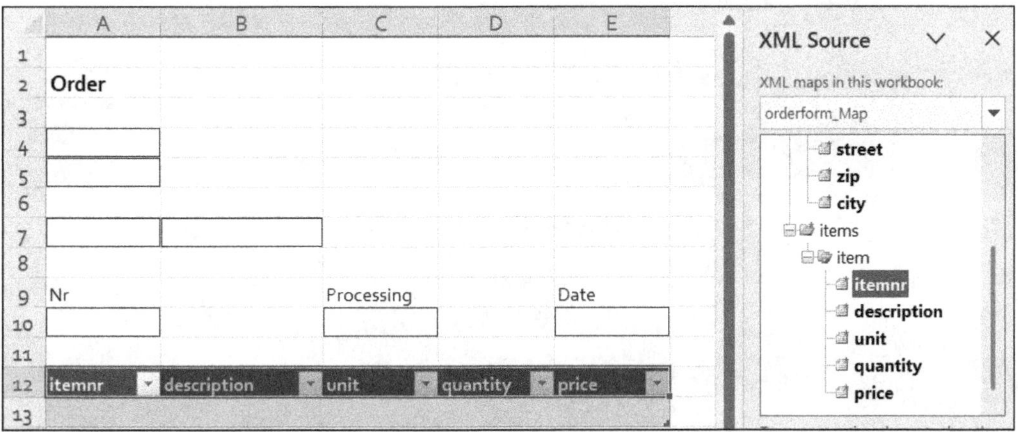

Figure 13.16 Separate and Repeating Elements

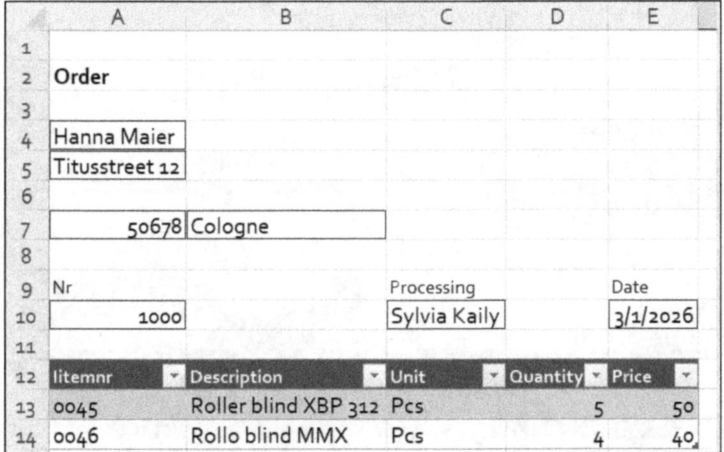

Figure 13.17 Form Filled with Source Data

Using defined XML schemas for various business processes, companies can establish standardized data structures for different workstations that can be flexibly applied to a wide range of analyses.

Use the **XML Maps** button in the **XML Source** task pane to add multiple data sources to a workbook if you need to. The mapping between the XML data source and worksheet cells is managed through special *XmlMaps* objects that can also be controlled through macros. The links to the data use XPath expressions, which are standard in XML applications.

Chapter 14
Collaborating on Workbooks

Teamwork is now a fundamental skill in the workplace, and many projects require collaboration among multiple people. For example, one person may compile data for one area, another person may compile data for a different one, and a third person may verify whether all that data is accurate.

Excel offers several ways to support collaboration on a document. This section covers teamwork within a local network and then discusses using OneDrive.

14.1 Teamwork in Local Networks

14

In Excel, the workbook-sharing feature known from earlier versions, which was available on the **Review** tab, is now hidden and maintained only as a legacy function. Microsoft aims to encourage users to organize teamwork in Excel via the cloud, as described in Section 14.2, but this form of collaboration requires a Windows 365 subscription. If you don't want to get a subscription, you can bring the hidden feature back into view and continue working with it. You can do this by making an appropriate extension of the ribbon menu, as follows:

1. Go to **File • Options** and then to the **Customize Ribbon** page.
2. Create a new group on the **Review** tab and name it, for example, **Teamwork.**
3. Under **Choose commands from**, select **All Commands** and add the **Share Workbook (Legacy)** command to the new group by using **Add**.
4. Add the **Track Changes (Legacy)** and **Compare and Merge Workbooks** commands.

Sharing lets you control the level of access others have for each workbook. Should the other person only be allowed to view the data? Should they have permission to edit the data? What if multiple people want to make different changes to the same cells? There are practical solutions for all of these scenarios.

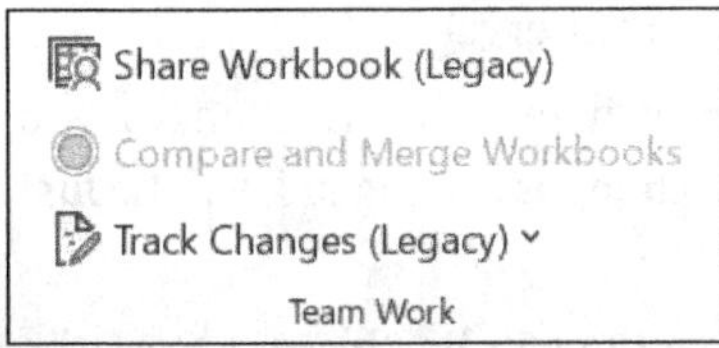

Figure 14.1 Commands for Sharing on Review Tab

14.1.1 Sharing a Workbook

For example, if you create a workbook for financial planning, you can choose whether to share it with others, such as HR staff who need to enter or review actual personnel cost data. Follow these steps:

1. Save the workbook in a shared network folder and allow editing, not just viewing.
2. On the **Advanced Review** tab, select the **Share Workbook** command.
3. In the dialog box, clear the first checkbox on the **Status** tab. This allows multiple users to edit the workbook.
4. Set detailed collaboration rules on the second **Advanced** tab. First, under **Track changes**, decide whether Excel should keep a log of changes and, if so, for how long. For example, to track changes for the last month, set it to **30 days**. This log lets you reconstruct who made which changes and when, if you need to.

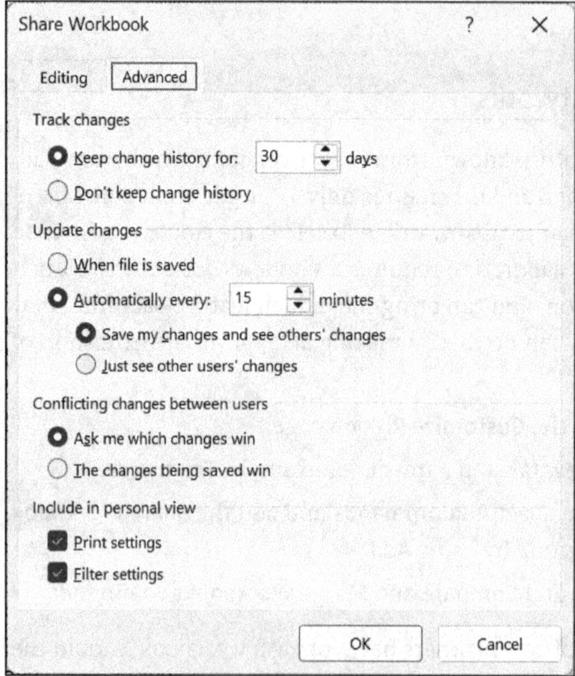

5. Under **Update changes**, choose whether changes will be applied only when the workbook is saved or automatically at specific intervals. If keeping the workbook's values as current as possible is important to you, select a short interval.
6. If you select a time interval, you can also choose to prioritize your own changes. You can either select the **Save my changes and see others' changes** setting or keep the **Just see other users' changes** setting.
7. If changes are allowed from different places, two people will be able to access the same cell. For example, colleague John will be able to enter personnel costs of

$40,000 even though colleague Anne just entered a corrected amount of $41,000. You'll need to prevent the correct value from being overwritten by an incorrect one if such conflicts occur, and you can do it by setting **Conflicting changes between users** to require manual resolution. You can also set it so that your own saved changes always take precedence.

8. Finally, under **Include in personal view**, you can allow each user to create their own workbook views with personalized print, sorting, and filtering settings.

When a shared workbook is opened, the title bar displays a note saying **Shared** and collaborative work on the workbook becomes fully enabled. An unlimited number of users can then work on the file, and each person who opens the shared file will initially work with a copy of the original. The names of active users will appear on the status tab.

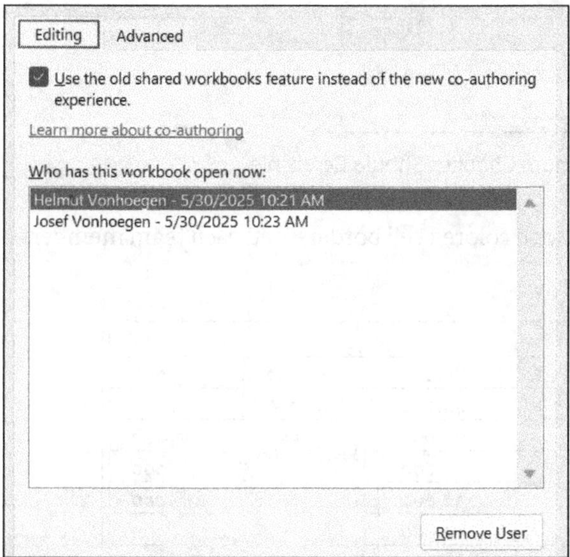

Figure 14.2 Display of Who Is Currently Working on Workbook

All changes are applied to the original file whenever it's saved or at the intervals set in step 4.

14.1.2 Showing Changes

To check whether cells in the workbook have been changed, use the **Track Changes • Highlight Changes** command. This command highlights changes directly on the worksheet as long as logging, which we mentioned in step 3, is enabled.

Keep the **Track changes while editing** option enabled. Under **When**, choose whether to show changes since the last save, show them since a specific date, or show all unreviewed changes. You can enter an exact date if you use the **Since date...** option. **Always** simply displays all changes.

To limit the review to changes made by specific people, check **Who** and select the name from the list box. Under **Where**, you can also restrict the review to specific ranges in the workbook. Changes can be shown directly on the affected worksheet or as a log on a separate sheet.

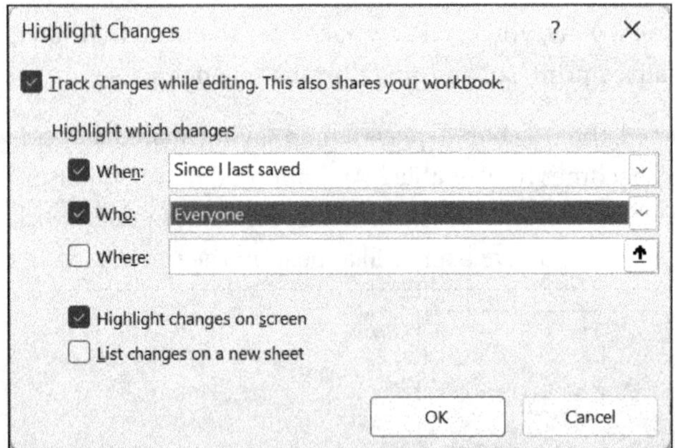

Figure 14.3 Selecting When and to Whom Changes Should Be Visible

Changes are highlighted on screen with colored cell borders, and each team member is assigned a unique color.

11	Expenditure			
12	Purchases of goods	80,000	83,000	3,000
13	Personnel costs	44,000	43,000	-1,000
14	Other costs	6,000	[illegible]	[illegible]
15	Loan repayments	4,000	[illegible]	[illegible]
16	Tax	30,000	[illegible]	[illegible]
17	Other expenditure	2,000	[illegible]	[illegible]
18	Total	166,000	[illegible]	[illegible]
19	Financing surplus/need	80,000	83,600	3,600

Josef Vonhoegen:
Value is too high

Figure 14.4 Table with Highlighted Changes

14.1.3 Reviewing Changes

If changes exist, you can choose whether to accept them by using the **Track Changes** and **Accept/Reject Changes** commands. The dialog box offers the same options to filter changes by age, person, or table range, and it's preset to review all unverified changes. When you click **OK**, a dialog box appears where you can accept or reject each change. You can also accept or reject changes individually or apply your decision to all changes at once.

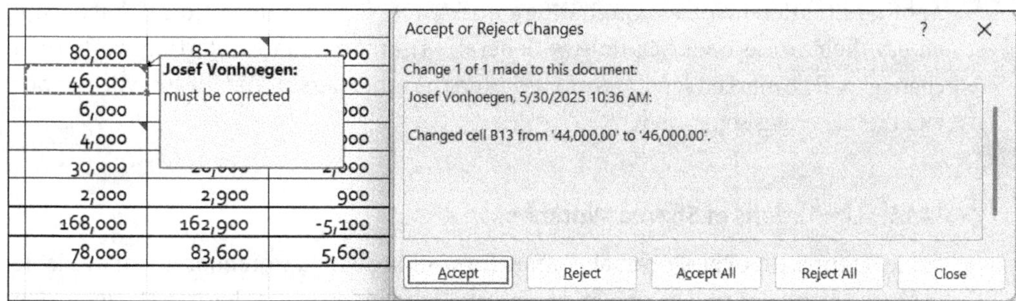

Figure 14.5 Decision to Accept or Reject a Change

If the change log output in the **Highlight Changes** dialog is placed on a separate sheet, you can closely track the sequence of changes, as Figure 14.6 displays.

	A	B	C	D	E	F	G	H	I	J	K
1	Action Number	Date	Time	Who	Change	Sheet	Range	New Value	Old Value	Action Type	Losing Action
2	1	5/30/2025	10:36 AM	Josef Vonhoegen	Cell Change	Plan2026	B13	46,000.00	44,000.00		
3	2	5/30/2025	10:45 AM	Helmut Vonhoegen	Cell Change	Plan2026	B7	6,000.00	5,000.00		
4	3	5/30/2025	10:47 AM	Helmut Vonhoegen	Cell Change	Plan2026	B13	47,000.00	46,000.00		
5	4	5/30/2025	10:53 AM	Helmut Vonhoegen	Cell Change	Plan2026	B13	46,000.00		Result of rejected action	3
6	5	5/30/2025	10:53 AM	Helmut Vonhoegen	Cell Change	Plan2026	B7	5000		Result of rejected action	2
7											
8	The history ends with the changes saved on 5/30/2025 at 10:53 AM.										

Figure 14.6 Log of Changes and How They Are Handled

If different people change certain values differently, a conflict will occur that you must resolve manually, based on the sharing settings you've previously chosen. A dialog box will appear showing the conflicts, and it will provide a button you can use to choose either option.

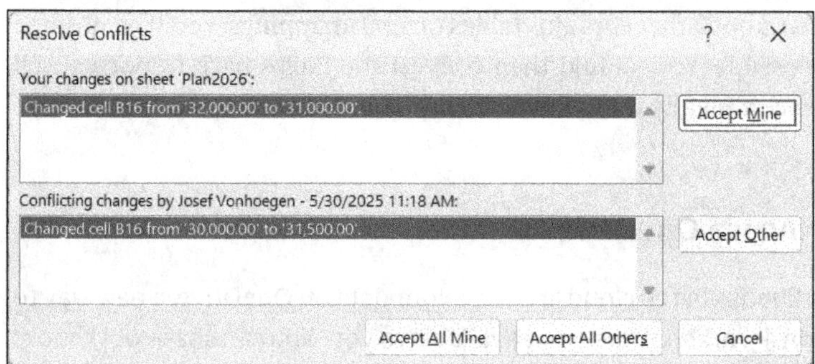

Figure 14.7 Resolve Conflicts Dialog Box

14.1.4 Sharing on a Single Workstation

Even if you're not working on a network, workbook sharing can be useful—for example, if you want to track specific changes you make to a calculation model over time.

Another situation involves you sharing a workspace with another person. If that person enters their name under **Options • General • Username** before starting work, then changes will be marked separately for you and the other person and the log will list your and their changes separately.

14.1.5 Limitations of Shared Workbooks

Some features that Excel normally offers for working with a workbook are unavailable in shared workbooks, so you should complete these tasks before sharing. The following actions are not possible:

- Deleting sheets
- Merging cells
- Defining conditional formats
- Setting validation rules
- Inserting and deleting cell ranges and excluding entire rows or columns
- Inserting or modifying charts, graphic objects, and hyperlinks
- Drawing with graphic tools
- Assigning or changing passwords
- Using or modifying scenarios
- Grouping and outlining
- Creating and pivoting tables
- Modifying menus and dialog boxes
- Developing new macros (existing macros can still be run)

If you try to share a workbook that contains tables or XML mappings, you'll be notified that sharing is not possible. You should then convert the tables back to normal cell ranges, and you must completely remove XML mappings.

14.2 Collaborating via OneDrive

If you store your workbooks in the cloud and save your data on OneDrive, a new way to organize teamwork on a workbook becomes available to you. You can share workbooks on OneDrive with others for editing, and many people can even work simultaneously on a shared workbook in a browser. No one has to wait for others to finish editing, and every change is instantly visible to everyone currently viewing the workbook online. No special locking commands are used to manage collaborative editing.

14.2.1 Sharing Workbooks

A prerequisite for this form of teamwork with Excel is that you have a Windows 365 subscription, and that's a primary reason why Microsoft recommends their subscriptions to users—who, hopefully, don't feel pressured too much.

When sharing workbooks, you choose whether the files can only be viewed or can also be edited. The steps to share a workbook are as follows:

1. If the workbook is open, use the Share button on the ribbon. Alternatively, you can use **File • Share** instead. In both cases, you'll see a **OneDrive** button that allows you to save the workbook in a OneDrive folder.

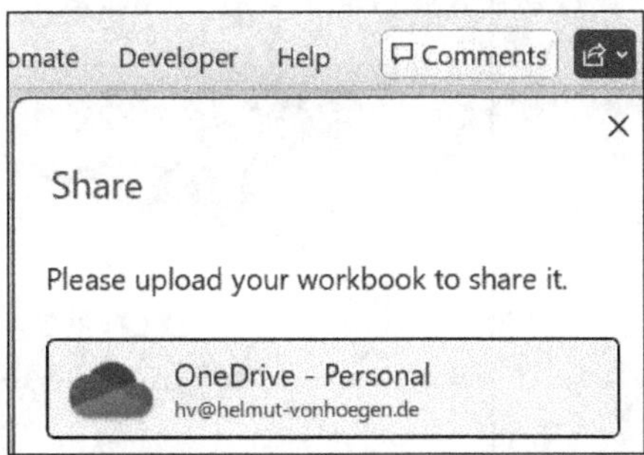

2. Share the workbook on OneDrive with the people who should view or edit it.

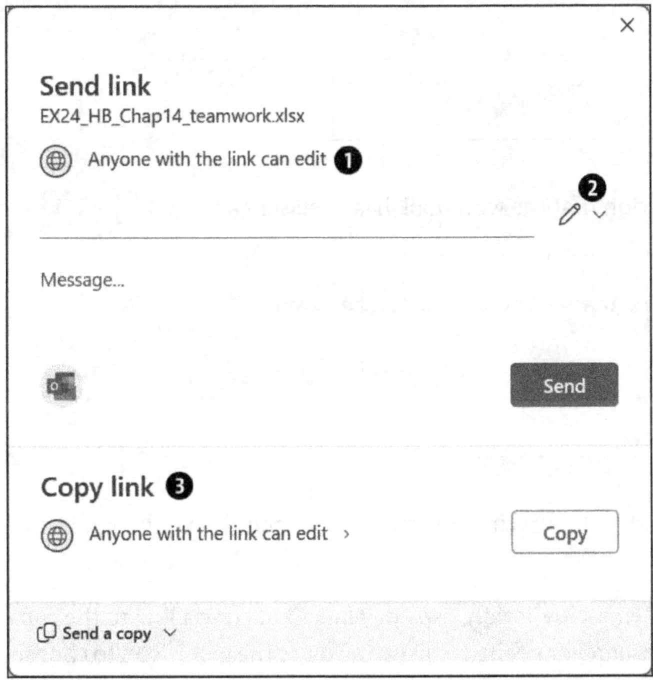

3. Under **Send Link**, specify who can access the workbook. The button with the sphere ❶ opens the link settings, where you can choose whether to allow collaborative editing. You can also set an expiration date or a password for access here.
4. Below that, enter one or more email addresses to receive a link to the workbook.
5. Use the menu or the pen ❷ to decide if the person can only view or also edit the workbook.
6. If you need to, write a message in the text box below. You can send the messages by clicking the **Send** button, but only if you have an email program installed.

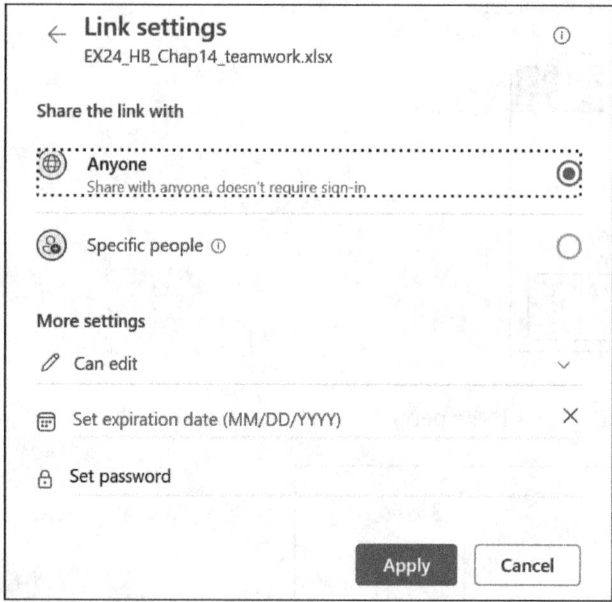

7. You'll receive a confirmation that the workbook has been sent.

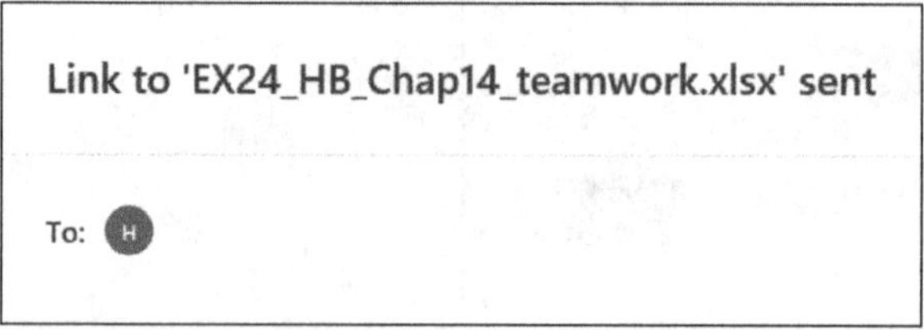

8. The sent email will include a link recipients can use to open the workbook in the online version of Excel.

As an alternative to a direct email invitation, use **Copy link** ❸ to copy a link to the clipboard, which you can paste wherever needed. You can also use the globe icon to choose whether the link will allow editing or view-only access.

	A	B	C	D
1	**Financial planning**		2026	Januar
2				
3		Plan	Is	Difference
4	Receipts			
5	Revenue	200.000	205.000	5.000
6	Extraordinary income	30.000	28.000	-2.000
7	Loan repayments	5.000	5.500	500
8	Tax refunds	5.000	500	-4.500
9	Other revenue	6.000	7.500	1.500

Plan2026

Figure 14.8 Workbook in Excel Online

14.2.2 Collaborate

If all team members have OneDrive access, collaborating on a workbook is particularly easy. When everyone opens the same workbook from OneDrive, the save buttons in the Quick Access Toolbar change. The first time, **AutoSave** is enabled, and the second time, a version of the **Save** button is enabled—one that has an added double arrow and not only saves your changes but also accepts visible changes from other team members.

Figure 14.9 Special Icons for Saving in Shared Files

While you're working on the workbook, each page instantly shows where external changes occur. Cells are outlined with different colored borders and display an abbreviation or the names when you hover over them (see Figure 14.10).

To track the sequence of changes, click or tap the document name in the title bar.

A	B	C	D
Financial planning		2026	Januar
	Plan	Is	Difference
Receipts			
Revenue	210,000	205,000	-5,000
Extraordinary income	30,000	28,000	-2,000
Loan repayments	5,000	5,500	500
Tax refunds	5,200	500	-4,700
Other revenue	6,000	7,500	1,500
Total	256,200	246,500	-9,700

Changes
This workbook
Helmut Vonhoegen 6 min ago
Edited Plan2026 B5
210,000
225,000

Figure 14.10 Navigating Changes in Selected Version; Changed Cells Are Highlighted

Use **Version History** to open the task pane with the same name. It lists the timestamp of each change, along with the **Open Version** option. Using **Open Version** lets you restore the workbook to the state it was in before a change. If you want to have a local copy of a OneDrive-stored workbook while editing the online version, use the **Download a Copy** command under **File**.

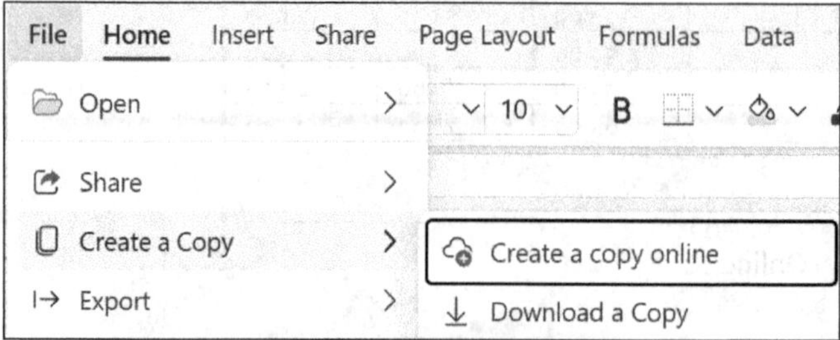

Figure 14.11 Information About Shared Document

The **Share** button also provides this menu:

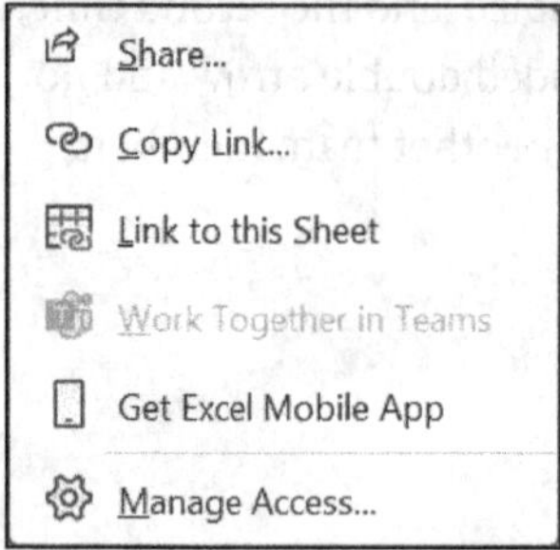

Figure 14.12 Share Button Menu

The **Manage Access** option always shows who can edit or at least view a shared workbook.

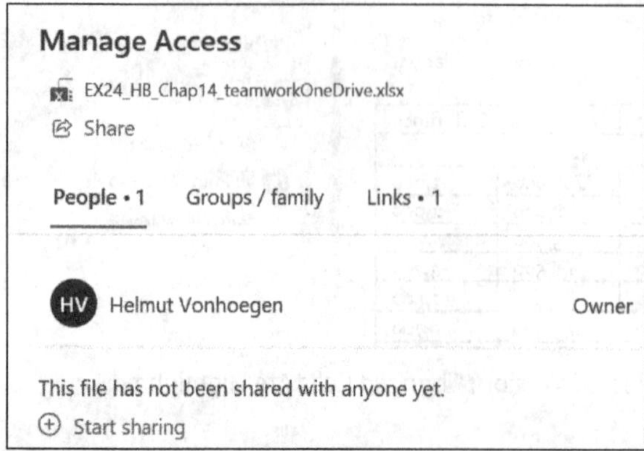

Figure 14.13 Managing Releases

Chapter 15
Table Functions

In previous chapters, you learned about several table functions that tell Excel what tasks to perform for you. Excel offers an overwhelming number of functions that derive new values from one or more inputs, sometimes in quite complex ways. Currently, there are over 475.

A *function* works like a black box: you put something in one side, and something else comes out the other. Without needing to understand the often complex formulas behind the function, the user receives the result after entering the required values.

Each function returns a result, and depending on the function, this result can be numeric—a specific calculation result—a string, text, or a logical value (TRUE or FALSE). Sometimes, the result is not a single value but an array of values, and you must consider the function's result type when referencing it in another formula. Otherwise, an error may occur if the data types don't match.

15

15.1 Functions Introduced in Excel 2013

Excel 2013 introduced several new mathematical functions—mainly to fill gaps related to documents saved in the ODS format. These include several trigonometric functions: ACOT(), ACOTH(), CSC(), CSCH(), COT(), and COTH(). New functions include BASE() and DECIMAL(), which convert between text strings and numbers, along with a combinatorics variant, COMBINA(), and MUNIT() for calculating the identity matrix.

The CEILING.MATH() and FLOOR.MATH() functions replaced the CEILING.PRECISE() and FLOOR.PRECISE() functions that had been introduced in 2010. The older CEILING() and FLOOR() functions have been moved to the **Compatibility** category and should generally be avoided going forward.

Financial functions now include PDURATION(), to calculate the number of periods until an investment reaches a target value. Additionally, RRI() was introduced to calculate the effective annual rate of return on an investment's growth.

The date and time functions added ISOWEEKNUM(), which finally lets you determine the correct week number according to standards without any hassle.

Some additional functions appeared in the **Statistics** category: BINOM.DIST.RANGE(), GAMMA(), GAUSS(), PHI(), SKEW.P(), and PERMUTATIONA().

The lookup functions introduced FORMULATEXT(), which displays a cell's formula as text—which is sometimes helpful for documentation.

The XOR() function, which was long overdue in the **Logic** category, simplifies working with exclusive or (XOR). The IFNA() function was added to handle missing values.

New text functions include UNICODE(), UNICHAR(), and NUMBERVALUE(), which converts text to numbers, regardless of the system's locale.

The info functions now include SHEET() and SHEETS(), which return the current sheet number and the total number of sheets in a workbook when called without arguments. If SHEETS() receives a reference, it returns the number of sheets for that reference; if SHEET() receives a word, it checks whether a sheet with that name exists and returns its sheet number if found, or it returns the #NA error value if not. The IS...() functions have been expanded with the addition of the ISFORMULA() function.

Five new functions related to bit operations have been added in the **Engineering** category: BITSHIFT(), BITOR(), BITRSHIFT(), BITAND(), and BITXOR().

A series of trigonometric functions has been added to the group of functions that work with imaginary numbers: IMCSC(), IMCSCH(), IMCOSH(), IMCOT(), IMSEC(), IMSECH(), IMSINH(), and IMTAN().

Three functions have been grouped under the **Web** category: ENCODEURL() converts strings into valid web addresses, WEBSERVICE() retrieves data from a webservice, and FILTERXML() extracts specific data from XML documents using an XPath expression.

15.2 Functions Introduced Since Excel 2016

Excel 2016 introduced only a few features that were not in Excel 2013. The previous FORECAST() function was moved to the **Compatibility** category, and five new functions were added: FORECAST.ETS(), FORECAST.ETS.CONFINT(), FORECAST.SEASONALITY(), FORECAST.ETS.STAT() and FORECAST.LINEAR(). *Exponential triple smoothing* (ETS) is a method of exponential smoothing of forecast values in time series analysis. You've already seen an example of these functions in Chapter 5, Section 5.6.

Excel 2019 adds a few more functions that have gradually become available to Excel 365 subscribers. These are the two text functions CONCAT() and TEXTJOIN(), the logical functions SWITCH() and IFS(), and the statistical functions MAXIFS() and MINIFS().

Later, Excel 365 introduced useful lookup functions: XLOOKUP() combines the features of HLOOKUP() and VLOOKUP() into one function. New functions included FILTER(), UNIQUE(), SORT(), SORTBY(), and XMATCH(). The GETPIVOTDATA() function was added for analyzing linked data types, and SEQUENCE() and RANDARRAY() were introduced as mathematical functions. The text functions now include LAMBDA(), LET(), and ARRAYTOTEXT().

The newly introduced STOCKHISTORY() financial function lets you output a sequence of stock values in an array.

15.3 Functions Introduced Since Excel 2021

Two groups of additional text functions enhance string handling: TEXTAFTER() and TEXTBEFORE() allow splitting of text after or before a delimiter, while TEXTSPLIT() distributes text across multiple cells. The second group uses regular expressions: REGEXEXTRACT(), REGEXREPLACE(), and REGEXTEST() enable string operations with regular expressions, as used in filter criteria or wildcards. The REGEX functions are initially available only in the paid version. Newly added are the function pair DETECTLANGUAGE() and TRANSLATE(), along with the PERCENTOF() and VALUETOTEXT() functions.

The updates are more extensive in the **Lookup and Reference** category: IMAGE(), EXPAND(), GROUPBY(), HSTACK(), PIVOTBY(), WRAPCOLS(), CHOOSECOLS(), TAKE(), VSTACK(), DROP(), WRAPROWS(), CHOOSEROWS(), TOCOL(), and TOROW(). These mostly involve working with data that's presented in matrix form.

The same applies to the new functions in the **Logical** category: MAP(), SCAN(), REDUCE(), MAKEARRAY(), BYCOL(), and BYROW(). This category now also includes the LAMBDA() and LET() functions, which were previously part of the text functions.

A function related to LAMBDA() that helps find missing parameters is still listed as ISOMITTED() in the **Information** category.

15

Functions introduced since Excel 2013 are not compatible with earlier Excel versions. If you plan to share workbooks with users who are running older versions, you should perform the compatibility check described in Chapter 11 and adjust calculation methods if necessary.

15.4 Structure and Use of Functions

Functions in Excel are instructions that perform operations for you. Table functions operate within a table, and you can enter them directly into it or include them in a macro that runs on the table. A function uses values that are provided as arguments to calculate other values. These can be simple or complex calculations, logical analyses, string manipulations, and similar tasks. The general structure of a function is this:

```
FUNCTION(Value...),
```

Here, Value... represents one or more arguments that the function requires.

15.4.1 Function Arguments

For many functions, arguments are divided into required and optional, meaning they don't always need to be specified. This doesn't mean the function will always work without the optional arguments; rather, if these arguments are omitted, the function will use predefined default values. Function arguments can include the following:

Arguments	Meanings
Constants	You can directly enter the values a function uses. For example, you can enter `=SUM(18, 15, 3)` in a cell to displays a result of 36.
References to cells or ranges	The values a function uses are already in cells or ranges in the table or will be entered there.
Range names	Suppose you've entered sales figures for several products in cells B3 through B15 of a table and want to calculate their total. If you've named the range B3:B15 *Sales*, simply enter this in the cell where you want the total: `=SUM(Sales)` Without a named range, you'd write: `=SUM(B3:B15)`.
Functions	Functions can take other functions as arguments, which provide the values they operate on. For example: `=SUM(SUM(2, 4), SUM(4, 6))` returns 16. You can nest functions up to 64 levels deep.

When only specific values are allowed for an argument, Excel displays them as drop-down lists in the formula bar when you're entering that argument. Figure 15.1 shows which functions are available for the first argument of the `AGGREGATE()` function once you enter the opening parenthesis.

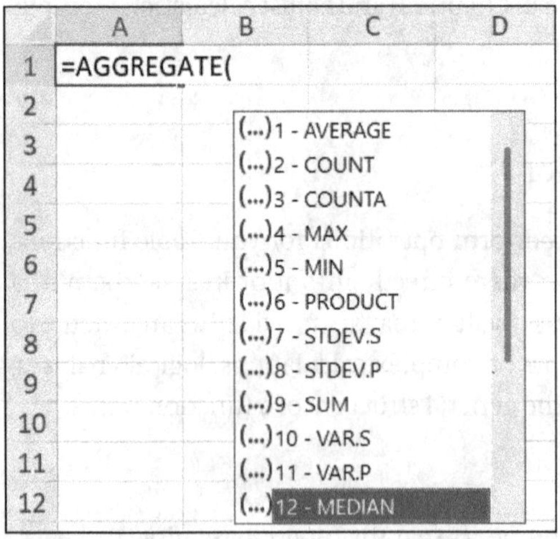

Figure 15.1 Offering Argument Values During Entry

Some functions (e.g., `PI()`, `TODAY()`, `NOW()`) don't require arguments. Still, you must include the parentheses for Excel to recognize the function.

15.4.2 Functions in Macros

When using an international version of Excel, such as the German edition, note that macro calls to worksheet functions typically use the English function names. Additionally, dots in function names are replaced by underscores. For more information, see Chapter 23.

15.5 Financial Mathematical Functions

Besides several functions covering general financial mathematics, three main groups stand out: annuity calculations, depreciation, and securities calculations.

15.5.1 Functions for Annuity Calculations

Here, *annuities* refer to privately agreed-upon regular payments. The simplest and most understandable model for grasping these functions is as follows: you deposit a certain sum of money in a bank, which accrues interest at an agreed-upon rate, and the bank then regularly pays you a fixed amount (the annuity) from this balance for as long as the funds last.

	A	B
1	**Annuity calculation**	
2		
3	**PV**	$160,000.00
4	**Nper**	13
5	**Interest**	1.40%
6	**Regular annual payout**	-$13,547.37

Figure 15.2 Example of Annuity Calculation

Several arguments frequently appear in these functions, so it's helpful to introduce them briefly.

Function Arguments	Explanations
`pv` [Present value]	This is the current calculated value of a series of regular, equal payments, such as an annuity.
`pmt` [Payment]	These are payments made or received regularly.
`rate` [Interest rate]	This is the interest rate for a payment period. It's expressed as a decimal (e.g., 0.08) or a percentage (e.g., 8%).
`per` [Period]	This is the individual payment period within the total time-frame.

Function Arguments	Explanations
fv [Future value]	This is the future value of an investment. If fv is used as an optional argument, then Excel defaults it to 0 when omitted.
nper [Number of payment periods]	This is the number of payment periods. The nper must match the units of the other arguments, meaning the function should consistently use years, months, or days. For example, if months are used, the monthly interest rate (annual interest rate / 12) must also be applied. Note that there are several conventions for dividing a year and month in financial contexts. One common method is the 30/360 approach, in which the interest year is divided into twelve months of 30 days each. Other methods may apply and can vary among countries and agreements.
f [Due date]	The f argument specifies whether the calculation is for payments made at the beginning or end of the period, such as annuities. If f is omitted or set to 0, the function calculates payments at the end of the period. If f is set to 1, it calculates payments at the beginning of the period.

15.5.2 Loan Calculations

Loan amortization is the counterpart to annuity calculations. It involves repaying a loan to the bank in fixed, periodic installments. The payment amount is calculated using the PMT() function, and each installment payment consists of an interest portion (calculated with IPMT()) and a principal portion that reduces the total debt. This portion is calculated using the PPMT() function. Figure 15.3 shows a simple example of loan repayment.

	A	B	C	D	E	F	G	H	I	J	K
1	**Repayment of loans**										
2											
3	Rate	3.00%	3.00%	3.00%	3.00%	3.00%	3.00%	3.00%	3.00%	3.00%	3.00%
4	Per	1	2	3	4	5	6	7	8	9	10
5	Nper	10	10	10	10	10	10	10	10	10	10
6	Pv	$20,000.00	$20,000.00	$20,000.00	$20,000.00	$20,000.00	$20,000.00	$20,000.00	$20,000.00	$20,000.00	$20,000.00
7											
8	PPMT()	-$1,744.61	-$1,796.95	-$1,850.86	-$1,906.38	-$1,963.57	-$2,022.48	-$2,083.16	-$2,145.65	-$2,210.02	-$2,276.32
9	IPMT()	-$600.00	-$547.66	-$493.75	-$438.23	-$381.04	-$322.13	-$261.45	-$198.96	-$134.59	-$68.29
10	PMT()	-$2,344.61	-$2,344.61	-$2,344.61	-$2,344.61	-$2,344.61	-$2,344.61	-$2,344.61	-$2,344.61	-$2,344.61	-$2,344.61

Figure 15.3 Loan Repayment Schedule

15.5.3 Depreciation Calculation

Depreciation involves considering an investment—such as in machinery, vehicles, or buildings—from two business perspectives. First, within the company, an acquisition will initially increase the value of the company's fixed assets by its purchase value, but this value will steadily decline thereafter due to wear and tear until only the scrap value remains.

On the other hand, such an investment can be deducted from taxes. The simplest method is to fully deduct the purchase amount in the year the purchase is made—and no further calculations will be required. For long-term investments, however, depreciation is typically spread over several years.

There are various methods you can use to do this. The simplest is *straight-line depreciation*, in which the depreciable amount is evenly spread over the entire depreciation period. Another method, which is supported by Excel functions, is called *declining balance depreciation*: The depreciation amount decreases each year, with a larger amount depreciated initially and smaller amounts in subsequent years. It's important to note that Excel cannot determine which method is most appropriate in each case or which one is currently allowed under tax law.

15.5.4 Example of a Depreciation Calculation

In this section, we'll create a worksheet that will help you calculate the depreciation rate for each year of an asset's useful life by using different depreciation methods. You must always verify which tax method is allowed under current tax law, as there have often been changes, especially concerning the various types of declining-balance depreciation.

Rows 3 and 4 initially store the data that's required as arguments for the different functions, and column B calculates straight-line depreciation for each year of the asset's useful life. The formula is the same for every year and uses absolute references:

```
=SLN($CS3, $C$4, $E$3)
```

The declining-balance depreciation formula is a bit more complex to allow copying downward:

```
=DDB($C$3, $C$4, $E$3, $E$3-($A$16-A7), $E$4)
```

Note that the reference to the year in each row is relative, while all other references are absolute. The sum-of-years' digits depreciation formula follows a similar approach:

```
=SYD($C$3, $C$4, $E$3, $E$3-($A$16-A7))
```

The formula for the fourth method is a bit more complex:

```
=VDB($C$3, $C$4, $E$3, $A7-$A$7, $A8-$A$7, $E$4, FALSE)
```

If you compare columns C and E, you'll notice that the declining-balance method yields lower amounts in the last two years than does the straight-line method. Therefore, switching to the straight-line method makes sense for these years, based on the current data.

To adjust the model for shorter or longer useful lives, insert or delete the necessary number of rows before the totals row and then copy the formulas down as needed. Instead of year 1, 2, 3 ..., you can use the actual years. The formulas in column E require the value currently in cell A17, which is a year beyond the final year of the depreciation period.

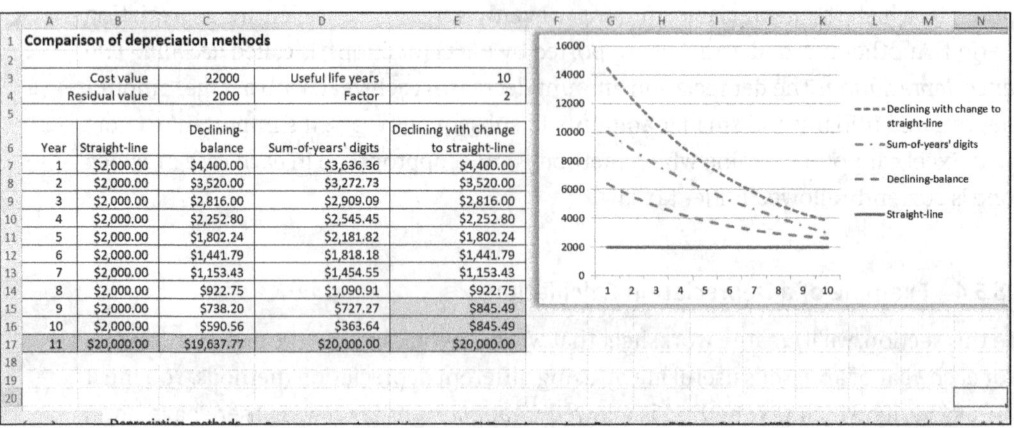

Comparison of depreciation methods

	Cost value	22000	Useful life years	10
	Residual value	2000	Factor	2

Year	Straight-line	Declining-balance	Sum-of-years' digits	Declining with change to straight-line
1	$2,000.00	$4,400.00	$3,636.36	$4,400.00
2	$2,000.00	$3,520.00	$3,272.73	$3,520.00
3	$2,000.00	$2,816.00	$2,909.09	$2,816.00
4	$2,000.00	$2,252.80	$2,545.45	$2,252.80
5	$2,000.00	$1,802.24	$2,181.82	$1,802.24
6	$2,000.00	$1,441.79	$1,818.18	$1,441.79
7	$2,000.00	$1,153.43	$1,454.55	$1,153.43
8	$2,000.00	$922.75	$1,090.91	$922.75
9	$2,000.00	$738.20	$727.27	$845.49
10	$2,000.00	$590.56	$363.64	$845.49
11	$20,000.00	$19,637.77	$20,000.00	$20,000.00

Figure 15.4 Impact of Different Depreciation Methods

15.5.5 Functions for Securities Calculations

A large group of functions focuses on fixed-rate securities and discount (or zero-coupon) bonds, and we'll cover those in this section.

Fixed-rate securities are valid for a specific period of time, and they pay interest regularly during that time. These include savings certificates, debentures, and corporate, federal, and municipal bonds. The buyer purchases the security for an amount called the *face value*, and the financial institution that sold the security makes monthly, semiannual, or annual interest payments to the buyer until the end of the term. Then, the buyer redeems the security for the face value. Sometimes, a discount will be applied at issuance, meaning a certain percentage—possibly based on the term—will be deducted from the face value.

Discount bonds—which are better known in the United States as *zero-coupon bonds*—are issued at a certain price and later redeemed at a higher price. They're called "zero-coupon" because they don't make periodic interest rate payments (which are called *coupons*)—but even so, you could calculate the annual rate of increase in such a bond's value and then express it as an annual interest rate. The buyer purchases the zero-coupon bond at a discount, meaning at a price that's below the face value, and they don't receive interest payments. Instead, they redeem the bond for the full face value at maturity. Alternatively, the buyer could purchase and redeem a zero-coupon bond at face value and receive the accrued, compounded interest all at once on the maturity date (rather than in the form of installment payments during the bond's term).

Based on a model where such a bond (fixed-rate or zero-coupon) is purchased from a financial institution and later sold, the following dates are important:

- **Date of issuance**
  This is the date when a bond is issued, meaning when it's introduced into the market and its interest rate payments or growth in value begins.
- **Settlement date**
  This is the date when a buyer acquires a bond. It can coincide with the issuance date, but it can also—and this is the more relevant case for calculations—occur at a later date.
- **Maturity date**
  This is the date when the final payment from a bond is due and the financial institution pays the owner the previously agreed-upon amount at maturity, which will include the principal plus the final interest payment (for fixed-rate bonds) or the face value (for zero-coupon bonds).

Since such bonds are sometimes traded during their term, questions will arise about their *current value*—the price at which they can be traded at any given time. This price should, of course, reflect the market value, meaning a potential buyer must be able to verify whether investing in the bond is worthwhile compared to other investment options. To do this, we use *yield*—the effective annual rate of return on the bond as a percentage of the capital invested—as a comparison metric. Choosing this metric is somewhat arbitrary: we calculate the price based on the yield that's achievable with other investments.

The `basis` argument appears in several functions. In interest calculations, it refers to the time base used. The available options are shown in the following table.

Time Bases	Meanings
0	This is a common US system: Months are counted as 30 days, and a year is counted as 360 days. An end date on the 31st of a month is treated as the 1st day of the following month if the start date is before the 30th of the month; otherwise, it is treated as the 30th.
1	The actual days in the months and years are counted.
2	The actual days of the month are counted, with the year counted as 360 days.
3	The actual days of the month are counted, with the year counted as 365 days.
4	This is a European system: Months are counted as 30 days, and a year is counted as 360 days. If a start or end date falls on the 31st of a month, it is treated as the 30th day.

Sale of a Fixed-Income Security

If you own fixed-income securities and want to sell some of them on a specific date, you can use financial functions to calculate the amount you'll receive. You'll need to know the price on the settlement date, which is the day the sale occurs. From the purchase settlement, you can find the other necessary details: the issue date, maturity date, nominal interest rate, face value, and number of interest payment dates per year. The first interest payment date is based on the number of payments.

Excel also requires you to specify the time base you'll use to calculate interest days. Here, the time base is 4, which means 360 days per year and 30 days per month. In cell B11, the current price is simply calculated with this formula:

```
=Nominal value * Price/100
```

Additionally, you can use this formula to calculate the number of interest days:

```
=COUPDAYBS(settlement_date, maturity, 1, 4)
```

However, this result is for informational purposes only and is not needed to calculate the accrued interest.

Besides the price, the amount of accrued interest matters when selling a bond. You can calculate this in E4 by using the ACCRINT() function. The formula is this:

```
=ACCRINT(issue, first_interest_date, settlement_Date,
nominal_rate, nominal_value, interest_periods_per_year,
base_for_calculating_interest_days)
```

Typically, a few additional costs apply: the bank charges a 0.125% commission on the price, the broker adds a fee of 0.0625% of the nominal value, and you'll also pay a fixed fee. The actual amount you pay will equal the price plus accrued interest, minus the three fees we just mentioned. Note that brokerage fees for foreign securities are usually different.

	A	B	C	D	E
1	**Sale of a fixed-income security**				
2					
3	Security:	Phantas			
4	Emission	6/3/2017		Accrued interest	$3,696.29
5	First interest date	6/3/2018		Provision	$39.08
6	Maturity	6/3/2029		Third-party expenses and expenses	$19.54
7	Settlement	1/6/2022		Exchange Fee	$3.50
8	Nominal interest rate	2.875%			
9	Face value	$28,000.00		Amount (credit)	$34,896.18
10	Course	111.65			
11	Market value	$31,262.00			
12	Interest dates per year		1		
13	Basis for calculating interest days		4		
14	Number of interest days		213		

Figure 15.5 Sale of a Fixed-Rate Security

Outputting Stock Market Data

The STOCKHISTORY() function lets you output stock market data for a specified period of time as a dynamic array. Microsoft provides the data through the Refinitiv information service, and you can find supported exchanges and delay times in Excel Help. Figure 15.6 shows a simple example of stock market data.

	A	B	C	D	E
1	**Example stock market history**				
2					
3			Meta share values		
4					
5		Close	Open	High	Low
6	1/1/2022	$313.26	$338.30	$343.09	$289.01
7	2/1/2022	$211.03	$314.56	$328.00	$190.22
8	3/1/2022	$222.36	$209.87	$231.15	$185.82
9	4/1/2022	$200.47	$224.55	$236.86	$169.00
10	5/1/2022	$193.64	$201.17	$224.30	$176.11
11	6/1/2022	$161.25	$196.51	$202.03	$154.25
12	7/1/2022	$159.10	$160.31	$183.85	$154.85
13	8/1/2022	$162.93	$157.25	$183.10	$155.23
14	9/1/2022	$135.68	$163.58	$171.39	$134.12
15	10/1/2022	$93.16	$137.14	$142.39	$92.60
16	11/1/2022	$118.10	$94.33	$118.74	$88.09

Figure 15.6 Example of Matrix with Stock Market Data

You can enter the function here in the empty cell B4. The syntax is as follows:

=STOCKHISTORY(**stock, start_date**, end_date, interval, headers, property1, property2,...)

The Stock argument requires a ticker symbol that's enclosed in quotation marks. If the data doesn't come from the specified national exchange, you must prefix it with the four-digit market identification code followed by a colon: for example, "XNAS:MSFT" for Microsoft stock. You can find the codes in the function's help section. The two date values specify the start and end of the period you want data for. Interval accepts the following values: 0 for daily (default), 1 for weekly, and 2 for monthly.

The number for headings indicates whether or not column headers appear in the first row of the result matrix: 0 = no and 1 = yes. The numbers for property values determine the column contents: 0 = Date, 1 = Close, 2 = Open, 3 = High, 4 = Low, and 5 = Volume.

15.5.6 List of Financial Math Functions

Functions	Descriptions
AMORLINC(**cost, date_purchased, first_period, salvage, period, rate**, basis)	Returns the straight-line depreciation amount based on the French accounting system

Functions	Descriptions
ACCRINT(**issue**, **first_interest**, **settlement**, **rate**, **par**, **frequency**, basis, ...)	Returns the accrued interest for securities
ACCRINTM(**issue**, **settlement**, **rate**, **par**, basis)	Returns the accrued interest on securities at maturity
RECEIVED(**settlement**, **maturity**, **investment**, **discount**, basis)	Returns the redemption amount of a fixed-rate security
STOCKHISTORY(**stock**, **start_date**, end_date, interval, headers, propertiesl, ...)	Retrieves historical data for a stock and returns it as a dynamic array
PV(**rate**, **nper**, **pmt**, fv, type)	Calculates the present value of an investment
SYD(**cost**, **salvage**, **life**, **period**)	Calculates depreciation using the sum-of-years' digits method
DISC(**settlement**, **maturity**, **price**, **redemption**, basis)	Calculates the discount on a security trade
DURATION(**settlement**, **maturity**, **coupon**, **yield**, **frequency**, basis)	Returns the hypothetical average capital tie-up period for a fixed-rate security
EFFECT(**nominal_rate**, **npery**)	Calculates the effective annual interest rate for an investment or loan
DDB(**cost**, **salvage**, **life**, **period**, factor)	Calculates the depreciation rate for a specific period using the double-declining balance method
DB(**cost**, **salvage**, **life**, **period**, month)	Calculates the depreciation rate for a specific period using the double-declining balance method
IRR(**values**, guess)	Calculates the internal rate of return for an investment
ISPMT(**rate**, **per**, **nper**, **pv**)	Calculates the interest paid during a specific period
PPMT(**rate**, **per**, **nper**, **pv**, fv, type)	Calculates the principal portion of a payment for an annuity loan
CUMPRINC(**rate**, **nper**, **pv**, **start_period**, **end_period**, **type**)	Calculates the total principal paid on an annuity loan over a period
CUMIPMT(**rate**, **nper**, **pv**, **start_period**, **end_period**, **type**)	Calculates the amount of interest paid
PRICE(**settlement**, **maturity**, **rate**, **yield**, **redemption**, **frequency**, basis)	Returns the price of a discount (zero-coupon) security

Functions	Descriptions
PRICEDISC(**settlement**, **maturity**, **discount**, **redemption**, basis)	Calculates the issue price of a discount (zero-coupon) security.
PRICEMAT(**settlement**, **maturity**, **issue**, **interest**, **yield**, basis)	Calculates the price of a fixed-rate security
SLN(**cost**, **salvage**, **life**)	Calculates the depreciation rate using the straight-line method
MDURATION(**settlement**, **maturity**, **coupon**, **yield**, **frequency**, basis)	Calculates the modified duration
NPV(**rate**, **value1**, value2, ...)	Calculates the net present value of periodic cash flows
NOMINAL(**effective_rate**, **npery**)	Calculates the annual nominal interest rate
DOLLARFR(**decimal_dollar**, **fraction**)	Calculates a decimal number expressed as a fraction
DOLLARDE(**fractional_dollar**, **fraction**)	Converts a fraction to a decimal number
PDURATION(**rate**, **pv**, **fv**)	Calculates the number of periods until an investment reaches the specified target value
MIRR(**values**, **finance_rate**, **reinvest_rate**)	Calculates the internal rate of return for a series of cash inflows and outflows
YIELD(**settlement**, **maturity**, **rate**, **price**, **redemption**, **frequency**, basis)	Calculates the annual yield of a security
YIELDDISC(**settlement**, **maturity**, **price**, **redemption**, basis)	Calculates the annual yield of a discount security
YIELDMAT(**settlement**, **maturity**, **issue**, **rate**, **price**, basis)	Calculates the annual yield of a security with interest paid at maturity
PMT(**rate**, **nper**, **pv**, fv, type)	Calculates the periodic payment for an annuity
TBILLEQ(**settlement**, **maturity**, **discount**)	Calculates the bond-equivalent yield of a treasury bill
TBILLPRICE(**settlement**, **maturity**, **discount**)	Calculates the price of a treasury bill
TBILLYIELD(**settlement**, **maturity**, **discount**)	Calculates the yield of a treasury bill
ODDFPRICE(**settlement**, **maturity**, **issue**, **first_interest_date**, **rate**, **yield**, **redemption**, **frequency**, basis)	Returns the price of a fixed-rate security with an irregular first interest period

15

Functions	Descriptions
ODDFYIELD(**settlement, maturity, issue, first_interest_date, rate, yield, redemption, frequency, basis**)	Returns the yield of a fixed-rate security with an irregular first interest period
ODDLPRICE(**settlement, maturity, last_interest_date, rate, yield, redemption, frequency**, basis)	Returns the price of a fixed-rate security with an irregular last interest period
ODDLYIELD(**settlement, maturity, last_interest_date, interest, price, redemption, frequency, basis**)	Returns the yield of a fixed-rate security with an irregular final interest period
VDB(cost, salvage, life, start_period, end_period, factor, no_switch)	Calculates depreciation by using the variable declining balance method
XIRR(values, dates, guess)	Calculates the internal rate of return for a series of irregular cash flows within a year
XNPV(**rate, values, dates**)	Calculates the net present value for a series of irregular cash flows within a subannual range
RATE(**nper, pmt, pv**, fv, type, guess)	Calculates the interest rate of an investment with regular payments
INTRATE(**settlement, maturity, investment, redemption**, dbasis)	Calculates the annual interest rate for an investment where no interest is paid between settlement and redemption
COUPNCD(**settlement, maturity, frequency**, basis)	Calculates the date of the first interest payment
COUPDAYS(**settlement, maturity, frequency**, basis)	Calculates the number of days in the interest period containing the settlement date
COUPDAYSNC(**settlement, maturity, frequency**, basis)	Calculates the number of days until the first interest payment
COUPDAYBS(**settlement, maturity, frequency**, basis)	Calculates the number of days from the last interest payment to the settlement date
COUPPCD(**settlement, maturity, frequency**, basis)	Calculates the date of the last interest payment
COUPNUM(**settlement, maturity, frequency**, basis)	Calculates the number of interest payments between the purchase date and the maturity date
IPMT(**rate, per, nper, pv**, fv, type)	Calculates the interest portion of a loan payment

Functions	Descriptions
RRI(**nper**, **pv**, **fv**)	Calculates the effective annual interest rate for an investment's growth
FV(**rate**, **nper**, **pmt**, pv, type)	Calculates the future value based on regular payments
FVSCHEDULE(**principal**, **schedule**)	Calculates the future value of an investment with varying interest rates
NPER(**rate**, **pmt**, **pv**, fv, type)	Calculates the number of payment periods for a loan

15.6 Date and Time Functions

Excel uses serial numbers for date and time functions. Dates typically start from January 1, 1900, which is serial number 1. The following day is serial number 2, and the serial numbers continue up to 2,958,465, which represents December 31, 9999. Excel also represents times by using serial numbers. These form the decimal part, where 0.00001 represents the first second and 0.5 corresponds to twelve noon. 15

Alternatively, you can use a different date system that matches the one commonly used on Macs to make it easier to exchange workbooks with Mac. This system starts from January 1, 1904, as day 0. In Excel, you can enable this date format by going to **File • Options • Advanced • When calculating this workbook** and checking **Use 1904 date system.**

By starting the count later, Apple avoided the problem that 1900 was not a leap year, which is assumed when counting from January 1, 1900. So, the incorrect date 2/29/1900 can be entered in this system, even though that day never existed. (This error is avoided only with the newer file format *Strict Open XML.*) Using the 1904 date system is also helpful on Windows when you're working with negative time values, as you'll often do when making work time calculations.

15.6.1 The Advantages of Using Serial Numbers

Converting dates and times into serial numbers makes calculations very simple. Since they are internally stored as regular numeric values, you can perform basic arithmetic operations on them—mainly subtraction and addition. Users don't need to see that date and time functions use serial numbers because the output usually appears in a date format. To display the result as a serial number, just change the format to **General** or **Number**.

When formatting the results of date and time functions, make sure not to assign a format that doesn't match the result's data type. For example, the formula =YEAR(TODAY()) returns the number 2024 in the year 2024. If this cell is mistakenly formatted as a date, it will display 07/16/1905.

15.6.2 Calculating Periodic Date Series

The following example demonstrates how to perform date calculations with the DATE() function. You can calculate periodic date series of any type. Just enter a starting date and the interval you want in days, weeks, or months. In the first row of the table with the date series, you can set the starting date by using an absolute reference.

The individual dates in the series are sequentially numbered in column A, and the formulas in the series reference these numbers. In the second row, formulas calculate the time interval. You only need to enter these formulas once, and then, you can copy them down.

The formula for a daily interval is this:

```
=DATE(YEAR($A$5), MONTH($A$5), DAY($A$5)+(A9-1)*$B$5)
```

Be sure to use absolute references if you want to follow the example.

The formulas for the monthly interval work similarly. A specific multiple of the given interval is always added to the corresponding part of the date. The weekly interval takes a bit more effort:

```
=DATE(YEAR($A$5), MONTH($A$5), DAY($A$5)+(A9-1)*$C$5*7)
```

Once you've set up the table, you can enter any interval values in cells B5 through D5 and instantly get the series you want.

	A	B	C	D	E
1	**Periodic data series**				
2					
3			Interval		
4	Start date	in days	in weeks	in months	
5	1/1/2026	14	4	3	
7	Date				
8	1	1/1/2026	1/1/2026	1/1/2026	
9	2	1/15/2026	1/29/2026	4/1/2026	
10	3	1/29/2026	2/26/2026	7/1/2026	
11	4	2/12/2026	3/26/2026	10/1/2026	
12	5	2/26/2026	4/23/2026	1/1/2027	
13	6	3/12/2026	5/21/2026	4/1/2027	
14	7	3/26/2026	6/18/2026	7/1/2027	

Figure 15.7 Calculating Date Series

15.6.3 Calculating Periodic Time Series

The next example shows you how to calculate periodic time series of any kind, as in the preceding date series example. You enter a starting time and then the interval you want in hours, minutes, or seconds. The first row of the time series table simply uses an absolute reference to the starting time. The individual dates in the time series are sequentially numbered in column A, and the formulas in the series reference these numbers. In the second row, formulas calculate the time interval, and you only need to enter these formulas once and then copy them down. The formula for an hourly interval is this:

```
=TIME(HOUR($A$5)+(A9-1)*$B$5, MINUTE($A$5), SECOND($A$5))
```

Be sure to use absolute references if you want to follow the example.

The formulas for minute and second intervals work similarly. A specific multiple of the given interval is added to the corresponding part of the time—here, to the minutes or seconds. Once you've set up the table, you can enter any interval values in cells B5 through D5 and instantly generate the series you want.

	A	B	C	D	E
1	**Periodic date series**				
2					
3			Interval		
4	Start date	in days	in weeks	in months	
5	1/1/2026	14	4	3	
7	Date				
8	1	1/1/2026	1/1/2026	1/1/2026	
9	2	1/15/2026	1/29/2026	4/1/2026	
10	3	1/29/2026	2/26/2026	7/1/2026	
11	4	2/12/2026	3/26/2026	10/1/2026	
12	5	2/26/2026	4/23/2026	1/1/2027	

Figure 15.8 Calculating Time Series

15.6.4 Calculating Working Hours

This example explains how to add times together. Excel's time functions typically return the time for a specific moment, which makes calculating time differences straightforward. Adding times can be tricky, though.

The following table shows working hours for a week. The start and end times are always entered in columns B and C, and three types of working hours are distinguished: regular, Saturday, and Sunday hours. You can use the time function to calculate the differences. The formula for the first day is this:

```
=TIME(HOUR(C5-B5)-1, MINUTE(C5-B5),)
```

It subtracts 1 from the hourly timesheets for a one-hour break.

15

When you're summing the three time values in the table while using the standard date format, Excel shows a time point instead of the amount of time. You can correct this with a custom format. Select the range from D5 to F12, open the **Format Cells** dialog, and choose the **Number** tab and then choose **Custom**. Enter the following format:

```
[h]:mm
```

The amount of time will now display correctly.

	A	B	C	D	E	F
1	**Calculation of working hours**				**week:**	**38**
2						
3		Beginning	End	Normal working hours	Saturday work	Sunday work
4	Monday, September 15, 2025	8:40	16:30	6:50		
5	Tuesday, September 16, 2025	8:30	16:30	7:00		
6	Wednesday, September 17, 2025	12:30	21:30	8:00		
7	Thursday, September 18, 2025	14:30	21:30	6:00		
8	Friday, September 19, 2025	8:30	16:30	7:00		
9	Saturday, September 20, 2025	8:30	12:30		4:00	
10	Sunday, September 21, 2025	8:30	10:30			2:00
11				10:50	4:00	2:00
12	one hour break					

Figure 15.9 Work Time Calculation with Incorrect Totals

	A	B	C	D	E	F
1	**Calculation of working hours**				**week**	**38**
2						
3		Beginning	End	Normal working hours	Saturday work	Sunday work
4	Monday, September 15, 2025	8:40	16:30	6:50		
5	Tuesday, September 16, 2025	8:30	16:30	7:00		
6	Wednesday, September 17, 2025	12:30	21:30	8:00		
7	Thursday, September 18, 2025	14:30	21:30	6:00		
8	Friday, September 19, 2025	8:30	16:30	7:00		
9	Saturday, September 20, 2025	8:30	12:30		4:00	
10	Sunday, September 21, 2025	8:30	10:30			2:00
11				34:50	4:00	2:00
12	one hour break					

Figure 15.10 Work Time Calculation with Correct Total Times

15.6.5 List of Date and Time Functions

Functions	Descriptions
WORKDAY(**start_date, days, holidays**)	Calculates a new date based on the specified workdays and holidays

Functions	Descriptions
WORKDAY.INTL(**start_date, days,** weekend, holidays)	Calculates a new date based on the specified work-days and holidays, with customizable weekend days
YEARFRAC(**start_date, end_date,** basis)	Calculates the fraction of the year between the start date and end date
DATEDIF(**start_date, end_date, unit**)	Calculates the difference between the start date and end date
DATE(**year, month, day**)	Returns the serial number for the specified date
DATEVALUE(**date_text**)	Converts a text string into a date value
EDATE(**start_date, months**)	Calculates a date shifted by a specified number of months
TODAY()	Returns the current system date
ISOWEEKNUM (**date**)	Returns the ISO standard week number
YEAR(**serial_umber**)	Returns the year of the specified date
NOW()	Returns the current system date and time
WEEKNUM(**serial_number**, return_type)	Returns the week number for the given date
MINUTE(**serial_umber**)	Calculates the minute component of a time value
MONTH(**serial_umber**)	Returns the month number from a date value
EOMONTH(**start_date, months**)	Returns the last day of the month, shifted by a specified number of months
NETWORKDAYS(**start_date, end_date,** holidays)	Calculates the number of working days between two dates
NETWORKDAYS.INTL(**start_date, end_date,** weekend, holidays)	Calculates the number of working days between two dates, allowing custom weekend days
SECOND(**serial_number**)	Returns the seconds portion of a time value
HOUR(**serial_number**)	Returns the hours portion of a time value
DAY(**serial_number**)	Returns the day number of a date value
DAYS(**end_date, start_date**)	Returns the number of days between the start date and the end date
DAYS360(**start_date, end_date, method**)	Returns the number of days between two dates, assuming a 360-day year

Functions	Descriptions
WEEKDAY(serial_number, return_type)	Returns the weekday number for a given date
TIME(**hour, minute, second**)	Calculates a serial number for the specified time
TIMEVALUE(**time_text**)	Converts a text string into a time value

15.7 Mathematical Functions

Given the wide range of functions in mathematics and trigonometry, we need to introduce you to those you'll likely use the most.

15.7.1 Sums and Conditional Sums

The most commonly used mathematical function is likely SUM(), which calculates the sum of the specified arguments. Since Excel 2007, it has supported up to 255 arguments, whereas older versions allow only 30. Keep this in mind when you're sharing workbooks with users who are running earlier versions of Excel. Arguments can be constants, array constants, cell references, or formula expressions that return values. Using this function is the most efficient way to sum cell ranges because you only need to specify the start and end of the range.

If the arguments include logical values or numbers entered as text, then they are treated as numbers. However, error values or text that can't be converted to numbers will cause the #VALUE! error. If the argument is a constant array or a cell reference, only numeric values are processed; empty cells, text, and logical values are ignored. The impact of different data types is shown Figure 15.11. Note the differences compared with formulas that use the + operator.

	A	B	C	D	E
1	**Data Types and the SUM() Function**				
2					
3	Area1	Area2	A+B	SUM(A,B)	SUM(A:B)
4	2	5	7	7	7
5	4	4	8	8	8
6	evening	morning	#VALUE!	0	0
7	TRUE	FALSE	1	0	0
8	3	apple	#VALUE!	3	3
9	TRUE	TRUE	2	0	0
12	SUM(Area1;Area2)		18		

Figure 15.11 SUM() Function Treats Certain Data Types Differently from + Operator

A very useful variant of the SUM() function is SUMIF(), which allows conditional addition of values. The syntax is as follows:

```
=SUMIF(range, criteria, sum_range)
```

The function compares the values in the cell range specified by range with the criteria given in the criteria argument. This can be a number, text string, an expression with a comparison operator like <10 or >John, or a reference to a cell containing a criterion.

If the values in the range meet the criteria, then the function looks for corresponding values in a second sum_range and sums them. If no value matches the criteria, then the function returns 0. If sum_range is not specified, then the function sums the part of the range that meets the search criteria.

Figure 15.12 displays a table showing multiple entries of how many red and blue products were sold. The function sums the numbers for the red product.

The SUMIFS() function expands on SUMIF() by allowing additional selection criteria. The syntax is as follows:

```
=SUMIFS(sum_range, criteria_range1, criteria1, criteria_range2, criteria2...)
```


	A	B
1	**Conditional Sum**	
2		
3	Area	Sum_Area
4	red	1
5	blue	3
6	red	5
7	blue	7
8	red	9
9	blue	11
10		SUMIF(Area;"red",Sum_Area)
11		15

Figure 15.12 Example of SUMIF()

The sum_range argument specifies the cell range that contains the values to sum. Below that is a list of range pairs, each of which specifies the range to which the following criterion applies. Up to 127 such range pairs are allowed. The criteria are additive, so summation occurs only if all criteria are met. All arguments for criteria_range(s) must specify ranges that are the same size as sum_range. The criteria expressions can also include wildcard characters. In the example in Figure 15.13, the student numbers are summed if they meet two criteria.

	A	B	C	D	E
1	**Multi-conditional sums**				
2					
3		Course 1	Course 2	Course 3	Course 4
4	Participant	80	60	70	110
5	under 20	30%	25%	15%	25%
6	20 - 50	60%	65%	45%	55%
7	over 50	10%	10%	40%	20%
8	Total number of participants in the courses with a share of under twenty years of age of less than 30% and a share of over fifty years of age of at least 20%.				180

Figure 15.13 Example of Summation in Which Two Conditions Must Be Met Simultaneously

15.7.2 Rounding Values

Many mathematical functions involve rounding values. This often means converting numbers with many decimal places into a form that's easier to handle in further calculations, but it also includes methods for adjusting values to specific interval limits. The ROUND() function uses this simple syntax:

```
=ROUND(number, num_digits)
```

It rounds the value given by `number` up or down to the number of digits specified by `num_digits`. If the decimal digit to be rounded is 5 or greater, then it rounds away from zero, so `=ROUND(3.45,1)` results in 3.5. If `num_digits = 0`, then `number` is rounded to an integer. On the other hand, the `INT()` function simply cuts off the decimal places.

The `ROUNDDOWN()` function, which uses the same syntax, behaves slightly differently. For example, this function returns –650:

```
=ROUNDDOWN(-657.65, -1)
```

It rounds the value given by `number` toward zero. Positive numbers round down, and negative numbers round up (toward zero). The `num_digits` argument specifies how many digits to consider when rounding. If `num_digits` is 1 or greater, then the number is truncated to that many decimal places; if it's 0, then the decimal places are removed, leaving an integer. If the argument is negative, it rounds left by the specified number of digits and toward zero. The opposite effect comes from the `ROUNDUP()` function, which always rounds away from zero. You can also specify the number of digits with the third argument.

The `INT()` and `TRUNC()` functions are somewhat related to the rounding functions we've described. While `INT()` removes all decimal places without rounding, `TRUNC()` truncates all decimal places beyond the specified number of digits. If `num_digits` is not specified, then it defaults to 0 and the integer part is returned. However, `TRUNC()` behaves differently than `INT()` with negative numbers.

This function results in –6:

```
=INT(-5.7)
```

And this function results in –5:

```
=TRUNC(-5.7)
```

Also worth mentioning is the ABS() function, which returns the absolute value of a number, effectively removing any minus sign.

Special rounding variants are provided by the function pair CEILING.MATH() and FLOOR.MATH(). The syntax is as follows:

```
=CEILING.MATH(number, significance, mode)
=FLOOR.MATH(number, significance, mode)
```

Both functions round the value given by number up or down to the nearest multiple of significance. For example, using 0.05 for significance ensures the hundredths place is always either 5 or 0 when rounding. A practical example is price calculations in which the smallest difference is 5 cents. The mode argument specifies whether to round away from zero or toward zero.

	A	B
1	**Upper limit**	
2		
3		Formula
4	5.00000	=CEILING.MATH(4.3)
5	-4.00000	=CEILING.MATH(-4.3)
6	6.00000	=CEILING.MATH(4.3,2)
7	6.00000	=CEILING.MATH(4.3,-2)

Figure 15.14 Examples of CEILING.MATH() Function

You can achieve a similar effect by using the MROUND() function. The syntax is as follows:

```
=MROUND(number, multiple)
```

This function rounds the value of number to the nearest multiple of multiple. If the remainder of number divided by multiple is less than half of multiple, it rounds down; otherwise, it rounds up.

Finally, you can round numeric values to the nearest even or odd number.

This function results in 7:

```
=ODD(5,3)
```

And this function results in 6:

```
=EVEN(5,3)
```

15.7.3 Basic Mathematical Functions

Many mathematical functions handle general math tasks, including exponentiation, root extraction, and logarithms. The PRODUCT() function lets you multiply up to 255 values at once.

This function returns 120:

```
=PRODUCT(4, 5, 6)
```

Several functions handle division:

This function returns the integer part of a division:

```
=QUOTIENT(numerator, denominator)
```

The remainder is discarded.

This function returns 2:

```
=QUOTIENT(15, 6
```

The inverse function is MOD().

This function returns 3:

```
=MOD(15, 6)
```

It returns the remainder—the modulo—of the division.

To calculate powers, use the POWER() function with this syntax:

```
=POWER(number, power)
```

It returns a number raised to a power, using the same calculation method as the ^ operator. Instead of =4^2, you can enter =POWER(4, 2).

Excel provides several functions for logarithms.

This function returns the natural logarithm of a number:

```
=LN(number)
```

That is, it returns the power to which Euler's number must be raised to produce the specified number.

This is the most general logarithm function:

```
=LOG(number, base)
```

It lets you calculate the logarithm of a number to any base, where the result is the value the base must be raised to in order to get that number. If you omit the base, Excel assumes base 10. This corresponds to the following function:

```
=LOG10(number)
```

Use the following function to extract the square root:

```
=SQRT(number)
```

It calculates the square root of a number. For higher roots, you can express the n-th root as raising to the power of 1/n. The cube root of a number is written as: `=number^(1/3)` or `=POWER(number,1/3)`.

15.7.4 Factorials and Combinations

It is common in mathematical calculations to need the factorial of a number. The function for this is as follows:

```
=FACT(number)
```

For example, the factorial of 4 is 24 (= `1*2*3*4`).

This is mainly used in *combinatorics*, which broadly explores how many ways elements from a set can be combined. Here's an example: How many ways can you choose 6 different numbers out of 59?

```
=FACT(59)/(FACT(6)*FACT(59-6))
```


The result is 45,057,474, meaning there are over 45 million ways to pick six numbers on a lottery ticket. To guarantee a "6 correct" lottery win, you would need to play every one of these combinations. Excel also offers the special function `COMBIN()` for these calculations. The syntax is as follows:

```
=COMBIN(n, k)
```

This function answers how many groups of size `k` can be formed from `n` elements when order doesn't matter. The function returns the binomial coefficient value. It calculates by using this formula:

```
n! / (k! * (n-k)!)
```

The lottery number example can be calculated even more simply like this:

```
=COMBIN(59, 6)
```

This also results in 45,057,474.

15.7.5 Generating Random Numbers and Sequences

The following functions that generate random values are especially useful during the testing phase of calculation models:

```
=RAND()
=RANDBETWEEN(lower_number, upper_number)
=RANDARRAY(rows, columns, min, max, integer)
```

The first function always generates values between 0 and 1. The second lets you specify the allowed value range with two arguments, and it always generates integers. The third function creates an array that has the specified number of rows and columns that contain values between `min` and `max`. The logical argument specifies whether only integers are allowed (`TRUE`) or not.

Enter the function in the first cell of the range and Excel will automatically fill the range defined by the `rows` and `columns` arguments.

A3 | =RANDARRAY(10,4,1,5,TRUE)

	A	B	C	D	E	F
1	**Random Array**					
2						
3	3	3	1	5		
4	2	1	5	4		
5	3	5	4	3		
6	3	4	3	2		
7	1	4	3	4		
8	4	5	2	2		
9	1	4	2	5		
10	2	3	3	2		
11	5	4	4	1		
12	3	5	3	4		

Figure 15.15 Matrix of Random Integers Between 1 and 5

Note that the random numbers change with each recalculation. To work with values that remain fixed, enter the function in the formula bar, press F9 after, and then press Enter.

Use the `SEQUENCE()` function to generate ascending number sequences with a specific interval for a cell range. Enter the function in the first cell of the range and Excel will fill the range specified by the arguments.

A3 | =SEQUENCE(12,4,100,10)

	A	B	C	D	E	F
1	**Example of the SEQUENCE() function**					
2						
3	100	110	120	130		
4	140	150	160	170		
5	180	190	200	210		
6	220	230	240	250		
7	260	270	280	290		
8	300	310	320	330		
9	340	350	360	370		
10	380	390	400	410		
11	420	430	440	450		
12	460	470	480	490		
13	500	510	520	530		
14	540	550	560	570		

Figure 15.16 Example of Sequence with Equal Intervals

15.7.6 Trigonometric Functions

Angle functions, which are essential not only in geometry but also in scientific fields involving oscillations (broadly defined, such as sound, light, electricity, and mechanics), originally derive from calculations with right triangles.

This work uses angle measurements expressed in radians, rather than degrees. A *radian* is the length of the arc that an angle subtends on the unit circle (a circle with a radius of 1). Since a circle's circumference is `2*r*PI`, the unit circle's circumference is `2*PI`, making the radian measure of a 360-degree angle `2*PI`.

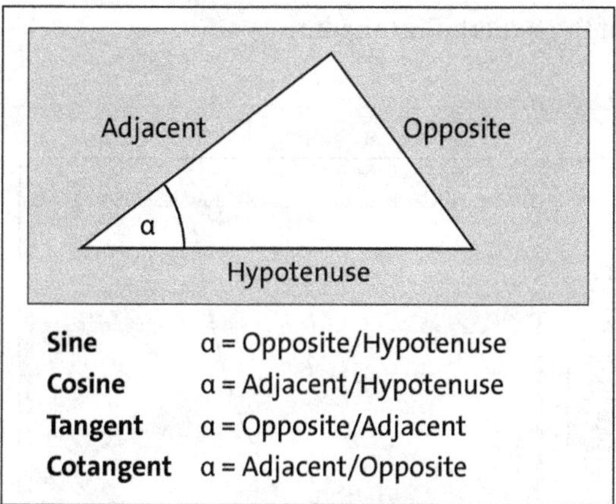

Figure 15.17 Trigonometric Functions

The conversion from degrees to radians is as follows:

```
Degrees = Radians * 180 / PI
Radians = Degrees * PI / 180
```

Trigonometric functions represent the ratios of specific sides in a right triangle relative to an angle:

```
SIN(x) = Opposite leg / Hypotenuse
COS(x) = Adjacent leg / Hypotenuse
TAN(x) = Opposite leg / Adjacent leg
COT(x) = Adjacent leg / Opposite leg
```

The cotangent is the reciprocal of the tangent:

```
COT(x) = 1/TAN(x)
```

Using the corresponding inverse functions, the angle in radians (the arc) is determined from a trigonometric function. All trigonometric functions follow a relationship of the form:

```
If y = SIN(x), then ASIN(y) = x
```

15.7.7 Hyperbolic Functions

Although their names resemble trigonometric functions, hyperbolic functions are not true trigonometric functions. This is clear because, unlike trigonometric functions, they are not periodic. The naming is justified by a strong formal similarity in the relationships between the individual functions and by mathematical connections between them. These functions are used, for example, in certain statistical approximation methods, static calculations, and analysis.

The inverse hyperbolic functions, which are often called *area functions*, begin with A and end with H. These functions yield a surface area in the geometric interpretation. The hyperbolic cotangent is the reciprocal of the hyperbolic tangent:

```
COTH(x) = 1/TANH(x)
```

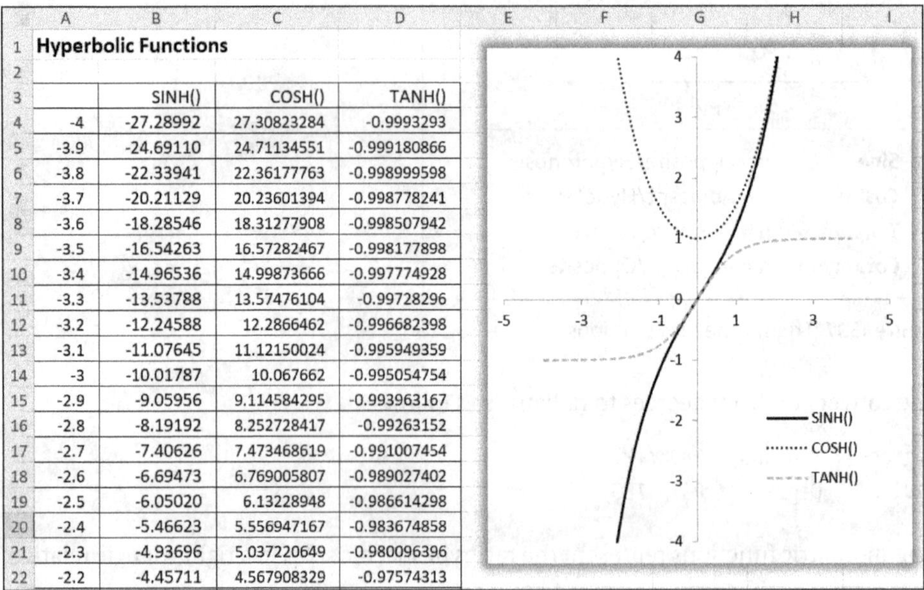

Hyperbolic Functions

	SINH()	COSH()	TANH()
-4	-27.28992	27.30823284	-0.9993293
-3.9	-24.69110	24.71134551	-0.999180866
-3.8	-22.33941	22.36177763	-0.998999598
-3.7	-20.21129	20.23601394	-0.998778241
-3.6	-18.28546	18.31277908	-0.998507942
-3.5	-16.54263	16.57282467	-0.998177898
-3.4	-14.96536	14.99873666	-0.997774928
-3.3	-13.53788	13.57476104	-0.99728296
-3.2	-12.24588	12.2866462	-0.996682398
-3.1	-11.07645	11.12150024	-0.995949359
-3	-10.01787	10.067662	-0.995054754
-2.9	-9.05956	9.114584295	-0.993963167
-2.8	-8.19192	8.252728417	-0.99263152
-2.7	-7.40626	7.473468619	-0.991007454
-2.6	-6.69473	6.769005807	-0.989027402
-2.5	-6.05020	6.13228948	-0.986614298
-2.4	-5.46623	5.556947167	-0.983674858
-2.3	-4.93696	5.037220649	-0.980096396
-2.2	-4.45711	4.567908329	-0.97574313

Figure 15.18 Hyperbolic Functions

For area functions, all hyperbolic functions follow this relationship (similar to trigonometric functions):

```
If y = SINH(x), then ASINH(y) = x
```

15.7.8 List of Mathematical Functions

Functions	Descriptions
ROUNDDOWN(**number, num_digits**)	Rounds a number toward zero on the number line
ABS(**number**)	Removes the sign from numeric values

Functions	Descriptions
AGGREGATE(**function, options, array, k**) AGGREGATE(**function, options, reference1**, reference2...)	Returns aggregates from data in cell ranges or arrays, letting you select the calculation method and how to handle specific values
ARABIC(**text**)	Converts a Roman numeral to an Arabic numeral
ACOS(**number**)	Returns the angle in radians for a given cosine value
ACOSH(**number**)	Returns the inverse hyperbolic cosine
ACOT(**number**)	Returns the arccotangent of a number in radians
ACOTH(**number**)	Returns the inverse hyperbolic cotangent
ASIN(**number**)	Returns the angle in radians for a given sine value
ASINH(**number**)	Inverse function of SINH()
ATAN(**number**)	Calculates the angle in radians for a given tangent value
ATAN2(**x_coordinate, y_coordinate**)	Calculates the arctangent directly from the xy coordinates
ATANH(**number**)	Inverse function of TANH()
ROUNDUP(**number, num_digits**)	Rounds a number away from zero
BASE(**number, base**, minimum_length)	Converts a number to text in the specified base
RADIANS(**angle**)	Converts an angle from degrees to radians
COS(**number**)	Calculates the cosine of the given angle
CSC(**number**)	Calculates the cosecant of an angle
CSCH(**number**)	Calculates the hyperbolic cosecant of an angle
COSH(**number**)	Returns the hyperbolic cosine
COT(**number**)	Returns the cotangent of an angle
COTH(**number**)	Returns the hyperbolic cotangent of an angle
DECIMAL(**text**, **base**)	Converts a text representation of a number in the specified base to a decimal number
EXP(**number**)	Raises Euler's number *e* (2.71828...) to the given power
FACT(**number**)	Returns the factorial of the specified *number*

15

Functions	Descriptions
INT(**number**)	Rounds a numeric expression down to the nearest integer
EVEN(**number**)	Rounds numbers up to the nearest even integer
GCD(**number1**, number2, ...)	Returns the greatest common divisor of integers
DEGREES(**angle**)	Converts radians to degrees
LCM(**number1**, number2, ...)	Returns the least common multiple
COMBIN(**n**, **k**)	Returns the value of the binomial coefficient
COMBINA(**number**, **number_chosen**)	Calculates the number of combinations with repetition for a given number of elements
TRUNC(**number**, number_digits)	Truncates to the specified number of digits
LN(**number**)	Returns the natural logarithm
LOG(**number**, base)	Returns the logarithm with the specified base
LOG10(**number**)	Returns the base-10 logarithm
MDETERM(**matrix**)	Returns the determinant of the specified matrix
MUNIT(**dimension**)	Returns the identity matrix of the specified size
MINVERSE(**matrix**)	Returns the inverse of a matrix
MMULT(**array1**, **array2**)	Returns the product of two matrices
CEILING.MATH(**number**, step, mode)	Rounds a number up to the nearest multiple of *Step*
PI()	Returns the numeric value of PI
MULTINOMIAL(**number1**, number2, ...)	Returns the factorial of the sum of the arguments divided by the product of their factorials
POWER(**number**, power)	Returns the specified power of a number
SERIESSUM(**x**, **n**, **m**, **coefficients**)	Returns the sum of powers of the number *x*
PRODUCT(**number1**, number2, ...)	Multiplies all arguments
SUMSQ(**number1**, number2, ...)	Calculates the sum of the squares of the values
QUOTIENT(**numerator**, **denominator**)	Returns the integer result of a division
MOD(**number**, **divisor**)	Returns the remainder (modulo) of a division
ROMAN(**number**, form)	Converts numbers to Roman numerals
ROUND(**number**, **num_digits**)	Rounds values up or down

Functions	Descriptions
SEC(**number**)	Returns the secant of an angle
SECH(**number**)	Returns the hyperbolic secant of an angle
SEQUENCE(**rows**, columns, start, step)	Returns a sequence of numbers within the specified range
SIN(**number**)	Calculates the sine of the given angle
SINH(**number**)	Returns the hyperbolic sine of a value.
SUM(**number1**, number2, ...)	Calculates the sum of the specified arguments
SUMPRODUCT(**array1**, array2, ...)	Multiplies corresponding elements in arrays and sums the results
SUMIF(**range**, **criteria**, sum_Range)	Calculates the sum of values that meet the specified criteria
SUMIFS(**sum_range**, **criteria_range1**, **criteria1**, criteria_range2, criteria2...)	Calculates the sum of values that meet multiple criteria
SUMX2MY2(**array_x**, **array_y**)	Subtracts the sums of the squared X values and the squared Y values
SUMX2PY2(**array_x**, **array_y**)	Adds the sums of the squared X values and the squared Y values
SUMXMY2(**array_x**, **array_y**)	Squares the difference between X and Y values, then sums the results
TAN(**number**)	Calculates the tangent of the specified angle
TANH(**number**)	Returns the hyperbolic tangent
SUBTOTAL(**function**, **reference1**, reference2, ...)	Returns a subtotal from a table or database
ODD(**number**)	Rounds numbers up to the next odd integer
FLOOR.MATH(**number**, significance, mode)	Rounds up to the next multiple of *significance*
SIGN(**number**)	Returns 1 for positive numbers, –1 for negative numbers, and 0 for zero
MROUND(**number**, multiple)	Rounds values to the nearest multiple
SQRT(**number**)	Calculates the square root of a number
SQRTPI(**number**)	Returns the square root of (*Number* × Pi)

15

Functions	Descriptions
RANDBETWEEN(**bottom_number**, **top_number**)	Generates random-number integers between *bottom_number* and *top_number*
RANDARRAY(**rows**, **columns**, **min**, **max**, **integer**)	Generates an array of random numbers between *min* and *max*
RAND()	Generates random numbers between 0 and 1
FACTDOUBLE(**number**)	Returns the double factorial of a positive number

15.8 Statistical Functions

In terms of function variety, Excel hardly needs to shy away from comparison with professional statistical software for statistical calculations. To help you navigate this area a bit more easily, here are at least a few brief preliminary remarks.

15.8.1 Overview of Statistical Functions

Statistical functions cover several areas: analyzing individual samples with one or more variables measured, analyzing and comparing multiple samples, comparing samples to a population, probability calculations and related topics, and probability distributions of random variables.

Descriptive and Inferential Statistical Methods

Statistical functions are initially divided into descriptive and inferential methods. *Descriptive statistical methods* summarize the original data set and serve as the starting point for any data analysis. Common functions include calculating frequency distributions, various averages, and measures of dispersion.

Inferential statistical methods are used to draw conclusions about a population based on data from a sample taken from that population. This is the typical area of opinion researchers who infer what millions of people think about a specific topic from a few hundred interviews. The statistical functions cover several areas: analyzing individual samples with one or more recorded characteristics, analyzing and comparing multiple samples, comparing samples to a population, and topics like probability theory and probability distributions of random variables.

Different Scales

To determine which statistical methods you can apply to a specific data set, you first need to consider the type of scaling involved. The primary types are nominal, ordinal, and metric scales.

For example, when recording the marital status of individuals, the possible categories represent values on a *nominal scale*. These are qualitative attributes where only differences can be identified, without any inherent order. However, they cannot be arranged in an order that assigns a higher rank to one characteristic over another, and nothing can be said about the distances between the characteristics.

With *ordinal-scaled values*, a specific ranking can be read from the data, as with school grades. Grade 1 is better than grade 2, and 4 is better than 5, but the distance between 1 and 2 is not necessarily the same as between 4 and 5.

When the exact distances between values provide useful information, as with data on weight and height, the data is based on a *metric scale*.

It's clear that certain statistical analyses don't make sense for some scales, such as calculating an average between the status categories *married* and *single*.

Another key distinction in data types is between *discrete values* and *continuous values*. While a person's marital status can only take a few values, height can be measured with any level of precision, including intermediate values.

15.8.2 Samples and Populations

Almost all statistical methods involve either samples or populations, often both directly and indirectly. A *sample* is a subgroup of elements that are randomly selected from a population. *Random* means the selection process does not favor any specific elements of the population, and the *population* is the set of elements from which the sample is drawn.

For example, when pollsters are conducting public opinion polls, simply randomly opening one or many phone books (which favors phone owners) or approaching people on the street (which disadvantages those who stay at home) is not enough to select a sample of eligible voters. Accordingly, the institutes that conduct opinion polls keep their methods for obtaining representative samples closely guarded.

15.8.3 Random Variables and Probability

Based on statistical surveys or sometimes theoretical considerations, it is often possible to assign a specific probability to the occurrence of certain values. For example, the probability of getting heads on a coin toss is 1 / 2 (0.5 or 50%), rolling a 6 on a six-sided die is 1 / 6 (0.166666), and drawing a specific card from a Poker deck is 1 / 52 (0.01923). Excel provides several functions to calculate probabilities that are less straightforward.

In other cases, you can determine probability by counting the total population. For example, if 51% of a country's population is female, the probability that a randomly selected person is female is 0.51. A quantity like this is called a *random variable*. If the variable is discrete, as in the examples given, you can directly specify the probability that it will take a specific value or one of several values.

It's different with continuous variables. For example, when you're measuring the physical height of people in a population, you'll encounter issues of both measurement accuracy and classification when specifying probabilities. You can't directly answer how likely people are to be 5 feet 8 inches tall until you answer the question of whether 5 feet 7.99 inches count as 5 foot 8. Therefore, you must group values into intervals or classes (e.g., from 5 feet 7.5 inches up to 5 feet 8.49 inches). For all such quantities, statistics uses probability distributions for continuous variables, several of which Excel offers.

15.8.4 Sample Analysis

When analyzing samples, two questions usually arise: What does the sample show, and what conclusions can be drawn about the population? To answer the first question, two key measures are used: the average and the spread. Typically, only two metrics are common here: the arithmetic mean and the standard deviation (or its square, which is called the *variance*). These measures allow you to estimate the corresponding population parameters, with estimates becoming more reliable as the sample size grows (see `AVERAGE()`, `VAR.S()`, and `STDEV.S()`).

Another important issue to address when analyzing a sample is whether the observed values follow a specific pattern. When you're examining whether there's a relationship between personal income and the size of living space used, it is reasonable to assume a connection: the higher the income, the larger the square footage. The *correlation coefficient* (`CORREL()`) and *covariance* (`COVARIANCE.P()`) measure whether and how strongly the data are related.

If the relationship is suspected to be linear or exponential, you can largely clarify it through regression by fitting the values to a line or an exponential curve. The powerful `LINEST()` and `LOGEST()` functions are available for this purpose.

15.8.5 Statistical Tests

Statistical tests generally aim to determine how confidently values from a sample can be used to infer values for the overall population. Excel provides two groups of functions for this purpose. One lets you perform tests directly on samples, while the other supplies values from probability distributions to verify parameters derived from those samples.

The t-test (`T.TEST()`) assesses whether two samples differ in their average value by chance (which indicates that both samples were randomly drawn from the same population) or whether they differ systematically (which indicates that they come from different populations or were not randomly drawn). The T.TEST() function returns a probability value directly.

The t-test can also determine whether the relative frequency of a characteristic in a sample deviates from its probability in the population by chance or not. Unfortunately, this case isn't covered by a function, so the t-distribution (T.DIST()) must be used.

The F-test (F.TEST()) determines whether two samples differ randomly in their variance. Again, using the F-distribution (F.DIST()) is a viable and sometimes necessary approach.

The chi-square test (CHISQ.TEST()) checks whether a sample with multiple values matches a population with known expected probabilities for those values. The distribution function is also available here.

15.8.6 Distribution Functions

We mentioned random variables earlier, and the question was about the probability of a specific event occurring. Excel provides several functions to answer this question and thus eliminates the need to consult extensive tables.

Based on the distinction between discrete and continuous random variables, their distributions can also be classified as discrete or continuous. Here is a brief overview with some application tips:

- **Binomial distribution: BINOM.DIST()**
  This models an event that can occur or not occur with a certain probability. Examples include coin tosses, dice rolls, and card draws (with replacement), plus male/female, employed/unemployed, and so on.
- **Hypergeometric distribution: HYPERGEOM.DIST()**
  When you draw cards without replacing them, the theoretical probability changes with each draw. In such cases, use the hypergeometric distribution.
- **Poisson distribution: POISSON.DIST()**
  This distribution is commonly used as an approximation of the binomial distribution when dealing with very large numbers and very small probabilities. However, since Excel can calculate the binomial distribution directly, this workaround isn't always necessary.
- **Normal distribution: NORM.DIST() and NORM.S.DIST()**
  When a random variable is based on a very large population (starting at 1,000) and measures a continuous variable, you can generally assume that it follows or approximates a normal distribution. This applies in many examples, making the normal distribution the most important continuous distribution. The special case is the standard normal distribution, where the mean is 0 and the standard deviation is 1.

Excel also provides some less common distributions that are useful for specialized applications.

To help simplify how you work with distribution functions, here's a final note: Probability distributions are always based on a density function, like the bell curve shown for the normal distribution. However, the Y-value alone doesn't indicate the probability of the corresponding X-value. Only the area between two X-values (the definite integral) represents the probability.

Based on discrete distributions, Excel uses a `cumulative` logical value to specify whether the function calculates the density (`FALSE`) or the probability (`TRUE`). The first case returns the density function value, while the second returns the probability distribution value.

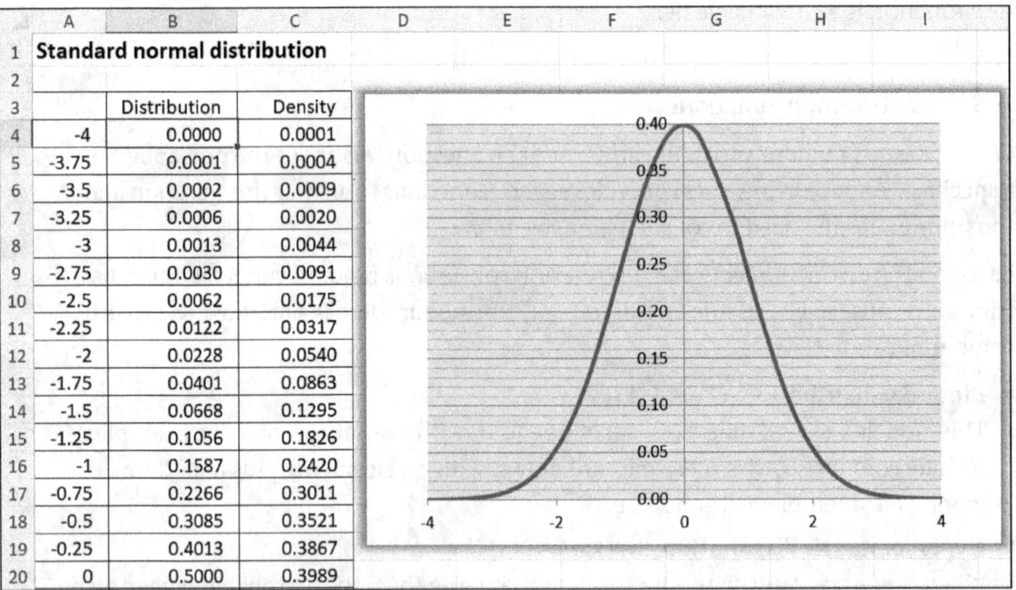

Standard normal distribution

	Distribution	Density
-4	0.0000	0.0001
-3.75	0.0001	0.0004
-3.5	0.0002	0.0009
-3.25	0.0006	0.0020
-3	0.0013	0.0044
-2.75	0.0030	0.0091
-2.5	0.0062	0.0175
-2.25	0.0122	0.0317
-2	0.0228	0.0540
-1.75	0.0401	0.0863
-1.5	0.0668	0.1295
-1.25	0.1056	0.1826
-1	0.1587	0.2420
-0.75	0.2266	0.3011
-0.5	0.3085	0.3521
-0.25	0.4013	0.3867
0	0.5000	0.3989

Figure 15.19 Graph of Standard Normal Distribution's Density Function

Most distribution functions also have an inverse function (`...INV()`), and the relationship between the two functions is as follows:

- The distribution function gives the probability that a random variable is less than or equal to a specified value (the argument `x`).
- The inverse function returns the value that corresponds to a given probability that is equal to or less than the random variable with that probability. Since this value is called the *quantile*, the relationship can also be expressed as follows:

```
...DIST(q) = p
...INV(p)  = q
```

  Here, `p` = probability and `q` = quantile.

15.8.7 Calculating the Standard Deviation of Test Results

When you're analyzing test results, you'll encounter questions about the average values you obtain and the degree of variation. In Figure 15.20, you see a simple example of this: a test taken by many people produced various scores, and the scores are listed in column B.

Once all values are available, calculating the arithmetic average is straightforward. You use the `AVERAGE()` function and provide the addresses of the range with the points as arguments. A related value is the `MEDIAN()`, which you can calculate in the same way. The result here is 1.208, which means that in this example, exactly half of the values are above 1.208 and the other half are below it.

However, the average itself does not reveal the spread of the results. With the same average, values can be clustered closely around the center or spread far apart. In column C, the differences from the average have been calculated for illustration by subtracting each individual value from the average. Calculating the average of these differences shows that the negative and positive deviations cancel each other out. Therefore, the result is not meaningful.

	A	B	C	D	E	F
1	**Standard Deviation and Variance in Test Results**					
2						
3	Test 1	Points	Deviation from the mean	Deviation absolute	Squares of Deviation	
4	Test person 1	970	153	153	23285	
5	Test person 2	1389	-267	267	71096	
6	Test person 3	878	244	244	59647	
7	Test person 4	673	450	450	202073	Standard deviation
8	Test person 5	700	423	423	178587	248
9	Test person 6	1258	-135	135	18226	
10	Test person 7	1108	15	15	227	
11	Test person 8	1302	-179	179	32084	Variance
12	Test person 9	1350	-227	227	51740	61727
13	Test person 10	1158	-35	35	1235	
14	Test person 11	1395	-272	272	74248	
15	Test person 12	1291	-168	168	28273	
16	Average	1123	0	214	61727	
17	Median	1208				

Figure 15.20 Calculating Standard Deviation of Test Results

The situation improves if, as in column D, the `ABS()` function is used to calculate the absolute difference from the average. The result in cell D16 is the average deviation. However, this simple calculation has the drawback that larger deviations do not have enough impact. To address this, variance and standard deviation functions work with the squares of the deviations, and the sum of the squared deviations is then divided by the count. The variance equals the average of the squared deviations. The formula is as follows:

```
=VAR.P(B4:B15)
```

The standard deviation is calculated as the square root of the variance:

```
=STDEV.N(B4:B15)
```

As you can see from the results, the standard deviation in this example is noticeably higher than the mean deviation.

15.8.8 Conditional Maximum and Minimum Values

Since Excel 2019, the `MAXIFS()` and `MINIFS()` functions have existed. They return the maximum or minimum in a cell range based on filter criteria, which you can specify in pairs by defining criteria ranges and their corresponding criteria. Up to 126 such pairs are allowed.

The size and shape of the criteria ranges must match those of the ranges containing the data to be evaluated; otherwise, the function will return an error. In the example in Figure 15.21, both formulas evaluate the weight data range and search the age range for values over 30 and under 40, respectively.

	A	B	C
1	**Conditional limits**		
2			
3	Person	Age	Weight
4	1	23	152
5	2	25	146
6	3	29	150
7	4	33	159
8	5	35	163
9	6	35	139
10	7	43	150
11	8	43	194
12	9	43	128
13	10	43	174
14	=MAXIFS(C4:C13,B4:B13,">30")		194
15	=MINIFS(C4:C13,B4:B13,"<40")		139

Figure 15.21 Conditional Maximum and Minimum Values

15.8.9 List of Statistical Functions

Functions	Descriptions
INTERCEPT(**y_values, x_values**)	Returns the y-intercept of the regression line where it crosses the y-axis
COUNT(**value1**, value2, ...)	Calculates the number of numeric values

Functions	Descriptions
COUNTA(**value1**, value2, ...)	Counts the number of values, including text and logical values
COUNTBLANK(**range**)	Counts the number of empty cells in the range
RSQ(**y_values, x_values**)	Returns the square of the Pearson correlation coefficient
BETA.INV(**probability, alpha, beta**, a, b)	Returns quantiles of a beta distribution
BETA.DIST(**x, alpha, beta, cumulative**, a, b)	Returns the probability distribution of a beta-distributed random variable
BINOM.INV(**trials, success_prob, alpha**)	Returns the smallest number of successful trials where the cumulative probability is at least the threshold probability
BINOM.DIST(**number_successes, trials, success_prob, cumulative**)	Returns the probability of outcomes in a binomial distribution
BINOM.DIST.RANGE(**trials, probability_s, number_successes**, number2_successes)	Returns the probability of success for a trial outcome using the binomial distribution
CHISQ.INV(**probability, degrees_freedom**)	Returns the quantile of the left-tailed chi-square distribution
CHISQ.INV.RT(**probability, degrees_freedom**)	Returns the quantile of the right-tailed chi-square distribution
CHISQ.TEST(**observed_values, expected_values**)	Returns the p-value for the chi-square test comparing observed and expected values
CHISQ.DIST(**x, degrees_freedom, cumulative**)	Returns values of the left-tailed distribution function for a chi-square distributed random variable
CHISQ.DIST.RT(**x, degrees_freedom**)	Returns values of the right-tailed distribution function for a chi-square distributed random variable
EXPON.DIST(**x, lambda, cumulative**)	Returns probabilities for an exponentially distributed random variable
F.INV(**probability**, degrees_freedom1, degrees_freedom2)	Returns quantiles of the left-tailed F distribution
F.INV.RT(probability, **degrees_freedom1, degrees_freedom2**)	Returns quantiles of the right-tailed F distribution

15

Functions	Descriptions
F.TEST(**matrix1, matrix2**)	Returns the probability that two samples have equal variances
F.DIST(**x, degrees_freedom1, degrees_freedom2, cumulative**)	Returns values of the distribution function (1-Alpha) for a left-tailed F-distributed variable
F.DIST.RT(**x, degrees_freedom1, degrees_freedom2**)	Returns values of the distribution function (1-Alpha) for a right-tailed F-distributed variable
FISHER(**x**)	Returns the Fisher transformation of a correlation coefficient
FISHERINV(**y**)	Inverse of FISHER()
Z.TEST(**matrix, x**, sigma)	Returns the one-sided probability for a Gaussian test
GAMMA(**x**)	Returns the value of the gamma function
GAMMA.INV(**probability, alpha, beta**)	Returns quantiles of the gamma distribution
GAMMA.DIST(**x, alpha, beta, cumulative**)	Returns probabilities for a gamma-distributed random variable
GAMMALN(**x**)	Returns the natural logarithm of the gamma function
GAMMALN.PRECISE(**x**)	Returns the natural logarithm of the gamma function
GAUSS(**x**)	Returns 0.5 less than the cumulative normal distribution
GEOMEAN(**number1**, number2, ...)	Calculates the geometric mean
TRIMMEAN(**array, percent**)	Calculates the arithmetic mean, excluding extreme values
HARMEAN(**number1**, number2, ...)	Calculates the harmonic mean
FREQUENCY(**data, classes**)	Returns the frequency within intervals defined by classes
HYPGEOM.DIST(**successes_s, sample_size_s, successes_g, population_size_g, cumulative**)	Calculates probabilities for a hypergeometrically distributed random variable
LARGE(**array, k**)	Returns the *k*-th largest value
SMALL(**array, k**)	Returns the *k*-th smallest value
CONFIDENCE.NORM(**alpha, standard_dev, size**)	Calculates the confidence interval

Functions	Descriptions
CONFIDENCE.T(**alpha, standard_dev, size**)	Calculates the confidence interval for a population with a t-distribution
CORREL(**matrix1, matrix2**)	Returns the correlation coefficient between two data sets
COVARIANCE.P(**array1, array2**)	Returns a measure of the relationship between two data sets
COVARIANCE.S(**array1, array2**)	Returns a measure of the relationship between two data sets for samples
KURT(**number1**, number2, ...)	Returns the kurtosis (peakedness) of a frequency distribution
LOGNORM.INV(**probability, mean, standard_dev**)	Returns the percentile of a lognormal distribution
LOGNORM.DIST(**x, average, standard_dev, cumulative**)	Returns the probability distribution value for a lognormal distribution
MAX(**number1**, number2, ...)	Returns the largest value
MAXA(**value1**, value2, ...)	Returns the largest value, including cells with text or logical values
MAXIFS(**max_range, criteria_range1, criteria1**, criteria_range2, criteria2, ...)	Returns the largest value in the specified range that meets the criteria in the criteria ranges
MEDIAN(**number1**, number2, ...)	Calculates the median of a dataset
MIN(**number1**, number2, ...)	Returns the smallest value
MINA(**value1**, value2, ...)	Returns the smallest value, including cells with text or logical values
MINIFS(**min_range, criteria_range1, criteria1**, criteria_range2, criteria2, ...)	Returns the smallest value in the specified range that meets the criteria in the criteria ranges
AVEDEV(**number1**, number2, ...)	Returns the average absolute deviation
AVERAGE(**number1**, number2, ...)	Returns the arithmetic mean
AVERAGEA(**value1**, value2, ...)	Returns the arithmetic mean, including cells with text or logical values
AVERAGEIF(**range, criteria**, average_range)	Returns the average of values that meet a specified criterion

Functions	Descriptions
AVERAGEIFS(**average_range, criteria_range1, criteria1**, criteria_range2, criteria2...)	Returns the arithmetic mean of values that meet multiple criteria
MODE.SNGL(**number1**, number2, ...)	Returns the most frequently occurring value in a dataset
MODE.MULT(**number1**, number2, ...)	Returns the most frequently occurring values in a dataset
NEGBINOM.DIST(**number_failures, number_successes, success_probability, cumulative**)	Calculates the probability of a compound event occurring
NORM.INV(**probability, mean, standard_dev**)	Returns quantiles of the normal distribution
NORM.S.INV(**probability**)	Returns quantiles of the standard normal distribution
NORM.S.DIST(**z, cumulative**)	Returns the probability that a random variable from the standard normal distribution is less than or equal to *z*
NORM.DIST(**x, average, standard_dev, cumulative**)	Returns values of a normal distribution
PEARSON(**matrix1, matrix2**)	Returns the Pearson correlation coefficient
PHI(**x**)	Returns the density function value for a standard normal distribution
POISSON.DIST(**x, average, cumulative**)	Returns probabilities for random variables following a Poisson distribution
FORECAST.ETS (**target_data, values, timeline, seasonality, data_completion, aggregation)**	Returns the forecasted value for a specific target date by using exponential smoothing
FORECAST.ETS.CONFINT(**target_date, values, timeline, seasonality, data_completion, aggregation)**	Returns a confidence interval for the estimated value at a specific target date
FORECAST.ETS.SEASONALITY(**values, timeline**, data_completeness, aggregation)	Returns the length of the repeating pattern that Excel detects in a series
FORECAST.ETS.STAT(**values, timeline, statistic_type**, seasonality, data_completeness, aggregation)	Returns statistical metrics for a forecast
FORECAST.LINEAR(**x, y_values, x_values**)	Returns an estimated *y* value for the given *x*, based on linear regression

Functions	Descriptions
PERCENTILE.EXC(**array**, **k**)	Returns the value in a data set below which a fraction k (between 0 and 1, excluding 0 and 1) of the data falls
PERCENTILE.INC(**array**, **k**)	Returns the value in a data set below which a given fraction k (ranging from 0 to 1) of the data falls
PERCENTRANK.EXC(**array**, **x**, accuracy)	Returns the proportion of data below the specified value (ranging between 0 and 1, excluding 0 and 1)
PERCENTRANK.INC(**array**, **x**, **accuracy**)	Returns the proportion of data below the specified value (ranging from 0 to 1)
QUARTILE.EXC(**array**, **quartile**)	Divides the data from *Array* into ranges with equal proportions based on percentile values (value range is between 0 and 1, excluding 0 and 1)
QUARTILE.INC(**array**, **quartile**)	Divides the data from *Array* into proportional ranges, based on percentile values (values range from 0 to 1)
RANK.EQ(**number**, **ref**, order)	Returns the rank of a value within a dataset based on its size
RANK.AVG(**number**, **ref**, order)	Returns the average rank of a value within a dataset based on its size
LINEST(**y_values**, x_values, const, stats)	Returns coefficients for linear regression
LOGEST(**y_values**, x_values, const, stats)	Returns coefficients for exponential regression
SKEW(**number1**, number2, ...)	Returns a measure of the asymmetry in the frequency distribution of a sample
SKEW.P(**number1**, number2, ...)	Returns a measure of the asymmetry in the frequency distribution of a population
STDEV.N(**number1**, number2, ...)	Calculates the standard deviation based on the population
STDEV.S(**number1**, number2, ...)	Estimates the standard deviation based on a sample
STDEV(**value1**, value2, ...)	Estimates the standard deviation based on a sample, including cells with text or logical values

15

Functions	Descriptions
STDEV.P(**value1**, value2, ...)	Calculates the standard deviation based on the population, including cells with text or logical values
STANDARDIZE(**x**, **average**, **std_deviation**)	Converts values from a normal distribution to the standard normal distribution
SLOPE(**y_values**, **x_values**)	Returns the slope of a regression line
STEYX(**y_values**, **x_values**)	Returns the standard error of the estimate for linear regression
DEVSQ(**number1**, number2, ...)	Returns the sum of squared deviations
T.INV(**probability**, **degrees_freedom**)	Returns left-tailed quantiles of the t-distribution
T.INV.2T(**probability**, **degrees_freedom**)	Returns two-tailed quantiles of the t-distribution
T.TEST(**array1**, **array2**, **pages**, **type**)	Returns values for comparing two samples
T.DIST(**x**, **degrees_freedom**, **cumulative**)	Returns left-tailed probability values for a t-distributed random variable
T.DIST.2T(**x**, **degrees_freedom**)	Returns two-tailed probability values for a t-distributed random variable
T.DIST.RT(**x**, **degrees_freedom**)	Returns right-tailed probability values for a t-distributed random variable
TREND(**y_values**, x_values, New_x_Values, Constant)	Calculates estimated Y-values based on linear regression
VAR.P(**number1**, number2, ...)	Calculates variance based on the entire population
VAR.S(**number1**, number2, ...)	Estimates population variance from a sample
VARA(**value1**, value2, ...)	Estimates population variance from a sample, including cells with text or logical values
VARPA(**value1**, value2, ...)	Calculates variance based on the entire population, including cells with text or logical values
GROWTH(**y_values**, x_values, new_x_values, constant)	Calculates values based on an exponential trend
PERMUT(**n**, **k**)	Calculates the number of ordered sequences that are possible with the given arguments

Functions	Descriptions
PERMUTATIONA(**number, chosen_number**)	Calculates the number of permutations for the specified number of objects (with repetitions) selected from the total set
PROB(**observed_values, observed_probabilities, floor,** ceiling)	Calculates the probability that an observation will fall within the specified interval
WEIBULL.DIST(**x, alpha, beta, cumulative**)	Returns the probabilities for a Weibull-distributed random variable
COUNTIF(**range, criteria**)	Returns the number of nonempty cells in the range that meet the specified criteria
COUNTIFS(**criteria_range1, criteria1**, criteria_range2, criteria2...)	Returns the number of nonempty cells in the range that meet multiple criteria

15.8.10 List of Compatible Functions

The following is a list of functions that remain valid as compatible functions in statistics. However, we recommend that you use the current functions in new workbooks. The help section for compatible functions always includes notes on the functions that replaced them.

Functions	Descriptions
BETAINV(**probability, alpha, beta**, a, b)	Returns quantiles of a beta distribution
BETADIST(**x, alpha, beta**, a, b)	Returns the probability distribution of a beta-distributed random variable
BINOMDIST(**number_successes, trials, success_prob, cumulative**)	Returns the probability of outcomes in a binomial distribution
CHIINV(**probability, degrees_freedom**)	Returns the quantiles of the chi-square distribution
CHITEST(**observed_values, expected_values**)	Returns the probability value for the chi-square test
CHIDIST(**x, degrees_freedom**)	Calculates the probability of exceeding the value for *x* and *degrees of freedom*, evaluating the fit between observed and expected values
EXPONDIST(**x, Lambda, cumulative**)	Returns probabilities for an exponentially distributed random variable
FINV(**probability, degrees_freedom1, degrees_freedom2**)	Returns the quantiles of the F distribution

Functions	Descriptions
FTEST(**matrix1, matrix2**)	Returns the probability that two samples have equal variances
FDIST(**x, degrees_freedom1, degrees_freedom2**)	Returns the values of the distribution function (1-Alpha) for an F-distributed random variable
GAMMAINV(**probability, alpha, beta**)	Returns quantiles of the gamma distribution
GAMMADIST(**x, alpha, beta, cumulative**)	Returns probabilities for a gamma-distributed random variable
ZTEST(**matrix, x**, sigma)	Returns the one-sided probability for a Gaussian test
HYPGEOMDIST(**successes_s, sample_size_s, successes_g, sample_size_g**)	Calculates probabilities for a hypergeometrically distributed random variable
CONFIDENCE(**alpha, standard_dev, sample_size_s**)	Calculates the confidence interval
COVAR(**matrix1, matrix2**)	Returns a measure of the relationship between two data sets
CRITBINOM(**trials, success_prob, alpha**)	Returns the smallest number of successful trials where the cumulative probability is at least the threshold probability
LOGINV(**probability, mean, standard_dev**)	Returns quantiles of a lognormal distribution
LOGNORMDIST(**x, average, standard_dev**)	Returns the probability distribution of a lognormal distribution
MODE(**number1**, number2, ...)	Returns the most frequently occurring value in a data set
NEGBINOMDIST(**number_failures, number_successes, success_prob**)	Calculates the probability of a compound event occurring
NORMINV(**probability, mean, standard_dev**)	Returns quantiles of the normal distribution
NORMDIST(**x, average, standard_dev, cumulative**)	Returns values of a normal distribution
CEILING(**number, step**)	Rounds a number up to the nearest multiple of *Step*
POISSON(**x, average, cumulative**)	Returns probabilities for random variables following a Poisson distribution

Functions	Descriptions
PERCENTILE(**array**, **alpha**)	Returns the value in a data set below which a fraction *Alpha* of the data falls
PERCENTRANK(**array**, **x**, accuracy)	Returns the percentage of data values below the specified value
QUARTILE(**array**, **quartile**)	Divides the data in *Array* into four ranges with equal proportions of data
RANK(**number**, **ref**, order)	Returns the rank of a value within a dataset. based on its size
FORECAST(**x**, **y_values**, **x_values**)	Returns an estimated *y* value for the given *x*, based on linear regression
STDEV(**number1**, number2, ...)	Estimates the standard deviation based on a sample
STDEVP(**number1**, number2, ...)	Calculates the standard deviation based on the population
NORMSINV(**probability**)	Returns quantiles of the standard normal distribution
NORMSDIST(**z**)	Returns the probability that a random variable from a standard normal distribution is less than or equal to *z*
TINV(**probability**, **degrees_freedom**)	Returns quantiles of the t-distribution.
TTEST(**array1**, **array2**, **pages**, **type**)	Returns values for comparing two samples
TDIST(**x**, **degrees_freedom**, **tails**)	Returns probabilities for t-distributed random variables
FLOOR(**number**, **step**)	Rounds a number down to the nearest multiple of *Step*
VAR(**number1**, number2, ...)	Estimates population variance from a sample
VARP(**number1**, number2, ...)	Calculates variance based on the entire population
CONCATENATE(**text1**, Text2, ...)	Joins text strings together
WEIBULL(**x**, **alpha**, **beta**, **cumulative**)	Returns probabilities for a Weibull-distributed random variable

15

15.9 Lookup and Reference Functions

The functions under the **Lookup and Reference** category primarily manage references, such as finding cell addresses, determining range sizes, and searching, sorting, and filtering ranges.

New functions include `IMAGE()` and several that are designed to work with matrix-style data: `EXPAND()`, `WRAPCOLS()`, `CHOOSECOLS()`, `TAKE()`, `VSTACK()`, `DROP()`, `HSTACK()`, `WRAPROWS()`, `CHOOSEROWS()`, `TOCOL()`, and `TOROW()`. Additionally, `GROUPBY()` and `PIVOTBY()` assist in summarizing table data.

15.9.1 Filtering, Sorting, and Reducing Data

If you want to filter values from a range or matrix based on specific criteria, in addition to the filter functions for data lists described in Chapter 17, you can use the `FILTER()` table function.

The syntax is as follows:

```
=FILTER(array, include, if_empty)
```

Here, `array` specifies the data range or array to filter, `include` is a logical expression that compares a value to values in a column of the array, and `if_empty` defines what to display if no match is found.

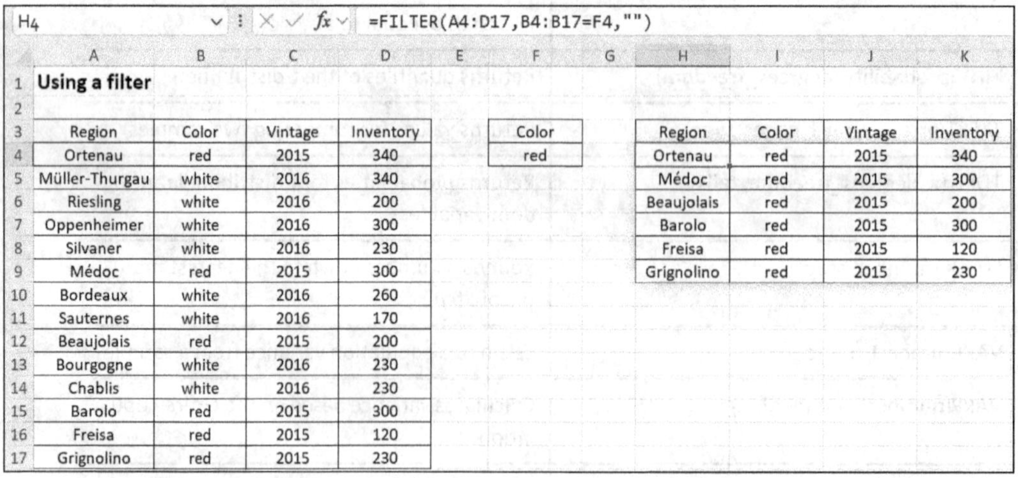

H4 =FILTER(A4:D17,B4:B17=F4,"")

	A	B	C	D	E	F	G	H	I	J	K
1	**Using a filter**										
2											
3	Region	Color	Vintage	Inventory		Color		Region	Color	Vintage	Inventory
4	Ortenau	red	2015	340		red		Ortenau	red	2015	340
5	Müller-Thurgau	white	2016	340				Médoc	red	2015	300
6	Riesling	white	2016	200				Beaujolais	red	2015	200
7	Oppenheimer	white	2016	300				Barolo	red	2015	300
8	Silvaner	white	2016	230				Freisa	red	2015	120
9	Médoc	red	2015	300				Grignolino	red	2015	230
10	Bordeaux	white	2016	260							
11	Sauternes	white	2016	170							
12	Beaujolais	red	2015	200							
13	Bourgogne	white	2016	230							
14	Chablis	white	2016	230							
15	Barolo	red	2015	300							
16	Freisa	red	2015	120							
17	Grignolino	red	2015	230							

Figure 15.22 Filtered Range at Right Shows Red Wines

The function in cell H4 returns a dynamic result array whose size depends on its arguments.

Besides the sorting functions introduced in Chapter 16, Excel offers the two SORT() and SORTBY() table functions for sorting. Their syntax is as follows:

```
=SORT(array, sort_index, sort_order, by_column)
=SORTBY(array, by_array1, [sort_order1],…)
```

Here, Array is the range or array of data to sort, and sort_index specifies the column to sort by. If sort_order = 1, then sorting is ascending; if –1, then sorting is descending. Also, by_column is a logical value, 1 sorts by columns, and 0 sorts by rows.

The second function lets you sort by multiple arrays, even with different sort orders, and both functions return a dynamic result array. Figure 15.23 shows two simple examples. The middle range shows the result of the SORT() function in cell E4, while the right range comes from the SORTBY() function in cell I4.

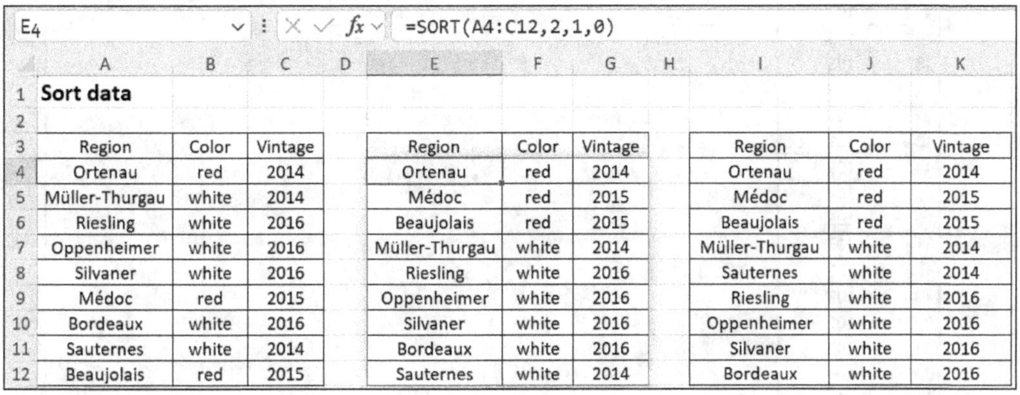

E4 =SORT(A4:C12,2,1,0)

	A	B	C	D	E	F	G	H	I	J	K
1	**Sort data**										
2											
3	Region	Color	Vintage		Region	Color	Vintage		Region	Color	Vintage
4	Ortenau	red	2014		Ortenau	red	2014		Ortenau	red	2014
5	Müller-Thurgau	white	2014		Médoc	red	2015		Médoc	red	2015
6	Riesling	white	2016		Beaujolais	red	2015		Beaujolais	red	2015
7	Oppenheimer	white	2016		Müller-Thurgau	white	2014		Müller-Thurgau	white	2014
8	Silvaner	white	2016		Riesling	white	2016		Sauternes	white	2014
9	Médoc	red	2015		Oppenheimer	white	2016		Riesling	white	2016
10	Bordeaux	white	2016		Silvaner	white	2016		Oppenheimer	white	2016
11	Sauternes	white	2014		Bordeaux	white	2016		Silvaner	white	2016
12	Beaujolais	red	2015		Sauternes	white	2014		Bordeaux	white	2016

Figure 15.23 Sorting with SORT and SORTBY

The FILTER() and SORT() functions can also be combined to sort filtered results immediately.

The UNIQUE() table function is also very useful for analyzing data in a range or array. It returns either numbers or text that appear only once in a range or array or those that appear multiple times. Figure 15.24 shows some simple examples.

The formula in C4 is as follows:

```
=UNIQUE(A4:A23, 0, 0)
```

The first argument specifies the cell range or array whose values the function evaluates. The second argument is a logical value. False or 0 compares values by row; 1 or TRUE compares by column. By default, the comparison is row-wise. If the exactly_once argument is 1 or TRUE, then the function returns values that appear only once; otherwise, it returns a single instance of values that appear multiple times. The function works equally well with numbers and text, and it also returns a dynamic result array. The number of output cells depends on the evaluation. If you enter the function in the first cell of the expected output range, you may get a #SPILL! error if the range needed for all results already contains entries that block the full array from spilling. Deleting these entries will display the desired array.

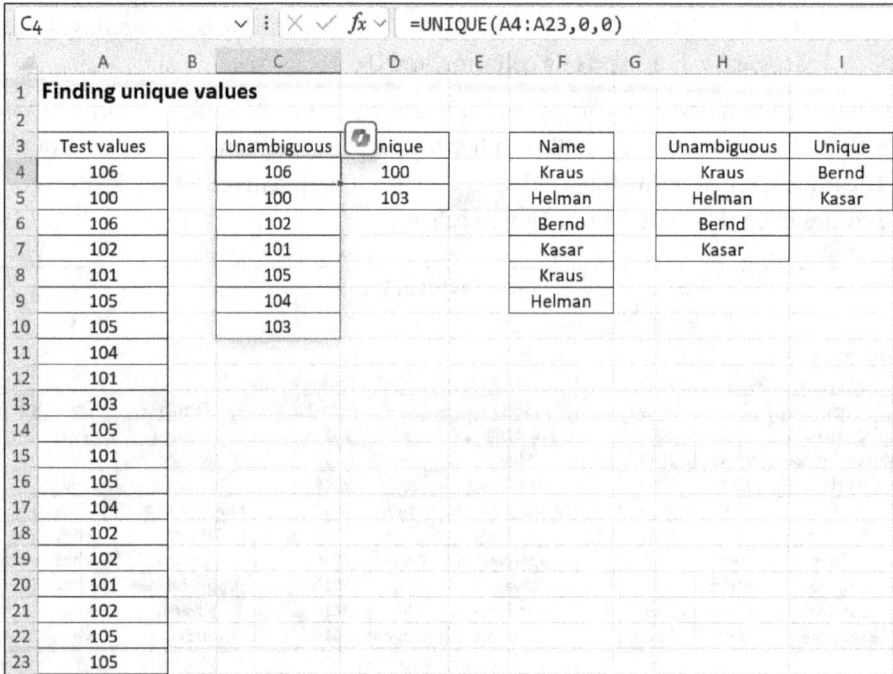

C4 =UNIQUE(A4:A23,0,0)

	A	B	C	D	E	F	G	H	I
1	**Finding unique values**								
2									
3	Test values		Unambiguous	nique		Name		Unambiguous	Unique
4	106		106	100		Kraus		Kraus	Bernd
5	100		100	103		Helman		Helman	Kasar
6	106		102			Bernd		Bernd	
7	102		101			Kasar		Kasar	
8	101		105			Kraus			
9	105		104			Helman			
10	105		103						
11	104								
12	101								
13	103								
14	105								
15	101								
16	105								
17	104								
18	102								
19	102								
20	101								
21	102								
22	105								
23	105								

Figure 15.24 Simple Examples of New UNIQUE() Function

All functions in this section return a dynamic array, as explained in Chapter 3, Section 3.5.

15.9.2 Querying Lookup Tables

Lookup functions like LOOKUP(), HLOOKUP(), VLOOKUP(), and XLOOKUP() are very useful in many situations. Use them whenever you need to extract specific information from a table whose first row or column is sorted in ascending order. Common examples include tax tables, inventory lists, directories, and catalogs.

Figure 15.25 shows an example of a horizontal lookup table that uses the HLOOKUP() function. It lists average values for apartments. The first row shows the number of rooms per apartment in ascending order, and the columns below list the specific data for apartments with different numbers of rooms. This means the data in cells B4 to B6 is for one-room apartments, the data in cells C4 to C6 is for to two-room apartments, and so on.

F7 =HLOOKUP(3,WOHNUNG,2)

	A	B	C	D	E	F
1	**Example: Horizontal reference table**					
2						
3	Rooms	1	2	3	4	5
4	Rent	300	600	800	1100	1400
5	Extra charges	80	150	200	220	240
6	Sq ft	470	846	1297	1711	2369
7	How much is the rent for 3-room apartments?					800

Figure 15.25 Horizontal Lookup Table

In this example, the entire data range is named *Apartment*. Now, if you want to retrieve the rent for the three-room apartments, the formula is as follows:

```
=HLOOKUP(3, Apartment, 2)
```

This function has three arguments:

- The lookup value (in this example, the number of rooms)
- The array, or table range (in this example, A3 to F6, which is named here as *Apartment*)
- The row index (in this example, 2 for the second row)

The *row index* indicates which row in the array provides the result. This depends on the information you want for the selected case. If you're interested in additional costs, the index would be 3. The function searches the first row from left to right for a value matching the lookup value, and once it finds the matching column, the cell pointer moves down to the row specified by the index. The function returns the cell content found there as its result.

This content can be either a numeric value or text. If no value in the row exactly matches the search criterion, then the function returns the value just before the next-higher one. For example, if the search criterion is 2.5 rooms, then the function returns the value for 2 rooms.

The `VLOOKUP()` function is used with vertically arranged tables. A typical example is a tax table, where the first column lists income levels in ascending order. For each income level, the tax amounts for various tax brackets appear in the columns to the right of the first column. To check what they owe, users first read the first column from top to bottom until they find an amount matching their income. Then, they move right along that row, column by column, until they reach the column for their tax bracket. That's exactly what the `VLOOKUP()` function does.

The simpler example depicted in Figure 15.26 is a currency exchange table. Typical euro amounts appear on the left, with corresponding dollar amounts in the columns to the right.

	A	B
1	**Vertical reference table**	
2		
3	Euro:	200.00
4	in Dollar:	223.22
5	Date	5/17/2025
6	EUR	Dollar
7	1	1.1161
8	10	11.1610
9	20	22.3220
10	50	55.8050
11	100	111.6100
12	200	223.2200
13	500	558.0500
14	1000	1116.1000

Figure 15.26 Vertical Lookup Table

In Figure 15.26, the entire data range is named *ExchangeRates*. Cell B3 contains a drop-down list for selecting euro amounts, and you set it up by using a data validation rule. This function can then answer in cell B4 how many dollars correspond to the selected euro amount. The formula is as follows:

```
=VLOOKUP(B3, ExchangeRates, 2)
```

As mentioned earlier, there's also a function that allows lookups in both directions: `XLOOKUP()`. It uses the following syntax:

`XLOOKUP(`**`lookup_value, lookup_array, return_array`**`, if_not_found, match_mode, search_mode)`

The first argument specifies the value to search for. Here, `lookup_array` is the range or array where the value should be found, and `return_array` provides the search results. All other arguments are optional: with `if_not_found`, you can display a message, and with `match_mode`, you can set the type of match. The various types of match are as follows:

- 0: This is an exact match; if none is found, the error value `#NA` appears (default).
- –1: This is an exact match or the next smaller item.
- 1: This is an exact match or the next larger item.
- 2: This is a wildcard match using the characters *, ?, and ~.

You can also set the search mode to determine how the search will be performed:

- 1: The search will start at the first item (default).
- –1: The search will start at the last item.
- 2: It will be a binary search, assuming the lookup array is sorted in ascending order.
- –2: It will be a binary search, assuming the lookup array is sorted in descending order.

The following example shows how, unlike with `VLOOKUP()`, you can search for a value in a column to the left of the lookup array. It also shows that the –1 `match type` finds the next smaller value when there's no exact match.

E4 =XLOOKUP(D4,B4:B8,A4:A8,"",-1,1)

	A	B	C	D	E	F	G
1	**Example of XLOOKUP() functions**						
2							
3	Gross wages	Tax		Tax	Gross wages		
4	3300	8.00%		10.00%	3700		
5	3500	9.00%					
6	3700	10.00%					
7	3900	11.00%					
8	4100	12.00%					

Figure 15.27 Example of the XLOOKUP() Function

15.9.3 Working with INDEX() Functions

Excel offers two versions of the `INDEX()` function:

- `=INDEX(`**`reference, row,`**` column, range)`
  We use this to query data tables.
- `=INDEX(`**`matrix, row,`**` column)`
  We use this to query a matrix.

The `INDEX()` function for data tables returns the value in the cell at the intersection of row n and column m within the `reference` range. The last argument, `range`, is used when `reference` contains a multiselection. For example, if three cell blocks are selected, a value of 2 for `range` means that the row and column numbers refer to the second cell block. If the `range` is not specified, the first block is always used.

The table in Figure 15.28 defines a range named *Production* that contains production results from several factories over four years. The cell in the third row and second column of this range (C6) contains the value 800,000.

E8 =INDEX(PRODUCTION,3,2)

	A	B	C	D	E
1	**Production results**				
2					
3		2021	2022	2023	2024
4	Factory 1	1,600,000	2,100,000	100,000	2,800,000
5	Factory 2	2,900,000	1,400,000	400,000	1,900,000
6	Factory 3	800,000	800,000	2,100,000	1,300,000
7	Factory 4	100,000	1,600,000	1,400,000	500,000
8			Production of Factory 3 in 2022:		800,000

Figure 15.28 Example of INDEX() Function

The INDEX() function works with the relative position of a cell within a defined range, not with the column and row labels of cell addresses. These column or row numbers can also be the results of a formula, and the function can also retrieve values from an array.

This can be a single value or an array, and you can enter the array argument as a range reference or an array constant. To retrieve not just a single value but an entire array, set the row and/or column value to 0 or leave it blank.

The following functions return the entire array:

```
=INDEX({3.5.9.7, 34.54.23.98}, 0, 0)
=INDEX({3.5.9.7, 34.54.23.98}, ,)
```

This function returns an array with the values from the second row:

```
=INDEX({3.5.9.7, 34.54.23.98}, 2)
```

And this function returns an array with the values from the second column:

```
=INDEX({3.5.9.7, 34,54.23.98}, , 2)
```

15.9.4 Example of the CHOOSE() Function

The CHOOSE() function returns a value from a list of up to 254 options. The syntax is as follows:

```
=CHOOSE(index, value1, value2, ... , value254)
```

The first argument specifies which element of the list to select. The index argument can be a number, reference, or formula, and the values can be numbers, cell references, names, text, or formulas. Range references aren't allowed because that's what the INDEX() function is designed to do.

Here's a practical example of using the CHOOSE() function: customers are divided into different discount levels. The discounts are listed in column E, a customer with discount level 1 receives the first discount from the list, a customer with discount level 2 gets the second discount, and so on. Discount levels are listed in column B, starting in cell B5. You start by creating the formula in cell C5, and you can write it as follows:

```
=CHOOSE(B5, E5, E6, E7)
```

Now, you need to determine the discount rates for the other customers. It makes sense to use the copy function, but you won't be satisfied with the result—an error will occur because the cell references for the discount rates were entered as relative. But the same discount list should apply to every customer, so the cells for the discounts must be referenced by absolute addresses.

You can correct the formula accordingly, like this:

```
=CHOOSE(B5, $E$5, $E$6, $E$7)
```

Then, you'll be able to copy the result without any problems.

Instead of cell references in the argument list, you can enter the percentages directly as constants in the formula, like this:

```
=CHOOSE(B56, 5%, 7%, 8%)
```

However, this would reduce flexibility. If the discount rate changes, you must update the formulas accordingly.

C5 =CHOOSE(B5,$E$5,$E$6,$E$7)

	A	B	C	D	E
1	**Example of the CHOOSE() function**				
3					
4	Customer	Discount level	Discount		Discounts
5	Customer A	2	7.00%		5.00%
6	Customer B	1	5.00%		7.00%
7	Customer C	3	8.00%		8.00%
8	Customer D	2	7.00%		
9	Customer E	1	5.00%		

Figure 15.29 Example of the CHOOSE() Function

The CHOOSE() function is especially useful for calculation alternatives. Different formulas can be used as arguments. Here's an example:

```
=CHOOSE(Dl, SUM(Items)*0.70, SUM(Items)*0.75, SUM(Items)*0.80+Shipping)
```

Depending on the value in Dl, the sum of the items is multiplied by a different factor, and in the third case, shipping costs are added. While the IF() function allows only a simple branch between two options—unless nested IF() functions are used—the CHOOSE() function lets you select from up to 254 options.

15.9.5 Lookup Functions

To find the position of a specific value or the closest match in a column or row, use the MATCH() or XMATCH() lookup function. The syntax is as follows:

```
=MATCH(lookup_value, lookup_array, match_type)
=XMATCH(lookup_value, lookup_array, match_mode, search_mode)
```

Here, lookup_value is the value you want to find, while lookup_array is the range or array to search. If exact matches aren't required, you should sort the array. The match_type and match_mode arguments use numbers from –1 to 2 to specify how the comparison is made, as follows:

- 0: This is an exact match; if none is found, the #NA error value appears (by default).
- –1: This is an exact match or the next smaller item.

- 1: This is an exact match or the next larger item.
- 2: This is a wildcard match using the characters *, ?, and ~ (only in `comparison` mode).

The `search_mode` argument determines the search sequence:

- 1: The search will start at the first item (default).
- –1: The search will start at the last item.
- 2: It will be a binary search, assuming the lookup array is sorted in ascending order.
- –2: It will be a binary search, assuming the lookup array is sorted in descending order.

The two examples in Figure 15.30 show the exact match case on the left, while the formula on the right shows the rank that would have been achieved with the specified time.

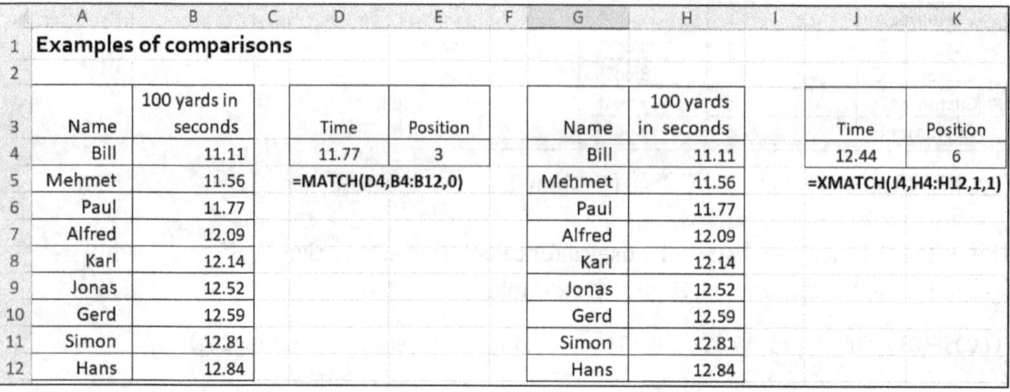

	A	B	C	D	E	F	G	H	I	J	K
1	**Examples of comparisons**										
2											
3	Name	100 yards in seconds		Time	Position		Name	100 yards in seconds		Time	Position
4	Bill	11.11		11.77	3		Bill	11.11		12.44	6
5	Mehmet	11.56		**=MATCH(D4,B4:B12,0)**			Mehmet	11.56		**=XMATCH(J4,H4:H12,1,1)**	
6	Paul	11.77					Paul	11.77			
7	Alfred	12.09					Alfred	12.09			
8	Karl	12.14					Karl	12.14			
9	Jonas	12.52					Jonas	12.52			
10	Gerd	12.59					Gerd	12.59			
11	Simon	12.81					Simon	12.81			
12	Hans	12.84					Hans	12.84			

Figure 15.30 Two MATCH Functions

15.9.6 Inserting Images as Cell Content

Besides the options to insert images into the worksheet that float above cells, the `IMAGE()` function allows you to fill cells directly with images. The `source` argument requires a URL pointing to the image file; the `alt_text`, `size`, `height`, and `width` arguments are optional. A value of `0` for size scales the image to fit the cell while preserving the aspect ratio. If the cell is enlarged, the image enlarges as well. A value of `1` also fills the cell but ignores the aspect ratio—while `2` maintains the image size, so parts of the image may be cut off. With `3`, the image size is adjusted using the height and width arguments. We already presented an example of this in Chapter 10, Section 10.5.3.

15.9.7 Additional Functions for Arrays

The `EXPAND()` function extends an existing array to a specified number of rows and columns, with the `pad` argument defining what fills the new cells. The default is `#NA`. Figure 15.31 indicates that the specified values for rows and columns represent the final size, not the number of additional elements. For example, if the row value is smaller than the existing row count, the function returns the #VALUE! error.

	A	B	C	D	E	F	G	H
1	**Expanding an Array**							
2								
3	2	4	8		=EXPAND(A3:C5,5,4)			
4	4	8	2		2	4	8	#N/A
5	3	5	7		4	8	2	#N/A
6					3	5	7	#N/A
7					#N/A	#N/A	#N/A	#N/A
8					#N/A	#N/A	#N/A	#N/A

Figure 15.31 Expanding an Array

Use the =HSTACK() and =VSTACK() functions to combine two matrices side by side or one below the other. If the arrays differ in height or width, empty cells in the resulting matrix will be filled with #NA.

	A	B	C	D	E	F	G	H	I	J	K	L	M	N	O	P	Q
1	**Stacking arrays, horizontal**																
2							=HSTACK(A3:B5,D3:E5)					=HSTACK(D3:E5,G3:J4)					
3	4	2		1	6		4	2	1	6		1	6	4	2	1	6
4	5	7		6	3		5	7	6	3		6	3	5	7	6	3
5	1	2		6	7		1	2	6	7		6	7	#N/A	#N/A	#N/A	#N/A
6																	
7	**Stacking arrays, vertical**																
8							=VSTACK(A9:B11,D9:E11)										
9	4	2		1	6		4	2									
10	5	7		6	3		5	7									
11	1	2		6	7		1	2									
12							1	6									
13							6	3									
14							6	7									

Figure 15.32 Combining Arrays

The =WRAPCOLS() and =WRAPROWS() functions fold a one-dimensional vector into a matrix with the specified number of elements per column or row.

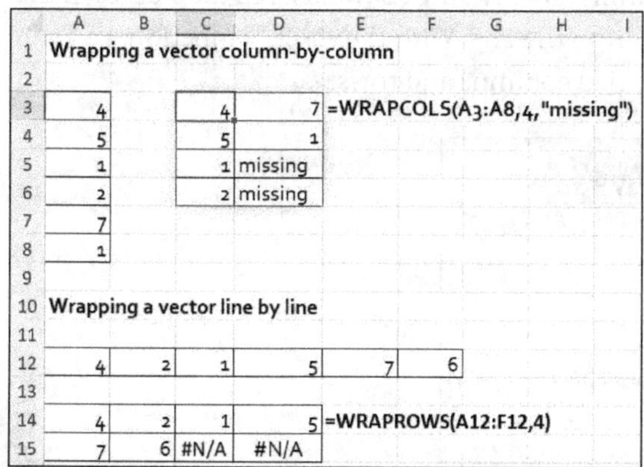

	A	B	C	D	E	F	G	H	I
1	**Wrapping a vector column-by-column**								
2									
3	4		4	7	=WRAPCOLS(A3:A8,4,"missing")				
4	5		5	1					
5	1		1	missing					
6	2		2	missing					
7	7								
8	1								
9									
10	**Wrapping a vector line by line**								
11									
12	4	2	1	5	7	6			
13									
14	4	2	1	5	=WRAPROWS(A12:F12,4)				
15	7	6	#N/A	#N/A					

Figure 15.33 Wrapping One-Dimensional Vector

The =CHOOSECOLS() and =CHOOSEROWS() functions let you extract a column or row from a matrix.

	A	B	C	D	E	F	G	H
1	Selects columns from an array							
2								
3	4	2		2	=CHOOSECOLS(A3:B5,2)			
4	5	7		7				
5	1	2		2				
6								
7	Selects rows of an array							
8					=CHOOSEROWS(A9:C10,2)			
9	4	2	1		5	7	6	
10	5	7	6					

Figure 15.34 Selecting Columns or Rows

With the =TOCOL() and =TOROW() functions, you can unfold a matrix into a column or row.

	A	B	C	D	E	F	G	H	I
1	Outputs an array as a column								
2									
3	4	2		4	=TOCOL(A3:B5)				
4	5	7		2					
5	1	2		5					
6				7					
7				1					
8				2					
9									
10	Outputs an array as a row								
11				=TOROW(A12:B14)					
12	4	2		4	2	5	7	1	2
13	5	7							
14	1	2							

Figure 15.35 Taking or Dropping Values from Matrix

With the =TAKE() function, you can extract the specified number of columns and rows from a matrix into a new matrix. Columns and rows are counted from the start or, if the function is preceded by a minus sign, from the end. With the =DROP() function, you can create a new matrix by dropping specified columns and rows.

	A	B	C	D	E	F	G	H
1	Returns columns or rows from an array							
2	4	2	1		4	2	=TAKE(A2:C4,2,2)	
3	5	7	6		5	7		
4	1	2	6					
5								
6	Omitting rows or columns of an array							
7	4	2	1		2	1	=DROP(A7:C9,-1,1)	
8	5	7	6		7	6		
9	1	2	6					

Figure 15.36 Taking or Dropping Values

15.9.8 Aggregating Values from Tables

With the GROUPBY() function, you can enable quick analysis of table values grouped together. The first argument specifies the column with the grouping criteria; the second specifies the cell range of values to aggregate; and the third specifies the summary type, such as SUM, AVERAGE, MAX, and so on. Optional arguments let you include the field description, display totals or subtotals, and apply sorting or filtering.

Group values

Artnr	Description	Category	Revenue
7777	Louvre Ccxs	Louvre	$120,000.00
5556	Roll BT 33	Rollo	$100,000.00
8444	Awning Blue Sk	Awning	$99,000.00
8443	Louvre VVx	Louvre	$88,000.00
8666	Roll Dark	Rollo	$110,000.00
6578	Louvre Louise	Louvre	$99,000.00
6666	Roll Top	Rollo	$70,000.00
5667	Awning Luxor	Awning	$60,000.00
5222	Roll XXs	Rollo	$80,000.00

=GROUPBY(C3:C12,D3:D12,SUM,,1)

Awning	$159,000.00
Louvre	$307,000.00
Rollo	$360,000.00
Total	$826,000.00

Figure 15.37 Summing Values Based on Grouping Criterion

15

The PIVOTBY() function aggregates values by rows and columns simultaneously, so you must first specify the relevant row and column fields.

Sales results

Representative	Product group	Region	2024
Hansen	Washing machines	East	120,000
Hansen	Refrigeration units	East	160,000
Gernot	Washing machines	East	180,000
Gernot	Refrigeration units	East	150,000
Schlier	Washing machines	West	120,000
Schlier	Refrigeration units	West	170,000
Gundar	Washing machines	West	140,000
Gundar	Refrigeration units	West	130,000
Seiffert	Washing machines	Nord	120,000
Seiffert	Refrigeration units	Nord	140,000
Adam	Washing machines	Nord	140,000
Adam	Refrigeration units	Nord	130,000
Karit	Washing machines	South	120,000
Karit	Refrigeration units	South	160,000
Lemgo	Washing machines	South	120,000
Lemgo	Refrigeration units	South	150,000

=PIVOTBY(sales_results[Product group],sales_results[Region],sales_results[2024],SUM)

	East	Nord	South	West	Total
Refrigeration units	310000	270000	310000	300000	1190000
Washing machines	300000	260000	240000	260000	1060000
Total	610000	530000	550000	560000	2250000

Figure 15.38 Example of PIVOTBY() Function

Both functions return results as dynamic arrays. The target range must be empty; otherwise, the functions will return the #SPILL! error.

15.9.9 List of Lookup and Reference Functions

Functions	Descriptions
ADDRESS(**row**, **column**, abs, a1, sheet_name)	Returns the address of the cell specified by *row* and *column*
OFFSET(**reference**, **rows**, **columns**, height, width)	Returns a range reference offset by a number of *rows* and *columns*
AREAS(**reference**)	Returns the number of ranges in *reference*
IMAGE(**source**, alt_text, size, height, width)	Inserts an image from the URL specified in the source
UNIQUE(**array**, by_column, exactly_once)	Returns the values in an array that occur only once
EXPAND(**matrix**, **rows**, columns, pad)	Expands an existing matrix to the specified number of rows and columns
FIELDVALUE(**value**, **field_name**)	Returns matching fields from the linked data type specified by *value*
FILTER(**matrix**, **include**, if_empty)	Filters data from a matrix based on the *include* criteria
FORMULATEXT(**reference**)	Returns the formula in the cell referenced by *reference* as a text string
GROUPBY(**row_fields**, **values**, **function**, field_headers, total_depth, sort_order, filter_array, field_relationship)	Groups data by specified criteria and summarizes it using the given aggregation function
HSTACK(**array1**, array2, ...)	Stacks arrays horizontally into a single array
HYPERLINK(**hyperlink_address**, friendly_name)	Creates a link to the specified address
INDEX(**reference**, row, column, area)	Returns a value from a range
INDEX(**array**, row, column)	Returns a value from an array
INDIRECT(**reference**, A1)	Returns the content of a cell referenced by another cell
TRANSPOSE(**array**)	Swaps the rows and columns in an array
GETPIVOTDATA(**data_field**, **pivot_table**, field1, item1, field2, item2, ...)	Returns values from a PivotTable

Functions	Descriptions
PIVOTBY(**row_fields**, **col_fields**, **values**, **function**, field_headers, row_total_depth, row_sort_order, col_total_depth, col_sort_order, filter_array, relative_to)	Aggregates values by rows and columns using the specified function
RTD(**prog_id**, **server**, **topic1**, topic2, ...)	Receives real-time data from a registered add-in
SORT(**array**, sort_index, sort_order, by_column)	Sorts data in a range or array
SORTBY(**array**, **by_array1**, sort_order1, by_array2, sort_order2,...)	Sorts data in a range or array based on data from other arrays
COLUMN(**reference**)	Returns the column number of a range
COLUMNS(**array**)	Returns the number of columns in a range or array
WRAPCOLS(**vector**, **wrap_count**, pad_with)	Splits a row or column vector into a new array with the specified number of values
CHOOSECOLS(**array**, **column_num1**, column_num2, ...)	Returns columns from an array or range
VLOOKUP(**lookup_value**, **table_array**, **col_index_num**, range_lookup)	Returns the value in the same row from another column, based on a cell in an array
TAKE(**array**, **rows**, columns)	Returns rows or columns from the start or end of an array
MATCH(**lookup_value**, **lookup_array**, match_type)	Searches a *lookup_array* or range for a *lookup_value* and returns its relative position
LOOKUP(**lookup_value**, **array**)	Searches the values in an array
LOOKUP(**lookup_value**, **lookup_vector**, **result_vector**)	Searches the values in a vector
VSTACK(**array1**, array2, ...)	Stacks arrays vertically into a single array
CHOOSE(**index**, **value1**, value2, ...)	Returns a value from a list of *values*
DROP(**array**, **rows**, columns)	Removes rows or columns from the start or end of an array
HLOOKUP(**lookup_value**, **array**, **row_index**, range_lookup)	Returns the content of the cell in the same column, based on a cell within an array
XMATCH(**lookup_value**, **lookup_array**, match_mode, search_mode)	Returns the relative position of an item within an array

Functions	Descriptions
XLOOKUP(**lookup_value**, **lookup_array**, return_array, if_not_found, match_mode)	Returns the corresponding cell content from a second range or array, based on a cell in the first range or array
ROW(**reference**)	Returns the row number of a reference
ROWS(**array**)	Returns the number of rows in a range or array
WRAPROWS(**vector**, **wrap_count**, pad_with)	Splits a row or column vector into a new array with the specified number of values
CHOOSEROWS(**array**, **column_num1**, column_num2, ...)	Returns rows from an array or range
TOCOL(array, ignore, scan_by_column)	Returns the array as a column
TOROW(**array**, ignore, scan_by_column)	Returns the array as a row

15.10 Database Functions

Besides convenient filtering options, Excel offers several functions for managing data tables, mainly for statistical analysis of datasets. The main difference between database functions and statistical functions is that in database functions, you can define specific criteria for the analysis.

15.10.1 Analyzing a Table

To simplify the explanation of each function, we'll use the small data table in Figure 15.39 as the "database."

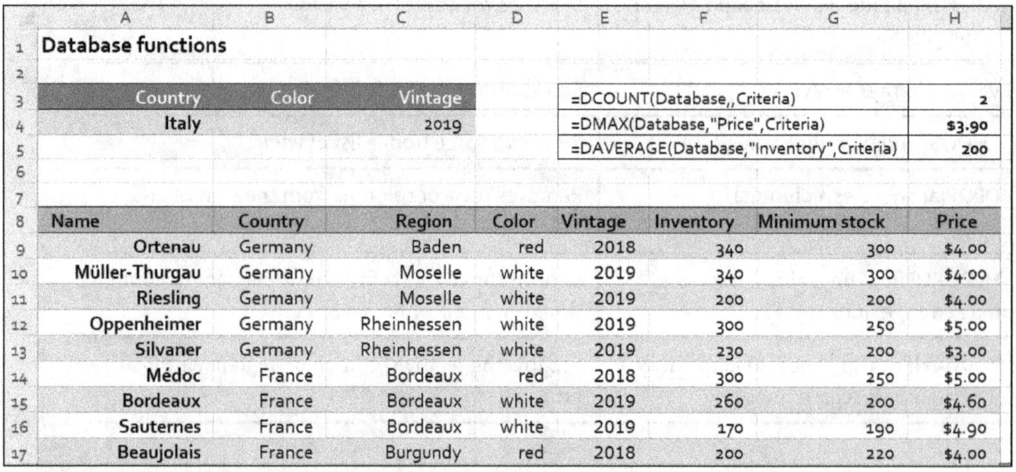

	A	B	C	D	E	F	G	H
1	**Database functions**							
2								
3	Country	Color	Vintage		=DCOUNT(Database,,Criteria)			2
4	Italy		2019		=DMAX(Database,"Price",Criteria)			$3.90
5					=DAVERAGE(Database,"Inventory",Criteria)			200
6								
7								
8	Name	Country	Region	Color	Vintage	Inventory	Minimum stock	Price
9	Ortenau	Germany	Baden	red	2018	340	300	$4.00
10	Müller-Thurgau	Germany	Moselle	white	2019	340	300	$4.00
11	Riesling	Germany	Moselle	white	2019	200	200	$4.00
12	Oppenheimer	Germany	Rheinhessen	white	2019	300	250	$5.00
13	Silvaner	Germany	Rheinhessen	white	2019	230	200	$3.00
14	Médoc	France	Bordeaux	red	2018	300	250	$5.00
15	Bordeaux	France	Bordeaux	white	2019	260	200	$4.60
16	Sauternes	France	Bordeaux	white	2019	170	190	$4.90
17	Beaujolais	France	Burgundy	red	2018	200	220	$4.00

Figure 15.39 Example of Querying a Data Table with Database Functions

The top three-row range from A3 to C5 is the area for the search criteria. If you name this range *SearchCriteria*, you can avoid specifying the range by cell addresses when entering functions. The range has two rows beneath the column names to show that you can enter search criteria in one or multiple rows. You'll need multiple rows when creating criteria linked by *or* (for example, records where the color is "red" or "white").

The large range that's formatted as a table is the database range. Like the criteria range, it contains the field names. At the top right are examples of analyses that use database functions, and these functions have the same meaning as the general statistical functions.

15.10.2 List of Database Functions

Functions	Descriptions
DCOUNT(**database, database_field, criteria**)	Counts the number of records matching the criteria
DCOUNTA(**database, database_field, criteria**)	Counts the number of records matching the criteria, including cells with text or logical values
DGET(**database, database_field, criteria**)	Returns the content of the specified field from the record matching the criteria
DMAX(**database, database_field, criteria**)	Returns the largest value in a database field from records that meet the specified criteria
DMIN(**database, database_field, criteria**)	Returns the smallest value in a database field from records that meet the specified criteria
DAVERAGE(**database, database_field, criteria**)	Returns the average of all values in a database field from records that meet the specified criteria
DPRODUCT(**database, database_field, criteria**)	Multiplies all values in a database field from records that meet the search criteria
DSTDEV(**database, database_field, criteria**)	Calculates the standard deviation of the database field for records that meet the search criteria (records treated as a sample)
DSTDEVP(**database, database_field, criteria**)	Calculates the standard deviation of the database field for records that meet the search criteria (records treated as the entire population)
DSUM(**database, database_field, criteria**)	Sums all values in a database field for records that meet the search criteria
DVAR(**database, database_field, criteria**)	Calculates the variance of values in a database field for records that meet the search criteria (records treated as a sample)

Functions	Descriptions
DVARP(**database, database_field, criteria**)	Calculates the variance of values in a database field for records that meet the search criteria (records treated as the entire population)

15.11 Cube Functions

Since version 7, Microsoft SQL Servers have supported online analytical processing (OLAP) through Server Analysis Services. This analysis method evaluates large volumes of corporate data by consolidating existing data from relational tables into a multidimensional data structure, called *cubes*, that are separate from regular SQL database transactions. These data cubes organize data across various dimensions (time, space, facts, etc.) to answer questions that are typically addressed in Excel with PivotTables.

Excel provides seven specialized functions you can use to analyze data offered by such a cube from a SQL Server. Alternatively, you can imported data offline from cube files that were previously generated from server data. Since Excel 2013, cube functions have also been applied to Excel data models. In this case, the functions use the predefined connection name "ThisWorkbookDataModel."

15.11.1 Special Features of Cube Functions

Unlike other Excel functions, these cube functions—except for `CUBESETCOUNT()`—return two results: one that's displayed in the cell range and an internal result that's used when a cube function is an argument for another cube function. For example, the `CUBESET()` function returns an internal value that the `CUBESETCOUNT()` function can then evaluate. Except for `CUBESETCOUNT()`, all cube functions use the text string that connects to the cube or cube file as their first argument. Data selection uses multidimensional expressions in MDX, which is the language for querying OLAP databases. You can find a reference in the Microsoft SQL Server documentation.

Figure 15.40 shows a simple example of an evaluation from a cube file that's available in the sample files as *Adventure Works DW2017.cub*.

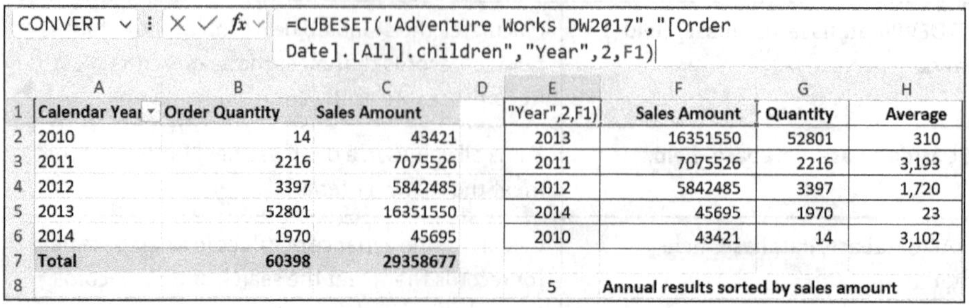

```
CONVERT  =CUBESET("Adventure Works DW2017","[Order Date].[All].children","Year",2,F1)
```

	A	B	C	D	E	F	G	H
1	Calendar Year	Order Quantity	Sales Amount		"Year",2,F1)	Sales Amount	Quantity	Average
2	2010	14	43421		2013	16351550	52801	310
3	2011	2216	7075526		2011	7075526	2216	3,193
4	2012	3397	5842485		2012	5842485	3397	1,720
5	2013	52801	16351550		2014	45695	1970	23
6	2014	1970	45695		2010	43421	14	3,102
7	Total	60398	29358677					
8					5	Annual results sorted by sales amount		

Figure 15.40 Example of Using Cube Functions

15.11.2 List of Cube Functions

Functions	Descriptions
CUBEMEMBER(**connection**, **member_expression**, caption)	Returns a member or tuple from a cube
CUBEMEMBERPROPERTY(**connection**, **member_expression**, **property**)	Returns the value of a property of a member in the cube
CUBEKPIMEMBER(**connection**, **kpi_name**, **kpi_property**, Caption)	Returns a property of a key performance indicator (KPI)
CUBESET(**connection**, **set_expression**, caption, sort_order, sort_by)	Sends a set expression for a specific group of members or tuples to the cube, which compiles the data and delivers it to Excel
CUBESETCOUNT(**set**)	Returns the count of members in a defined set
CUBERANKEDMEMBER(**connection**, **set_expression**, **rank**, caption)	Returns the item in a set specified by *rank*
CUBEVALUE(**connection**, **member_expression1**, member_expression2 ...)	Returns an aggregated value from a cube

15

15.12 Text Functions

Text functions are those that are unrelated to calculations. For example, you can use them to combine labels from multiple cell values or change their formatting. Alternatively, you can use them to modify text or convert between text and numeric values.

15.12.1 Extracting Parts of Strings

A small group of additional text functions enhances string handling. The new functions TEXTAFTER() and TEXTBEFORE() allow you to extract parts of text strings. Figure 15.41 shows simple examples of =TEXTBEFORE() and =TEXTAFTER().

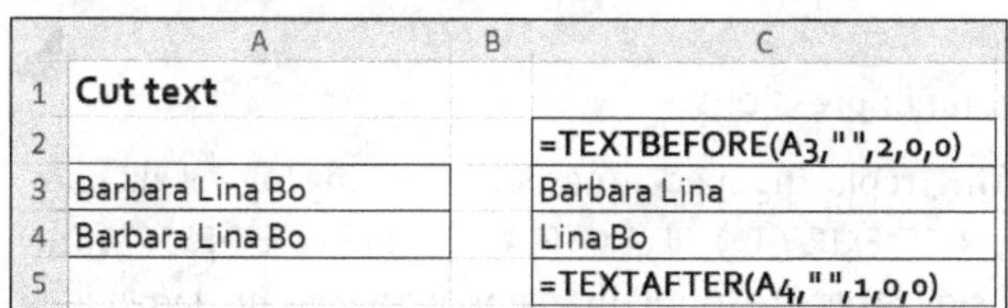

A
B
C
1 Cut text
2 =TEXTBEFORE(A3," ",2,0,0)
3 Barbara Lina Bo
Barbara Lina
4 Barbara Lina Bo
Lina Bo
5 =TEXTAFTER(A4," ",1,0,0)

Figure 15.41 Examples of TEXTBEFORE() and TEXTAFTER()

The TEXTBEFORE() and TEXTAFTER() functions require a reference to the text string, followed by the delimiter before or after which the extraction should occur. Optionally, you can specify the instance. The default is 1, and it refers to the first occurrence of the delimiter, while -1 starts the search from the end. The third argument specifies whether to consider case sensitivity (0 = default) or not (1). The third, optional element determines whether the end of the text is treated as a delimiter (0 = default) or not (1). The last argument, if_not_found, lets you specify a string to return if nothing is found.

The =TEXTSPLIT() function splits parts of a text string across columns in a row, down rows in a column, or both at once. It's an alternative to the **Text to Columns** command on the **Data** tab under **Data Tools**. The function returns a dynamic array.

The first argument is the text to split, and the second or third argument specifies the character where overflow to the next column or row (which is optional) occurs. If you want to split at multiple delimiters, you provide them as an array constant. Also, =TEXTSPLIT(A1,{",","."}) splits at commas and periods.

The logical value TRUE for the fourth argument ignores consecutive delimiters. The default, however, is FALSE, which returns an empty cell if there is nothing between two delimiters. The fifth argument specifies whether to consider case sensitivity (0 = default) or not (1). The last argument specifies a value to fill in if a value is missing.

To spread data among multiple columns and rows simultaneously, two delimiters are used, as shown in the second example in Figure 15.42.

	A	B	C	D	E	F
1	**Split strings**					
2			**=TEXTSPLIT(A3,{",";" "},,TRUE)**			
3	Boener, Rosa 7/6/2010 Buffalo		Boener	Rosa	7/6/2010	Buffalo
4	Hansen, Sabine 12/12/2012 Paris		Hansen	Sabine	12/12/2012	Paris
5	Duerr, Max 10/8/1999 Berlin		Duerr	Max	10/8/1999	Berlin
6						
7						
8			**=TEXTSPLIT(A9,":",",",TRUE,1,"fehlt")**			
9	Test1: 2000,Test2:1900,Test3:,Test4:2100		Test1	2000		
10			Test2	1900		
11			Test3	fehlt		
12			Test4	2100		

Figure 15.42 Examples of Splitting into Rows and Columns

15.12.2 Operations with Regular Expressions

Extracting characters from a string, replacing specific parts, and testing for certain characters is now possible with the REGEXEXTRACT(), REGEXREPLACE(), and REGEXTEST() functions, which use regular expressions that are found in many programming languages. The first argument, text, specifies the string to process, while the key argument, pattern, is a pattern expression that follows the rules for regular expressions. Commonly used elements of regular expressions include the following:

- `"[0-9]"`
  This is for any numeric digit.
- `"[a-z]"`
  This is for a character in the range a to z.
- `"[A-Z]"`
  This is for a character in the range A to Z.
- `"."`
  This is for any character.
- `"a"`
  This is for the “a” character
- `"a*"`
  This is for zero or more “a” characters.
- `"a+"`
  This is for one or more “a” characters.
- `"\d"`
  This is for any numeric digit.
- `"\w"`
  This is for a word character (i.e., any digit, letter, or underscore).

The possible expressions are defined in the Perl Compatible Regular Expressions, Version 2 (PCRE2) open-source library. Details are available at *https://github.com/PCRE2Project/pcre2*.

The `return_mode` argument specifies which strings to extract. The default is 0, but the following values are possible:

- 0: This returns the first text string that matches the pattern.
- 1: This returns all text strings matching the pattern as a dynamic array.
- 2: This returns capture groups as a dynamic array. *Capture groups* come from the first match of the specified patterns, and they're parts of a regular expression pattern enclosed in quotation marks and parentheses [`"(...)"`]. This allows individual parts of a match to be returned separately.

The `case_sensitive` argument determines whether case sensitivity is applied (0 is the default) or not (1).

Figure 15.43 shows examples of text extraction. In the first case, a substring with one uppercase letter followed by several lowercase letters is extracted. In the second case, the matches found are distributed across multiple cells. In the third case, only the numbers are extracted, and in the fourth case, three capture groups are distributed across three cells.

	A	B	C	D
1	**Extract Strings**			
2				
3		=REGEXEXTRACT(A4,"[A-X][a-x]+",0,0)		
4	Martha Gast-Brehm	Martha		
5		=REGEXEXTRACT(A4,"[A-X][a-x]+",1,0)		
6		Martha	Gast	Brehm
7				
8		=REGEXEXTRACT(A9,"[0-9()]+ [0-9-]+ [0-9-]+",0)		
9	Tom Heese (44) 6767 545643	(44) 6767 545643		
10	Tony Krass (41) 6464 75754	(41) 6464 75754		
11				
12		=REGEXEXTRACT(A13, "(\d{3})-(\d{3})-(\d{4})",2)		
13	021-444-7722	021	444	7722

Figure 15.43 Substrings Can Be Extracted or Spread Across Multiple Cells

When you're using the REGEXREPLACE() function, the third required argument specifies what to replace the found pattern match with. Specifying the occurrence is optional and applies when the pattern match appears multiple times.

	A	B	C	D	E
1	**Replace Strings**				
2					
3	785339-8777	***-8777	=REGEXREPLACE(A3,"[0-9]+-","***-")		
4	66232-79567	***-79567			

Figure 15.44 Example of How Parts of Data Can Be Hidden

The REGEXTEST() function simply checks whether the string matches the specified pattern and returns the logical values TRUE or FALSE accordingly.

	A	B	C	D	E
1	**Testing Strings**				
2					
3	Uppercase	TRUE	=REGEXTEST(A3,"[A-Z]+",0)		

Figure 15.45 This Checks Whether Uppercase Letters Are Present

15.12.3 Concatenating Strings

Sometimes, it's useful to create a string in a cell by using a formula that concatenates different characters or strings. Suppose you want to expand item numbers by two characters that represent the product group. If you enter "=C5&C8" in cell C9, it concatenates the contents of cells C5 and C8. If C5 contains the product group code PX and C8 contains the item number 3370086, then the result in cell C9 will be PX3370086. If the existing item number in C8 is entered as a number instead of text, Excel automatically converts the number to text. To insert a space between the two text elements, enter "=C5&" "&C8" in the cell.

You can achieve the same result by using the `=CONCATENATE(C5," ",C8)` function, which combines text strings provided as arguments into one string. It accepts up to 255 arguments. Constants must be enclosed in quotation marks, and if an argument is a cell reference to a number, the number's format resets to the default. For example, 1,000 EUR becomes 1000.

To preserve the format, apply the function to the currency amount first. Use `TEXT()` to convert a value into a string that preserves the format: `=TEXT(1000,"#.##0 [$EUR]")`. The same applies when combining text with date or time values. If cell C7 contains the word *Appointment* and C8 contains the date 12/12/2025, then `=CONCATENATE(C7," ",C8)` returns `Appointment` 46003, showing only the date's serial number. To fix this, enter "=TEXT(DATEVALUE("12/12/2025"),"MM/DD/YYYY")" in cell C8.

Starting with Excel 2019, `CONCATENATE()` has been replaced by the `CONCAT()` function. It uses the same syntax as `CONCATENATE()`. The `CONCATENATE()` function is still available under the **Compatibility** category.

F4 =CONCAT(A4,"-",B4,"-",C4,"-",D4,"-",E4)

Concatenating Strings

Prrfix	Country/language code	Publisher number	Title number	Check digit	ISBN
978	3	89842	736	6	978-3-89842-736-6
978	0	471	40261	3	978-0-471-40261-3
978	0	471	45380	2	978-0-471-45380-2
978	0	596	00382	X	978-0-596-00382-X
978	3	8274	1631	5	978-3-8274-1631-5

Figure 15.46 Example of Text String

The `TEXTJOIN()` function was also introduced with Excel 2019. The first argument is a delimiter, and the second is a logical value for *ignore_empty* that, if `TRUE`, causes empty cells to be ignored. You can then specify up to 252 text arguments—strings or cell references. The formula from the last example that used `CONCAT()` can be simplified with this function as follows:

```
=TEXTJOIN("-", TRUE, A2:E2)
```

Figure 15.47 shows an example in which country names from a column are concatenated with commas.

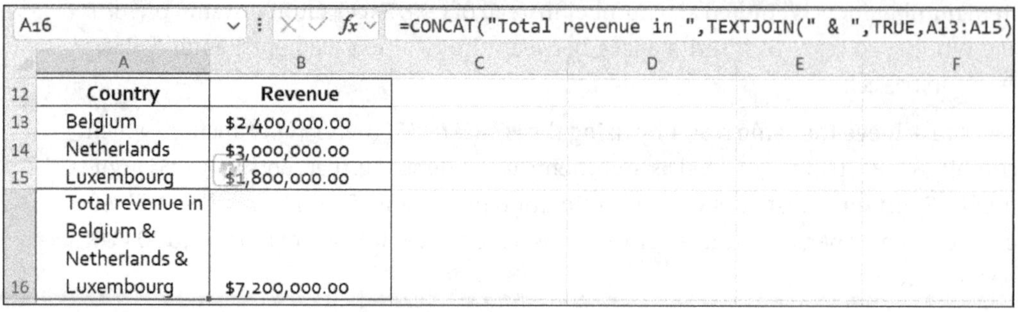

	A	B	C	D	E	F
12	Country	Revenue				
13	Belgium	$2,400,000.00				
14	Netherlands	$3,000,000.00				
15	Luxembourg	$1,800,000.00				
16	Total revenue in Belgium & Netherlands & Luxembourg	$7,200,000.00				

A16 =CONCAT("Total revenue in ",TEXTJOIN(" & ",TRUE,A13:A15)

Figure 15.47 Concatenating Country Names from Range

15.12.4 Sorting with Text Functions

If product descriptions always include a product group code in the first or last two characters, you can extract those characters with a text function to sort a list accordingly.

This function returns the last two characters:

```
=RIGHT(C9, 2)
```

And this function returns the first two characters:

```
=LEFT(C9, 2)
```

15.12.5 Including Logical Values in Text

A cell that contains a logical value can also be concatenated with another cell's content into a text string.

This function produces the phrase *The statement ... is FALSE* if the condition in C9 is not met:

```
="The claim that sales are increasing is " & C9
```

15.12.6 Combining Text with a Date

Combining text with a date is a bit more challenging. When you reference a date field, Excel returns the serial number that represents the date internally, instead of the actual date. You can resolve this issue by using other date functions. For example, if you use a daily form with the header *Billing from ...*, you can enter the following formula in a cell:

```
="Billing from "&MONTH(TODAY())&"/"&DAY(TODAY())&"/"&YEAR(TODAY())
```

The form will then print automatically with the current date.

15.12.7 Detecting Languages and Translating

Excel's linked translation services let you use two additional text functions: DETECTLANGUAGE() identifies the language of a text and returns the country code based on the ISO 3166-1 standard, while the TRANSLATE() function translates it. For the latter function, you must provide the text argument along with the country codes for the source and target languages. Possible codes are suggested as you enter the function.

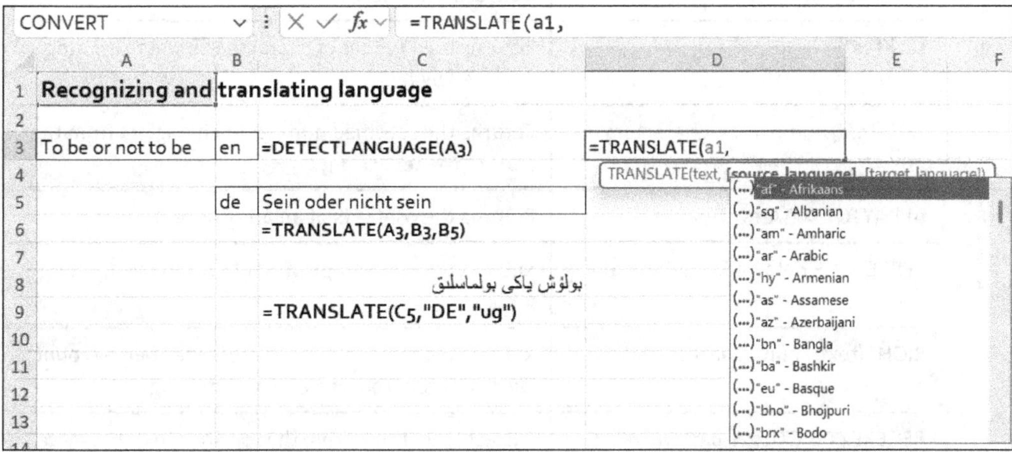

Figure 15.48 Detecting and Translating

15.12.8 List of Text Functions

Functions	Descriptions
BAHTEXT(**number**)	Converts a number to text in the Thai currency format
CODE(**text**)	Returns the code of the first character in the string
DOLLAR(**number**, decimal_places)	Converts the value to a text string in the current currency format
REPLACE(**Oold_text**, **start_character**, **number_of_characters**, **new_text**)	Replaces a specified number of characters in a text string, starting at a given position, with new characters
FIXED(**number**, decimal_places, no_commas)	Converts a numeric value to a text string
FIND(**find_text**, **within_text**, start_character)	Checks whether the text string specified by *find_text* appears within another text string
TRIM(**text**)	Removes leading and extra spaces
UPPER(**text**)	Converts all letters in the text string specified by *text* to uppercase

Functions	Descriptions
PROPER(**text**)	Converts the first letter to uppercase and the rest to lowercase
EXACT(**text1,text2**)	Compares two text strings to determine if they match exactly
LOWER(**text**)	Converts all letters to lowercase
LEN(**text**)	Returns the number of characters in the text string specified by *text*
LEFT(**text**, number_of_characters)	Returns the specified number of characters from the start of the text string
ARRAYTOTEXT(**Array**, Format)	Returns the contents of an array as text
PERCENTOF(data_subset, data_all)	Returns the percentage of a subset relative to the whole
RIGHT(**text**, count_char)	Returns the number of characters specified by *count_char*, counted from the end of the string
REGEXEXTRACT(**text**, **pattern**, return_mode, case_sensitivity)	Extracts substrings from the specified text that match the pattern
REGEXREPLACE(**text**, **pattern**, **replacement**, occurrence, case_sensitivity)	Replaces substrings in the specified text that match the pattern
REGEXTEST(**text**, **pattern**, case_sensitivity)	Checks whether the string matches the pattern
CLEAN(**text**)	Removes all nonprintable characters from the string provided in the *text* argument
DETECTLANGUAGE(**text**)	Returns the two-letter country code for the specified text
SEARCH(**find_text**, **within_text**, start_num)	Checks whether the text string specified by *find_text* appears within another text string
T(**Value**)	Checks whether the *value* argument returns a number or a text string
MID(**text**, **start_num**, **num_chars**)	Extracts a substring from the text string
TEXT(**value**, **Format_text**)	Converts a numeric value to text
CONCAT(**text1**, text2, ...)	Joins text strings together

Functions	Descriptions
TEXTAFTER(**text**, **Delimiter**, Instance_num, Match_mode, Match_end, If_not_found)	Returns the text following the delimiter
TEXTSPLIT(**text**, **col_delimiter**, row_delimiter, ignore_empty, match_mode, pad_with)	Splits text into rows or columns using delimiters
TEXTJOIN(**delimiter, ignore_empty, text1, ...**)	Joins text strings, inserting the specified delimiter between them
TEXTBEFORE(**text**, **delimiter**, instance_num, match_mode, match_end, if_not_found)	Returns the text preceding the delimiter
TRANSLATE(**text**, source_language, target_language)	Translates the specified text from the source language to the target language
UNICODE(**text**)	Returns the code for the first character in *text*
UNICHAR(**number**)	Returns the Unicode character for the specified code
SUBSTITUTE(**text**, **old_text**, **new_text**, nth_occurrence)	Replaces old text with new text within the string
VALUE(**text**)	Converts the string to a numeric value if it represents a valid number
VALUETOTEXT(**value**, format)	Returns any value as text
REPT(**text**, **Multiplier**)	Repeats the specified string in the cell n times
NUMBERVALUE(**text**, decimal_separator, group_separator)	Converts text to numbers regardless of the current locale
CHAR(**number**)	Returns the character corresponding to the code number

15.13 Logical Functions

Logical functions check whether specific facts or conditions are true or false. A logical formula can include multiple conditions simultaneously. These conditions can be combined in two ways: alternatively or conjunctively. *Alternative combinations* use the `OR()` or `XOR()` functions, which require at least or exactly one condition to be true. *Conjunctive combinations* use the `AND()` function, which requires all conditions to be true. Here are some examples:

```
=AND(B6>B7, B9>B10)
=AND(A10>50, A10<100)
```

In these cases, the condition is TRUE only if both subconditions are met.

```
=OR(MONTH1="May", MONTH2="June", MONTH3="Oct")
=XOR(MONTH1="May", MONTH2="June", MONTH3="Oct")
```

The first formula returns TRUE if at least one of the three month cells contains the specified month; the second formula returns TRUE only if at most one of the month cells contains the specified month.

	A	B	C	D
1	**The difference between OR() and XOR()**			
2				
3	Month1	April	TRUE	=OR(Month1="April",Month2="Juni",Month3="Okt")
4	Month2	Juni	FALSE	=XOR(Month1="April",Month2="Juni",Month3="Okt")
5	Month3	July		

Figure 15.49 Difference Between OR() and XOR(), Illustrated with Example

The NOT() function negates a comparison. For example, the following condition is true whenever A1 <> 100:

```
=NOT(A1=100)
```

Therefore, the following formulas are interchangeable:

```
=A1<>100
=NOT(A1=100)
```

Logical comparisons are often performed with the IF() function, but sometimes, it's more practical to first evaluate a complex condition in a cell and then use the IF() function to reference that cell's value, rather than embedding the complex condition directly in the IF() function. This is especially helpful when the condition must be checked in multiple places.

15.13.1 TRUE or FALSE as Arguments

To indicate whether an IF() function's condition is met, you can use the TRUE() and FALSE() functions as arguments. This approach is useful when it's not immediately clear whether a condition is met. Here's an example:

```
=IF(Profit01>500000, TRUE(), FALSE())
```

The result is FALSE if Profit01 = 300,000.

15.13.2 Checking Conditions

The IF() function checks whether a specific condition is TRUE. If it is, the TRUE expression is evaluated and determines the formula's result. If not, the function evaluates the FALSE expression. The function requires three arguments to be entered in the correct order to avoid unwanted results:

- **logical_test**
  This is a condition expressed as a logical expression.
- **value_if_true**
  This is the expression you use to evaluate whether the condition is met.
- **value_if_false**
  This is an expression that's evaluated when the condition is not met.

The IF() function is generally used for two distinct purposes:

- It lets you perform checks. For example, it can determine whether certain limits have been exceeded, there are deviations from a standard, or specific target values have been achieved.
- Excel uses the IF() function to perform different operations, depending on whether a condition is true or false. This is similar to branching, which is a fundamental operation in all programming languages.

  However, the branching only affects the result within the cell itself. You cannot use the IF() function to directly enter a new value into another cell. For example, you might write the following into cell B3 because you are used to using the = as an assignment to a variable in programming languages:

  ```
  =IF(B1>1000, B2=100, B2=200)
  ```

  But if B1 is greater than 1,000, this formula only checks whether cell B2 contains the value 100. If it does, the result in cell B3 will be TRUE.

You can only assign a value directly to a cell, potentially overwriting existing data, by using macros. This doesn't prevent another cell from referencing the value in the cell with the IF() function. For example, say you enter this into cell B3:

```
=IF(B1>1000, 100, 200)
```

And say you also write this into cell B2:

```
=B3
```

Then, you'll get the result you intended with the first formula.

15

15.13.3 Checking Multiple Conditions

Excel provides the IFS() function, which allows multiple branching. It takes pairs of logical tests and the values to return if the test is true. Up to 127 such pairs are allowed. For demonstration purposes, Figure 15.50 includes a small table that determines which meteorological season a column of date values falls into. The formula evaluates the month of the date:

```
=IFS(AND(MONTH(A5)>2,(MONTH(A5)<6)),"Spring",
     AND(MONTH(A5)>5,(MONTH(A5)<9)),"Summer",
     AND(MONTH(A5)>8,(MONTH(A5)<12)),"Autumn",
      OR(MONTH(A5)>11,(MONTH(A5)<3)),"Winter")
```

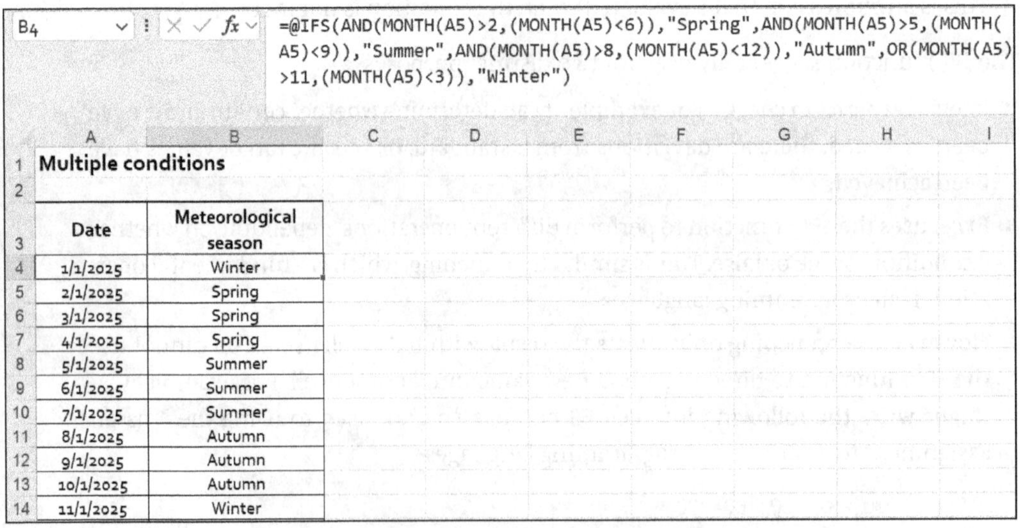

B4 =@IFS(AND(MONTH(A5)>2,(MONTH(A5)<6)),"Spring",AND(MONTH(A5)>5,(MONTH(A5)<9)),"Summer",AND(MONTH(A5)>8,(MONTH(A5)<12)),"Autumn",OR(MONTH(A5)>11,(MONTH(A5)<3)),"Winter")

Multiple conditions

Date	Meteorological season
1/1/2025	Winter
2/1/2025	Spring
3/1/2025	Spring
4/1/2025	Spring
5/1/2025	Summer
6/1/2025	Summer
7/1/2025	Summer
8/1/2025	Autumn
9/1/2025	Autumn
10/1/2025	Autumn
11/1/2025	Winter

Figure 15.50 Example of Using IFS() Function

The SWITCH() function works in a very similar way, but the syntax is slightly different. First, an expression is evaluated, and then, its result is compared to a series of values. If the result matches one of these values, the corresponding result is returned:

```
=SWITCH(expression, value1, result1, default_or_value2, result2, … default_or_value126, result126)
```

If no match is found, you can specify a default value as the last argument; otherwise, the function returns the error value #NA. In the example shown in Figure 15.51, the month value of the date in column A is evaluated, and month numbers are assigned their corresponding abbreviated month names.

	A	B	C	D	E	F	G	H	I	J
B5	=@SWITCH(MONTH(A5),1,"Jan",2,"Feb",3,"Mar",4,"Apr",5,"May",6,"Jun",7,"Jul",8,"Aug",9,"Sep",10,"Oct",11,"Nov",12,"Dec","no correct									
1	**Multiple choice**									
2										
3	Date	Month								
4	1/1/2022	Jan								
5	2/1/2022	Feb								
6	3/1/2022	Mar								
7	4/1/2022	Apr								
8	5/1/2022	May								
9	6/1/2022	Jun								
10	7/1/2022	Jul								
11	8/1/2022	Aug								
12	9/1/2022	Sep								
13	10/1/2022	Oct								
14	11/1/2022	Nov								
15	12/1/2022	Dec								

Figure 15.51 SWITCH() Function Returns Abbreviated Month Names

15.13.4 Automatically Adjusting Text

The following table shows customer sales for a company over two years, with 2023 sales in column B and 2024 sales in column C. Customers whose sales have dropped more than 2% compared to the previous year should be marked in column D for quick identification.

Enter the formula in cell D7 and copy it down the column. The formula could be as follows:

```
=IF(C7<(B7*0.98), "significant decline", "")
```

For customers whose sales dropped by more than 2% in 2023, the note *significant decline* appears in column D. For all other customers, the cell in column D remains empty as a result of the formula.

	A	B	C	D
1	**Checking conditions**			
2				
3		2023	2024	
4	Customer A	$100,000	$98,800	
5	Customer B	$120,000	$110,000	significant decline
6	Customer C	$87,000	$89,000	
7	Customer D	$250,000	$240,000	significant decline
8				=IF(C4<(B4*0.98),"significant decline","")

Figure 15.52 Checking Conditions

While the previous example focused on analyzing customer sales trends, the next formula demonstrates how the IF() function can respond differently to various conditions. The company might consider tying the year-end bonuses to sales performance:

Customers with sales over $250,000 would receive a 0.5% bonus, while others would receive nothing. The formula for this case is as follows:

```
=IF(C5>250000, C5*0.05, 0)
```

When writing conditions and instructions for `TRUE` and `FALSE`, you can use any valid combination of operators and functions.

15.13.5 Conditional Text Display

Is it possible to print text in a worksheet only under specific conditions? The `IF()` function works well here. Imagine that the worksheet is used to prepare and print invoices. On December invoices, a note about a special Christmas offer should appear in the invoice footer. The solution could be this formula:

```
=IF(MONTH(NOW())=12, "Special 20% discount on all Oriental rugs", "")
```

15.13.6 Text Checking

Of course, you can use the `IF()` function to check more than just numeric cells. You can evaluate cells that contain text or strings, and the text may be entered directly, referenced from another cell, or returned as the result of a formula that outputs a string. Here's a simple example:

```
=IF(B10="Berlin", "Capital", "")
```

However, string comparisons can sometimes lead to unexpected results. The condition that cell B10 contains the name *Berlin* is only true if it contains exactly the word *Berlin*. A common, hard-to-spot error in Excel is the presence of extra spaces. If the spacebar was accidentally pressed again after the word *Berlin*, Excel will treat the content of cell B10 as if it were not *Berlin*. Strictly speaking, that's true, but the difference only appears when you press [End] in the formula bar. However, you can catch this error by using the `=TRIM(B10)` function. The following formula is protected against this specific potential error:

```
=IF(TRIM(B10)="Berlin"...
```

The `TRIM()` function is a text function that removes extra spaces.

15.13.7 Checks with Complex Conditions

The `logical_test` argument can consist of multiple individual conditions. In many cases, an operation depends on multiple conditions. For example, a product component should only be ordered from a specific supplier if the price is acceptable and the delivery time is no more than one month. The formula for this could be as follows:

```
=IF(AND(C12<12500, D12<4), "order", "do not order")
```

Cell C12 contains the price, cell D12 contains the delivery time in weeks, and the condition is met only if both parts are true simultaneously.

In other cases, a decision depends on whether one condition or another is met. In this example, a higher price might be acceptable if the delivery time is short. A small change to the formula accounts for this. Instead of the AND() function, OR() is used:

```
=IF(OR(C12<12500, D12<4), "order", "do not order")
```

The new condition is TRUE in three cases:

- The first part of the condition is met.
- The second part of the condition is met.
- Both conditions are met.

The OR() function is not an exclusive *or*, which would exclude the third case. Since Excel 2016, an *either-or*—but not *both*—has been available via the XOR() function:

```
=IF(XOR(B10=5, C10=7), "ok", "check")
Before Excel provided the XOR() function, the expression would have been rather complex:
=IF(AND(OR(B10=5, C10=7), NOT(AND(B10=5, C10=7))), "ok", "check")
```

In both cases, if cell B10 equals 5 and cell C10 equals 7 at the same time, the result is check, so the condition is not met.

15.13.8 Multiple Branching

Both the value_if_true and value_if_false arguments can contain an IF() function that enables multiple branching.

	A	B	C
1	**Nested conditions**		
2			
3	Customer	Revenue	Gift Category
4	Berger	600,000	Gift A
5	Wehner	430,000	Gift B
6	Brech	30,000	Gift B
7	Schub	3,290	no gift
8	Scheimer	24,000	Gift B
9	Vosken	1,200	no gift
10	Erber	530,000	Gift A
11	Kosinsky	23,000	Gift B
12	Marchat	12,000	Gift B
13	**=IF(B4>=10000,IF(B4>=500000,"Gift A","Gift B"),"no gift")**		

Figure 15.53 Nested Conditions

Say a company's Christmas gifts are divided into two groups based on value. Group A gifts go to customers with sales over $500,000, while Group B requires at least $10,000 to exclude mini-customers entirely. The first formula filters out mini-customers and then distinguishes between Group A and Group B:

- `IF(B4>=10000, "Gift", "No gift")`
  This excludes the mini-customers.
- `=IF(B4>10000, IF(B4>=50000, "Gift A", "Gift B"), "no gift")`
  The `value_if_true` is now the complete `IF()` function from the first step, which includes its own `logical_test`, a `value_if_true`, and a `value_if_false`.

15.13.9 Array Functions

The new `MAP()`, `SCAN()`, `REDUCE()`, `MAKEARRAY()`, `BYCOL()`, and `BYROW()` functions all work with arrays, and nearly all return results as dynamic arrays. Each uses an expression with a `LAMBDA()` function as a key argument to compute values, and `MAKEARRAY()` creates a new matrix with the specified number of rows and columns, calculating values using the `LAMBDA()` function with the `r` and `c` row and column indices.

	A	B	C	D	E	F	G	H	I
1	**Create array**								
2									
3	1	2	3						
4	2	4	6						
5	3	6	9						
6	=MAKEARRAY(3, 3, LAMBDA(r,c, r*c))								
7									
8	**Creating an new array**								
9									
10	1	4	2	1		1	16	4	1
11	3	8	6	3		9	64	36	9
12	7	4	2	1		49	16	4	1
13	1	4	7	4		1	16	49	16
14	=MAP(A10:D13,LAMBDA(a,IF(a>1,a^2,a)))								
15									
16	**Generate intermediate values**								
17									
18	20	4	12		20	80	960		
19	12	8	13		11520	92160	1E+06		
20	=SCAN(1, A18:C19, LAMBDA(a,b,a*b))								

Figure 15.54 Examples of MAKEARRAY(), MAP(), and SCAN()

The `MAP()` function creates a second matrix from an existing one and calculates each value with a `LAMBDA()` function. With `SCAN()`, you can also generate a second matrix in which each value uses the previous calculation's result as an intermediate step for the next.

Using the BYCOL() function and the BYROW() function, you can, for example, compile the highest or lowest values per row or column into a separate matrix. The REDUCE() function combines the matrix values into a single value by using a LAMBDA() calculation.

	A	B	C	D	E	F	G
1	**Evaluate array column-by-column**						
2							
3	5	15	30		33	44	30
4	12	23	8				
5	11	44	2				
6	33	2	26				
7	=BYCOL(A3:C6, LAMBDA(array, MAX(array)))						
8							
9	**Evaluate array row-by-row**						
10							
11	5	15	30		30		
12	12	23	8		23		
13	11	44	2		44		
14	33	2	26		33		
15	=BYROW(A11:C14, LAMBDA(array, MAX(array)))						
16							
17	**Reduce an array to an accumulated value**						
18							
19	1	2	3		91		
20	4	5	6				
21	=REDUCE(, A19:C20, LAMBDA(a,b,a+b^2))						

Figure 15.55 Examples of BYCOL(), BYROW(), and REDUCE()

15.13.10 LET() and LAMBDA()

Both of these functions simplify calculations. The LET() function combines step-by-step calculations within a function, passing a result to the next step. The syntax is very flexible. You start with at least one variable name and its value, and up to 126 pairs are allowed.

After these pairs, you define a calculation that uses the variables. The variables can store intermediate results used in the next calculation step, and Figure 15.56 shows an example of such a sequence of steps. First, values from the table are assigned to the Price, Costs, and Sales variables. Next, the Totalprofit variable is assigned the result of a calculation using those variables. Then, the sales are summed, and finally, the average profit for each product is calculated.

The LAMBDA() function lets you encapsulate complex calculations within a custom worksheet function. Unlike user-defined functions created in VBA (as explained in Chapter 23), this occurs directly in the worksheet (see Figure 15.57).

	A	B	C	D	E	F	G
1	**Example of the LET() function**						
2							
3	**Category**	Price	**Costs**	**Sales**	**Revenue**		**Profit**
4	WG1	150.00	100.00	3000	450,000.00		150,000.00
5	WG2	200.00	114.00	2000	400,000.00		172,000.00
6	WG3	60.00	39.00	4000	240,000.00		84,000.00
7	WG4	50.00	27.00	5000	250,000.00		115,000.00
8				14000	1,340,000.00		521,000.00
9							
10				Average profit per item:			37.21
11				**=LET(Price,B4:B7,**			
12				**Costs,C4:C7,**			
13				**Sales,D4:D7,**			
14				**Totalprofit,SUM((Price-Costs)*Sales),**			
15				**Totalsales,SUM(Sales),**			
16				**Profitmean,Totalprofit/Totalsales**			
17				**,Profitmean)**			

Figure 15.56 Example of LET() Function

	A	B	C	D	E	F	G
1	**Minimization of material consumption**						
4							
5	Radius	Height	Material				
6	5.42	10.84	553.58				
7		Can volume:	**1000.00**	**=A6^2*PI()*B6**			
8							
9			**1000.00**	**=LAMBDA(Radius,Height,Radius^2*PI()*Height)(Radius,Height)**			
10							
11			**1000.00**	**=ZYVOL(Radius,Height)**			

Figure 15.57 Example of LAMBDA() Function

The syntax first requires you to specify the variables used in the calculation, and you can include up to 253 parameters. The last argument is the calculation performed with those variables, and after the closing parenthesis, you provide the values for the variables. In this example, the values come from named cells A6 and B6. Constants and simple cell references also work.

To simplify, you can assign a name to the function. Use **Define Name** and choose a name that briefly describes the calculation, like `ZYVOL` for cylinder volume. First, copy the formula from the formula bar and paste it under **Refers to**, leaving out the part with the variable values. Then, add a brief note under **Comment**, explaining what the function does. After that, you can use `ZYVOL(Radius, Height)` throughout the workbook. If you want to use it in another workbook, copy a sheet where the name is defined into the new workbook.

15.13.11 List of Logical Functions

Functions	Descriptions
SWITCH(**expression**, **value1**, **result1**, default_or_value2, result2, ...)	Compares the result of an expression to a list of values and returns the matching result (returns the default value if none match)
FALSE()	Stores the logical value FALSE in the cell
LAMBDA(parameter1, parameter2,..., **calculation**)	Lets you define custom functions that perform frequently used calculations using the specified parameters
LET(**name**, **Name_value1**, ...)	Assigns names to the specified calculations
MAP(**array**, lambda_or_array2, ...)	Returns a new array of data created by mapping existing array values using Lambda functions
MAKEARRAY(**rows**, **columns**, **function**)	Returns an array calculated by a Lambda function with the specified number of rows and columns
BYCOL(**array**, function)	Applies a Lambda function to each column and returns an array of results
BYROW(**array**, function)	Applies a Lambda function to each row and returns an array of results
NOT(**logical**)	Reverses the value of *logical*
OR(**logical1**, logical2, ...)	Returns TRUE if any argument is TRUE; otherwise, returns FALSE
REDUCE(**initial_value**, **array**, **function**)	Reduces an array to a single accumulated value by applying a Lambda function to all values
SCAN(**initial_value**, **array**, **function**)	Scans an array, applies a Lambda function to each value, and returns the results as a new array
AND(**logical1**, Logical2, ...)	Returns TRUE only if all arguments are TRUE; otherwise, returns FALSE
TRUE()	Inserts the logical value TRUE into a cell
IF(**logical_test**, **value_if_true**, Value_if_false)	Returns value_if_true if the logical test is TRUE; otherwise, returns value_if_false
IFERROR(**value**, **value_if_error**)	Allows custom error messages to be displayed when an error occurs
IFNA(**value**, **value_if_na**)	Returns the first value if the expression does not return the #N/A error; returns the second value otherwise

15

Functions	Descriptions
IFS(**logical_test1**, **value_if_true1**, Logical_test2, Value_if_true2, ...)	Checks if at least one condition is true and returns the corresponding value; returns the #N/A error otherwise
XOR(**logical_value1**, logical_value2, ...)	Returns the exclusive OR of all arguments

15.14 Information Functions

Information functions are mainly used to make calculation results—together with the IF() function—dependent on the contents of specific cells. Additionally, information functions work well in formulas that are used for validation rules or conditional formatting.

15.14.1 Example: Preventing Errors

Consider this example: In a table, columns A and B from row 3 contain both numbers and text entries, and you want to display the product of A and B in column C. If you enter the formula =A3*B3 in C3 and copy it down, you'll get the #VALUE! error whenever column A or B doesn't contain numbers. To avoid this, enter the following formula in C3:

```
=IF(AND(ISNUMBER(A3), ISNUMBER(B3)), A3*B3, "")
```

This formula calculates the product only if both cells contain numbers; otherwise, it returns an empty string. Copying this formula down will prevent unpleasant error messages.

15.14.2 List of Information Functions

Functions	Descriptions
SHEET (value)	Returns the sheet number of a reference
SHEETS(reference)	Returns the number of sheets in a reference
ERROR.TYPE(**error value**)	Returns the error number
INFO(**type**)	Returns information about the system environment
ISREF(**value**)	Tests whether *value* is a valid name or reference
ISERR(**value**)	Tests whether *value* is an error value other than #NA

Functions	Descriptions
ISERROR(**value**)	Tests whether *value* is an error value
ISFORMULA(**reference**)	Tests whether the specified cell contains a formula
ISEVEN(**number**)	Tests whether *number* is evenly divisible by 2
ISNONTEXT(**value**)	Tests whether a cell contains no text
ISBLANK(**value**)	Tests whether a cell is empty
ISLOGICAL(**value**)	Checks whether *value* returns a logical value
ISNA(**value**)	Checks whether a cell contains the #NA error value
ISTEXT(**value**)	Checks whether *value* returns a text string
ISODD(**number**)	Tests whether *number* is evenly divisible by 2
ISNUMBER(**value**)	Checks whether *value* is a number
N(**value**)	Returns *value* converted to a number
NA()	Inserts the #NA error value into the cell
TYPE(**value**)	Returns the data type of *value*
ISOMITTED(**argument**)	Checks whether an argument is omitted from a Lambda function
CELL(**info_type**, reference)	Returns various information about a cell

15

15.15 Technical Functions

This group contains different variants of Bessel functions, which are particularly used in vibration and wave calculations. It also includes a variety of conversion functions to translate values between different number systems. Another group involves complex numbers. Since Excel 2013, functions for bit operations like left and right shifts have been added.

15.15.1 Converting Units of Measure

The CONVERT() function is very useful for converting between different units of measure, and you can easily build a table that provides conversions for the units you use most often. Figure 15.58 lists common unit abbreviations, and you must enter any abbreviation into a function with quotation marks unless you use a cell reference.

	A	B	C	D	E	F	G	H
1	**Names of units**							
2								
3		Mass			Magnetism			Energy
4	Gram	g		Tesla	T		Joule	J
5	Pound (Trade Weight)	lbm		Gauss	ga		Erg	e
6	U (atomic mass unit)	u					Thermodynamic Calorie	c
7	Ounce (Commercial Weight)	ozm			Achievement		Calories	cal
8				Horsepower (HP)	HP		Electron volt	eV
9		Length		Watt	W		Horsepower times the hour	HPh
10	Metre	m					Watts times Hour	Wh
11	British Mile	mi			Pressure		BTU	BTU
12	Nautical Mile	Nmi		Pascal	Pa			
13	Inch	in		Atmosphere	atm			Dimensions for liquids
14	Foot	ft		mm Mercury	mmHg		Teaspoon	tsp
15	Yard	yd					Tablespoon	tbs
16	Ångström	ang			Power		Fluid Ounce	oz
17	Pica (1/72 inch)	Pica		Newton	N		Cup	cup
18				Dyn	dyn		Pint	pt
19		Time					Quarter	qt
20	Year	yr			Temperature		Gallon	gal
21	Day	day		Degrees centigrade	C		Litre	l
22	Hour	hr		Degrees Fahrenheit	F			
23	Minute	mn		Degrees Kelvin	K			
24	Second	sec						

Figure 15.58 Abbreviations of Units of Measure

In the table from Figure 15.59, you only need to enter the values you want to convert in the cells of column A. The formulas refer to the names of the units of measure.

	A	B	C	D	E
1	**Converting Units of Measure**				
2					
3	1000	Gram	conform:	2.205	Pound (Trade Weight)
4		g			lbm
5	1000	Metre	conform:	0.621	UK Miles
6		m			mi
7				39370.08	Inch
8					in
9				3280.84	Foot
10					ft
11				1093.613	Yard
12					yd
13	32	Degrees centigrade	conform:	89.6	Degrees Fahrenheit
14		C			F
15	1000	Joule	conform:	238.85	Calories
16		J			cal
17	10	Horsepower (HP)	conform:	7457.00	Watt
18		HP			W

Figure 15.59 Conversion of Common Units of Measure

15.15.2 Bessel Functions

Bessel functions (also known as *cylinder functions*) are several related functions used in physics and engineering, especially for vibration and wave calculations. In Figure 15.60, the BESSELJ() function is graphically displayed for several orders, clearly illustrating that the function is well suited to calculating dampened oscillations.

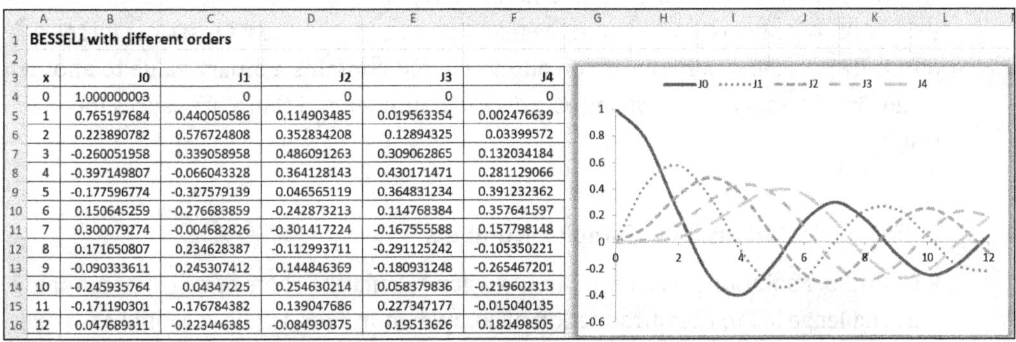

BESSELJ with different orders

x	J0	J1	J2	J3	J4
0	1.000000003	0	0	0	0
1	0.765197684	0.440050586	0.114903485	0.019563354	0.002476639
2	0.223890782	0.576724808	0.352834208	0.12894325	0.03399572
3	-0.260051958	0.339058958	0.486091263	0.309062865	0.132034184
4	-0.397149807	-0.066043328	0.364128143	0.430171471	0.281129066
5	-0.177596774	-0.327579139	0.046565119	0.364831234	0.391232362
6	0.150645259	-0.276683859	-0.242873213	0.114768384	0.357641597
7	0.300079274	-0.004682826	-0.301417224	-0.167555588	0.157798148
8	0.171650807	0.234628387	-0.112993711	-0.291125242	-0.105350221
9	-0.090333611	0.245307412	0.144846369	-0.180931248	-0.265467201
10	-0.245935764	0.04347225	0.254630214	0.058379836	-0.219602313
11	-0.171190301	-0.176784382	0.139047686	0.227347177	-0.015040135
12	0.047689311	-0.223446385	-0.084930375	0.19513626	0.182498505

Figure 15.60 Graph of BESSELJ() Function

15.15.3 Conversions Between Number Systems

Twelve functions in this category convert numbers from one number system to another. Besides the decimal system, the binary, octal, and hexadecimal systems are included. Each system includes three functions that convert numbers into the other systems.

Number systems differ by the number of digits they use, which is known as the *base*. The decimal system, as you know, uses ten digits (0 through 9). When counting beyond 9, the first digit increases and a 0 is added. The binary system, also called base-2, uses only the digits 0 and 1. You count 0, 1, 10, 11, 100, and so on. The octal system uses eight digits, counting 0, 1, 2, ..., 7, 10, 11, ..., 17, and 20. The hexadecimal system uses the letters A through F along with the digits 0 through 9 to represent sixteen characters.

Number Systems

Decimal	Octal	Binary	Hexadecimal
0	0	0	0
1	1	1	1
2	2	10	2
3	3	11	3
4	4	100	4
5	5	101	5
6	6	110	6
7	7	111	7
8	10	1000	8
9	11	1001	9
10	12	1010	A
11	13	1011	B
12	14	1100	C
13	15	1101	D
14	16	1110	E
15	17	1111	F

Figure 15.61 First Sixteen Numbers in Various Number Systems

Binary values can quickly become very long because only two digits are available, so the octal and hexadecimal systems are used to represent binary values more compactly. This works well because bases eight and sixteen are powers of 2. However, Excel outputs the results of conversion functions as text strings, not numbers, unless the target system is decimal. For example, if you add a cell that contains a binary value to another value, Excel treats the text string as a decimal number and therefore causes incorrect results.

15.15.4 Calculations with Complex Numbers

Real number solutions reach their limits when calculating roots of negative numbers. This challenge led to the introduction of complex numbers, which combine a real part and an imaginary part. To address this, the imaginary unit `i` was introduced as a new number. It's defined as `i = √-1`.

When an Excel function requires a complex number as an argument, enter it as a string in the form `x+yi` or `x+yj`. The `j` notation is used in electrical engineering to avoid confusion with current. Lowercase `i` or `j` is required. The `x` and `y` can be any real numbers, and `yi` or `yj` is shorthand for `y*i` and `y*j`. This means a real number is multiplied by an imaginary number. If you enter a complex number directly as a function argument, enclose it in quotation marks, as with `=IMABS("3+3i")`. If you use a cell reference as an argument, the string appears without quotation marks in the referenced cell.

You can work with imaginary numbers just like with regular variables in arithmetic—as if `i` were simply a variable—provided you keep the following points in mind:

`i^2 = -1; i^3 = -i; i^4 = 1; i^7 = -i` and so on:

```
4i + 2i = 6i
4i - 4i = 0
3i * 4i = -12
10i/2i  = 5
```

Complex numbers play a key role in physics and engineering, especially because they simplify solving differential equations that involve oscillations. In electrical engineering, complex numbers are used to calculate reactance in AC circuit analysis. In pure mathematics, they appear in function theory.

15.15.5 List of Technical Functions

Functions	Descriptions
BESSELI(**x**, **n**)	Returns the modified `In(x)` Bessel function
BESSELJ(**x**, **n**)	Returns the `Jn(x)` Bessel function

Functions	Descriptions
BESSELK(**x, n**)	Returns the modified Kn(x) Bessel function
BESSELY(**x, n**)	Returns the Yn(x) Bessel function
BIN2DEC(**number**)	Converts a binary value to a decimal number
BIN2HEX(**number**, digits)	Converts a binary value to a hexadecimal number
BIN2OCT(**number**, digits)	Converts a binary value to an octal number
BITSHIFT(**number, shift_amount**)	Returns the number resulting from shifting bits left by the shift amount
BITOR(**number1, number2**)	Returns the bitwise OR of two numbers
BITRSHIFT(**number, Shift amount**)	Returns the number obtained by shifting bits to the right by the specified amount
BITAND(**number1, number2**)	Returns the bitwise AND of two numbers
BITXOR(**number1, number2**)	Returns the bitwise exclusive OR of two numbers
DELTA(**number1**, number2)	Checks whether values are equal
DEC2BIN(**number**, digits)	Returns the binary representation of a decimal number
DEC2HEX(**number**, digits)	Returns the hexadecimal representation of a decimal number
DEC2OCT(**number**, digits)	Returns the octal representation of a decimal number
ERF.PRECISE(**x**)	Returns values of the Gaussian error function
ERF(**lower_limit**, upper_limit)	Returns values of the Gaussian error function
ERFC(**lower_limit**)	Returns the complementary values of the Gaussian error integral
ERFC.PRECISE(**x**)	Returns the complementary values of the Gaussian error integral
GESTEP(**number**, step)	Checks whether the value specified by *number* meets or exceeds the threshold set by *step*
HEX2BIN(**number**, digits)	Returns the binary value of a hexadecimal number
HEX2DEC(**number**)	Returns the decimal value of a hexadecimal number
HEX2OCT(**number**, digits)	Returns the octal value of a hexadecimal number
IMABS(**complex_number**)	Returns the absolute value (magnitude) of a complex number

Functions	Descriptions
IMAGINARY(**complex_number**)	Returns the imaginary part of a complex number
IMPOWER(**complex_number, Power**)	Returns the power of a complex number
IMARGUMENT(**complex_number**)	Returns the argument of a complex number
IMCOS(**complex_number**)	Returns the cosine of a complex number
IMCSC(**complex_number**)	Returns the cosecant of a complex number
IMCSCH(**complex_number**)	Returns the hyperbolic cosecant of a complex number
IMCOSH(**complex_number**)	Returns the hyperbolic cosine of a complex number
IMCOT(**complex_number**)	Returns the cotangent of a complex number
IMDIV(**complex_number1, complex_number2**)	Divides complex numbers
IMEXP(**complex_number**)	Raises *e* (Euler's number) to the power of a complex number
IMCONJUGATE(**complex_number**)	Returns the conjugate of a complex number
IMLN(**complex_number**)	Returns the natural logarithm of a complex number
IMLOG10(**complex_number**)	Returns the common logarithm of a complex number
IMLOG2(**complex_number**)	Returns the binary logarithm of a complex number
IMPRODUCT(**complex_number1**, complex_number2, ...)	Calculates the product of complex numbers
IMREAL(**complex_number**)	Returns the real part of a complex number
IMSEC(**complex_number**)	Returns the secant of a complex number
IMSECH(**complex_number**)	Returns the hyperbolic secant of a complex number
IMSIN(**complex_number**)	Returns the sine of a complex number
IMSINH(**complex_number**)	Returns the hyperbolic sine of a complex number.
IMSUB(**complex_number1, complex_number2**)	Subtracts two complex numbers
IMSUM(**complex_number1**, complex_number2, ...)	Adds complex numbers
IMTAN(**complex_number**)	Returns the tangent of a complex number

Functions	Descriptions
IMSQRT(**complex_number**)	Returns the square root of a complex number
COMPLEX(**real part**, **imaginary_ part**, suffix)	Creates a complex number from two real numbers
OCT2BIN(**number**, digits)	Returns the binary value of an octal number
OCT2DEC(**number**)	Returns the decimal value of an octal number
OCT2HEX(**number**, digits)	Returns the hexadecimal value of an octal number
CONVERT(**number**, **from_unit**, **to_unit**)	Converts between different units of measure

15.16 Web Functions

The **Web** category in Excel 2016 introduced three functions to simplify accessing web data. We cover them in this section.

The `ENCODEURL()` function converts text strings into properly formatted web addresses or parts of them. The `FILTERXML()` function extracts from XML content can be read using the appropriate XPath expression.

15.16.1 Web Queries

The `WEBSERVICE()` function retrieves data from a web service that's specified by its URL. The URL must include the credentials and all parameters for the web service, and only simple requests (like GET) are supported.

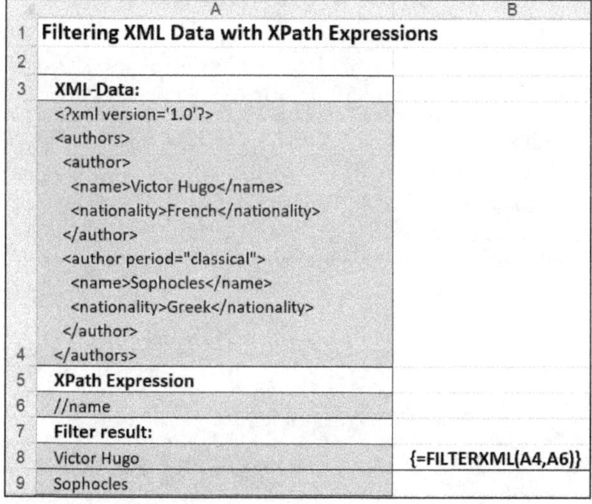

	A	B
1	**Filtering XML Data with XPath Expressions**	
2		
3	**XML-Data:**	
4	<?xml version='1.0'?> <authors> <author> <name>Victor Hugo</name> <nationality>French</nationality> </author> <author period="classical"> <name>Sophocles</name> <nationality>Greek</nationality> </author> </authors>	
5	**XPath Expression**	
6	//name	
7	**Filter result:**	
8	Victor Hugo	**{=FILTERXML(A4,A6)}**
9	Sophocles	

Figure 15.62 Example of Querying Data in XML Format

15.16.2 List of Web Functions

Functions	Descriptions
ENCODEURL(**text**)	Converts text into a URL-encoded string
WEBSERVICE(**url**)	Returns data from a web service
FILTERXML(**xml, xpath**)	Delivers data from XML content specified by the *XPath* expression

Chapter 16

Organizing and Managing Information as Tables

Everyday office tasks include compiling data in tabular lists, such as phone directories, address lists, product lists, and order lists. Excel offers powerful features you can use to create, maintain, and analyze these uniformly structured tables, and some are comparable to database programs. Additionally, those table ranges come equipped with special features that are designed to facilitate the maintenance, expansion, and formatting of the tables themselves.

16.1 Tables, Data Lists, and Data Tables

Those who are familiar with even older versions of Excel know that Microsoft has shown some inconsistency in the terminology used for *table*, *data list*, and *data table*. Up to version 2003, the terms *data lists* and *databases* were commonly used; since then, the terms *tables* and *Excel tables* have been used instead—to distinguish them from simple cell ranges on worksheets that do not offer the special functionality of an Excel table.

The term *data tables* now has a clearer meaning. They were described in Chapter 5, Section 5.4, in relation to multiple operations. To avoid confusion, the special functions for analyzing such Excel tables remain categorized under **Database**. For simplicity's sake, this chapter refers to them simply as *tables*.

16.2 A Table for an Inventory List

The following sections explain how to work with these tables by using an inventory list of wines from various countries as an example. Within each country, the wines are organized by growing region and then by color: red, white, and rosé. These sorting criteria help demonstrate how to group, summarize, and analyze data in tables.

16.3 Applications for Using Tables

You'll use tables whenever you're working with information that can be arranged evenly in rows. Each cell in a table row contains data for a specific unit, such as a person, group, location, time period, or subject. For example, a table row holds various details about a specific person, while a table column lists the same type of information for each person, such as their name.

The table's contents are indicated by the field name in the column header of the table's first row. Placing such tables in an Excel worksheet is good practice since the sheet already provides a tabular structure.

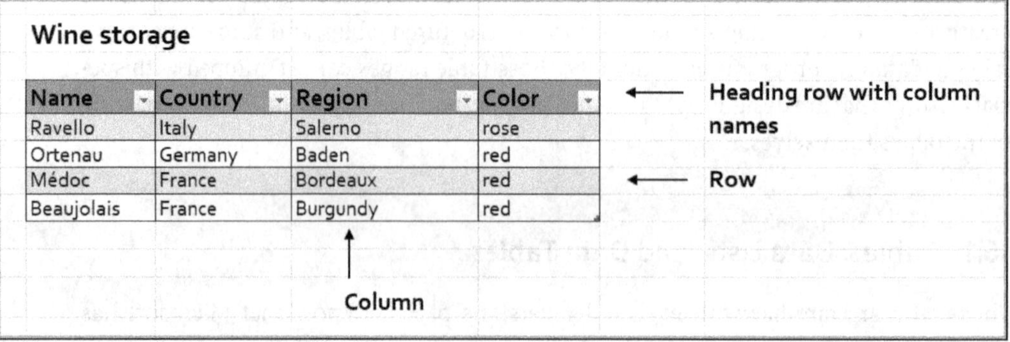

Wine storage
Name
Country
Region
Color
Ravello
Italy
Salerno
rose
Ortenau
Germany
Baden
red
Médoc
France
Bordeaux
red
Beaujolais
France
Burgundy
red
Heading row with column names
Row
Column

Figure 16.1 Table Structure

Excel's table features do not aim to replace specialized database programs. Microsoft Access serves as a dedicated database program within the Office suite, and Microsoft SQL Server versions offer even broader capabilities. The free Express edition is a practical solution for handling moderate amounts of data.

Data exchange with external databases is a key strength of Excel. The program is well prepared to import external data into a worksheet, and special drivers enable direct access to data from numerous databases. This is covered in Chapter 19.

When you're managing very large data volumes, we generally recommend that you use specialized database programs because the data volumes these programs can handle are not limited by main memory size. On the other hand, while Excel can access files that are not loaded into main memory, it typically works with all data loaded into main memory. For moderate amounts of data, the main-memory sizes that are common today are sufficient to run applications effectively, so the main advantage of using these applications is that you can directly evaluate data sets with formulas in the table.

16.4 Defining Table Structure

Working with tables or data lists has been simplified step by step since the first Excel versions. In most cases, defining specific ranges before using table tools is no longer necessary. Excel usually identifies the data belonging to the table automatically, but exceptions occur in special cases where criteria ranges must be defined for data evaluation.

Excel offers numerous very useful functions for analyzing such tables. They include creating result rows, extensive filter functions, and especially interactive PivotTables, which let you rearrange and summarize data from different perspectives. You'll learn more about them in Chapter 17 and Chapter 18.

[«]

Backward Compatibility

Excel can continue using older data list applications that were created with defined database ranges without changes. This also applies to importing Lotus 1-2-3 models.

16.4.1 An Inventory Table for a Wine Warehouse

The first step in building the intended table for fictional wine warehouse is to determine which information is necessary to manage the inventory effectively. In practice, this consideration first appears in the table's header row. Which column names are necessary depends on the table's subject and, more importantly, on the questions the table aims to answer.

Whether you should organize your thoughts mentally, on paper, or directly in a table depends on your personal work style. The table is useful for experimenting because you can easily correct initial attempts by using drag-and-drop without re-entering data.

Defining the Subject

The first thing to determine is the subject of the table. What unit should each row of the table represent? The unit can be a person in a specific role, such as a customer or employee, or it can be an object, like an item, a product kit, a process such as a service, or a fact like assigning people to projects. In our example, the unit is an individual wine, identified by a specific name.

Information Elements

What information is available for each unit or needs to be obtained and recorded? Which distinguishing features and sorting criteria are necessary or useful? Answering these questions depends on the practical decisions you'll make based on the information in the table. For example, if you want to know which wines need to be reordered from inventory, you should include a column for minimum stock and a column for current inventory.

As a general rule, you should include only the information that's necessary for the item and the expected questions. Excess information takes up space, makes the table confusing, and slows down all processes. Often, it is better to use multiple tables than a list with unnecessary repetitions. In an order list, only the supplier number should appear, not the full supplier address, if a supplier list assigns the address to the supplier number.

The order of columns is basically arbitrary, but it makes sense to arrange related information side by side. If the table contains only a few columns, then this is not a major issue, but in tables with many columns, you should place the information that uniquely identifies the subject or person in the first columns.

Key Columns for Access

Often, it makes sense to designate one or more column names as *key fields* that uniquely identify a specific data record. In our example, this data is the wine's name—though instead of the name, you could use a unique item number. You can easily generate such numbers with a starting number by using **Start • Home • Editing • Fill • Data series.**

16.4.2 Data Types and Field Lengths

Nothing stops you from entering a number in a field in one row and a text string in the next. However, this flexibility can cause issues during sorting, since Excel always places numbers before text. If you sometimes enter an item number as a number and sometimes as text, then the order will become mixed. Problems also occur when defining search criteria. For example, if you entered the zip code sometimes as a number and other times as text, a criterion comparing the zip code as a number might not display all records.

For each field, you need to decide which data type to use in the respective column. Is it a numeric value, a date or time value, a Boolean value, or a text string? Unlike a dedicated database program like Access, Excel does not initially validate data type or input length when you're entering data into a table, unless you control data entry with a macro or use validation rules. We already covered this feature in Chapter 2, Section 2.5. It offers excellent options for verifying correct data entry but is not mandatory.

Although you don't need to make a final decision when defining the table structure, it is usually helpful to estimate how much space each piece of information will require to set the column width accordingly.

Estimating Data Amount

You should also consider the expected total amount of data during planning. However, since Excel has greatly expanded worksheet size, limiting the table to a single sheet is usually no longer an issue.

Is the main memory sufficient to handle the planned table, or will you need to split the table? Only rough estimates are possible here. If you work with multiple tables, decide whether to link them. Is there a common column in both tables that is suitable for linking?

16.4.3 Rules for Choosing Column Names

Follow these rules when choosing column names:

- Column names should generally be unique. You can use column names multiple times, but when you use the **Advanced Filter** function, this will cause issues because it will only search the first column with that name, not subsequent ones.
- Avoid commas, periods, spaces, hyphens, and semicolons, and don't use names that look like cell references. If a name has multiple parts, use underscores instead of spaces.
- For column name length, balance clarity and brevity. Long names increase user effort in writing queries and the risk of typos.

After you entering the header row with the column names, the table structure will be established. Figure 16.2 shows the header row for the wine warehouse table.

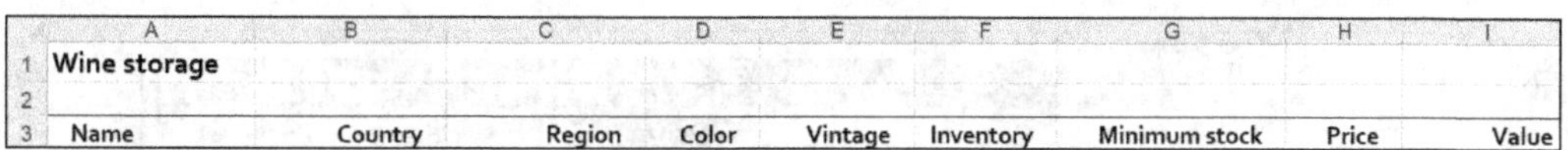

A
B
C
D
E
F
G
H
I
1
Wine storage
2
3
Name
Country
Region
Color
Vintage
Inventory
Minimum stock
Price
Value

Figure 16.2 Header Row for Wine Storage Table

16.5 Table Ranges

To simplify working with tables, especially when adding rows or columns, you can explicitly convert a structured cell range into a table range. This is not required to use features like filtering functions, but it is highly recommended because it helps prevent errors when you're expanding and defining named cell ranges.

16.5.1 Converting Cell Ranges to Table Ranges

You can convert a cell range into a table either at the start with a label row or later with a range that already contains data below the label row. In the first case, place the cell pointer in a cell of the label row or below it, then select **Insert • Tables • Table**. You can also use the `Ctrl`+`T` keyboard shortcut instead of the command.

If Excel detects a table range, it asks you to confirm it in the small **Create Table** dialog. If the range is empty, you can define it by dragging with the mouse or entering the range addresses, and you can also select **My table has headers** if applicable. If no header exists, Excel assigns the default name *ColumnN*. Excel creates a table range and displays filter buttons for each column by default, and we'll explain them later. Excel also applies to the created range a default table style, which you can change at any time.

You can access the same dialog by placing the cursor in an existing cell range and clicking the **Format as Table** button under **Home • Styles**. This method immediately applies the selected style from the palette to the range.

Figure 16.3 Converting Cell Range into Table

Another option is to create a table this way: When you select the entire data range, the **Quick Analysis** button appears at the far right. If you select **Tables** in the palette, a **Table** icon will appear and instantly convert the cell range into a table with the default style.

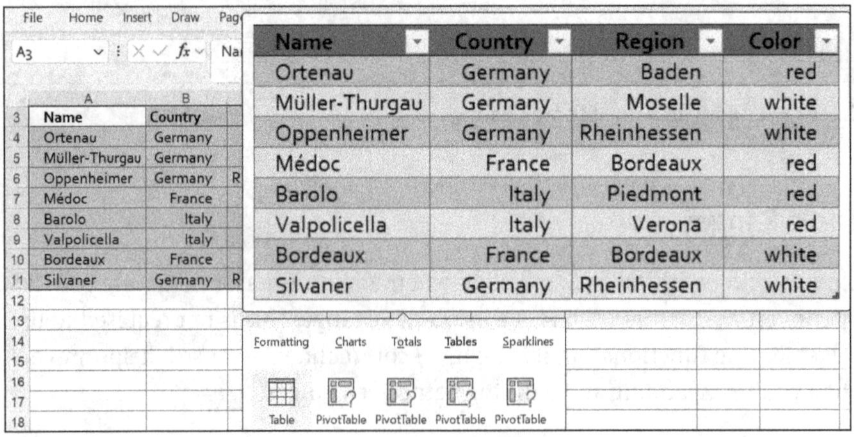

Figure 16.4 Creating Table with Quick Analysis Button

The **Table Design** tab appears for every new table and offers five groups of options for table design. Under **Properties • Table Name**, you can overwrite the default numbered name ❶. You can use this name for structured references in formulas, as we'll explain later. It works like a defined range name.

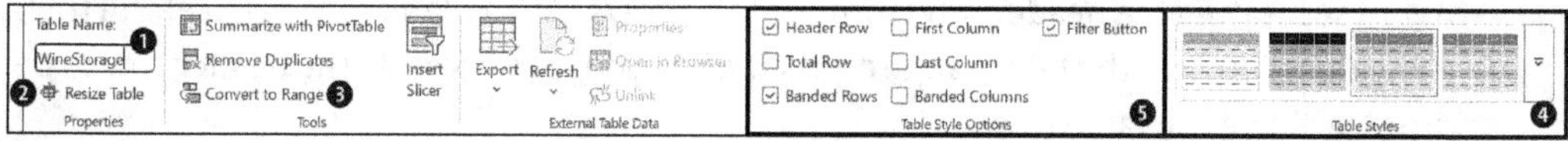

Figure 16.5 Option Groups for Designing the Table Range on the Table Design Tab

You can also expand the table range later via **Table Design • Properties • Resize Table ❷**. The new range must overlap the existing one, and the header row must stay in the same row.

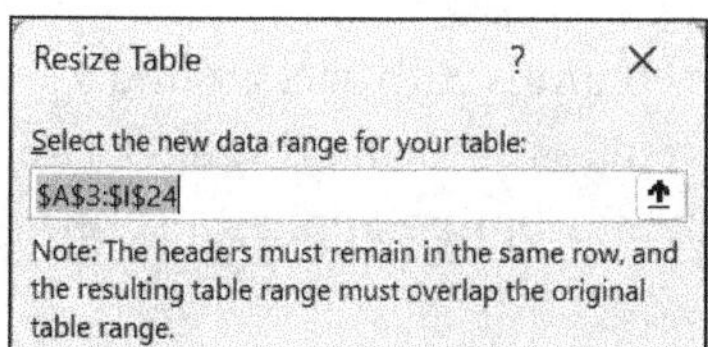

Figure 16.6 Dialog for Resizing Table

If you need to, you can convert the table range back to a normal cell range by using **Table Design • Tools • Convert to Range ❸** as well. Strangely, this is necessary even when working with subtotals, which we'll explain further later.

16.5.2 Formatting Tables

To distinguish the header row from the table rows and keep each row clear, you apply an appropriate table style. If you haven't selected a style before, select a cell in the table and choose **Table Design • Table Styles ❹**. From the extensive range of **Quick Styles**, you can assign the desired pattern with a click or tap.

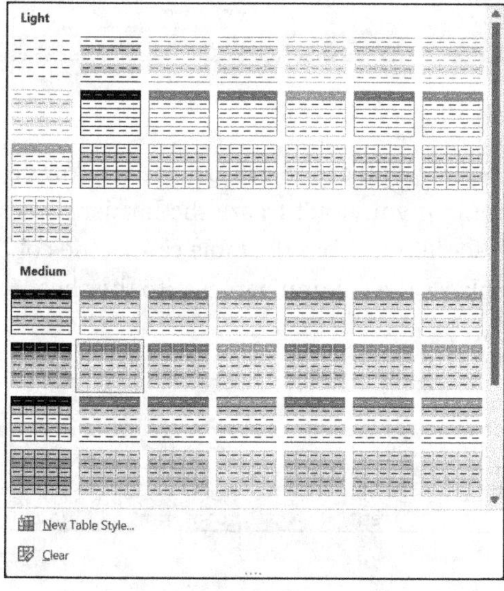

Figure 16.7 Assigning Table Style

Avoid separating the header row from the table rows with a blank row, as Excel will then struggle to identify the table range correctly. The program expects the header row to be directly above the table rows.

16.5.3 Table Options

In the **Table Style Options** group ❺ on the **Table Design** tab, you'll find useful settings related to working with tables. If you need to, you can hide the **Header Row** with the column names.

Activating **First Column** or **Last Column** highlights these columns with background formatting from the table styles. The **Banded Rows** option makes the formatting alternate in pairs by creating a striped pattern. The **Banded Columns** option applies a similar effect to the columns.

16.5.4 Freezing Labels

For long tables, you can keep the header and possibly the first column visible by using **View • Window • Freeze Panes**. This keeps them on the screen even when you're scrolling to distant rows or columns. Place the cell pointer on the first row below the field names and to the right of the first column.

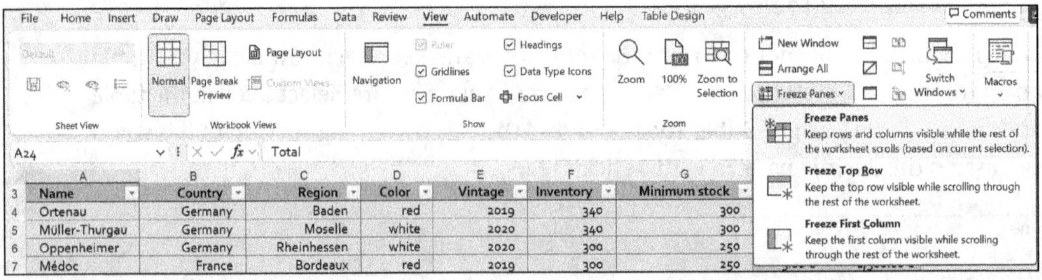

	Name	Country	Region	Color	Vintage	Inventory	Minimum stock
4	Ortenau	Germany	Baden	red	2019	340	300
5	Müller-Thurgau	Germany	Moselle	white	2020	340	300
6	Oppenheimer	Germany	Rheinhessen	white	2020	300	250
7	Médoc	France	Bordeaux	red	2019	300	250

Figure 16.8 List with Frozen Header

Instead of using the cell pointer to control freezing, you can freeze just the top row or first column via the **Freeze Panes** button menu. If you don't freeze the header rows, Excel still provides a useful reference. When scrolling within the table causes the column names to disappear, Excel temporarily places the column names into the sheet's column headers.

	Name	Country	Region	Color	Vintage	Inventory
4	Ortenau	Germany	Baden	red	2019	340
5	Müller-Thurgau	Germany	Moselle	white	2020	340
6	Oppenheimer	Germany	Rheinhessen	white	2020	300
7	Médoc	France	Bordeaux	red	2019	300

Figure 16.9 Column Labels in Column Headers

16.5.5 Entering Data

Once the table structure is fixed, you can start entering data. The simplest method is to enter data directly below the header row. Since you usually enter data row by row, you can end each cell entry by pressing [Tab], which moves you immediately from the end of one record to the start of the next. Apart from that, you work as with a normal table by inserting rows or deleting them if data becomes unnecessary.

[+]

Make Sure Spelling is Consistent!

When entering text, you need to make sure your spelling is consistent. Company names and different spacing often cause issues. For example, "A B C" will be sorted before "AAS" in ascending alphabetical order. In such cases, you can insert a second column to standardize all names and thus prevent sorting problems.

16.5.6 Uniqueness and Duplicates

If you use a key field like item number, customer number, or invoice number, the program won't respond if a key appears multiple times. You must ensure key uniqueness yourself, so to verify uniqueness later, use **Remove Duplicates** from the **Data • Data Tools** group. In the dialog, select the column to check for duplicates, and once they're confirmed, Excel will remove the duplicates and only keep the first occurrence of each item.

16

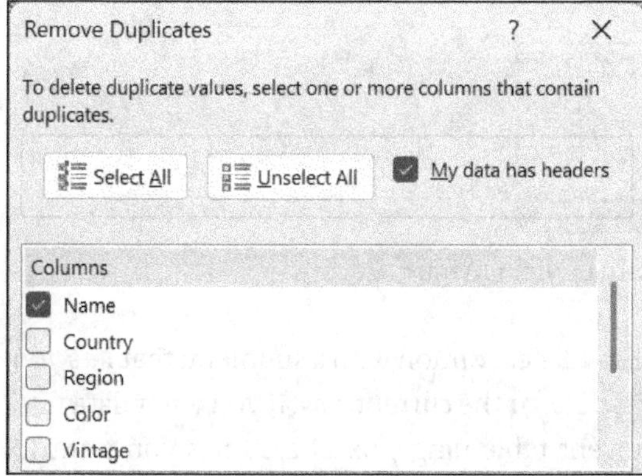

Figure 16.10 Checking for Duplicates

This method is rather crude and offers no further options for user intervention. Often, it is better to mark duplicates with conditional formatting and then clean the table selectively. You can do this via **Home • Styles • Conditional Formatting • Highlight Cells Rules • Duplicate Values.**

16.5.7 Expanding and Formatting Ranges

When you expand a normal cell range by filling additional table rows, you need to ensure that the formatting will carry over and any formulas will copy down. To do this, on the **Advanced** page in the **Excel Options** dialog box, enable **Extend data range formats and formulas, which will make** Excel automatically apply the format and formulas from the last row to any new row where you enter data within a cell range. Excel will also recognize contiguous cell ranges when you place the cell pointer in any cell within the range and use a command from the **Data • Sort and filter** group.

In tables, formats and formulas automatically apply to new rows, regardless of the **Expand data range formats and formulas** option you've selected, which only affects normal cell ranges. However, the new row must not be inserted below a result row.

16.5.8 Expanding Tables

The context menu of any cell you select in the table will provide commands under **Insert row/column** that you can use to add rows or columns within the table without affecting the rest of the worksheet. Similarly, a delete command is available for rows or columns you select within the table range. This lets you easily create multiple tables on the same sheet, either side by side or stacked.

To add a new row at the end of a table, finish entering data in the last cell and press Tab or start typing directly in the new row. In both cases, Excel adds the new row to the table range and applies the formatting from the last row.

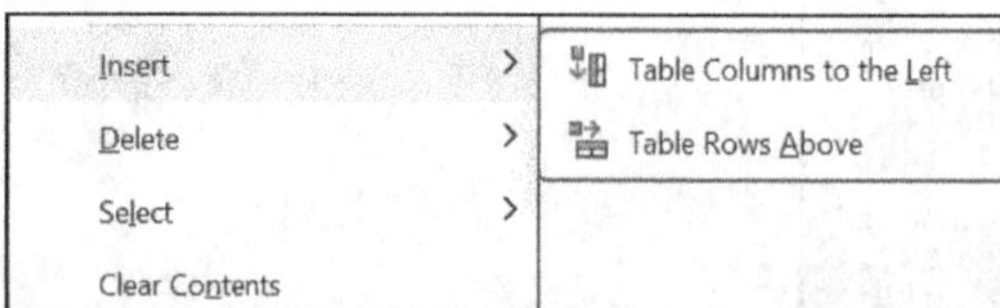

Figure 16.11 Inserting Rows or Columns with Context Menu

The context menu for tables also includes a **Select** option with a submenu that lets you quickly highlight the current column, its data, or the current row. If you enter data into a cell directly next to or below the current table range, Excel assumes you want to include that data in the table range. If you don't, you can undo the automatic table expansion via the **AutoCorrect options** button menu.

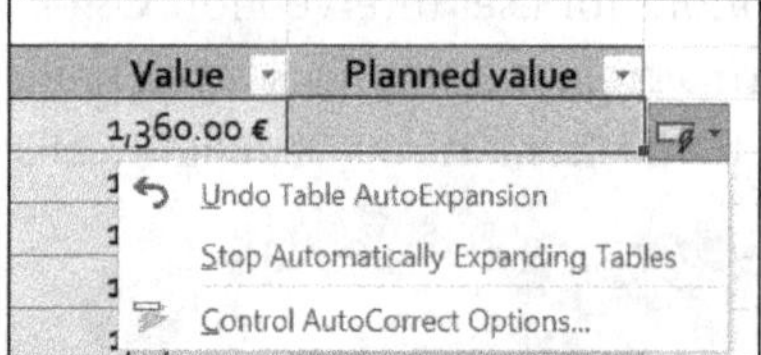

Figure 16.12 Expanding Table by One Column

16.5.9 Inserting Totals Rows

Another option for the table range is to insert a totals row at the end of the table, which moves automatically with the expanding range and adjusts its formulas. To do this, use **Table Design • Table Style Options • Total row** or **Table • Totals row** in the context menu. Use the small arrow buttons to select the calculation type individually for each column. Besides standard calculations like **Sum**, **Count**, and **Average**, you can access additional functions through the **More Functions** option.

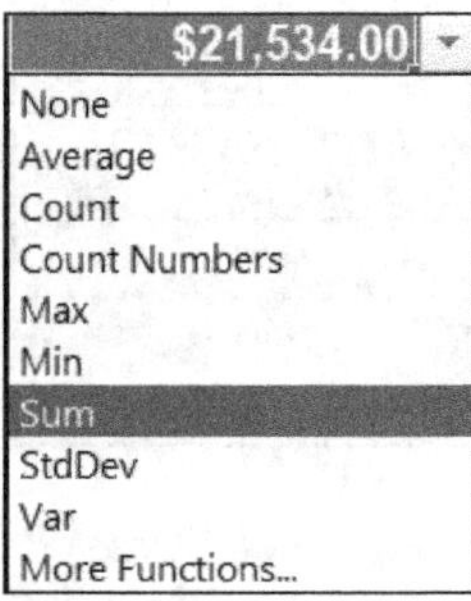

Figure 16.13 Totals Row with List Box

It's especially useful to display column sums in the totals row by using the `SUBTOTAL()` function instead of the `SUM()` function and activating it via **QuickAnalysis • Totals • Running**. For example, say you use these formulas in the `Inventory` and `Value` columns:

```
=SUBTOTAL(109,[Inventory])
=SUBTOTAL(109,[Value])
```

Only the values for visible rows will be shown, and when you change the filter for the column, the result will update automatically.

16.5.10 Calculated Columns

In addition to columns where you enter data, a table can include columns with values that are calculated from values in other columns or from values outside the table. For example, you can calculate the sales price by applying a percentage surcharge to the entered purchase price. To do that, you enter a formula into the cell instead of a value and copy it down for each new data record.

For formulas in such tables, Excel offers a simplified reference format that's called a *structured reference*. For example, if you want to multiply the price in the **Value** column by the inventory, Excel automatically shows the respective column name in square brackets when you're selecting cells during formula entry, without explicitly naming the ranges. Excel also places an implicit @ intersection operator before each name.

Therefore, this is the formula:

```
=[@Inventory]*[@Price]
```

It's applied unchanged to all table rows, but it evaluates the correct values in each row. After you enter the first formula, it will automatically fill the entire column, and the **AutoCorrect Options** button will appear so you can undo this action if you need to.

Using the **Control AutoCorrect Options** menu, you can deselect the **Fill formulas in tables to create calculated columns** option on the **AutoFormat as you type** tab if needed.

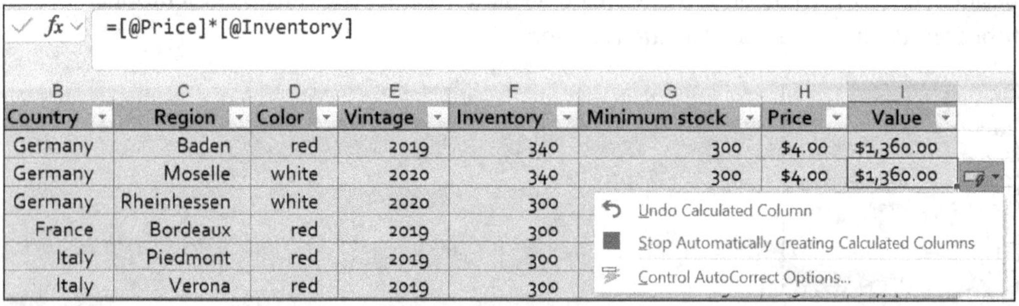

fx =[@Price]*[@Inventory]

B	C	D	E	F	G	H	I
Country	Region	Color	Vintage	Inventory	Minimum stock	Price	Value
Germany	Baden	red	2019	340	300	$4.00	$1,360.00
Germany	Moselle	white	2020	340	300	$4.00	$1,360.00
Germany	Rheinhessen	white	2020	300			
France	Bordeaux	red	2019	300			
Italy	Piedmont	red	2019	300			
Italy	Verona	red	2019	300			

Undo Calculated Column
Stop Automatically Creating Calculated Columns
Control AutoCorrect Options...

Figure 16.14 Automatically Creating Calculated Column

16.5.11 Working with Structured References

Using structured references instead of those in normal cell ranges simplifies working with table data. These references can point to different ranges within a table, and this applies to formulas both inside and outside the table. Outside formulas must always include the table name, while inside formulas can omit it. The assigned table name refers to all table data except the header and result rows.

Column identifiers are derived from the column names in the table, and they're enclosed in square brackets. They refer exclusively to the column data itself, excluding the column name and any potential totals row. When references are nested, double brackets are required. Additionally, the table supports special identifiers for certain table elements.

Identifiers	Meanings
`#All`	Refers to the entire table, including column names, data, and totals rows
`#Data`	Refers to the data without the totals row
`#Headers`	Refers to the header with the column names
`#Totals`	Refers to the totals row and returns 0 if none exists
`@`	Refers to the data in the current row

If the table is named *Wine storage* as in the example, then the following formula returns the count of blank cells in the current table:

```
=COUNTBLANK(Wine_warehouse[#All])
```

To calculate the difference between inventory and minimum stock for each row, use this formula outside the table:

```
=Wine_warehouse[Inventory]-Wine_warehouse[Minimum_stock]
```

Excel helps you enter such references as soon as you enter an open square bracket, as Figure 16.15 shows.

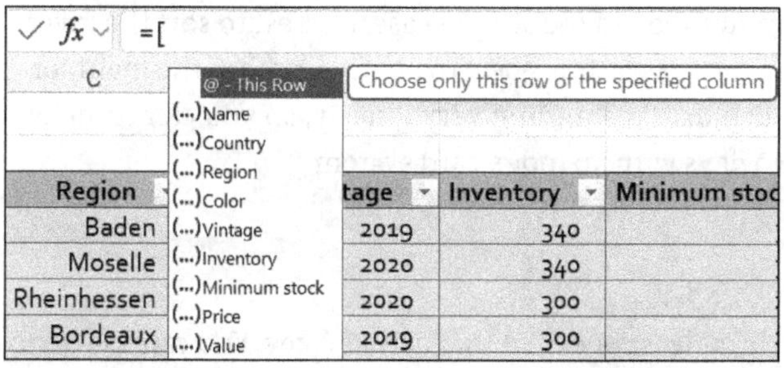

C				
Region		tage	Inventory	Minimum stoc
Baden		2019	340	
Moselle		2020	340	
Rheinhessen		2020	300	
Bordeaux		2019	300	

Figure 16.15 Availability of Structured References When Entering Formulas

To use structured references in this form, enable **Use table names in formulas** under **Excel Options** on the **Formulas** page, within **Working with formulas.** If you later convert the table back to a normal cell range, the structured references will automatically revert to standard references. However, no automatic conversion to structured references occurs when you convert a normal cell range into a table. For example, entering the first formula with structured references in one column is enough; it will automatically copy down to the column's end.

Note that all column names are treated as text strings, even if they include a date or year. If a column name contains special characters—such as spaces, commas, semicolons, periods, and similar characters—use double brackets.

16.6 Sorting Data

Sorting data can be useful in any cell range, but with tables, the need to sort arises especially often. We cover that topic in this section. You can enter data into a table randomly or in somewhat organized fashion. Often, it makes sense to reorder data after the initial entry. For the wine warehouse list, it's logical to sort data by country, region within each country, and wine name within each region. You can also sort by number of the wines.

Any field in the table can serve as the base for sorting. For example, to find which wines have the most bottles in inventory, sort the table by the *Inventory* field. Sorting by selecting **Descending** places the wines with the highest inventory first. You can also sort chronologically by a date or time field, such as the year. Descending order lists the youngest vintages first and the oldest last.

16.6.1 Sorting for Different Purposes

Different scenarios often call for different sorting methods. Before sorting your data, consider which arrangement is the default and which sorting is temporary to answer a specific question. Choose the default sorting based on how you primarily access the data.

For the inventory list of our wines, the question is whether customers usually order by wine name or primarily by item number. In the second case, it's best to sort the inventory list first by item number; in the first case, it's best to use the name as the main sorting criterion. Excel performs such data reshuffling very quickly, so you can complete tasks that might have taken two days with an index card system.

16.6.2 The Sort Key

The program always sorts data by at least one field that acts as a key. You can use up to sixty-four keys at once, and ranges can be sorted not only by values but also by formatting like cell color, font color, and cell icons.

You can't sort directly by part of a field or by combining multiple fields. Instead, you can create additional columns and use formulas to extract, for example, the first three characters of an item number.

Here's an example of a formula that enables sorting by the first eight characters of the name:

```
=LEFT(Name, 8)
```

Preserving the Original Order

In some tables, the order in which data was originally entered is important, and if you sort such a table by a field, then the original order may no longer be recoverable thereafter. This occurs when Excel cannot identify the old order based on the values in a field.

If chronological order matters, you should either use an additional field with time stamps or add a column with sequential numbers before the first sort. The best method is to use the **Home • Editing • Fill • Data Series** command or the fill handle. This sequence number lets you restore the table's original order anytime.

Avoiding Duplicates

If the first key can contain duplicates, you can specify an additional key that determines the order of duplicates. For example, if wines are first sorted by country of origin, then you can use the growing region within each country as the second key, which will only be used if the first key cannot definitively determine the order. If duplicates remain in the second key, you can specify a third key, which would be the wine name in this case.

Sort Order Direction

The sort order can be ascending or descending, and Excel determines the sort direction separately for each key. For example, you can sort storage locations from Z to A, in descending order, while sorting item numbers within those locations in ascending order. By default, Excel sorts rows based on keys in one column or up to sixty-four columns. However, you can also reorder columns based on keys in one or multiple rows.

16.6.3 The Sort Order

Excel sorts data in a range according to the sort order option you choose. You can use Excel's default sort order or define a custom one. The default sort order is set by the language settings that are configured in Windows' Control Panel.

When you're creating tables for other countries, keep in mind that the sort order may change accordingly. The default sort order, unless you specify otherwise, is ascending, as follows:

1. Numeric values are sorted numerically, from the smallest negative number to the largest positive number.
2. Spaces.
3. Character strings that start with special characters are sorted in Unicode order.
4. Strings that start with numbers are sorted in numerical order.
5. Strings that start with letters are sorted alphabetically, with lowercase letters placed before uppercase.

Selecting descending sort reverses the order exactly. The following rules also apply to sorting:

- For Boolean values, FALSE precedes TRUE.
- All error values are considered equal.
- When sorting by a column's content, rows with identical entries in that column remain in order (this is called *stable sorting*). When sorting by a row's content, the same applies to columns.
- Rows with empty cells in the key column always appear at the end of the sorted table, while columns with empty cells in the key row move to the right edge of the table. Sorting order, whether ascending or descending, does not affect this.
- Hidden rows or columns remain unaffected by sorting. This rule excludes rows or columns that are temporarily hidden within an outline.

Excel saves the sorting settings you use in the dialog box until you change them.

Use Caution with Formulas During Sorting

Before sorting, you should verify whether any formulas will be affected. Formulas with relative or mixed references may cause issues, but formulas that reference addresses in the same row or column pose no problem. Since the entire row shifts during sorting, all addresses in the formula shift accordingly. If formulas refer to cells outside the sort range, you must use absolute addresses.

If the formula refers to cells in other rows, sorting may cause undesirable effects. The same applies to references to other columns when sorting by columns. There are two ways you can avoid this situation:

- Sort the data first and then create the formulas.
- Convert the formulas to their results by using **Copy** and **Paste • Paste Values**, then sort.

Sorting by up to Sixty-Four Criteria

Sorting is simplest when a table is sorted by only one key. To do this, place the cell pointer in any cell of the column you want to sort the table rows by, and Excel will automatically expand the selection to the entire table range, excluding the row with the field names. If a total row exists, Excel will exclude it from the sort range. This applies only if the data is a table. For example, if the table has been converted back to a simple cell range, Excel will include a totals row in the sort range. In this case, you must select the correct range before running the command.

On the **Home** tab, in the **Editing** group, you can use the **Sort & Filter** button and then click one of the two sort icons in its menu. The range will sort in seconds. **Sort & Filter** commands also appear separately on the **Data** tab in the **Sort & Filter** group.

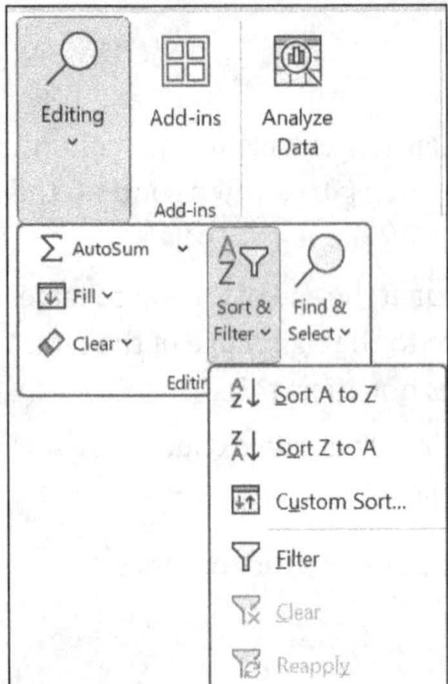

Figure 16.16 Menu of Sort & Filter Button

Depending on whether the active cell is in a text or value column, the **Sort A to Z** or **Sort by Size** command may appear. For date or time entries, the **Sort by Date** or **Sort by Size** labels will appear, provided that the column uses valid date and time formats. Alternatively, you can use the **Sort** command in the context menu of a cell in the relevant table column. Always check the sorting result immediately, and if something goes wrong, you can quickly restore the previous state by using the **Undo** icon.

To sort only a specific column or a set of columns within a table, you should explicitly select the range first. The active cell's position within the range you select determines which column will be sorted, so use the [Tab] key to move the active cell within the range you select if you need to.

	A	B	C	D
1	**Wine storage**			
2				
3	Name	Country	Region	Color
4	Barolo	Italy	Piedmont	red
5	Beaujolais	France	Burgundy	red
6	Bordeaux	France	Bordeaux	white
7	Bourgogne	France	Burgundy	white
8	Brolio	Italy	Tuscany	red
9	Chablis	France	Burgundy	white
10	Chianti	Italy	Tuscany	red
11	Frascati	Italy	Rome	white
12	Freisa	Italy	Piedmont	red
13	Grignolino	Italy	Piedmont	red

Figure 16.17 The Wine List Sorted Ascending by Name

16.6.4 Custom Sorting

Sorting with the icon buttons always sorts rows, so if you want more control, you can click the **Sort** button on the **Data** tab in the **Sort & Filter** group. Usually, you don't need to select a sort range; just place the cell pointer anywhere in the table. The program must be able to identify the correct range, and this always applies when the data forms a closed block with a label row at the top. Excel will then automatically exclude that header row from the sort range.

In the **Sort** dialog box, under **Column**, select the column to sort by first. You also select the column label you want from the list box. If Excel should not treat the first row as headers during automatic selection, clear the **My data has headers** option and Excel will then list column letters under **Column**. This option is only available for a normal cell range; it's disabled for a table range.

Under **Sort by** ❶, select the sorting type, and under **Order** ❷, choose the direction. For example, to sort the wine list first by country, select the **Country** column. To add another sorting criterion, click the **Add Level** ❸ button each time. You can set the order of rows within the same country under **Then by** ❹. This could be the **Region**, for example.

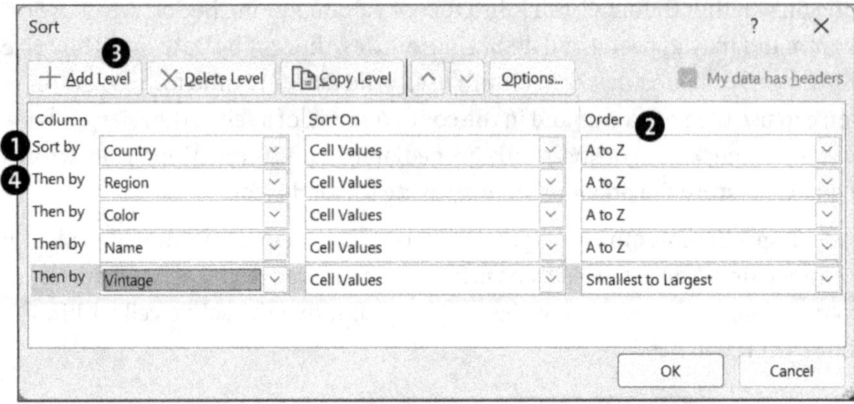

Figure 16.18 Sort Dialog Box

You can then determine the order within the *Region* at the next **Then by** level. If you select **Color**, Excel sorts the wines within the growing region by color. The next level could be the **Name**, with the **Year** as the final level. Figure 16.19 shows the result of sorting by five keys, all in ascending order.

	A	B	C	D	E
1	**Wine storage**				
2					
3	**Name**	**Country**	**Region**	**Color**	**Vintage**
4	Médoc	France	Bordeaux	red	2019
5	Bordeaux	France	Bordeaux	white	2020
6	Sauternes	France	Bordeaux	white	2020
7	Beaujolais	France	Burgundy	red	2019
8	Bourgogne	France	Burgundy	white	2020
9	Chablis	France	Burgundy	white	2020
10	Ortenau	Germany	Baden	red	2019
11	Müller-Thurgau	Germany	Moselle	white	2020
12	Riesling	Germany	Moselle	white	2020
13	Oppenheimer	Germany	Rheinhessen	white	2020

Figure 16.19 Sorting with Five Keys

Sort Options

The **Options** button in the **Sort** dialog box offers additional command features, and we've already covered the options under **Orientation**. Instead of the more common **Sort top to bottom**, you can also choose **Sort left to right** ❺.

Here is a simple example of where this is useful. Say you entered test results in a table, with multiple measurements per test in separate columns, and now, you want to arrange the test columns so the values for a specific measurement are sorted in ascending order. Select the range B3:G11 to exclude column A from sorting, and then, under **Options**, select **Sort left to right** and confirm with **OK**. Next, under **Sort by**, choose the appropriate row and sort direction. Figure 16.20 shows the result when values are sorted by the first test value.

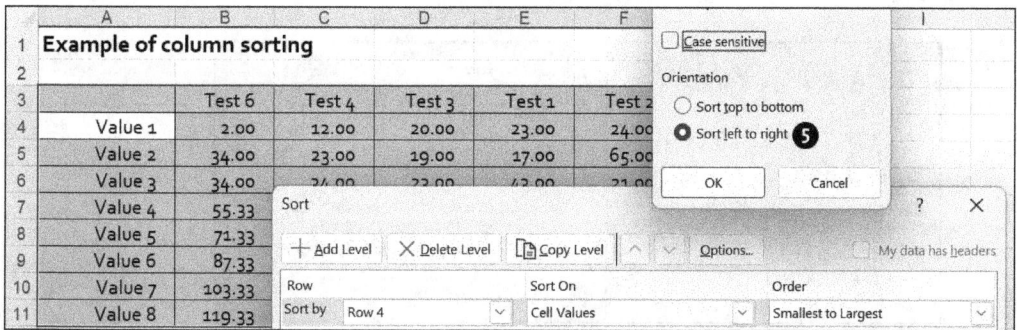

Figure 16.20 Sorting Columns

16.6.5 Sorting by Formatting

You can also sort ranges and tables based on specific formatting criteria, and these functions apply when ranges are manually or conditionally formatted with different cell or font colors. Another option is to sort by symbols that are assigned to cell contents. You can choose sorting criteria in the **Sort** dialog box we just described by using the **Sort On** list box. Then, under **Order**, select the background color, text color, or symbol to appear at the beginning or end—specifying this in the adjacent list box for row sorting as **On Top** or **On Bottom** or for column sorting as **On Left** or **On Right**. You may need to add multiple levels to achieve a specific order.

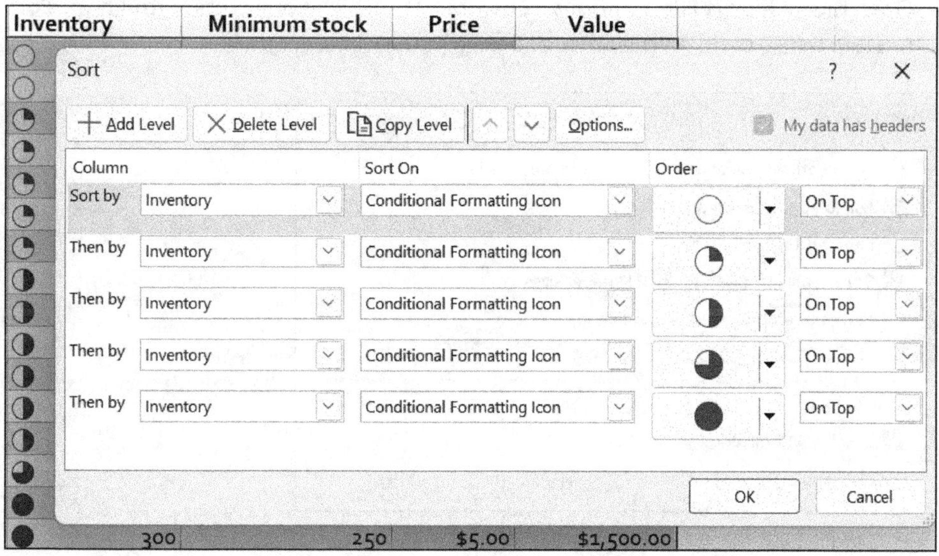

Figure 16.21 Sorting by Cell Icons

Instead of the dialog box, you can use a quick method via the context menu of the cell that contains the color, text color, or symbol that you want to appear first (see Figure 16.22).

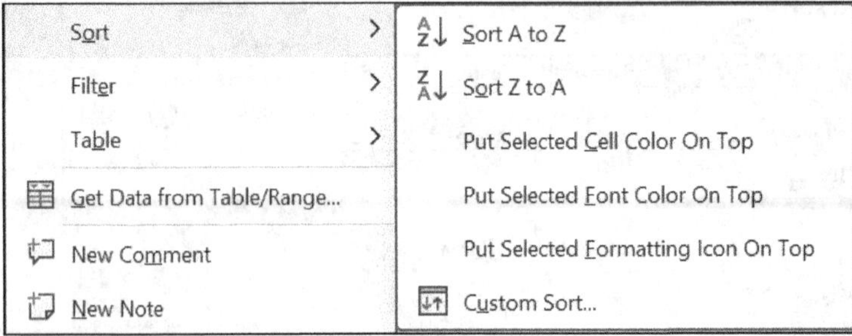

Figure 16.22 Context Menu Options for Sorting

16.6.6 Sorting by Using a Custom Order

In Chapter 2, Section 2.4, you learned how to define custom label sequences and repeatedly create them by using the fill handle or the **Home • Editing • Fill • Data Series** command with the **AutoFill** option. However, you can use custom lists not only to generate labels but also to sort data records in a table.

Using our wine warehouse list as an example, it might be more practical to sort countries of origin by relevance rather than alphabetically. Suppose that in order of trading volume, you mainly trade wines from France, Italy, and Germany. Setting up this sorting is simple:

1. Open the dialog via **File • Options • Custom Lists**, which you access through the **Edit Custom Lists** button on the **Advanced** page in the **General** group.

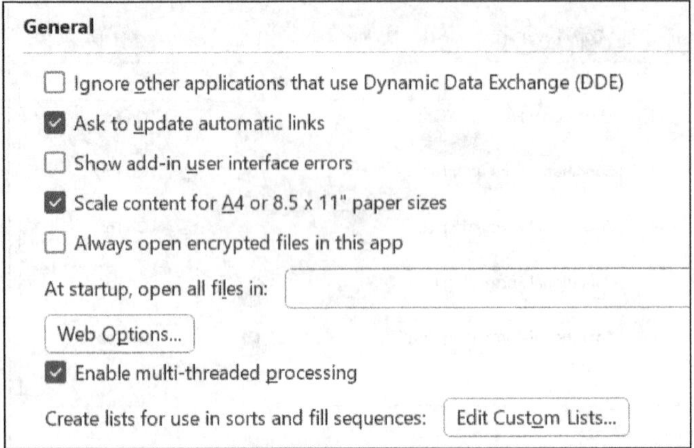

2. Under **New List**, enter the list items for the series: France, Italy, Germany.

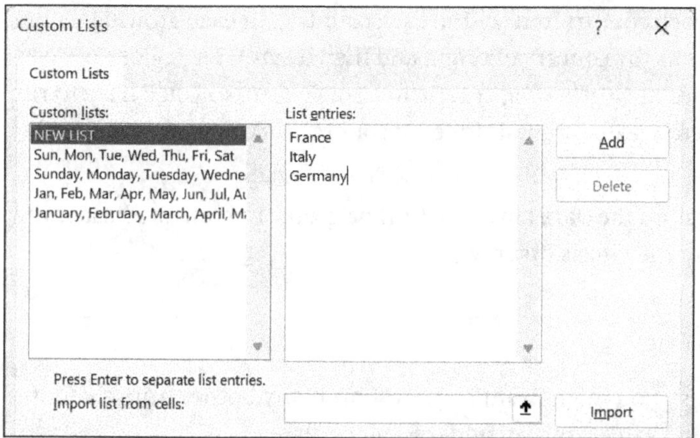

3. Next, place the cell pointer in the inventory list and open the **Sort • Custom Sort** command from the context menu.
4. Select the **Country** column, and under **Order**, choose the previously defined country order.
5. When you confirm with **OK**, the wine list will display the custom order.

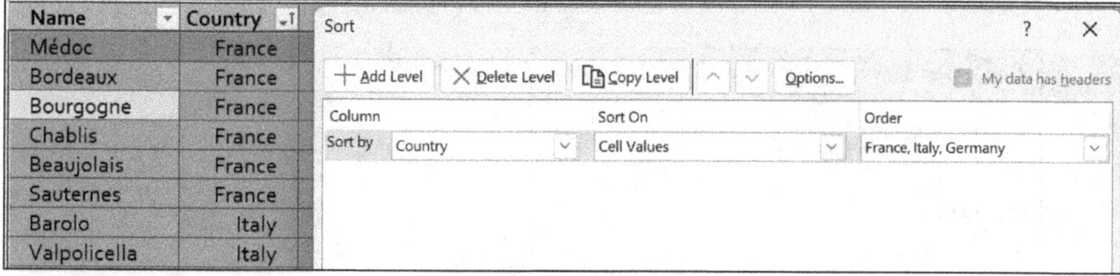

16.7 Group Data

If the wines in the wine warehouse are sorted by country of origin, you can easily create subtotals for each country, and if they are sorted by growing region, you can generate subtotals for each growing region. This function works only with lists that represent a normal range, so you must first convert a table range back by using **Table Design • Tools • Convert to range**.

The prerequisite for automatic subtotals in a table is that several data records share at least one common attribute. For example, multiple wines must come from the same country or the same growing region. The second requirement is that the table data must be sorted exactly by the column that defines the grouping. The **Subtotal** command is usually preceded by the **Sort** command, unless the data was initially entered in a specific order.

To create subtotals for each country and, within each country, for each growing region, you must first sort the table by country of origin and then by growing region. You cannot generate subtotals for both grouping criteria in a single step, so you'll have to run the function twice: first to create subtotals for each country and then for each growing region.

The **Subtotal** command is on the **Data** tab in the **Outline** group. It requires a header row with column headings in the table's first row.

16.7.1 Inserting Subtotals

Returning to the wine list, you might want to know how many bottles from a country are in stock and their total value. You can find this out easily:

1. If the data is sorted by country of origin, place the cell pointer anywhere in the list and choose **Data • Outline • Subtotal**. Excel will automatically select the entire table range, using the top row as the header row.
2. In the dialog box, you'll need to make three main decisions. First, you need to decide which column to use for grouping. Select the column to apply and choose the field name for grouping. In the first example, wines are grouped by country, so **Country** is the grouping criterion you should select. You can only use one criterion at a time.

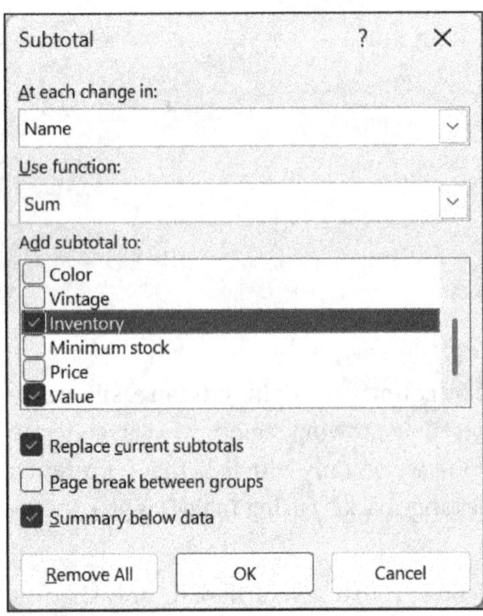

3. Second, you need to make a decision about the content and data type of the next selected fields, specifically about the types of subtotals Excel should calculate. Should it calculate the sum of values, just the count of cases, or another statistical evaluation? The list for **Use function** offers the following evaluation methods:

Methods	Meanings
Sum	Sum of values in the group
Count	Count of cases in the group
Average	Average of values in the group
Max	Highest value in the group
Min	Lowest value in the group
Product	Product of all values in the group (Value 1 × Value 2 × ... Value n)
Count Numbers	Count of data records with a numeric value in the referenced field
StdDev	Estimate of the standard deviation, treating the group data as a sample
StdDevp	Calculation of the standard deviation, treating the group data as the population
Var	Estimate of the variance, treating the group data as a sample
Varp	Calculation of the variance of the data within the group

4. Third, you need to decide which columns or fields to include in the group evaluation. Under **Add subtotal to** in the **Subtotal** dialog, all column headings or field names are listed again. You can check multiple boxes simultaneously if the same type of evaluation is needed. If columns should display different subtotals (sometimes sum and sometimes count), apply the **Data • Subtotal** command multiple times.

A grand total always appears at the end of the table, and it's always calculated from individual data records—not from subtotals, as shown in the formula. To find how many bottles are in stock per country and their value, check the **Inventory** and **Value** fields and select **Sum** as the function. The result will then appear as shown in Figure 16.23.

To determine how many different wines from a country are in stock, select the **Name** and **Count** fields under **Add subtotal to** with **Use function**.

Since the command—as mentioned—can and must be applied multiple times to the same list in some cases, the dialog includes a checkbox that controls how Excel handles existing subtotals. Selecting **Replace current subtotals** deletes the previous subtotals, and that's useful when you need to summarize data by a completely different criterion.

	A	B	C	D	E	F	G	H	I
3	Name	Country	Region	Color	Vintage	Inventory	Minimum stock	Price	Value
4	Ortenau	Germany	Baden	red	2019	340	300	$4.00	$1,360.00
5	Müller-Thurgau	Germany	Moselle	white	2020	340	300	$4.00	$1,360.00
6	Riesling	Germany	Moselle	white	2020	200	200	$4.00	$800.00
7	Oppenheimer	Germany	Rheinhessen	white	2020	300	250	$5.00	$1,500.00
8	Silvaner	Germany	Rheinhessen	white	2020	230	200	$3.00	$690.00
9		Germany Total				1410			$5,710.00
10	Médoc	France	Bordeaux	red	2019	300	250	$5.00	$1,500.00
11	Bordeaux	France	Bordeaux	white	2020	260	200	$4.60	$1,196.00
12	Sauternes	France	Bordeaux	white	2020	170	190	$4.90	$833.00
13	Beaujolais	France	Burgundy	red	2019	200	220	$4.00	$800.00
14	Bourgogne	France	Burgundy	white	2020	230	170	$5.30	$1,219.00
15	Chablis	France	Burgundy	white	2020	230	150	$6.80	$1,564.00
16		France Total				1390			$7,112.00
17	Barolo	Italy	Piedmont	red	2019	300	200	$3.80	$1,140.00
18	Freisa	Italy	Piedmont	red	2019	120	100	$3.90	$468.00
19	Grignolino	Italy	Piedmont	red	2019	230	250	$4.20	$966.00
20	Frascati	Italy	Rome	white	2020	230	280	$3.80	$874.00
21	Ravello	Italy	Salerno	rosé	2019	200	150	$5.90	$1,180.00
22	Brolio	Italy	Tuscany	red	2019	230	200	$4.70	$1,081.00
23	Chianti	Italy	Tuscany	red	2019	120	100	$5.00	$600.00
24	Valpolicella	Italy	Verona	red	2019	300	250	$5.80	$1,740.00
25	Soave	Italy	Verona	white	2020	170	200	$3.90	$663.00
26		Italy Total				1900			$8,712.00
27		Total				4700			$21,534.00

Figure 16.23 Subtotals per Country for Inventory and Value

16.7.2 Calculations for Subgroups

To add a subgroup by growing region alongside grouping by country, apply the **Subtotal** command again and clear the **Replace current subtotals** checkbox; otherwise, only one group summary will remain, by growing region. You must sort the list accordingly beforehand—first by country and then by growing region.

Check the **Page breaks between groups** box to force a page break after each subtotal, which is useful when groups are very large. **Summary below data** is selected by default, and if you uncheck it, subtotals will appear above each group.

Excel automatically structures the table so you can hide individual items in each group. You can also hide subtotals, leaving only the grand total visible. You create subtotals with the `SUBTOTAL()` table function. The first argument specifies the calculation type, and the second defines the range to evaluate. The function excludes cells in the column from being counted again if they already contain a `SUBTOTAL()` function.

To remove the subtotals and outline from the table sheet, reopen the **Subtotals** dialog and click the **Remove All** button.

Chapter 17
Data Queries and Data Extracts

Anyone who takes the trouble to gather a large amount of information in a table and organize it neatly does not want to end up with a useless data dump. The information should help support decisions and make work in the applicable area more effective, so it's essential for such a data collection to provide exactly the information users need in a wide variety of situations without them having to expend much effort.

17.1 What's the Best Way to Formulate Queries?

Excel only "understands" questions when they're formulated in a specific way. When you ask questions of a table, you need to specify certain restrictions or conditions, such as *Release_Year = 2020* in a DVD list. This lets you extract the exact data records that are important for a decision or process from the large amount of available data. In Excel, this process is called *filtering*. Sometimes, you need to use multiple filters to find the data that answers your question. Two basic methods are available: simple **Filter** and **Advanced** filter. Simple **Filter** is designed for quick, basic queries—while the **Advanced** filter allows more complex queries but is somewhat more cumbersome, and it partly follows procedures from older Excel versions.

17.2 Filtering Relevant Data

Using the wine warehouse as an example, if you want to extract a specific subset of wine, such as only wines from France, the **Sort & Filter** group on the **Data** tab offers the **Filter** option, which you can use to quickly achieve this:

1. Place the cell pointer in any cell within the table. (The function requires the table to have a header row with column names.)
2. Select the **Data • Sort & Filter • Filter** command and filter arrow buttons will appear next to each column name in the table.

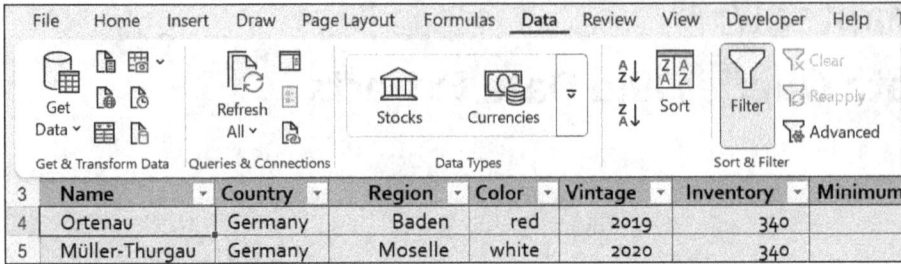

3	Name	Country	Region	Color	Vintage	Inventory	Minimum
4	Ortenau	Germany	Baden	red	2019	340	
5	Müller-Thurgau	Germany	Moselle	white	2020	340	

3. Click or tap one of the arrows to display all unique values in that column as choices, along with special selection criteria. The second field shows the **Country** of origin, and the filter menu lists the names of different countries. If the list does not fit, you can resize the menu by dragging the handles at the bottom right.
4. Select one of these countries, such as **France**, to display only the records from that country.

5. The **Total** row will then show only the sum of the values for that country.

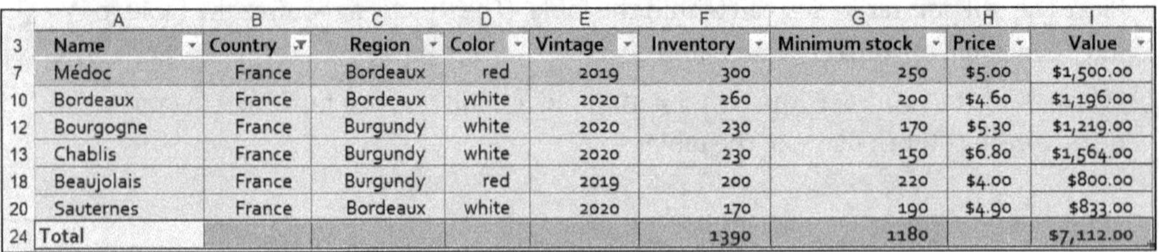

	A	B	C	D	E	F	G	H	I
3	Name	Country	Region	Color	Vintage	Inventory	Minimum stock	Price	Value
7	Médoc	France	Bordeaux	red	2019	300	250	$5.00	$1,500.00
10	Bordeaux	France	Bordeaux	white	2020	260	200	$4.60	$1,196.00
12	Bourgogne	France	Burgundy	white	2020	230	170	$5.30	$1,219.00
13	Chablis	France	Burgundy	white	2020	230	150	$6.80	$1,564.00
18	Beaujolais	France	Burgundy	red	2019	200	220	$4.00	$800.00
20	Sauternes	France	Bordeaux	white	2020	170	190	$4.90	$833.00
24	Total					1390	1180		$7,112.00

All data that does not meet the filter criterion will be temporarily hidden, and you'll be able to tell because the row numbers of the displayed records will remain unchanged.

The button in the filtered column will show a filter icon, so you won't have to remember which column has a filter applied. Hover the mouse pointer over it to display the current filter criterion.

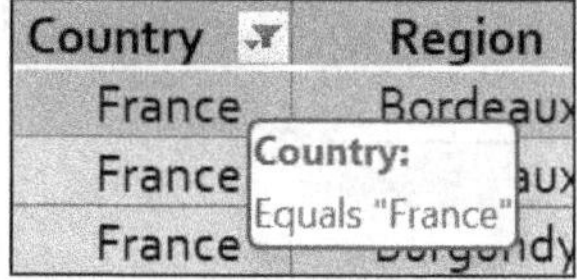

Figure 17.1 Display of Column's Filter

17.2.1 Location-Independent Filtering and Sorting

The previous chapter mentioned that Excel moves column labels into the row with preset column headers for tables that extend beyond the screen height, so you can always see the content in each column. This also applies to the filter buttons when they're enabled. The key advantage is that no matter how far you navigate into a table, you always have access to the filter button menu to apply filters or sorting. You no longer need to navigate to the table header each time.

	Name	Country	Region	Color
4	Ortenau	Germany	Baden	red
5	Müller-Thurgau	Germany	Moselle	white

Figure 17.2 Filter Button in Workbook's Column Headers

17.2.2 Text Filters

The filter button menu offers additional filter options beyond the column values; these depend on the column's data type. For text, as in this case, **Text filters** are available with the following options:

Options	Effects
Equals	Displays all records with the selected value
Does not equal	Displays all data records that don't match the selected value
Begins with	Filters by the first characters of the value
Ends with	Filters by the last characters of the value
Contains	Shows data records where the specified text appears
Does not contain	Shows data records where the specified text does not appear
Custom filter	Allows entering up to two criteria for a single column

To see all countries except Italy, use **Text filter • Does not equal**, and then, in the **Custom Autofilter** dialog, choose **Italy** in the second list box as the value. If the comparison value is text, do not enter it in quotation marks. You can also use wildcard characters: ? stands for exactly one character and * stands for any number of characters. The expression matches r* filters: both **Red** and **Rosé**, for example.

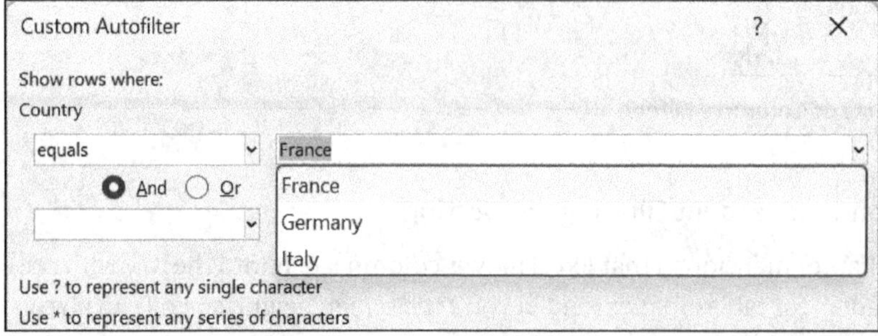

Figure 17.3 Custom Autofilter

17.2.3 Number Filters

Your filter options expand when you're working with numeric values. To filter by years, for example, in addition to the existing year values under **Number filter**, you can use the following options:

Options	Effects
Equals	Displays all records with the selected value
Does not equal	Displays all data records that do not match the selected value
Greater than	Shows all values greater than the specified value
Greater than or equal to	Shows all values greater than or equal to the specified value
Less than	Shows all values less than the specified value
Less than or equal to	Shows all values less than or equal to the specified value
Between	Shows all values between the two specified values
Top 10 ...	Instantly delivers the top values, either absolute or percentage-based
Above average	Shows all values above the average of the column
Below average	Shows all values below the average of the column
Custom filter	Allows entering up to two criteria for a single column

17.2.4 Date Filters

Date filters are very useful for data containing dates. When you open one, a tree structure appears first, and it allows you to select specific years or months.

Figure 17.4 Selecting Date Values as Filters

The menu for **Date filter** also offers the following options:

Options	Effects
Equals	Displays all data records with the selected date
Before	Displays all values before the specified date
After	Displays all values after the specified date
Between	Displays all values between two specified dates
Tomorrow	Displays all values dated the next day
Today	Displays all values dated today
Yesterday	Displays all values dated the previous day
Next Week	Displays all values dated next week
This Week	Displays all values dated within the current week
Last Week	Displays all values dated within last week
Next Month	Displays all values dated within next month
This Month	Displays all values dated within the current month

Options	Effects
Last Month	Displays all values dated within last month
Next Quarter	Displays all values dated within the next quarter
This Quarter	Displays all values dated within the current quarter
Last Quarter	Displays all values dated within the last quarter
Next Year	Displays all values dated within next year
This Year	Shows all values with a date from the current year
Last Year	Shows all values with a date from last year
Year to Date	Shows all values in the current year before the current date
All Dates in the Period	Offers the four quarters and individual months of the current year for selection
Custom filter	Allows entering up to two criteria for a single column, with calendar controls that assist in selecting values

In Figure 17.5, data for a specific quarter is filtered from values like page views for a website.

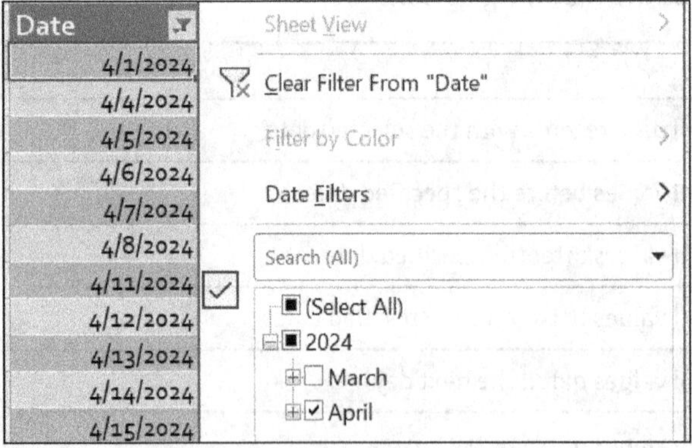

Figure 17.5 Filter with Data for Second Quarter

17.2.5 Color Filters

Another option in the filter button menu is **Filter by Color**. If the column contains conditional formatting with different cell colors, font colors, or cell icons, all will appear in the submenu of this option for selection. The pattern you select will determine which data will be displayed. In Figure 17.6, for the wine warehouse inventory, three symbols indicate stock status.

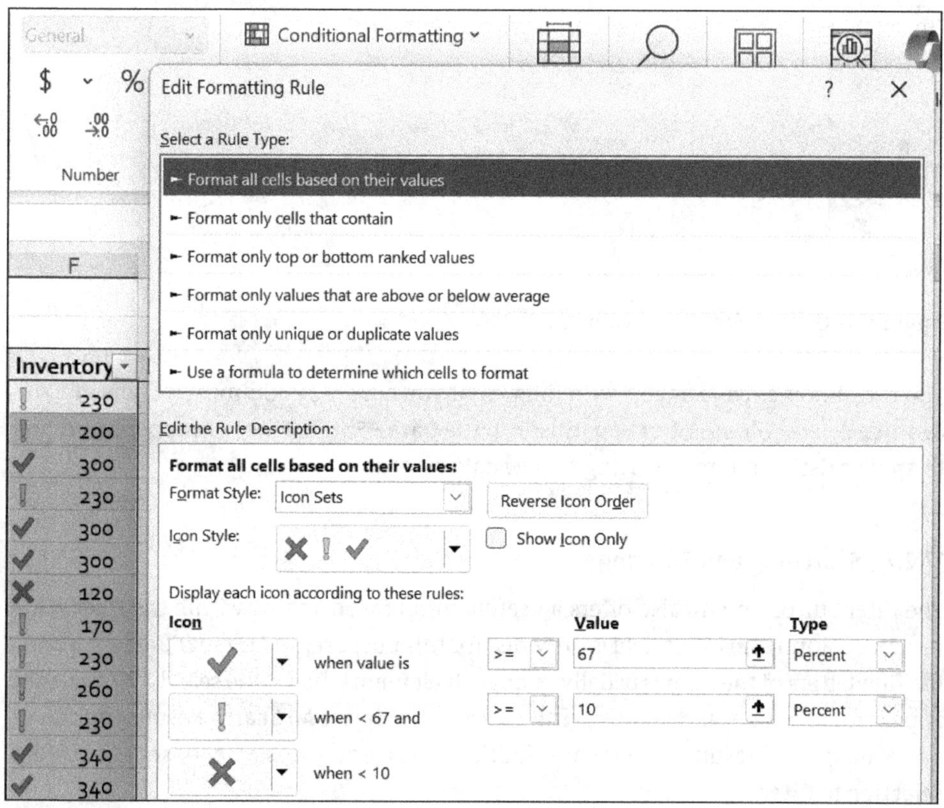

Figure 17.6 Conditional Formatting for Inventory Assessment

When you select the color filter for this column, the symbols for possible ratings appear. To view critically low inventory at a glance, select the **X** symbol.

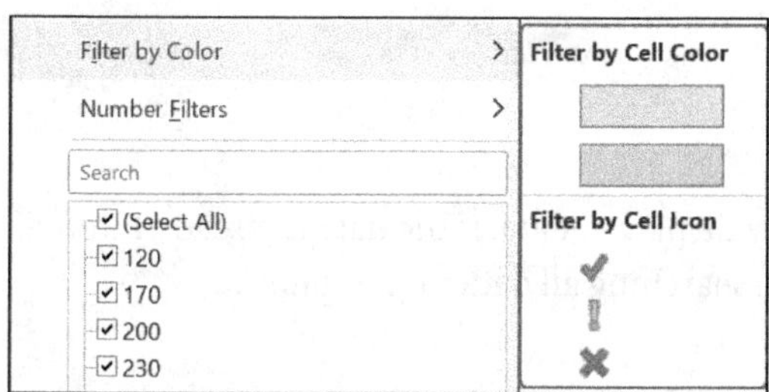

Figure 17.7 Selecting Cell Icon as Filter

17.2.6 Sorting

All filter button menus also offer options to sort data in ascending or descending order. Sorting also includes the **Sort by Color** option. If the column contains conditional formatting with different cell colors, font colors, or cell icons, then they appear as options in the submenu. The pattern you select determines the starting point of the sorting.

Inventory
✔ 300
✔ 300
✔ 300
✔ 340
✔ 340
✔ 300
! 230
! 200
! 230

Figure 17.8 Column Sorted by Symbols

You can always expand a table with filters that you enable by adding new columns. You can also delete columns, but you must run the **Data • Sort & Filter • Reapply** command to apply existing filters or sorting to new data records.

17.2.7 Searching and Filtering

The filter buttons menu also offers a useful option for searching within the filter criterion in a column. An input field is provided for this purpose, and for text fields, entering the first letters of the item is usually enough. If elements match the search criteria, they appear for selection, and you can either accept the **(Select All Search Results)** option or select individual results if there are multiple ones. You can also choose **Add current selection to filter.**

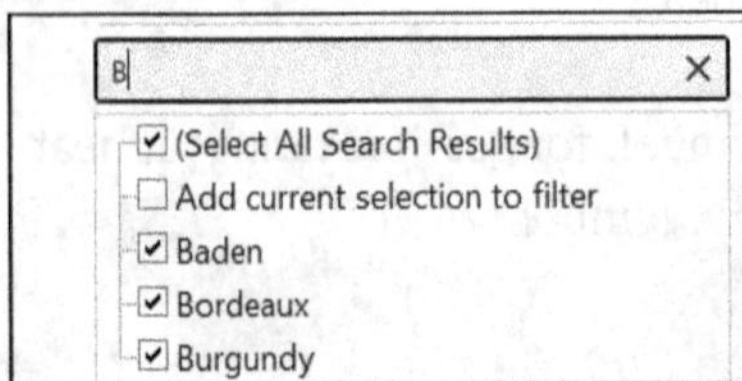

Figure 17.9 Filtering by Search Criterion

The search always applies to the currently displayed data. If the data is already filtered, you must first deactivate the filter before searching all data in the column.

Filtering Out Empty Fields

If the column contains empty fields, the **(Blanks)** filter option is available. It shows data records where this column is still empty, and it's useful for identifying records that are missing specific information.

Multiple Filtering

You can chain multiple filters as needed. For example, to see only wines from a specific growing region among the filtered French wines, you can apply an additional filter to refine your selection. To do this, open the **Filter** menu at **Region** and select, for example, **Bordeaux**.

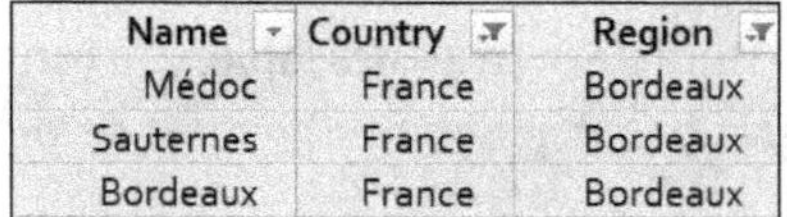

Name	Country	Region
Médoc	France	Bordeaux
Sauternes	France	Bordeaux
Bordeaux	France	Bordeaux

Figure 17.10 Filter Showing Only Bordeaux Wines from French Selection

Undoing Filtering

You can undo table filtering in three steps, depending on your next action:

1. Select **(Select All)** in the list box to remove the filter from that column. Filters in other columns remain unaffected. If you select *Bordeaux* as the growing region within France, the **(Select All)** option under **Growing region** will display all French growing regions again. Alternatively, you can use the **Clear Filter from "Column name"** option.
2. To keep the filter arrows but clear only the current filter, use the **Clear** command in the **Sort & Filter** group.
3. To remove all filters and display all data records at once, click or tap the **Filter** button again in the **Sort & Filter** group.

The First and the Last

To quickly find the ten most expensive wines in stock, choose **Number filter • Top 10** in the **Price** column. A small dialog box will appear, and you'll just need to confirm it.

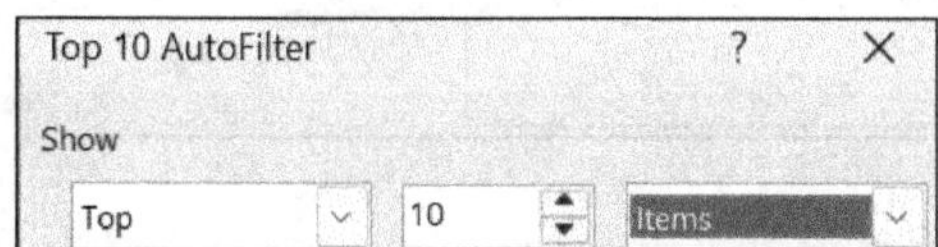

Figure 17.11 Top 10 AutoFilter Dialog Box

The result may not be fully satisfactory if the ten most expensive wines are found but not shown in sorted order. A descending sort by price is appropriate here.

	A	B	C	D	E	F	G	H	I
3	Name	Country	Region	Color	Vintage	Inventory	Minimum stock	Price	Value
6	Oppenheimer	Germany	Rheinhessen	white	2020	300	250	$5.00	$1,500.00
7	Médoc	France	Bordeaux	red	2019	300	250	$5.00	$1,500.00
9	Valpolicella	Italy	Verona	red	2019	300	250	$5.80	$1,740.00
10	Bordeaux	France	Bordeaux	white	2020	260	200	$4.60	$1,196.00
12	Bourgogne	France	Burgundy	white	2020	230	170	$5.30	$1,219.00
13	Chablis	France	Burgundy	white	2020	230	150	$6.80	$1,564.00
16	Brolio	Italy	Tuscany	red	2019	230	200	$4.70	$1,081.00
19	Ravello	Italy	Salerno	rose	2019	200	150	$5.90	$1,180.00
20	Sauternes	France	Bordeaux	white	2020	170	190	$4.90	$833.00
23	Chianti	Italy	Tuscany	red	2019	120	100	$5.00	$600.00
24	Total					2340	1910		$12,413.00

Figure 17.12 Ten Most Expensive Wines in Sorted Order

The dialog box also offers other options. Under **Show**, you can select **Bottom** instead of **Top**, and the number of items doesn't have to be ten; you can limit the selection to three or expand it to twenty. Also, instead of the top ten items in the table, you can query the top 10% by choosing **Percent** instead of **Items** in the third field. In a sales table, you can quickly find which products belong to the top 5% group that generates the most revenue.

17.2.8 Filtering and Sorting by Cell Values

To filter by the value of a selected cell in a column, you can use the context menu of that cell. For example, to show only wines from the Piedmont growing region, select **Filter by Selected Cell's Value** from the context menu of a cell displaying that value. The context menu also shows a delete command for this filter. You can filter by values, cell color, font color, or an assigned icon. For example, if a column with conditional formatting has different manually assigned cell colors, you can select the cell with the color you want to filter by and then choose **Filter by Selected Cell's Color** from the context menu.

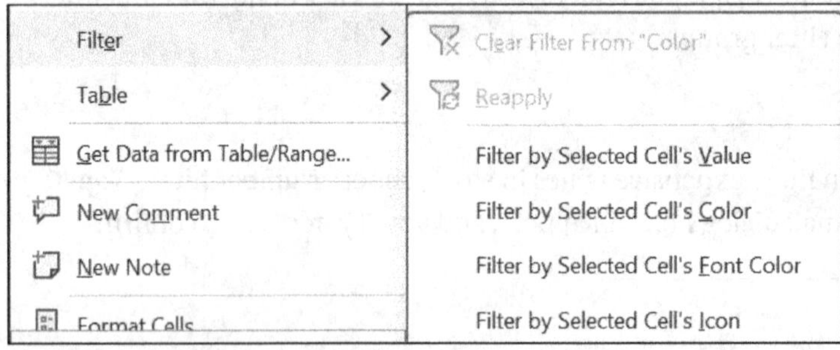

Figure 17.13 Filter Menu for Cell

The context menu also offers **Sort** options you can use to move the selected cell color, font color, or cell icon to the top of the column.

17.2.9 Combining Filters

You can create combined or alternative filters for each column instead of single filter conditions. If you want to find wines with more than 200 but fewer than 300 units in inventory, follow these steps:

1. Select the **Number Filters • Custom Filter** option in the **Inventory** column.
2. Excel opens a dialog box that allows you to enter up to two criteria related to the column. Enter them as a logical comparison in which each cell's content in the column is compared to a specific value.

3. The dialog box already shows the field name of the selected column. Choose the comparison operator in the first list box. Operators are translated into text: = is translated into *equals*, > is translated into *is greater than*, and so on. Here, you select **is greater than**.
4. In the next input field, select the comparison value from the list or enter it directly—here, you enter 200.
5. To enter the second condition, select **And**, choose the **is less than** operator, and enter 300 as the value.

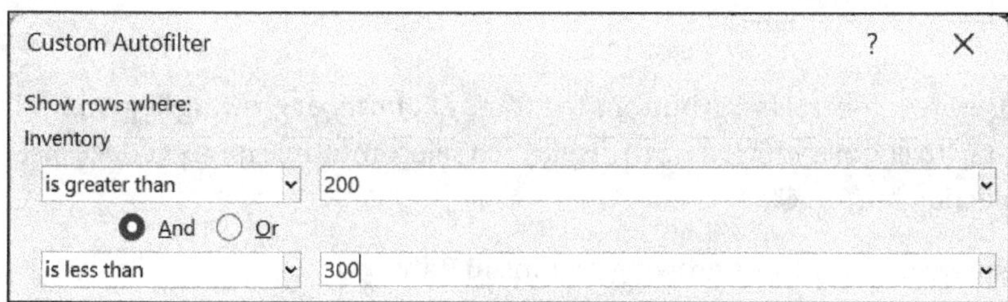

Only Simple Comparisons Are Allowed

The **Filter** dialog box does not support comparisons with values in other columns. You cannot enter a filter like *Inventory* < *Minimum* stock here, so for such comparisons, use the **Advanced Filter** function.

Applying Multiple Filters

You can define filter criteria for multiple fields simultaneously, and only data that meet all criteria will be displayed. For example, if you select **red** under **Color** and **Bordeaux** under **Growing region**, only red wines from Bordeaux will be displayed.

17.2.10 Filtering with Slicers

Since Excel 2013, you've been able to apply the slicer feature to tables. (It was previously available only for PivotTables and charts.) Because we cover this convenient filtering tool in detail for PivotTables, we refer you to Chapter 18 for that and provide only a simple example here. To quickly filter wines by specific countries and growing regions, follow these steps:

1. Select a cell in the wine warehouse table and click the **Slicer** button in the **Filters** group on the **Insert** tab.
2. Select the columns where you want slicers to appear.

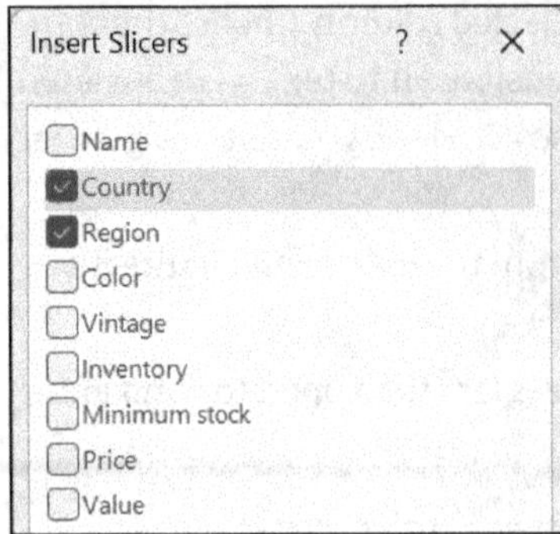

3. Excel will insert two controls with buttons to filter by **Country** or growing **Region**, and you click or tap to choose which data to display. To select multiple items at once, use the **Multiselection** button ❶.

Figure 17.14 displays data for three growing regions in Italy.

	A	B	C	D	E	F	G	H	I
1	**Wine storage**								
2									
3	Name	Country	Region	Color	Vintage	Inventory	Minimum stock	Price	Value
8	Barolo	Italy	Piedmont	red	2019				
14	Grignolino	Italy	Piedmont	red	2019				
15	Frascati	Italy	Rome	white	2020				
19	Ravello	Italy	Salerno	rose	2019				
22	Freisa	Italy	Piedmont	red	2019				

Country: Italy, France, Germany, (blank)

Region ❶: Piedmont, Rome, Salerno, Tuscany, Verona, Baden, Bordeaux, Burgundy

Figure 17.14 Two Slicers for Convenient Filtering

17.3 Complex Queries with Advanced Filters

Using the filter function to reduce a table to data that's needed for further processing with simple criteria is straightforward. You can complete most table queries quickly this way, or if this process is insufficient, as it is when you're comparing inventory with minimum stock in our example, you must use the **Advanced Filter** dialog. You access it from the **Data** group • **Sort & Filter** by using **Advanced**.

17.3.1 Table and Criteria Ranges

A special cell range defines the query criteria, and this criteria range must include at least two rows. The first row contains column names, and the second and any additional rows below are reserved for criteria entry.

Usually, you don't need to manually select or name the table range for this function. Excel highlights the appropriate range based on the previously described rules when the cell pointer is inside the table, but the specified table range is required to store the criteria that determine which data records are selected. Column names must exactly match the table's column names, except for additional names used for calculated criteria, which we'll explain later. The easiest way to accomplish this is to copy the header row from the table into the first row of the criteria range.

Not all field names need to appear in the criteria range; you can drop individual field names. For combined criteria, you can use a column name multiple times. You can also place the criteria range on the worksheet to prevent accidental changes from inserting or deleting columns or rows.

If the criteria range is created within the sheet containing the table, you should insert new rows above the table for safety. This allows the table to grow downward without risking a collision with the criteria range. Another safe option is to place the criteria range on a separate sheet. When you set the criteria range in the **Advanced Filter** dialog box, Excel automatically names it *Criteria*. You can quickly access this range by using [F5] or the name box, and you can also assign a different name.

	A	B	C	D	E	F	G	H	I
1	Name	Country	Region	Color	Vintage	Inventory	Minimum Stock	Price	Value
2				red		>=200			
3				rose					
4									
5	Name	Country	Region	Color	Vintage	Inventory	Minimur		
7	Ravello	Italy	Salerno	rose	2019	200			
8	Valpolicella	Italy	Verona	red	2019	300			
10	Médoc	France	Bordeaux	red	2019	300			
14	Brolio	Italy	Tuscany	red	2019	230			
16	Grignolino	Italy	Piedmont	red	2019	230			
18	Ortenau	Germany	Baden	red	2019	340			
19	Beaujolais	France	Burgundy	red	2019	200			
23	Barolo	Italy	Piedmont	red	2019	300			
26	Total								

Figure 17.15 Criteria Range Above Table Range

17.3.2 Data Extracts in the Output Range

With the **Advanced Filter** function, you can copy filtered data to another area of the sheet instead of filtering within the table. Use this whenever you want to work with a filter extract over time, independent of the original table.

You cannot filter the extract directly into another sheet or workbook; instead, you must work in reverse. Activate the sheet for the output range and then select **Data • Sort & Filter • Advanced**. You must explicitly select the table range on the other sheet and the criteria range.

Defining the Output Range

If you want to use an output range, you may need to prepare before opening the **Advanced Filter** dialog. You have three options:

- Specify only the top-left cell of the output range, and Excel will copy all columns, including headers, into the output range.
- Specify a row with column names as the output range.
- Select a range for the entire output, with column names in the first row.

If you want to specify column names for the output range, the easiest method is to copy the column names or the entire column name row from the table into the output range. Use a free range below the table for this. If you always use the same output range, assign a name to it. You can also drop individual column names or change their order.

	A	B	C	D	E	F	G	H	I
1	Name	Country	Region	Color	Vintage	Inventory	Minimum Stock	Price	Value
2				red		>=200			
3				rose					
4									
5	Name	Country	Region	Color	Vintage	Inventory	Minimur		
7	Ravello	Italy	Salerno	rose	2019	200			
8	Valpolicella	Italy	Verona	red	2019	300			
10	Médoc	France	Bordeaux	red	2019	300			
14	Brolio	Italy	Tuscany	red	2019	230			
16	Grignolino	Italy	Piedmont	red	2019	230			
18	Ortenau	Germany	Baden	red	2019	340			
19	Beaujolais	France	Burgundy	red	2019	200			
23	Barolo	Italy	Piedmont	red	2019	300			
26	Total								
27									
28	Name	Country	Region	Color	Vintage	Inventory	Price		
29	Ravello	Italy	Salerno	rose	2019	200	$5.90		
30	Valpolicella	Italy	Verona	red	2019	300	$5.80		
31	Médoc	France	Bordeaux	red	2019	300	$5.00		
32	Brolio	Italy	Tuscany	red	2019	230	$4.70		
33	Grignolino	Italy	Piedmont	red	2019	230	$4.20		
34	Ortenau	Germany	Baden	red	2019	340	$4.00		
35	Beaujolais	France	Burgundy	red	2019	200	$4.00		

Figure 17.16 Example of Filtering with Output Range

Swapping Columns

To change the order of columns in a table, you can use an output range where the column names are rearranged accordingly and then use the **Advanced Filter** dialog without criteria. Leave the **Criteria range** field empty so all data from the original table copies to the output range.

17.3.3 Checking Inventory with the Advanced Filter

Now, we'll test the operation of the advanced filter on a previously mentioned problem. The program should filter wine records in which inventory falls below the minimum stock to inform you which wines you need to reorder. Follow these steps:

1. Create a criteria range with a column name and a criterion below it. The following figure shows two rows that have been inserted for this purpose. Here, the column name must be new and must not match any column name in the table, since the criterion compares values from two fields rather than a field with a constant value. This is a calculated criterion. Therefore, you enter the column name **Requirement** into cell A1. Enter the following formula in A2: `=F5<G5`

   F5 is the address of the first cell under the **Inventory** column name, and G5 is the address of the first cell under **Minimum Stock**. Make sure to enter the cell references as relative. Excel will return `TRUE` if the condition is met in the first data row of the table; otherwise, it will return `FALSE`.
2. After you enter the necessary data in the criteria range, move the cell pointer back to the table. Then, access the **Data • Sort & Filter • Advanced** command option.
3. Select under **Action** how Excel should filter the data. The default setting is **Filter the list, in-place**, which means Excel will filter normally and temporarily hide the filtered-out data.
4. Excel usually fills the **List range** field automatically when you place the cell pointer correctly within the table. Alternatively, you can select the range manually or insert a range name by pressing [F5].
5. Accurate referencing of the criteria range is essential, so select the range or enter the reference or name if it's assigned.
6. Check the **Unique records only** box to hide or exclude duplicate data records from the extract. Duplicates occur when data records are mistakenly entered twice.
7. After you confirm the dialog box, Excel will filter out all data records from the entire table that meet the criteria in the criteria range. The following image shows the data records of wines that need to be reordered.

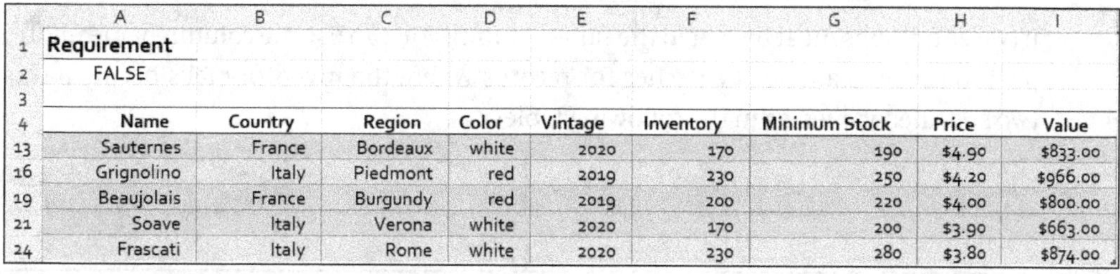

	A	B	C	D	E	F	G	H	I
1	Requirement								
2	FALSE								
3									
4	Name	Country	Region	Color	Vintage	Inventory	Minimum Stock	Price	Value
13	Sauternes	France	Bordeaux	white	2020	170	190	$4.90	$833.00
16	Grignolino	Italy	Piedmont	red	2019	230	250	$4.20	$966.00
19	Beaujolais	France	Burgundy	red	2019	200	220	$4.00	$800.00
21	Soave	Italy	Verona	white	2020	170	200	$3.90	$663.00
24	Frascati	Italy	Rome	white	2020	230	280	$3.80	$874.00

Excel retains the range entries in the **Advanced Filter** dialog box until you change them. If you use the command multiple times, you only need to update it if the ranges in the table sheet have changed.

17.3.4 What Selection Criteria Are Available?

Use a selection criterion to specify which data Excel should filter from a table. The criterion is always stated positively as the condition a data record must meet to pass the filter. For example, the **Color = red** criterion lets all red wines pass through the filter while filtering out all non-red wines.

The simplest form of a criterion in the criteria range applies to a single column, and you either enter a constant value under the column name as a criterion or combine a logical operator with a constant value. This matches the syntax you saw in the custom filter dialog box.

Combined Criteria

If you enter values in multiple fields of the first row in the criteria range, the program will treat it as a combined criterion. It will search for data records where all specified fields exactly match. For example, if you enter "Piedmont" under **Region** and "red" under **Color** in the criteria range, the program will search for red wines from Piedmont—that is, for records that meet both criteria simultaneously. This matches the logical `AND()` function:

```
=AND(Region="Piedmont", Color="red")
```

Before you enter a new criterion, always check that no entries from previous queries remain, as they may act as unintended additional criteria.

Region	Color
Piedmont	red

Figure 17.17 Multiple Criteria to Be Searched for Simultaneously

If two conditions must be met in the same column, you can use the column name multiple times. For example, it searches for records where the inventory falls between the two specified values, as in the following table:

Inventory	Inventory
>200	<300

Alternative Criteria

You can also use criteria can as alternatives by entering the criteria one below the other in a column of the criteria range. For example, under the column name **Color** in the criteria range, you can enter "red" and "rosé," one below the other. Then, all wines with either of the two colors will be found. In this case, you've simply expanded the criteria range by one row.

Region	Color
	red
	rose

Figure 17.18 Alternative Criteria

To query wines from either Italy or Rheinhessen, place the criteria on separate rows; otherwise, Excel will treat them as `AND()` conditions:

Country	Region
Italy	
	Rheinhessen

You can also mix `AND()` and `OR()` combinations. Here, wines must be red but can come from either Italy or Rheinhessen:

Country	Region	Color
Italy		red
	Rheinhessen	red

For comparative search criteria, a cell's content in the table can be compared to a string, a number, a Boolean, or an error value. If you only enter an equal sign as a criterion, Excel will search all records with no entry in the critical field—that is, empty cells. If you enter <> as a criterion, Excel will search all records with any entry in the relevant column, regardless of type.

Text Strings as Criteria

When you're using only a text string as a criterion, Excel searches for records where the table field matches the entry in the criteria range. Consider this criterion:

Color
red

It's a shortened form of this criterion:

Color
="red"

Exact Text Comparisons

Text comparisons usually ignore case. Also, a criterion like *red* finds not only *red* but also *reddish*. To match exactly, ="=red" is what you should enter. Instead of matching a text string, you can use the position in alphabetical order as a criterion.

This criterion excludes white wines, while it includes red and rosé wines.

Color
<white

This condition includes red, rosé, and white wines.

Color
>r

And this condition selects all wines except French ones:

Country
<>France

When working with text strings, names, and labels, users often face the problem of not knowing the exact spelling. In such cases, wildcard characters can help.

Wildcard Characters	Effects
*	This represents any number of characters in a text string.
?	This represents exactly one character in a text string.
~	Place this character before a question mark or asterisk to search for wildcard characters themselves. The tilde prevents these characters from acting as wildcards.

The following table shows examples of such approximate search criteria.

Examples	Results
*wand	Finds *band, land,* and *shorthand*
M*er	Finds *Maker, Mixer,* and *Mother*
w??en	Finds *woven* and *widen*
~*(?,~)-character	Finds *-*character* (or ?, or ~) when they appear in the cell content

Numeric Values as Criteria

If you enter a number in a criteria range field, the value must exactly match the value in the data record you are searching for. The format does not have to match. For example, if you enter a number in standard format in the criteria range but the corresponding column in the table is formatted with a currency symbol, it does not affect the search.

Problems can occur if the number in the table is rounded for display but stored internally with multiple decimal places. This often happens when the number results from division. If the criteria range shows 3.33 but the table displays the result of 10/3, the data record will not be found. In such cases, it helps to use a calculated search criterion with a formula like this:

```
=ROUND(F9,2)=3.33
```

Otherwise, you can use the comparison operators we've already mentioned when comparing numbers.

17.3.5 Searching with Calculated Criteria

The simple comparisons we covered earlier do not always suffice to filter data that's relevant to a specific problem. Often, you'll need to compare values from different columns, or you'll need to multiply values by a factor and then the compare the result with a value outside the table. You can use calculated criteria in Excel for all these cases. Earlier, you saw the example of calculating the difference between inventory and minimum stock. For calculated criteria some special rules apply:

- The calculated criterion requires a unique column name that does not appear elsewhere in the table.
- The formula must always be logical and return `TRUE` or `FALSE`.
- The formula must reference at least one field in the table, either by the address of the first cell in the relevant column or by a column name.

When referencing the first value in a column, the formula in the criteria range displays the corresponding truth value. If the criterion is met in the first row, it shows `TRUE`; otherwise, it shows `FALSE`. Note that the cell reference must be relative in this case. With a calculated criterion, you can create complex logical comparisons by using `AND()` or `OR()`, or you can use other table functions.

This function searches for wines with an above-average number of bottles in stock:

```
=F6>AVERAGE($F$6:$F$25)
```

If a calculated criterion refers to a value outside the table, such as a constant, the reference must be absolute.

[+]

If the Data Search Fails...

Suppose you tried to query a table using an **Advanced Filter**, and despite the criterion, all data records appeared. You'd need to check whether the criteria range used an incorrect column name. If it didn't, you'd need to verify whether the criteria range still contained a blank row from a previous query or if a field accidentally included a space. Empty fields allow any values, and when cells are listed vertically in the criteria range, Excel treats them as alternative criteria and displays all data because the query means either this specific value or any value. This condition is always true.

17.4 Further Processing of Filtered Data

When you hide records by using the **Filter** or **Advanced Filter** functions, Excel shows in the status bar how many records were found from the total in the table. When data is filtered, many commands that follow apply only to the filtered data, not the entire table. This includes printing, sorting, and creating subtotals.

This also makes it easy to create a chart from filtered data. To do this, select the intended columns in the entire table and create a chart from them. When you apply a filter, the chart immediately updates to reflect the new selection.

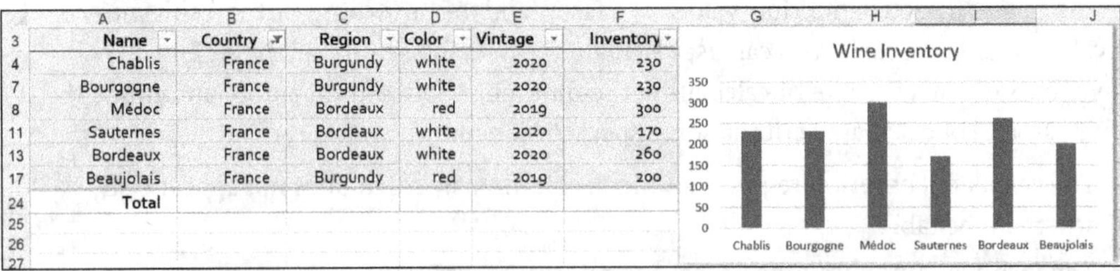

	A	B	C	D	E	F
3	Name	Country	Region	Color	Vintage	Inventory
4	Chablis	France	Burgundy	white	2020	230
7	Bourgogne	France	Burgundy	white	2020	230
8	Médoc	France	Bordeaux	red	2019	300
11	Sauternes	France	Bordeaux	white	2020	170
13	Bordeaux	France	Bordeaux	white	2020	260
17	Beaujolais	France	Burgundy	red	2019	200
24	Total					
25						
26						
27						

Figure 17.19 Chart for French Wines

Copying Filtered Results

When you copy the range of filtered data to the clipboard, you can paste the extract into any workbook or worksheet without including hidden records.

Some commands behave differently when applied to filtered data.

Commands	Effects
Sum symbol	Sums only visible cells
Cell formatting commands	Formats only visible cells
Clear contents	Clears only the contents of visible cells

This feature lets you apply a specific background color to all rows with French wines. To do this, filter out all non-French wines and then format the remaining rows accordingly.

17.5 Calculations with Database Functions

Besides calculating subtotals for selected groups in a table, Excel offers several statistical functions that are designed specifically for tables. The main difference from standard statistical functions is that you can set specific criteria for the evaluation. In the wine warehouse, this lets you determine how many varieties of Italian red wine are available, the total number of those bottles, the price of the most expensive or cheapest bottle, and the overall value of these wines.

In the following example, cells A2 to B3 define a range for search criteria, and a double criterion is set: the wines must be from Italy and red. Cells D2 to H2 use five different database functions. To simplify argument entry, the table range—including the row with column names—was named *WineWarehouseTotal*. The criteria range is automatically named *Criteria*.

```
=DCOUNT(WineWarehouseTotal,,Criteria)
=DSUM(WineWarehouseTotal,"Inventory",Criteria)
=DMAX(WineWarehouseTotal,"Price",Criteria)
=DMIN(WineWarehouseTotal,"Price",Criteria)
=DSUM(WineWarehouseTotal,"Value",Criteria)
```


All functions use three arguments: The first is the range of the entire table, the second is the name or number of the column to evaluate—with DCOUNT the field can be omitted—and the third is the range containing the search criteria. If you use a field number, the first column of the table counts as number 1 (note that the column name must be in quotation marks).

Database functions are especially useful when you need specific statistical analysis for changing criteria. You can simply update the values in the criteria range.

	A	B	C	D	E	F	G	H	I
3	Country	Color			=DCOUNT(Database,,Criteria)			6	
4	Italy	red			=DMAX(Database,"Price",Criteria)			$5.80	
5					=DMIN(Database,"Price",Criteria)			$3.80	
6					=DSUM(Database,"Value",Criteria)			$5,995.00	
7									
8	Name	Country	Region	Color	Vintage	Inventory	Minimum stock	Price	Value
9	Ortenau	Germany	Baden	red	2018	340	300	$4.00	$1,360.00
10	Müller-Thurgau	Germany	Moselle	white	2019	340	300	$4.00	$1,360.00
11	Riesling	Germany	Moselle	white	2019	200	200	$4.00	$800.00
12	Oppenheimer	Germany	Rheinhessen	white	2019	300	250	$5.00	$1,500.00
13	Silvaner	Germany	Rheinhessen	white	2019	230	200	$3.00	$690.00
14	Médoc	France	Bordeaux	red	2018	300	250	$5.00	$1,500.00
15	Bordeaux	France	Bordeaux	white	2019	260	200	$4.60	$1,196.00
16	Sauternes	France	Bordeaux	white	2019	170	190	$4.90	$833.00
17	Beaujolais	France	Burgundy	red	2018	200	220	$4.00	$800.00
18	Bourgogne	France	Burgundy	white	2019	230	170	$5.30	$1,219.00
19	Chablis	France	Burgundy	white	2019	230	150	$6.80	$1,564.00

Figure 17.20 Example of Statistical Analysis Using Database Functions

Chapter 18
PivotTables and Charts

Companies often collect and process large amounts of data they need to manage various tasks. A typical example is the data generated during invoicing. Companies that work with sales representatives create tables that assign sales figures to each representative, and sales are usually divided by product groups to show which representative sold greater or lesser amounts of which products. Figure 18.1 shows a simple example of such a list.

	A	B	C	D	E	F
1	**Sales results**					
2						
3	**Representative**	**Product group**	**Region**	**2022**	**2023**	**2024**
4	Hansen	Washing machines	East	200,022	140,000	120,000
5	Hansen	Refrigeration units	East	160,000	160,000	160,000
6	Gernot	Washing machines	East	110,000	110,000	180,000
7	Gernot	Refrigeration units	East	150,000	230,000	150,000
8	Schlier	Washing machines	West	120,000	180,000	120,000
9	Schlier	Refrigeration units	West	170,000	170,000	170,000
10	Gundar	Washing machines	West	120,000	120,000	140,000
11	Gundar	Refrigeration units	West	130,000	130,000	130,000
12	Seiffert	Washing machines	North	120,000	120,000	120,000
13	Seiffert	Refrigeration units	North	140,000	160,000	140,000
14	Adam	Washing machines	North	120,000	120,000	140,000
15	Adam	Refrigeration units	North	130,000	130,000	130,000
16	Karit	Washing machines	South	120,000	120,000	120,000
17	Karit	Refrigeration units	South	160,000	160,000	160,000
18	Lemgo	Washing machines	South	120,000	120,000	120,000
19	Lemgo	Refrigeration units	South	150,000	150,000	150,000

Figure 18.1 Sales Table as Source Data

This list links together quite different dimensions. One aspect shows sales growth over time, another shows the spatial distribution across regions—and a third is the breakdown of sales by product groups, which is a categorical dimension. This last dimension could be further subdivided by product type or sales representative.

In the version in Figure 18.1, the list is not very informative. You can quickly check how much sales representative Hansen made with washing machines in 2024, but the share of total sales that washing machine sales represented in the North region is not immediately visible. Also, whether sales growth in the East region has been faster than in the

West region over the last three years is not immediately clear. Nevertheless, the list contains everything you need to answer these questions, and the key is to transform raw data into useful information.

18.1 Interactive Tables and Charts

Excel provides PivotTables and PivotCharts for this purpose. Unlike the tables described earlier, a PivotTable is an *interactive* table. In this context, a *pivot* is a pivot point, like a door hinge. This emphasizes that a PivotTable lets you arrange, summarize, and analyze data from different perspectives.

You do this by creating buttons from the column names, which you can move between different areas, with the related data moving accordingly. You can also rearrange the row and column labels in the PivotTable. These elements and their associated data are internally linked, summary calculations are automatically generated and adjusted whenever the data grouping changes, and you can add more calculated fields to the table. Since Excel 2013, the **Recommended PivotTables** feature has suggested the best way to analyze your data based on the dataset. The program supports creating PivotTables from multiple tables by using data models that define relationships among them. Useful features include slicers for interactive filtering and timelines you can employ to analyze data over time.

The number of PivotTables is not limited per sheet or workbook—you can create multiple PivotTables from a single source table or from different source tables within a sheet or workbook. However, this feature is memory-intensive, so actual limits depend on your main memory. Excel stores source data in a special PivotTable cache that's created from a source range.

Since Excel 2010, PivotTable capacity has expanded significantly. Each field can contain up to 1,048,576 unique items, while earlier versions supported only 32,500. The number of row and column fields is now limited only by available memory, while the number of report filters and value fields is capped at 256.

You can also create interactive charts from PivotTables, and you can expand, collapse, or rearrange these charts by using the buttons for each field in the underlying list. The PivotTable and the associated PivotChart are linked so that changes to the chart also affect the table and vice versa.

Excel delivers faster performance with PivotTables and charts when tasks run on multiple threads. To enable this, you need to activate multithreaded calculation (see Chapter 3). In addition to the built-in pivot functions, there's a powerful add-in called *Power Pivot* that we'll briefly introduce at the end of this chapter.

18.2 Applications

The **PivotTable** and **PivotChart** functions generate reports and charts that can help you perform many different kinds of analysis of existing datasets. Here are some of the report types they can produce:

- Sales management reports
- Inventory analyses
- Assortment planning reports
- Personnel statistics
- Project planning and control reports
- Statistical analyses
- Error logs and material test evaluations

With PivotTables and PivotCharts, Excel provides a powerful, dynamic way to analyze and summarize large data sets. These tools let you easily create different reports from the same data set by changing the focus. You can also rearrange, reassemble, or filter data without changing any cell in the original table. Excel offers two comprehensive tabs for these tasks: **PivotTable Analyze** and **Design**.

18.3 Suitable Data

PivotTables and PivotCharts use data from regular cell ranges (with a suitable structure) or from tables. Converting the data range into an Excel table as previously described is not required but is always recommended.

A PivotTable initially contains no formulas. It only contains the results of formulas from the underlying original table and additional calculations performed by the pivot function itself, such as creating groups or totals. Data from external sources, like Access or SQL databases, can also be analyzed in a PivotTable. Existing PivotTables and PivotCharts can serve as the base for additional tables and charts.

18.4 Data Analysis with PivotTables

You can only use the pivot function when suitable data is available. First, we'll address how to use it when data is stored as a table in a worksheet.

18.4.1 From Source Data to PivotTable

In this section, we'll revisit the sales control example. To use a table as the basis for a PivotTable, you start by just selecting any cell within the table. If the source table has a header row with column headings and a contiguous data block with individual records, the PivotTable feature will be able to correctly identify the entire table range.

Remove Filters or Subtotals First

If the table contains subtotals or filters, remove them before proceeding.

Next, open the dialog to create a PivotTable from the **Insert** tab by using the **PivotTable** button in the **Tables** group. This group also includes the **Recommended PivotTables** button, which you'll learn about later.

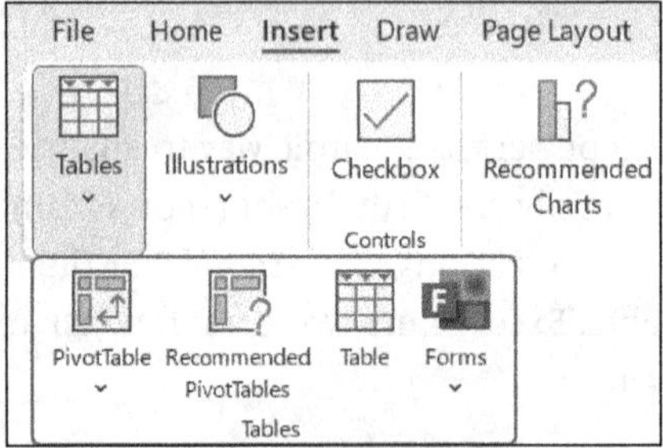

Figure 18.2 PivotTable Button in Tables Group

Next, identify the data source. If the cell pointer is inside the data block to be analyzed, Excel will show the detected range under **Table/Range**, and you can adjust it if necessary. For regular cell ranges, make sure the row with the column headers is included. If it's an Excel table, Excel will suggest the assigned table name.

	A	B	C	D	E	F
1	**Sales results**					
2						
3	**Representative**	**Product group**	**Region**	**2022**	**2023**	**2024**
4	Hansen	Washing machines	East	200,022	140,000	120,000
5	Hansen	Refrigeration units	East	160,000	160,000	160,000
6	Gernot	Washing machines	East	110,000	110,000	180,000
7	Gernot					
8	Schlier					
9	Schlier					
10	Gundar					
11	Gundar					
12	Seiffert					
13	Seiffert					
14	Adam					
15	Adam					
16	Karit					
17	Karit					
18	Lemgo					
19	Lemgo					

PivotTable from table or range
Select a table or range
Table/Range: SalesResults
Choose where you want the PivotTable to be placed
New Worksheet
Existing Worksheet
Location:
Choose whether you want to analyze multiple tables
Add this data to the Data Model
OK
Cancel

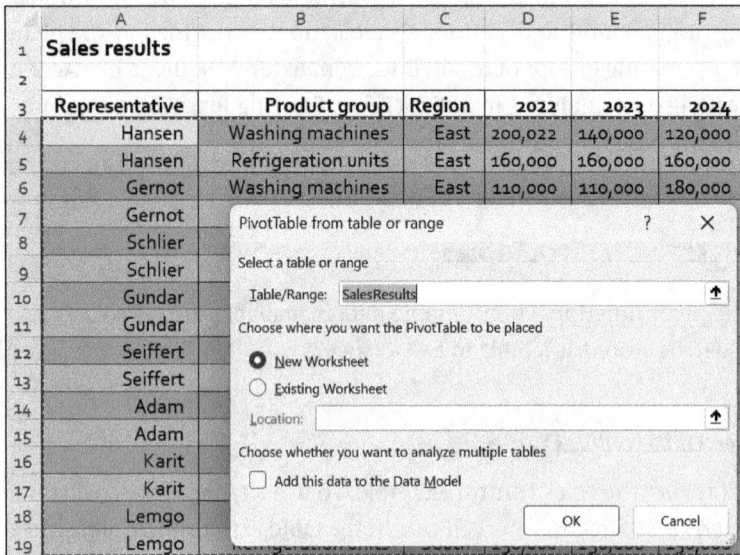

Figure 18.3 PivotTable from Table or Range Dialog

At the bottom of the dialog, choose where to place the PivotTable: on a new sheet or in a specific location on the active sheet. In the latter case, enter or select the address of the upper-left cell for the PivotTable under **Location**. Note that the new table will overwrite any existing data within the specified range, so you'll see a warning and should select a different range if you want to keep the existing data.

It is usually better to use a separate sheet in the workbook. When you confirm the dialog, Excel creates an initially empty PivotTable range ❶ and opens the **PivotTable Fields** task pane on the right.

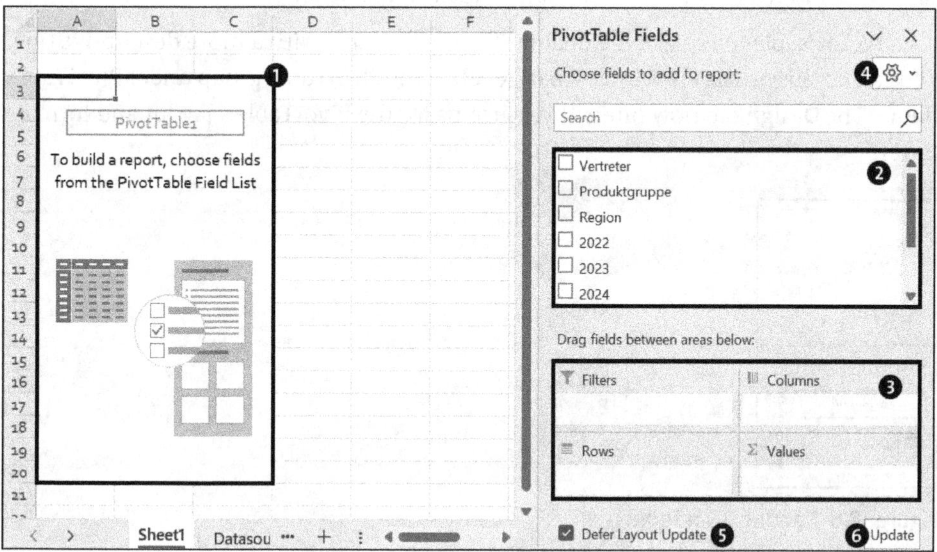

Figure 18.4 Empty PivotTable Range and Editing Tools

The task pane contains the list of fields ❷ and simultaneously displays the areas ❸ to which the fields can be assigned. You can split the task pane layout in various ways using the button palette at the top right ❹.

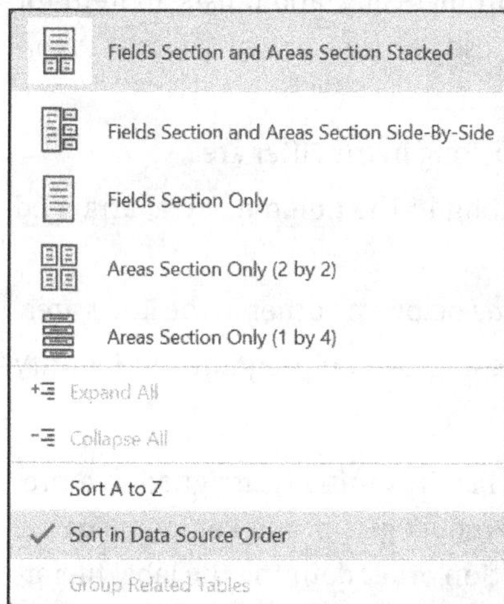

Figure 18.5 Palette for Splitting Task Pane

At the bottom of the PivotTable **Fields** task pane, you'll find the **Defer Layout Update** ❺ option, and if you check it, data linked to a field won't update immediately in the Pivot-Table when you change field assignments to ranges. Instead, the data will update only after you finish the layout and click the **Update** ❻ button. This is useful when fields link to very large data sets.

The task pane will be hidden as soon as you select a cell outside the PivotTable, and you can select any cell inside the PivotTable to make it reappear.

It's also advisable to replace the default PivotTable name with a more descriptive one. You can do this on the **PivotTable Analyze** tab in the **PivotTable** group under **PivotTable Name**. The **Design** tab now offers three groups for the PivotTable's layout and formatting.

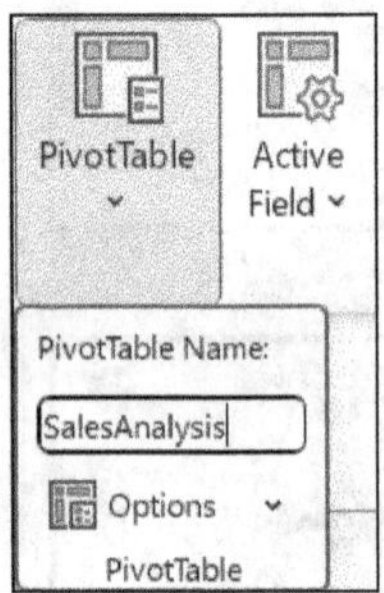

Figure 18.6 Naming PivotTable

18.4.2 PivotTable Layout

Next, you need to determine the main perspective for viewing the data. The new table should arrange and summarize the data to show how the share of the two product groups has changed over the years, both overall and across the four regions. Excel provides a PivotTable **Fields** task pane that appears automatically, and it lists an item for each column in the source table. You can assign these items to one of the four Pivot-Table areas:

- Fields that determine the order of report pages belong in the **Filter** area.
- Fields that determine the order of columns belong in the **Columns** area, arranged from left to right.
- Fields that determine the order of rows belong one below the other in the **Rows** area.
- You must insert at least one field in the **Values** area so the program can identify which values to display in the table.

When you select a field by checking it in the field list, it is initially assigned to a predefined area. In this example, the **Representative**, **Product group**, and **Region** fields are added sequentially to the **Rows** area, with the selection order defining the label hierarchy.

The fields with the years are inserted by default into the **Values** area, and the **Columns** area then automatically splits based on the values in the **Values** area; in this case, a separate column is created for each year with summed values. After you select all the fields, the report shown in Figure 18.7 is generated.

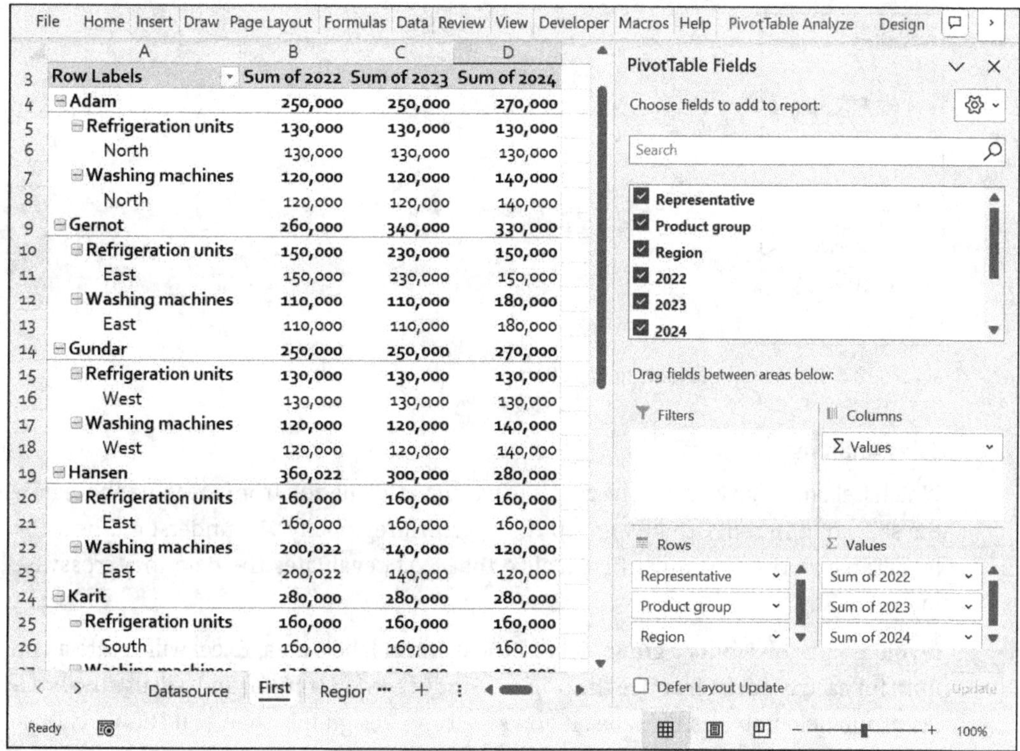

Row Labels	Sum of 2022	Sum of 2023	Sum of 2024
Adam	250,000	250,000	270,000
Refrigeration units	130,000	130,000	130,000
North	130,000	130,000	130,000
Washing machines	120,000	120,000	140,000
North	120,000	120,000	140,000
Gernot	260,000	340,000	330,000
Refrigeration units	150,000	230,000	150,000
East	150,000	230,000	150,000
Washing machines	110,000	110,000	180,000
East	110,000	110,000	180,000
Gundar	250,000	250,000	270,000
Refrigeration units	130,000	130,000	130,000
West	130,000	130,000	130,000
Washing machines	120,000	120,000	140,000
West	120,000	120,000	140,000
Hansen	360,022	300,000	280,000
Refrigeration units	160,000	160,000	160,000
East	160,000	160,000	160,000
Washing machines	200,022	140,000	120,000
East	200,022	140,000	120,000
Karit	280,000	280,000	280,000
Refrigeration units	160,000	160,000	160,000
South	160,000	160,000	160,000

Figure 18.7 First Analysis with Predefined Area Assignments

To insert a field into a different area from the default area, you select the target area from the field name's context menu.

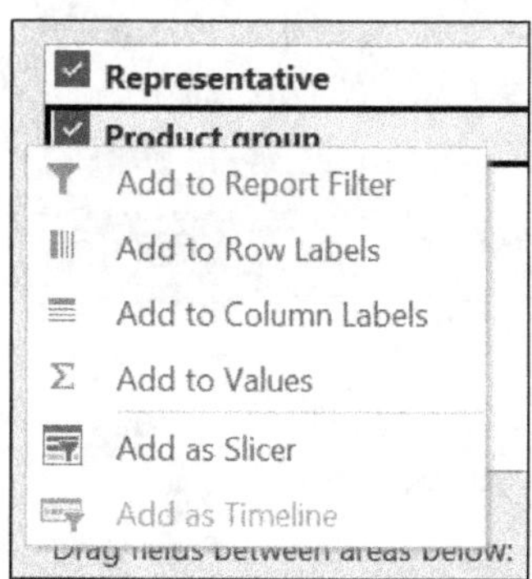

Figure 18.8 Context Menu for Field in Field List

Alternatively, you can arrange the distribution of fields across the four possible ranges by dragging them directly, and the task pane will show the four areas with their currently assigned fields. For example, you can move the **Region** field from the **Rows** area to the **Filter** area.

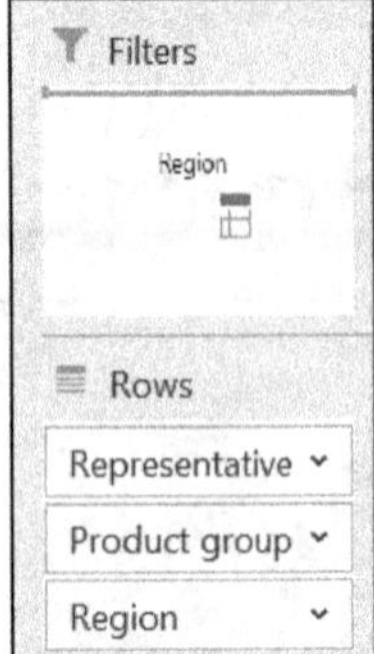

Figure 18.9 Rearranging Fields in Task Pane

Sales Analysis

Which field must go where if the goal is to extract information from the table about how the sales of different product groups have grown regionally? To understand how the PivotTable works, you must first realize that Excel evaluates the data in a crosswise manner.

If you assign the **Product group** field to the column labels area, Excel will create a column for each product group in the original **Product group** column and will use the *Product group* field name as the label. If you select the **Design** tab, then you should choose **Show in Outline Form** or **Show in Tabular Form** in the **Layout** group on the **Report Layout** button menu. The alternative compact format shows general labels like **Column Headers** instead of field names.

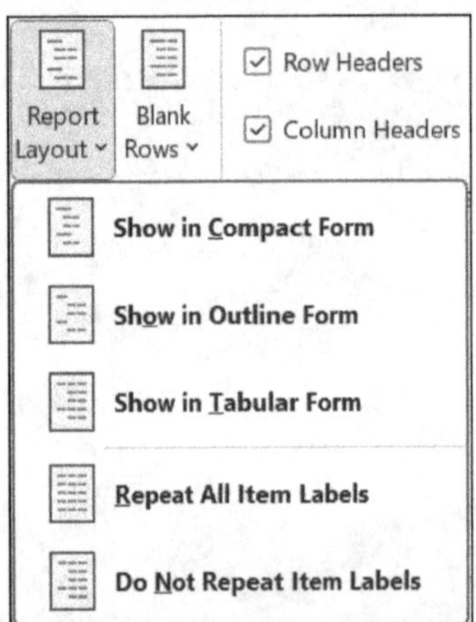

Figure 18.10 Selecting Report Layout

Here, only two product groups exist, *Washing Machines* and *Refrigerators*, so two columns are created per year. An extra column for the grand total will also be added unless you disable this setting in the **PivotTable Options** dialog. More details follow.

You can deselect the **Representative** field here, and when you assign the **Region** field to the **Rows** area, Excel will create a row element for each of the four sales regions. This forms a table with four row elements and two column elements. Each cell in the data range represents the intersection of a product group and a region element, and the column and row with their respective grand totals are also included. After you place the field for the first year in the values area, the PivotTable will populate with data from the original table. Clicking the button for the **2022** field displays the **Sum of 2022** default label. Figure 18.11 displays the analysis for the first year.

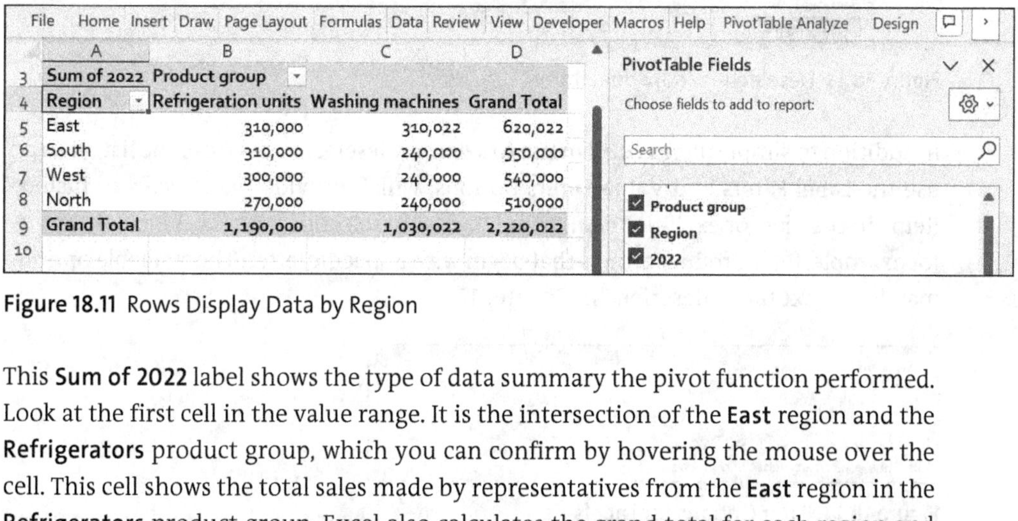

File Home Insert Draw Page Layout Formulas Data Review View Developer Macros Help PivotTable Analyze Design

	A	B	C	D
3	Sum of 2022	Product group		
4	Region	Refrigeration units	Washing machines	Grand Total
5	East	310,000	310,022	620,022
6	South	310,000	240,000	550,000
7	West	300,000	240,000	540,000
8	North	270,000	240,000	510,000
9	Grand Total	1,190,000	1,030,022	2,220,022
10				

PivotTable Fields
Choose fields to add to report:
Search
Product group
Region
2022

Figure 18.11 Rows Display Data by Region

This **Sum of 2022** label shows the type of data summary the pivot function performed. Look at the first cell in the value range. It is the intersection of the **East** region and the **Refrigerators** product group, which you can confirm by hovering the mouse over the cell. This cell shows the total sales made by representatives from the **East** region in the **Refrigerators** product group. Excel also calculates the grand total for each region and product group.

	A	B	C	D
3	Sum of 2022	Product group		
4	Region	Refrigeration units	Washing machines	Grand Total
5	East	310,000	310,022	620,022
6	South	31[illegible]	[illegible]00	550,000
7	West	30[illegible]	[illegible]00	540,000
8	North	27[illegible]	[illegible]00	510,000
9	Grand Total	1,19[illegible]	[illegible]022	2,220,022

Sum of 2022
Value: 310,000
Row: East
Column: Refrigeration units

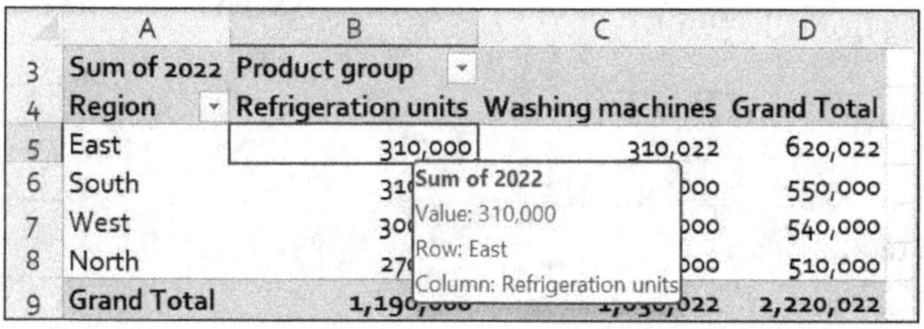

Figure 18.12 Tool Tip for Value

Item Filters

Filter buttons appear next to **Region** and **Product group**—the two field names we're using as sorting criteria—as you learned in Chapter 17 for tables. Clicking or tapping one of these buttons opens a detailed filter menu with the list of items. For example, clearing the checkmark for *Washing Machines* in the **Product group** column removes the data for washing machines from the table. In the **PivotTable Fields** pane and next to the field name in the PivotTable, filter icons indicate that a filter is applied to that field.

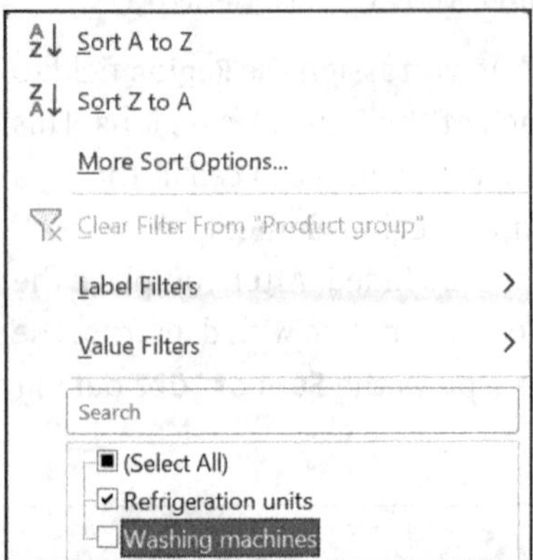

Figure 18.13 Deselecting Individual Items

In addition to simple filters you can use to select or deselect items from the list, you can use the **Label Filters** and **Value Filters** options, which provide many ways to include items in the view or exclude them from it by using selection criteria. A label filter can, for example, filter product groups that begin with a specific letter. The available options match the text filters described in Chapter 17.

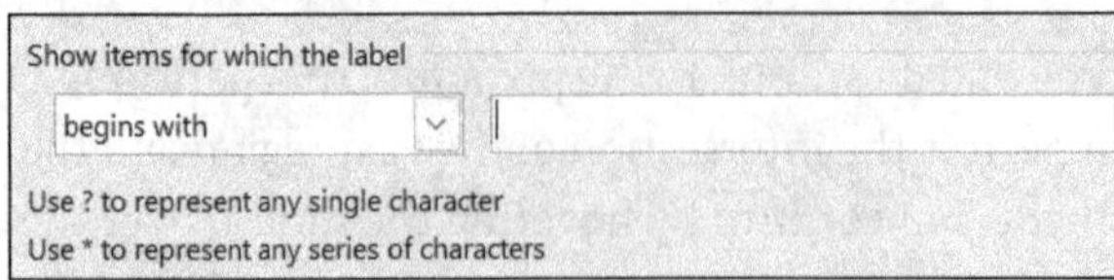

Figure 18.14 Filter Options for Labels

A value filter can filter product groups with the highest sales, and value filters correspond to the number filters from Chapter 17.

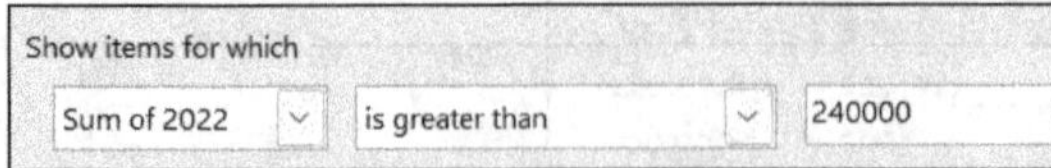

Figure 18.15 Value-Based Filter Options

If a filter is no longer needed, you can remove it by using the now active **Clear Filter from** command.

Item Search

The filter button menus for items also include the search function described in Chapter 17. This is especially useful for fields with many items. With the item search, you can quickly find relevant elements even when the PivotTable contains thousands of rows.

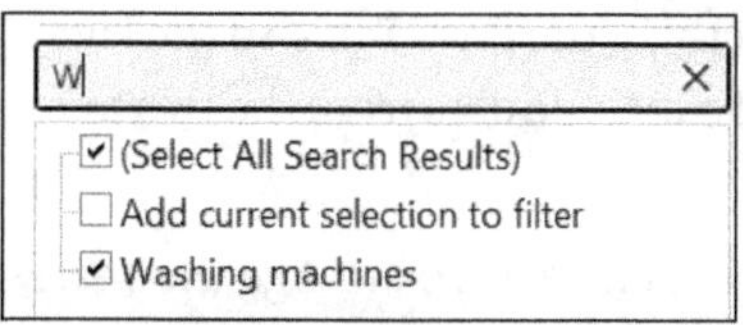

Figure 18.16 Searching for Items in Field

18.4.3 Recommended PivotTables

Here, we briefly explore a feature that helps you create PivotTables. To use it, instead of selecting the **Insert • Tables • PivotTable** command, you select the **Recommended PivotTables** button in the same group.

The task pane will recommend and show previews of various types of tables, based on the source table's structure and selected data. You can review different options at your convenience, and after you choose and confirm a layout, Excel will immediately create a new sheet with the corresponding PivotTable.

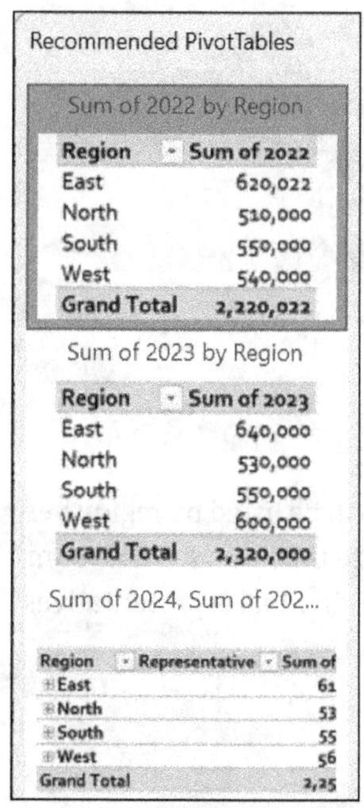

Figure 18.17 Suggestions for PivotTables Based on Current Data

18.4.4 Adding Value Columns

The solution we're aiming for in the example we're describing here has deliberately not been fully completed yet, to keep the way the pivot function works as transparent as possible. Compared to the initial goal, this first version still lacks temporal development since it only shows data for one year.

You can easily correct this. Place the cell pointer directly in the PivotTable so the **PivotTable Fields** task pane shows the field buttons again. To include data from the two missing years, check the **2023** and **2024** fields as well.

Additional columns will appear in the values area for each year. To arrange the yearly values vertically, use the **Move to Row Labels** option in the menu for the **Values** button under the **Columns** area. Alternatively, you can use the **Move Values to • Move Values to Rows** option in the context menu of the table cell labeled **Values**.

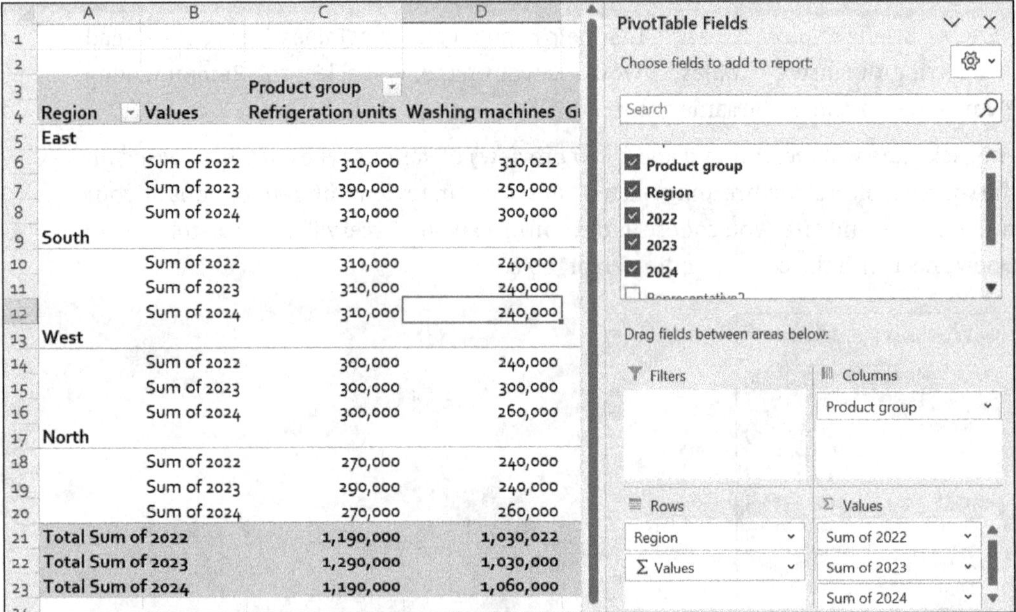

Figure 18.18 PivotTable Showing the Evaluation for Three Years

After that, the results per product group for the three years will be listed by region vertically, with the three total sums at the end. This will complete the intended data summary. Details that aren't relevant at that moment, such as individual representatives' results, will not be included in the calculated totals.

18.4.5 Changing the PivotTable Layout

You may find the column-wise arrangement of the time progression somewhat unsatisfactory. We usually expect to display and read timelines from left to right, so you can reorganize the data by swapping the **Values** and **Product group** fields as follows:

1. In the range section of the **PivotTable Fields** pane, drag the **Values** item, which was previously used as a row label, into the **Column Headings** area.
2. Do the opposite with **Product group**. Drag the field from the **Columns** area and place it below **Region** in the **Rows** area. The rows will then be grouped into two levels. The

primary criterion is **Region**, and the secondary one is **Product group**. The yearly results will appear side by side.

3. To remove the awkward **Sum of ...** field label, click the small arrow next to each field in the **Values** range, select **Value Field Settings**, and replace the label under **Custom Name** with "Sales" and the corresponding year.

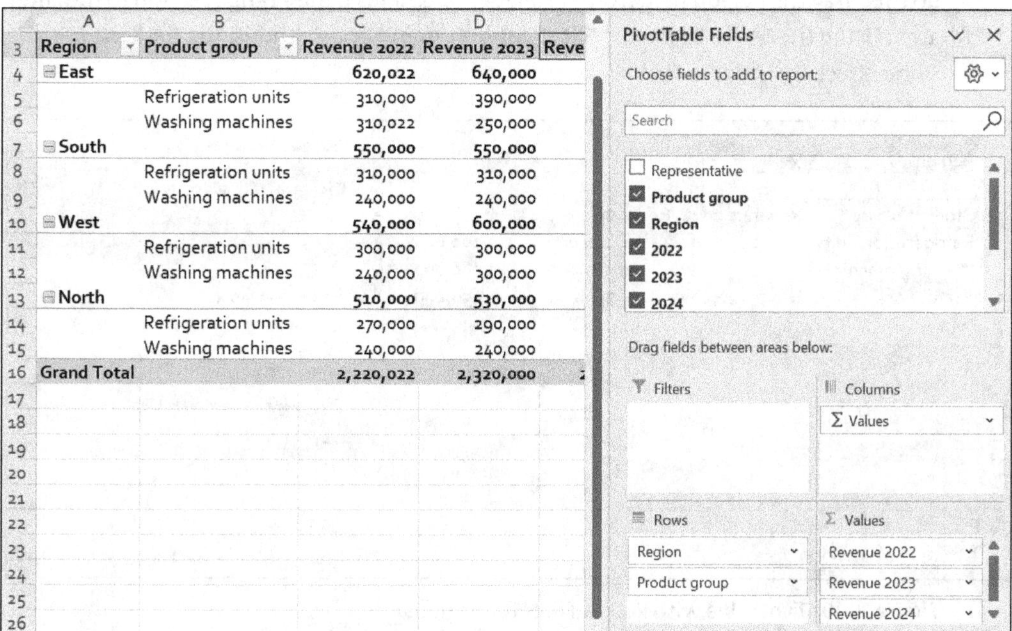

4. If the year columns are out of order, select the fields and use the options in the first group to rearrange them as needed.

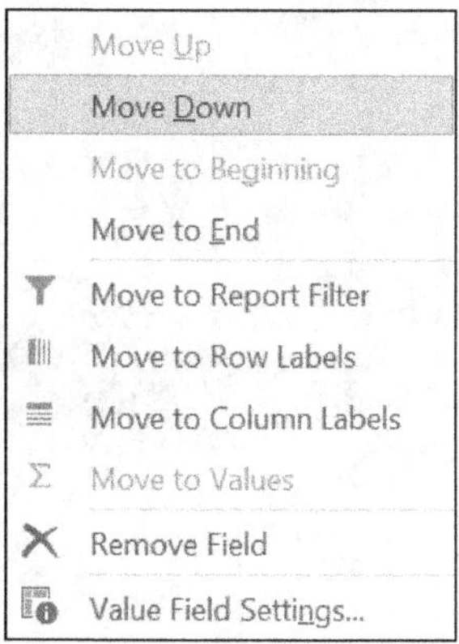

When you have multiple row labels, their order matters. For example, if you list the product group first and the region second, the data will group first by product group and then by region within each group.

To achieve this arrangement later, drag the **Region** field below the **Product group** field in the **Rows** area.

Report Filter

You might consider compressing the data further to display only individual regions. To do this, drag the **Region** field from the **PivotTable Fields** pane's range section to the **Filter** area. Using the **Move to Report Filter** context menu command in the field list has the same effect.

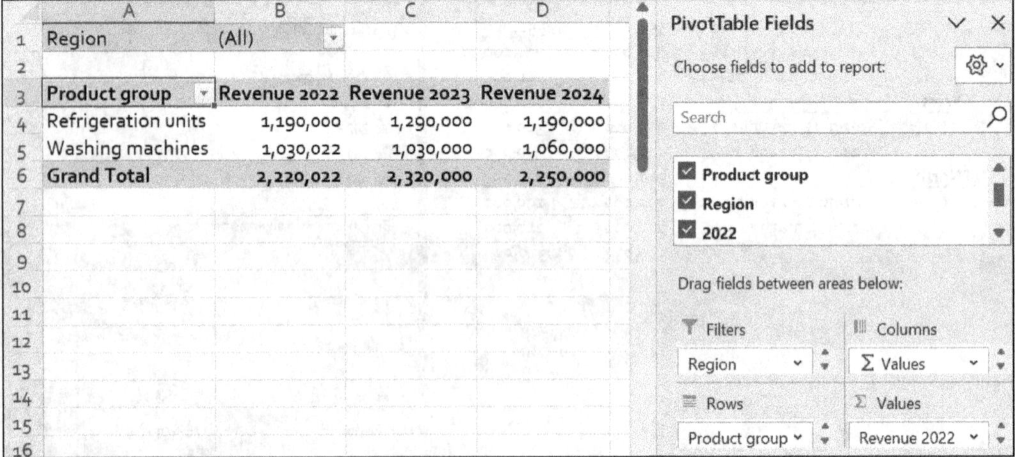

Figure 18.19 PivotTable with Report Filter for Regions

The initially displayed values show the grand total for all areas, but you can use the filter button to display data for regions individually or in any selection. To do this, check the **Select Multiple Items** box.

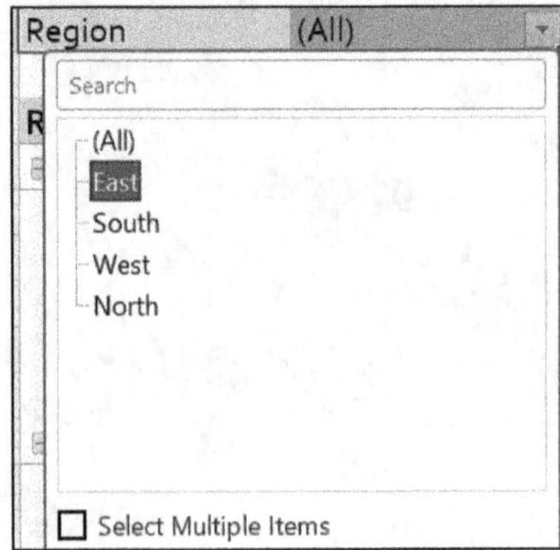

Figure 18.20 Selecting Individual Regions

You may also want to have a separate table for each region, and you can quickly create them with the current PivotTable. In the **PivotTable Analyze • PivotTable** group, open the **Options** menu and then use the **Show Report Filter Pages** command to insert a new sheet for each page (for each region in this example) and label the tabs with the item names. All new tables will be fully functional PivotTables, and you can easily copy or move them to other workbooks to share them.

	A	B	C
2	Region	North	
3			
4	Representative	Revenue 2022	Revenue 2023
5	Adam	250,000	250,000
6	Seiffert	260,000	280,000
7	Grand Total	510,000	530,000

Figure 18.21 Separate Sheets for Separate Regions

Further Compressing Data

You can completely remove the **Region** sorting criterion from the PivotTable to summarize the analysis further. Simply deselect the field in the field list, and the sales figures for each year will be grouped only by **Product group**.

	A	B	C	D
3	Product group	Revenue 2022	Revenue 2023	Revenue 2024
4	Refrigeration units	1,190,000	1,290,000	1,190,000
5	Washing machines	1,030,022	1,030,000	1,060,000
6	Grand Total	2,220,022	2,320,000	2,250,000

Figure 18.22 Reduced View

18.4.6 Options for the PivotTable Report

When you select a cell in a PivotTable, the **Options** button appears on the **PivotTable Analyze** tab in the **PivotTable** group. You can also open the PivotTable **Options** dialog directly from the context menu of any PivotTable cell.

In the first field, you can rename the PivotTable. The **Layout & Format** tab provides options that control the display of labels and values, and the **For error values show** option lets you enter text to display instead of an error message. You can also specify text for empty cells by entering something like "Value missing." Checking the **Preserve cell formatting on update** box ensures that a specific format, like a background color, stays applied to certain data elements even when the data is regrouped.

On the **Totals & Filters** tab, you can choose whether to show row or column totals. The **Display** tab offers options to **Show expand/collapse buttons** and **Show contextual tooltips**. Additional options affect field labels and their order in the field list. The **Printing** tab lets you control how a PivotTable prints, which especially applies to multipage prints, in which you can enable repeating row labels or print titles. The **Data** tab allows you to configure how PivotTable data is handled, and checking **Save source data with file** means that copies of external data will be saved with the PivotTable.

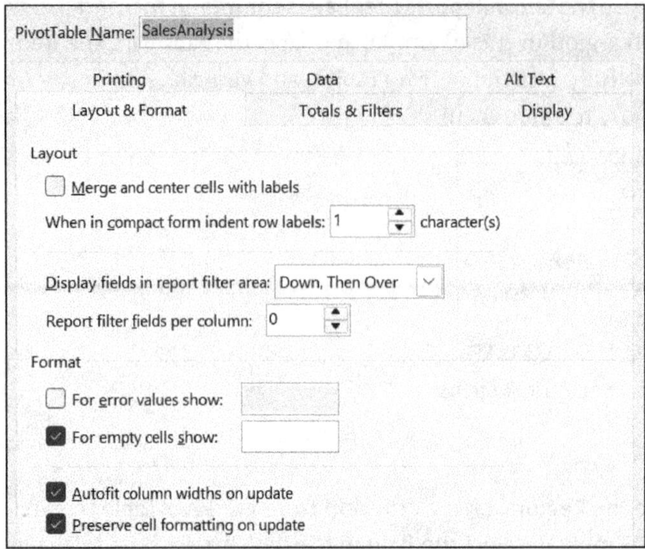

Figure 18.23 PivotTable Options Dialog Box

18.4.7 Adding Fields

The individual representatives have been omitted in the PivotTables shown so far, but if you need their results, you can easily rearrange the PivotTable, create a new one, or copy a sheet within the workbook and then create a new table layout in the copy.

To do this, place the cell pointer back in the PivotTable and assign the **Representative** field under **PivotTable Fields** to either the **Rows** or **Columns** area via the context menu, depending on how you want to organize the data.

	A	B	C	D	E
3	Product group	Representative	Revenue 2022	Revenue 2023	Revenue 2024
4	Refrigeration units		1,190,000	1,290,000	1,190,000
5		Adam	130,000	130,000	130,000
6		Gernot	150,000	230,000	150,000
7		Gundar	130,000	130,000	130,000
8		Hansen	160,000	160,000	160,000
9		Karit	160,000	160,000	160,000
10		Lemgo	150,000	150,000	150,000
11		Schlier	170,000	170,000	170,000
12		Seiffert	140,000	160,000	140,000
13	Washing machines		1,030,022	1,030,000	1,060,000
14		Adam	120,000	120,000	140,000
15		Gernot	110,000	110,000	180,000
16		Gundar	120,000	120,000	140,000
17		Hansen	200,022	140,000	120,000
18		Karit	120,000	120,000	120,000
19		Lemgo	120,000	120,000	120,000
20		Schlier	120,000	180,000	120,000
21		Seiffert	120,000	120,000	120,000
22	Grand Total		2,220,022	2,320,000	2,250,000

PivotTable Fields
Choose fields to add to report:
Search
Representative
Product group
Region
2022
2023
Drag fields between areas below:
Filters
Columns
Σ Values
Rows
Σ Values
Product group
Representative
Revenue 2022
Revenue 2023

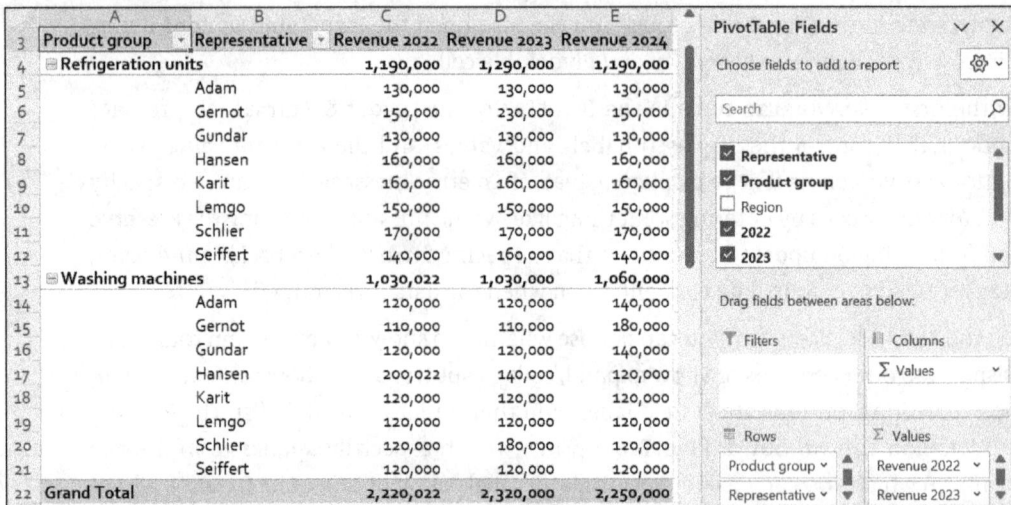

Figure 18.24 Product Group Results Broken Down by Representative

If you drag the **Representative** field into the **Rows** area, you can choose to position it above or below the **Product group** field. In the first case, the data will be sorted first by representatives and then by product groups for each representative. In the second case, the order will be reversed. If you add the **Representative** field to the **Columns** range instead, you'll see three columns with annual results for each representative.

If the Values in the List Change

If you try to change any values in the PivotTable's data range, you'll find it is not possible. A message will inform you that this part of the PivotTable cannot be edited, so you should always make data changes in the source table, which contains the original data the PivotTable analyzes.

However, Excel doesn't automatically update the PivotTable(s) when a value in the sales list changes—and unfortunately, you'll receive no indication that the source data and the analyses in the PivotTable no longer match. To update the PivotTable, select a cell within it and click the **Refresh** button in the **Data** group on the **PivotTable Analyze** tab. Alternatively, you can press the keyboard shortcut [Alt]+[F5].

You can also set a PivotTable to update automatically when the workbook opens. To do this, enable the **Refresh data when opening the file** option on the **Data** tab in the **PivotTable Options** dialog box.

18.4.8 Sorting in the PivotTable

Just as the overall PivotTable can have different sorting criteria depending on how its various areas are arranged, the elements belonging to each of these fields be arranged in different ways. There are two distinct methods you can use for this:

- You can sort manually by using the standard Excel sorting commands.
- You can set a dynamic sort for each field that updates automatically whenever the source table data changes and you refresh the PivotTable with the **Refresh** command.

You can also choose to sort by label order or by a value.

Manual Sorting

For example, consider a PivotTable showing sales per representative. To sort the table by representative names used as row labels, select the field name—which is **Representative** here—or one of the representatives' names and then use the sort icons in the **Data • Sort & Filter** group. The data linked to the items will be assigned automatically. Alternatively, you can access the **Sort** command via the context menu of the button with the field name or the context menu of an item in the field.

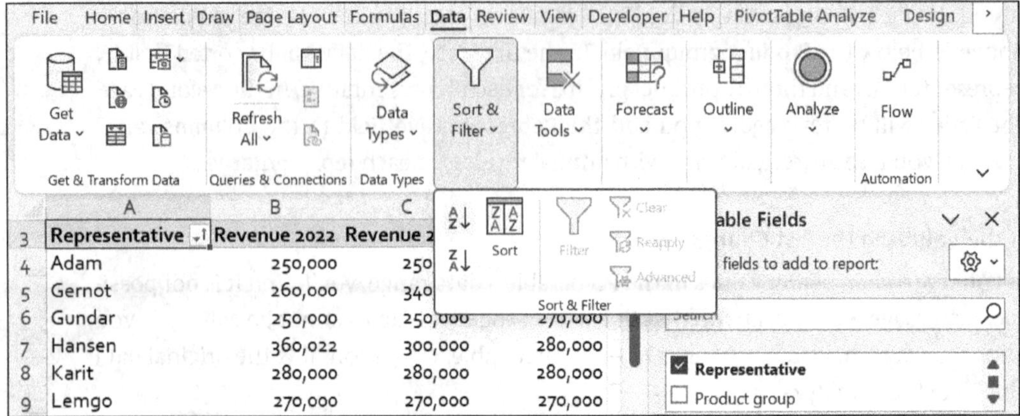

Figure 18.25 Representatives Table Sorted by Name

To sort by the ranking of sales results for 2024, select a sales value in the 2024 column and then use the sort icon or the corresponding context menu option, which are both labeled **Sort Largest to Smallest** (descending).

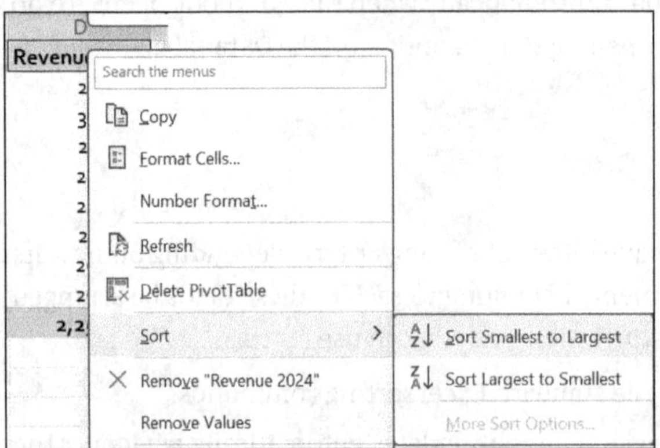

Figure 18.26 Sorting by Values

Dynamic Sorting

You can set a specific sort order for each PivotTable field that restores automatically with each table update. Use the sales table analysis as an example, where representatives are arranged in the row field. The table should always show the representative with the highest total sales first.

To dynamically sort representatives by sales revenue, select the **Representative** field name, click the **Sort** button in the **Data • Sort & Filter** group, select **Descending (Z to A) by**, and choose **Revenue 2024** from the list box.

The table will always show the representative with the highest grand total for the selected year first, and you can test the automatic sorting by changing specific values in the source table. Refreshing the PivotTable restores the desired sorting order.

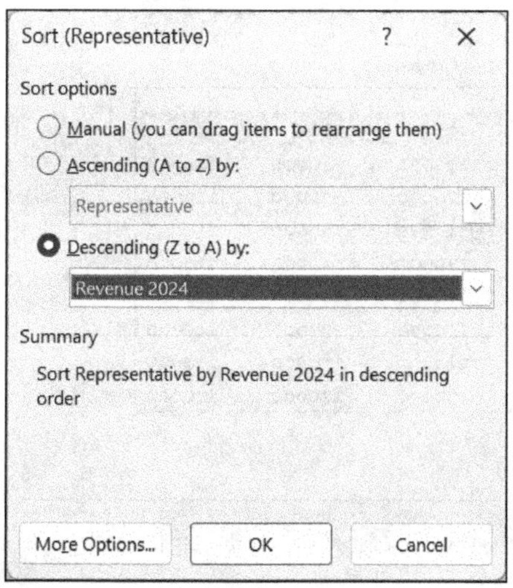

Figure 18.27 Options for Dynamic Sorting

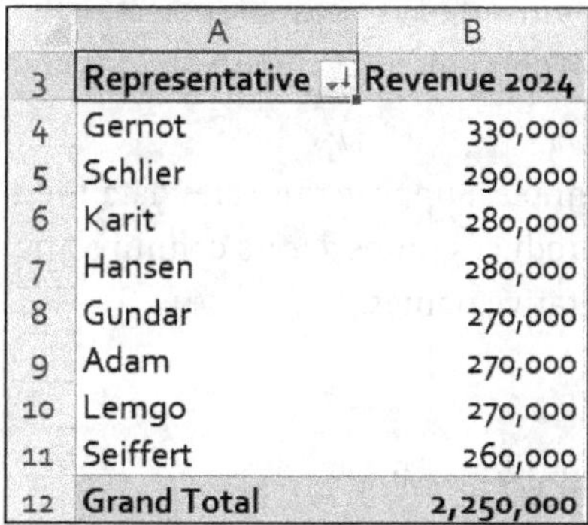

	A	B
3	Representative	Revenue 2024
4	Gernot	330,000
5	Schlier	290,000
6	Karit	280,000
7	Hansen	280,000
8	Gundar	270,000
9	Adam	270,000
10	Lemgo	270,000
11	Seiffert	260,000
12	Grand Total	2,250,000

Figure 18.28 Results of Dynamic Sorting

18.4.9 Quick Data Extracts for Individual Values

The link between the PivotTable data and the source table has another practical benefit: you can quickly extract specific data from the entire dataset. For example, if you double-click or tap a summary value, like a sales total in the PivotTable, all data from the original table that contributed to this value will appear on a separate sheet. If you select cell B5 in the compressed PivotTable described earlier, you get a detailed list of the washing machines. This new table is not a PivotTable and can be treated like a regular table. Excel applies one of the predefined table formats to the list.

You can disable this feature, which breaks down a value into its contributing detail values, to prevent others from viewing the detail data. This is useful when you copy a PivotTable into a separate workbook to share. To do this, when creating the PivotTable, use the **PivotTable Options** dialog box on the **Data** tab to enable or disable the **Show Details** option.

	A	B	C	D	E	F
1	Details for Revenue 2024 - Product group: Washing machines					
2						
3	Representative	Product group	Region	2022	2023	2024
4	Hansen	Washing machines	East	200022	140000	120000
5	Lemgo	Washing machines	South	120000	120000	120000
6	Gernot	Washing machines	East	110000	110000	180000
7	Karit	Washing machines	South	120000	120000	120000
8	Schlier	Washing machines	West	120000	180000	120000
9	Adam	Washing machines	North	120000	120000	140000
10	Gundar	Washing machines	West	120000	120000	140000
11	Seiffert	Washing machines	North	120000	120000	120000

Figure 18.29 Details for Summarized Value

18.4.10 Slicers

The **Slicers** feature is especially useful when working with PivotTables. It provides an interactive filter component that is easy to use and lets you quickly switch between different filters. This feature uses a special control that's inserted as a graphic object in the worksheet, which enables functions under the **Slicer Tools** tab in the ribbon, such as layering, aligning, grouping, rotating, and resizing.

This function is best understood with a simple example. Suppose our sales data for a year is summarized in a PivotTable that arranges product groups across columns and organizes rows first by region and then by representative names.

Setting Up a Slicer

For the preceding sales data example, set up three slicers as follows:

1. Select any cell in the PivotTable.
2. In the **PivotTable Analyze** group, click the **Insert Slicer** button under **Filter**.
3. In the **Insert Slicers** dialog, select the fields to create filter components ❶.

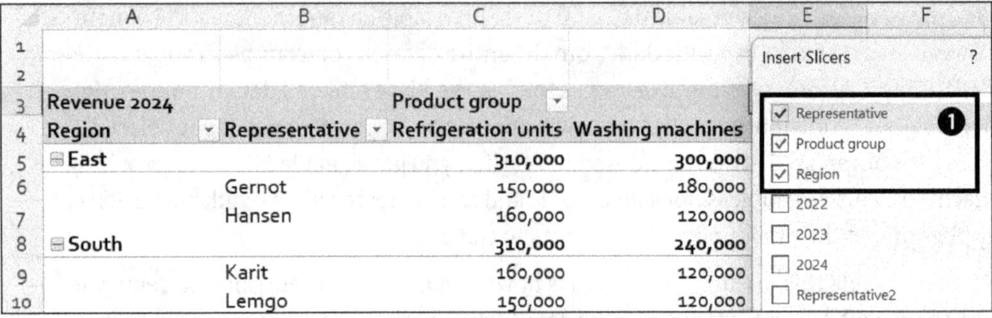

	A	B	C	D
1				
2				
3	Revenue 2024		Product group	
4	Region	Representative	Refrigeration units	Washing machines
5	⊟East		310,000	300,000
6		Gernot	150,000	180,000
7		Hansen	160,000	120,000
8	⊟South		310,000	240,000
9		Karit	160,000	120,000
10		Lemgo	150,000	120,000

4. If you select multiple fields, Excel creates a stack of filter components ❷.

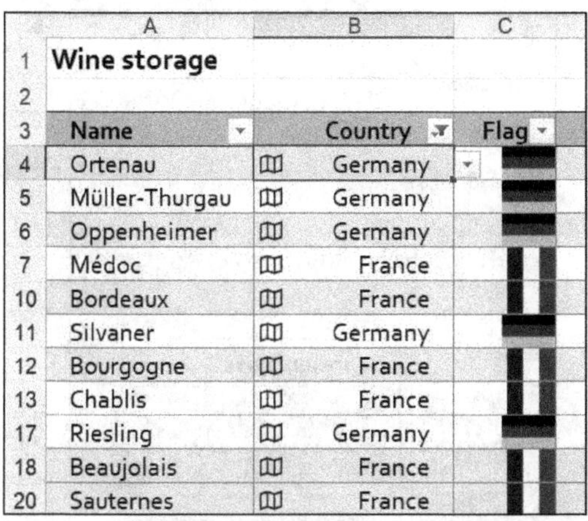

Wine storage
Name
Country
Flag
Ortenau
Germany
Müller-Thurgau
Germany
Oppenheimer
Germany
Médoc
France
Bordeaux
France
Silvaner
Germany
Bourgogne
France
Chablis
France
Riesling
Germany
Beaujolais
France
Sauternes
France

5. You can resize and move the components like any graphic object.

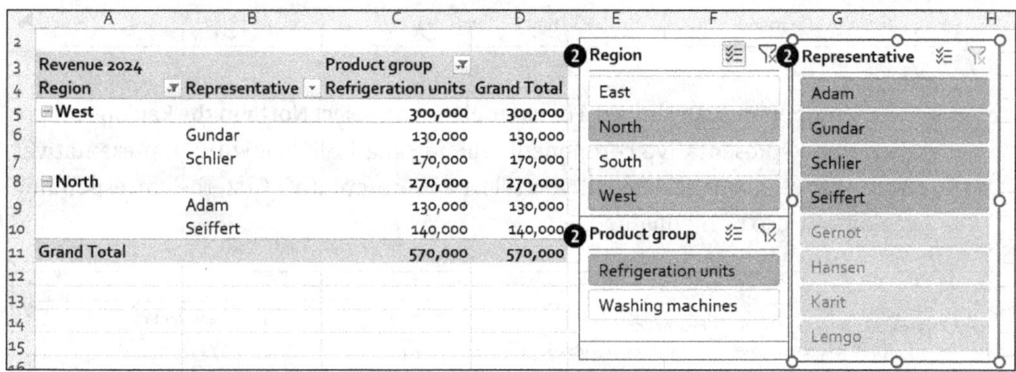

Revenue 2024
Product group
Region
Representative
Refrigeration units
Grand Total
West
300,000
300,000
Gundar
130,000
130,000
Schlier
170,000
170,000
North
270,000
270,000
Adam
130,000
130,000
Seiffert
140,000
140,000
Grand Total
570,000
570,000
2 Region
East
North
South
West
2 Product group
Refrigeration units
Washing machines
2 Representative
Adam
Gundar
Schlier
Seiffert
Gernot
Hansen
Karit
Lemgo

6. Excel shows the **Slicer** tab. Use the options in the **Arrange** group ❸ to align the components.
7. If you select a component, you can assign one of the styles from the **Slicer Styles** palette ❹.
8. Each component displays the related field in its title, and clicking the icon ❺ at the top right clears the current filter. Below, buttons list all items in the field, and clicking or tapping an item adds it to the filter. Only data for that item will appear; data for others will be hidden. To select multiple items for the filter, hold down [Ctrl] while clicking or use the **Multi-selection** icon ❻ in the title bar. Each item you select will show a different color to clearly indicate the active filter.

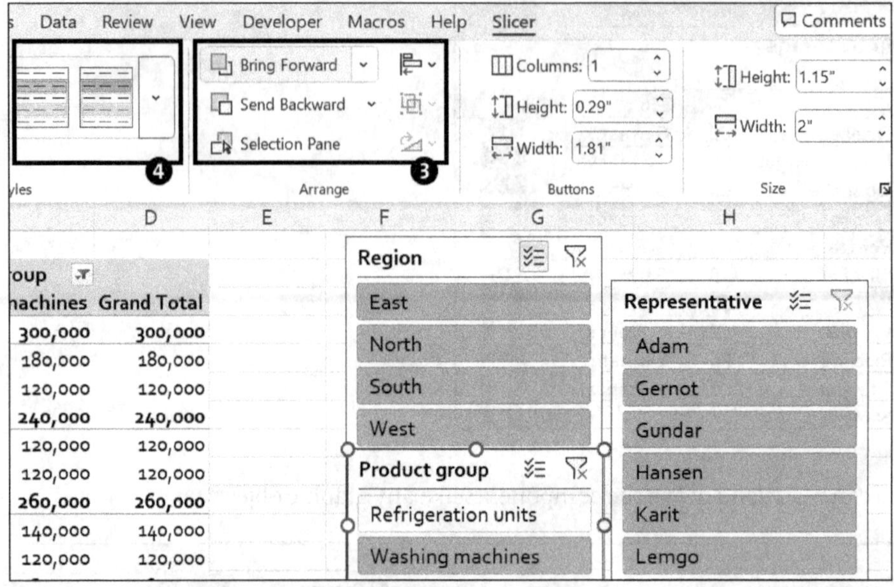

9. Filter components stay linked. For example, if you select **North** in the **Region** component, the **Representative** component will automatically show the representatives from that region as selected. If you select a representative first, the corresponding region will also be highlighted.

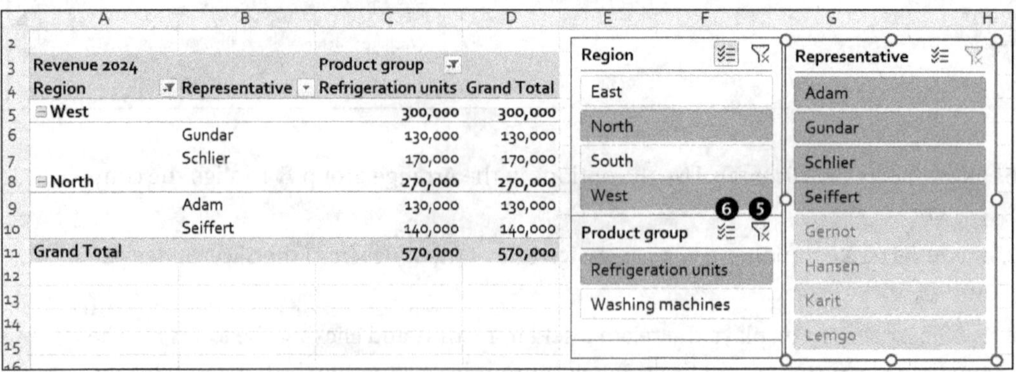

If the component contains many elements, a scrollbar will appear automatically. The **Slicer • Slicer • Slicer Settings** command opens a dialog where you can adjust how elements display in the component.

Since slicer components can be referenced by macros, a field lets you replace the default name—which is the field name. You can overwrite or hide the label, which uses the field name by default. A more useful option is to sort item names in **Ascending** or **Descending** order. You can also use custom lists, as explained in Chapter 16. You can visually mark items without data or move them to the end of the list.

Slicer Settings
Source Name: Representative
Name to use in formulas: Slicer_Representative
Name: Representative
Header
Display header
Caption: Representative
Item Sorting and Filtering
Ascending (A to Z)
Descending (Z to A)
Use Custom Lists when sorting
Hide items with no data
Visually indicate items with no data
Show items with no data last
Show items deleted from the data source

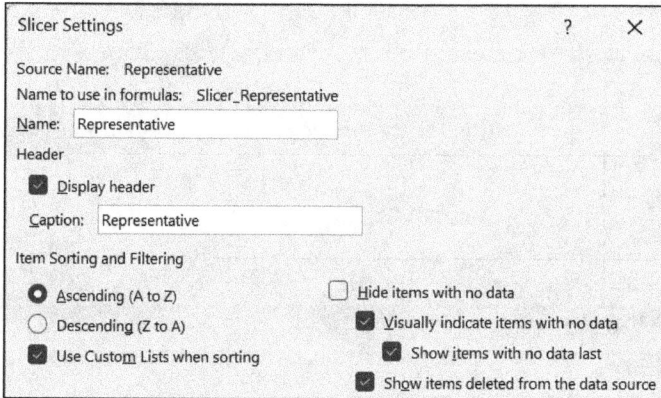

Figure 18.30 Dialog for Slicer Settings

You can also detach slicers from one PivotTable and apply them to others. Use the **Report Connections** button in the **Slicer** group to do this. In the dialog box, you can delete the active connection and select another. All PivotTables in the workbook that use the same source data will be listed. You can also link the slicer to multiple PivotTables at once.

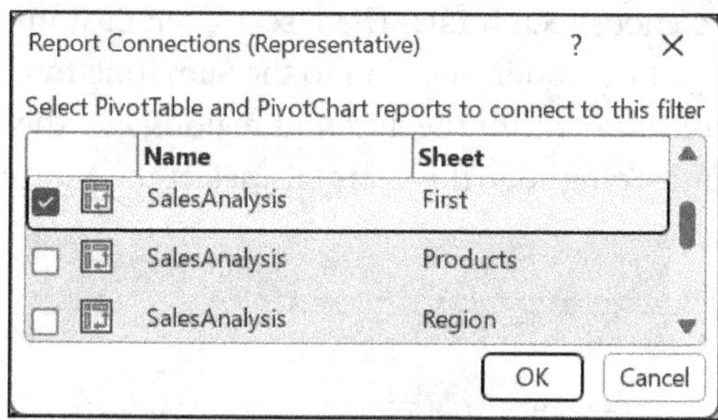

Figure 18.31 New Slicer Assignment

18.4.11 Changing Settings for Individual Fields

You can set specific options for each field in the PivotTable. To do this, select the field name in the PivotTable and then choose the **Field Settings** option in the **PivotTable Analyze • Active Field** group. The **Field Settings** dialog box will open and offer two tabs. Then, under **Custom Name**, you can rename the field to create a clearer label.

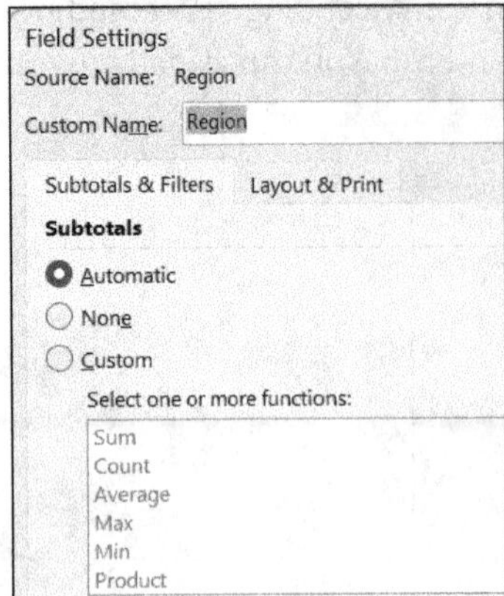

Figure 18.32 Field Settings Dialog Box

Selecting the Evaluation Type

The options on the **Subtotals & Filter** tab offer more possibilities. Excel usually shows subtotals only for the top column or row fields when multiple fields exist. If the values are numeric, sums appear, and for text fields, Excel shows the count of entries. You can add more subtotals for column and row fields or remove all subtotals.

For example, assume the PivotTable is initially sorted by regions in the rows area and then by product groups within each region. Excel will first provide a subtotal for each sales region, and to display the average along with the subtotal, you select the field name, open the **Field Settings** dialog box, choose **Subtotals**. Then, select the **Custom** option, and from the list, select the **Average** function in addition to the **Sum** function. Figure 18.33 shows a table with two types of subtotals for the different regions, and the average refers to the individual sales results. To remove the subtotals, select the **None** option.

	A	B	C
3	Region	Product group	Revenue 2022
4	⊟East		
5		Refrigeration units	310,000
6		Washing machines	310,022
7	East Sum		620,022
8	East Average		155,006
9	⊟South		
10		Refrigeration units	310,000
11		Washing machines	240,000
12	South Sum		550,000
13	South Average		137,500
14	⊟West		
15		Refrigeration units	300,000
16		Washing machines	240,000
17	West Sum		540,000
18	West Average		135,000
19	⊟North		

Field Settings
Source Name: Region
Custom Name: Region
Subtotals & Filters
Layout & Print
Subtotals
Automatic
None
Custom
Select one or more functions:
Sum
Count
Average
Max
Min
Product
Filter
Include new items in manual filter

Figure 18.33 Table with Multiple Subtotals

18.4.12 Showing and Hiding Subtotals and Grand Totals

You can temporarily hide subtotals or grand totals at any time after setting them up. On the **Design** tab, the **Layout** group offers the **Subtotals** and **Grand Totals** buttons, which open menus with options to display results.

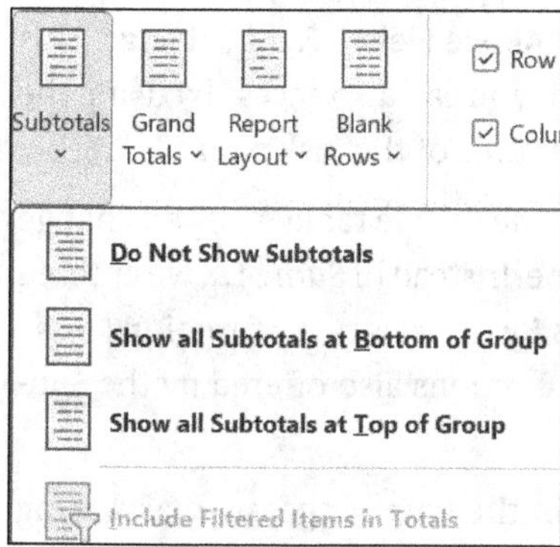

Figure 18.34 Options for Showing and Hiding Subtotals and Grand Totals

18.4.13 Options for Layout and Printing

The second tab in the **Field Settings** dialog lets you adjust how item names appear in the PivotTable. You can choose between tabular form and outline form, and in the latter case, you can also set subtotals to appear above each group.

You can also display the next field's label in the same column in a more compact form. This matches the short format, which the next section will cover. To visually separate item groups more clearly, you can use the **Insert blank row after each item name** option. For printing, you can set a page break after each item.

The **Repeat item labels** option ensures that region names repeat in every row of the example table.

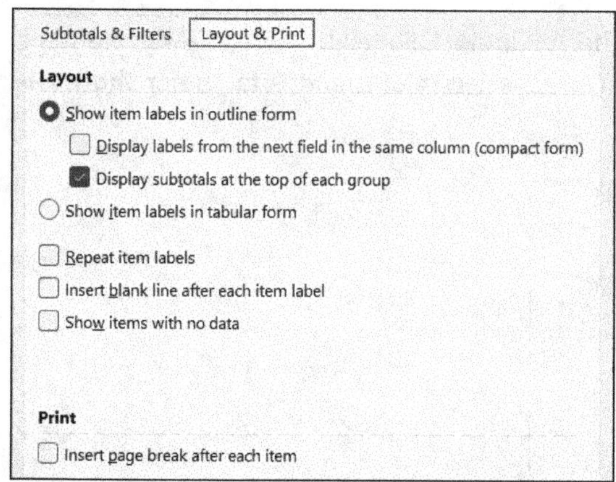

Figure 18.35 Options for Table Layout

18

18.4.14 Changing the Calculation Method

If you don't change anything, Excel sums related values in the value range cells, such as sales by different representatives for a product group. For text fields, it shows the count, like the number of representatives in a region. You can also select other calculation methods to aggregate data, and this change applies to a specific value field. To do this, select a value field and then use **PivotTable Analyze • Active Field • Field Settings**. This command opens the **Value Field Settings** dialog, which you can also access by using the menu of the corresponding value button in the **Values** area of the task pane.

Next, you'll see the source name, which is the label in the source table. You can change the automatically generated name under **Custom Name**. Instead of *Sum of...*, we entered *Sales ...* in earlier examples. On the **Summarize Values By** tab, select the function to use for the calculation. Here you will find the statistical functions also offered by the **Subtotals** command.

The **Number Format** button lets you format values in the data range, such as adding thousand separators and hiding decimal places.

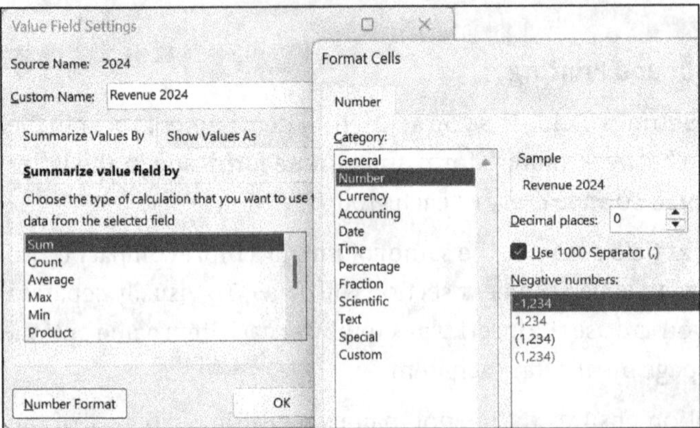

Figure 18.36 Value Field Settings

The second tab, **Show Values As**, offers useful options. For example, to view percentage shares of summarized values instead of sums, select **% of Grand Total** under **Show values as**.

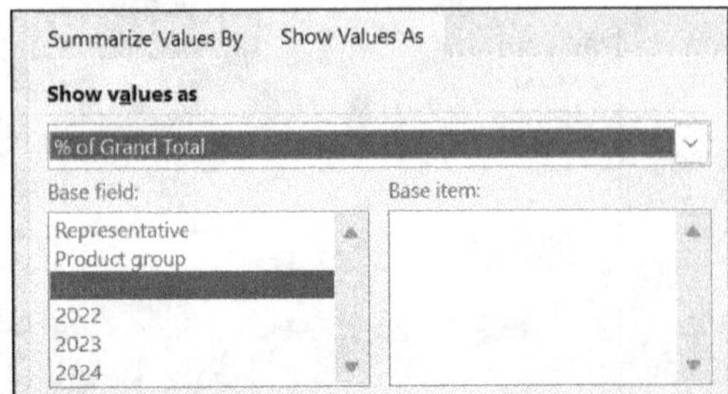

Figure 18.37 Choosing Percentage Display Relative to Grand Total

You can also calculate differences with other columns. To do this, select the appropriate entry under **Base field** or **Base item**. Figure 18.38 shows a table with percentage shares.

	A	B	C	D	E
3	Region	Representative	Revenue 2024	Sum of 2023	Sum of 2022
4	**East**		**27.11%**	**640,000**	**620,022**
5		Gernot	14.67%	340,000	260,000
6		Hansen	12.44%	300,000	360,022
7	**South**		**24.44%**	**550,000**	**550,000**
8		Karit	12.44%	280,000	280,000
9		Lemgo	12.00%	270,000	270,000
10	**West**		**24.89%**	**600,000**	**540,000**
11		Gundar	12.00%	250,000	250,000
12		Schlier	12.89%	350,000	290,000
13	**North**		**23.56%**	**530,000**	**510,000**
14		Adam	12.00%	250,000	250,000
15		Seiffert	11.56%	280,000	260,000
16	**Grand Total**		**100.00%**	**2,320,000**	**2,220,022**

Figure 18.38 Table with Percentage Shares

You can also calculate differences with other columns. To calculate the difference of other regions compared to the **West** region, select **% Difference From** on the list and then choose **Region** under **Base field** and **West** under **Base item.**

To speed up switching between analyses, the context menu for a value field includes **Summarize Values By** and **Show Values As** options, which offer many calculation alternatives that you can use directly for the selected value field. You can enter the values for the **Base field** or **Base item** as needed in a small dialog box. 18

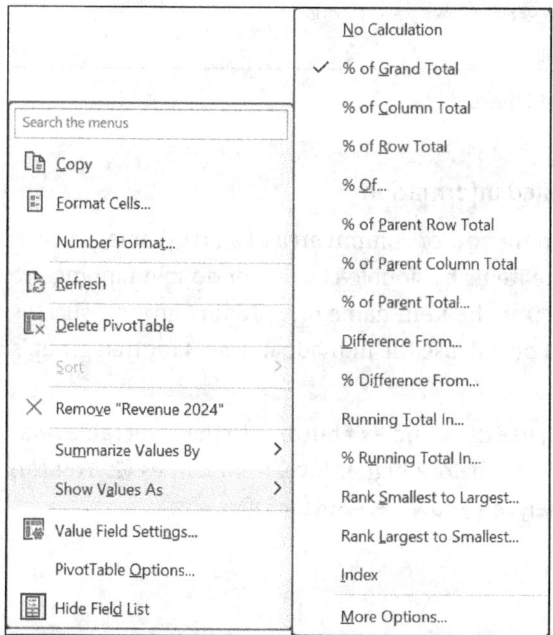

Figure 18.39 Context Menu for Value Field

18.4.15 Special Options for Report Filters

Excel offers additional layout options for a PivotTable with report filters. Like row or column fields, you can work with multiple report filter fields simultaneously. In this example, the first filter is **Region**, and the second is **Product group**. When using multiple report filters, you can arrange them vertically in a column or horizontally in a row. With more than two page fields, you can arrange them by rows and columns.

To choose an arrangement, you can use the previously described **PivotTable Options** dialog box. For example, to arrange report filters side by side, select **Over, Then Down** under **Display fields in report filter area** on the **Layout & Format** tab. If there are four page fields, set **Report filter fields per row** to 2 so as to create two rows with two filters each, arranged hierarchically from top to bottom.

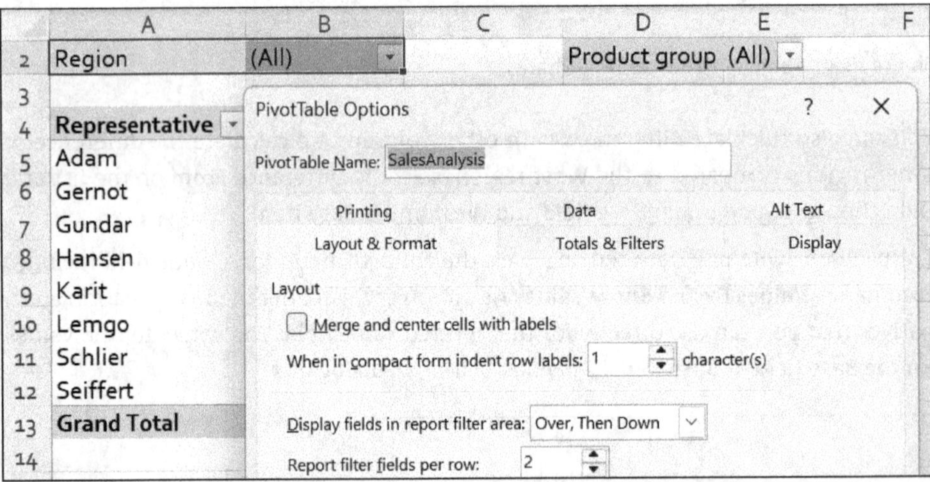

Figure 18.40 Choosing Layout for Report Filters

18.4.16 Showing and Hiding Detailed Information

When working with multiple fields in the row or column area of a PivotTable, you can easily expand or collapse subordinate items by double-clicking or double-tapping the higher-level item. For example, selecting the item name of a product group—such as **Washing machines**—lets you expand or collapse the individual results for that group's representatives.

Instead of double-clicking or tapping, use the **+** and **–** ❶ buttons in the **PivotTable Analyze • Active Field** group or the **+/–** icons before higher-level item names ❷. You can show or hide these via **PivotTable Analyze • Show • +/- Buttons** ❸.

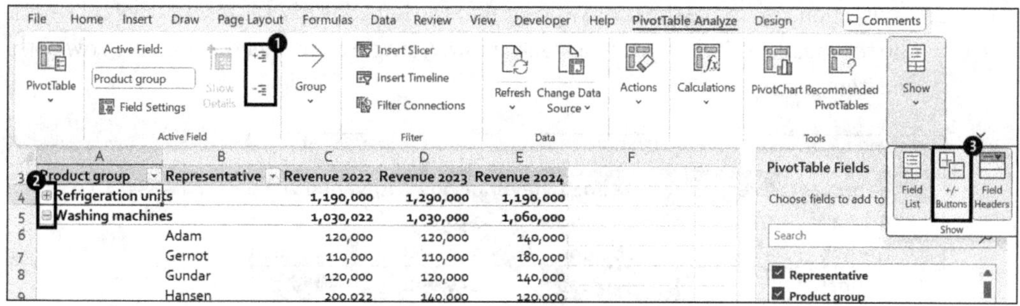

Figure 18.41 Hidden Individual Results for Refrigerator Product Group

18.4.17 Creating New Groups

If a field has many items, you can create subgroups and calculate results for them. For text items, do this manually; for number or date items, you can also group automatically by intervals.

If a company has multiple representatives, it doesn't mean they are all treated equally. The commission may be higher for some than for others. Suppose you have two commission levels and want to assign representatives to one of the two groups. The following example assumes a PivotTable where only representatives appear as row labels.

Follow these steps to create new groups:

1. If you haven't already, activate the special selection cursor that makes selecting rows or columns easier. In the **PivotTable Analyze** group, choose **Actions**, click the **Select** button, and select **Enable Selection**.

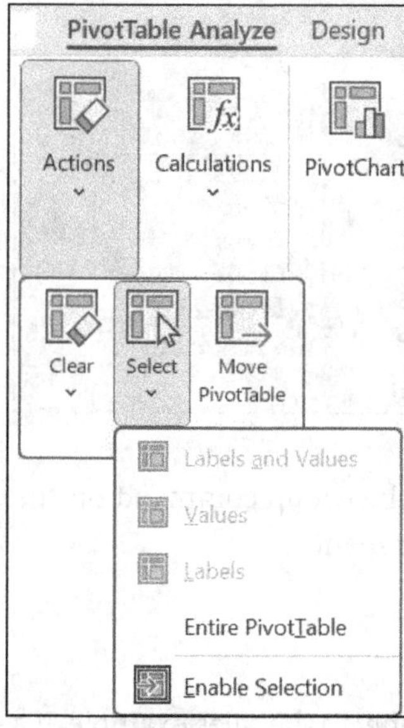

2. Now, when you move the mouse pointer to the left edge of a representative cell—not the row header—it changes to a black arrow. A single click selects the entire row with the item name and data.

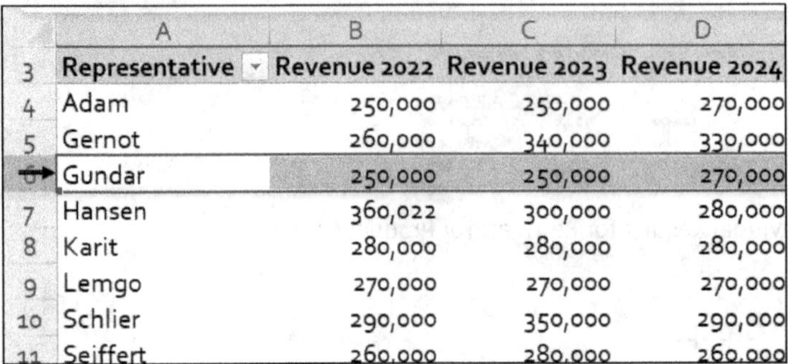

	A	B	C	D
3	Representative	Revenue 2022	Revenue 2023	Revenue 2024
4	Adam	250,000	250,000	270,000
5	Gernot	260,000	340,000	330,000
6	Gundar	250,000	250,000	270,000
7	Hansen	360,022	300,000	280,000
8	Karit	280,000	280,000	280,000
9	Lemgo	270,000	270,000	270,000
10	Schlier	290,000	350,000	290,000
11	Seiffert	260,000	280,000	260,000

3. To split the column with representatives into two groups, select the representatives for the first group with the mouse. You can select multiple rows by dragging, and the items do not need to be adjacent. Hold down `Ctrl` and click each desired item. On a touchscreen, you can select with the pen.
4. Use the **Group Selection** command in the **PivotTable Analyze • Group** group. Excel will group the items you've selected and insert a field and a group item with temporary names. You can rename both in the formula bar after selecting them.
5. Select the remaining representatives the same way and click the **Group Selection** button again. Rename the new group field in the field settings, and the new field will appear in the **PivotTable Fields** pane.

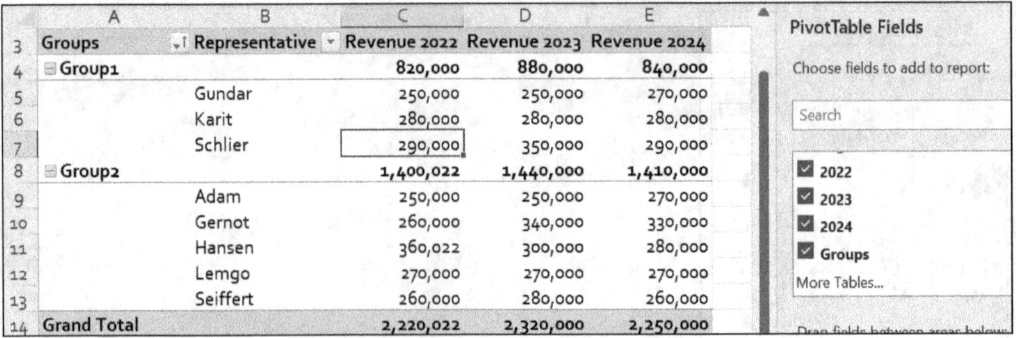

	A	B	C	D	E
3	Groups	Representative	Revenue 2022	Revenue 2023	Revenue 2024
4	Group1		820,000	880,000	840,000
5		Gundar	250,000	250,000	270,000
6		Karit	280,000	280,000	280,000
7		Schlier	290,000	350,000	290,000
8	Group2		1,400,022	1,440,000	1,410,000
9		Adam	250,000	250,000	270,000
10		Gernot	260,000	340,000	330,000
11		Hansen	360,022	300,000	280,000
12		Lemgo	270,000	270,000	270,000
13		Seiffert	260,000	280,000	260,000
14	Grand Total		2,220,022	2,320,000	2,250,000

To ungroup, select the new group field button and use the **Group** command on the **PivotTable Analyze** tab or select **Ungroup** from the context menu.

18.4.18 Organizing Numerical Data

At this point, we'll cover some other options that don't apply to the sales example. If a column contains numbers, you can group them into specific intervals. This example uses a small table to display test results. The second column shows each test subject's age, and the third column shows their test results.

You might initially doubt that a PivotTable can provide any insight here, and the first attempt is rather underwhelming. You can create a PivotTable that uses age as the row label and test results as the value field. The PivotTable will then essentially replicate the original table, except for the sum displayed at the end. However, the possibilities don't end there. For example, if you want to analyze average test results for a specific age group, the PivotTable can assist.

	A	B	C
1	Test person	Age	Test result
2	1	22	5036
3	2	38	4873
4	3	46	4762
5	4	14	4683
6	5	11	4563
7	6	28	4481
8	7	44	4381
9	8	10	4284
10	9	13	4168
11	10	23	3942
12	11	31	3895
13	12	13	3745
14	13	30	3494
15	14	47	3400
16	15	33	3330
17	16	49	3308
18	17	42	3218
19	18	48	3099
20	19	19	2939
21	20	43	2879
22	21	28	2833

Figure 18.42 Source Table

Select any item in the **Age** column, then use the **Group Field** command in the **PivotTable Analyze • Group** group. Since these are numeric values, a small dialog box will appear and allow you to set specific age intervals (see Figure 18.43).

The dialog box shows the lowest and highest age values and suggests a ten-year interval under **By**. You can accept or adjust these values and then click **OK** to display the Pivot-Table with the age groups.

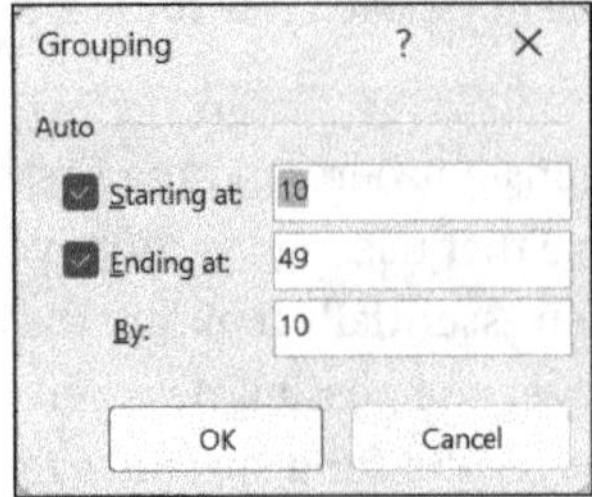

Figure 18.43 Choosing Intervals

However, summing the test results is not meaningful in this case, but obtaining the average test result per age group is important. This is easy to do:

1. Select a test result.
2. In the context menu under **Summarize Values By**, select **Average**.
3. To remove extra decimal places, use **Number Format** in the context menu and choose a format without decimals.

The table in Figure 18.44 shows the distribution results across different age groups.

E	F
Age	**Test_result**
10-19	25467
20-29	20702
30-39	23419
40-49	28747
Grand Total	**98335**

Figure 18.44 Grouping by Age Group

Creating Groups by Time or Date

If a field contains date or time elements, you can proceed similarly. The dialog box offers grouping by days, months, quarters, or years. Selecting **Days** lets you group multiple days together, such as by week.

This example begins with a small list recording daily visitor numbers for an exhibition over one month. To see how visitor numbers distribute across the weeks of the month, follow a similar process as in the previous example:

1. Create a PivotTable with the date as the row label and visitor numbers as the value field.
2. Select any date and use the **PivotTable Analyze • Group • Group Field** command.
3. Under **By**, select **Days** and enter "7" under **Number of days**. Use **Starting at** to specify when the first week begins.

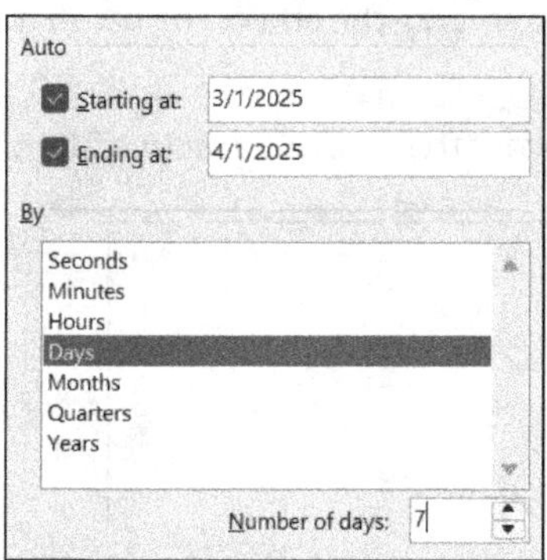

You will then see the table sorted into 7-day intervals, as in Figure 18.45.

D	E
Days (Day)	Visitors
3/1/2025 - 3/7/2025	1696
3/8/2025 - 3/14/2025	2032
3/15/2025 - 3/21/2025	1233
3/22/2025 - 3/28/2025	1665
3/29/2025 - 4/1/2025	948
Grand Total	7575

Figure 18.45 Grouped Table

18.4.19 Inserting Timelines

Another feature that, like slicers, simplifies navigating compiled data is the timeline. This is also a graphical element that you can add, and it lets you easily browse data organized by time. The table from the last section showing visitor numbers is a good candidate for this solution.

If the PivotTable includes a time dimension, adding a timeline is useful. The prerequisite is at least one field with a valid date format. The procedure is as follows:

1. Select any cell in the PivotTable. Use the **Insert Timeline** button in the **PivotTable Analyze • Filter** group.
2. The dialog offers fields with a date format for selection, and confirming the dialog creates a timeline that covers at least the time span in the selected field. The division is based on existing data, but you can change it by using the list box in the upper right corner.
3. As with any graphic element, adjust size and position by dragging the frame or handles.
4. Clicking or tapping a section of the scale activates the corresponding filter functions, and the table will then show only values for the selected day, month, quarter, or year.

5. To display data for a specific period, such as several days, drag the end of the bar that marks the current selection, as Figure 18.46 shows.
6. Click or tap the **Clear Filter** icon in the upper right corner to remove the active filter.

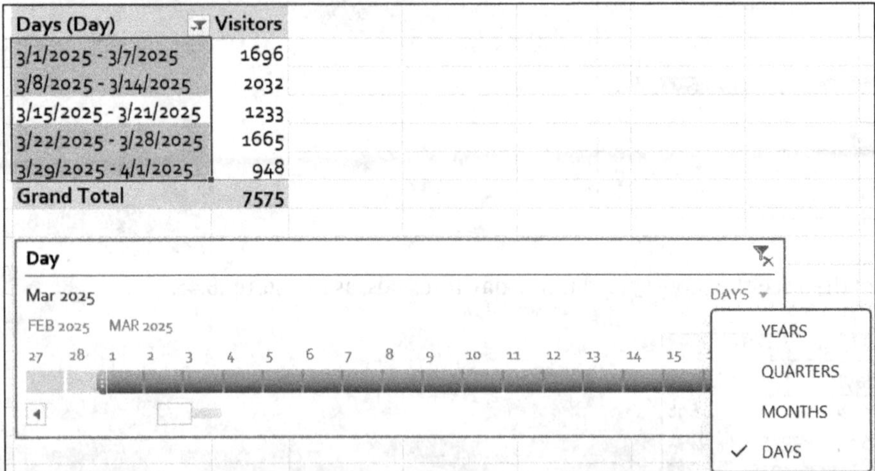

Figure 18.46 Timeline for PivotTable with Date Field

You can access more formatting options through the context menu by selecting **Size and Properties** to open the **Format Timeline** pane.

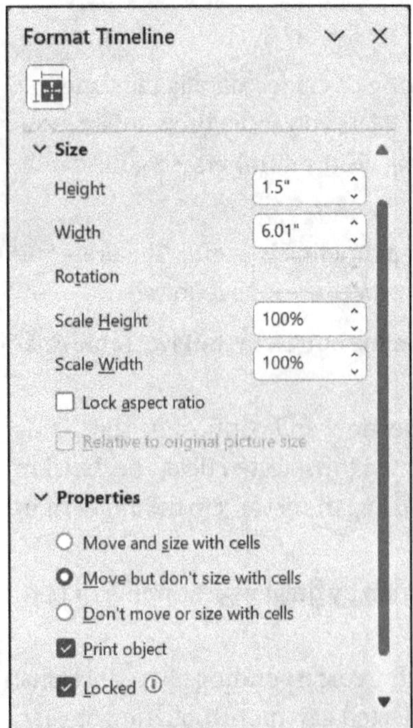

Figure 18.47 Pane for Formatting Timeline

18.4.20 Calculated Fields and Items in PivotTables

PivotTables allow you to analyze existing tables and experiment with the displayed data. While a PivotTable shows only the results of formulas in the source table—not the formulas themselves—you can extend it with calculated fields. These formulas refer to either a field or a single element within a field.

However, formulas can only include references to fields within the PivotTable itself and constant values, not references to other cells in the table sheet. To refer to PivotTable elements, use the element names or field names directly.

Inserting Calculated Fields

These fields allow what-if analyses, such as determining the values for individual items when sales increase by 20%. Consider the source table in Figure 18.48.

	A	B	C	D
3	Revenue 2024	Product group		
4	Representative	Refrigeration units	Washing machines	Grand Total
5	Adam	130,000	140,000	270,000
6	Gernot	150,000	180,000	330,000
7	Gundar	130,000	140,000	270,000
8	Hansen	160,000	120,000	280,000
9	Karit	160,000	120,000	280,000
10	Lemgo	150,000	120,000	270,000
11	Schlier	170,000	120,000	290,000
12	Seiffert	140,000	120,000	260,000
13	Grand Total	1,190,000	1,060,000	2,250,000

Figure 18.48 PivotTable Before Inserting Calculated Field

To insert a calculated field, follow these steps:

1. Select the **Product group** column label field, and on the **PivotTable Analyze** tab, in the **Calculations** group, select the **Fields, Items, & Sets** command and the **Calculated Field** submenu.
2. In the dialog box, replace the suggested field name with a suitable one, such as "*target_2025*."
3. Enter the formula in the next text box. The formula should reference the **2023** value field, so select this field from the **Fields** list and insert it into the formula by using **Insert Field**.
4. The planned value is set to exceed the 2024 values by 20%, so the formula appends `* 1.2`.
5. If you don't want to define additional calculated fields, close the dialog box by clicking **OK**; otherwise, select **Add** and specify the next calculated field. Remove unnecessary calculated fields with **Delete** when they're selected under **Name**.

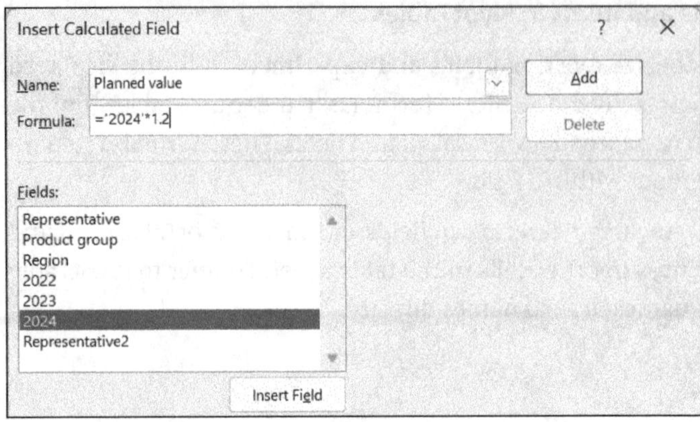

6. The table adds extra columns with planned values for each product group for the new field, and you can adjust the field names accordingly.

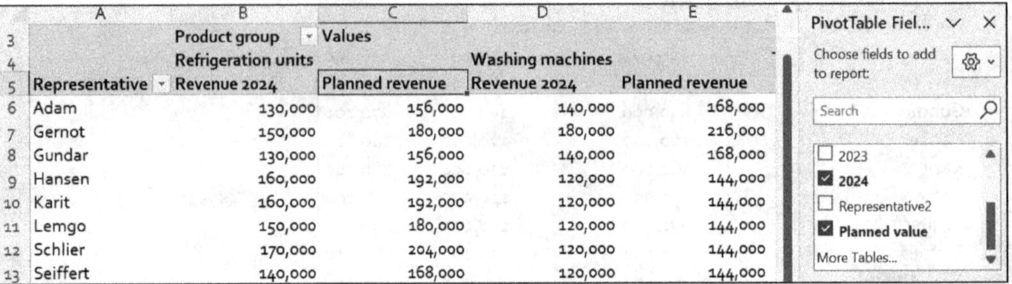

	Product group	Values		
	Refrigeration units		Washing machines	
Representative	Revenue 2024	Planned revenue	Revenue 2024	Planned revenue
Adam	130,000	156,000	140,000	168,000
Gernot	150,000	180,000	180,000	216,000
Gundar	130,000	156,000	140,000	168,000
Hansen	160,000	192,000	120,000	144,000
Karit	160,000	192,000	120,000	144,000
Lemgo	150,000	180,000	120,000	144,000
Schlier	170,000	204,000	120,000	144,000
Seiffert	140,000	168,000	120,000	144,000

When you add calculated fields to a PivotTable, they also appear in the field list in the task pane.

Formulas for Individual Items

In this case, the preceding field formula calculates planned values for all entries in the **Product group** field. You can also use formulas that return values for a specific item in a field. For example, to calculate how sales for washing machines compare percentage-wise to refrigerators, follow this procedure:

1. Select the item name **Washing machines**, and then, from the **PivotTable Analyze • Calculations • Fields, Items & Sets** button menu, choose the **Calculated Item** command.
2. Enter a name for the new field, such as "R/W."
3. In the **Formula** field, select the item names **Washing machines** and **Refrigeration units** from the **Items** list and insert the division sign between them.

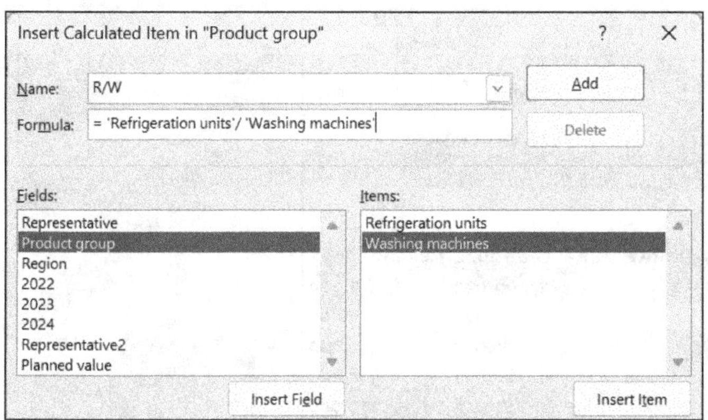

4. The table will add an extra row below the washing machine sales column. Format the new values as percentages.

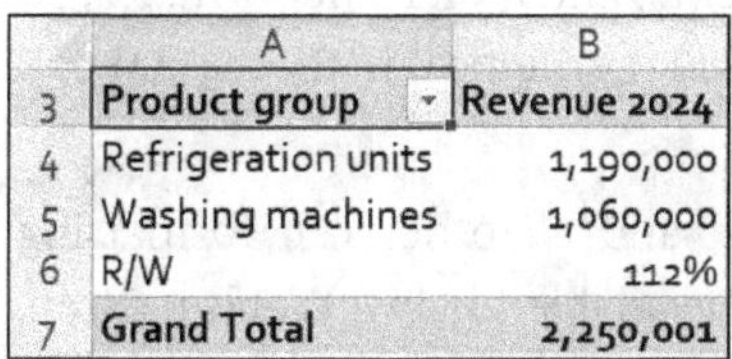

	A	B
3	Product group	Revenue 2024
4	Refrigeration units	1,190,000
5	Washing machines	1,060,000
6	R/W	112%
7	Grand Total	2,250,001

Calculated items do not appear in the field list like calculated fields because they are treated as additional items within an existing field. If PivotTables from the same source were previously created in the workbook, calculated items may also appear in earlier tables. To hide them, apply the appropriate filter.

Setting the Calculation Order

Different formulas can sometimes affect the values in a PivotTable cell, so you should use the **PivotTable Analyze • Calculations • Fields, Items, & Sets • Solve Order** command to specify which formula determines the cell's value. In the dialog box, you can change the priority of the selected formulas by using the **Move Up** or **Move Down** buttons. The last formula in the series determines the current value of the cell. Since formulas are not documented directly in the table or formula bar, the **Fields, Items, & Sets** menu includes a command that lists formulas on a separate sheet.

18.4.21 Formatting PivotTables

Using predefined patterns in the **Design • PivotTable Styles** group is especially helpful when formatting PivotTables, as the patterns largely match table formats. In the **PivotTable Style Options** group, you can apply special formatting to row and column headers, as you do with table styles, and set alternate row or column formatting.

Subtotals | Grand Totals | Report Layout | Blank Rows — Layout
☑ Row Headers ☑ Banded Rows ☑ Column Headers ☐ Banded Columns — PivotTable Style Options

A4 | fx | Group2

	A	B	C	D
3	Representative2	Representative	Revenue 2022	Revenue 2023
4	⊟ Group2		1,400,022	1,440,000
5		Adam	250,000	250,000
6		Gernot	260,000	340,000
7		Hansen	360,022	300,000
8		Lemgo	270,000	270,000

Figure 18.49 PivotTable Style with Banded Rows

To create custom templates for PivotTables, open the dialog via the **New PivotTable Style** palette command, select individual table elements, and assign the formats you want by using the **Format** button. Your custom styles will then appear in the **Format** palette in a separate section.

We've already covered setting the number format, and all other cell formatting options—font, background, borders, and more—are available for PivotTables. Often, users want to highlight specific data groups, for example, by applying different background colors. Excel ensures that color assignments remain intact when data is regrouped.

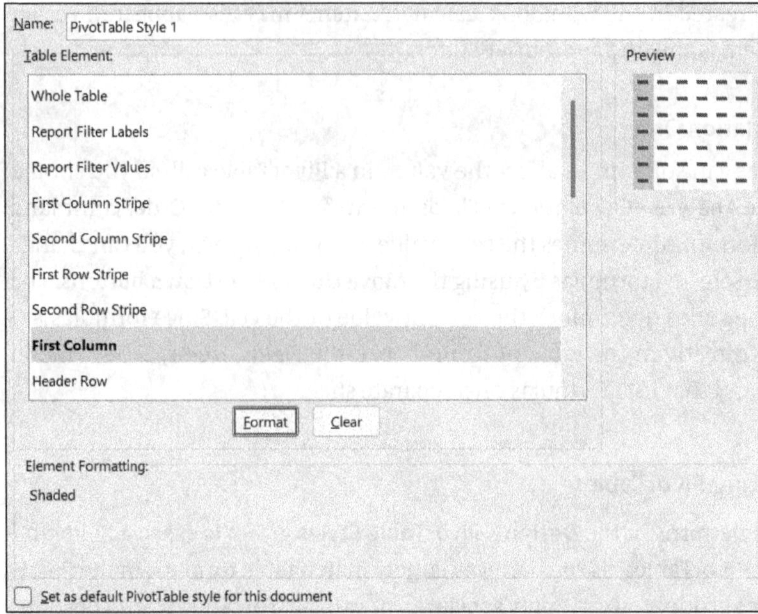

Figure 18.50 Creating Custom PivotTable Formats

18.4.22 Quickly Selecting Data Groups

To simplify selecting data for formatting or other tasks like copying, Excel offers commands under the **Select** button in the **PivotTable Analyze • Actions** group. These commands let you select specific parts or the entire PivotTable. The first three options become available only after you select the **Entire PivotTable** or activate the selection cursor with **Enable Selection**, which you can use to mark individual items. The **Values** selection highlights all data in the value range, including existing analyses. You can also select labels separately to assign different fonts.

Figure 18.51 Commands for Selecting Data Groups

18.4.23 Conditional Formatting in PivotTables

You can use conditional formatting options very effectively in PivotTables. Figure 18.52 shows a table with data bars, and when you assign this format to a cell selection in a PivotTable, the **Formatting Options** button appears at the end of the range. It lets you make targeted assignments to specific PivotTable values.

	A	B	C	D
3	Representative	Revenue 2022	Revenue 2023	Revenue 2024
4	Adam	250,000	250,000	270,000
5	Gernot	260,000	340,000	330,000
6	Gundar	250,000	250,000	270,000
7	Hansen	360,022	300,000	280,000
8	Karit	280,000	280,000	280,000
9	Lemgo	270,000	270,000	270,000
10	Schlier	290,000	350,000	290,000
11	Seiffert	260,000	280,000	260,000
			2,320,000	2,250,000

Apply formatting rule to ...

- Selected cells
- All cells showing "Revenue 2022" values
- All cells showing "Revenue 2022" values for "Representative"

Figure 18.52 PivotTable with Conditional Formats

Instead of selecting all relevant cells first, you can assign the format to a single cell initially. The second option also formats the grand total, while the third option formats only the individual values of the representatives in the respective columns.

	A	B	C	D
3	**Representative**	**Revenue 2022**	**Revenue 2023**	**Revenue 2024**
4	Adam	250,000	250,000	270,000
5	Gernot	260,000	340,000	330,000
6	Gundar	250,000	250,000	270,000
7	Hansen	360,022	300,000	280,000
8	Karit	280,000	280,000	280,000
9	Lemgo	270,000	270,000	270,000
10	Schlier	290,000	350,000	290,000
11	Seiffert	260,000	280,000	260,000
12	**Grand Total**	**2,220,022**	**2,320,000**	**2,250,000**

Figure 18.53 Data Bars for Values of Representatives

18.4.24 Changing, Moving, and Deleting a Data Source

After you create a complex analysis as a PivotTable, you may find that you can analyze other data sources with the same method. In such cases, you can change the data source afterward. To do this, use the **Change Data Source** button in the **Data** group on the **PivotTable Analyze** tab and then select the new data source in the dialog.

If the new data source contains additional fields, they will be added to the field list, and you can select them to include the corresponding data in the PivotTable. If the new data source has fewer fields, the missing fields will be re removed from the field list and the table.

To move a PivotTable, click the **Move PivotTable** button in the group **PivotTable Analyze • Actions** and specify the new location in the dialog. The **Actions** menu also includes the **Clear** command. **Clear All** removes the entire PivotTable, while **Clear Filters** removes only the applied filters.

18.4.25 Data Types and Images in PivotTables

PivotTables can now handle data types and images, and product tables can be enhanced with images that are inserted into table cells, as described in Chapter 10. A simple example based on the wine storage from Chapter 16 shows what is possible.

First, select the column with country names and overwrite it by using the **Geography** option in the **Data • Data Types** group. A symbol indicating the name as part of the data type will appear before the country name, and you can use the small icon next to the flag to insert a new column for it. Then, drag the fill handle down to the last entry.

When you create a PivotTable from the table, the flag column is treated like any other data column. Figure 18.56 shows a simple summary of the source table.

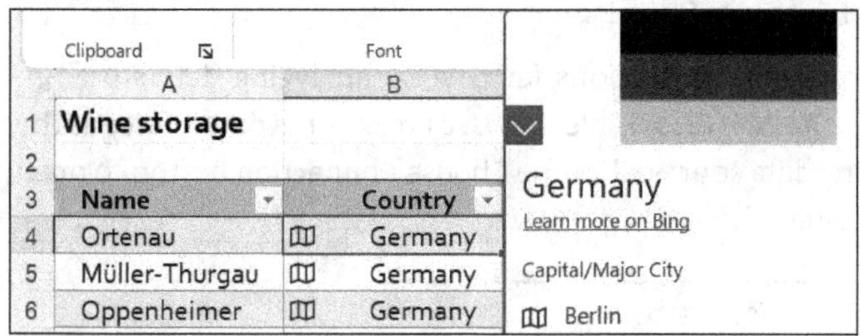

Figure 18.54 Details About Selected Data Type

	A	B	C
1	**Wine storage**		
2			
3	**Name**	**Country**	**Flag**
4	Ortenau	Germany	
5	Müller-Thurgau	Germany	
6	Oppenheimer	Germany	
7	Médoc	France	
10	Bordeaux	France	
11	Silvaner	Germany	
12	Bourgogne	France	
13	Chablis	France	
17	Riesling	Germany	
18	Beaujolais	France	
20	Sauternes	France	

Figure 18.55 Geography Data Type Provides Corresponding Flag

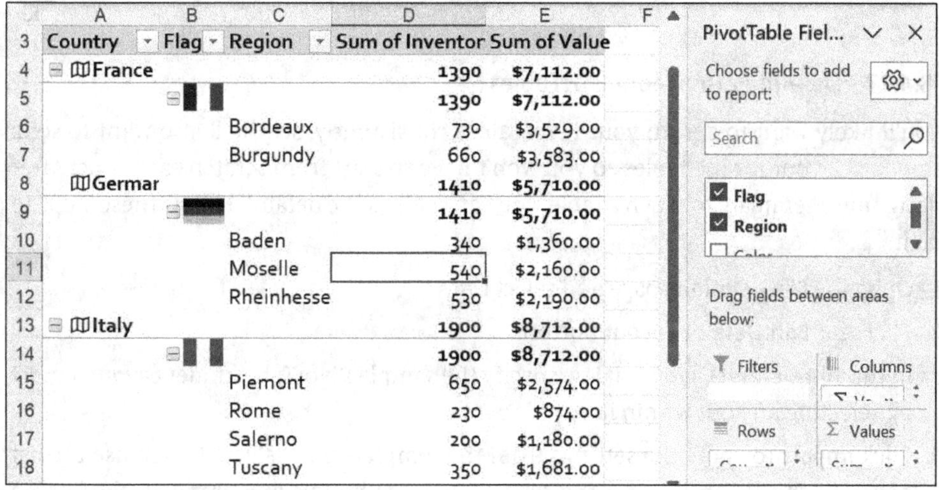

	A	B	C	D	E
3	Country	Flag	Region	Sum of Inventor	Sum of Value
4	France			**1390**	**$7,112.00**
5				**1390**	**$7,112.00**
6			Bordeaux	730	$3,529.00
7			Burgundy	660	$3,583.00
8	Germar			**1410**	**$5,710.00**
9				**1410**	**$5,710.00**
10			Baden	340	$1,360.00
11			Moselle	540	$2,160.00
12			Rheinhesse	530	$2,190.00
13	Italy			**1900**	**$8,712.00**
14				**1900**	**$8,712.00**
15			Piemont	650	$2,574.00
16			Rome	230	$874.00
17			Salerno	200	$1,180.00
18			Tuscany	350	$1,681.00

Figure 18.56 PivotTable with Data Types and Images

18.4.26 PivotTables from External Data

PivotTables and PivotCharts offer applications far beyond analyzing data stored in workbooks. The menu under **Insert • PivotTable** lists **Use an external data source** as the second option. To specify the data source, click the **Choose Connection** button to open the **Existing Connections** dialog.

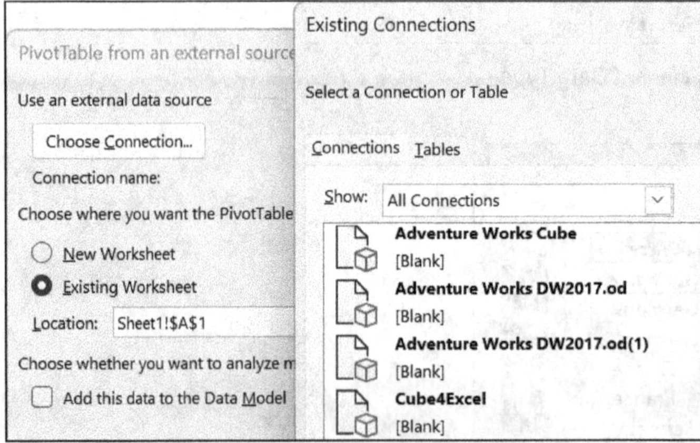

Figure 18.57 Creating a PivotTable from External Data

Setting up connections for pivot analysis, such as database access, is explained in the next chapter.

When selecting an existing connection, you may need to enter credentials for server-based data sources.

18.4.27 Default Format for PivotTables

You'll likely want to design your PivotTables consistently, so you'll also want to set a standard layout for the tables so you won't have to start from scratch each time. After designing a template for such a table, you can set it as the default. Follow these steps to do it:

1. Open the **File • Options** page and select **Data**.
2. Click the **Edit Default Layout** button.
3. In the **Edit Default Layout** dialog, specify the sample PivotTable under **Layout Import** by selecting any cell within it.
4. Click **Import** to apply the settings from the template you selected, then use the list boxes to choose options for the report layout, subtotals and grand totals, filtered items, item labels, and inserting blank lines between items.

After you confirm the dialog, the layout you selected will become the template for your next PivotTable.

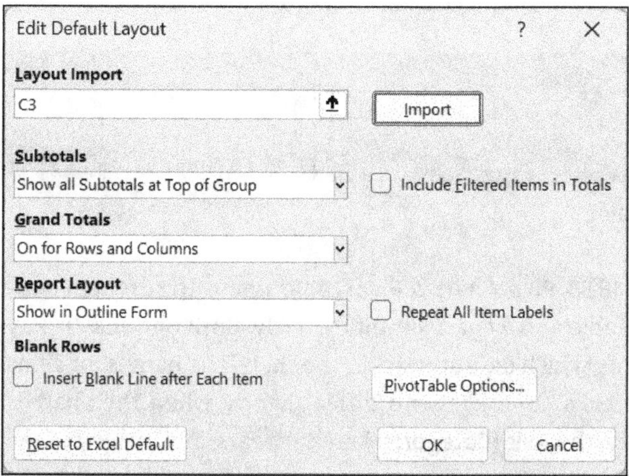

18.5 Dynamic Charts from PivotTables

Charts based on PivotTables also allow you to interactively summarize or group data. The **Insert** tab includes a dedicated **PivotChart** button for this, and the menu offers the **PivotChart** and **PivotChart & PivotTable** options.

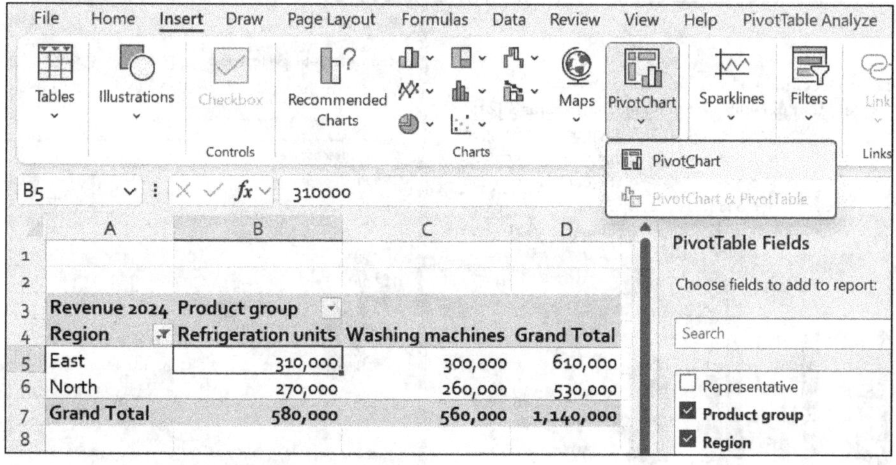

Figure 18.58 Inserting a PivotChart

You use the first option to add a chart to an existing PivotTable. If the cell pointer is inside the PivotTable, Excel will open the **Insert Chart** dialog, where you can select the chart type.

If the cell pointer is not in a PivotTable, then Excel will open the **Create PivotTable** dialog. Except for the title bar, it initially matches the dialog used to create a PivotTable. Once you confirm the dialog, an empty PivotTable and a linked chart area will be created.

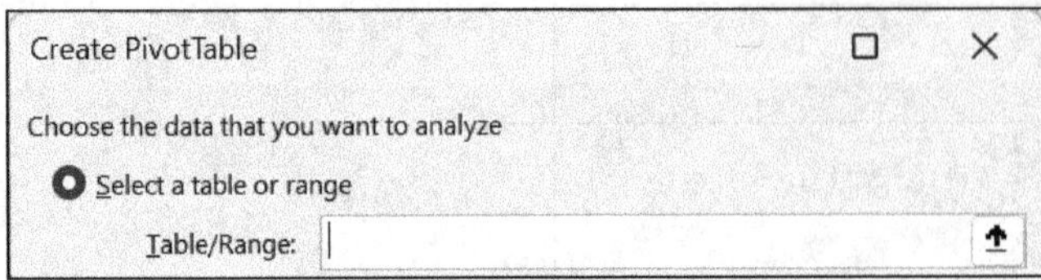

Figure 18.59 Create PivotTable Dialog

The second option from Figure 18.58, **PivotChart & PivotTable**, opens the **Create PivotTable** dialog but creates both a table and a chart simultaneously. Both options display the task pane with fields to specify which data appears in the chart. The names for areas and assignment options in a field's context menu differ slightly when the chart is selected. Instead of the **Rows** area, the **Axis (Categories)** area appears, and instead of the **Columns** area, the **Legend (Series)** area is shown.

To analyze the sales table as a chart, use the **Product group** field as the legend and the region as the axis. The sales fields remain assigned to the values area. The result is a default column chart showing total sales by region, broken down by product group. The chart matches the table, which is created automatically. To simplify use, filter buttons appear on the chart, and their list boxes let you remove items from the chart, such as a specific product group.

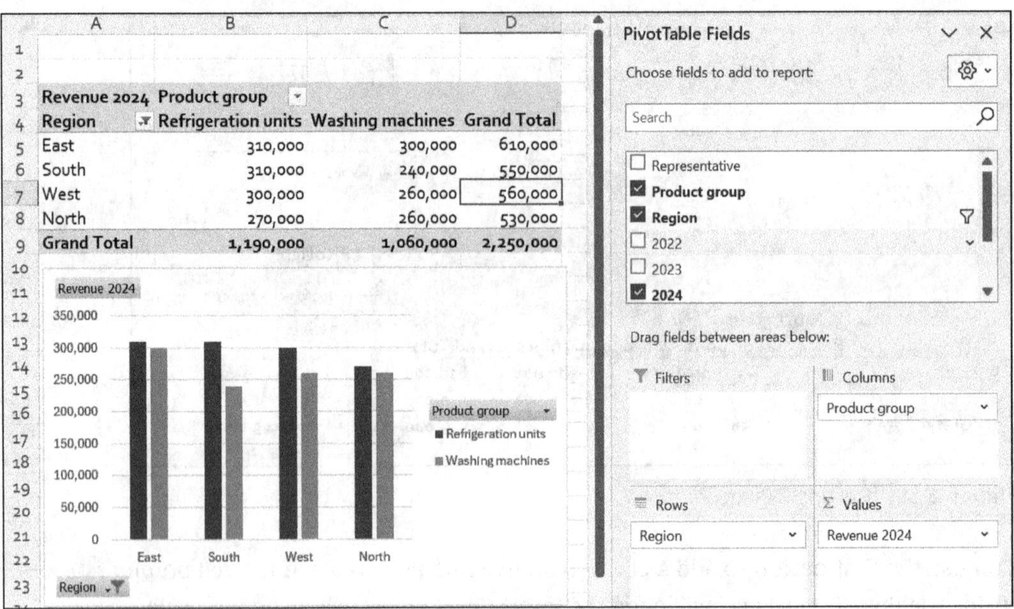

Figure 18.60 PivotChart and PivotChart Fields Task Pane

If these filter buttons are not visible, use the **Field Buttons** menu on the **PivotTable Analyze** tab in the **Show/Hide** group, where you can show or hide the field buttons individually.

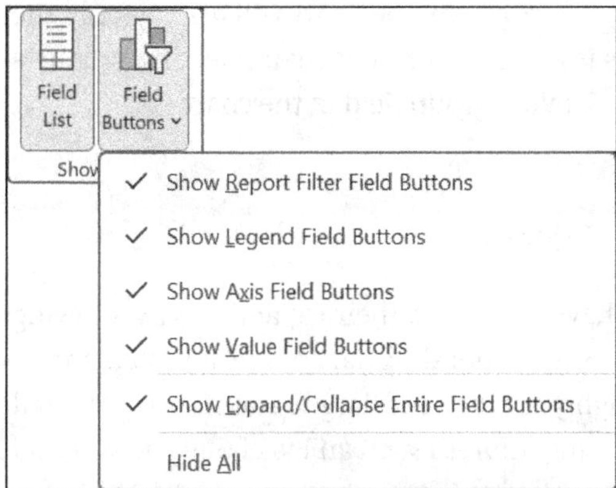

Figure 18.61 Menu for Field Buttons in Chart

You can drag fields between areas in the task pane to rearrange the chart, and the linked table will update automatically. For example, you can drag the **Region** field to the **Filters** area and then use the **Representative** field as the axis field to display columns separately for each representative.

Use the options on the **Design** tab to edit the chart when it's selected on the worksheet. Changing the chart type or applying a chart style works as described in Chapter 7. Use **Design • Location • Move Chart** to move the chart to its own chart sheet or back into a worksheet. For selected PivotCharts, the editing tab is **PivotChart Analyze.**

18

18.5.1 Slicers for a PivotChart

The slicers we described previously also work with PivotCharts. When you select a PivotChart, the **Slicers** button appears in the **Filter** group on the **PivotChart Analyze** tab. The procedure for using slicers on PivotCharts matches that for using slicers on PivotTables, as described previously.

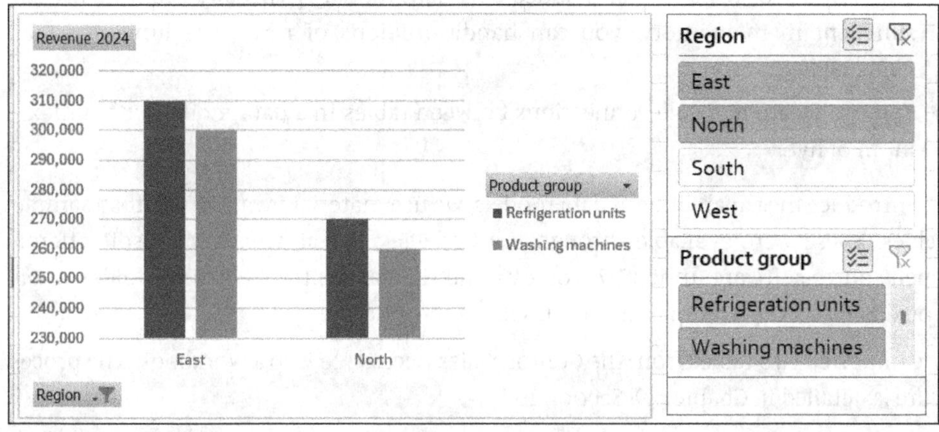

Figure 18.62 Chart with Slicer

For example, if a slicer is set up for the product group, the chart will display only the sales data for the product you select. This is more convenient than selecting products each time through the filter button of the **Product group** field in the chart.

18.6 Data Models with Multiple Tables

Since Excel 2013, PivotTables and charts have expanded their capabilities by allowing multiple tables as source data for analysis. A corresponding data model is built in memory for this purpose. Because the data is highly compressed, large datasets can be stored in Excel files and easily shared. When you uncompress such an Excel file, you will find a file with the *.data* extension that contains this data inside. The limit for the 32-bit version is between 500 and 700 MB; for the 64-bit version, only available memory size determines the limit.

18.6.1 Building a Model for a PivotTable

To set up a data model for a PivotTable, select **Add this data to the Data Model** in the **Create PivotTable** dialog when you create the PivotTable from the first data source.

The data model concept allows Excel to create an internal relational data source that links multiple tables through defined relationships and supports analysis in both PivotTables and Power Pivot. This matches the joins that are commonly used in database applications. The following rules apply:

- Only one data model is allowed per workbook.
- You can add each table in the workbook to the data model.
- You can also add data from external sources to the data model. You can process data from leading database systems, as well as from the web (via data feeds), text files, or existing Excel tables.
- The number of rows in tables added to a data model is virtually unlimited. With sufficient main memory, you can handle millions of rows. The limit is set at 1,999,999,997.
- You can create multiple connections between tables in a data model, with a maximum of five.

To introduce the capabilities of data models, we use material from the Contoso sample tables. These were available through the CodePlex portal, where Microsoft offered open-source software until 2017. You can find the data in the *ContosoV2* folder when you download the examples for this book.

We imported two tables from the ContosoSales Access file into a workbook. The procedure is detailed in Chapter 19, Section 19.1.

The first table contains a product list, with products identified by **ProductKey** and **ProductSubcategoryKey**. Only the **ProductName**, **Manufacturer**, and **UnitPrice** fields should be included. You can delete other fields with the Power Query Editor before inserting the data into the table sheet. Rename the imported table as "Products" by using **Table Design • Properties • Table name**. The second table is a short category list.

Then, add the product list to the data model as the first table and select **Add this data to the Data Model** in the **PivotTable from table or range** dialog.

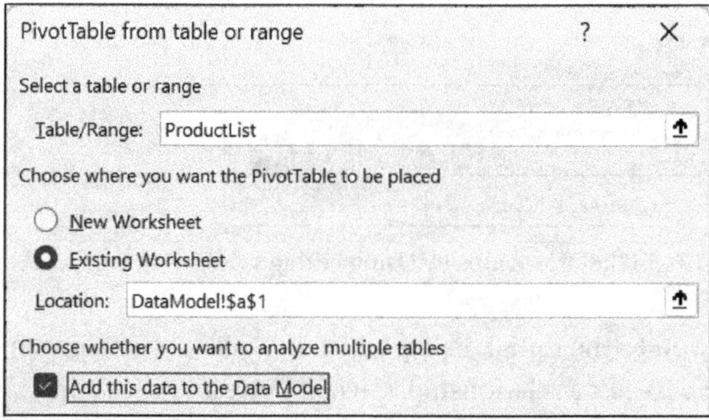

Figure 18.63 Adding Table to Data Model

The **PivotTable Fields** pane will display two links, **Active** and **All**, below the title. Under **Active**, you'll initially see only the product list's field list. Under **All**, you'll find both tables. All tables in the workbook are automatically included in the data model once it is created with any table.

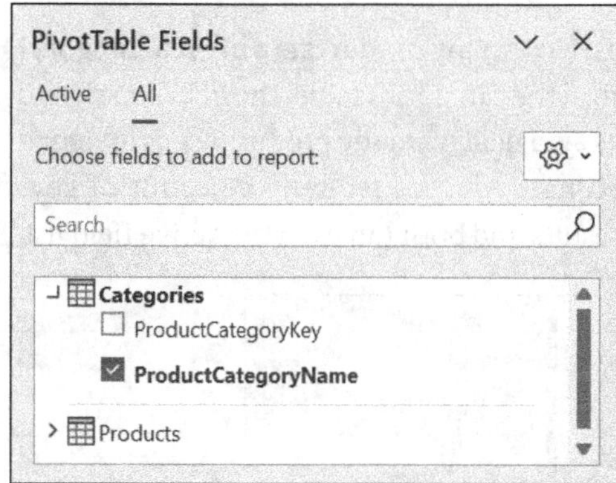

Figure 18.64 Active and All Links for Managing Data Model

To select fields from the category list for the PivotTable, choose the **Show on the Active Tab** context menu option.

18.6.2 Defining Relationships

The next step is to create a relationship between the two tables. Access the dialog via **PivotTable Analyze • Calculations • Relationships** and then click **New** to define a relationship. In this case, the relationship is based on the two keys assigned to the product category; in the product list, this field is **ProductSubcategoryKey**, and in the category list, it's **ProductCategoryKey**.

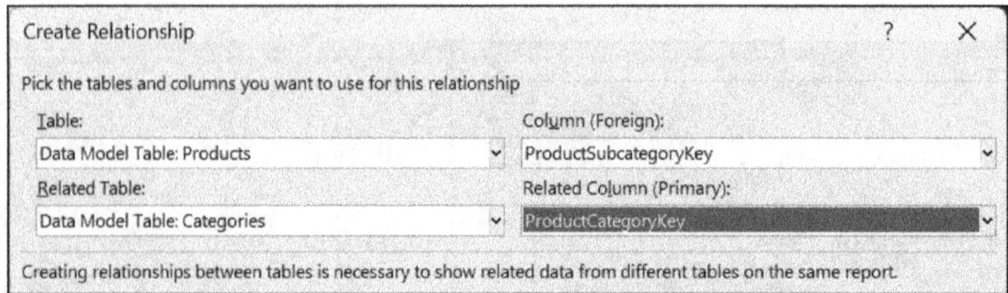

Figure 18.65 Dialog That Lets You Create Relationships by Using Foreign Keys

Clicking the **OK** button returns you to the initial dialog, which lists all relationships in the data model (see Figure 18.66). To edit a relationship, select it there and click **Edit**.

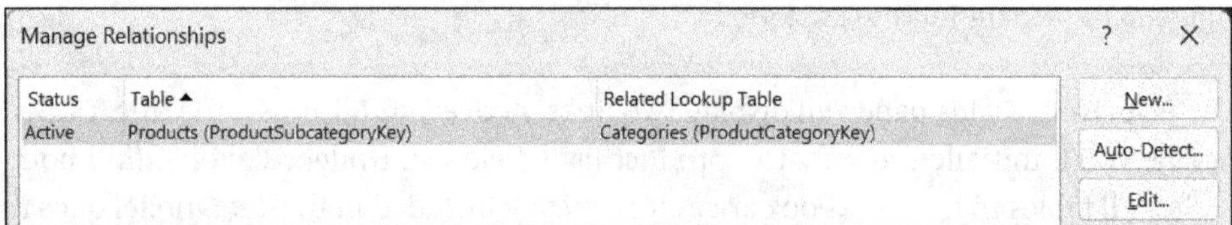

Figure 18.66 Dialog That Lets You Manage All Relationships

The simple question for the PivotTable is this: How many products are in each category? The table uses the product category name from the categories table; the field is assigned to the row labels, while **ProductSubcategoryKey**, a field from the product list, is assigned to the **Values** range. Using **Value Field Settings**, you can request the count of keys instead of the sum. Figure 18.67 shows the results and both tables in the active field list.

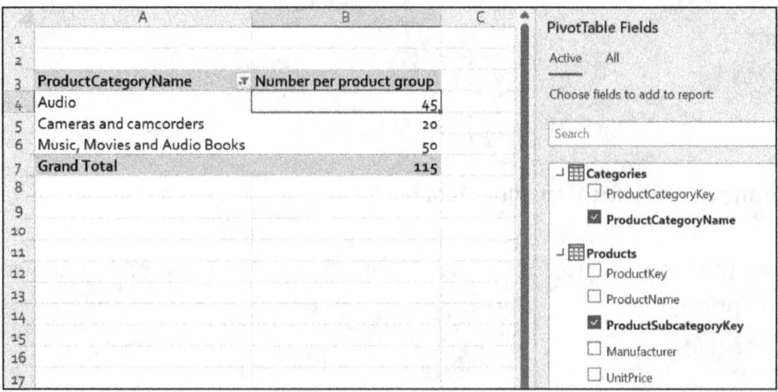

Figure 18.67 Analysis Based on Two Tables

18.7 Power Pivot

If Excel's analysis functions for PivotTables and charts are insufficient, you can use the Power Pivot add-in. Power Pivot connects to the data model we described previously but offers an advanced modeling environment. When you use **Power Pivot • Data Model • Manage** to open the Power Pivot management window in the workbook we described previously, you will find the tables of this model.

You can build the data model in Excel as described earlier, continue working in Power Pivot, and add more data to the model at any time. Alternatively, you can compile the data for the model directly with Power Pivot's built-in tools—especially the filter functions, which let you reduce the data volume for further processing in advance.

Power Pivot supports building hierarchies to organize data and thus enables the use of detailed or summarized views as needed. Defining calculated fields or key performance indicators is essential, and you can create these formulas with the special *Data Analysis Expressions* (DAX) language. Data prepared in the Power Pivot window is passed to Excel's pivot functions as the source for analysis.

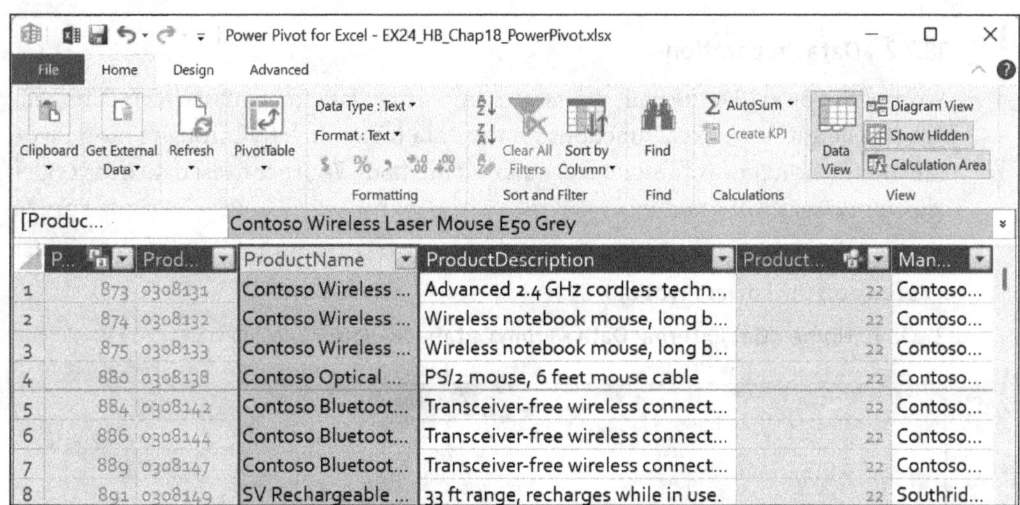

Figure 18.68 Tables of Data Model Described in Previous Section Within Power Pivot Window

18.7.1 Activating the Add-In

To activate the add-in, go to **File • Options • Add-Ins**, select **COM Add-Ins** under **Manage**, and click the **Go** button. Then, in the **COM Add-Ins** dialog, check the **Microsoft Power Pivot for Excel** option (see Figure 18.69).

Alternatively, you can use the **COM Add-Ins** button on the **Developer** tab. When Power Pivot is enabled, Excel shows an additional **Power Pivot** tab (see Figure 18.70).

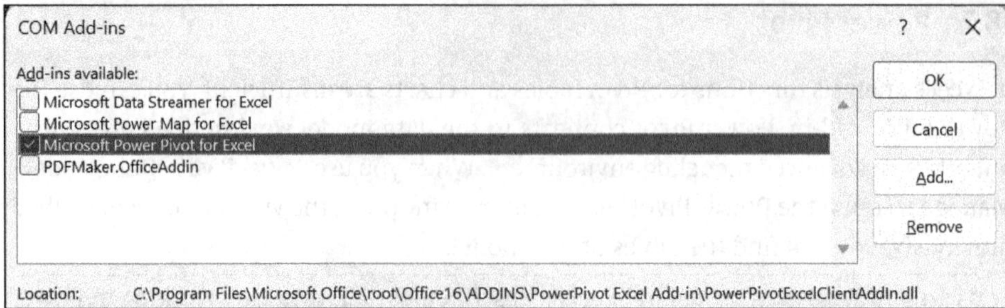

Figure 18.69 Dialog for Activating COM Add-Ins

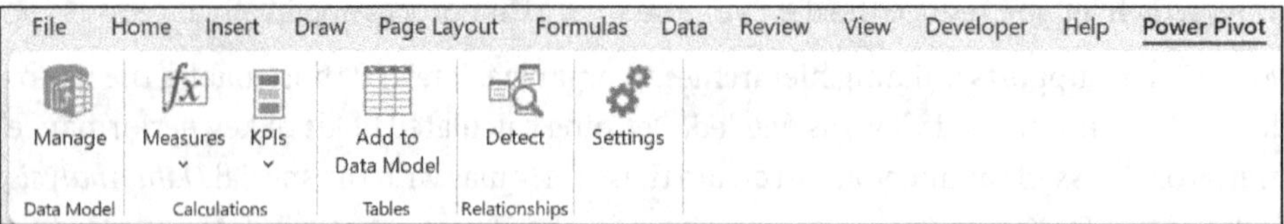

Figure 18.70 Power Pivot Tab

18.7.2 Data Preparation

Work with Power Pivot usually starts with importing data for analysis. As mentioned, you can do this with Excel functions on the **Data** tab or directly through Power Pivot functions. The following describes the second method. We use the test data from CodePlex for Power Pivot again.

1. Open the Power Pivot window by clicking the **Manage** button in the **Data Model** group on the **Power Pivot** tab.
2. Go to **Home • Get External Data • From Database • From Access.**

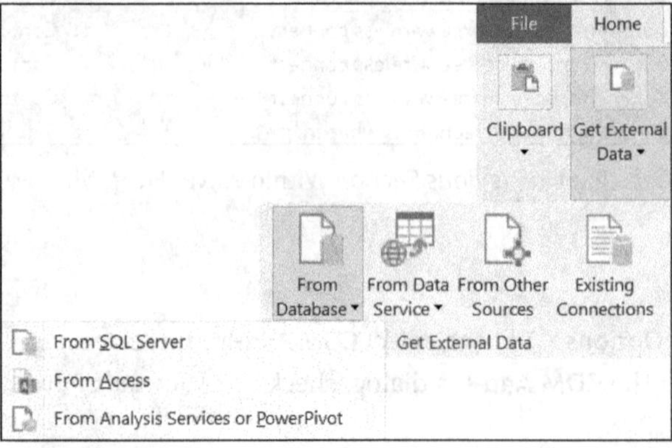

3. Excel will launch a table import wizard that will request the name of the Access database. We use the sample ContosoSales database here.
4. Choose whether to import from a list of tables and views or by using an SQL query.

5. If you use the first option, it will open a table selection dialog, and you can check he first box before **Source Table** to select all tables.
6. To reduce the data volume included in the pivot, apply filters to each selected table and click the **Preview & Filter** button.

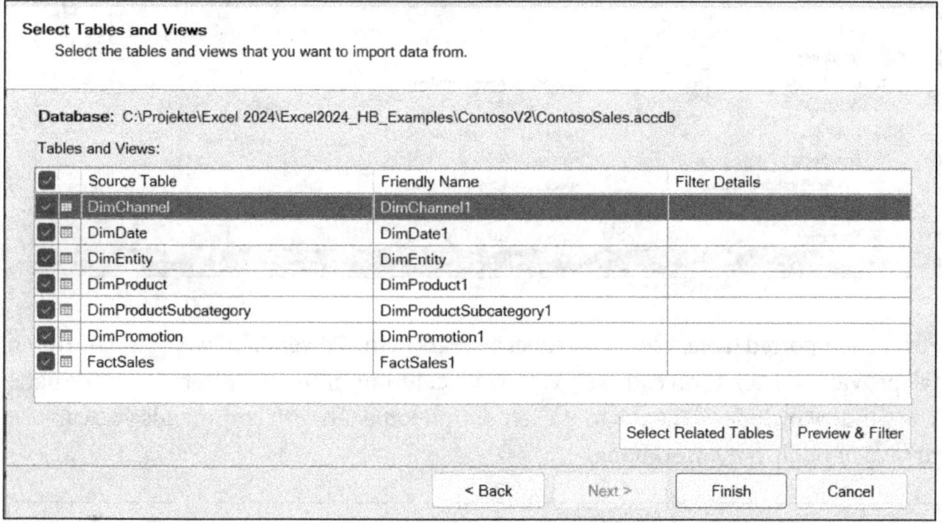

7. The table will appear in this preview with filter buttons, which—like those in Excel tables—limit the data volume. Apply this to each column individually until you identify the exact data needed for the planned analysis. To ignore a column entirely, uncheck the box next to its field name.

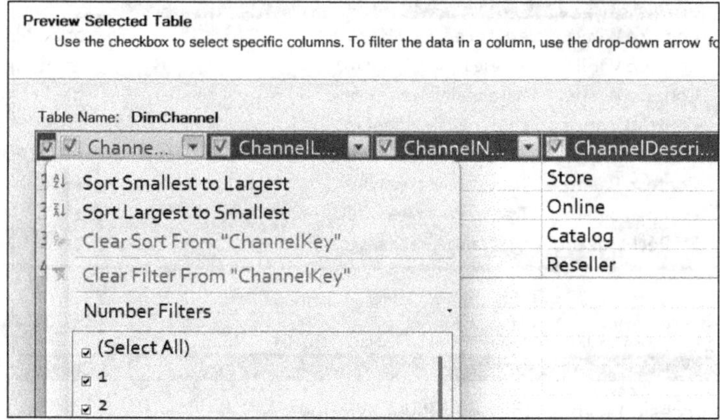

8. After the import completes, you'll receive an overview of the imported data volumes. You can also click the **Details** link to display the relationships that were recognized and established during import. The following figure shows that despite the filter applied previously, the largest table still contains 2.2 million rows.

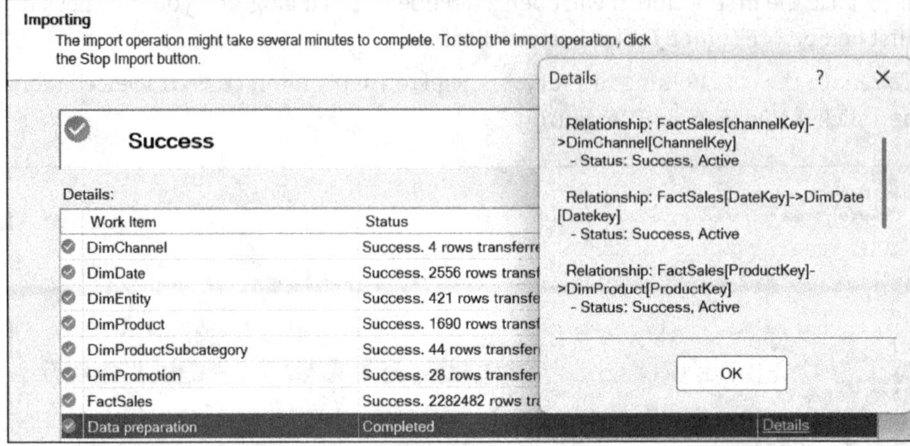

For each imported table, a separate sheet appears in the Power Pivot window. The **Home** tab provides buttons you can use to reformat columns or further filter or sort the data. The window includes a **File** tab to the left of the **Home** tab with commands you can use to save or publish the workbook.

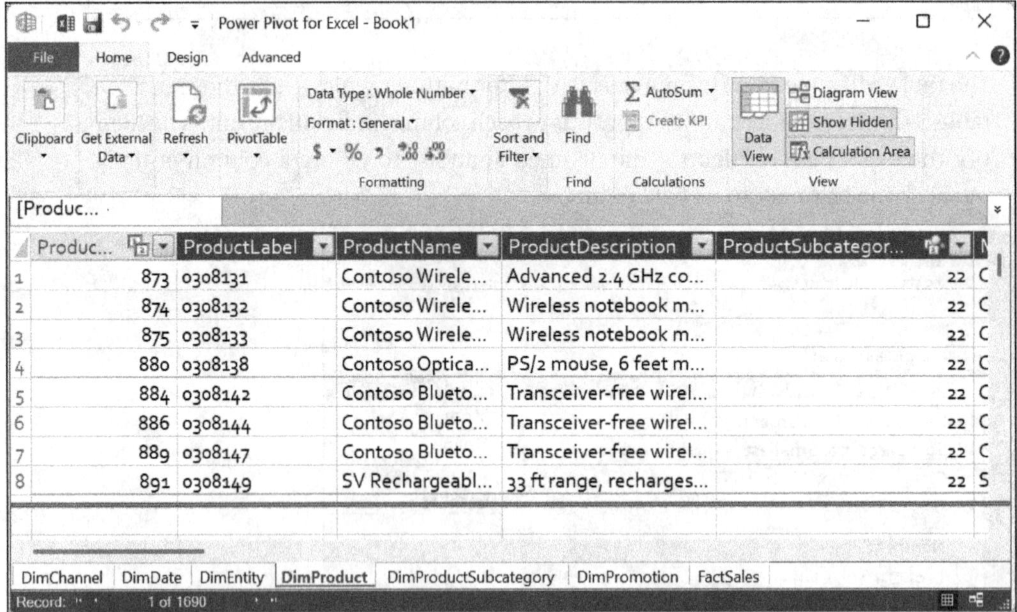

Figure 18.71 Tables from Access Database in Power Pivot Window

Initial Analysis

To perform an initial analysis, use the **PivotTable** button on the **Home** tab.

For the initial analysis, assign the **Sum of SalesAmount** field from the **FactSales** table to the **Values** area, the **Dates** field from the **DimDate** table to the **Columns** area, the

ProductSubcategory field from **DimProduct** to the **Rows** area, and add the **BrandName** field below it in the **Rows** area.

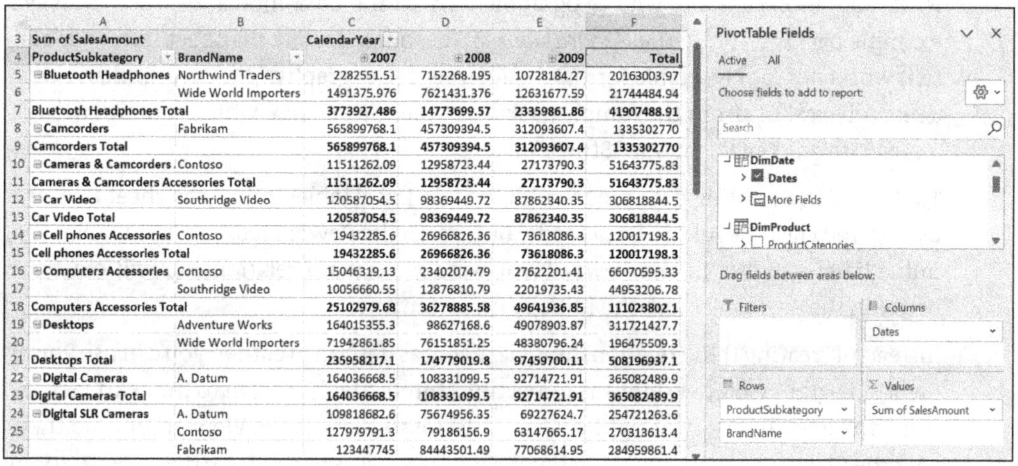

	A	B	C	D	E	F
3	Sum of SalesAmount		CalendarYear			
4	ProductSubkategory	BrandName	2007	2008	2009	Total
5	Bluetooth Headphones	Northwind Traders	2282551.51	7152268.195	10728184.27	20163003.97
6		Wide World Importers	1491375.976	7621431.376	12631677.59	21744484.94
7	Bluetooth Headphones Total		3773927.486	14773699.57	23359861.86	41907488.91
8	Camcorders	Fabrikam	565899768.1	457309394.5	312093607.4	1335302770
9	Camcorders Total		565899768.1	457309394.5	312093607.4	1335302770
10	Cameras & Camcorders .	Contoso	11511262.09	12958723.44	27173790.3	51643775.83
11	Cameras & Camcorders Accessories Total		11511262.09	12958723.44	27173790.3	51643775.83
12	Car Video	Southridge Video	120587054.5	98369449.72	87862340.35	306818844.5
13	Car Video Total		120587054.5	98369449.72	87862340.35	306818844.5
14	Cell phones Accessories	Contoso	19432285.6	26966826.36	73618086.3	120017198.3
15	Cell phones Accessories Total		19432285.6	26966826.36	73618086.3	120017198.3
16	Computers Accessories	Contoso	15046319.13	23402074.79	27622201.41	66070595.33
17		Southridge Video	10056660.55	12876810.79	22019735.43	44953206.78
18	Computers Accessories Total		25102979.68	36278885.58	49641936.85	111023802.1
19	Desktops	Adventure Works	164015355.3	98627168.6	49078903.87	311721427.7
20		Wide World Importers	71942861.85	76151851.25	48380796.24	196475509.3
21	Desktops Total		235958217.1	174779019.8	97459700.11	508196937.1
22	Digital Cameras	A. Datum	164036668.5	108331099.5	92714721.91	365082489.9
23	Digital Cameras Total		164036668.5	108331099.5	92714721.91	365082489.9
24	Digital SLR Cameras	A. Datum	109818682.6	75674956.35	69227624.7	254721263.6
25		Contoso	127979791.3	79186156.9	63147665.17	270313613.4
26		Fabrikam	123447745	84443501.49	77068614.95	284959861.4

PivotTable Fields
Active All
Choose fields to add to report:
Search
DimDate
Dates
More Fields
DimProduct
ProductCategories
Drag fields between areas below:
Filters
Columns
Dates
Rows
Values
ProductSubkategory
Sum of SalesAmount
BrandName

Figure 18.72 Initial Analysis from Multiple Tables in Database

18.7.3 Interactive Linking of Tables

When you import multiple tables from a database—whether Access or Microsoft SQL Server—Power Pivot usually registers and maintains the relationships between these tables automatically. You can switch to the Excel window using the button in the title bar, and you can find the automatically registered relationships under **PivotTable Analyze • Calculations • Relationships** in the **Manage Relationships** dialog.

Data that's imported via Power Pivot automatically becomes part of the current workbook's data model, but if data from different sources is loaded into the Power Pivot window, you must set up any relationships among these tables manually.

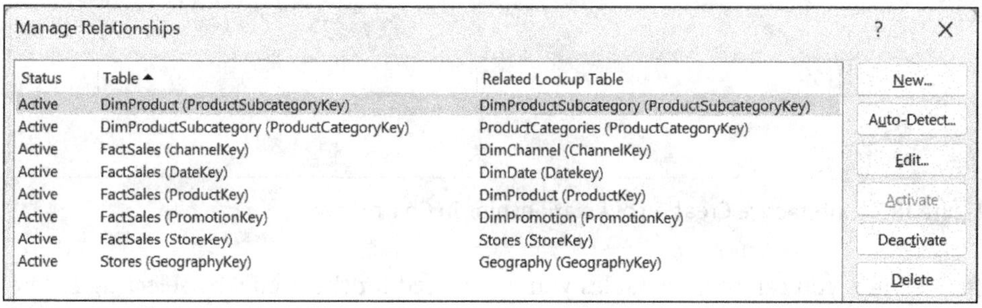

Manage Relationships

Status	Table ▲	Related Lookup Table
Active	DimProduct (ProductSubcategoryKey)	DimProductSubcategory (ProductSubcategoryKey)
Active	DimProductSubcategory (ProductCategoryKey)	ProductCategories (ProductCategoryKey)
Active	FactSales (channelKey)	DimChannel (ChannelKey)
Active	FactSales (DateKey)	DimDate (Datekey)
Active	FactSales (ProductKey)	DimProduct (ProductKey)
Active	FactSales (PromotionKey)	DimPromotion (PromotionKey)
Active	FactSales (StoreKey)	Stores (StoreKey)
Active	Stores (GeographyKey)	Geography (GeographyKey)

New...
Auto-Detect...
Edit...
Activate
Deactivate
Delete

Figure 18.73 List of Relationships

To demonstrate the procedure, we use a second Access database included in the CodePlex example data package. You should repeat the previous steps, naming the database file "ProductCategories" this time. This database contains only a small table named

18

DimProductCategory. Filtering is not required here, and the new table appears in the Power Pivot window.

Next, add an Excel table with geographic data to the data model. In the **ContosoV2** example package, locate the *Geography.xlsx* file, open it, copy the sheet into your current workbook, apply a table format, and name it "Geography." Proceed similarly with a second Excel file from the sample package named *Stores.xlsx*, which contains a list of stores. Assign the table name "Stores" to it.

To expand the existing data model with the two Excel tables, select the sheet tabs one by one and click the **Add to Data Model** button on the **Power Pivot** tab. Both tables will immediately appear in the **Power Pivot** window, but no relationship will exist yet between the new tables and the initially imported ones.

Instead of creating these relationships via dialog as described earlier, you can use Power Pivot's diagram view to establish connections interactively. Switch to the chart view by using the icon in the lower right corner or via **Chart View** in the **View** group, and then, click the **Chart** button in the Power Pivot window's status bar and use the slider to zoom out until all tables are visible. The most recently added tables initially have no connections to the other tables.

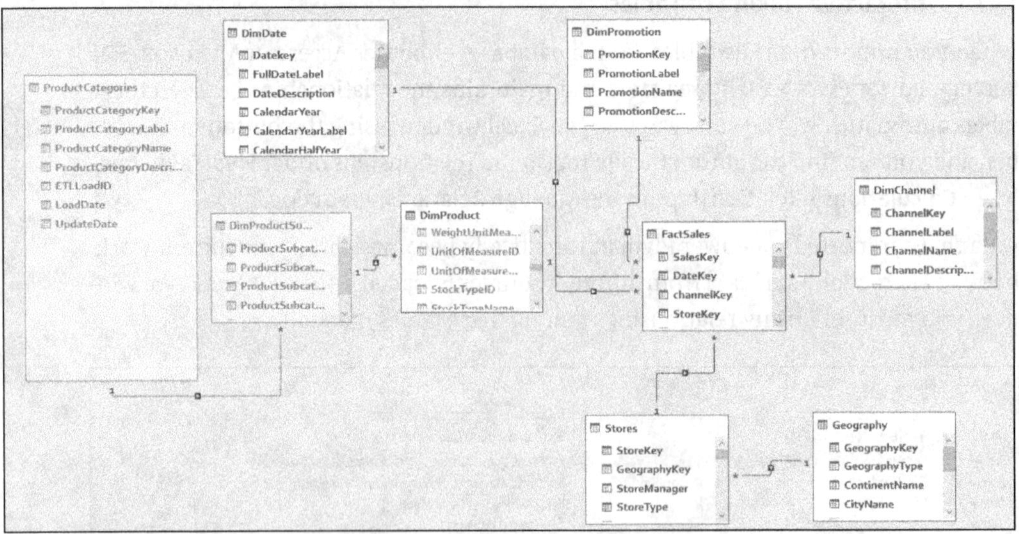

Figure 18.74 Interactive Creation of Relationships in Chart View

In this view, you can remove tables you don't need for the model by selecting Delete from the context menu. Then, create a relationship between the **FactSales** table and the Stores table. **FactSales** contains sales and cost data for the fictional company Contoso, along with keys to other tables, such as store codes used in the **Stores** table.

Then, in the **FactSales** table's field list, select the **StoreKey** field, drag it to the **StoreKey** field in the **Stores** table, and arrange the tables as needed to enable easy connection. An arrow will appear in the diagram, pointing from the primary table to the related table. Use the same process to create a relationship between the **ProductCategoryKey** fields in the **DimProductSubcategory** and **DimProductCategory** tables.

Advanced Analysis

The new relationships already enable additional analyses in the PivotTable. Use the **Switch to Workbook** button in the Power Pivot window's title bar to return to the table sheet with the existing PivotTable. Click the **All** link to display all tables in the data model. Expand the **Stores** table and drag **StoreName** to the **Filters** area. Remove the **BrandName** field from the **Rows** area.

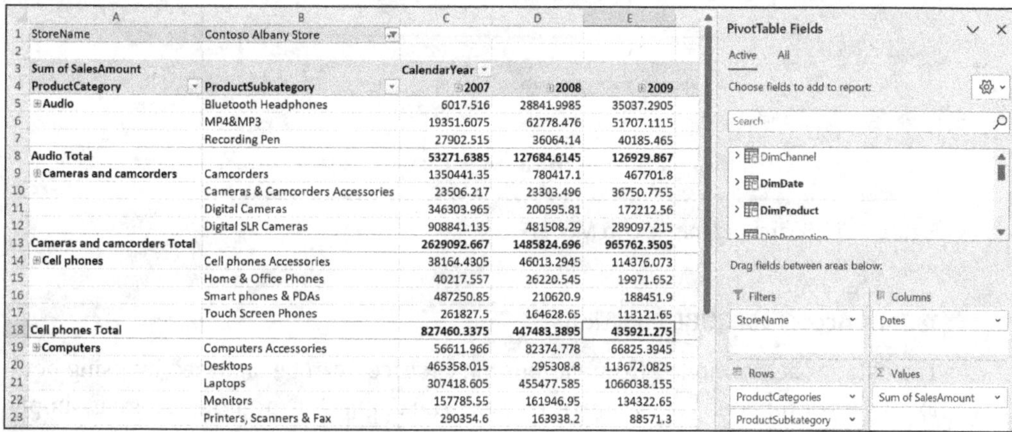

	A	B	C	D	E
1	StoreName	Contoso Albany Store			
2					
3	Sum of SalesAmount		CalendarYear		
4	ProductCategory	ProductSubkategory	2007	2008	2009
5	Audio	Bluetooth Headphones	6017.516	28841.9985	35037.2905
6		MP4&MP3	19351.6075	62778.476	51707.1115
7		Recording Pen	27902.515	36064.14	40185.465
8	Audio Total		53271.6385	127684.6145	126929.867
9	Cameras and camcorders	Camcorders	1350441.35	780417.1	467701.8
10		Cameras & Camcorders Accessories	23506.217	23303.496	36750.7755
11		Digital Cameras	346303.965	200595.81	172212.56
12		Digital SLR Cameras	908841.135	481508.29	289097.215
13	Cameras and camcorders Total		2629092.667	1485824.696	965762.3505
14	Cell phones	Cell phones Accessories	38164.4305	46013.2945	114376.073
15		Home & Office Phones	40217.557	26220.545	19971.652
16		Smart phones & PDAs	487250.85	210620.9	188451.9
17		Touch Screen Phones	261827.5	164628.65	113121.65
18	Cell phones Total		827460.3375	447483.3895	435921.275
19	Computers	Computers Accessories	56611.966	82374.778	66825.3945
20		Desktops	465358.015	295308.8	113672.0825
21		Laptops	307418.605	455477.585	1066038.155
22		Monitors	157785.55	161946.95	134322.65
23		Printers, Scanners & Fax	290354.6	163938.2	88571.3

Figure 18.75 Advanced PivotTable

Then, expand the **DimProduct** table. Drag **ProductCategories** to the **Rows** area and place it above **ProductSubcategory**, and when you click the **Active** link in the field list, you'll find the new fields there as well.

The last relationship should integrate the **Geography** table into the model. Here, we introduce the search function that's available in the diagram view.

	A	B	C
1	GeographyKey	GeographyType	ContinentName
2	1	Continent	Asia
3	2	Continent	Europe
4	3	Continent	North America
5	269	Country/Region	Asia
6	270	Country/Region	Asia
7	271	Country/Region	Asia

Figure 18.76 Table with Geographical Data

18

Search Metadata

The **Geography** table contains a **GeographyKey** field that uniquely identifies its rows. To check whether this key is used by other tables in the model, go to the **Home** tab and click the **Find** button to open the **Find Metadata** function.

Then, use the **Find Next** button to cycle through matches in the model's tables. Here, the **Stores** table is the target. As described earlier, you should then create a relationship between the key fields in the **Geography** and **Stores** tables. This links all tables in the model in some way.

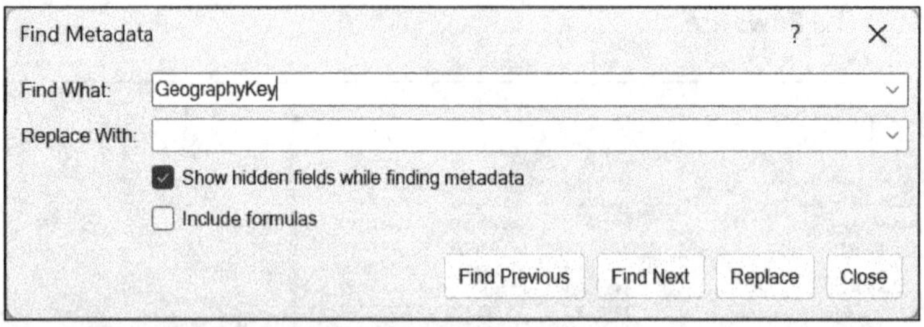

Find Metadata

Find What: GeographyKey

Replace With:

Show hidden fields while finding metadata

Include formulas

Find Previous | Find Next | Replace | Close

Figure 18.77 Searching for Field in Model

18.7.4 Access via CUBE Functions

The data model, compiled from various data sources, can be analyzed by using both Excel's Pivot functions and CUBE functions. The *Connection* argument here is simply `"ThisWorkbookDataModel"`.

For example, this stores the count of product categories in cell A40, where only the label appears:

```
=CUBESET("ThisWorkbookDataModel", "[ProductCategoryKey].[All].children",
"Product categories")
```

Then, this returns the actual count:

```
=CUBESETCOUNT(A40)
```

18.7.5 Creating Calculated Columns

The second major way Power Pivot extends Excel is through a special expression language called DAX, which is a simplified version of *Multidimensional Expressions* (MDX), the language used on SQL Server to analyze OLAP cubes. DAX is used in Power Pivot to add calculated columns and fields to the model.

This complex topic is only briefly covered in this book. A simple example is a profit calculation assigned to an additional column in the **FactSales** table. Follow these steps:

1. In the Power Pivot window, use the **Grid View** button to switch back to grid view.
2. Select the **FactSales** table from the tab.
3. Follow the **Design • Columns • Add** path.
4. Enter the formula for the column in the formula bar above the table. The **AutoComplete** feature helps you enter exact column and table names easily. References to the table's columns are enclosed in square brackets, and after the first square bracket following the equal sign, possible values appear for selection. The required formula is this:

   ```
   = [SalesAmount] - [TotalCost] - [ReturnAmount]
   ```

   It subtracts the other two values from the **SalesAmount** value in each table row.
5. The new column will fill up immediately with the calculated values. Replace only the default dummy name, open its context menu, and click **Rename Column**.

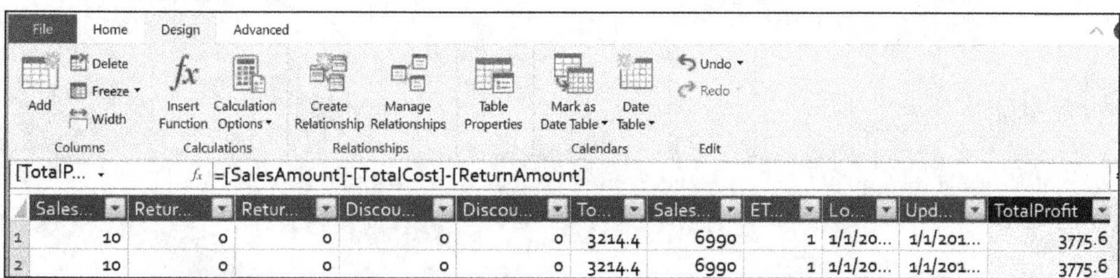

DAX consists of functions, operators, and constants. You can find the complete syntax at *msdn.microsoft.com/en-us/library/library/ee634396.aspx*.

Chapter 19
Working with External Data

Although you can now store tables with hundreds of thousands of rows on a single sheet, Excel is not primarily designed to store and manage large datasets. Excel's strengths lie mainly in data analysis, but its ability to access data from other sources has steadily improved. However, this book cannot cover all variations in detail.

A key area is importing data from external databases, whether local or over the network, so you need to have easy-to-use linking options to use Excel's features with this data smoothly. Excel uses connection definitions, which you create once and reuse repeatedly.

A *connection definition* specifies how to establish the connection to the data source and which parameters to use. A workbook can use multiple such connections without issue, and you can either fully import external data sets from specific sources or filter them by query criteria before importing. This often helps reduce data to only what is needed for the desired analysis.

Excel provides many tools you can use to access external data sources. The first group, **Get & Transform Data**, contains the tools you need to import external data. The second group, **Queries & Connections**, manages existing queries and connections. 19

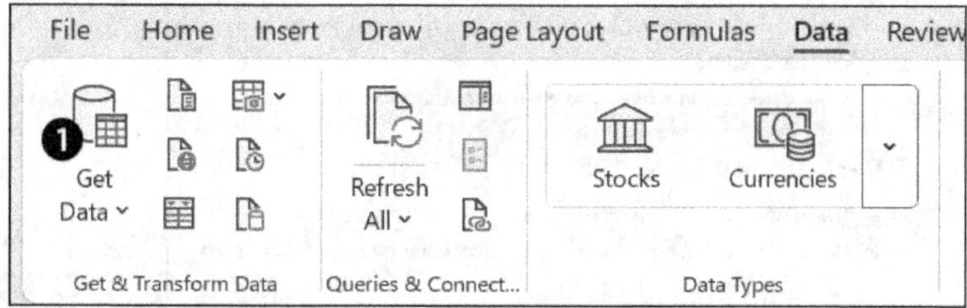

Figure 19.1 Tools for Accessing External Data

In the first group, you'll find a **Get Data** button ❶ that opens commands you can use to import data from files, databases like Access or SQL Server, Azure, online services, and other sources. The menu for query formulation also provides the Power Query Editor ❷. **Data Source Settings** ❸ lets you manage data source permissions.

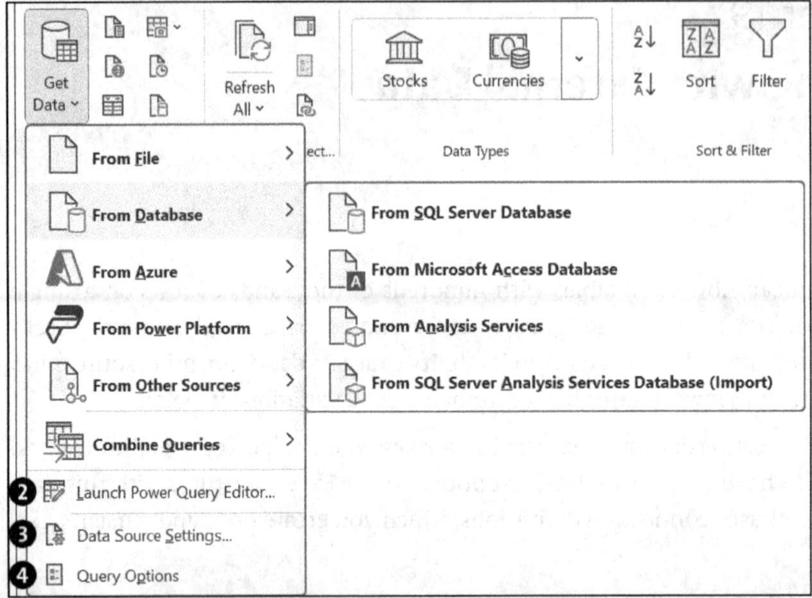

Figure 19.2 Options for Get Data Button

Query Options ❹ cover global and workbook-specific settings for loading and caching data, data security, privacy rules, and more.

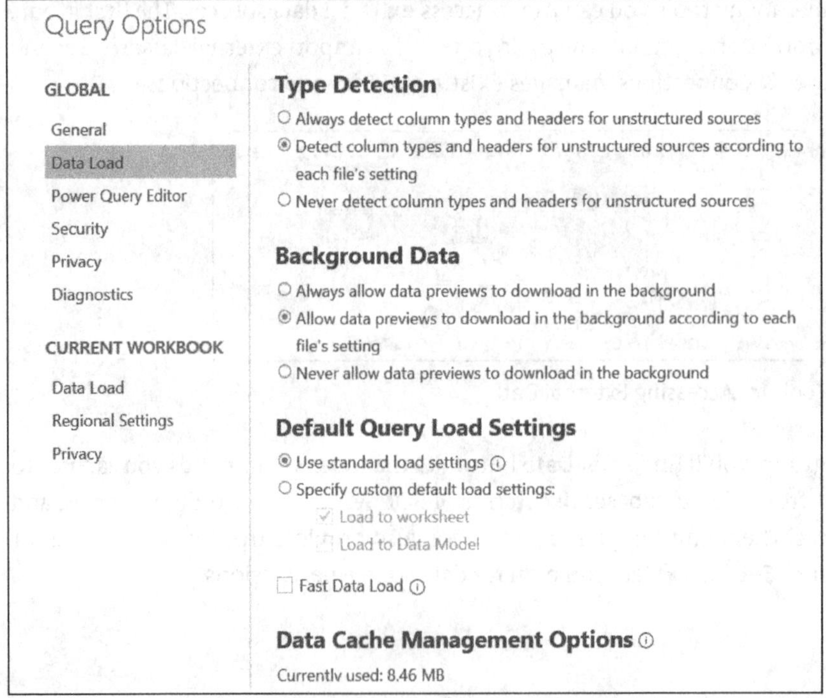

Figure 19.3 The Query Options Dialog

The same group includes icons for frequently used query sources that link directly to the following: **From Text/CSV**, **From Web**, and **From Table/Range**. You can also access the last option in the context menu of a selected cell range or table as **From Table/ Range**. In the first case, the cell range is converted into a table, then the Power Query Editor opens for that table. In the second case, the editor opens directly with the table's data.

To reuse a recently used source, use the **Recent Sources** button.

Use **Existing Connections** to recall all previously defined connections. This applies to connections defined in the workbook, as well as local or network connection files when you select the appropriate option under **Show**.

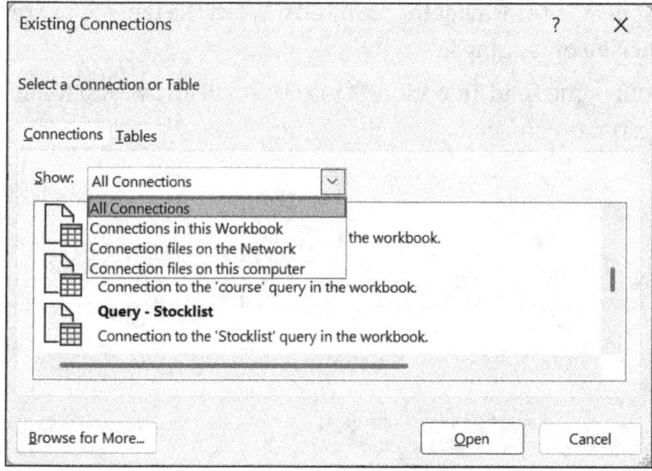

Figure 19.4 Existing Connections Dialog

In the **Queries & Connections** group, find the button with the same name, which opens the **Queries & Connections** task pane (see Figure 19.5). **Queries** and **Connections** appear on separate tabs there.

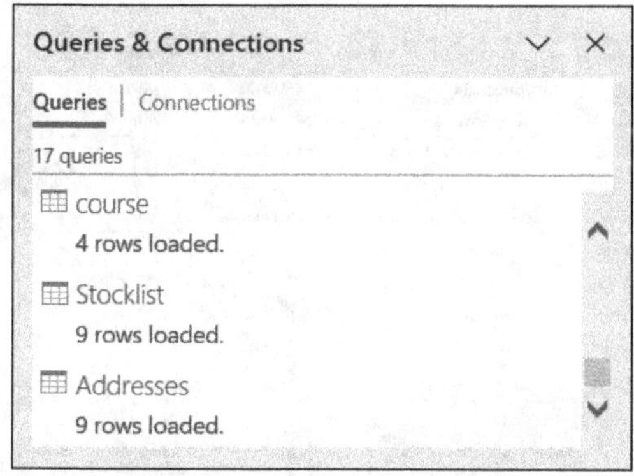

Figure 19.5 Queries and Connections Task Pane

19.1 Importing Access Data

Access is a user-friendly, well-equipped database program that serves users with moderate data needs effectively. To demonstrate how to import external data from an available source, we'll first describe an example of importing data from an Access database. Follow these steps:

1. In a blank worksheet, select the **Data • Get Data • From Database • From Microsoft Access Database**. Excel will open the **Import Data** dialog box. Access databases are preset as a data type, and they use the*.mdb* file extension for older versions and *.accdb* for files from version 2007 onward.
2. After you select the desired Access file, confirm the dialog. Then, the tables and queries in this database will appear in the **Navigator** dialog box. Select the table or query you're interested in by clicking or tapping it.
3. The **Load** button ❶ also offers the **Load To** option ❷ via the small arrow, and it lets you specify the import destination format.

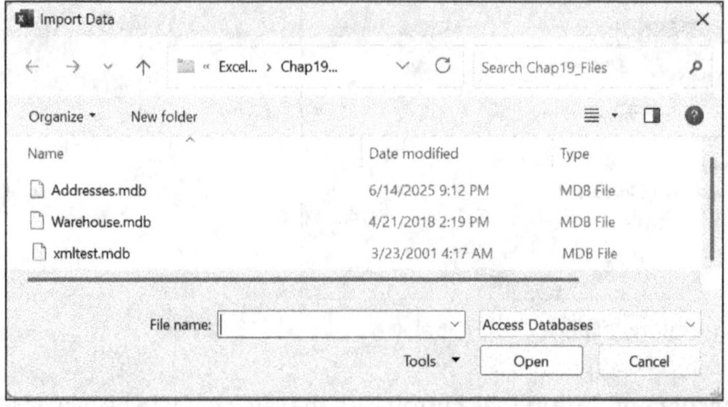

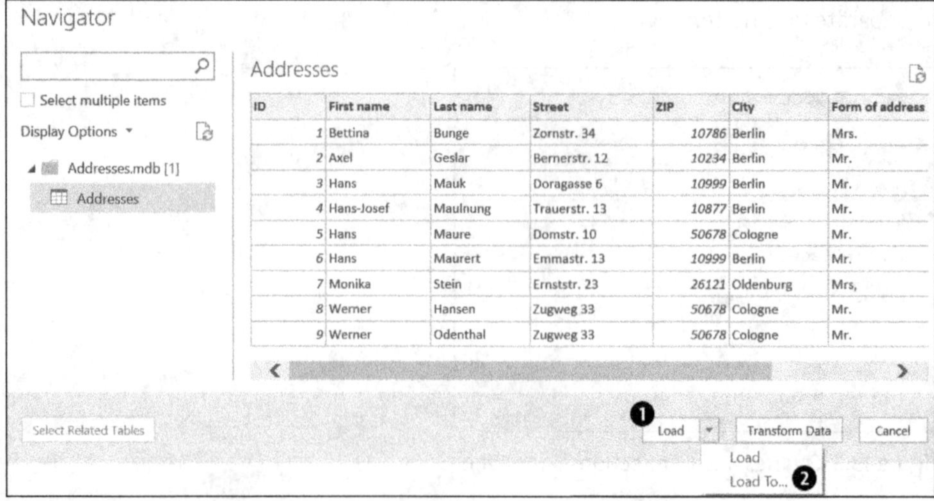

4. Besides using the **Table** ❸ option, you can import directly into a PivotTable report or a PivotChart here. Finally, you should specify where to place the data. Either select **Existing worksheet** and enter the address of the upper-left cell or place the table on a new sheet.

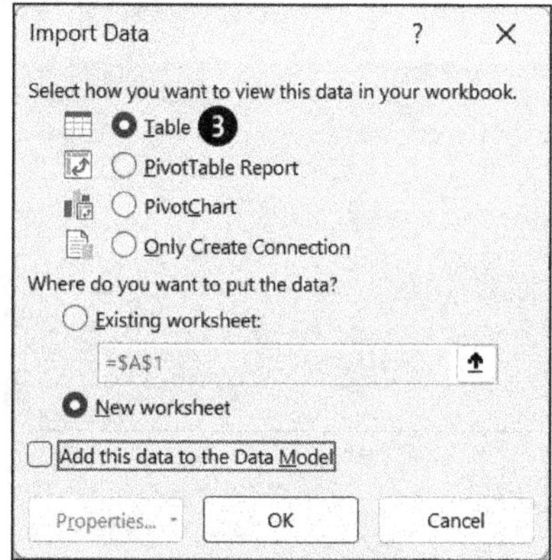

5. The database data will be inserted at the specified location in the chosen format. If you select **Table** as the format, Excel will create a table with a predefined style.

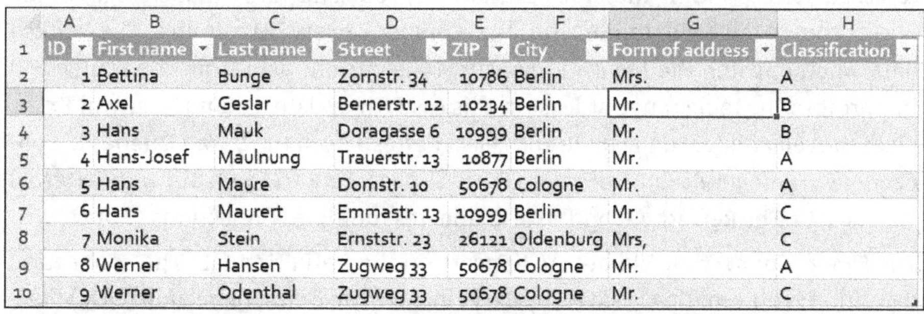

ID	First name	Last name	Street	ZIP	City	Form of address	Classification
1	Bettina	Bunge	Zornstr. 34	10786	Berlin	Mrs.	A
2	Axel	Geslar	Bernerstr. 12	10234	Berlin	Mr.	B
3	Hans	Mauk	Doragasse 6	10999	Berlin	Mr.	B
4	Hans-Josef	Maulnung	Trauerstr. 13	10877	Berlin	Mr.	A
5	Hans	Maure	Domstr. 10	50678	Cologne	Mr.	A
6	Hans	Maurert	Emmastr. 13	10999	Berlin	Mr.	C
7	Monika	Stein	Ernststr. 23	26121	Oldenburg	Mrs,	C
8	Werner	Hansen	Zugweg 33	50678	Cologne	Mr.	A
9	Werner	Odenthal	Zugweg 33	50678	Cologne	Mr.	C

6. When you select a cell in the imported table, Excel displays the **Table Design** and **Query** tabs.

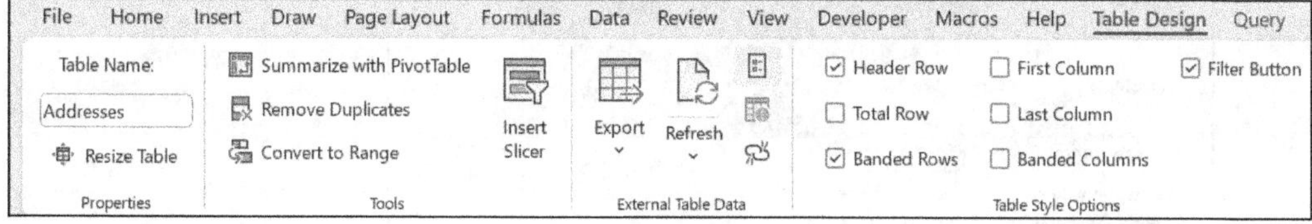

7. Excel automatically connects to the Access database during this process. You can view and edit the data range properties under **Table Design • External Table Data • Properties**. This applies to both formatting and updating when the source data changes. It also specifies how to handle changes in the number of rows.

8. The **Include row numbers** option adds a column with an index before the imported data records, which can be useful. When **Adjust column width** is enabled, the column width adapts to the data. The **Preserve column sort/filter/layout** and **Preserve cell formatting** options prevent these settings from being overwritten.

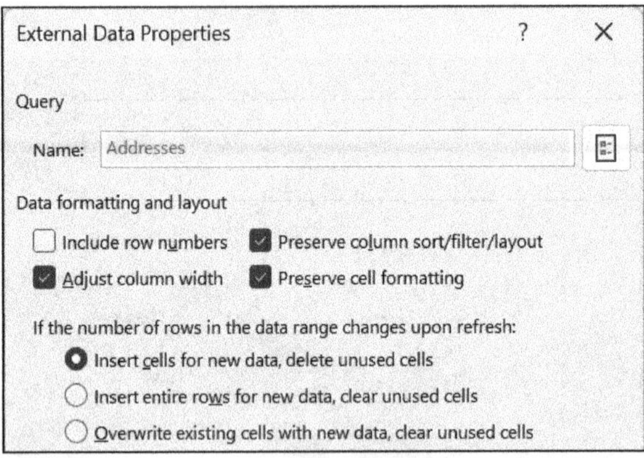

9. The **Query • Query Properties** button opens the **Query Properties** dialog, where you can configure update settings on the **Usage** tab. If the data changes frequently, you can set a corresponding interval. For example, you can specify that changes in the Access source table will import into the workbook every 60 minutes. The **Refresh data when opening the file** option ensures the current data from the source will import into the table range at least in this case. To update data manually in Excel, place the cell pointer in the table and select **Refresh** from the **Data • Queries & Connections • Refresh All** button menu. This applies only to the table with external Access data. The **Refresh All** option also updates multiple connections at once.
10. The **Export Connection File** button appears on the **Definition** tab. The connection definition is automatically saved by Excel in an *Office Data Connection* (ODC) file.

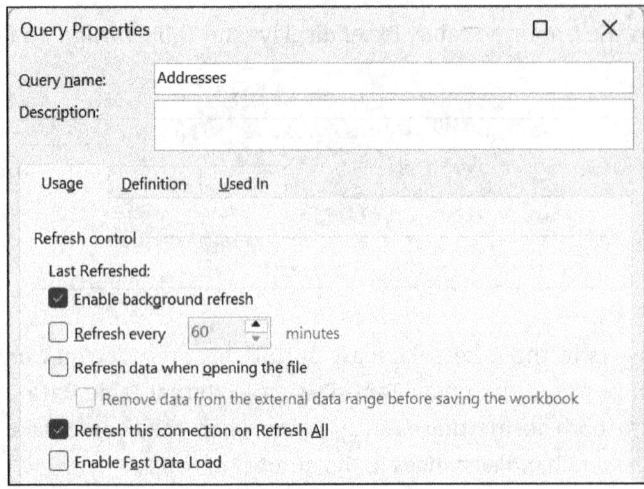

By default, Excel saves these files in a user-specific folder called *My Data Sources*. From then on, the query will be available under **Existing Connections,** and you can use it from any workbook to import data. The new query also appears in the **Queries & Connections** task pane.

Ranges you can create in the table sheet when importing from an external database include features that simplify further editing, and data you import from external sources appears in the Excel table sheet in the format you select. The query definition saves with the workbook. When you select a cell in the range, a special context menu appears, and you can use the **Edit Query** command to view and edit the credentials used during data import.

To disconnect the range from the source file in Access, choose **Unlink from Data Source** from the context menu. The range will then become a normal table range, and the connection will be removed from the workbook. The external data range will convert into an Excel table that preserves the displayed data.

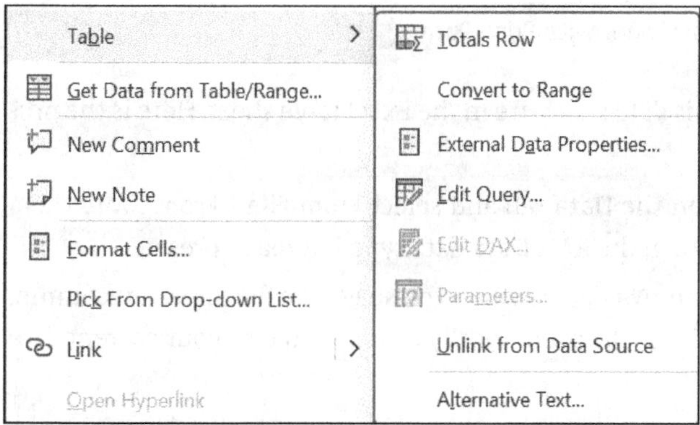

Figure 19.6 Context Menu for External Data Range

19.2 Querying XML Files

Today, large volumes of data are increasingly stored not only in databases but also in XML documents. Also, XML documents are sometimes stored within databases using special fields. Excel offers several functions for handling XML data. This section covers the simple method, while the more complex approach is discussed in Chapter 13, Section 13.4.

Existing XML files can be easily imported into a worksheet, and we'll use an XML document that contains course data as an example. Here is an excerpt:

```
<?xml version="1.0" encoding="UTF-8"?>
<courseCatalog>
  <course>
```

```
    <title>XML Basics
    </title>
    <duration>2 days</duration>
    <instructor>Hans Fromm</instructor>
    <date>2025-12-11</date>
    <description>Introduction to programming</description>
  </course>
  <course>
    <title>XML Practice</title>
    <duration>3 days</duration>
    <instructor>Hans Fromm</instructor>
    <date>2025-09-09</date>
    <description>Introduction to programming</description>
  </course>
  ...
</courseCatalog>
```

Listing 19.1 Excerpt from XML Document with Price Data

The question is how to store this data structure in the Excel table sheet. Here is the procedure:

1. Click the **Get Data** button on the **Data** tab and select **From File • From XML**. Then, select the XML file in the dialog; the **XML Data** data type is already preset.
2. Excel will show a preview in the **Navigator** when you select the file in the left column. The date format in the XML file will automatically be converted to your current date format.
3. To add an extra column that creates an abbreviation from the speakers' names, open the Power Query Editor using the **Transform Data** button.

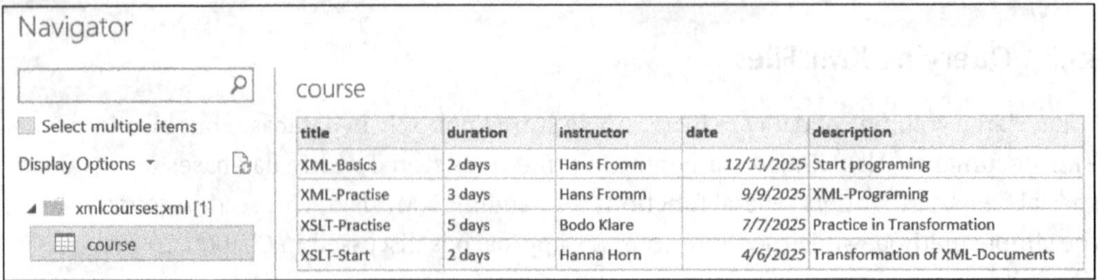

Navigator
Select multiple items
Display Options
xmlcourses.xml [1]
course
course
title | duration | instructor | date | description
XML-Basics | 2 days | Hans Fromm | 12/11/2025 | Start programing
XML-Practise | 3 days | Hans Fromm | 9/9/2025 | XML-Programing
XSLT-Practise | 5 days | Bodo Klare | 7/7/2025 | Practice in Transformation
XSLT-Start | 2 days | Hanna Horn | 4/6/2025 | Transformation of XML-Documents

4. Type the column **instructor**. On the **Add Column** tab, select the **Column from Examples** icon and choose **From Selection**, which is helpful here.

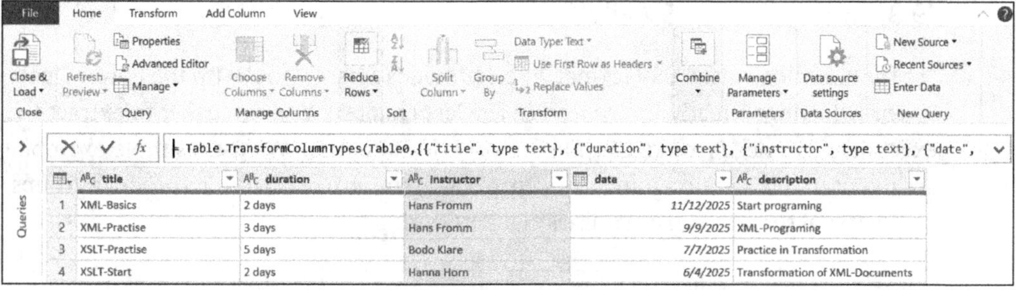

5. With the **instructor** column selected, define an additional column by entering an example in the first row. Here, you create the abbreviation from the first letters of the first and last names. Finish the first entry by pressing Ctrl + Enter.
6. If the first example isn't unique for Excel, enter another example and then confirm the transformation by clicking **OK**.

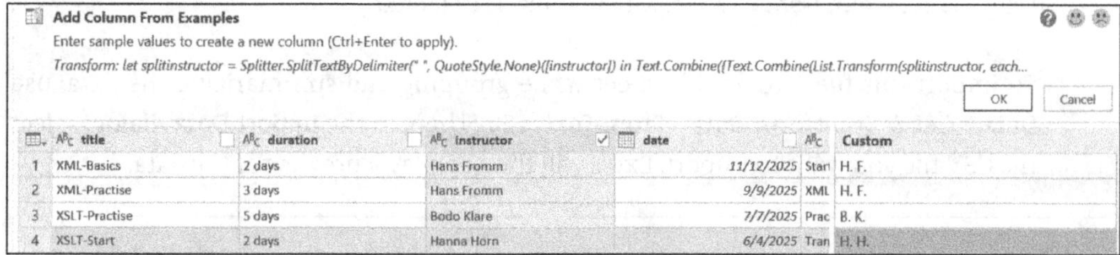

7. Rename the column "Abbreviation" and close the editor via **File • Close & Load**.
8. Excel inserts the XML data as a table at the selected location and creates a link between this range and the XML source. This is called an *XML mapping*.

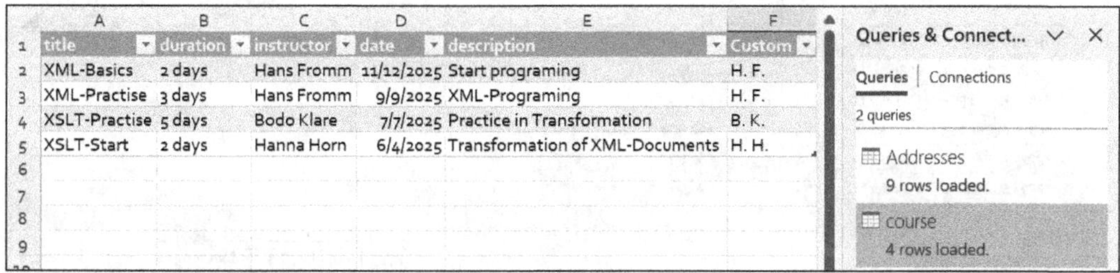

title	duration	instructor	date	description	Custom
XML-Basics	2 days	Hans Fromm	11/12/2025	Start programing	H. F.
XML-Practise	3 days	Hans Fromm	9/9/2025	XML-Programing	H. F.
XSLT-Practise	5 days	Bodo Klare	7/7/2025	Practice in Transformation	B. K.
XSLT-Start	2 days	Hanna Horn	6/4/2025	Transformation of XML-Documents	H. H.

The table only shows the lowest elements of the XML document as column headings; parent element names like `courseCatalog` or `course` do not appear.

19.3 Importing a CSV File

For the still widely used CSV format, Excel provides a handy tool with the Power Query Editor (which we already introduced in the last chapter). You can use it to extract the exact data you need from a CSV file for a specific situation. We use a simple example: an inventory list with ongoing sales data. The items are divided into product groups, and material serves as an organizational criterion.

```
Artnr,Name,Category,Material,Revenue
7777,Louvre Ccxs,Louvre,Plastic,"$23,801 "
8443,Louvre VVx,Louvre,Metal,"$40,000 "
7774,Louvre Vvxx,Louvre,Plastic,"$87,690 "
5667,Awning Luxor,Awning,Natural fiber,"$20,000 "
8666,Roller Dark,Roller blind,Plastic,"$7,000 "
5554,Roller PCx,Roller blind,Natural fiber,"$19,000 "
6666,Roller Top,Roller blind,Plastic,"$10,000 "
5222,Roller XXs,Roller blind,Plastic,"$6,000 "
7999,Sunset,Roller blind,Natural fiber,"$24,000 "
```

Figure 19.7 Excerpt from CSV File, with Comma as Delimiter

To import this file into a table sheet while grouping and summarizing the data, use **Data • Get & Transform Data • From Text/CSV**. Then, in the **Import Data** dialog, select the CSV file you want to import. Excel will then display a preview of the data.

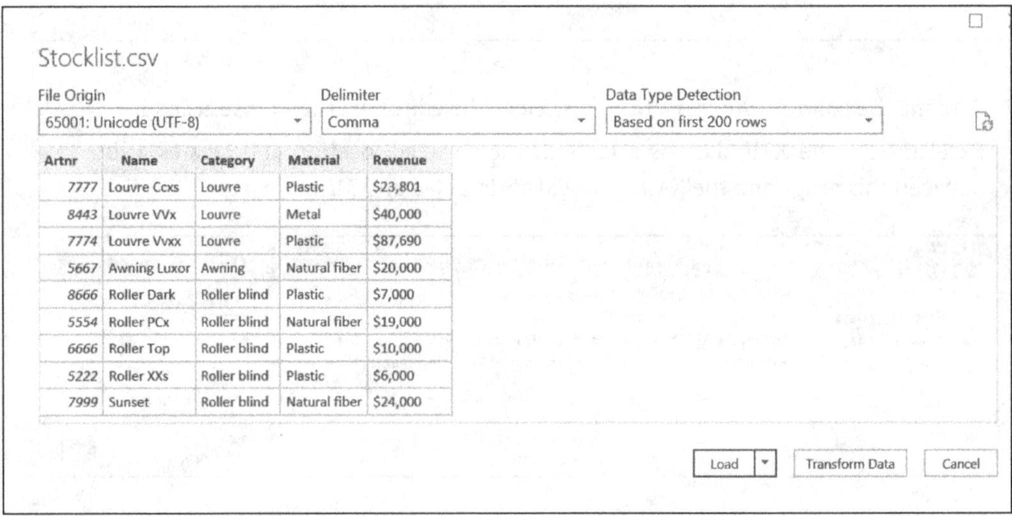

Figure 19.8 CSV Data Preview

Use the **Transform Data** button to review the table in the Power Query Editor window. If the data in the first row does not appear as column headers, enable this by using the **Use First Row as Headers** ❶ command in the **Home • Transform** group. Click the **Group By** button ❷ to summarize the list by product groups.

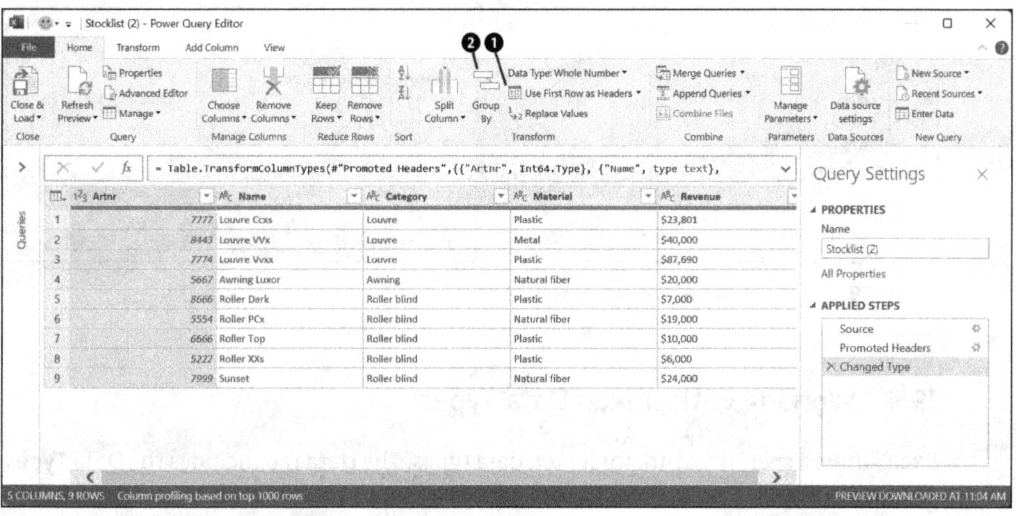

Figure 19.9 The CSV File in the Query Editor

Under **Group By**, specify the **Category** column, and for **New column name**, enter **Count**. Under **Operation**, select **Count Rows**, and keep **Column** blank.

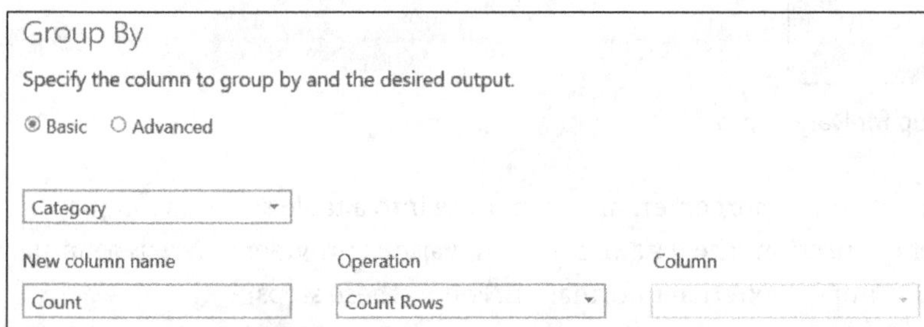

Figure 19.10 Grouping the Data

After you confirm, the summarized data will appear in the editor. Use the menu for the data field to change the selected column's format and use the sort icons to set the desired order.

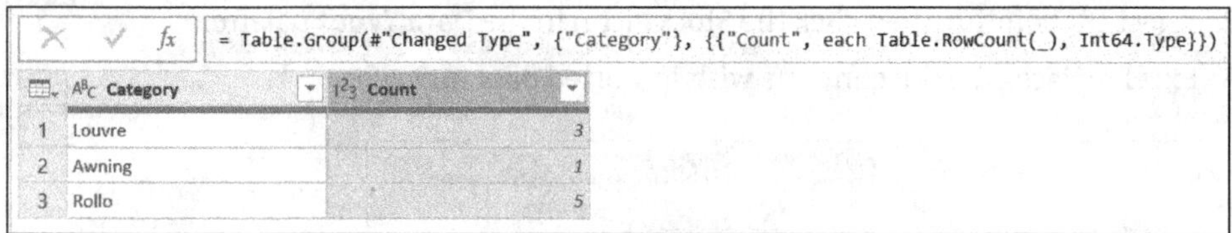

Figure 19.11 Results of Grouping by Product Categories

Finally, transfer the query result to the selected table sheet by using the **Close & Load** button. Excel will automatically open the **Queries & Connections** task pane.

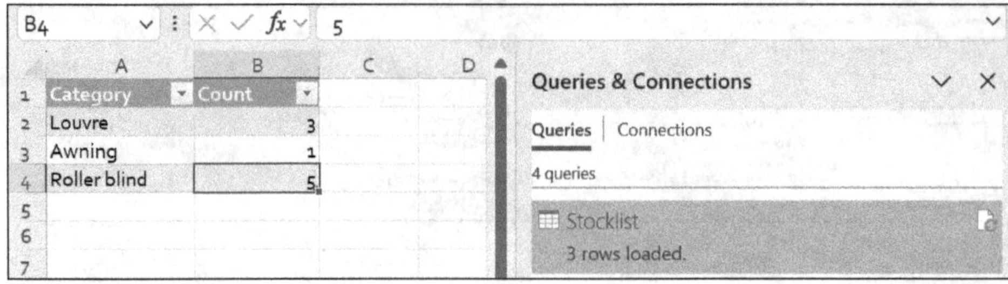

Figure 19.12 Results of CSV Query in Table Sheet

19.4 Working with Linked Data Types

Excel offers a special feature for linked data types. The **Data** tab includes the **Data Types** group for this.

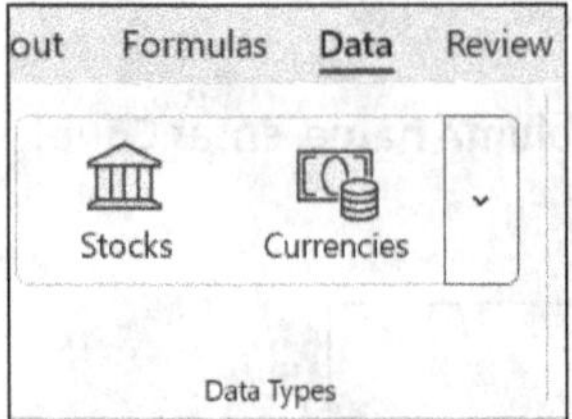

Figure 19.13 Group for Data Types

These data types let you import external information into a table sheet, as long as you have an internet connection. The data comes from various providers. Here is a simple example of how to import external information. Follow these steps:

1. Enter a series of stock codes into a table column.

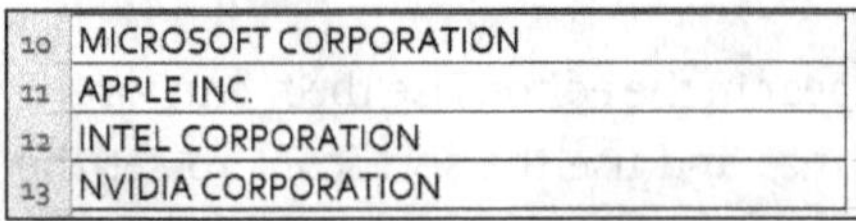

2. Select the column, then click the **Stocks** icon in the **Data Types** palette.
3. Excel replaces the cell contents with imported data for each stock.

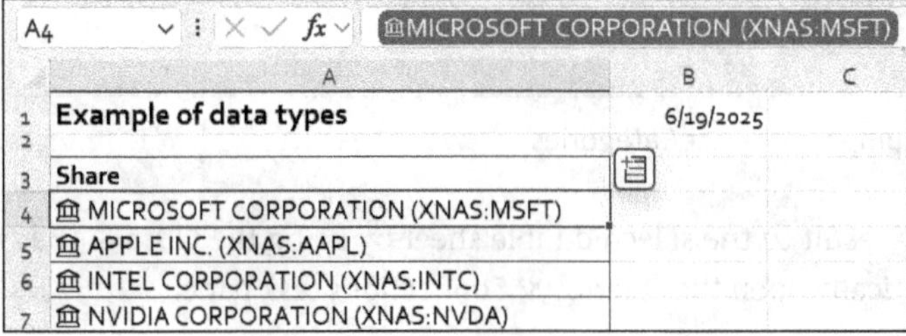

4. Click or tap the icon at the start of the cell to view details about the security on a data type card.

5. The button above the first row on the right lets you fill adjacent columns with specific data about the securities. It opens a menu with many individual data options.

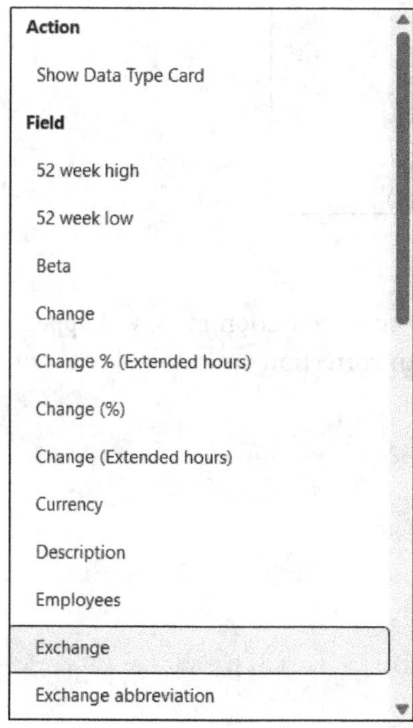

19

6. You can select the data you want individually, and each time you select a data item, a new column will be created.

	A	B	C	D
1	**Example of data types**			6/19/2025
2				
3	Share	High	Deep	Market capitalization
4	MICROSOFT CORPORATION (XNAS:MSFT)	$481.00	$474.46	$3,569,404,930,560
5	APPLE INC. (XNAS:AAPL)	$197.57	$195.07	$2,936,085,000,000
6	INTEL CORPORATION (XNAS:INTC)	$21.60	$20.66	$93,739,380,000
7	NVIDIA CORPORATION (XNAS:NVDA)	$145.65	$143.12	$3,549,711,999,999

Opening the data type palette reveals additional data types from various sources (see Figure 19.14). Earlier Wolfram data types are no longer supported.

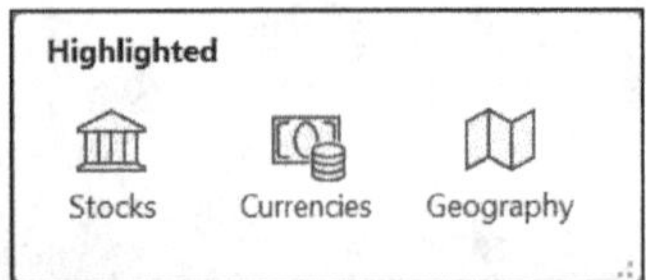

Figure 19.14 Palette of Various Data Types

The example in Figure 19.15 uses the **Currencies** data type to fetch the current dollar/EUR rate and convert the dollar price into euros. It uses an accepted currency pair like **USD/EUR** and enters it into a cell.

	A	F	G
1	**Example of data types**		
2		USD/EUR	€ 0.98
3	Share	High in EUR	
4	MICROSOFT CORPORATION (XNAS:MSFT)	€ 472.53	
5	APPLE INC. (XNAS:AAPL)	€ 194.09	
6	INTEL CORPORATION (XNAS:INTC)	€ 21.22	
7	NVIDIA CORPORATION (XNAS:NVDA)	€ 143.09	

Figure 19.15 Converted American Price Values

If you enter a stock name that Excel cannot recognize, a question mark will appear before it) and the **Data Selector** pane will open with correction options and relevant hints (see Figure 19.16).

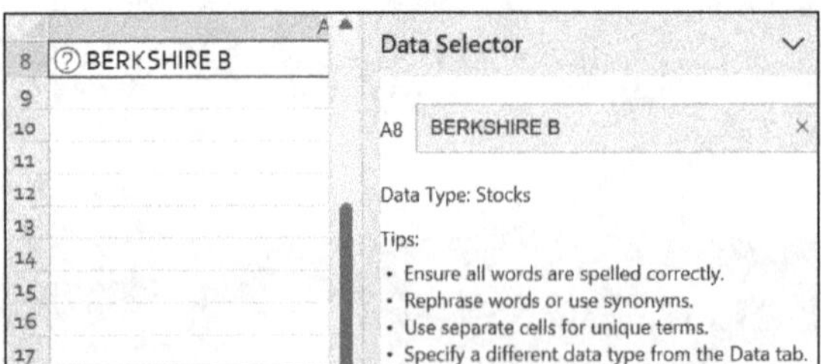

Figure 19.16 Notes on Unrecognized Stock

If you change the name, Excel will search the task pane and may offer a solution to apply.

The cell range with data types provides special commands for linked data types in its context menu, including converting back to a simple text range (see Figure 19.17).

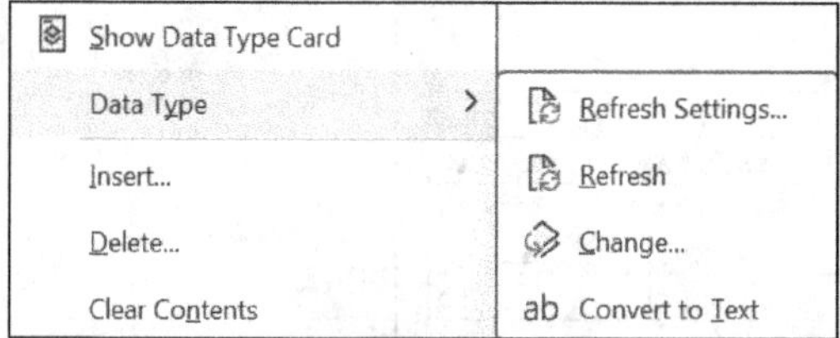

Figure 19.17 Context Menu for Linked Data Types

Clicking **Update Settings** opens a task pane where you can set the update frequency, which is especially important for stocks. **Update** always fetches the latest values immediately.

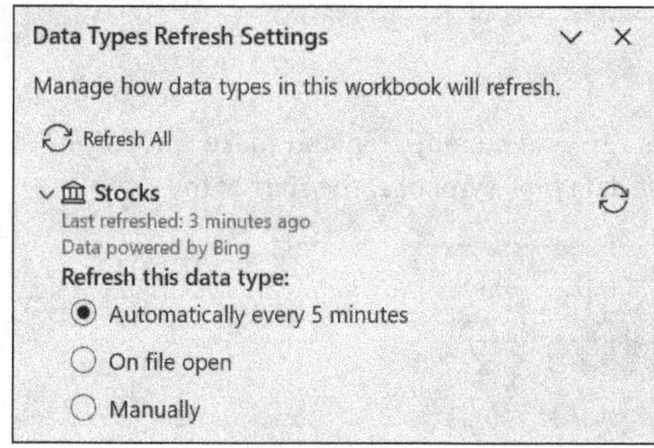

Figure 19.18 Updating Settings for Data Types

19.5 Importing Data from Pictures

Sometimes, it helps to extract text and numbers directly from an image into cells. On the **Data** tab, you'll find a **Picture From File** button that lets you import data from a file or the clipboard.

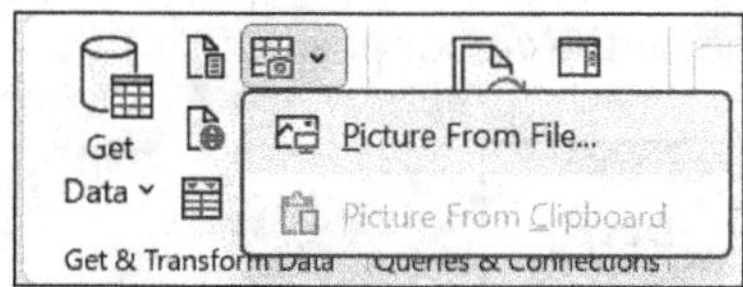

Figure 19.19 Options for Importing Data from Pictures

This feature works best when the image shows a clear table layout, such as a screenshot of tabular data. In Figure 19.20, a graphic with price data uses the stock data type. Excel opens a **Data from Picture** task pane where you can import and initially review the data. The result appears as a table.

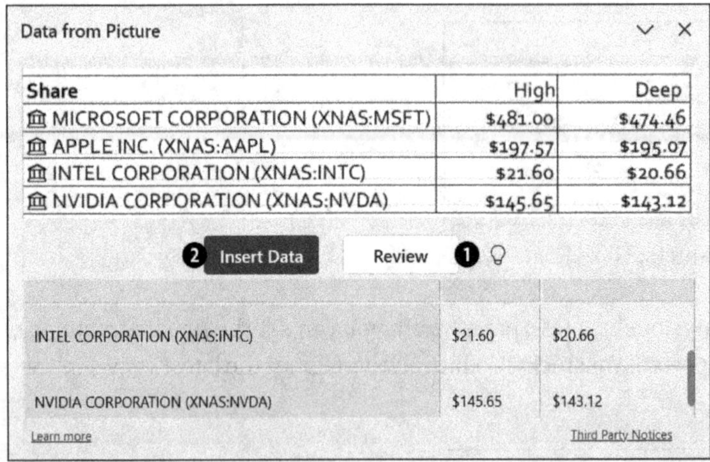

Figure 19.20 Importing from Screenshot

Use the **Review** button ❶ to examine the data step-by-step and make corrections if needed. In this case, always remove the data type symbol at the start of the row.

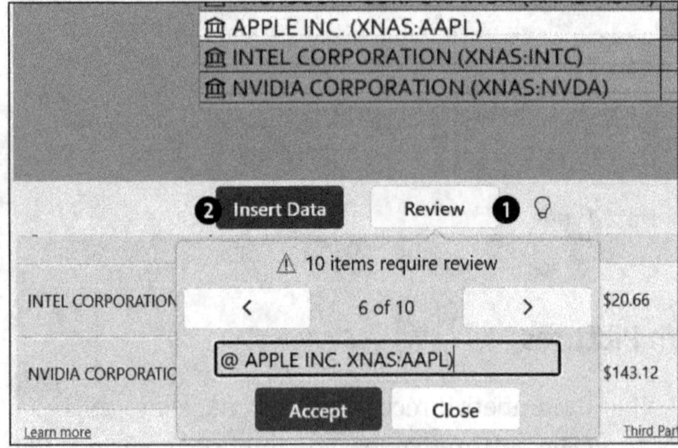

Figure 19.21 Correcting Entry

Using **Insert Data** ❷ inserts the data into the table sheet at the cursor position.

	F	G	H
27	Share	High	Deep
28	INTEL CORPORATION (XNAS:INTC)	$21.60	$20.66
29	NVIDIA CORPORATION (XNAS:NVDA)	$145.65	$143.12

Figure 19.22 Imported Data

Chapter 20
Export and Import of Files

Company employees who want to collaborate must be able to freely exchange data. The situation is similar with customers and suppliers, and individual users must also be able to transfer data between applications with minimal loss.

20.1 Data Formats and Filters

Since office work often involves various programs, error-free translation of data between formats is essential, especially when data comes from external sources. Conversely, data often needs further processing in other applications or at different workstations, and re-entering existing data simply because a different program was used initially is unacceptable.

This section briefly covers some traditional data exchange methods that remain relevant in practice. First, we cover importing and exporting files in other formats. This typically involves copying entire files and converting them to a different data format, and special converters or filters handle this. Such conversion methods are also required to process files in the current Excel version that were created with older Excel versions—or vice versa, when data must be processed on a workstation running a much older Excel version.

20.2 Supported File Formats

Excel can create formats from older versions and process files in those formats. However, newer Excel features are not compatible with older versions. Back-translation involves some loss.

There were three major version jumps in the past. Excel 5 and Excel 95 used the same file format, but the jump between formats 5.0/95 and 97–2003 was significant. Some features of the 97–2003 format cannot be reproduced in the 5.0/95 format. The next jump came with Excel 2007, and these file formats are still current. Excel continues to support the 5.0/95 and 97–2003 formats, and no fundamental changes have occurred since Excel 2010. However, the **Strict Open XML Workbook** variant was added, as described in Chapter 1, Section 1.5.

If you work in an office that mostly uses older Excel versions, you can set an older format as the default file type if you need to. You can find this setting under the **File • Options** command, on the **Save** page, under **Save files in this format**.

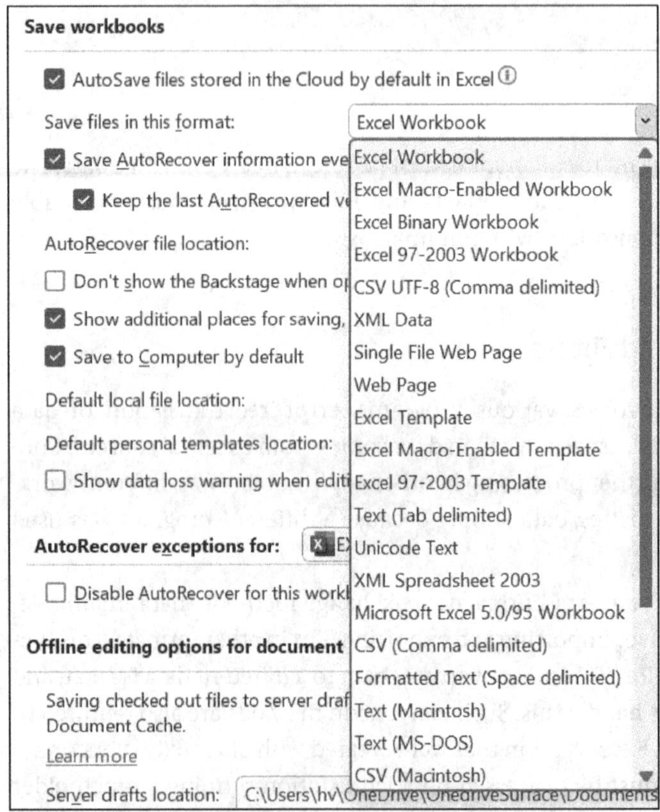

Figure 20.1 Setting Older Format as Default

In Excel, import and export functions are usually integrated into the **Open** and **Save As** dialogs. When importing, the program usually detects the required action automatically after reading the external file's header. When exporting, the file type selection controls the translation.

20.2.1 Output Formats

The following table lists the formats Excel offers for saving files.

Formats	Types	Descriptions
Excel Workbook	*XLSX*	XML-based standard file format
Strict Open XML-Workbook	*XLSX*	Variant of the XML-based standard format

Formats	Types	Descriptions
Excel Workbook with Macros	*XLSM*	XML-based format for macro-enabled files that saves VBA macro code and Excel 4.0 macro templates (XLM)
Excel Binary Workbook	*XLSB*	Excel binary file format
Excel Template	*XLTX*	Excel standard file format for templates
Excel template with macros	*XLTXM*	Excel standard file format for templates with macros
Excel 97–2003 Workbook	*XLS*	Excel 97–2003 binary file format
Excel 97–2003 Template	*XLTT*	Excel 97–2003 binary file format for templates
Microsoft Excel 5.0/95 workbook	*XLS*	Excel 5.0/95 binary file format
XML spreadsheet 2003	*XML*	XML spreadsheet 2003 File format (XMLSS)
XML data	*XML*	XML data format
Excel add-in	*XLAM*	XML-based standard format for Excel add-ins with VBA support
Excel 97–2003 add-in	*XLA*	Excel 97–2003 add-in with VBA support
Excel 4.0 workbook	*XLW*	Version 4.0 file format (open only)
OpenDocument spreadsheet	*ODS*	Standard format for spreadsheets in programs like OpenOffice or LibreOffice
Single web archive	*MHT, MHTML*	Output format for a web archive
Web page	*HTM, HTML*	Output format for a webpage
PDF	*PDF*	Output format for a PDF file
XPS document	*XPS*	Output format for an XPS file

20.2.2 Import Formats

The list of file formats Excel can open differs slightly from the list that's available for saving. In some cases, a file can be opened but not saved again in its original format. This applies to dBASE III and IV formats. The dBase II, Lotus 1-2-3, Microsoft Works, and Quattro Pro formats are no longer supported.

20.2.3 Working in Compatibility Mode

When you open a file in the 97–2003 format, Excel automatically switches to a special compatibility mode that's indicated in the title bar. In this mode, some Excel features are disabled to prevent data loss when you save in the older format.

If a file you edit in this mode should later use the new features, you can convert it through the **File** tab. Select the **Information** page and then click the **Convert** button in the **Compatibility Mode** section.

If you want to save a current file in an older format, run a compatibility check first to identify potential data loss. Do this via **File • Information • Check for Issues • Check Compatibility**. This was already covered in Chapter 11.

20.2.4 Text Formats

When data cannot be exchanged directly between applications, intermediate formats like text formats, in which individual data elements are separated by specific characters, can often help. Although some formatting features are usually lost, the data is preserved.

Formats	Types	Descriptions
Formatted text	*PRN*	Lotus format with spaces as delimiters
Text (tab-delimited)	*TXT*	Tab-delimited text file
Text (Macintosh)	*TXT*	Tab-delimited text file for the MacOS
Text (MS-DOS)	*TXT*	Tab-delimited text file for MS-DOS
Unicode Text	*TXT*	Saves a workbook as Unicode text
CSV-UTF-8 (delimiter-separated)	*CSV*	Delimiter-separated text file—using commas or semicolons—that supports the UTF-8 character set
CSV (Macintosh)	*CSV*	Delimiter-separated text file—using commas or semicolons—for the MacOS
CSV (MS-DOS)	*CSV*	Delimiter-separated text file—using commas or semicolons—for MS-DOS that supports only the ANSI character set
DIF	*DIF*	Data Interchange Format
SYLK	*SLK*	Multiplan SYLK files

20.3 Importing Text Files

The situation is more complex when structured data, such as an address list, is stored as a simple text file. If the information in the text file is separated by tabs, Excel can neatly distribute the data across columns when you use the clipboard. However, if you open the text file through the **Open** dialog, Excel behaves differently.

20.3.1 Importing an Address List

Here, as our example, we use an ASCII file that contains addresses. The data are arranged in separate rows that are separated only by tabs.

```
First name	Last name	Street	ZIP	CITY
Bettina	Bunge	Zornstr. 34	10786	Berlin
Axel	Goslar	Bernerstr. 12	10234	Berlin
Hans	Mauk	Doragasse 6	10999	Berlin
```

Figure 20.2 Small Address List in Text Format, Separated by Tabs

To import the data, follow these steps:

1. In the **Open** dialog box, select **Text Files** under **File Type** and then choose the appropriate file.
2. If you confirm with **Open**, the **Text Import Wizard** will start. It helps you distribute data across different columns. In the lower part of the dialog box for step 1, Excel displays a preview of the text file's records.
3. Under **Original data type,** you must first decide which data type applies to the file. There are two file types: **Delimited** and **Fixed width**. The preceding example is **Delimited** because the data fields are separated by a tab. Other delimiters can include semicolons, spaces, and commas, and more.
4. In the **Start import at row** field, specify whether to import data from the first text row or a later one. This is useful if the first row contains formatting characters that are unsuitable for the table.
5. In the **File origin** field, specify the data source. This refers to the character codes used in the file. Windows text files use the ANSI code, while DOS or OS/2 text files use the PC-8 code. For MacOS files, select the matching setting. If accented characters like umlauts don't display correctly in the preview, try a different option here.

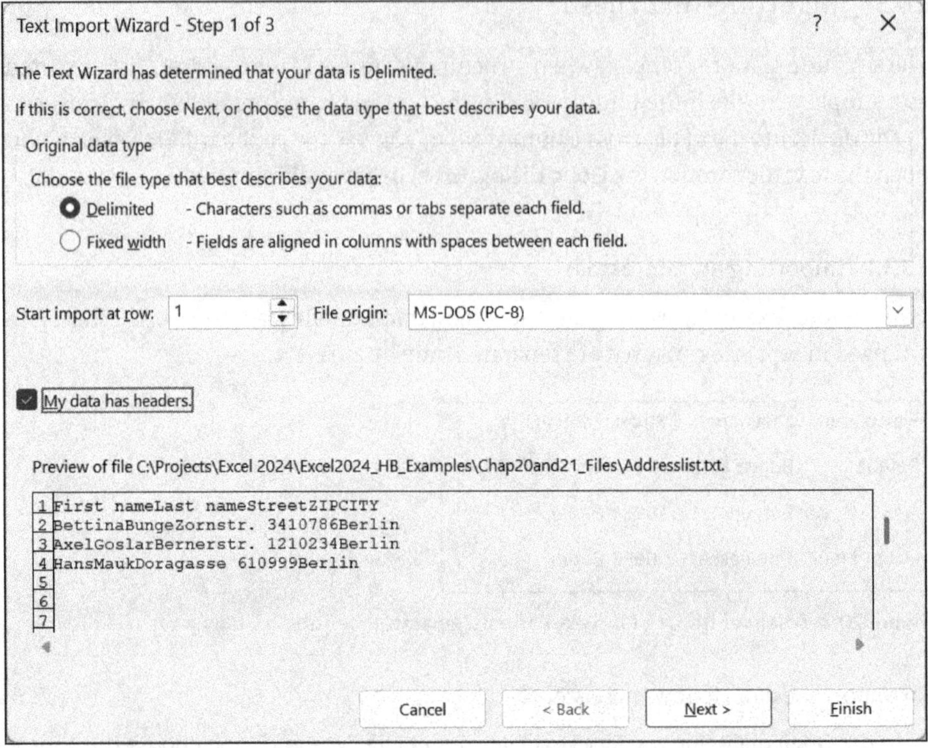

6. After you choose the settings, the next dialog shows which character the Text Import Wizard assumes to be the delimiter. In this case, the tab character is correctly recognized, but if the assumption is incorrect, you can correct it by selecting which **delimiter** to use. If none of the offered characters were used, select **Other** and enter the actual character in the small text field. Once you choose the correct character, the text under **Data Preview** will appear in a neat, tabular format.
7. A special problem arises when a character that's normally used as a delimiter—such as a space—is needed as a text character. The solution is to use a different character, such as quotation marks, as a **Text qualifier**. An entry like *Trajanstr.12* occupies only one column. You can also select multiple delimiters to catch input errors or separate data later. For example, if a text file uses semicolons to separate data but first and last names are not separated, you can use the space as a second delimiter. However, you should check the box for **Treat consecutive delimiters as one** because it prevents unnecessary empty columns from being created.

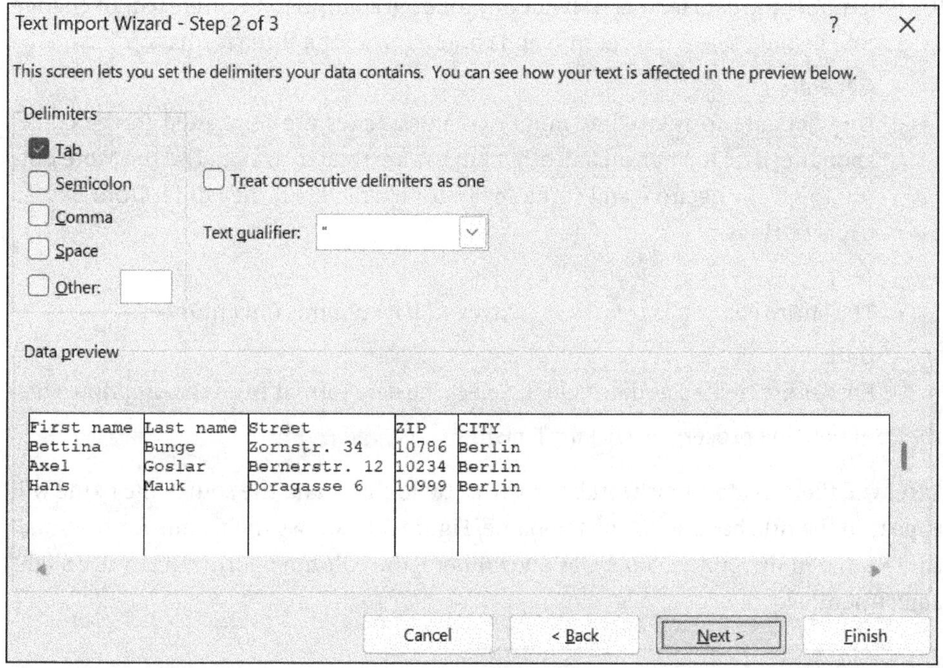

8. If the selection in the dialog box shows a correct preview of the text file data, you can review the individual columns created during import in the next dialog box. First, you can exclude columns from import by clicking or tapping anywhere in the column shown under **Data Preview**. Then, select **Do not import column (skip)** under **Column data format**.

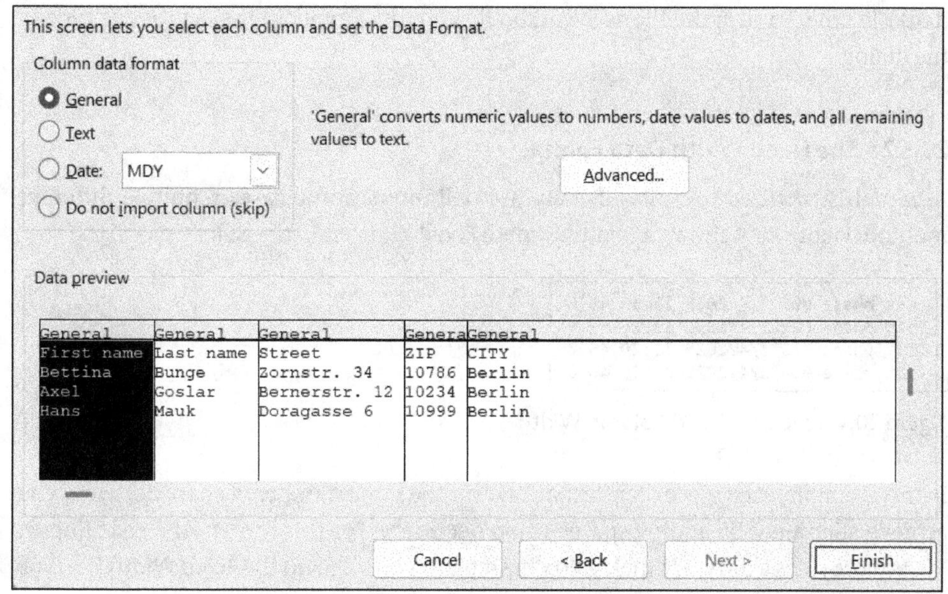

9. You can set the data format individually for each column to be imported. To change Excel's default, select the column and choose one of the three format options:
   - **General**
     Numbers are formatted as numeric values, dates are formatted in the corresponding date format, and all other entries are treated as text. Use the **More** button to set the decimal and coma separator characters if they differ from Excel's usual settings.
   - **Text**
     The entire column is treated as text, even if it contains only numbers.
   - **Date**
     Entries are treated as date values. Select the date format from the dropdown list.
10. Complete the process by clicking **Finish**.

Excel will then create a new workbook with a table sheet, and the source file name will appear in the title bar and as the tab name. Figure 20.3 shows the results for the small address file in the table. Usually, as shown here, the column width will require slight adjustment.

	A	B	C	D	E
1	First name	Last name	Street	ZIP	CITY
2	Bettina	Bunge	Zornstr. 34	10786	Berlin
3	Axel	Goslar	Bernerstr. 12	10234	Berlin
4	Hans	Mauk	Doragasse 6	10999	Berlin

Figure 20.3 Imported Text Data

When you save the new file, Excel will initially suggest that you use the **Text (Tab delimited)** file type. To save the file as a workbook, you must explicitly select the file type in the dialog.

20.3.2 The Fixed Width Data Format

Although structured text files usually use delimiters, some cases require a different method. Figure 20.4 shows a text file where **Fixed Width** must be selected.

```
Test phase 001 1/4/2025 12:10 45.5
Test phase 002 1/4/2025 12:15 45.6
Test phase 003 1/4/2025 12:20 45.7
Test phase 004 1/4/2025 12:25 45.8
```

Figure 20.4 Text with Fixed Column Width

In this example, each piece of information uses the same number of characters, and you achieve alignment by adding spaces when necessary. For this format, the Text Import Wizard provides a different dialog box in step 2. If you selected the **Fixed Width** file type in step 1, Excel will try to detect columns and their widths based on spaces, and it will display the assumed layout under **Preview of selected data**. You can then adjust this assumption by moving the arrow lines that indicate the suggested column breaks, using a mouse, finger, or pen. If Excel does not detect a delimiter, you can set the break lines manually. To do this, insert an arrow at the desired position with a click or tap. You can also delete any unnecessary column breaks by double-clicking or double-tapping the corresponding arrows. Then, proceed as described previously.

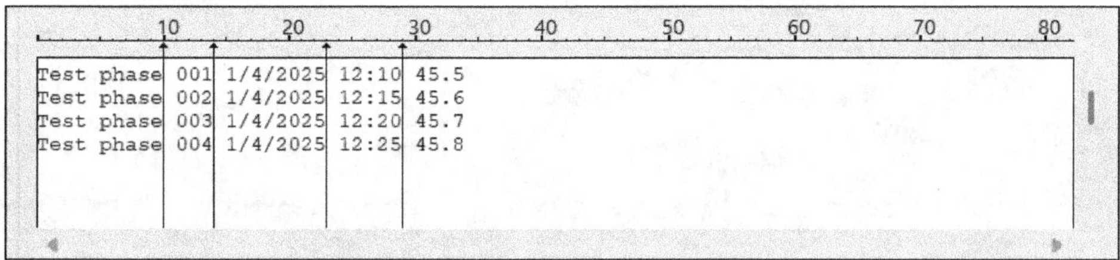

Figure 20.5 Splitting Text by Setting Break Lines

20.4 Distributing Text Across Columns

You can also use the text conversion wizard when text you've pasted into a single column of a table sheet needs to be formatted into a table, or when you need to split a column that contains full names into separate first- and last-name columns. In this case, use the **Data • Data Tools • Text to Columns** command and select the column that contains the text you want to split into multiple columns.

Alternatively, you can proceed the same way you would when importing a text file. The only difference is that in step 3, you can specify the destination range where the processed data will appear in the sheet. By default, the upper-left cell of the selected range is set, and if you don't enter another reference, the data will be split into columns in place. The adjacent columns must be empty; otherwise, they will be overwritten. If not, the original data remains unchanged and is copied to the destination range.

A new alternative to this method is the `=TEXTSPLIT()` table function, which we demonstrated with an example in Chapter 15, Section 15.12.

Chapter 21
Exchanging Data with Other Applications

This chapter explains methods for exchanging data between different documents and applications, focusing on parts of documents or objects within them. These methods range from simple copying to embedding or linking objects. Unlike importing or exporting entire files, the methods described here require both applications to be installed on the PC during the exchange.

The emphasis is on mixing different types of information within a document, such as tables with images or image sequences.

21.1 Exchanging Data via the Clipboard

The simplest way to exchange data between applications is through the clipboard. What does that mean? The key difference here isn't that the data exchange uses the clipboard, since that also happens with other data exchange methods. "Simple" means that specific data or objects are transferred only once, and no link remains between the target document and the source document afterward.

Either data is cut from the source file and placed into the target file, or the source file remains unchanged while the target file receives a one-time copy to complete the data exchange.

The handy temporary storage of multiple data sets at once via the Office clipboard, which supplements the system clipboard, was already covered in Chapter 2. It can also be used especially for exchanging data with other Office programs.

21.1.1 Word Imports Data from Excel

First, we'll explain the clipboard exchange in more detail by using Excel and Word as an example. Naturally, the exchange works both ways. A table or chart from a workbook can be inserted or copied into a Word document, and a list created in Word can be transferred into a worksheet in a workbook.

	A	B	C	D
1	**Production plan week 21**			
2				
3		Product A	Product B	Product C
4	Monday	100	100	60
5	Tuesday	50	60	70
6	Wednesday	80	90	60
7	Thursday	80	90	60
8	Friday	80	90	60

Figure 21.1 Source Table in Excel

21.1.2 Exporting Data via the Clipboard

Let's start with the first case. To directly transfer a small production plan for one week from an Excel workbook into a Word document, follow these steps:

1. In Excel, open the file from which the table will be transferred.
2. Select the appropriate table range.
3. Go to **Home • Clipboard • Copy**.
4. If Word and the target document are open, click the Word button on the task bar to switch to it.
5. In the Word document, place the cursor where you want to insert the table.
6. Select **Home • Clipboard • Paste**, and the table will be pasted from the clipboard into the selected spot. The following figure shows the table in the Word window.
7. After pasting, the data you selected in Excel will remain outlined with a moving border. Press [Esc] to exit this mode in Excel when you return.

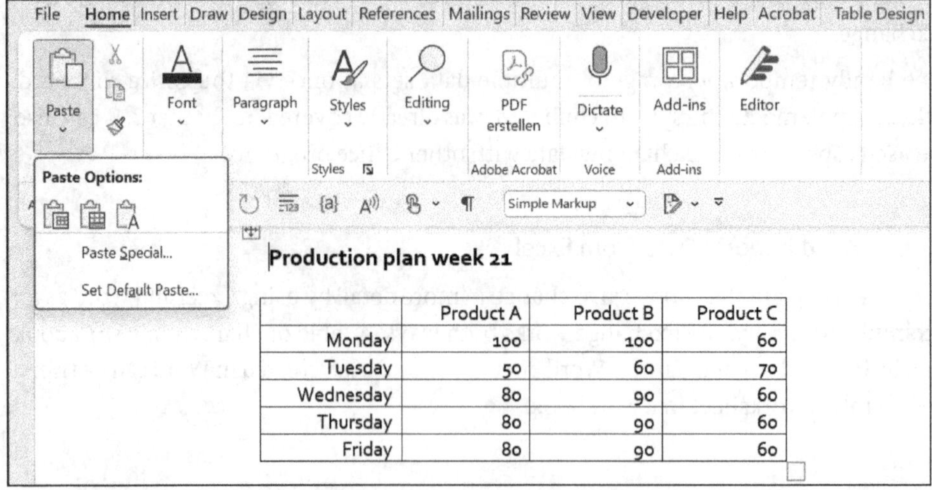

As you can see in the figure, the table structure in Word remains intact, and the formatting of text and numbers is preserved. If you click the second button instead of the first,

Use destination styles, Excel's formatting will be replaced by Word's predefined table style.

21.1.3 How the Clipboard Works

This example shows that the clipboard is more than just passive temporary storage. Data isn't moved back and forth arbitrarily. Before pasting data into another application, the program determines the best format to use, and it always selects the format the target application handles best. In the example we just described, Word can create tables on its own. Therefore, unless you specify otherwise, Excel data will be pasted as a table. Since Excel 2000, the default format has been HTML.

If you want to paste data in a different format than the one the target application automatically uses, you can use one of the buttons under **Paste Options**, or you can select **Paste** and then **Paste Special** from the menu. There, you can choose from various formats and methods for exchanging data.

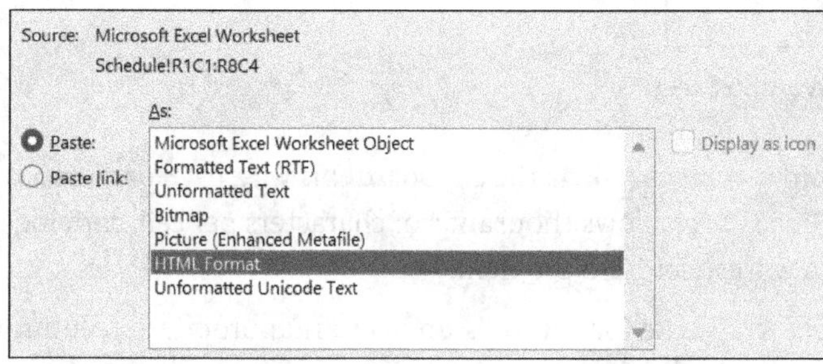

Figure 21.2 Formats for Pasting via Clipboard into Word

In Word, you can set the default for how copies from other applications are pasted. This occurs in the **Word Options** dialog under **Advanced** in the **Cut, Copy, and Paste** section under **Paste from other programs.**

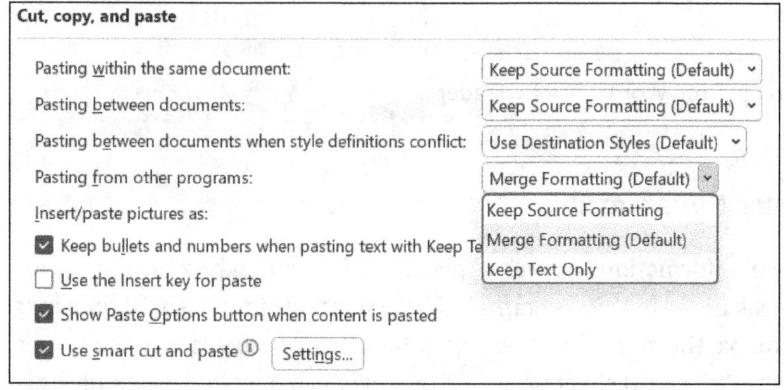

Figure 21.3 Cut, Copy, and Paste Settings in Word

21.1.4 Importing Text from Word

When you're importing text from an existing Word document into an Excel workbook, the outcome depends on what is currently selected in the workbook. For example, it can be useful to insert a longer, complex technical term from an open Word document directly into a specific spot in the formula bar or into the cell you've opened for editing.

If text is pasted directly into a selected cell, it overwrites the cell's previous content, and by default, the pasted text keeps the formatting from Word. This can also apply to an entire column, as Figure 21.4 shows.

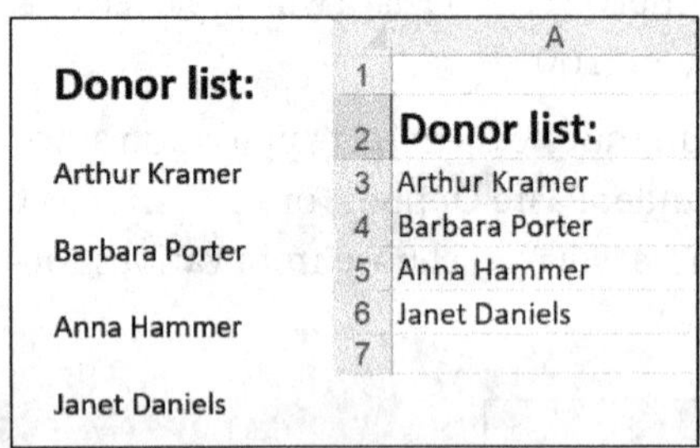

Figure 21.4 List Pasted from Word to Excel

When flowing text from Word is transferred via the clipboard this way, each paragraph is copied into a single cell. Since Excel allows thousands of characters per cell, copying large amounts of text onto a worksheet is no problem.

However, you should enable wrap text for the cells and set an appropriate column width. Merging cells is also useful in this case. Figure 21.5 shows an example in which a longer text was transferred into a merged cell range.

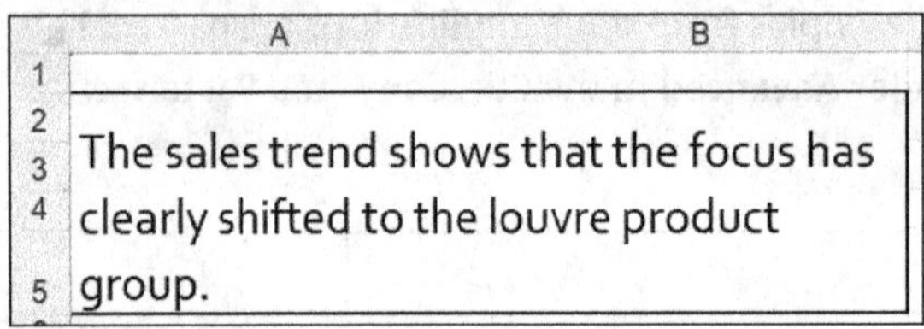

Figure 21.5 Text Passage from Word in Table Header

21.2 Linking Files Dynamically

A one-time transfer of information between applications isn't enough when it's important to keep data consistent in both. For example, if an Excel table imports a list of names from another workbook, the question arises: What happens if the names in the source catalog change? If the Excel worksheet needs to track every change in the catalog, linking both documents is recommended.

Although you can create such a link via the clipboard, continuous data updates from the source to the target file happen independently of the clipboard. This happens via the opening of a direct channel between a specific element in the source document and the corresponding element in the target document, which allows for data exchange.

21.2.1 Linking Between Documents

For example, consider a list of labels in an Excel document that will be used to label table rows in Excel. Follow these steps to link the documents:

1. Copy the selected data from the source workbook to the clipboard and then switch to the second Excel workbook.
2. Select the first cell of the range where you want to place the data. This time, use the **Paste special** command from the context menu. Then, in the dialog box, select the **Paste link** option and, under **As**, choose **Text**.

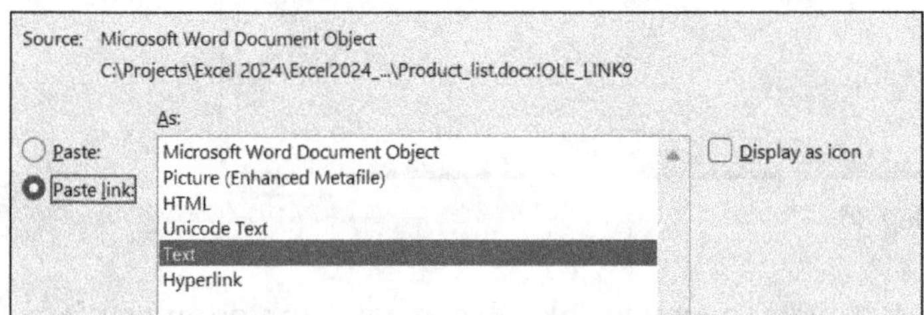

3. When you confirm, Excel will insert the data. If you check the Formula Bar, you'll see the same array formula in each cell that creates the link. The formula will look something like this:

```
{=Word.Document.12|'C:\Projects\Excel 2024...\Product_list.docx'!'!OLE_LINK9'
```

Once the link is established, the data in the Excel worksheet will update whenever it changes in the Word document.

21.2.2 Update Control

Whether changes in a source file immediately appear in the Excel worksheet depends on whether automatic or manual updating is enabled. You can manage this in the task pane (or dialog) of the **Edit Links** command, which you can use to manage all links in a workbook. This command appears in the **Data • Queries & Connections** group only if the workbook contains links. Since this isn't a link to another workbook, the task pane offers the **Edit Other Links** option. It opens the **Edit Links** dialog box, which you may recognize from earlier Excel versions.

If, under **Update,** it is not set to **A** for **Automatic**, which is the default, you can update a selected link in the list to the latest version by clicking the **Update Values** button.

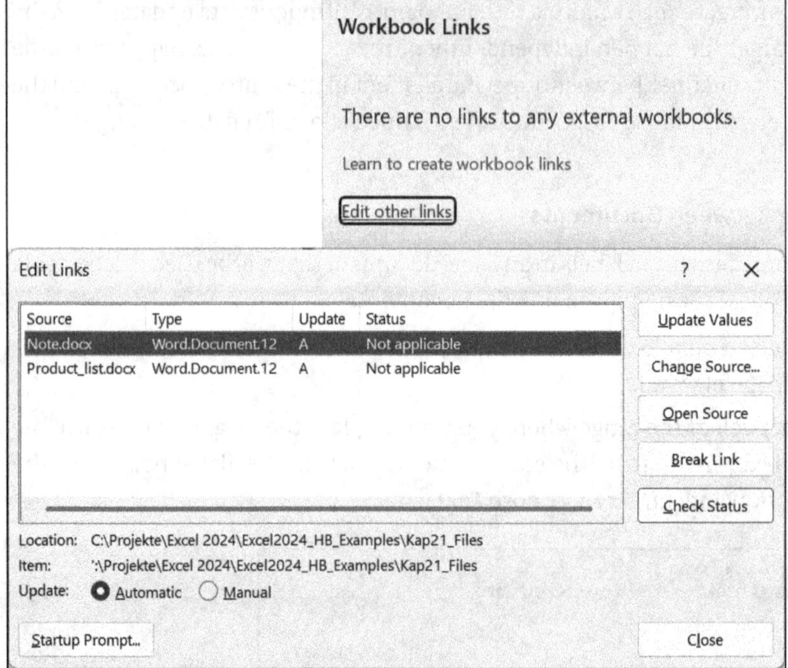

Figure 21.6 Edit Links Dialog Box

When you open a file that contains a remote link—that is, a link to a document from another application—you are usually asked whether you want to update the links "to an external source."

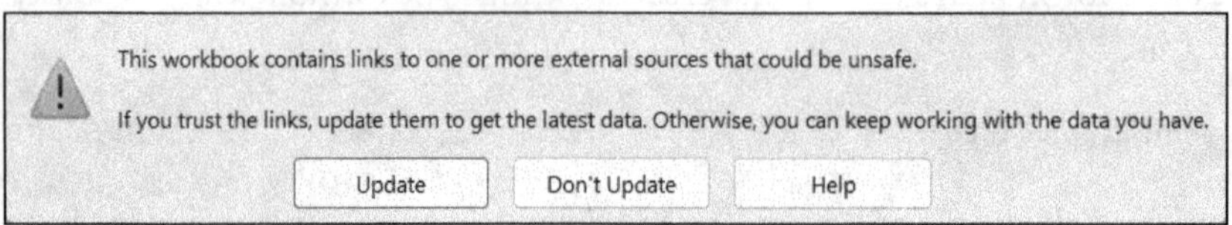

Figure 21.7 Note on Links

If you confirm, the data in the table will update to match the source file; otherwise, changes in the source file won't be applied. You can disable this query by using the **Startup Prompt** button in the **Edit links** dialog. The alternatives are automatic updating without warning or completely blocking the update.

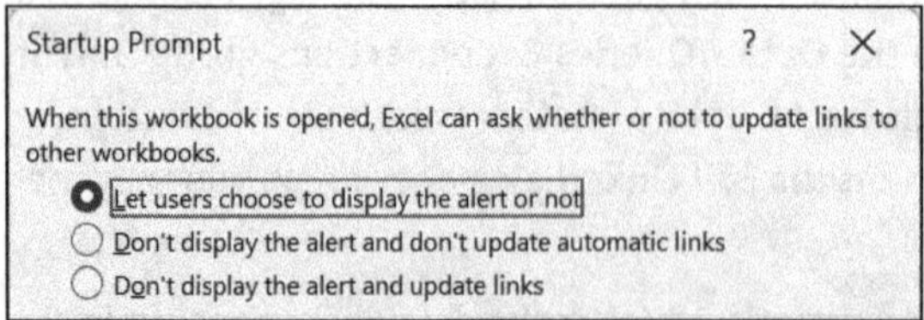

Figure 21.8 Options When Opening Linked Documents

If the file is in a location that's not yet marked as trusted (see Chapter 1, Section 1.8.24), Excel will notify you that automatic updating has been disabled when you open the file. You can override this restriction by clicking **Enable Content**.

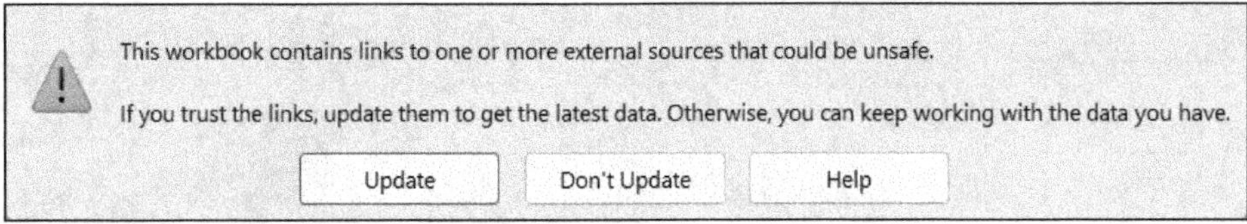

Figure 21.9 Notice of Disabled Updating

Chapter 22
Automating Routine Tasks with Macros

Much of the office work in Excel is routine. You often need tables with the same layout, and you repeatedly fill out and print certain forms. Since the earliest versions of Excel, you've been able to automate such actions by recording them as macros. Excel includes the Visual Basic for Applications (VBA) programming language, which is a variant of Visual Basic (VB), which is one of the key programming languages on Windows.

While VB is used as a standalone language to create complete applications, VBA runs as a guest inside a running application. If you only record macros, you barely need to know this language because Excel generates the code automatically during recording. An introduction to VBA is in the next chapter.

22.1 Recording Macros

The first example for macro recording here is a project schedule that always covers the next 30 days. If the schedule always appeared in the same spot in a table, having a template would be useful. If you want to use the schedule anywhere in a workbook, it's best to record a macro for it.

	A	B	C
1	**Schedule**		
2		Task	Status
3	Tuesday, June 24, 2025		
4	Wednesday, June 25, 2025		
5	Thursday, June 26, 2025		
6	Friday, June 27, 2025		
7	Saturday, June 28, 2025		
8	Sunday, June 29, 2025		
9	Monday, June 30, 2025		
10	Tuesday, July 1, 2025		
11	Wednesday, July 2, 2025		
12	Thursday, July 3, 2025		
13	Friday, July 4, 2025		
14	Saturday, July 5, 2025		
15	Sunday, July 6, 2025		

Figure 22.1 Schedule with Weekdays

Before you start recording the macro, run through the workflow manually and note each step to avoid getting stuck during recording and having to start over.

22.1.1 Preparations

First, ensure the **Developer** tab is visible. To do this, go to **File • Options** and select **Customize Ribbon**. Then, enable **Developer** under **Main Tabs** in the right pane. The toolbar includes these groups: **Code**, **Add-Ins**, **Controls**, and **XML** (see Figure 22.2).

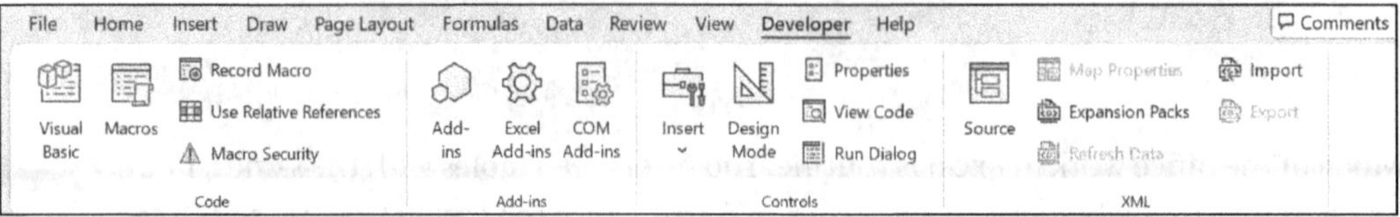

Figure 22.2 Developer Tools Tab

You cannot deselect individual groups here (see Figure 22.3).

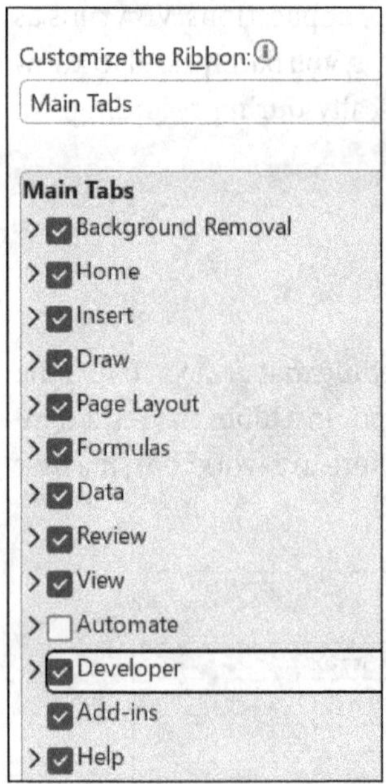

Figure 22.3 Displaying Developer Tools Tab

Also, make sure the **Record Macro** icon appears on the left side of the status bar. If it isn't, open the status bar's context menu and enable the button.

Figure 22.4 Button to Start Macro Recording in Status Bar

If you only want to use recorded macros, you can keep the **Developer** tab hidden, which is the default. You can record and run macros from the **Macros** group on the **View** tab, and the **Macros** button provides a small menu for this.

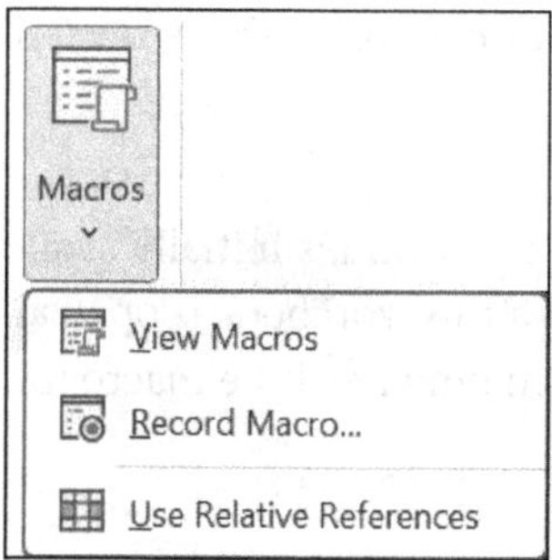

Figure 22.5 Macro Commands on View Tab

Follow these steps to prepare the actual recording:

1. Select a blank sheet in a new workbook and use the **Record Macro** icon or the **Record Macro** command in the **Code** group on the **Developer** tab.
2. In the dialog box, replace the default name with a descriptive one—something you'll still recognize after six months. If the name has multiple words, connect them with underscores. Periods are not allowed in macro names, and the first character must be a letter.
3. Adding a brief **Description** can be helpful. For example, it might explain exactly what the macro does and what to watch for when using it.
4. You can start a macro in several ways. One option is to assign a keyboard shortcut here to launch the macro. Under **Shortcut key**, enter a letter to use while holding down `Ctrl` and `⇧`. In this case, entering "S" might be a good choice to help you remember *Schedule*. When you confirm the dialog, Excel checks if the shortcut is available. If it warns you that it's not, you should choose a different shortcut.

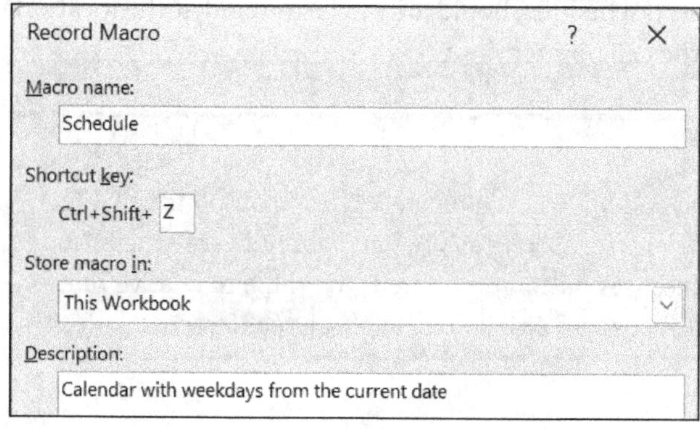

Unauthorized Keys

Keyboard shortcuts cannot include numbers or special characters. Keyboard shortcuts in a workbook override other predefined shortcuts while the workbook is open.

5. Under **Store macro in**, you must decide where to save your macro. You have several options. The choice depends on what the macro is for:
   - **This Workbook**
     This means the macro is recorded in the current workbook and is initially available only there. It's available to other workbooks only if this workbook is open at the same time. Usually, you'll choose this option if you only need the macro for this workbook.
   - **New Workbook**
     This means a separate workbook is created for the macro. The macro is available whenever this workbook is open. This option is recommended if you use macros often, but not always, across different workbooks.
   - **Personal Macro Workbook**
     Recording into this means the macro is saved in a special workbook named *Personal.xlsb*, which Excel automatically stores in the *XLSTART folder* in a hidden state. This means two things: First, this file opens automatically every time you start Excel, so the macros are always available. Second, you don't see this file; it runs behind the scenes. This option is ideal for macros that you use frequently and want to be accessible in every Excel session. For the planned macro, this choice makes sense if the schedule is used repeatedly. For simplicity, the data will initially be saved in the current workbook.
6. After you confirm the dialog, recording can begin. The **Record Macro** icon will change slightly, and you'll be able to use it to stop the recording. The same applies to the **Record Macro** command in the **Code** group, which then offers **Stop Recording**.
7. The **Developer • Code** group also contains another button labeled **Use Relative References**. Click it if the recording should use relative cell addresses. This recording method is necessary because the schedule should always be entered at the location you previously marked with the cell pointer.

[»]

Mixing References

By toggling relative recording on and off, you can mix relative and absolute cell references in a macro whenever you need to. As long as you need absolute references, you must keep the **Use Relative References** button unselected. To switch to relative references, click this button, and click it again to switch back to absolute references.

22.1.2 Recording a Schedule

Now, you can start recording. Perform the steps for the schedule in order:

1. Enter the "Schedule" heading into the current cell and format it with the **Bold** button and font size **12**.

2. Select a cell two rows below and enter the current date there with the `TODAY()` function. Since a fixed value is needed, copy the cell and then paste the value the formula returns into the same cell by using **Paste • Values**.
3. Select a range of twenty cells, starting from the cell with the first date.
4. Go to **Home • Editing • Fill • Series** to open the **Series** dialog. From **Series in,** choose **Columns, Type: Date, Date unit: Weekday**, and **Step value: 1** to create a date series that includes only weekdays.
5. The range you select will then be formatted in the **Date, long** format, with the day name appearing before the date.
6. Adjust the column to fit the new entries by double-clicking or double-tapping the column boundary.
7. Label the cell above the empty adjacent **Task** column and the **Status** cell next to it. Widen the task column slightly to make room for the entries.
8. Select the entire range and highlight it with a border grid and a different cell background.
9. Place the cell pointer in the first input cell and then stop the recording with **Stop Recording**.

[+]

Copy the Macro

If you later decide to transfer a macro from one workbook to another or to the personal workbook, you don't need to record it again. Use the clipboard to copy the text from the module window to the new location. You must temporarily display the personal workbook for this, so do it via **View • Window • Unhide**.

22.1.3 What Does the Recording Look Like?

22

For this macro, which you can reuse as is, you don't really need to know how Excel recorded the commands. The macro works, and that's what counts. Still, you should take a look at the recording's results. Macros appear and are edited in their own window—the VBA Editor window. If you want to view and possibly edit the results, follow these steps:

1. Select **Developer • Code • Macros** or press `Alt`+`F8` to open the **Macro** dialog.
2. In the list under **Macro name**, you'll find the entry **Schedule** if you've completed the previous steps correctly. When you select the macro, a brief description will appear below. Select the macro name and click the **Edit** button.

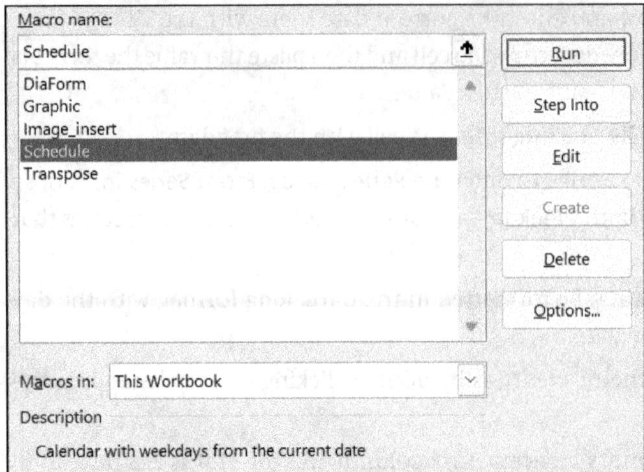
Macro name:
Schedule
DiaForm
Graphic
Image_insert
Schedule
Transpose
Run
Step Into
Edit
Create
Delete
Options...
Macros in:
This Workbook
Description
Calendar with weekdays from the current date

3. Excel will open the VBA window and show the macro in its own module window for viewing and editing.

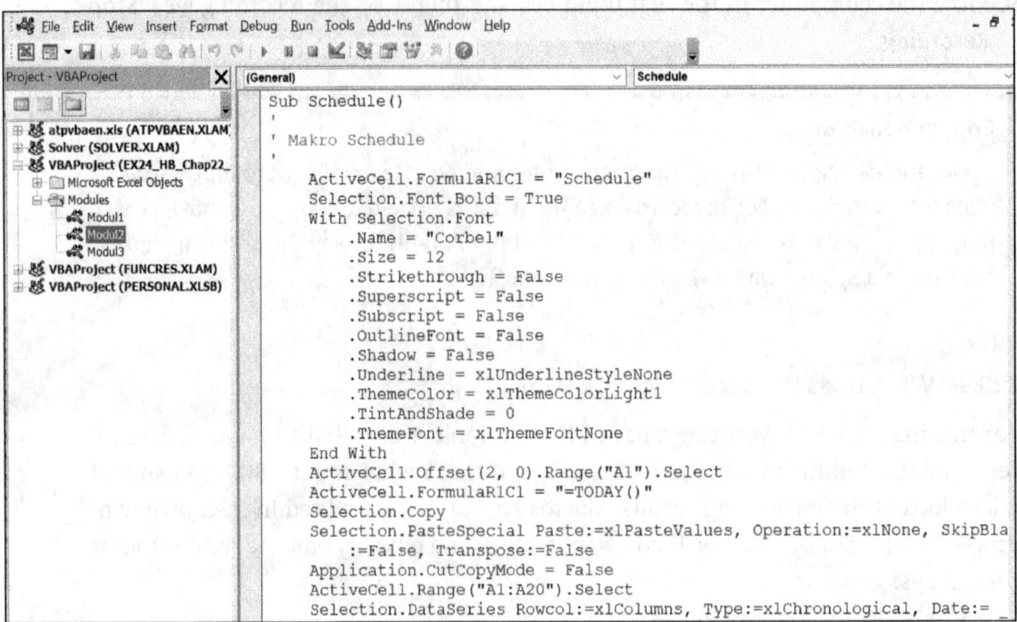
File Edit View Insert Format Debug Run Tools Add-Ins Window Help
Project - VBAProject
atpvbaen.xls (ATPVBAEN.XLAM)
Solver (SOLVER.XLAM)
VBAProject (EX24_HB_Chap22_
Microsoft Excel Objects
Modules
Modul1
Modul2
Modul3
VBAProject (FUNCRES.XLAM)
VBAProject (PERSONAL.XLSB)
(General)
Schedule
Sub Schedule()
'
' Makro Schedule
'
ActiveCell.FormulaR1C1 = "Schedule"
Selection.Font.Bold = True
With Selection.Font
.Name = "Corbel"
.Size = 12
.Strikethrough = False
.Superscript = False
.Subscript = False
.OutlineFont = False
.Shadow = False
.Underline = xlUnderlineStyleNone
.ThemeColor = xlThemeColorLight1
.TintAndShade = 0
.ThemeFont = xlThemeFontNone
End With
ActiveCell.Offset(2, 0).Range("A1").Select
ActiveCell.FormulaR1C1 = "=TODAY()"
Selection.Copy
Selection.PasteSpecial Paste:=xlPasteValues, Operation:=xlNone, SkipBla
:=False, Transpose:=False
Application.CutCopyMode = False
ActiveCell.Range("A1:A20").Select
Selection.DataSeries Rowcol:=xlColumns, Type:=xlChronological, Date:= _

As you can see, Excel has recorded a `Sub` procedure. Most lines begin with `ActiveCell` or `Selection` (which refer to the active cell or selected range) and set specific properties for these cells (such as `Size = 12` for the font size), or they apply the `DataSeries` method, which creates a series. You can learn more about this in Chapter 23. To exit the VBA window, use **File • Close and Return to Microsoft Excel** or press `Alt`+`Q`.

[«]

What Excel Records and What It Doesn't

Excel only records completed actions. For example, if you select several cells in a row before entering data, Excel only records the selection of the cell where you actually enter something. The same applies to selections in a dialog box. Only the final confirmed state is recorded; canceled commands are not. So, it's no problem if you accidentally open the wrong dialog box and exit it by pressing [Esc].

22.1.4 Saving the Workbook with the Macro

Excel uses a special file type for workbooks that contain macros when using the XML-based standard format. This takes some getting used to. When you're saving the workbook, be sure to select the correct file type—**Excel Macro-Enabled Workbook (*.xlsm)**—in the **Save As** dialog. Otherwise, you'll get a warning.

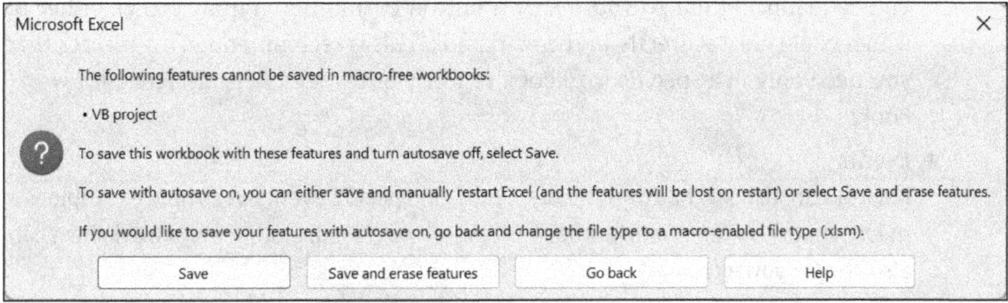
Microsoft Excel
The following features cannot be saved in macro-free workbooks:
• VB project
To save this workbook with these features and turn autosave off, select Save.
To save with autosave on, you can either save and manually restart Excel (and the features will be lost on restart) or select Save and erase features.
If you would like to save your features with autosave on, go back and change the file type to a macro-enabled file type (.xlsm).
Save
Save and erase features
Go back
Help

Figure 22.6 Warning When Saving with Wrong File Type

If you save the workbook as an Excel binary workbook with the*.xlsb* file extension, it doesn't matter whether it contains macros.

22

22.1.5 Running the Macro

You can run the recorded macro in two ways:

- The fastest way is to use the keyboard shortcut you set when you started recording. After placing the cell pointer in the cell where the schedule should begin, press [Ctrl]+[Shift]+the specified key—in our example, this is [S]. The entire schedule will be entered into the table instantly.
- Another way is via **Developer • Code • Macros** or **View • Macros • View Macros.** If you select the macro name in the dialog box, you can start the macro by clicking the **Run** button. Under **Macros in**, you can choose to display only macros from the current workbook or from all open workbooks.

You can assign a keyboard shortcut later as well. To do this, click the **Options** button while the macro is selected.

22.2 Integrating Macros into the Workflow

What you've learned so far is the simplest way to record and run a macro. For some macros, this method works well—but if you work with macros frequently, you'll find this approach limiting. Using keyboard shortcuts is fast, but the number of possible combinations is limited. Another issue is that most people can't remember many combinations.

Starting through **Developer • Code • Macros** is tedious, especially when macros are meant to speed up tasks. For frequently used macros, this method is too cumbersome. In such cases, other options are available. Here's a brief overview:

- **Icons**
  You can link a macro to a button icon and add it to the Quick Access Toolbar. This works well for macros that handle general tasks and should always be accessible.
- **Buttons or graphic objects**
  You can launch a macro with its own button or another graphic object, either of which you'd create directly in the workbook. This is recommended for macros that you need only in a specific workbook or form, or for templates designed for workbooks.
- **Events**
  Macros can also start automatically, based on specific events. A simple example is a macro that runs when its workbook is opened or closed. Changes to a worksheet can also trigger a macro.

The following sections offer examples of how to integrate your macros as effectively as possible into your work environment.

22.2.1 Quick Start with Icons

It's convenient to start a frequently used macro with a single click. Follow these steps to create a custom icon that launches a macro:

1. Go to **File • Options** and open the **Quick Access Toolbar** page.
2. Under **Choose commands from**, select **Macros**. In the list box on the right, choose whether the customization applies to all documents or only the current workbook.
3. Select the macro you want, click the **Add** button ❶, and use the small arrow buttons to adjust the position of the new icon.
4. Excel will assign a default icon to the selected macro. To use a different button, click **Modify** ❷.

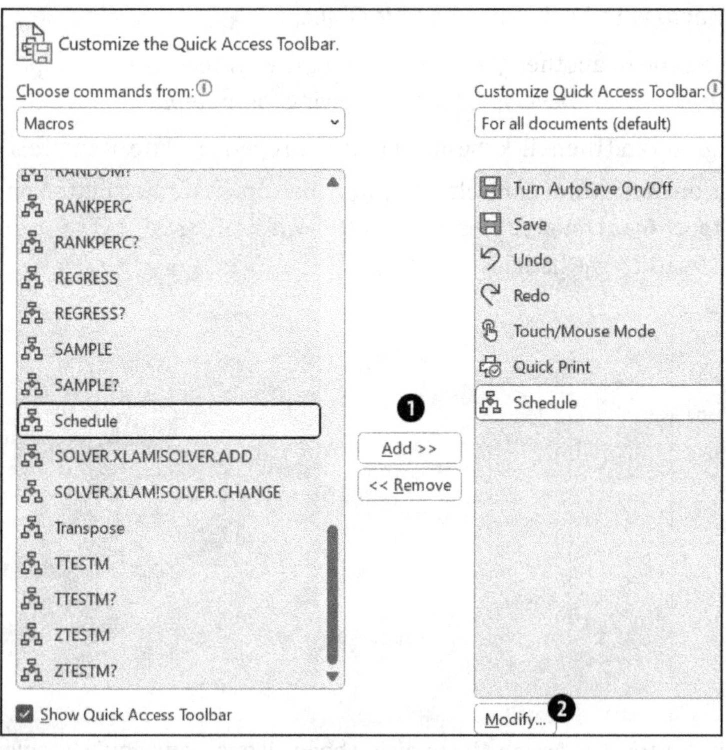

5. The dialog offers a wide selection of icons. The display name defaults to the macro name, but you can change it here.

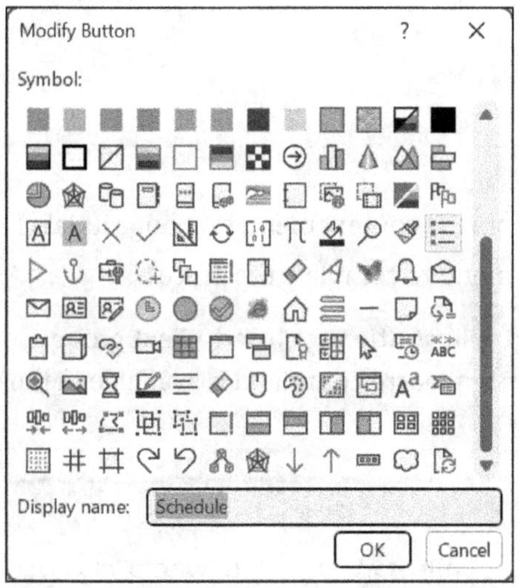

Depending on the setting you choose, the Quick Access Toolbar extension will apply either to all workbooks or only to workbook you select.

22.2.2 Starting a Macro with Buttons or Graphic Objects

Linking a macro to a button or another graphic object on the worksheet is very simple. You can insert the button via the **Developer** tab by following these steps:

1. Click the **Controls** group and then click the **Insert** button to open a palette of controls.
2. Select the **Button** icon in the **Form Controls** group and then drag to draw a button on the worksheet. In **touch input** mode, a preset-size button will be inserted, and you can move and resize it by using the handles.

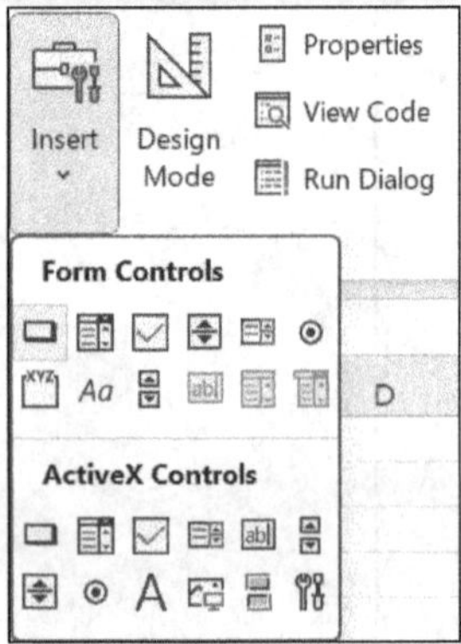

3. After you insert the button, the **Assign Macro** dialog box will open, and you'll be able to select the macro to link.

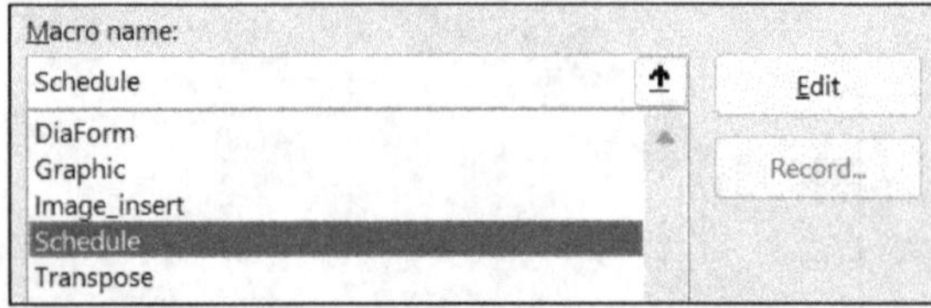

4. Right-click the button and select **Edit Text** from the context menu to change its label.
5. Select the **Exit Edit Text** option in the same menu to complete the process.

For objects you created with tools in the **Insert • Illustrations** or **Insert • Text** groups—such as a graphic element or WordArt—the **Assign Macro** command is available in the object's context menu.

Buttons or Icons?

Although these two options look very similar—both start the macro with a mouse click—they represent two distinct types: Icons are the most global option, while buttons are the most local. This affects where the associated macros should be stored. Macros for buttons belong in the same workbook as the sheets with the buttons, while macros for icons should preferably be stored in the personal macro workbook.

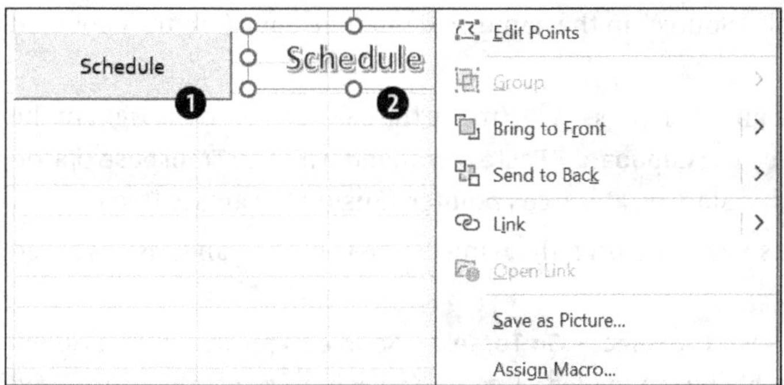

Figure 22.7 Starting Macro via Button ❶ or Graphic Object ❷

22.3 Flipping a Table with a Macro

Sometimes, you need to adjust a recording slightly so it can be used in many different situations. We'll demonstrate this with a macro that swaps (transposes) columns and rows in a table. This is often useful when tables grow in the wrong direction. For clarity, it's usually better for a table to grow vertically rather than horizontally.

For example, when you're analyzing a test over time, you'll want to continuously add new columns with test results, so it would be helpful to flip the table. To do this manually, you'd copy the table, paste it by using **Transpose** outside the old table, delete the old table, and move the new transposed table to that spot. However, simply recording these steps is not enough. The macro would only work with tables that are always the same size, so its usefulness would be limited.

	A	B	C	D	E	F	G	H	I
1		Test 1	Test 2	Test 3	Test 4	Test 5	Test 6	Test 7	Test 8
2	Test person 1	51	26	34	73	61	75	94	62
3	Test person 2	22	104	23	68	49	110	111	28
4	Test person 3	74	20	93	40	66	98	47	107
5	Test person 4	60	93	55	27	52	107	101	71

Figure 22.8 Original Table to Be Flipped

22.3.1 Transposing with a Macro

The goal is for the macro to remember the old table's location so it can paste the transposed table back there at the end. You can also record the necessary steps, which are as follows:

1. Select the table you want to transpose, and to the right of the table, leave a large enough range free to temporarily hold the transposed table.
2. Start recording by clicking the **Record macro** button and choosing **Use relative references.**

3. Name the selected table "Source" in the name box and then copy it to the clipboard by using **Copy**.
4. Press [Ctrl]+[?] twice and then press [?] to move the cell pointer to the right of the selected table. Then, use the **Clipboard • Paste** command with the **Transpose** option to paste the rotated table starting at the cell pointer. (Ensure the area is free.)
5. Without changing the selection, assign the name "Offset" to this range as described previously.
6. Go to **Home • Editing • Find & Select • Go To**, select **Source** to return to the original table, and delete the table by using **Clear**. Use **Go To** again to select the transposed table named **Offset** and cut it to the clipboard.
7. Use **Go To** to select the **Source** range, move the cell pointer with the [?] key to the top-left cell of this range, and clear the range. This is important because otherwise, the target range size won't be usable when you're pasting. Place the transposed table here by using **Paste**.
8. Finally, open the **Name Manager** dialog box again via **Formulas • Defined Names • Name Manager,** select the two **Source** and **Offset** names there, and delete them with **Delete**. After closing the dialog box, stop the recording using the **Stop Recording** icon in the status bar.

	K	L	M	N	O
1		Test person 1	Test person 2	Test person 3	Test person 4
2	Test 1	51	22	74	60
3	Test 2	26	104	20	93
4	Test 3	34	23	93	55
5	Test 4	73	68	40	27
6	Test 5	61	49	66	52
7	Test 6	75	110	98	107
8	Test 7	94	111	47	101
9	Test 8	62	28	107	71

Figure 22.9 Flipped Table

Figure 22.10 shows the macro. However, it still has a weakness that becomes clear when you review the macro recording.

In both cases, fixed table ranges are specified for the ranges that the macro temporarily names, and that doesn't work here because the macro should work regardless of the table's location or size. You'll need to perform manual adjustment, so in the **Macro** dialog box, select the new macro and open the VBA module window by clicking the **Edit** button.

```
(General)                                    Transpose
Sub Transpose()
    ActiveWorkbook.Names.Add Name:="Source", RefersToR1C1:= _
    "=Transpose!R2C1:R6C9"
    Selection.Copy
    Selection.End(xlToRight).Select
    Selection.End(xlToRight).Select
    ActiveCell.Offset(0, 1).Range("A1").Select
    Selection.PasteSpecial Paste:=xlPasteAll, Operation:=xlNone, SkipBlanks
        False, Transpose:=True
    Application.CutCopyMode = False
    ActiveWorkbook.Names.Add Name:="Offset", RefersToR1C1:= _
    "=Transpose!R2C10:R10C14"
    Application.Goto Reference:="Source"
    Selection.ClearContents
    Application.Goto Reference:="Offset"
    Selection.Cut
    Application.Goto Reference:="Source"
    ActiveCell.Select
    ActiveSheet.Paste
    ActiveWorkbook.Names("Source").Delete
    ActiveWorkbook.Names("Offset").Delete
End Sub
```

Figure 22.10 Recorded Macro for Transposing a Table

In the lines with `Names.Add Name`, replace the previous range addresses with `RefersTo=Selection` after each name, without quotation marks. Then, you'll be able to use the corrected macro for any range. After that, in the VBA window, select **File • Save** and **Close and Return to Microsoft Excel** to switch back to the normal Excel window.

```
(General)                                    Transpose
Sub Transpose()
    ActiveWorkbook.Names.Add Name:="Source", RefersTo:=Selection
    Selection.Copy
    Selection.End(xlToRight).Select
    Selection.End(xlToRight).Select
    ActiveCell.Offset(0, 1).Range("A1").Select
    Selection.PasteSpecial Paste:=xlPasteAll, Operation:=xlNone, SkipBlanks
        False, Transpose:=True
    Application.CutCopyMode = False
    ActiveWorkbook.Names.Add Name:="Offset", RefersTo:=Selection
    Application.Goto Reference:="Source"
    Selection.ClearContents
    Application.Goto Reference:="Offset"
    Selection.Cut
    Application.Goto Reference:="Source"
    ActiveCell.Select
    ActiveSheet.Paste
    ActiveWorkbook.Names("Source").Delete
    ActiveWorkbook.Names("Offset").Delete
End Sub
```

Figure 22.11 Corrected Macro

Cell References in Macros

Visual Basic uses two types of addressing: *R1C1 notation*and *A1 notation*. The latter type is common when working with spreadsheets, while the less common R1C1 notation names cells by their row and column positions (R stands for *Row*, and C stands for *Column*). R3C2 means "row 3, column 2," which corresponds to B3.

Since A1 addresses are strings, you can modify them with string operators: instead of *A5*, you could write *A & 5*. The R1C1 notation allows you to elegantly display relative references. RC always refers to the current cell: R(1)C is one row down, R(-1)C is one row up, RC(1) is one column to the right, R(2)C(3) is two rows down and three columns to the right, and so on.

22.4 Macros for Chart Formatting

Since Excel 2010, you've been able to record chart formatting. The object model underlying Excel has been expanded accordingly, and this is a valuable enhancement, especially when you're applying a consistent design across a series of charts.

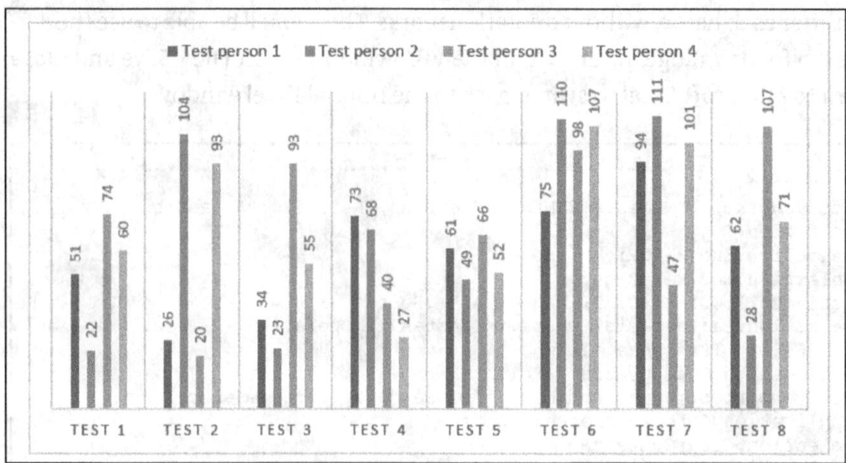

Figure 22.12 Original Format

The following code shows an example in which an existing chart on a worksheet is resized to specific dimensions and a gradient fill is applied to the chart area. The macro assumes there is exactly one chart on a worksheet. For this, the chart name automatically assigned by Excel, as in this:

```
ActiveSheet.ChartObjects("Chart 1").Activate
```

This is replaced by an index:

```
ActiveSheet.ChartObjects(1).Activate
```

```
(General)                                    DiaForm
Sub DiaForm()

    ActiveSheet.ChartObjects(1).Activate
    ActiveSheet.Shapes(1).Width = 500
    ActiveSheet.Shapes(1).Fill.Visible = msoTrue

    With ActiveSheet.Shapes(1).Fill
        .Visible = msoTrue
        .ForeColor.ObjectThemeColor = msoThemeColorAccent1
        .ForeColor.TintAndShade = 0.3399999738
        .ForeColor.Brightness = 0
        .BackColor.ObjectThemeColor = msoThemeColorAccent1
        .BackColor.TintAndShade = 0.7649999857
        .BackColor.Brightness = 0
        .TwoColorGradient msoGradientHorizontal, 2
    End With

End Sub
```

Figure 22.13 Formatting Chart with Macro

As a result, the chart looks like the one shown in Figure 22.14.

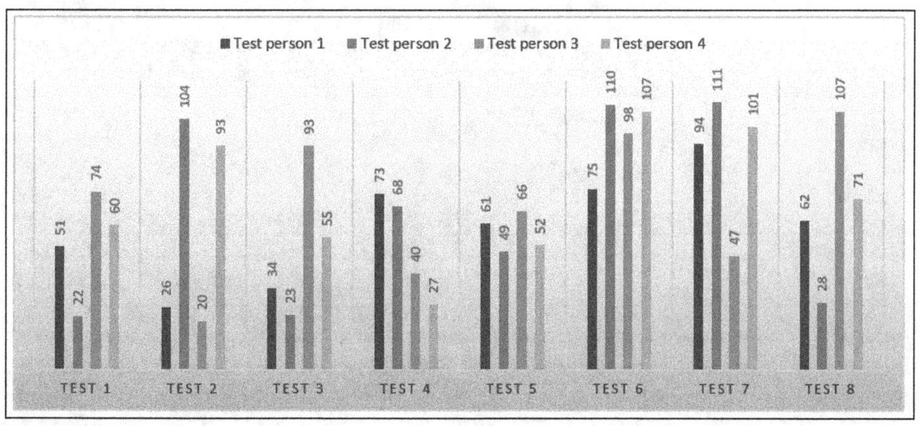

Figure 22.14 Newly Formatted Chart

22.5 Macros from Older Excel Versions

Macros in the Excel 97/2000 file format usually run without issues in the current Excel version. Macros in Excel 5 and 7 were also programmed in VBA, but they used a different programming environment. Specifically, macros are no longer displayed or edited in their own worksheets but in the VBA Editor window. Reverse translation is generally possible without major problems, but only if no objects, methods, or properties are used that the older VBA version did not support.

For example, this is not possible in Excel 95:

```
View = ActiveWindow.View
```

It can be annoying that the following does not work because it's often used in error routines:

```
Error = Err.Description
```

If you open a file with macros from Excel version 5 or 7, the module sheets disappear from the workbook. However, the modules only change location, and they're saved with the file and shown and edited in the VBA Editor's special module windows. If the macros were developed in the German VBA version, they are automatically translated into English. Although forms replaced dialog sheets starting with Excel 97, dialog boxes defined by dialog sheets remain functional.

Chapter 23
Visual Basic for Applications

Recording macros lets you condense workflows that you repeatedly perform manually in Excel into a single keystroke, mouse click, or tap. However, Excel's options in this area go far beyond that. With Visual Basic for Applications (VBA), you can program complete applications and thus create functions that Excel itself doesn't offer. VBA programs can even extend beyond Excel to integrate other applications.

23.1 Basics of VBA

VBA, as mentioned earlier, is a special version of Visual Basic (VB), which is a programming language that Microsoft developed for Windows and modeled on earlier Basic languages. The VBA development environment includes powerful tools for creating forms. Naturally, this book can only provide an introduction to the capabilities of this powerful language, so it provides a brief overview of the language, introduces the integrated programming environment, and shows some examples of using VBA within Excel.

23.1.1 The Excel Object Model

VB and VBA share with other object-oriented programming languages the concept that applications work with objects and are themselves made up of elements that can be addressed as objects. The definition of a specific object type is called a *class*, and each object has a specific set of properties and a set of procedures to work with it, which are called *methods*. For example, an object can be a cell range, a PivotTable, or a chart. A cell range has properties such as being bold formatted. You can query or modify properties and when you copy a range, the Copy method is applied to it. 23

The objects that an application provides and their relationships are defined by the object model—which is a predefined hierarchy where the *application* is the top-level object. You can view the object mapping through the object catalog interface, and different icons in the catalog indicate whether an item is an object, a method of the object, a property of the object, or an event associated with the object (see Figure 23.1).

To work confidently with Excel's objects or those of another Office application that you access from Excel, you need to understand and navigate the object model that underlies each application. The VBA object model has expanded with each new Excel version to support the program's enhanced features. For example, new elements have been added for new chart types, table functions, or working with data models.

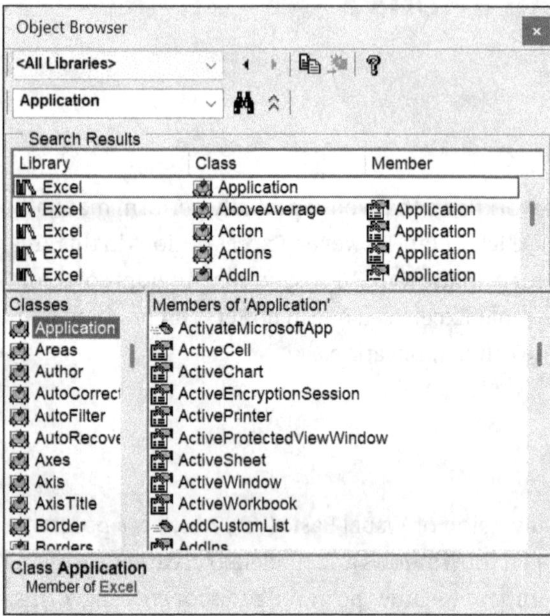

Figure 23.1 Object Catalog for Excel

The object models resemble Russian nesting dolls. The top-level object is the application itself, and all other objects are contained within this main object. The next level down in Excel is the workbooks, and within workbooks, different collections of peer objects exist: worksheets, charts, and modules. You can access individual items in these collections by index or assigned name. `Worksheets(3)` and `Worksheets("Forecast")` are examples of how to reference an element in a collection of objects.

It can be confusing that in VBA, the same name is sometimes used for both an object and a property. For example, `Legend` is a property of a `Chart` object, and this property returns another object—the `Legend` object, which is a subobject of the `Chart` object. That's why a property that returns an object is often used in code instead of the object itself.

Three questions will frequently arise when you're working with objects:

- How do you correctly reference an object or subobject?
- What properties does an object have?
- Which methods can be used under which conditions?

For example, to fill the interior of a cell range with a different color in an Excel procedure, use this line of code:

```
Worksheets("Table1").Range("A1").Interior.Colorindex = 3
```

In plain language, this statement means, "Set the property `Colorindex` to `3` for the object `Interior` (the cell background) within the `Range A1` object, which is part of the object named `Table1` in the `Worksheets` collection."

The development environment offers handy tools—especially the object catalog and various editing aids that are available during program entry—to help you work with objects. Controls are also treated as objects, and you can easily drag them into a form. For example, if an application needs a list box to select a specific value—like a country name—you can add the entire list box to the appropriate form. The only remaining question is how to link the list box to the country names and where they should appear.

23.1.2 Events Control the Program Flow

The second key feature of VBA programs is that an application's flow usually doesn't follow a fixed plan but is driven by events. When you click a button, something happens; when a worksheet is activated, certain calculations run automatically. A file can be opened, saved, or closed. A worksheet can be activated, changed, or recalculated, and a window can be activated, deactivated, or resized.

Programming with VBA usually follows these main steps:

1. Designing forms that users employ for data exchange and interaction with the application
2. Selecting the properties of the controls used in those forms
3. Programming the actions the program performs when specific events occur with these controls or other objects

23.1.3 Variables and Constants in VBA

In VBA, as in any programming language, variables store specific values that are created during program flow to allow their reuse elsewhere as needed. To differentiate values, you give variables names. You also specify the type of information a variable holds—such as numbers or text—by using a data type that defines the maximum size or length of the stored data, as well as the available methods.

You can declare variables implicitly or explicitly. To do it implicitly, you use the data type `Variant`, which is especially flexible because it accepts any kind of data. However, you must be careful about what data is actually assigned to a variable of this type before working with it.

Declaring Variables Implicitly

You can implicitly declare a variable directly within a procedure by assigning it a value. Take these two lines as an example:

```
value1 = 3
value2 = "xyz"
```

They define two variables named `value1` and `value2` and assign them the values `3` and `xyz`, which are a number and a string. This simple way of defining variables is very convenient. You declare a variable exactly when you need it, which lets you work quite intuitively. The data type (number, text, etc.) for the variable is not fixed, and that makes the definition very flexible. You can use the same variable within a procedure as a number at one point and as a string at another.

However, this convenience has two drawbacks. First, a variable defined this way is only valid within the procedure where it's defined, so you can't access its values from other procedures. Second, in larger programs, it's easy to lose track of where variables are used. Still, this method is useful for quickly developing smaller programs.

Declaring Variables Explicitly

You can explicitly declare variables at the start of a procedure or module—under *declarations*—by using statements like `Dim`. In this case, you explicitly set the data type unless you're using the `Variant` type. This form of variable declaration is the cleaner programming solution for larger projects because it allows clear work that you can understand later. VBA provides the variable types listed in the following table.

Variable Types	Meanings
`Boolean`	Boolean values (TRUE or FALSE) or numbers, where 0 means FALSE and any other number means TRUE
`Byte`	0–255
`Integer`	Whole numbers from –32,768 to 32,767
`Long`	Whole numbers from about –2 billion to +2 billion
`Single`	Floating-point numbers from 1.4E–45 to 3.4E38 for positive values and from –1.4E–45 to –3.4E38 for negative values
`Double`	Floating-point numbers from 4.9E–324 to 1.79E308 for positive values and from –4.9E–324 to –1.79E308 for negative values
`Decimal`	Twenty-eight digits before and after the decimal point
`Currency`	Numbers from –9.22E15 to 9.22E15 rounded to four decimal places
`String`	Character string (up to about 2 billion characters)
`String * Length`	Character string with fixed length, with longer strings truncated to the specified length, from 1 to 65,400

Variable Types	Meanings
Date	Serial numbers (as used in table functions for date and time) or date and time values written between #...#
Object	References to an object
Variant	Numeric values (such as *double*) or strings
LongLong	Stores signed 64-bit numbers; the value range is between –9,223,372,036,854,775,808 and 9,223,372,036,854,775,807

The following procedure demonstrates a simple exercise that assigns variables to different data types:

```
Sub variable_chain()
  Dim logical_value As Boolean
  Dim amount As Currency
  Dim note As String
  Dim dateValue As Date
  logical_value = 77 * 13 < 12 * 84
  amount = 10
  note = "Playing around with data types"
  dateValue = #7/30/2024#
  MsgBox note & " on " & dateValue & ": " & " Bet " & amount  _
       & " Euros, that this " & logical_value & " is."
End Sub
```

The procedure displays a message box in which the values of different variables are concatenated into a text string using the & operator.

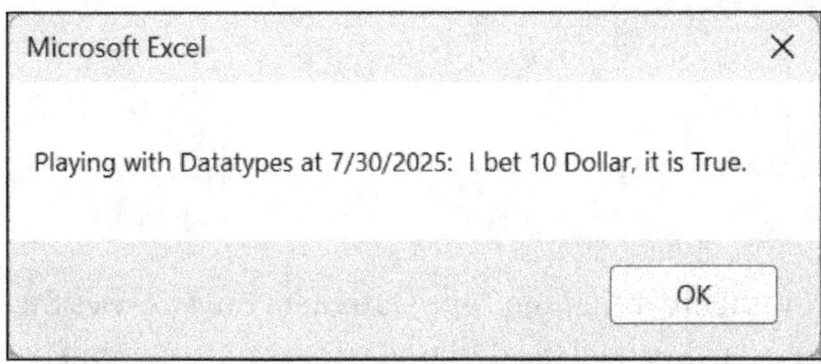

Figure 23.2 All Variables Concatenated into Text String

This concatenation shows the flexibility of variable definitions in VBA: since the & operator normally concatenates only strings, VBA apparently treats all variable types as strings here.

Arrays with Multiple Dimensions

Variables can also be defined as *arrays* that are accessed by indexes. This lets you define many similar variables with a single variable declaration. However, this only makes sense if the arrays actually contain similar variables, like a matrix in Excel. You access individual array elements by their indexes:

```
Dim regions() As String
regions = Array("North","East","South","West")
region = regions(0)
```

You can set whether indexes start at 0 or 1 by using the `Option Base` command. The default is 0, and you must place the `Option Base 1` statement at the start of a module, before you declare any data arrays.

Custom Variables

Defining your own complex variable types—which are commonly called *records*—is especially useful because it lets you assign an entire data set (such as an Excel table) to a single variable name. Here's how you do it:

```
Option Base 1

Private Type contacts
    name As String
    email As String
    webaddress As String
End Type

Sub dataset()
  Dim contact As contacts
  contact.name = "Jan Willem"
  contact.email = "jw@net.de"
  contact.webaddress = "www.jw.net"
  MsgBox contact.name & " " & contact.email
End Sub
```

When defining custom data types, remember that the Type statement only defines the data type itself, not a variable. The actual variable is declared with the `Dim` statement, and this declaration can occur at both the module level and the procedure level. The procedure shows how to access individual elements of variables.

Variable Scope

All variables have a scope that defines which parts of the program they apply to and when they are removed from memory. A VBA application usually consists of a collection of modules and forms, and the modules are divided into individual procedures. A variable that's declared within a procedure is typically usable only within that procedure, and you cannot use such variables to pass values to another procedure or module. The exception is variables that are declared with the `Static` keyword. This allows you to use the variable beyond its normal scope:

```
Static variableName As dataType
```

To use a variable in all procedures of a module, declare it within the module's *declaration section*. This section is automatically available in the module's **Module** window. To use a variable throughout the entire application, declare it as a global variable. Use the `Public` keyword, for example:

```
Public variableName As dataType
```

User-defined variable types must be declared at the module level (that is, not within procedures); their scope can be limited to the module, or it can extend across all modules.

Special Case: Object Variables

VBA code primarily involves manipulating objects, as mentioned earlier, and to reference objects, you assign object variables to them. Like other variables, object variables have different types, and for object variables, these types correspond to different object classes. In Excel, for example, the `Workbook`, `Worksheet`, and `Range` classes exist. If you don't know the object type beforehand, you can use the generic `Object` object type.

To assign or retrieve specific properties of an object, you can use an object variable instead of the object itself. The same applies when you're using methods of an object. You assign an object variable to an object with the `Set` statement. Here is a simple example:

```
Dim topRange As Object
Set topRange = Worksheets(1).Range("Toplist")
topRange.Copy
```

First, declare the `topRange` object variable, and then, assign the variable a `Range` object from a worksheet. The last command applies the `Copy` method to the `Range` object that's represented by the `topRange` object variable. Using object variables simplifies references to frequently used objects and thus reduces the code you need to write.

Choosing the Variable Name

VBA gives you great freedom when naming variables. The name must start with a letter and cannot include periods, and within limits, you can even use terms from the VBA vocabulary as variable names.

Despite this freedom, you should avoid certain practices. Using VB keywords only causes confusion because you'll soon lose track of which variables are your own. For the same reason, avoid naming variables so they resemble VB keywords: `MyText` might look like a nice name, but typographically, it's not immediately clear that it's a variable. For these reasons, it's best to write all variable names consistently.

Like other Basic dialects, VBA lets you add a type indicator at the end of a variable name to specify its type. For example, a dollar sign indicates that a variable is a string. With the `Dim` statement, you must not write ... `As` ... but only the variable name.

Use Descriptive Names

While you can keep variable names short—`fn` and `ln` for first and last names are quicker to type than long names—such abbreviations often become unclear over time.

Comments

It's best to add comments to longer code so it remains understandable later. Text following an apostrophe is always treated as comments, which the program ignores.

User-Defined and Built-In Constants

When certain values are repeatedly needed in an application, it makes sense to declare them as constants. Here, you must consider the scope, just like with variables. If constants should apply to the entire application, use the `Public` keyword. A simple example is the two VAT rates:

```
Public Const VAT1 = 7%
Public Const VAT2 = 5%
```

This lets you use these two values throughout the entire application. If the VAT rate changes, you should update only the value in the constant declaration—not every calculation involving the VAT rates.

Besides user-defined constants, many built-in constants are available and are assigned to individual Office objects, mainly for setting properties. You can see which ones are predefined in the object catalog. Otherwise, they are used like other constants.

23.1.4 Basic Units and Language Elements

Larger VBA applications usually consist of multiple modules and often some related forms. The modules themselves consist of individual procedures. The art of programming lies in skillfully breaking down an application's tasks into small subtasks so procedures can be reused whenever possible to reduce overall programming effort.

Procedures and Functions

The smallest unit of programs in VBA is the procedure. There are three types: Sub procedures, Function procedures, and Property procedures.

```
Sub ProcedureName()
   ... statements
End Sub

FunctionName (Arguments)
   ... statements
End Function

Property Get ... Let ... Set
   ... statements
End Property
```

You'll only use Property procedures when you want to define properties while programming classes. Get reads the property's value, Let assigns a value to a property, and Set also assigns a value to a property but includes an object reference. Classes act as templates for objects, and objects are instances of classes. Since Excel 97, users have been able to program their own classes and work with object instances of those classes. However, this topic goes beyond the scope of an introduction, so we won't cover it here.

Unlike Sub procedures, Function procedures return a value that can be used later in the program flow. Within such a Function-procedure, not all statements allowed in a Sub procedure can be used.

Procedures and functions can be called from various points within the program flow. For Sub procedures, the procedure name and any necessary arguments are sufficient. If the procedure is in a different module, the module name must be prefixed:

```
ModuleName.ProcedureName Argument1, Argument2, ...
```

The process is slightly different for functions. The value returned by a function is assigned to a variable so the result can be used later in the program.

```
VariableName = ModuleName.FunctionName(Argument1, Argument2, ...)
```

You usually define a function along with the arguments passed to it. You can also specify the data type of the value the function returns. If a function requires no arguments, you still need to include the parentheses. When there are multiple arguments, separate them with commas inside the parentheses. Note that not all VB keywords can be used in functions. For example, you cannot assign a specific font to a cell through a function. You can employ user-defined functions not only within programs but also directly in the worksheet.

The following two listings show a function that calculates the circumference of a circle and its call within a procedure that first prompts for the radius:

```
Public Function Circumference(Radius)
   Circumference = Radius * 2 * Application.WorksheetFunction.Pi
End Function
Sub calculateCircumference()
  Radius = InputBox("Radius")
  MsgBox("The circumference of the circle is " & Circumference(Radius))
End Sub
```

Program Structure

Based on what we've said so far, we can roughly distinguish three main parts of a VB program:

- **General section**
  This includes general settings (like `Option Base`), variable declarations (`Dim ...`), and definitions for custom variable types (`Type ...`).
- **Sub procedures**
  These form the main body of the program, which contains the instructions the program executes.
- **Functions**
  Functions are defined to calculate specific values needed in the program.

A VBA module (programs spanning multiple modules are beyond the scope of this introduction) generally looks like this: it begins with the general section, which includes the program's title, description, and general notes. This section also contains the declaration of variables that apply to the entire module and the definition of custom data types.

Next come the functions, and procedures within the module access them. These functions could be placed anywhere, but grouping them in one block makes the module easier to navigate. Short comments (starting with an apostrophe) improve clarity.

The module's procedures come last, and short comments again help maintain clarity.

Objects and Their Properties

Working with objects plays a central role in VBA programming. In general, the program specifies what should happen with which objects under what conditions over time. For example, a program might select a specific worksheet from a list of sheets (an object), modify the cell contents (properties), activate a dialog box (an object), and read and process inputs (properties) in certain dialog box elements.

If you need to get the properties of an object in a program line, you usually assign the result to a variable so you can work with it further. For example, to find out the value of a cell in a worksheet of a workbook, you can write this:

```
value1 = Range("F7").Value
```

Excel treats the value entered in a cell as a property of the `Range` object, and the object name and the property are separated by a period. Instead of assigning it to a variable, you can also assign it to a dialog box, like this:

```
MsgBox Range("F7").Value
```

To assign a different value to the cell—that is, to change the `Value` property—the syntax you use is as follows:

```
Range("F7").Value = "John Haenks"
```

To assign the value of cell F17 to cell F7, write this:

```
Range("F7").Value = Range("F17").Value
```

Most properties can be read and modified, while some are read-only. Many properties use built-in constants to set their values.

Using Excel Constants

Excel internal constants often determine properties. They usually start with `xl` (for *Excel*), `mso` (for *MS Office*), or `vb` (for *Visual Basic*). These constants can be used in many contexts, and they typically represent codes that are otherwise difficult to understand. When recording macros, they are used automatically. Using them in programming makes a program clearer but is somewhat more complex.

Using Methods

When using a method that's applicable to a specific object, you can use different notations. Most methods require certain arguments that specify how the method should be applied. Usually, some of these arguments are necessary, while others are optional. Some methods, however, work without any arguments.

For example, the following method applies the `AutoFit` method, to the three columns of the first sheet in a workbook and automatically adjusts the column width:

```
Worksheets(1).Columns("D:F").AutoFit
```

Some methods use default settings when no arguments are provided.

Named Arguments

If a method has multiple arguments, you have two options. The first is to provide all arguments in the correct order or at least include placeholders for any omitted arguments. For example, the SaveAs method, which saves workbooks, has nearly a dozen arguments. You could use it like this:

```
Workbook.SaveAs(FileName, FileFormat, , WritePassword ...)
```

However, this input method is cumbersome and prone to errors. The alternative is to use named arguments, where only the arguments with specified values are listed and each is named individually and separated from others by commas. The order doesn't matter, which is a big relief!

This would be an example:

```
Workbook.SaveAs Filename:="Plan2022.xlsx", WriteResPassword:="secret"
```

The VBA Editor helps you select argument names, as you'll see in the following sections.

Methods with Return Values

Methods can provide feedback when you use them in your program. For example, if you apply the CheckSpelling method to a cell range to perform a spell check, you can check the result. If no error is found, the method returns True, which you can assign to a variable. Here's how you use it:

```
correct = Worksheets("Plan_2022").CheckSpelling(IgnoreUppercase:=True)
```

As you can see in the line, the method's arguments are enclosed in parentheses this time. This is necessary when you want to use a method's return value.

Methods that Return Objects

Certain methods return a new object themselves. In the following statement, for example, the method Offset returns a new object, which has a subobject (Range), to which the method Select is then applied:

```
ActiveCell.Offset(1,0).Range("A1").Select
```

This method also returns a new object, namely Selection, to which another method like Copy can be applied.

Because many methods return objects that can be handled like objects within a program, you can assign these returned objects to object variables and continue working with them in your code. Here is an example with the following lines:

```
Dim SelectionRange As Object
Set SelectionRange = Selection.Offset(5, 2)
SelectionRange.Copy
```

An object is assigned to the object variable, after which the Copy method can be applied to the variable.

Range Object

Ranges are among the most commonly used objects in a worksheet. This object has over 170 properties and methods available in the current object model. Here's an example of how they're used:

```
Sub rangeColoring()
    Sheets("Table1").Select
    Range("A1:B12").Select
    With Selection.Interior
        .Pattern = xlSolid
        .PatternColorIndex = xlAutomatic
        .Color = 49407
        .TintAndShade = 0
        .PatternTintAndShade = 0
    End With
End Sub
```

First, the `Select` method is applied to `Sheets()`, which is the list of sheets, and then to `Range()`, which is a range. Both are objects, with `Sheets` being a subobject of a workbook (e.g., *ActiveWorkbook*) that contains subobjects—which are the individual sheets of the workbook. The selected `Range` is also a subobject of the selected sheet. The interior of the selected range is then filled with color.

Using Operators

In many previous examples, operators were used without further explanation. This was fine because the operators we've used so far closely matched those you've already seen when creating formulas in Excel worksheets. However, since they aren't exactly the same and operators are often needed when setting conditions for program flow, here's a brief overview. VBA distinguishes between arithmetic, logical, comparison, and concatenation operators:

- **Arithmetic operators**
  These are `+`, `-`, `*`, `/`, and `^`, and they're the same as those used in tables. The `\` operator is the one for integer division (`5\3` returns `1`), and `Mod` returns the remainder of such a division (`5 Mod 3` returns `2`).

23

- **Logical operators**
  These (except `Not`) link expressions that return Boolean values and also return Boolean values. Their function is best understood through truth tables.

	A	B	C	D	E	F	G	H
1	**Truth Values**							
2								
3	a	b	not a	a Eqv b	a Imp b	a Or b	a And b	a XOr b
4	TRUE	TRUE	FALSE	TRUE	TRUE	TRUE	TRUE	FALSE
5	TRUE	FALSE	FALSE	FALSE	FALSE	TRUE	FALSE	TRUE
6	FALSE	TRUE	TRUE	FALSE	TRUE	TRUE	FALSE	TRUE
7	FALSE	FALSE	TRUE	TRUE	TRUE	FALSE	FALSE	FALSE

Figure 23.3 Truth Table for Logical Operators

  In this table, `a` and `b` are expressions that return Boolean values. Assigning these expressions a value of null (i.e., no value) is not considered here. The operators follow these rules:

  - The `Not a` operator is true if `a` is false and false if `a` is true.
  - The `a Eqv b` (a is equivalent to b) operator is true if `a` and `b` have the same Boolean value.
  - The `a Imp b` (a implies b) operator is true if `b` is true or if `a` is false.
  - The `a Or b` operator is true if at least one of the two expressions is true.
  - The `a And b` operator is true if both expressions are true.
  - The `a XOr b` (a exclusive or b) operator is true if exactly one of the two expressions is true.

  When logical operators are combined, related elements must be enclosed in parentheses, just like with arithmetic operators. If you don't want to memorize the rules, you can simply enclose everything by default.

- **Comparison operators**
  These compare values. They are exactly the same as those used in tables: `<`, `>`, `<=`, `>=`, `=`, and `<>`. They represent "less than," "greater than," "less than or equal to," "greater than or equal to," "equal to," and "not equal to." With numbers, the meaning is clear; with strings, `>` means "later in alphabetical order" (so `a > b` is false).

- **Concatenation operators**
  These are `&` and `+`. You can use `&` to join strings (as in tables). However, you should not use `+` as a concatenation operator.

Statements and VBA Functions

Statements generally control program flow. Declaring and defining variables, setting the start and end of a procedure or function, and constructing branches and loops—all of this is done through statements. Besides the functions you can program yourself,

VBA offers many built-in functions that also serve to return or convert values. Generally, values are passed to a function, and the function returns values. Some of these functions perform tasks similar to their corresponding table functions, such as the mathematical functions: Abs(), Cos(), Exp(), Log(), and others. In addition to mathematical functions, there are built-in functions that handle date and time values, manipulate or evaluate strings, and more.

Controlling Processes with Branches and Loops

Regardless of the programming approach you use, many work scenarios within a program can be seen as branches or loops for you to consider. For example, a program should respond differently to various inputs: If the input is this, then take this action; if it's different, then take that action, and so on. A program should perform a specific activity until the user takes a certain action or a specific event occurs. This means that as long as something is not true, this activity occurs; once it becomes true, a different activity takes place.

Alternatively, a program should perform a certain activity at a specific frequency, meaning it should count how many times the activity has been performed, and once the specified number is reached, it should proceed with other (or no) activities.

If-Then

A common task is responding to the occurrence of a situation. In VBA, this task is usually handled with a simple If-Then structure. For example, if a function includes an optional argument, this branching controls what happens when the argument is missing, such as assigning a default value:

```
Function Week(Optional d As Date)
  If d = 0 Then
    d = Date
  ...
End If
```

23

You can extend such structures with an Else branch. Here's another example from a function we'll introduce later:

```
If d7 <= 31 Then
  Date = CDate(CStr(d7) & ". 3. " & CStr(Year))
Else
  Date = DateValue(CStr(d7 - 31) & ". 4. " & CStr(Year))
End If
```

Since sub-branches can also use Else-If, you can create arbitrarily complex decision structures in a program, but this will quickly cause you to lose track. For complex branches, check if the problem can be solved by using the next structure (Select Case).

Select Case

If branching depends on the value of an expression, you can use a structure that provides different responses for different values. In the following example, the user can select different periods for a calculation. The function uses a Select Case branch that is very clear:

```
Select Case periode
      Case "Years"
         fn = 1
      Case "Months"
         fn = 1 / 12
      Case "Days"
         fn = 1 / 360
End Select
```

This type of branching always stays clear. Of course, clarity can be lost through nesting within this structure, but that won't be shown here.

While

A common programming task is to run a section of code repeatedly until a certain condition is met. For example, this occurs when user inputs are read into a list or an array variable until the user ends the process, or when an operation continues until a specific value is reached. The following loop runs as long as the variable diff has a value less than 0:

```
While diff < 0
    attempt = NomIrate(m, erg, bab)
        diff = attempt - nz
    erg = erg + 0.1
Wend
```

Inside the loop are all the instructions needed to execute in this case.

Do-Loop

These tasks can be solved especially elegantly with a similar but more flexible structure. You can place the loop condition at the beginning or the end of the loop (in the previous example, the condition must be at the beginning). You can also set exit conditions within the loop. The example can be rewritten as follows:

```
Do While diff < 0
    attempt = nomrzins(m, erg, bab)
        diff = attempt - nz
    erg = erg + 0.1
Loop
```

For-Next

Use a `For-Next` loop when you need to repeat a statement or a series of statements a specific number of times. Here's an example that reads sheet names into an array:

```
count = Sheets.Count
Dim blist() As String
ReDim blist(count)
For i = 1 To count
    blist(i) = Sheets(i).Name
Next i
```

First, the number of sheets is assigned to the `count` variable, which allows the initially variable `blist` array to be resized. The loop can then run up to the determined number to write the sheet names into the array.

With this overview of VBA program structures, the general introduction to VBA is now complete. The next section introduces the VBA development environment.

23.2 The Development Environment

VBA programming takes place in a separate window with its own menus and toolbars. Within the VBA Editor window, additional windows open as needed. This happens partly automatically when you select certain objects or when you select the **View** menu or click the corresponding toolbar icons. Excel continues to use the traditional interface with menus and freely movable toolbars. Clicking the **Help** icon opens the developer reference and provides access to detailed support for VBA development.

[«]

Save Often

A project and its related elements are automatically saved when you save the workbook it belongs to. However, since software development can be time-consuming, it's wise to save your work regularly to avoid losing valuable progress due to, for example, a power outage. Use **File • ... save** in the editor window or press Ctrl + S.

23.2.1 Project Explorer and Module Window

The Project Explorer is the control center where you manage your project development. It shows the files and objects in a project and lets you select them, view the selected object or its code, or open a new window for it. The Project Explorer window provides a quick overview of the project's progress and access to all its elements.

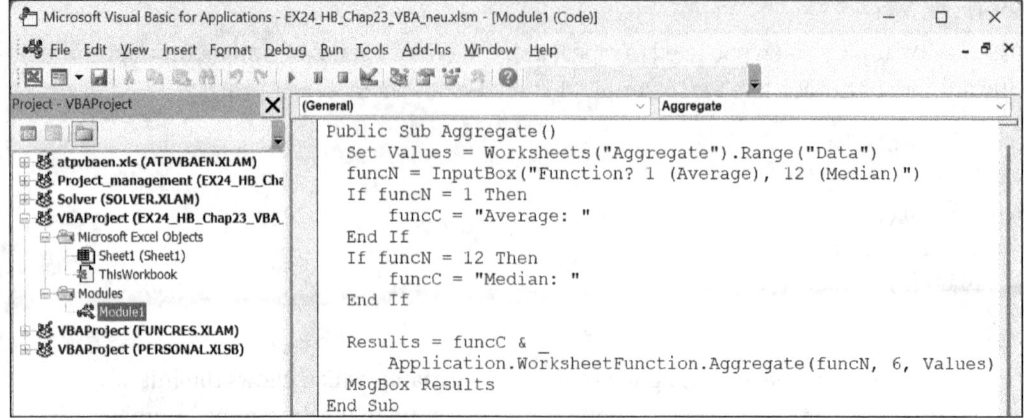
Microsoft Visual Basic for Applications - EX24_HB_Chap23_VBA_neu.xlsm - [Module1 (Code)]
File Edit View Insert Format Debug Run Tools Add-Ins Window Help
Project - VBAProject
atpvbaen.xls (ATPVBAEN.XLAM)
Project_management (EX24_HB_Cha
Solver (SOLVER.XLAM)
VBAProject (EX24_HB_Chap23_VBA_
Microsoft Excel Objects
Sheet1 (Sheet1)
ThisWorkbook
Modules
Module1
VBAProject (FUNCRES.XLAM)
VBAProject (PERSONAL.XLSB)
(General)
Aggregate
Public Sub Aggregate()
Set Values = Worksheets("Aggregate").Range("Data")
funcN = InputBox("Function? 1 (Average), 12 (Median)")
If funcN = 1 Then
funcC = "Average: "
End If
If funcN = 12 Then
funcC = "Median: "
End If
Results = funcC & _
Application.WorksheetFunction.Aggregate(funcN, 6, Values)
MsgBox Results
End Sub

Figure 23.4 Project Explorer and Module Window Within VBA Window

A separate tree is created for each project, and it lets you work on multiple projects simultaneously without losing track. You can create a project for each open workbook in Excel, and the project will then belong to that file. You can select code elements and move or copy them within the project or to another project. If the file contains elements that are linked to program code, such as worksheet or chart sheets, icons for these will appear in the tree structure. As the project progresses, icons for custom forms, program code, class modules, and references to other Office documents are added.

The **Forms** and **Modules** subfolders are created only when you insert the corresponding elements. When you record macros, they are initially inserted as procedures in **Module1**. If you later distribute macros across different modules to organize them by specific criteria, it won't affect how the macros function. To change the default project name, select **Properties of** from the context menu and enter a suitable name.

Export – Import: Program Code for Multiple Projects

To make forms or program code available for other projects, you can save the data in separate files. Use **File • Export File** to create files with the *.frm* extension for selected forms, *.bas* for Basic code files, and *.cls* for classes. To import this data into other projects, use **File • Import File** and select the location in the Project Explorer where you want to insert the file. To insert existing code lines at a specific spot in a module, use **Paste • File**.

Window for the Program Code

Code is written in special windows, and VBA provides separate windows for each workbook, sheet, module, and project form. For example, if you double-click the workbook icon in the Project Explorer, the VBA Editor will open a module window. The right list box will show a procedure skeleton for every event that's available for the `Workbook` object when you select **Workbook** in the left list box.

The module window always has two list boxes below the title bar, which displays the module name and its associated document. The left list box lists the objects ❶, and the right list box lists the individual procedures ❷ belonging to the module (see Figure 23.5). In the bottom left corner (not shown), you can use the two small buttons to choose whether to display only one procedure or all procedures of the module sequentially in the window.

For example, if you want to specify what should happen each time the active document opens—such as selecting certain settings—then enter the commands in the `Open` event procedure. To color the gridlines red, these lines are sufficient:

```
Private Sub Workbook_Open()
    ActiveWindow.GridlineColorIndex = 3
End Sub
```

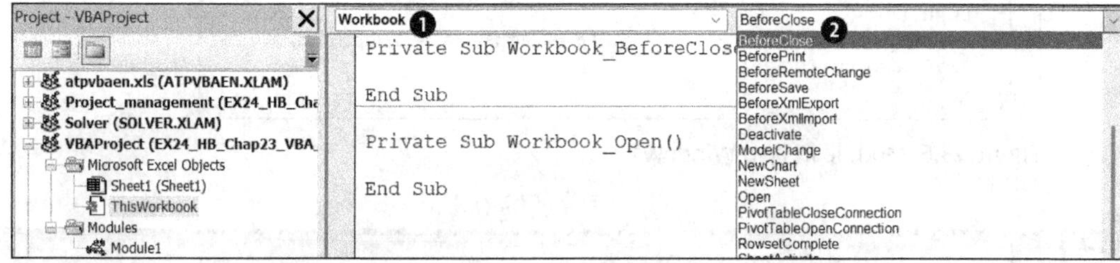

Figure 23.5 Procedure Skeletons for Events

Editing in the Module Window

Working in a module window is much like using a simple text editor. You can write, paste, copy code, etc. But two differences are immediately clear: there is no automatic line wrap for longer lines, and when you complete a row by pressing `Enter`, its typographic appearance may change. Colors might shift, some lowercase letters may become uppercase, spaces may be added, and occasionally, a message may appear. This occurs because automatic syntax checking is enabled by default.

Within a module, you can position the insertion point by clicking or tapping. By pressing the `Insert` key, you can toggle back and forth between insert mode (which is the default) and overwrite mode. You can also navigate easily within a module window by using the arrow keys, and combined with pressing `Ctrl`, you can use them to move word by word left and right and paragraph by paragraph up and down. To select text while moving, hold down the `Shift` key. With the mouse, you can select text by dragging over it while holding down the left button. You can cut or copy selected text by using the familiar icons or menu commands and then paste it elsewhere.

You'll appreciate the copy option when you realize that VBA often requires writing many long lines with only slight variations. Copying and then making small corrections saves a lot of work.

Fortunately, you can split the module window to view different ranges of a larger module at once. This makes copying and pasting program text, comparing procedures, and checking variable definitions easier. Using **Window • Split** splits the module window horizontally in the middle; you can drag the split line up or down with the mouse or finger.

```
(General)                                        Aggregate
Public Sub Aggregate()
  Set Values = Worksheets("Aggregate").Range("Data")
  funcN = InputBox("Function? 1 (Average), 12 (Median)")
  If funcN = 1 Then
      funcC = "Average: "
  End If
-------------------------------------------------------------
  If funcN = 12 Then
      funcC = "Median: "
  End If

  results = funcC & _
      Application.WorksheetFunction.Aggregate(funcN, 6, Valu
  MsgBox results
```

Figure 23.6 Module in Two Windows

Splitting Long Lines

Since VBA commands can quickly become quite long, it's important to know how to manage them. It's confusing when parts of a command aren't visible in the window. You can split a line into multiple lines by adding a space and an underscore at the end of each line. These two characters tell Excel to treat the next line as part of the same command.

Font Formatting as a Programming Aid

You can freely choose the module's typographic design, especially the colors assigned to different text elements. You can also specify the font type and size. You access these options in the VBA window via **Tools • Options** on the **Editor Format** tab.

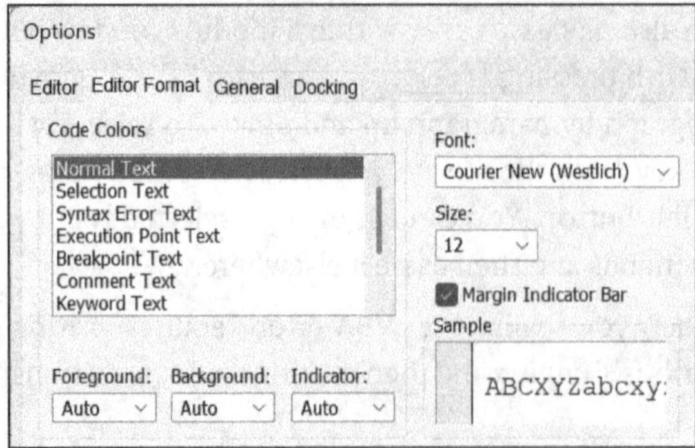

Figure 23.7 Dialog Box for Editor Formatting

Options for Code Input

More key options are found under the **Editor** tab., where the upper section partly relates to typographic display. You can turn automatic indentation on or off via **Auto Indent**, which works by setting tabs at the start of a line and then indenting all following lines accordingly. This makes individual program sections easier to read.

You can set the **Tab Width** separately; for deeply nested code, a setting of 2 is ideal to avoid lines being indented too far to the right. You can undo indentation by pressing `Shift`+`Tab`.

You can also uncheck **Auto Syntax Check** to prevent distracting error messages. This is useful if you're working with incomplete code lines or copying snippets from different sources.

The preset option for **Require Variable Declaration** is important, because it removes the ability to define variables simply by assigning a fixed value without the `Dim` statement when needed. This promotes good programming style but makes programming somewhat more tedious. You can achieve the same effect with the `Option Explicit` statement at the start of a module, but this setting will only apply to that specific module.

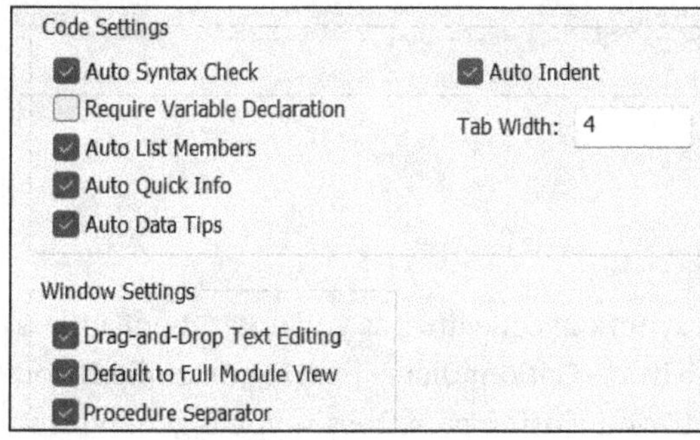

Figure 23.8 Settings for VBA Editor

23

Window Optimization

A not entirely unimportant skill when using the VBA Editor window is knowing how to manage the many sub-windows that appear. You can freely move each window around the workspace, and as a helpful feature, Excel allows you to dock these windows to specific sides of the main window. This is usually a good idea, as having many open windows can quickly lead to a cluttered workspace. You can specify which windows are dockable on the **Docking** tab in the **Options** dialog box (see Figure 23.9).

You can drag each window by its title bar to the edge of the window where you want it to be. For example, it's convenient to dock the windows for projects and watch expressions so you always have all important data in view. Drag the borders between windows to adjust the layout as you wish; the other windows will resize automatically. Use the **Close** button to hide individual windows, and the remaining windows will fill the freed-up space (see Figure 23.10).

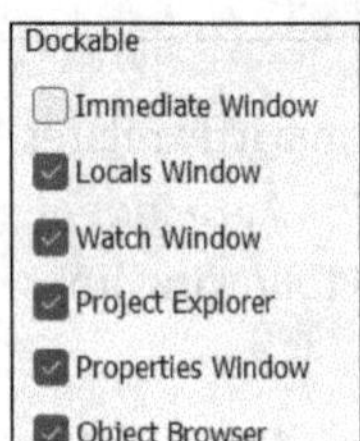

Figure 23.9 Windows That Can Be Docked

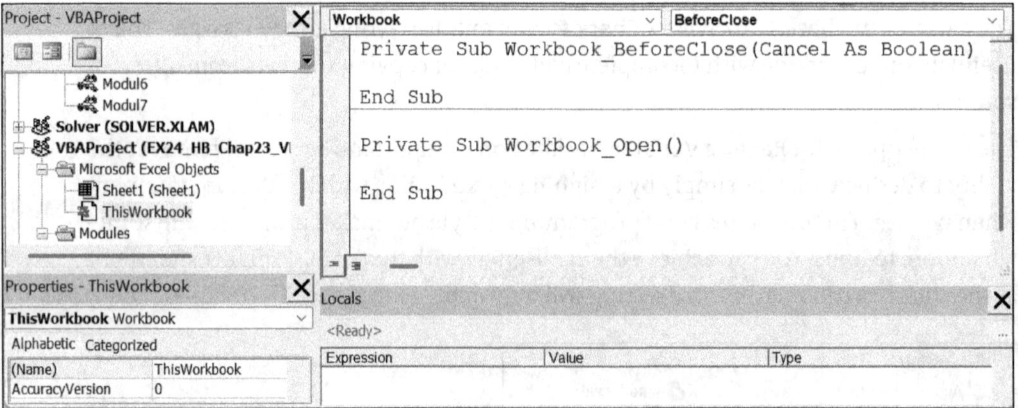

Figure 23.10 Example of Anchored Windows

23.2.2 Editing Aids

The VBA Editor immediately checks the syntax of code lines as you enter them, unless this behavior is disabled on the **Editor** tab in the **Options** dialog box. For example, if you forget a comma between two arguments or a closing parenthesis, you'll get an error message.

Use Lowercase Letters When Typing

Since reserved keywords in VBA always start with a capital letter, you should enter keywords entirely in lowercase. If a word doesn't appear with a capital letter at the start after you press Enter, you may have made a typo or used the wrong word.

VBA offers convenient tools for working in a module window. The **Edit** toolbar contains several icons for this, and you can show it by selecting **View • Toolbars**. When you've entered enough characters for VBA to clearly identify a word, VBA can complete it automatically.

For example, if you type "App" and then click the **Complete Word** icon, a small list of possible words will appear and you can double-click or tap the word you want to complete it.

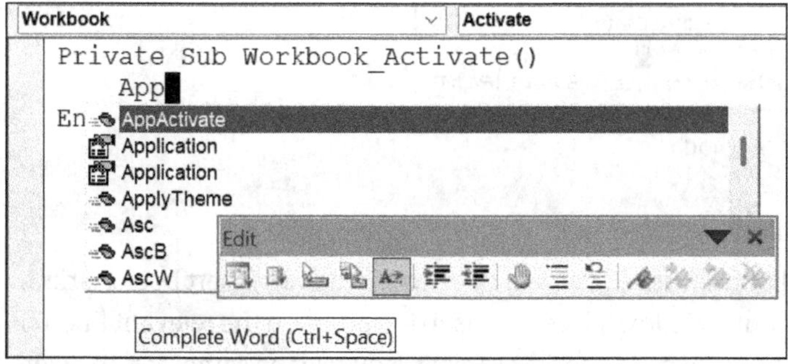

Figure 23.11 Edit Toolbar and Complete Word Icon

Selecting Properties and Methods

When you enter an object's name followed by a period, a list of available methods and properties will appear. You can then type another letter to filter the list to show only properties and methods that start with that letter or a later letter alphabetically. Properties are marked with a hand icon; methods are shown with a flying package icon. You can select an entry by double-clicking or double-tapping it.

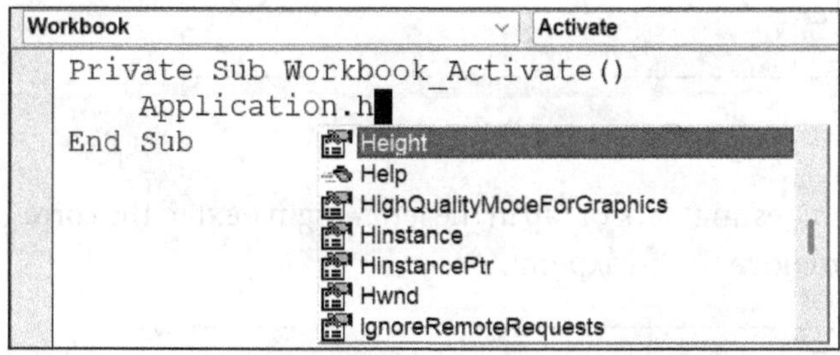

Figure 23.12 List of Object's Methods and Properties

When you assign a property to an object, possible constants appear after the equal sign.

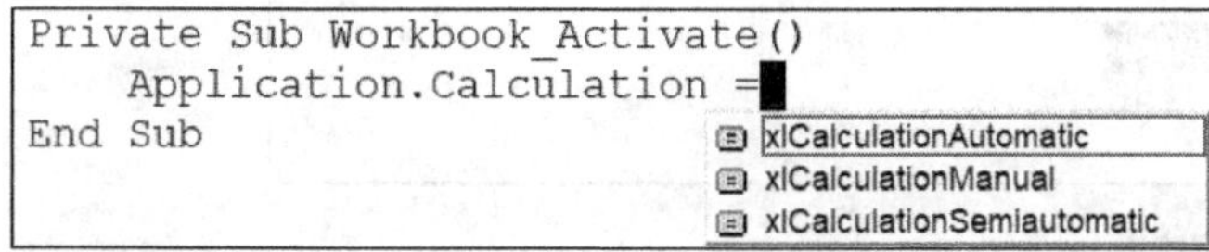

Figure 23.13 Possible Constants for Property

QuickInfo

QuickInfo for the syntax of functions, methods, or procedures is very helpful. For example, when you enter a method and then type the next space or an opening parenthesis, a brief reference with required and optional arguments will appear directly below the insertion point.

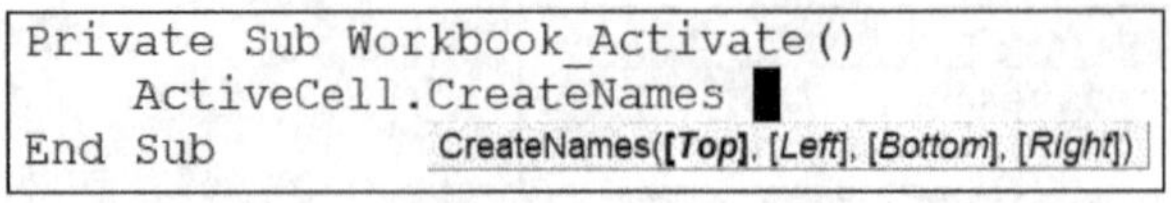

```
Private Sub Workbook_Activate()
    ActiveCell.CreateNames
End Sub         CreateNames([Top], [Left], [Bottom], [Right])
```

Figure 23.14 QuickInfo for Method

Bookmarks and Breakpoints

To quickly access critical or unfinished parts of your code, you can insert bookmarks in the left margin of the module window. Place the insertion point on the relevant line and click the **Toggle Bookmark** flag icon in the **Edit** toolbar. You use the same icon to delete individual bookmarks, and the other two flag icons let you jump forward or backward to a bookmark.

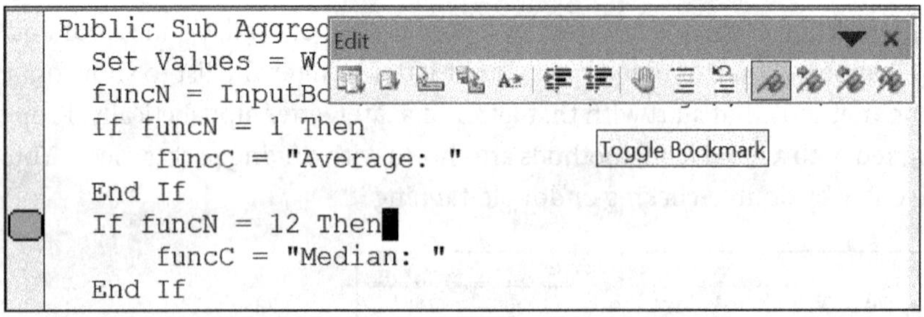

```
Public Sub Aggreg
  Set Values = Wo
  funcN = InputBo
  If funcN = 1 Then
      funcC = "Average: "
  End If
  If funcN = 12 Then
      funcC = "Median: "
  End If
```


Figure 23.15 Setting Bookmark

To set a breakpoint for program testing, click or tap in the left margin next to the corresponding row. Click again to remove the breakpoint.

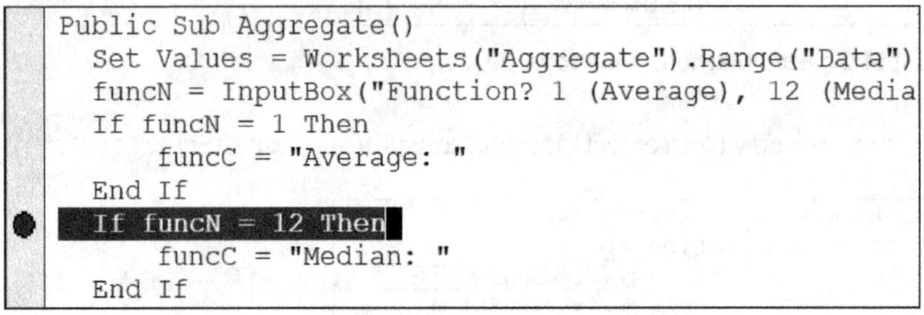

```
Public Sub Aggregate()
  Set Values = Worksheets("Aggregate").Range("Data")
  funcN = InputBox("Function? 1 (Average), 12 (Media
  If funcN = 1 Then
      funcC = "Average: "
  End If
  If funcN = 12 Then
      funcC = "Median: "
  End If
```

Figure 23.16 Breakpoint in Margin

The two comment icons are very useful. The one on the left turns all selected rows into comments, while the one on the right converts comments back into code. This lets you temporarily exclude parts of the code from testing.

There's also a pair of icons for indenting code lines. While not required, it's good practice to indent lines inside loops or branches. To do this, simply select the appropriate rows and click the **Indent** icon.

```
Public Sub Aggreg
  Set Values = Wc
  funcN = InputBc
  If funcN = 1 Then
     funcC = "Average: "
  End If
'  If funcN = 12 Then
'     funcC = "Median: "
'  End If
```

Edit

Comment Block

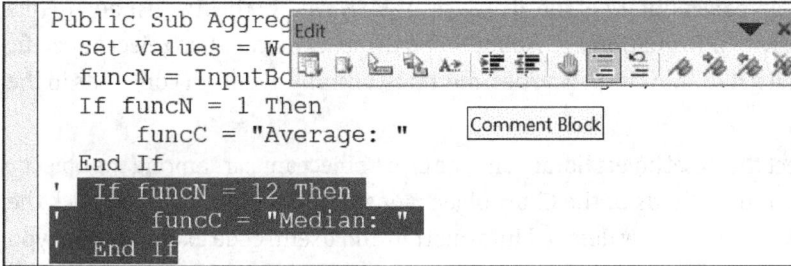

Figure 23.17 Commented-out Procedure Section

Working with the Object Catalog

The vast number of objects Excel offers, along with countless methods and properties, would challenge any developer's memory unless they have access to helpful tools. A key aid in programming is the object cataloG, and you can open it using the **Object Catalog** icon or pressing F2. The window can stay open while you're doing your work.

What the object catalog displays depends on what you select in the first list box. When you choose a project, the left side of the window lists the workbook's objects, including worksheets, chart sheets, and modules. When you select a module on the left, its individual procedures appear on the right. With the **View • Code** command, the F7 key, or a double-click or double-tap, you can open the module window for that procedure.

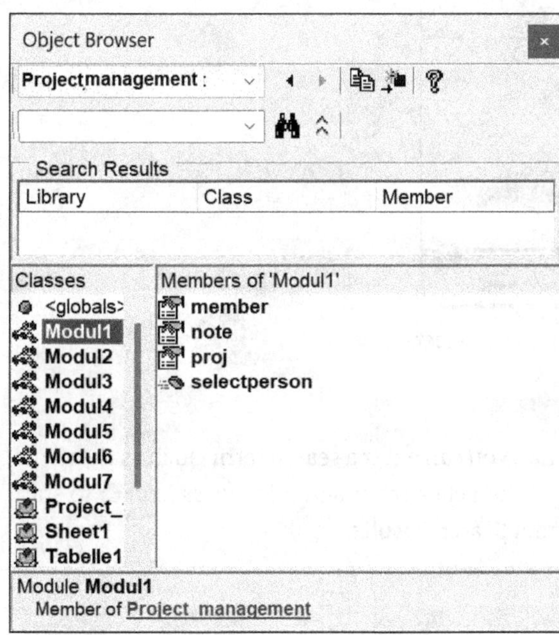

Figure 23.18 List of Elements for Module in Project Management Project

You can select an object library name, like *Excel* or *MSForms*, in the first list box to display all objects with information in that library on the left. You can then select a specific object from that list to show its properties, methods, and any events or constants in the right pane.

When you select the Excel object library and the `Chart` object appears among the objects, the properties and methods of the Chart object appear on the right. You can click the question mark icon ❶ to view detailed information and useful code examples that you can copy into your own procedure. If you select the `PrintOut` method, its arguments will appear at the bottom of the object catalog ❷ and the object hierarchy will also be displayed. You can click a link for any object in this hierarchy to select that object in the left pane.

The object catalog isn't just for reference; you can also use it as an input aid. For example, if the `PrintOut` method is selected for the `Chart` object, you can click the copy icon ❸ and then **Paste** in the module window.

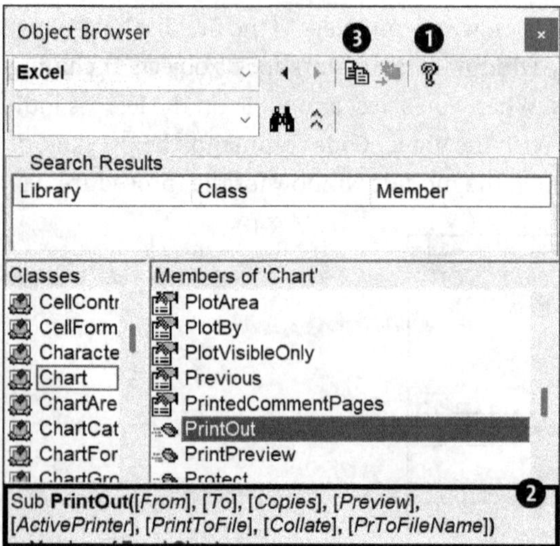

Figure 23.19 Properties and Methods of Chart Object

Search Elements

In the second list box below the title bar, you can enter a search term (such as the name of an object, method, or property) and then click the magnifying glass icon ❹ to start the search. The results will appear under **Search Results**.

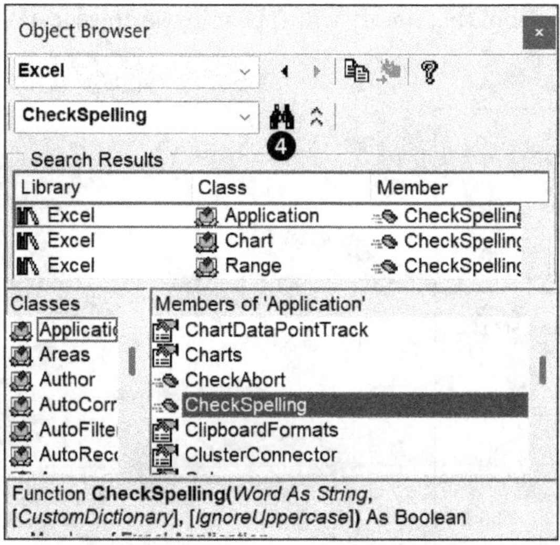

Figure 23.20 Objects to Which CheckSpelling Method Can Be Applied

Adding More Object Libraries

For some projects, you may want to add more object libraries to the existing ones so you can use the objects they define. You can do this by setting a reference to the library by selecting **Tools • References** in the VBA Editor window. Then, select the object library you want to reference, such as the **Microsoft XML, v6.0 parser**. If a library isn't listed, you can locate the file by clicking **Browse**.

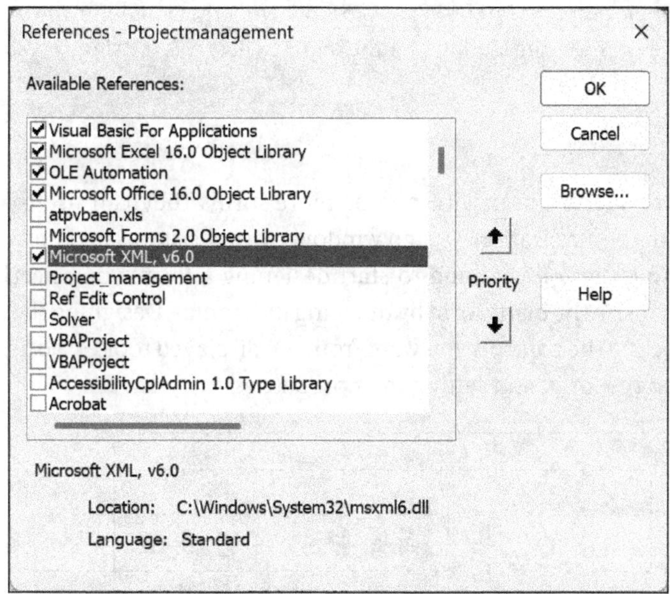

Figure 23.21 Reference to XML Parser

After you create a reference, the classes from that library will appear in the object catalog.

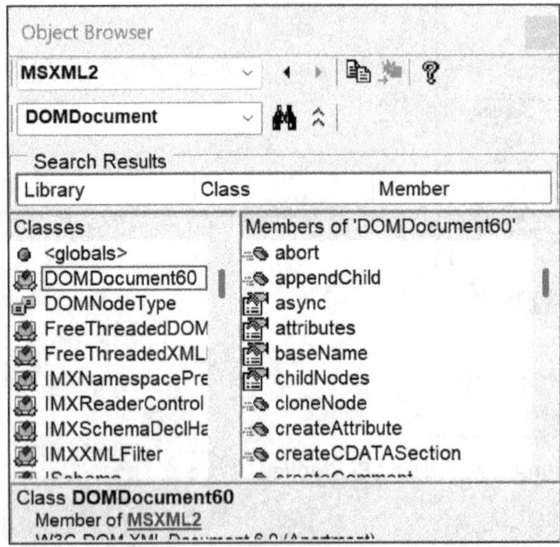

Figure 23.22 List of Classes in XML Library

Be Careful with References

References apply only to the specific project where they are added. When sharing macro applications, remember that any libraries you reference—such as in a workbook—must also be installed on the other computer for the macros to work properly. Otherwise, the other user will receive an error message stating that the required library is missing.

Form Design Window

Most VBA applications provide the user with one or more forms they can use to exchange information with the application. Design windows with the associated toolset are available for creating these forms, and to start designing a form, you select **Insert • UserForm**. You can resize the blank form by dragging its handles. Designing the form mainly involves selecting the controls you want from the displayed toolset, positioning and sizing them on the form, and setting their properties.

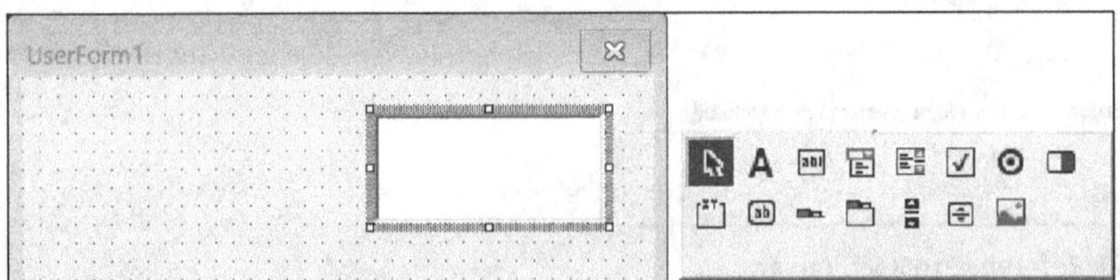

Figure 23.23 Drawing Text Box in Form Design

Properties Window

Each control object has specific properties that you can view or modify. When a control is selected in the form, you can open its properties window by clicking the **Properties window** icon or pressing F4.

This window displays the properties of the selected object. On the **Alphabetic** tab, properties are listed alphabetically. On the **Categorized** tab, properties are grouped by criteria such as **Appearance**, **Position**, or **Behavior**. The property is listed on the left, and the current value is shown on the right. When you select a property, small buttons often appear to help you choose values. Otherwise, you can enter the value you want directly.

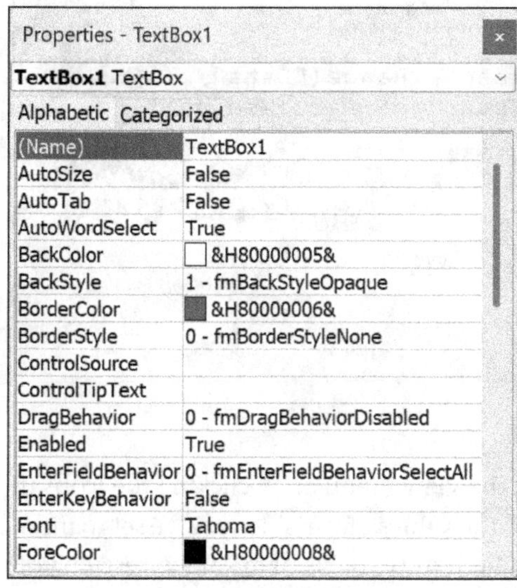

Figure 23.24 Properties Window for Selected Text Box

23.2.3 Testing Programs

VBA program code must be compiled before it can be used. If you choose the **Run • Run Sub/UserForm** command or press F5 and the code hasn't been compiled since the last changes, it will usually compile automatically. However, it's often useful to compile the code beforehand to catch errors that only the compiler can detect. Select **Debug • Compile VBAProject** to start the compilation. If the compiler finds an error, it will mark it in the code and show a corresponding error message.

```
Public Sub Aggregate()
  Set Values = Workshe
  funcN = InputBox("Fu
  If funcN = 1 Than
     funcC = "Average
  End If
```

Microsoft Visual Basic for Applications

Compile error:

Expected: Then or GoTo

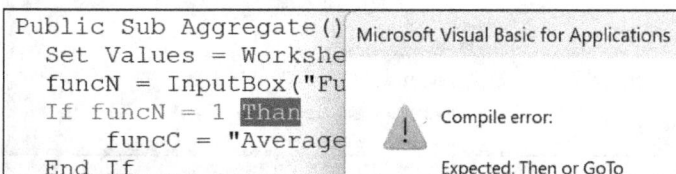

Figure 23.25 Compiler Immediately Detects Typo in Selected Method

The **Help** button provides detailed explanations whenever possible. Otherwise, the VBA Editor offers three special windows for program testing, and the **Debug** toolbar is also available. You can run the test step-by-step, and the best way to do it is by clicking the **Step Into** icon or pressing F8, either stepping through the procedure or up to the insertion point. These commands are also in the **Debug** menu. Individual program lines are highlighted in color so you can closely monitor the program flow.

By selecting **Run • Run Sub/UserForm** or pressing F5, you can execute the finished program in the editor window. If breakpoints are set, you can press F5 to run the program up to the first breakpoint, which allows you to check specific values there.

(General) | Aggregate

```
Public Sub Aggregate()
  Set Values = Worksheets("Aggregate").Range("Data")
  funcN = InputBox("Function? 1 (Average), 12 (Median)")
  If funcN = 1 Then
      funcC = "Average: "
  End If
  If funcN = 12 Then
      funcC = "Median: "
  End If
```

Debug

Step Into (F8)

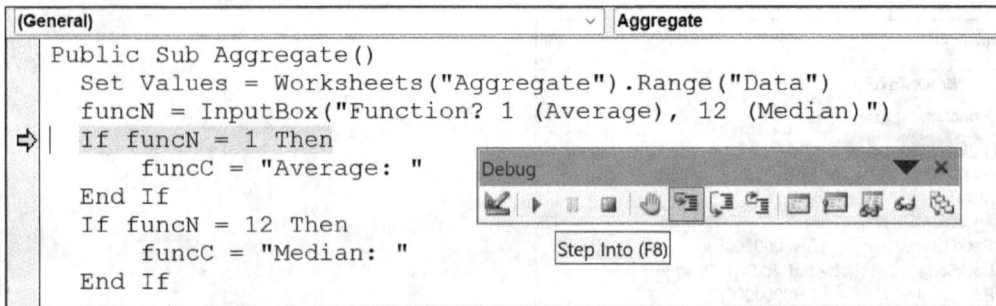

Figure 23.26 Single-Step Test

Checking Values

You can open additional test windows from the **View** menu or by clicking the icons in the **Debug** toolbar. The **Watch window** shows the values of variables you've set to monitor, and the easiest way to set them is to highlight the variable name or expression you want to watch in the procedure. For variable names, right-click or press and hold them briefly and then choose **Add Watch** from the context menu.

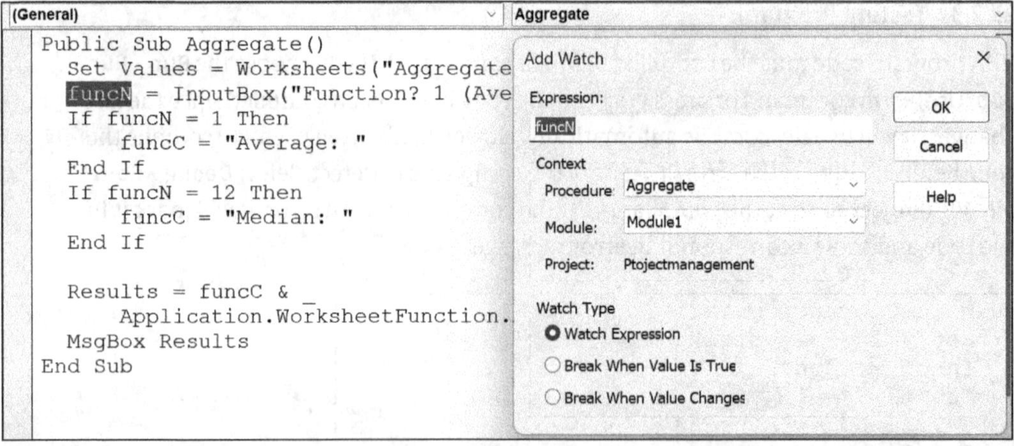

Figure 23.27 Expression in Add Watch Dialog Box

The editor will then open a dialog box that displays the variable name or expression for your review. Under **Watch Type**, you can select how the watch will be performed. The **Watch Expression** option displays only the value of the variable or expression, and for the other two options, program execution pauses as soon as the expression becomes `True` or changes.

After you confirm, the **Watch** window will open and show the watch expression you entered with its current value.

```
    If funcN = 1 Then
        funcC = "Average: "
    End If
    If funcN = 12 Then
        funcC = "Median: "
    End If

    Results = funcC & _
        Application.WorksheetFunction.Aggregate(funcN, 6, Values)
    MsgBox Results
End Sub
```

Watches

Expression	Value	Type	Context
funcN	"1"	Variant/String	Module1.Aggregate

Figure 23.28 Watch Expression in Watch Window

Locals and Immediate Windows

To temporarily test a different value for a variable, you can display the **Locals** window via **View • Locals window**. If you overwrite a displayed value there, the procedure will use that value when you continue testing by selecting **Debug • Step Into** or pressing [F8].

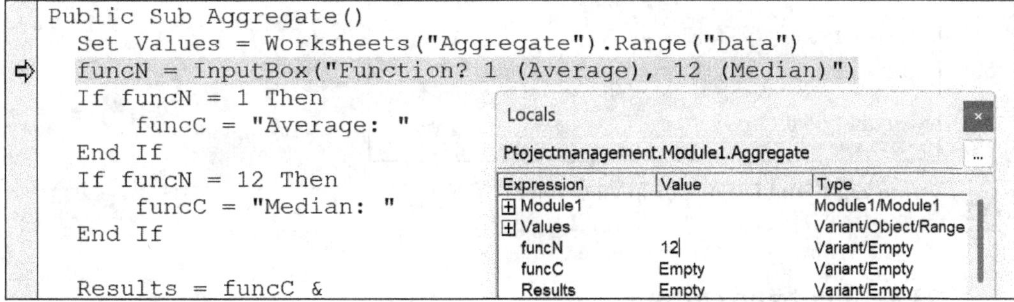
Public Sub Aggregate()
Set Values = Worksheets("Aggregate").Range("Data")
funcN = InputBox("Function? 1 (Average), 12 (Median)")
If funcN = 1 Then
funcC = "Average: "
End If
If funcN = 12 Then
funcC = "Median: "
End If
Results = funcC &
Locals
Ptojectmanagement.Module1.Aggregate
Expression
Value
Type
Module1
Module1/Module1
Values
Variant/Object/Range
funcN
12
Variant/Empty
funcC
Empty
Variant/Empty
Results
Empty
Variant/Empty

Figure 23.29 Changed Value in Locals Window

During step-by-step testing, you can also check variables or expressions directly in the **Module** window by hovering the mouse over the variable name or expression. If a variable has already been assigned a value during program execution, it will appear below the row.

```
If funcN = 12 Then
  funcN = "12" = "Median: "
End If
```

Figure 23.30 Value Display for a Selected Variable

In the **Immediate Window**, you can use the `Print` and `MsgBox` commands to check the value of variables or the results of functions. To reset a module to its initial state, click the **Reset** icon. This matches the **Run • Reset** command.

[+]

> **Switching between the VBA and Application Windows**
>
> To quickly switch to the VBA Editor window and back, use the Alt+F11 keyboard shortcut. Since the VBA Editor has its own button in the Excel Windows group on the taskbar, you can quickly switch between windows by using these buttons.

23.2.4 Printing Code and Forms

To document programming work, it usually makes sense to print the program code and forms at least once. You can do this by using the **File • Print** command. In the dialog box, you can choose to print selected ranges, the current module, or the entire project. If you select a form, you can print both the form design and its code.

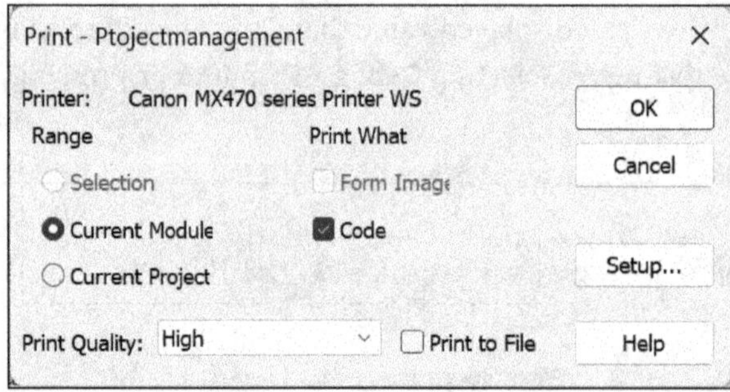

Figure 23.31 Print Dialog Box in VBA Project

23.3 Input and Output

A key part of programming is designing how a program interacts with its users. This interaction occurs on two levels. First, there is direct dialogue between the program and the user, where the user enters values or triggers events (such as mouse clicks) that the program responds to. Second, the program can be set to read from and write to tables.

23.3.1 Simple Input Dialog

The simplest way to get user input in a program is through the input dialog box. You can call this standard dialog box in VBA by using the `InputBox()` function. The output counterpart to the input dialog is the message dialog, which uses the `MsgBox()` function.

For programs requiring only a few inputs, the input dialog box is very practical; for more complex inputs, however, you should use the forms we discuss in the following paragraphs. The dialog box already includes **OK** and **Cancel** buttons. When **OK** is clicked, the dialog box returns the input to a variable. In its simplest form, the `InputBox()` function requires only one argument, like this one:

```
InputBox("Amount")
```

To customize the dialog box, you can use additional arguments: a prompt, a custom window title, or a default value that's used if nothing is entered.

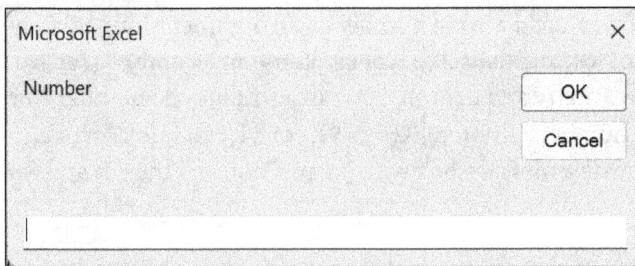

Figure 23.32 Example of Input Window

The code for this dialog looks like this:

```
Sub inputTestValue()
    Dim input As Variant
    input = _
      InputBox("Test value: ", "Input Test Values", "1000", 200, 150)
    MsgBox input
End Sub
```

The key point of the input window is capturing the entered value in a variable. The dialog always returns its results as a string. If you want more control over the data type, use the `InputBox` method instead of the `InputBox()` function. This method belongs directly to the `Application` object, and it accepts the same arguments as the `InputBox()` function plus an additional `Type` argument that sets the data type of the result. For example, Type:=1 requires a number, 2 requires text, and 8 requires a cell reference. You can also use combinations:

```
zt = InputBox(Prompt:= "Please enter numbers or text:",Type:=3)
```

23.3.2 Message Dialogs

The counterpart to the input dialog is the message dialog. You can specify the text to display either as a quoted string or as a variable:

```
Sub report()
    Dim message As String
    message = "Still raining in August 2015"
    MsgBox "Unpleasant: "
    MsgBox message
End Sub
```

At first glance, the message box appears to be just an output tool. However, with additional arguments, you can set it up to consider the user's response in the program's flow, though the user's reaction is limited to clicking buttons. To select which buttons the message box displays, you can add a code number as an argument immediately after the message text: 0 (**OK**), 1 (**OK** and **Cancel**), 2 (**Cancel**, **Retry**, and **Ignore**), 3 (**Yes**, **No**, and **Cancel**), 4 (**Yes** and **No**), or 5 (**Retry** and **Cancel**). You can also add code numbers for displayed icons to the button numbers: 16 (Stop), 32 (?), 48 (!), or 64 (i for Info). For example, you can add code 68 to create a message box with an info icon and the **Yes** and **No** buttons.

As with methods, there are two ways to write functions:

```
message = MsgBox(expression)
MsgBox expression
```

To display a simple message, using the second form is enough. But to capture the result of the message dialog (which button was clicked) in a variable, you must use the first form.

23.3.3 Selecting Ranges in Worksheets

When working with VBA in Excel, the core of programming is working with worksheets: selecting cells or ranges, reading data from them, and writing data to them. To select a cell in a workbook, use the Range property of the Worksheets object and the Select method. In VBA, a single cell is treated as a range containing just one element.

For example, this selects cell A3 on Sheet1:

```
Worksheets("Sheet1").Range("A3").Select
```

You'll get the same result by splitting the selection into two statements:

```
Worksheets("Sheet1").Select
Range("A3").Select
```

The second statement refers to the previously selected sheet. When a procedure is called from a specific worksheet, you can work directly with the Range object and the Select method:

```
Range("A3").Select
```

To select a cell range, specify the corresponding addresses:

```
Range("A3:F3").Select
```

If the cell range has a name, you can use that name here instead of the cell addresses. The name must be enclosed in quotation marks:

```
Range("Range1").Select
```

Navigating Within the Worksheet

If a macro refers to the cell that's active when called, you can move anywhere on the sheet from that point by using the `Offset()` method.

This selects a cell that's three rows down and five columns to the left of the active cell:

```
ActiveCell.Offset(3, 5).Select
```

Note that the argument order—row number and then column number—is the exact reverse of a cell address, where the column letter comes first and the row number comes second. You can extend such a statement with an additional range reference:

```
ActiveCell.Offset(3, 5).Range("A1:A10").Select
```

23

This moves the selection away from the active cell and selects a range of ten cells in one column starting from this new point. The A1:10 range address is not treated as an absolute address but an address that's relative to the cell reached by `Offset(3, 5)`. For example, if the active cell is A1, then the F4:F13 range will be selected.

Variable Cell Selection

The following code retrieves the row and column numbers of the active cell and then calculates new row and column numbers from these values to use for the next selection:

```
Sub calculatedSelection()
  Dim row, column, newRow, newColumn
  row = ActiveCell.Row
```

```
  column = ActiveCell.Column
  newRow = row + 1
  newColumn = column + 2
  Cells(newRow, newColumn).Select
  With Selection.Interior
    .Pattern = xlSolid
    .PatternColorIndex = xlAutomatic
    .Color = 65535
  End With
End Sub
```

The Cells Property

Another way to select cells or ranges is by using the `Cells` property. Using this property from `Application` returns a `Range` object that covers all cells on the active worksheet. The following procedure demonstrates how to use this property to fill the entire worksheet background with color.

```
Sub accessCells()
    Application.Cells.Select
    With Selection.Interior
        .Pattern = xlSolid
        .PatternColorIndex = xlAutomatic
        .Color = 65535
    End With
End Sub
```

The second object where you can use the `Cells` property is `Range`. With the following statement, you can select the A1:E8 cell range on the active worksheet:

```
Range(Cells(1,1), Cells(8, 5).Select
```

23.3.4 Assigning Values to Cells

To enter a value or formula into a cell, you can also use properties. For values—text, numbers, dates, or times—you use the `Value` property. Unlike manual work on the worksheet, a program doesn't need to make a cell active before writing to it. VBA lets you write directly into cells using both absolute and relative references:

```
Range("A3").Value = "Title"
Cells(3, 2).Value = Range("A3").Value & "ID"
Range("A3:F3").Value = "Sale"
```

The last statement shows that you can write a value to multiple cells at once.

23.3.5 Entering Formulas

There are several properties you can use to enter a formula into a cell or range. The simplest is Formula, and here are some examples:

```
Range("A3").Value = "Title"
Range("A4").Formula = "=Right(A3,2)"
Cells(5, 1).Formula = Range("A4").Formula
```

The second statement uses the English function name and a comma as the argument separator. However, cell A4 automatically displays the formula with the German function name.

The alternative property here is FormulaLocal. It lets you specify the function in the system's local language, which in this case could be the following (for a German locale):

```
Range("A4").FormulaLocal = "=Rechts(A3;2)"
```

Other properties you can use are FormulaR1C1 and FormulaR1C1Local. As the names suggest, both use the alternative R1C1 notation for ranges. The statements look like this:

```
Range("A4").FormulaR1C1 = "=Right(R3C1,2)"
Range("A4").FormulaR1C1Local = "=Rechts(R3C1;2)"
```

The R1C1 notation in VBA makes managing relative references easier. The following table clearly illustrates this.

Addresses	Meanings
R1C1	The A1 cell (row 1, column 1)
RC(1)	One column to the right
RC(-1)	One column to the left
R(1)C(1)	One row down and one column to the right

In the next example, two nested For-Next loops fill a cell range with random numbers. The row and column numbers iterate through specific values:

```
Sub writeMultipleTimes()
   Dim i As Integer, j As Integer
   For i = 1 To 8
      For j = 1 To 8
         Cells(i, j).Formula = _
            "=RANDBETWEEN(0,9)"
      Next j
   Next i
End Sub
```

23

23.3.6 Reading Data from Worksheets

When reading cell contents from a worksheet into a variable, you can use either the Value property or the Text property. This works even if the cell contains a numeric value. The difference is that the Text property always returns a string, even if the cell holds a number or a formula. The following example shows how to assign cell values to three variables and write them back to adjacent cells:

```
Sub readCellValues()
    Dim value1, value2, value3
    value1 = Range("A2").Value
    value2 = Range("A3").Value
    value3 = Range("A4").Value

    Range("B2").Value = value1
    Range("B3").Value = value2
    Range("B4").Value = value3

    value1 = Range("A2").Text
    value2 = Range("A3").Text
    value3 = Range("A4").Text

    Range("D2").Value = value1
    Range("D3").Value = value2
    Range("D4").Value = value3

End Sub
```

To read a formula from a cell in the table, use the Formula or FormulaR1C1 property:

```
Dim formula1 As Variant
Range("A4").Formula = "=SQRT(16)"
formula1 = Range("A4").Formula
Range("A5").Formula = formula1
```

If the cell contains no formula or is empty, these properties simply return the cell's content—which is a constant or an empty string.

23.3.7 A Macro for a Sheet List

In workbooks with many sheets, navigating via tabs is often inefficient because only a part of the tabs is visible at a time. A practical solution is a sheet list on a front-inserted worksheet that links to the other sheets, and you can easily create such a list with a small macro. The following example shows a simple solution that reads the sheet names in a loop and arranges them in a multicolumn table if needed. The procedure first inserts a new sheet named *sheetslist* at the start of the workbook.

The count variable holds the current number of sheets in the workbook, and this value controls the loop that enters the other sheet names into the cell range. To help you quickly jump from the sheet list to the corresponding sheets, hyperlinks are created using the sheet names. The sheet names provide both the subaddress and the link label shown in the cell. To display the list in multiple columns, the macro checks the current row number and moves to the next column after filling 25 rows:

```
Public Sub sheetsList()
  Dim i, s, r, number As Integer
  i = 2
  r = 2
  s = 1
  If Sheets(1).Name = "SheetList" Then Sheets(1).Delete
  End If
  Sheets.Add Before:=Sheets(1)
  Sheets(1).Name = "SheetList"
  count = Sheets.Count
  title = ActiveWorkbook.Name
  Cells(2, s).Value = title

  Do While i <= number
    Cells(r + 2, s).Hyperlinks.Add _
    Anchor:=Cells(r + 2, s), _
    Address:="", _
    SubAddress:=Sheets(i).Name & "!A1", _
    TextToDisplay:=Sheets(i).Name
    i = i + 1
    r = r + 1
    If r > 25 Then
      r = 2
      s = s + 1
    End If
  Loop
End Sub
```

23.4 Designing Forms

The message and input dialogs you learned about in the previous section are only suitable for single inputs and outputs. When larger amounts of input are needed, it becomes frustrating if a separate dialog box appears for each entry. Moreover, input windows offer very limited options: they can't include list boxes for selecting entries, nor can they include option buttons or checkboxes that can be simply clicked. The primary tool for creating complete dialog boxes in VBA is the form, which you can create in the VBA Editor window.

23.4.1 Developing an Input Form

Here's how you create a form in the VBA Editor window:

1. Open the workbook that will contain the program with the form. Switch to the VBA window by pressing `Alt`+`F11`.
2. Open a new form with **Insert • UserForm.** This opens a window for designing the form, and you can resize it as needed by dragging the handles.
3. Insert a label that will serve as a caption for the next control, which is the list box for employees. From the toolbox—which appears automatically—select the **Label** ❶ icon and then draw a suitable field to display the list caption.
4. You can move the field freely or align it to the grid points. Drag the handles to change the size and proportions.
5. A preset label appears first in the design window, which you should replace. Open the properties window for the first control by pressing `F4` and then select the **By Category** tab ❷. Under **Appearance**, you'll find properties that are important to the project. Under **(Name)** ❸, you'll see the VBA name assigned automatically to the control. This name is used to reference the control in the program code. We recommend that you to enter a descriptive name here, but you don't have to.
6. For **Caption** ❹, the previously mentioned name is initially set as temporary text. This text is replaced by the label of the intended list, which will use a combo box.
7. To change the font, go to the **Font** row (not shown) and click the button with three dots. Then, in the dialog box, select the desired font.

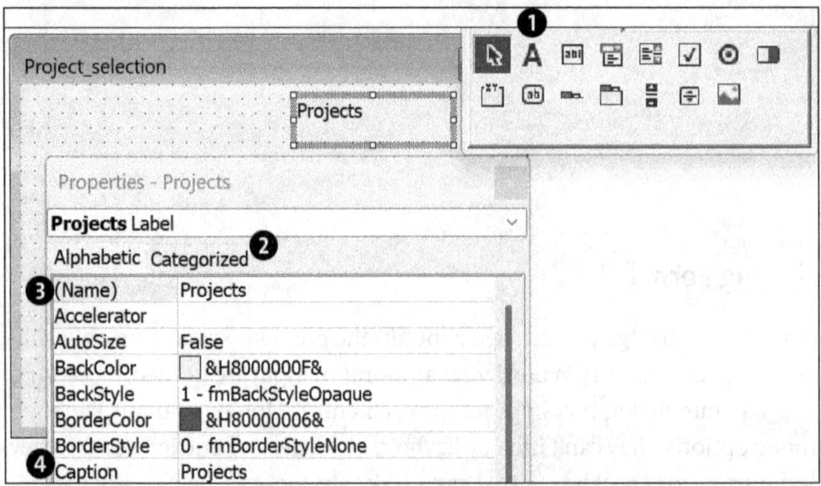

8. At this point, you should set a different title for the form's title bar. Select the form while the **Properties** window is still open, and under **Caption**, enter a different title, such as "Selection Form." You can also enter a different name under **Name** for the form, such as "Project Form," which the program will use to reference it.

[+]

Snap to Grid

To place controls precisely on a form, it's helpful to enable snap to grid. You can find this in the **Tools • Options** dialog box under **General**. To create a finer grid, reduce the point value for grid spacing there.

23.4.2 Inserting Input Controls

You can keep the **Properties** window open while adding other controls. If the toolbox isn't visible, click the **Toolbox** icon on the default toolbar and then add input controls as follows:

1. Add a combo box for the employee list to select employee names. Select the combo box icon from the toolbox and draw a rectangle for the field on the form. Enter a descriptive name for the control: "EmployeeList."
2. Add a second label field with the `Caption` property set to "Projects." You can also enter the text directly into the field without opening the properties window. To do this, select the field and overwrite the default text.
3. You'll also need a combo box for projects, and ideally, you should name it "Person-List." Otherwise, proceed as you did with the employee field.

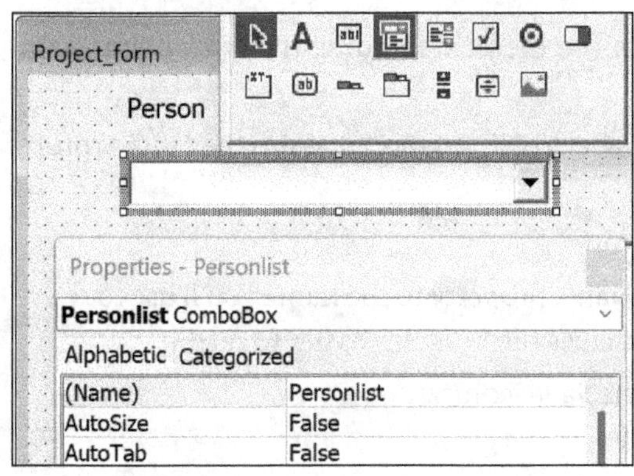

4. Use a text box to enter notes. Select the **TextBox** control from the toolbox and draw a larger field on the form.
5. For a text box, these properties are important: To allow longer entries, add scroll bars. Under **Scroll** and **ScrollBars**, choose the value 3 to add both vertical and horizontal scroll bars.

23

6. Under **Behavior**, set **MultiLine** to `True` to enable multiline entries.

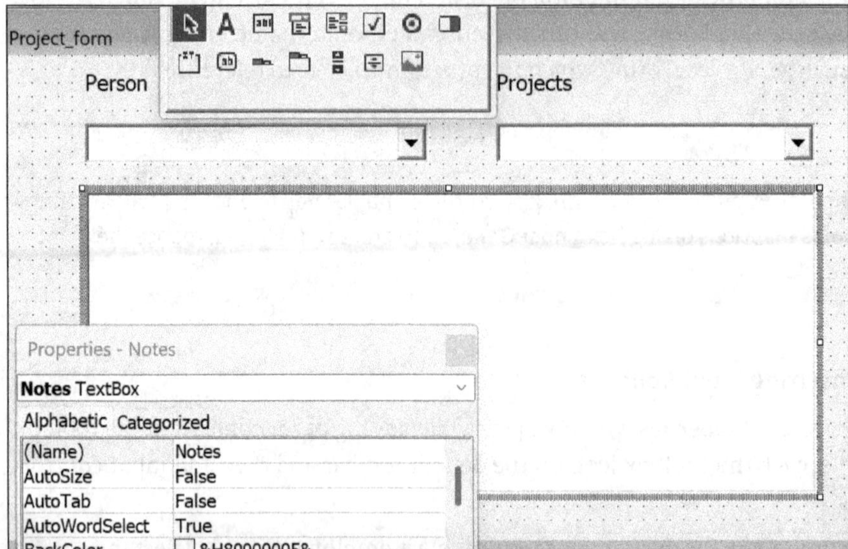

This configures the data entry controls and sets their properties.

23.4.3 Adding Buttons

Two buttons are still needed: one to transfer the selected and entered data into the workbook and another to cancel the dialog box if you don't want to transfer the data. Here's how you create them:

1. Select the **CommandButton** icon from the toolbox and drag the button you want onto the form.
2. Select the new button and change the label to **OK**. This also updates the `Caption` property.
3. Repeat the process with the other button.
4. In the properties window, change the **Name** property to the name you'll use to reference the button in the program. This makes the program easier to read.
5. Save your progress again by clicking the **Save** icon.

The order in which controls are selected when you press the Tab key is controlled by the **TabIndex**. Excel follows the order in which the controls were created, but if this doesn't fit, you can adjust the **TabIndex** values for the controls in the **Properties** window.

23.4.4 Entering the Procedures

After designing the form for user-program communication, you can start programming the procedures that call the form and transfer the data into the workbook:

1. Open the Project Explorer window. In the project tree, the developed form will appear under **Forms**.

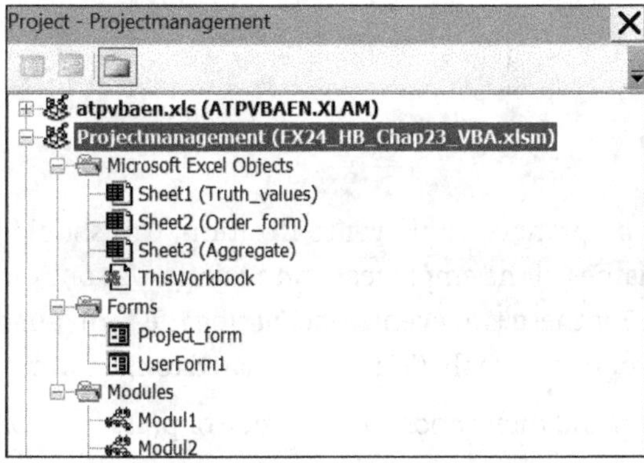

2. Select **Insert • Module** to enter the program. Excel will open a module window and will expect variable declarations. On the left is **General**, and on the right is **Declarations**. This case requires declaring three variables.
3. Enter the declarations for `note`, `member`, and `proj` as strings in the text area.

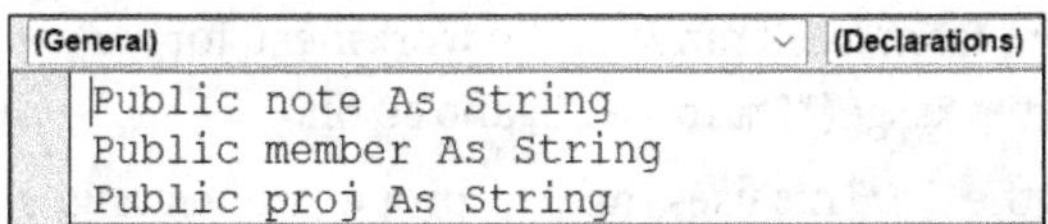

4. Select the **Paste • Procedure** command, enter a name for the procedure (such as "selectperson"), select **Sub** as the type, and click **OK**.

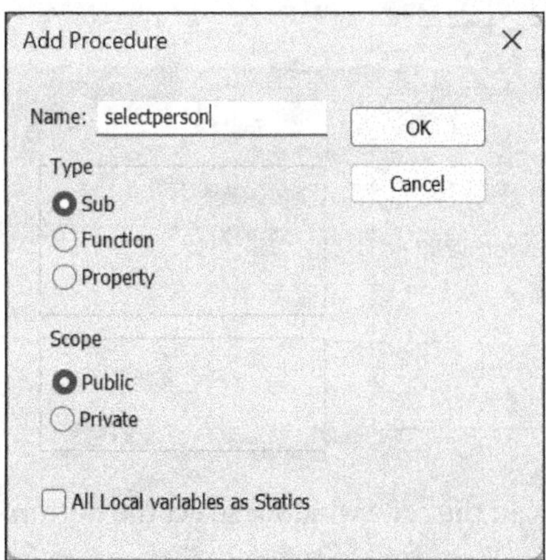

5. At the start of the new procedure, the declared variables are assigned empty strings as initial values. You can display the form by applying the `Show` method to the `Formular` object.

```
(General)                                    selectperson
Public Sub selectperson()
  note = ""
  Person = ""
  proj = ""
  Project_form.Show
```

The Code for the Form

The `Show` command temporarily hands program control over to the form. This should not appear empty; it should show the names of the employees and projects. Use one of the two buttons to complete the form. For these two events—activating the form and clicking a button—you can enter the program code that's directly related to the form:

1. In the Project Explorer, select the form and then choose **View • Code** or press F7 to open the code window.
2. In the object list of the module window, `UserForm`, meaning the form, is selected. From the right list, select the `Activate` event, which occurs when the form is activated by the previously entered `Show` command.
3. A procedure skeleton for this event will appear immediately, and you should use the `AddItem()` method to fill both lists with entries. Here, you can simply enter the list values directly. You can also pull them from a cell range in the worksheet, for example, by using a range name with `AddItem Range("Employees")`, and so on.
4. The last line selects the first entry in the list. This uses the `Text` property and assigns it the first list entry, which has an index of 0 since these lists are zero-based.

```
(General)                                    (Declarations)
Private Sub UserForm_Activate()
  Project_form.Personlist.AddItem "Hamann"
  Project_form.Personlist.AddItem "Stein"
  Project_form.Personlist.AddItem "Werle"
  Project_form.Personlist.AddItem "Stohaus"
  Project_form.Personlist.Text = _
    Project_form.Personlist.List(0)
  Project_form.Projectlist.AddItem "Berlin"
  Project_form.Projectlist.AddItem "Bonn"
  Project_form.Projectlist.AddItem "Hamburg"
  Project_form.Projectlist.Text = _
    Project_form.Projectlist.List(0)
End Sub
```

5. Add the procedures for the two buttons. Stay in the code window, select the button name in the left list box, and then choose the `Click` event in the right list box.

6. In the procedure for the **OK** button, assign the entries you've selected from the employee and project lists to the member and proj variables. Use the Text property to get the selected entry.
7. Assign the note you've entered into the text box to the note variable, which is declared as Private. You do this by retrieving the Value property of the note control.

```
OK                                    Click
Private Sub OK_Click()
  member = Project_form.Personlist.Text
  proj = Project_form.Projectlist.Text
  note = Project_form.Notes.Value
  Project_form.Notes.Value = ""
  Project_form.Hide
End Sub
```

8. Reset this property for the next input and hide the form with the Hide method.
9. The procedure for the **Cancel** button requires only this last command.

```
Cancel                                Click
Private Sub Cancel_Click()
    Project_form.Hide
End Sub
```

This completes the work on the form. It is clear what should happen when the user interacts with the form.

The End of the Main Procedure

Now, control returns to the main procedure. Next, you should pass the variables to the worksheet and double-click the module in the project window. Then, simply use the Value property of the Range objects three times. After that, you can write the selection and inputs from the form directly into the specified cells without selecting them first (see Figure 23.33). 23

```
(General)                             selectperson
Public Sub selectperson()
  note = ""
  Person = ""
  proj = ""
  Project_form.Show
  Range("B3").Value = member
  Range("B4").Value = proj
  Range("B6").Value = note
End Sub
```

Figure 23.33 Finished Main Procedure

Save the project by using **File • Save As**, then switch back to the workbook window with **File • Close and Return to Microsoft Excel**. When you start the macro—via **Developer**

Tools • Code • Macros or a button assigned to the macro—you can fill out the form (see Figure 23.34).

Project_form
Person
Projects
Hamann
Berlin
OK
Start is May 2026
Cancel

Figure 23.34 Input into Form

This example has been intentionally simple, and our main goal has been to show you how to enter data into a form and then transfer it to a workbook. You can easily expand the project. For example, you can write the form values directly into a table template—meaning a work order—and then print it. Using a loop, you can fill out several work orders at once. You can also use named ranges to store employee and project names.

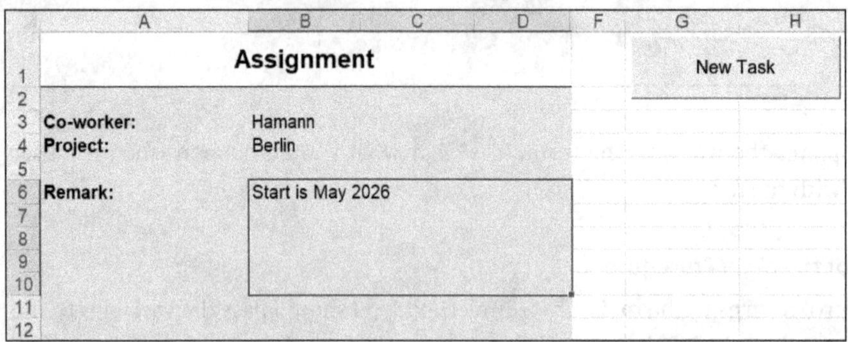

Figure 23.35 Data in Worksheet

To transfer the names into the form lists, you can use a For-Each loop like this one:

```
For Each c In Worksheets("Table1").Range(Employees).Cells
  ProjectForm.EmployeeList.AddItem c.Value
Next c
```

23.4.5 Expandable Tool Collection

The VBA tool collection includes a set of predefined elements, and we'll briefly introduce the most common ones here. Except for the **Select Objects** icon, handling the elements is always the same. You select them in the tool collection and then draw a rectangle on the form to define the element's position. The elements are always initially assigned numbered English default names.

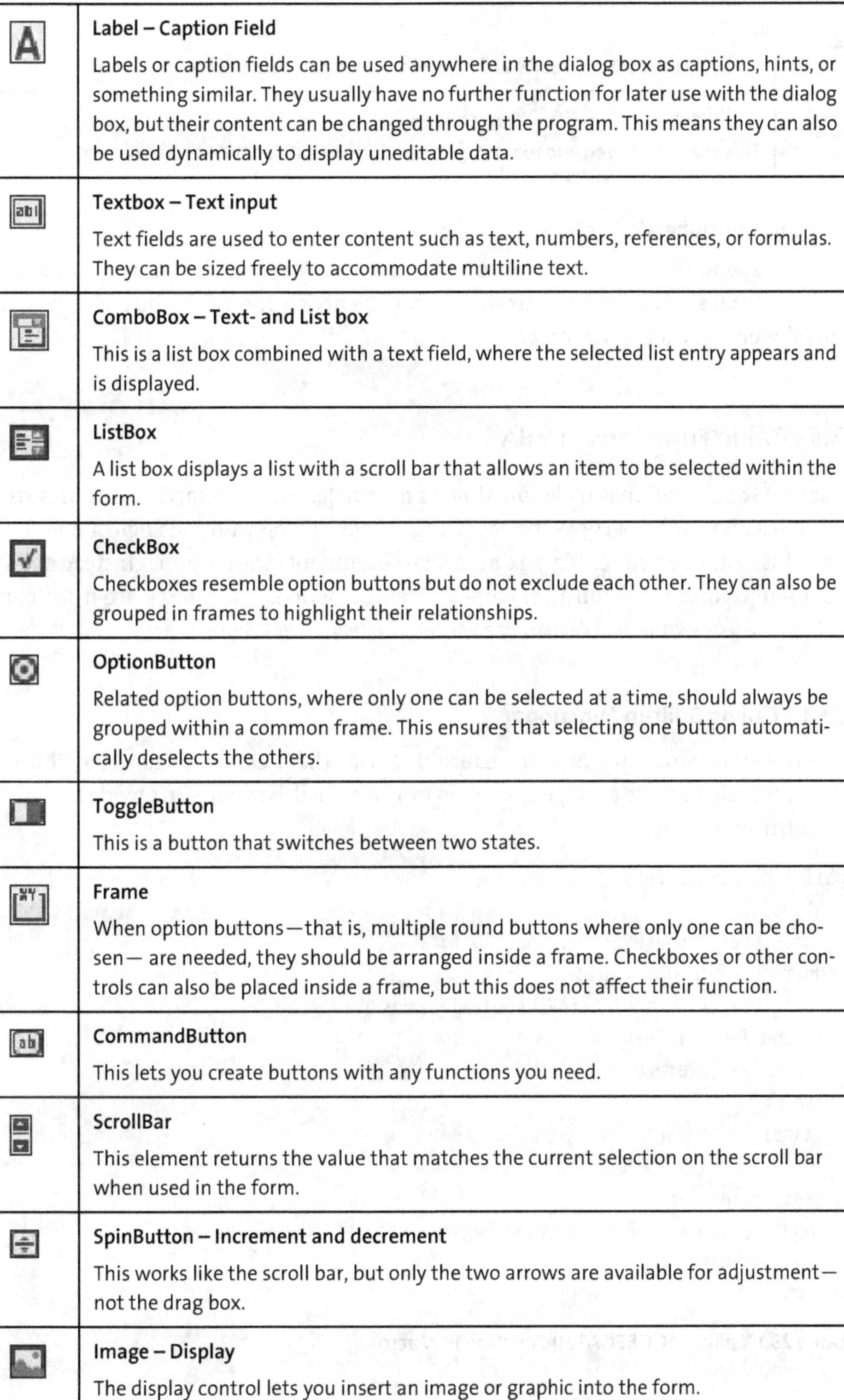

	Label – Caption Field Labels or caption fields can be used anywhere in the dialog box as captions, hints, or something similar. They usually have no further function for later use with the dialog box, but their content can be changed through the program. This means they can also be used dynamically to display uneditable data.
	Textbox – Text input Text fields are used to enter content such as text, numbers, references, or formulas. They can be sized freely to accommodate multiline text.
	ComboBox – Text- and List box This is a list box combined with a text field, where the selected list entry appears and is displayed.
	ListBox A list box displays a list with a scroll bar that allows an item to be selected within the form.
	CheckBox Checkboxes resemble option buttons but do not exclude each other. They can also be grouped in frames to highlight their relationships.
	OptionButton Related option buttons, where only one can be selected at a time, should always be grouped within a common frame. This ensures that selecting one button automatically deselects the others.
	ToggleButton This is a button that switches between two states.
	Frame When option buttons—that is, multiple round buttons where only one can be chosen— are needed, they should be arranged inside a frame. Checkboxes or other controls can also be placed inside a frame, but this does not affect their function.
	CommandButton This lets you create buttons with any functions you need.
	ScrollBar This element returns the value that matches the current selection on the scroll bar when used in the form.
	SpinButton – Increment and decrement This works like the scroll bar, but only the two arrows are available for adjustment—not the drag box.
	Image – Display The display control lets you insert an image or graphic into the form.

	TabStrip This control lets you create tabs in the form that can hold additional controls.
	MultiPage This control lets you add extra pages to the form that can hold additional controls.

Inserting Additional Controls

You can expand the number of controls that are available for design as needed via the command **Tools • Additional Controls**. When you check a control in the dialog box, it immediately appears in the toolbar.

23.5 Table Functions in VBA

Chapter 15 explained that table functions can be called within a macro. This uses the `WorksheetFunction` property of the `Application` object along with the English function name. Inside the parentheses, you can separate arguments with commas instead of the usual semicolons. If the function name contains a period, you can replace it with an underscore. For example, you can use `VAR_P()` instead of `VAR.P()`.

23.5.1 Calling Built-In Functions

Here is a simple example of a procedure that uses the `AGGREGATE()` function. It asks whether to calculate the average or median from a named `Data` range in a worksheet and then displays the result.

```
Public Sub Summarize()
  Set values = _
    Worksheets("Aggregate").Range("Data")
  function = _
    InputBox("Function? 1 (AVERAGE), 12 (MEDIAN)")
  If function = 1 Then
     label = "Average: "
  Else
    label = "Median: "
  End If
  result = label & _
    Application.WorksheetFunction.Aggregate(function, 6, values)
  MsgBox result
End Sub
```

Listing 23.1 Calling AGGREGATE() Function in Macro

Figure 23.36 shows the result when Average is selected as the function argument for the AGGREGATE() function.

	A	B	C	D
1	**Data**			
2				
3	128	Stein	553	834
4	212	Köln	139	689
5	397	370	101	465
6	764	590	553	640
7	226	400	650	951
8	395	899	320	845
9	723	816	623	268

Microsoft Excel
Median: 553
OK

Figure 23.36 Calling AGGREGATE() Function in Macro

23.5.2 Custom Table Functions

With nearly 500 table functions built into Excel, you don't have to settle for less than what you need. If you often repeat specific calculations, you can create your own functions using VBA that return a calculated value from given input data.

In this section, we'll show you how to define a table function, using a simple circumference calculation as an example:

1. To use the function in all workbooks, it's best to add it to the usually hidden personal workbook *Personal.xlsb* file.
2. Switch to the VBA environment by pressing Alt+F11 and select **VBAProject(PERSONAL.XLSB)** in the **Project** window. To collect the new functions in a separate module, choose **Insert • Module**; otherwise, open an existing module by double-clicking it in the Project Explorer.
3. Select **Insert • Procedure** and enter the function name. Under **Type**, choose **Function**, and select **Public** for the scope so the function can be called universally.

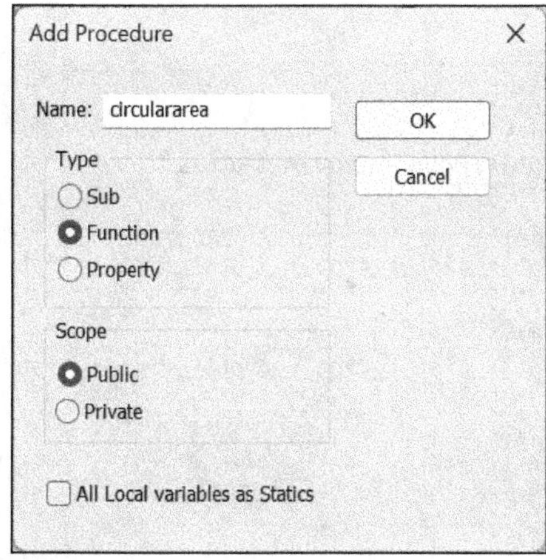

4. A skeleton for the function will appear in the module window. Add the argument names inside the parentheses after the function name. In this case, a single argument called Radius is enough.
5. Write the instruction for the calculation between the two lines:

```
Public Function Circumference(Radius)
 Circumference = Radius * 2 * _
   Application.WorksheetFunction.Pi
End Function
```

6. Save the VBA project by using the **Save** button. Switch back to the workbook window by using the **View Microsoft Excel** icon.

If you want to call this function within a procedure to calculate the circumference, follow the example in this listing:

```
Sub calculateCircumference()
  radius = InputBox("Radius")
  MsgBox ("The circumference is " & Circumference(radius))
End Sub
```

To use the function directly on the worksheet, use the **Insert Function** dialog as usual. Under **Select Category**, choose **User Defined** and then select the new function.

Inserting a Function Description

To perfect the function further, add a description of the function and make sure it appears in a specific category in the **Insert Function** dialog. You need a procedure that only runs once in the workbook that contains the function, and for our circle function, it might look like this:

```
Sub DescribeFunction()
  Dim functionName As String
  Dim functionDescription As String
  Dim functionCategory As String
  Dim argumentDescription(1 To 3) As String
  functionName = "Circumference"
  functionDescription = "Calculates the circumference from the radius"
  functionCategory = 3 'Math

  argumentDescription(1) = "Radius of the circle"
  Application.MacroOptions Macro:=functionName, _
    description:=functionDescription, _
    Category:=functionCategory, _
    ArgumentDescriptions:=argumentDescription
End Sub
```

When you're pasting via the **Insert Function** dialog, the function appears in the category you selected and the **Function Arguments** dialog displays the description of the required argument. This description is saved with the workbook where the function is defined.

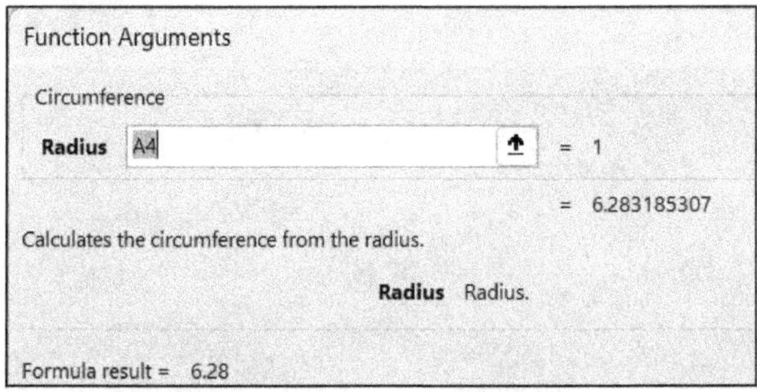

Figure 23.37 Custom Function to Calculate Circumference

23.5.3 Retrieving the Sheet Name

If you want to list the names of sheets in a workbook in a table, a function that retrieves the **Name** property of the sheets can help:

```
Public Function SheetName(id As Integer) As Variant
    SheetName = Sheets(id).Name
End Function
```

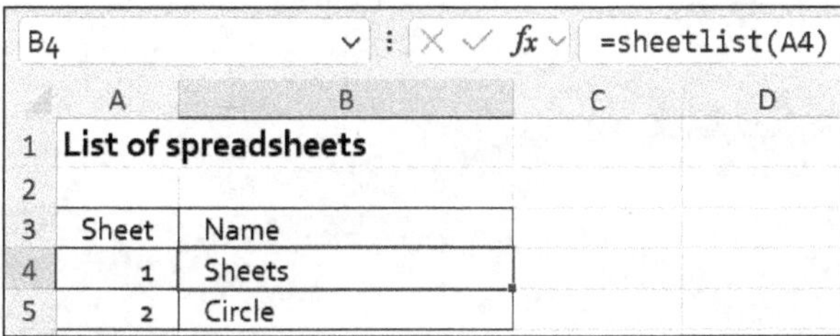

Figure 23.38 List of Sheets in Workbook

Note that Excel requires the workbook name that contains the function to be prefixed before the function name for custom functions that are not saved in the active workbook. For example, you should use `=Personal.xlsb!SheetName()` instead of `=SheetName()`.

If you copy the function into the active workbook, you only need to specify the function name. It's often practical to add custom functions to templates so they're automatically available in all workbooks that are created from that template.

Chapter 24
Preview

As a bonus, I want to highlight two features that are only available in basic form in the current version of Excel. One is an alternative to VBA solutions: Instead of recording scripts with VBA, the JavaScript variant TypeScript is used. The other is support for working with Excel through the Copilot AI solution.

24.1 Office Scripts: The Alternative to Macros

Office Scripts are saved by default in your OneDrive storage, with the *.osts* file type. The workbook is first saved to OneDrive, so an internet connection is required. Before you can start recording, the **Automate** tab must be visible in the menu bar. You can click the **Record Actions** button to start the process and open the task pane, and you can click the stop icon there to end the recording. In the example in Figure 24.1, you should sort the table by the data in the **Category** column.

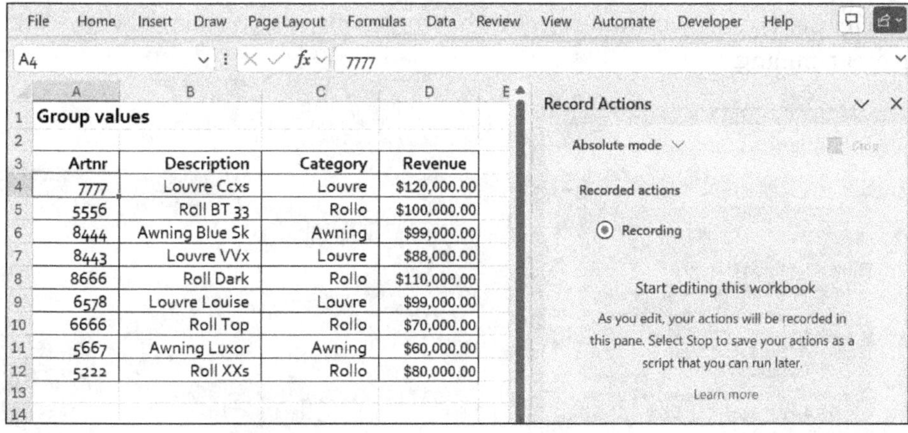

Figure 24.1 Start Recording of Office Script

In this case, you should sort the table by product groups, so simply select the table, trigger the appropriate sort on the **Data** tab, and apply a style.

The recording is documented as a sequence of individual steps that are displayed one after another in the task pane (see Figure 24.2) while the underlying code remains hidden at first. You should replace the default script name with a suitable one, and you can also add a description of your script.

Within the task pane, you can later change, for example, the addresses of ranges. In the menu under the three dots, you'll find the **Advanced edit** command, which opens the generated code for corrections.

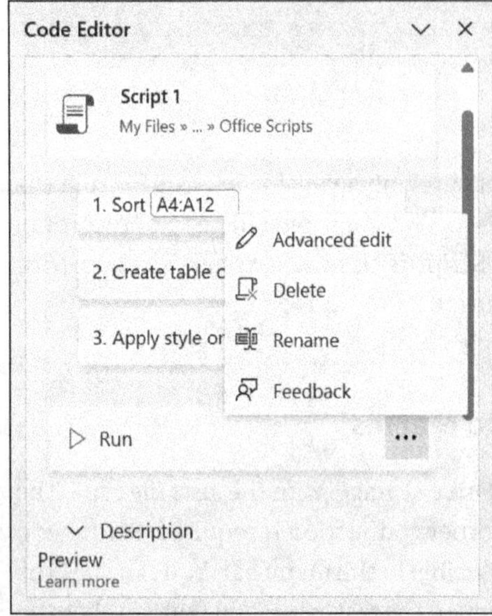

Figure 24.2 First Result of Recording

Every script starts with a `main` function that includes the workbook type as the first parameter (see Figure 24.3). However, as with recorded macros, you don't have to work with programming.

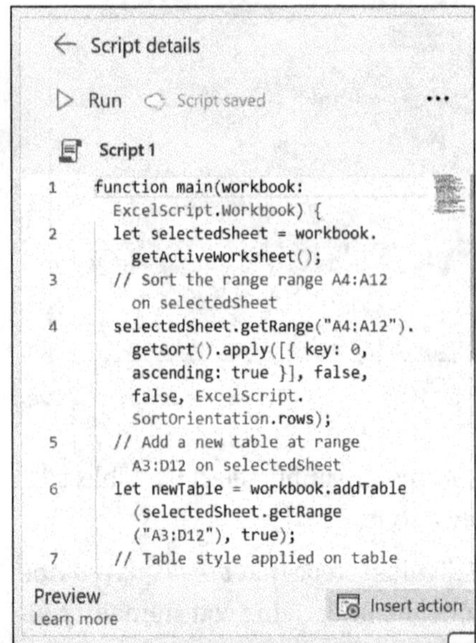

Figure 24.3 Script's Source Code

Use the **Insert action** button to record an additional step if you need to and click the small triangle in the task pane to run the script. The tab lists existing scripts.

The TypeScript object model includes counterparts for many Excel objects, but development is still in progress. If you record something that can't yet be reproduced as a script, like the function for subsets, the task pane will display an error message.

	A	B	C	D
1	**Group values**			
2				
3	Column:	Descriptio	Categor	Revenu
4	5222	Roll XXs	Rollo	$80,000.00
5	5556	Roll BT 33	Rollo	$100,000.00
6	5667	Awning Luxor	Awning	$60,000.00
7	6578	Louvre Louise	Louvre	$99,000.00
8	6666	Roll Top	Rollo	$70,000.00
9	7777	Louvre Ccxs	Louvre	$120,000.00
10	8443	Louvre VVx	Louvre	$88,000.00
11	8444	Awning Blue Sk	Awning	$99,000.00
12	8666	Roll Dark	Rollo	$110,000.00

Figure 24.4 Sorted Table

24.2 Copilot: The Helpful Assistant

Copilot, which is Microsoft's AI solution, has a fitting name since it supports the person in control. Microsoft offers *Copilot Pro* as a paid add-on to its Windows 365 subscription. If you have it, you'll find a **Copilot** button in Excel's **Home** tab that opens the task pane.

If the workbook you want to use Copilot with isn't saved on OneDrive and automatic saving isn't enabled, a button will appear to help you complete these steps. Once you've done this, the **Copilot** task pane will show initial tips and support options (see Figure 24.5).

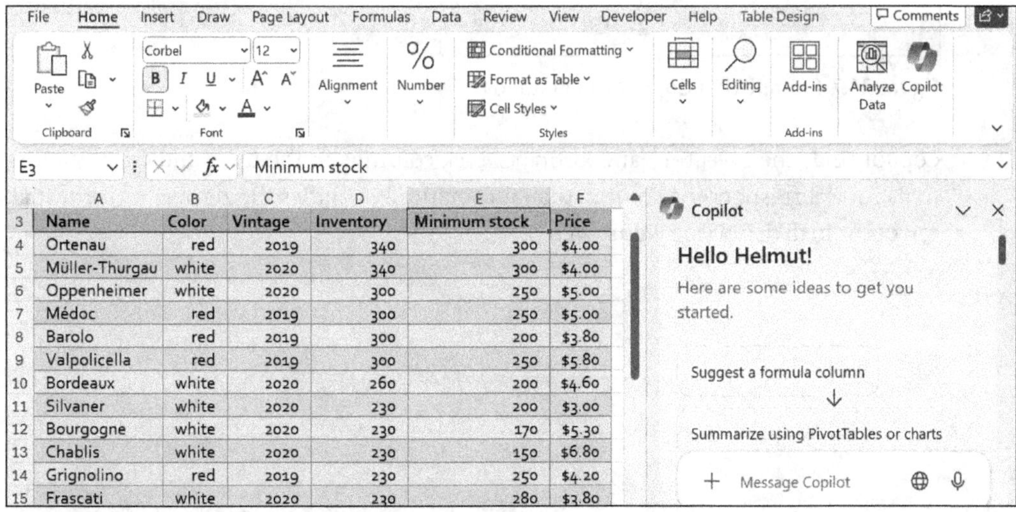

	A	B	C	D	E	F
3	Name	Color	Vintage	Inventory	Minimum stock	Price
4	Ortenau	red	2019	340	300	$4.00
5	Müller-Thurgau	white	2020	340	300	$4.00
6	Oppenheimer	white	2020	300	250	$5.00
7	Médoc	red	2019	300	250	$5.00
8	Barolo	red	2019	300	200	$3.80
9	Valpolicella	red	2019	300	250	$5.80
10	Bordeaux	white	2020	260	200	$4.60
11	Silvaner	white	2020	230	200	$3.00
12	Bourgogne	white	2020	230	170	$5.30
13	Chablis	white	2020	230	150	$6.80
14	Grignolino	red	2019	230	250	$4.20
15	Frascati	white	2020	230	280	$3.80

Figure 24.5 Sample Table and Copilot Task Pane

If you click the **Suggest a formula column** option, Copilot will analyze the data in the current worksheet and offer an initial suggestion (see Figure 24.6). Here, it makes sense to calculate the inventory value by multiplying the unit price by the stock quantity. If you agree, click the **Insert column** button.

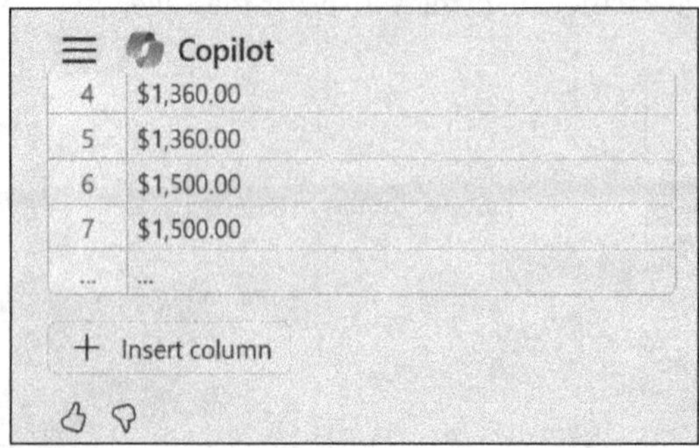

Figure 24.6 Suggestion for Formula and New Column

If the result isn't usable, click **Undo**. This simple example shows that the entire function works like a conversation between you and your assistant. It should also remind you that the helper can sometimes be wrong: the AI always operates on the probability that something is true, but it can be false.

Copilot's greatest feature is in the input field, where you can type or speak prompts. Copilot understands much of what you want to do in Excel, and you can simply say what you want to calculate (see Figure 24.7).

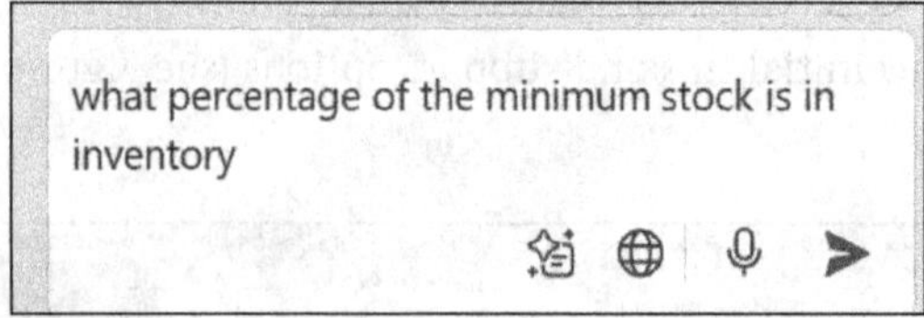

Figure 24.7 Prompt That Copilot Understands

Copilot finds the `Inventory` and `Minimum_stock` columns in the table and creates a formula for a corresponding column that automatically handles the zero-value error that can occur in division (see Figure 24.8).

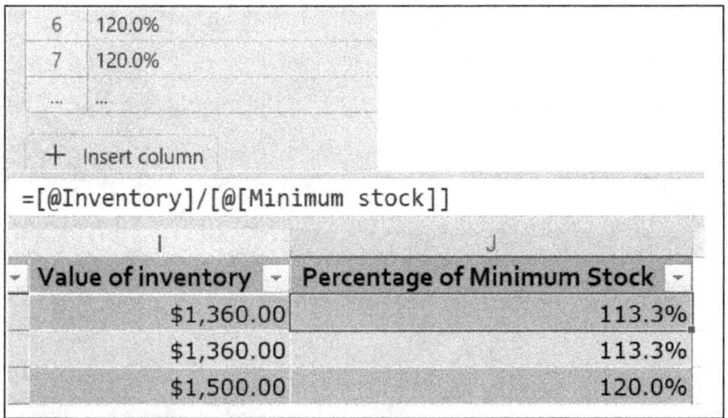

6 120.0%
7 120.0%
+ Insert column
=[@Inventory]/[@[Minimum stock]]
I
J
Value of inventory
Percentage of Minimum Stock
$1,360.00
113.3%
$1,360.00
113.3%
$1,500.00
120.0%

Figure 24.8 New Value of Inventory Column

When writing your questions, you can access a prompt gallery by clicking the first button (see Figure 24.9).

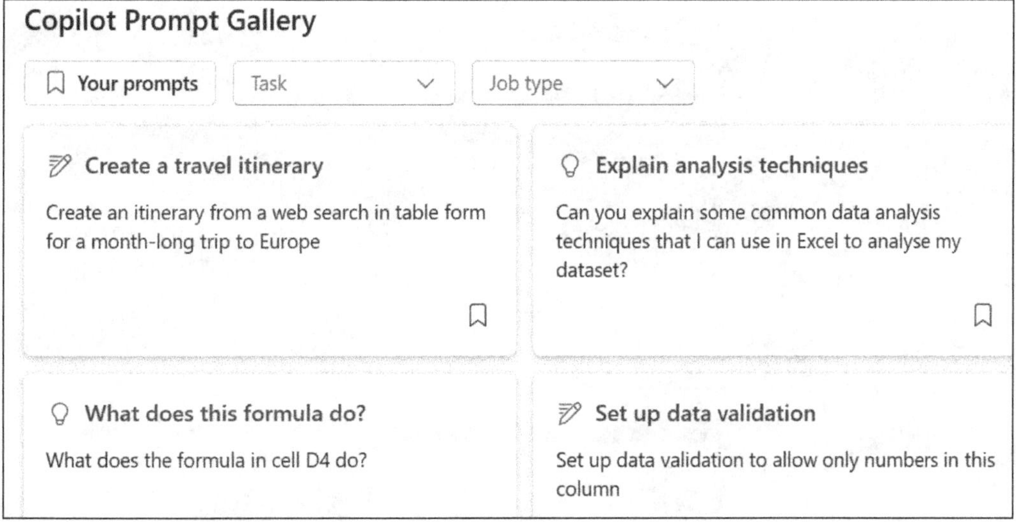

Your prompts

Create a travel itinerary
Create an itinerary from a web search in table form for a month-long trip to Europe
Explain analysis techniques
Can you explain some common data analysis techniques that I can use in Excel to analyse my dataset?
What does this formula do?
What does the formula in cell D4 do?
Set up data validation
Set up data validation to allow only numbers in this column

Figure 24.9 Prompt Gallery

Useful versions are organized by topic. Although many things aren't possible yet, remember that AI learns quickly, so check back often to see what Copilot can offer you.

Appendix A
Function Keys, Keys, and Keyboard Shortcuts

A.1 Overview of General Key Functions

The following table lists individual keys, function keys, and keyboard shortcuts that let you perform many tasks in Excel without using the ribbon or quick access toolbar. This is especially helpful for users who have developed tendon sensitivities from excessive mouse use.

Keyboard shortcuts involve pressing multiple keys either individually or combined: [Ctrl], [Shift], [Alt], and sometimes two of these at once. In all these cases, all keys in the combination must be pressed simultaneously, unlike key sequences that also trigger ribbon commands.

Keyboard Shortcut	Description
[Alt]	Shows key tips for accessing commands on the ribbon.
[Alt]+[=]	Inserts the function =SUM().
[Alt]+[F1]	Creates an embedded chart from the selected data using the default chart type.
[Alt]+[F4]	Closes Excel.
[Alt]+[F8]	Displays the **Macro** dialog box.
[Alt]+[F11]	Opens the VBA editor window.
[Alt]+[F12]	Opens the Power Query Editor.
[Alt]+[Spacebar]	Switches between keyboard layouts when multiple are enabled.
[Alt]+[Shift]+[F1]	Inserts a new worksheet into the workbook.
[Enter]	Confirms an entry or command.
[Esc]	Cancels an entry or command.
[F1]	Opens help.
[F2]	Opens the cell for editing.
[F5]	Opens the **Go To** dialog.

Keyboard Shortcut	Description
F6	Switches between windows, the ribbon, and the task pane, or moves to the next pane if the window is split.
F7	Activates spell check.
F8	Turns extend mode on or off.
F9	Starts recalculation in all open workbooks if automatic recalculation is off.
F10	Shows key tips for accessing commands on the ribbon.
F11	Creates a chart from the selected table data on a separate sheet using the default chart type.
F12	Opens the **Save As** dialog.
Ctrl + :	Inserts the current system date into the cell or input field.
Ctrl + Shift + ;	Inserts the current system time into the cell or input field.
Ctrl + ,	Copies the formula from the cell above into the active cell.
Ctrl + -	Equivalent to the **Delete Cells** command.
Ctrl + Shift + +	Opens the **Insert Cells** dialog.
Ctrl + `	Corresponds to the command **Show Formulas**.
Ctrl + 1	Opens the **Format Cells** dialog.
Ctrl + 2	Toggles bold formatting on the selection.
Ctrl + 3	Toggles italic formatting on the selection.
Ctrl + 4	Toggles underline formatting on the selection.
Ctrl + 5	Toggles strikethrough formatting on the selection.
Ctrl + 6	Shows or hides graphic objects.
Ctrl + 7	Shows or hides outline symbols.
Ctrl + 8	Hides the columns containing the currently selected cells, supporting multiselection.
Ctrl + 9	Hides the rows containing the currently selected cells, supporting multiselection.
Ctrl + Alt + V	Opens the **Paste Contents** dialog.
Ctrl + Page ↑	Previous sheet
Ctrl + Page ↓	Next sheet

Keyboard Shortcut	Description
Ctrl + A	Selects the entire worksheet or, if the sheet contains data, the current range first.
Ctrl + B	Emphasizes selected text bold or removes bold.
Ctrl + C	Copies the current selection to the clipboard.
Ctrl + F	Opens the **Find and Replace** dialog with the **Find** tab selected.
Ctrl + F1	Toggles the ribbon on and off.
Ctrl + F10	Toggles the maximized view for the active workbook.
Ctrl + F11	Inserts a new sheet for Excel 4 macros into the workbook.
Ctrl + F2	Shows the print preview.
Ctrl + F3	Opens the **Name Manager** dialog.
Ctrl + F4	Equivalent to the command **File • Close**.
Ctrl + F5	Restores the workbook window to its previous size if it was maximized.
Ctrl + F6	Switches to the next open workbook if multiple workbooks are open.
Ctrl + F7	Allows moving the workbook window if it's not maximized.
Ctrl + F8	Allows resizing the workbook window if it's not maximized.
Ctrl + F9	Minimizes the workbook window to an icon.
Ctrl + G	Opens the **Go To** dialog.
Ctrl + H	Opens the **Find & Replace** dialog with the **Replace** tab selected.
Ctrl + I	Emphasizes selected text italics or removes italics
Ctrl + K	Opens the **Insert Hyperlink** dialog.
Ctrl + L	Opens the **Create Table** dialog.
Ctrl + Spacebar	Selects the column of the active cell.
Ctrl + N	Inserts a new workbook.
Ctrl + O	Opens the **Open** dialog.
Ctrl + P	Opens the **File** tab with the **Print** page.

Keyboard Shortcut	Description
Ctrl + R	Copies the content and format of the first column in a multicolumn range to the adjacent cells on the right, matching the **Fill Right** command.
Ctrl + S	Opens the **Save As** dialog.
Ctrl + Tab	Switches to the next workbook when multiple workbooks are open.
Ctrl + D	Copies the content and format of the first row in a multirow range to the cells below, matching the **Fill Down** command.
Ctrl + Shift + +	Selects the entire data block containing the active cell.
Ctrl + Shift + ;	Copies the value from the cell above the active cell into the active cell; for formulas, this copies the result, not the formula.
Ctrl + Shift + /	Selects the entire array containing the active cell.
Ctrl + Shift + 0	Unhides columns.
Ctrl + Shift + 9	Unhides rows.
Ctrl + Shift + F12	Opens the **File** tab with the **Print** page.
Ctrl + Shift + F3	Opens the **Create Names from Selection** dialog to name values using adjacent labels.
Ctrl + Shift + Spacebar	Selects the entire worksheet or, if the sheet already contains data, the current range.
Ctrl + Shift + O	Selects all cells that contain notes.
Ctrl + Shift + Tab	Switches to the previous workbook when multiple workbooks are open.
Ctrl + Shift + $	Applies a two-decimal currency format to the selection.
Ctrl + Shift + %	Applies a percentage format to the selection.
Ctrl + Shift + !	Applies a two-decimal number format to the selection.
Ctrl + Shift + *	Selects the current cell range around the active cell.
Ctrl + U	Underlines the selection or removes the underline.
Ctrl + Shift + V	Pastes an unformatted value previously copied.
Ctrl + V	Pastes the current selection from the clipboard.
Ctrl + W	Closes the selected workbook window.

Keyboard Shortcut	Description
Ctrl + X	Cuts the current selection and places it on the clipboard.
Ctrl + Y	Repeats an action or command.
Ctrl + Z	Undoes an action or command.
Alt + F10	Shows the context menu for the current selection.
Alt + F11	Inserts a new worksheet into the workbook.
Alt + F2	Adds a new note or opens one for editing.
Alt + F3	Opens the **Insert Function** dialog.
Alt + F5	Opens the **Find and Replace** dialog.
Alt + F6	Switches between the window, ribbon, and task pane, or moves to the next pane if the window is split.
Alt + F8	Toggles **Add** mode to expand an existing selection.
Shift + F9	Recalculates only the active worksheet if automatic recalculation is off.
Shift + Spacebar	Selects the row of the selected cells.

A.2 Edit Keys and Keyboard Shortcuts

These keys and shortcuts work only when editing formulas or other cell contents.

Keyboard Shortcut	Description
Alt + Enter	Starts a new line in the formula bar.
End	Go to the end of the input field
F2	Switches between **Edit** and Enter modes.
F3	Opens the **Insert Name** dialog when entering formulas.
F4	Toggles between relative and absolute addresses
F9	Converts a formula to its result.
→ / ←	Move one character to the right/left
Home	Go to the start of the input field
Ctrl + → / ←	Go to the start of the next/previous word or to the end/start of a number

Keyboard Shortcut	Description
Ctrl + Shift + → / ←	Extends the selection right/left to the end or start of a data block.
Shift + → / ←	Extends the selection right/left.

A.3 Navigation and Selection Keys

Navigation keys speed up movement and selection in the worksheet.

Keyboard Shortcut	Description
Page ↑	Move one window up
Page ↓	Move one window down
Alt + Page ↑	Move one screen left
Alt + Page ↓	Move one screen right
← or Shift + Tab	Move one column left
↑	Move one row up
→ or Tab	Move one column right
↓	Move one row down
Home	Jump to the beginning of the row
Ctrl + Page ↑	Jump to the previous sheet
Ctrl + Page ↓	Jump to the next sheet
Ctrl + End	Jump to the end of the used range
Ctrl + Backspace	Makes the active cell visible again if the view was previously moved using a scroll bar or by scrolling.
Ctrl + ←	Jump left in the row to the start of a data block or to the end of the next data block.
Ctrl + ↑	Jump up in the column to the start of a data block or to the end of the next data block.
Ctrl + →	Jump right in the row to the end of a data block or to the start of the next data block.
Ctrl + ↓	Jump down in the column to the end of a data block or to the start of the next data block.

Keyboard Shortcut	Description
Ctrl + Home	Jump to the beginning of the worksheet.

You can quickly and easily select ranges using the arrow keys.

Keyboard Shortcut	Description
Ctrl + Shift + ← ↓ ↑ →	Extend the selection to the end or beginning of the block in the direction of the arrow key, or to the start or end of the next data block
Ctrl + Shift + End	Extend the selection to the end of the table
Ctrl + Shift + Spacebar	Select the entire data block
Ctrl + Shift + Home	Extend the selection to the beginning of the table
Shift + ← ↓ ↑ →	Extend the selection by one row or column
Shift + Page ↑	Extend the selection up by one window
Shift + Page ↓	Extend the selection down by one window
⇧ + Spacebar	Select the entire row
Shift + Backspace	Reduce the selection to the active cell
Shift + Home	Extend the selection to the beginning of the row

Appendix B
The Author

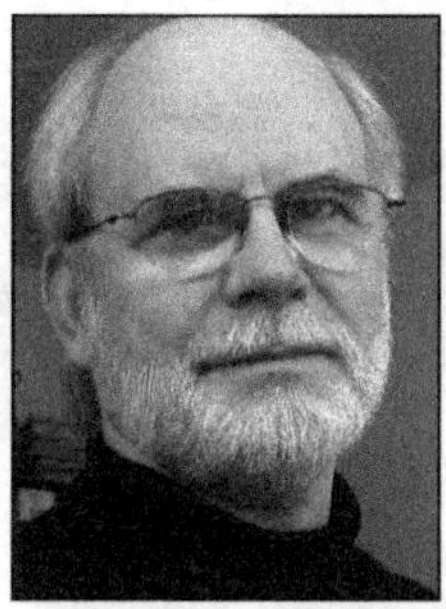

Helmut Vonhoegen is a freelance author and IT consultant. He has published more than 80 books since 1992 and written numerous articles in specialist journals. His focus is Microsoft Office, Windows, web programming, and XML.

Index

B

C

E

G

H

I

J

K

L

M

N

O

Q

T

U

V

W

X

Y

Z

- A practical guide to Python for nonprogrammers
- Work with NumPy, SymPy, SciPy, Matplotlib, and VPython
- Automate numerical calculations, create simulations and visualizations, perform statistical analysis, and more

Veit Steinkamp

Python for Engineering and Scientific Computing

It's finally here—your guide to Python for engineers and scientists, by an engineer and scientist! Get to know your development environments and the key Python modules you'll need: NumPy, SymPy, SciPy, Matplotlib, and VPython. Understand basic Python program structures and walk through practical exercises that start simple and increase in complexity as you work your way through the book. With information on statistical calculations, Boolean algebra, and interactive programming with Tkinter, this Python guide belongs on every scientist's shelf!

511 pages, pub. 03/2024
E-Book: $54.99 | **Print:** $59.95 | **Bundle:** $69.99
www.rheinwerk-computing.com/5852